ZOOLOGY

ALFRED M. ELLIOTT
Professor Emeritus of Zoology
University of Michigan

DARRYLL E. OUTKA
Associate Professor of Biochemistry and Biophysics
Iowa State University

ZOOLOGY

FIFTH EDITION

Prentice-Hall, Inc., Englewood Cliffs, New Jersey

Library of Congress Cataloging in Publication Data

ELLIOTT, ALFRED MARLYN (date)
  Zoology.

  Includes bibliographies and index.
  1. Zoology.   I. Outka, Darryll E., joint author
II. Title.
  QL47.2.E4   1976   591   75-26964
  ISBN   0-13-984021-4

ZOOLOGY, *Fifth Edition*

Alfred M. Elliott and Darryll E. Outka

Previous editions were under the sole
authorship of Alfred M. Elliott.

10 9 8 7 6 5 4 3 2 1

Printed in the United States of America

Prentice-Hall International, Inc., *London*
Prentice-Hall of Australia, Pty. Ltd., *Sydney*
Prentice-Hall of Canada, Ltd., *Toronto*
Prentice-Hall of India Private Limited, *New Delhi*
Prentice-Hall of Japan, Inc., *Tokyo*
Prentice-Hall of Southeast Asia (Pte.) Ltd., *Singapore*

CONTENTS

APPENDICES 653

PREFACE

The study of living things is becoming more exciting each year owing to the advances being made in many areas, particularly at the levels of molecules, cells, and whole animals (behavior). New and improved instruments are advancing our knowledge on many fronts. Physicists and chemists, employing the tools and conceptual methods of their disciplines, are contributing mightily to our understanding of life processes at the molecular level. One of the significant contributions made in this field of research is the regulation of gene activity, which has given us a new approach toward understanding how genes function in development. With the continual improvement of electron microscopes, together with refined techniques of fractionation and purification of cell particulates, the ultrastructure and function of the various parts of cells are rapidly being clarified. The physiologist has at his disposal highly complex recording devices for gaining more information about how animals and their parts function. Remarkable accomplishments are being made in the study of animal behavior using miniaturized electronic equipment, some of which are designed for space travel.

It is our conviction that as much of this new information as possible should be included in a modern textbook of zoology. Students need to be aware of what is going on at the frontiers, but at the same time they must also understand that the newer knowledge is built on a firm foundation established by many brilliant biologists who labored in the past. Consequently, we have tried to integrate the new with the traditional in order that the student may understand that what we are learning today is merely the extension of questions that were asked many years ago.

As we are firmly convinced that organic evolution is the most reasonable approach to the study of zoology, it is the central theme in this edition, as it was in earlier editions. The newer knowledge from laboratory and field studies supplements our understanding of this unifying concept. Molecular biology is adding much to our understanding of evolution at the macromolecular level. It is becoming clearly evident that the way animals behave has evolved as truly as has their morphology and physiological processes. For these reasons, organic evolution is emphasized in every chapter with the hope that there is no question about its importance.

The simplest way to tell the story of organic evolution is, we believe, from the historical approach. Consequently, we have started with the physical laws that were responsible for the formation of our planet and the major events that led eventually to the first forms of life. In order to understand this, a certain amount of elementary chemistry and physics has been included. Much of modern biology requires this background and it seems logical to place it at the beginning.

The subsequent evolution of animals is portrayed, essentially, in the sequence in which it occurred. Throughout these chapters other important principles are included. It is our hope that students will recognize some of the major unsolved problems facing biologists today and the methods used in attacking them. It should become apparent that most of the important problems remain unsolved, and we hope that this information will stimulate students to appreciate, and indeed take part in the solution of, the many mysteries of life that challenge scientists today.

The goal in writing this edition is the same as in earlier editions: to present salient features of animals without overwhelming the student with details. Some of the subject matter is complex, but much effort has gone into writing it as simply as possible. We have tried to present it in such a fashion as to challenge the pre-professional student and to stimulate the general education student.

No effort has been made to slant the material toward either group since we feel that all that is included in this book is of equal value to both.

The book is organized the same as in the fourth edition. All chapters have been reevaluated in the light of recent knowledge. Some have been extensively rewritten, others have been supplemented. Some remain essentially unchanged. Considerable new material has been included in the chapters on cell structure and function and genetics. Most of the earlier illustrations have been redrawn and many new ones added. In order to portray the true beauty of animals, a number of colored photographs have been included.

The book consists of eight parts. Following a brief discussion of biology as a science, there follows—in Part I—a description of the scientist, his methods, and the part he plays in society. It is important to recognize that biologists employ the same techniques and disciplines used by all scientists.

The story of evolution begins—in Part II —with a brief discussion of the physical evolution of our planet and the inception of the earliest forms of life. The elementary principles of chemistry and physics are introduced; these must be mastered in order that later discussions of molecular and cellular biology will be meaningful. The cell is also considered in this Part and continued in Part III, where the organization of cells into many-celled organisms is described. The principles of taxonomy are also included in this part. The long evolutionary sequence of animals is covered in Part IV. Representatives of each phylum are treated in some detail, with constant emphasis on how each higher phylum evolved from the next lower group.

The organ systems of man are covered in Part V. Considerable attention should be devoted to this section because of its importance, particularly by those students who take no more courses in zoology.

The continuity of life is discussed in Part VI, beginning with the replication of DNA and protein synthesis and continuing with cell duplication, continuity of the individual, and of the species. Considerable new material has been added to these chapters.

Organic evolution, its meanings, theories, and mechanism are taken up in Part VII. The book is concluded with Part VIII, which considers animal behavior and the relationship of animals to their environment.

Suggested Supplementary Readings are listed at the end of each chapter. They include books and articles primarily from *Scientific American*. Books that are published as paperbacks are so indicated. These titles have been selected for differing reasons. Some contain excellent illustrations; others have presented the chapter contents in a different and more inclusive fashion. A few books are included which treat the topic-in-question in depth. We regard the books and articles as not necessarily the "best," but certainly worthwhile reading. They provide a guide to biological literature which may be very attractive to some students.

Colleagues at the University of Michigan and elsewhere have been most generous in supplying us with illustrative materials and checking the accuracy of chapters devoted to specialized areas of zoology. We should, again, like to express our appreciation to those specialists who read certain chapters in previous editions: Professors L. C. Stuart (Chapters 1–6), H. F. Hogg (Chapters 2–3), J. Hendricks (Parasites), H. vanderSchalie (Mollusks), E. T. Hooper (Mammals), and A. C. Clement (Chordates). The following Professors critically examined specific sections of this revision: E. E. Steiner (Part II—Continuity of Life), B. E. Frye (Chapter 15—Endocrinology), N. E. Kemp (Chapter 22—Embryology), R. D. Alexander (Chapter 27—Animal Behavior), F. F. Hooper (Chapter 28—Ecology). We are grateful to these specialists for their expert opinions. The entire manuscript was read by Professor I. J. Cantrall, whose criticism we greatly appreciate.

As has been mentioned previously, special attention has been given to the illustrations in this revision. The new drawings and those that have been modified or completely redrawn, were done by Jan Powers whose clear lucid style has greatly enhanced the attractiveness of this edition.

In addition to the acknowledgements listed in the back of the book for photographs, we should like to mention our special appreciation to Professors Keith R. Porter, Hugh Huxley, and George Rose for the excellent electron and light micrographs which they so kindly permitted us to use.

We are deeply grateful to the staff of the large elementary zoology course at the University of Michigan for their constant assistance in bringing errors to our attention and for their helpful suggestions regarding the presentations of certain topics. Moreover, we wish to thank the many teachers who have used earlier editions and shared their experiences with us in order to improve this one. We hope that those who read this edition will let us have the benefit of their criticisms.

A. M. E.   D. E. O.

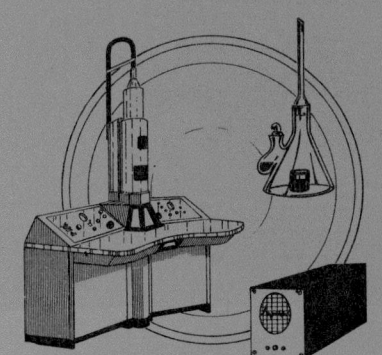

ZOOLOGY
AS A SCIENCE

1

THE SCIENCE OF ZOOLOGY

We are told by astronomers that it is likely that there are a great many planets in the universe that possess the necessary physical requirements to support living things. Proof of this, one way or another, may come within this century. For the present we must be satisfied with what we can learn about living things as they exist on planet Earth where they are extremely successful in kinds and numbers. The extent of their success is indicated by the fact that if all of the physical world were swept away, the outline of its crust would be maintained in essentially its present form consisting only of the body framework of living organisms.

Possibly the most challenging problems faced by scientists in the past, and even more so today, are those centered around the origin and subsequent history of these living things. The physical world appeared first and life was derived from it at some later time; hence the most obvious classification of objects which make up the world is into non-living and living. The physical sciences represent the accumulated knowledge of the former, whereas all that we know about the latter comprises **biology**, the science of life. The living world has been arbitrarily divided into **botany**, the study of plants, and **zoology**, the study of animals, the subject of this book.

Physical scientists have made great strides toward understanding many natural laws, impressively so in recent decades when the atom has been torn apart and its components examined. We now know that living things operate within the framework of these same physical laws, but that they have added some remarkable features which help distinguish the living from the nonliving world. Thus, cells and organisms can assemble matter and energy to make additional cells and organisms, developing high levels of order and complexity. Perhaps the most exciting discoveries made today are those at the level where the knowledge of the chemist, physicist, and biologist converge on the problem of the composition of living matter. There is no greater intellectual achievement than the understanding of the laws of nature that initiated and fostered the development of living things on the earth. This becomes more apparent when one considers that these same laws are responsible for the origin of man himself. Few would argue the relative merits of achieving a thorough understanding of the composition and functioning of this one species.

Are the mysteries of the universe being understood? Some of those concerned with our physical world are yielding to our constant probing, and today we are fairly certain of the workings of many natural laws. The atom is gradually yielding its secrets and we are approaching an understanding of the nature of the universe. But what do we know about life? In spite of all of the great accomplishments in chemistry, we have not yet been able to identify the exact chemical composition of a living organism. We do not understand how an animal grows, how a fertilized egg develops into an individual, and how the brain of man works. These and hundreds of other problems are some of the most complex man has so far encountered, but without their solution there can be no understanding of life itself.

In spite of this dearth of knowledge concerning the actual nature of life, there has accumulated a mass of information about animals, particularly about their structure, physiology, and interrelationships and all of it is included in the science of **zoology**. However, as with all scientific knowledge, zoology has become more and more specialized. It has divided into many compartments, such as, among others, **anatomy** (study of gross structure), **physiology** (study of function), **embryology** (study of early development), and **genetics** (study of heredity). In rendering a broad view of the entire field, therefore, it is necessary to select the salient parts of each of these segments and fit them into a unified whole. The purpose of this book is to lay these bits of information before you with the hope that you will grasp the unified picture and as a result gain an understanding

of animal life as it has come to be what it is today. More importantly, if you can come to a clearer comprehension of your own origin and the place, you occupy in this world of living things, you may be able to formulate a more satisfying philosophy of life. If you do, the effort you put forth in gaining this knowledge will be well worthwhile and the purpose of this book will have been accomplished.

WHAT SCIENCE IS

Many attempts have been made to define science, and much confusion exists in the minds of most people about just what science is. Science is actually a systematic approach to the solution of problems. It involves observation, the collection of facts, and the drawing of verifiable conclusions based on these facts.

Science has come to mean **organized knowledge** which includes all that man has learned about his world. In volume and scope this has become so tremendous that no one person can even read the titles of research papers that appear each day. Indeed, it requires much reading for a specialist to become familiar with the works of others in his own field. One of the critical problems in modern research that must be solved is some way of condensing the current literature that pours from laboratories all over the world and reducing it to such form that it can be read and understood by those working in the field. Many techniques have been devised, such as microfilming and microcarding, but none have been entirely satisfactory.

Science is more than the mere gathering of facts. A fact in itself has meaning only when it is related to a whole body of information. Counting the number of leaves on a tree or the number of hairs on a dog would result in facts, but such information is relatively useless and anyone doing it might be considered queer. However, even this information could be useful if one was studying the value of certain compounds to increase foliation in trees or fur quality in animals. Indeed, factual information is essential to the solution of many problems. It is meaningless only when out of the context of related information. In other words, gathering of facts is an essential part of all research and should not be ridiculed. The annals of science are replete with instances where certain isolated facts, dug out of the old literature, have been the key to the solution of some new problem.

The highest level of achievement in scientific thought is the formulation of **theories** which explain observed facts. Theories are particularly gratifying if they can be reduced to simple terms such as mathematical formulas. The physical scientists and astronomers have been able to do this to a remarkably advanced degree, but in biology very few processes or principles can be fitted neatly into a formula. The movement of the planets and forces controlling a falling body obey laws which can be stated in mathematical form, but the factors controlling the growth of an embryo cannot be so clearly stated. Biological processes certainly do obey laws, some of which are quite clear today, and a constant struggle is made by biologists to reduce the multitude of complex data that accumulates each day to some sort of working formula.

A good theory is one which explains all of the facts that have been observed but, more important, it makes possible **predictions** of what one might expect in future explorations. In other words, it points the way for future work and takes the "guesswork" out of one's efforts.

In recent years scientists have been in the public eye more than ever before for reasons that are quite obvious. Atomic bombs, intercontinental missiles, space travel, and computerization have commanded the attention of everyone. The people responsible for these achievements are the scientists. What kind of people are they? Scientists are no different from any other group of intelligent, highly motivated people, with one possible exception; they possess a spirit of inquiry which stimulates them to investigate the unknown.

Open-mindedness is another characteristic of a good scientist. Preformed ideas about the ultimate outcome of any investigation are inevitable, but one must be certain that such thoughts do not flavor in any way the facts that are obtained. In biology this tendency is a much greater hazard than in the physical sciences because living things are more complex and thus seemingly less amenable to, for example precise mathematical analysis; oftentimes there is considerable room for interpretation and even speculation, both of which are important in all scientific research. Indeed, perhaps the highest level of achievement in science is in the realm of speculation where sweeping ideas, only partially proven, have spurred further investigation. Out of such mental efforts have come our greatest theories and laws, Newton's Law of Gravity and Darwin's Theory of Natural Selection, to give two examples.

A scientist must be critical of all scientific information, particularly his own. It is only by checking and rechecking one another's work that the ultimate truth can be known. Scientists are their own severest critics, and so they should be. Conclusions must be drawn only after the most rigorous tests have been made, and such conclusions must allow for future discoveries which may alter or extend present knowledge.

COMMUNICATION AMONG SCIENTISTS

In the early days of science, men often worked alone and in secrecy. This was possible because their tools were simple and for the most part could be made by the worker himself. As the problems became more complex and more and more scientists were working on the same or related problems, it became necessary for scientists to depend on one another both for tools, chemicals, and for exchange of ideas. Communication then became a very important aspect of science. Groups of scientists with a common interest formed societies where they could meet at regular intervals to "read" papers, that is, describe their achievements to their peers.

Publication, as understood by scientists, has a meaning somewhat different from that usually assigned to the word. To the scientists, it means putting down on paper the details of his work in such a fashion that anyone properly trained can repeat his observations at any time in the future. **Repeatability** is the very heart of science, and without publication there would be no opportunity to repeat one another's work and scientific progress would be slow indeed.

It has also become traditional that scientific publications be open to the public. One might ask what purpose does open publication serve? When a great many men in all parts of the world are working on a common problem, the knowledge gained by each is available to all if open publication is practiced. This greatly speeds up the solution to problems and is largely responsible for the meteoric advance of science in the past 200 years.

WHAT SCIENCE HAS DONE AND CAN DO

Consider for a moment what science has done for people of the world, what it is doing today, and what it could or might do in the future. One example is the world population. Approximately 275 years ago there was a world population of some 600 millions; that number has increased five-fold to the present time and is now increasing at an unprecedented figure. This means that within this relatively short period there has been a tremendous increase in population compared to the preceding period of approximately 500,000 years which produced the number of people alive in 1700. What has been responsible for this sudden burst of reproductive powers in man? Certainly there has been no physical evolution in man himself in so short a time. It is generally agreed that food is one of the limiting factors in the growth of any population, be it fish or man. Populations always encroach upon the food supply just as closely as they can, often with widespread starvation. This increase in population, then, is due in part to increased food production; increased food production has come about through the application of the scientific method to food production problems. This

might be all changed again by a single important discovery, for example, an economical method of producing food from inorganic sources, such as sugar from carbon dioxide and water. A discovery such as this would change the entire food problem of the world overnight. Feeding the half-starved world of today is within the realm of possibility now, although if populations increase at the present rate, it is doubtful that even the best efforts of scientists can keep pace in supplying food.

In colonial days, the average life span of a man was under 40 years; today it is well over 65. It is not that men are any better physically today than they were then. It is due almost entirely to the progress made by science in understanding the cause and prevention of infectious diseases. All of the advancements that have been made in medicine have been accomplished through the agency of science. Before the scientific method was employed in the study of medicine, medical knowledge was largely based on superstition and mysticism. Since the discovery of the germ theory of disease, aseptic surgery, and anesthesia, many of man's ills have been partially or wholly conquered. There is no doubt that many of the infectious as well as organic diseases that plague man today will eventually be eliminated from civilized societies.

It seems clear that the application of the scientific method to the solution of man's physical betterment has been good and bad. It has lifted many burdens from his shoulders by simplifying the work essential for his physical needs; it has extended his average life span also, but, at the same time, has provided him with deadly weapons, such as the atomic and hydrogen bombs. For the first time he has an instrument within his grasp that can annihilate the whole of the civilized world as we know it. Such a situation in an uneasy world certainly is not good when considered from the point of view of survival of a race. Perhaps the application of the scientific method to man's social ills might have some of the success it has had in the alleviation of his physical ills.

SUBDIVISIONS OF ZOOLOGY

Botany and zoology are the two traditional branches of biology which encompass the study of all living things in the universe. Both botany and zoology have been arbitrarily divided into various divisions or disciplines. Since we are dealing only with animals in this book, let us consider the subdivisions of this great branch of biology. For convenience these various subdivisions can be regarded as falling into two principal groups, each inextricably linked to the other. The first centers around the organism, that is, it is **organism-oriented**; the second is oriented toward the **approach** taken to studying the subject matter. For example, **protozoology** is the study of protozoa and **entomology** is the study of insects. Zoologists working in these fields are oriented toward the study of a group of organisms. On the other hand, the approach-oriented zoologist may study physiology, for example, which involves function. He may study the physiology of digestion in many different groups of animals or he may investigate the endocrine glands (**endocrinology**) of only one, man, for example. The principal disciplines or subdivisions of zoology are summarized in Table 1–1.

The science of zoology is so vast and so specialized that a zoologist may spend his

TABLE 1–1.   Subdivisions of Zoology

ORGANISM-ORIENTED	
1. Protozoology: single-celled animals	5. Herpetology: reptiles
2. Malacology: mollusks	6. Ornithology: birds
3. Entomology: insects	7. Mammalogy: mammals
4. Ichthyology: fishes	8. Anthropology: man

APPROACH-ORIENTED
1. Morphology: structure and form
a. Anatomy: gross structures visible to the naked eye
b. Embryology: formation and development of embryos
c. Microanatomy: visible only under the microscope
1. Histology: tissue and organ structure
2. Cytology: cell structure
2. Physiology: functional mechanisms
a. Cellular: function of single cells
b. Comparative: similarities and differences of mechanisms between animals
c. Endocrinology: endocrine glands and their secretions
d. Neurophysiology: nervous systems
3. Animal behavior: response of animals to one another and their environment
4. Biochemistry and Biophysics: structure and function at the molecular level
5. Genetics: mechanisms of inheritance
6. Evolution: origin and history
7. Taxonomy: evolutionary relationships as they relate to classification
8. Ecology: interrelationships of plants and animals to each other and their nonliving environment.
9. Parasitology: animal parasites in health and disease
10. Zoogeography: distribution of animals
11. Paleontology: fossil animals

entire life effort on one small phase of the subject. For example, the **entomologist** concentrates on insects and may be fairly well acquainted with a large number of them, but a **dipterologist** is an expert on flies, one order of insects. Likewise, a **physiologist** may be familiar with the general functioning of many organ systems in a large number of animals, but a **neurophysiologist** is an expert on coordinating mechanisms, indeed, he may be a specialist on the function of the nerves of invertebrates and may be known as an **invertebrate neurophysiologist.** It is obvious that as the frontiers of zoology are pushed back, each person who concentrates in a discipline becomes extremely narrow in his research. This is essential because so much information has accumulated in all areas of zoology that individuals can attain near-perfection in only a very limited area of their science. This is true of all science.

Where are the frontiers of zoology today? During the past few decades the frontiers have been pushed farther and farther toward the molecular level. Whole animals are still being used in laboratories all over the world, and a great deal of information is coming from these studies, but for a complete understanding of the way an animal functions, zoologists have been led to the chemicals of which the animal is composed. At this point zoology merges with chemistry and, as a result, a young discipline has come into existence: **biochemistry.** It is already becoming clear that the next frontier will involve atoms of which the chemicals are composed. The infant science of **biophysics** has already emerged and is making great strides in the field of zoology. Obviously, in order to understand the universe we must seek the answers at the most elementary level and that inevitably must be in the atom of which the universe is made. How atoms are put together to form molecules which, in turn, form animals will be the task for biologists, chemists, and physicists in the future.

SUGGESTED SUPPLEMENTARY READINGS

Books

*Beveridge, W. I. B., *The Art of Scientific Investigation.* New York: Norton, 1957.

Cohen, I. B., *Science, Servant of Man.* Boston: Little, Brown, 1948.

Conant, J. B., *Modern Science and Modern Man.* New York: Columbia University Press, 1952.

Fulton, J., *Selected Readings in the History of Physiology.* Springfield, Ill.: Charles C. Thomas, 1930.

*Gabriel, M. L., and Fogel, S., *Great Experiments in Biology.* Englewood Cliffs, N.J.: Prentice-Hall, 1955.

Gardner, E. J., *History of Biology.* Minneapolis, Minn.: Burgess, 1965.

*Hall, T. S., *A Source Book in Animal Biology.* New York: McGraw-Hill, 1951.

Harvey, W., *Anatomical Studies on the Motion of the Heart and Blood,* trans. by C. D. Leake. Springfield, Ill.: Charles C. Thomas, 1931.

*Johnson, W. H., and Steere, W. C., *This is Life.* New York: Holt, Rinehart & Winston, 1962.

Nordenskiold, E., *The History of Biology.* New York: Alfred A. Knopf, 1933.

Singer, C., *A History of Biology.* New York: Schuman, 1950.

* Available in paperback.

Articles

Butterfield, H., "The Scientific Revolution." *Scientific American,* September, 1960.

Terman, L. M., "Are Scientists Different?" *Scientific American,* January, 1955.

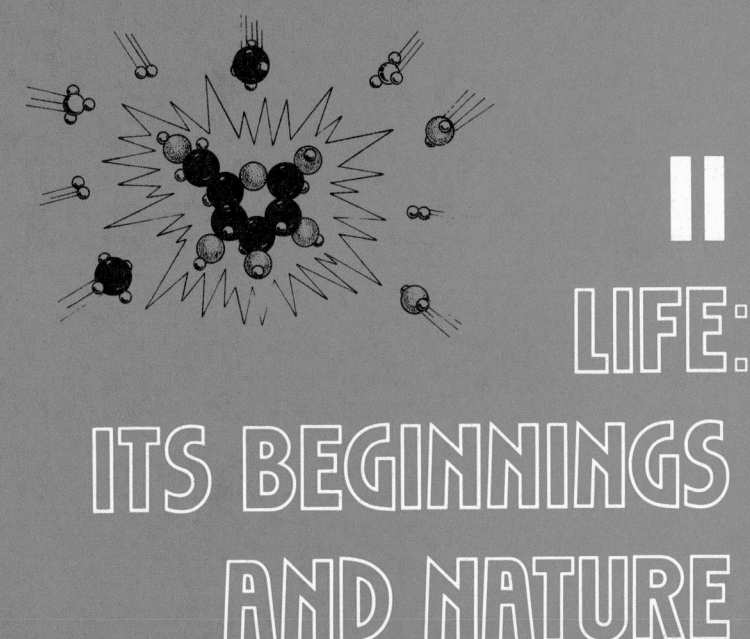

II

LIFE:
ITS BEGINNINGS
AND NATURE

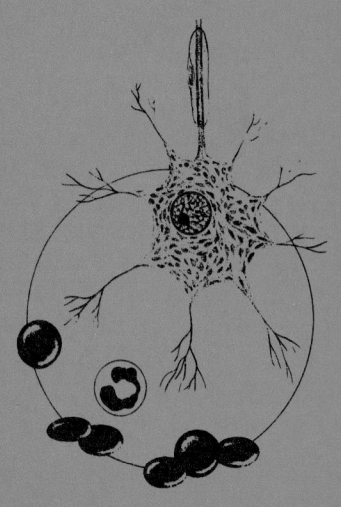

2

EARLY HISTORY OF LIFE

There is general agreement among biologists that life originated on the earth and that all of the elements that compose living things have come from the earth and its enveloping atmosphere. Speculations as to how this came about is considered a little later, but at this point attention is given to the components of the earth and the forces operating which, together, have made life possible. The crust of the earth, where all living things exist, is made up of many elements and compounds, some of which have become an integral part of all living things. The water and the gases which occur abundantly at the surface of the earth are also important, not only in the composition of organisms, but also in influencing their activity. All of the nonliving world obeys certain physical laws, most of which are well known, and these laws also operate within living things. It is important, then, for us to understand how the physical world is put together before we attempt to learn something about the far more complex living objects that exist in such great profusion at its surface.

ORIGIN OF THE EARTH

Astronomers generally agree that our planet was born at least five billion years ago, probably not long after the entire universe came into existence, possibly as a result of one terrific explosion. Along with millions of other solar systems, our particular sun was formed out of a very large, flattened, and highly rarefied mass of gas and dust. Smaller masses of these gases separated from the sun and became the planets which rotated about it. As the rotation proceeded, with the regions nearest the sun moving faster than the outer regions, whirling eddies must have formed in different parts of this tenuous nebula. Although most of the gases and dust probably escaped the solar system, the larger and denser of these whirlpools were able to coalesce by the forces of gravitation and to attract additional matter into them (Fig. 2–1). After some millions of years, these planetary nuclei had grown slowly and steadily by accretion (addition of particles from the outside) into the solid planets that we observe today.

Planet Earth, like all planets and the sun as well, probably started as a glowing mass of various gases, the most abundant of which was hydrogen, the lightest of all atoms. All of the gases existed only in the form of atoms, owing to the tremendously high temperatures that they had attained when the sun and planets were first formed. The sun today is still primarily atomic hydrogen gas

and is extremely hot. As a result of gravitational forces the heavier atoms moved toward the center of the sphere and the lighter ones formed shells at succeeding distances from the center. Geophysicists today believe the earth to be composed of three main concentric spheres (Fig. 2–1). The core consists of molten iron, with some nickel, of about 4,000 miles in diameter. This is surrounded by a shell, known as the "mantle" which is 2,000 miles thick and reaches almost to the surface. It is composed of an iron-magnesium silicate, a heavy green crystalline substance, and even today remains hot throughout. The next solid layer is very thin (10–30 miles in thickness), about the same proportion as the skin of the apple is to the fruit, but this is the only part of the earth with which man has some first-hand information. Its inner

Fig. 2–1. Planet Earth is one of several planets formed by the condensation of dust clouds encircling the sun. Later it cooled and became a rigid ball. Its inner structure today may be revealed by cutting an imaginary section as indicated. The outer shell is 10–30 miles thick and the mantle about 2,000 miles. The core consists of molten iron.

Iron core

Mantle

Outer shell
basalt and
granite

portion is composed of basalt (black rock) which provides the floors of our deepest oceans, and the surface consists of granite. The density of these materials decreases as the surface is approached, which means that our continents are essentially floating on the material below.

Even before the earth became a solid sphere, the lightest atoms, such as hydrogen, nitrogen, oxygen, and carbon, remained at the greatest distance from the center. With gradual cooling these atoms reacted among themselves to form molecules. Since hydrogen is the most active it was more likely to combine with the others than they were with one another. Consequently, the first atmosphere was probably composed primarily of water (H_2O), ammonia (NH_3), and methane (CH_4). The energy which was necessary to cause atoms to combine to form molecules probably came as radiant energy from the sun, although the earth itself was still sufficiently hot to provide part of the energy for this fusion.

The early atmosphere probably consisted primarily of water in the form of steam which existed as a thick gaseous layer covering the entire planet. As the steam pushed upward into the cooler regions, condensation occurred, causing dense fog and torrential rains. Eventually, after still further cooling, the raindrops penetrated the heat and reached the hot rocks below, only to evaporate as steam once more. Finally, however, with further cooling, water stayed in the depressions of the earth's extremely wrinkled surface, forming the infant oceans of the world. Hot rivers formed, of course, which were forever rushing to fill the oceans, carrying with them any minerals of the substratum that would dissolve. These substances were deposited in the young oceans, resulting in a constant increase in the chemical composition there. Water evaporated from their great surfaces, just as it does today, leaving the heavier salt particles behind and thus continually raising their total salt content. That is why the ocean water of today is salty. The importance of these great bodies of water lies in the fact that undoubtedly life originated in them sometime during this early period.

The physical and chemical forces operating in the atmosphere and oceans at this early period made it possible for life to appear later. These warm seas contained an abundance of minerals, but more important they contained dissolved ammonia and methane, two key compounds that, together with water, were the starting molecules for the first living things.

Here, then, we see a spinning new world, sufficiently far from the sun to be warm in most of its parts and fixed in an orbit that insures evenly spaced seasons. Most of its surface is covered with water and that which is exposed consists of rocks, gravel, and sand. No soil has been produced at this early period. A homogeneous gas, rich in nitrogen, envelops its surface, with clouds of water vapor moving slowly with the air currents. This is the setting in which life started, and once started, it spread over most of the earth's surface, becoming ever more diversified as time passed. Among the tremendous numbers of living things that arose, only one, man, evolved who can contemplate his own origin.

THE PHYSICAL NATURE OF THE EARTH

To understand the origin and nature of life, it is essential that we discuss its chemical composition and the physical laws that operate in the universe. Knowledge of these must precede any discussion of the beginning and subsequent development of life because all living things were derived from the inanimate world, and to study living things without this knowledge would be meaningless. The following discussion involves only the most elementary information in the fields of chemistry and physics. For some students this will be a review, and for others it may be the first contact with these fields. In the latter case some extra effort will be required to master the elementary principles set forth in the following pages. For both groups it is important that these principles be mastered before biological principles are considered.

Matter

Everything in the universe is composed of **matter,** which occupies **space** and has **mass.** Mass may be defined as the amount of matter in a body which can be measured in terms of resistance to change of motion. This is more meaningful when it is considered in the light of attraction between bodies. Bodies attract one another according to their respective masses. For example, the earth has a greater attraction (gravity) for a man than does the moon because the earth has greater mass. We have a convenient method of measuring this attractive force between bodies; we call it **weight.** Weight, of course, changes, depending on where it is taken. A person weighing 130 pounds on the earth would weigh 30 pounds on the moon, although his mass would be the same. Weight simply measures the pull of **gravity.**

Animal bodies are constructed to compensate for the pull of gravity. For example, small animals, such as mice, have relatively light skeletons in comparison to their weights, whereas larger animals, such as the elephant, have much heavier skeletons with respect to weight. This fact limits the size of animals, for should they go on increasing in bulk, they would reach a point where the skeleton alone could not bear its own weight. If life, as we know it, occurs on other planets, it too would show relation to weight. Animals on Jupiter, for example, would have to be constructed on an entirely different plan from those on the earth, because the pull of gravity is so much greater. They would probably be heavily boned animals and greatly flattened.

One cannot mention the motion of matter without referring to another force that operates on bodies, namely, **inertia.** When a swift elevator starts up, one is conscious of a sudden increase in weight; likewise, when it comes to rest, one seems suddenly and momentarily lighter than usual. There may be a simultaneous peculiar feeling in the mid-section as the internal organs respond to the effects of inertia. This force is the resistance of a body to change in its rate of motion. If standing still, it resists movement; if moving at a certain rate of speed it resists any change in this rate. That is why we need low gear and good brakes on our cars—it requires more power to get started or to stop than to keep rolling.

Most animals are little affected by inertia except when they are suddenly stopped, such as when a bird flies into the side of a building. The sudden cessation of forward motion can be fatal to bird or man, as attested by car accidents. Another modern machine that brings the effects of inertia into prominence is the airplane. Pilots often "black out" because their forward motion is suddenly changed, as in coming out of a power dive. The blood, being fluid, tends to follow the forward motion it has attained and is thus pulled away from the head, causing the "black-out" or faint. The normal movements of animals are little affected by inertia, and the problem arises in man only when he steps into one of his mechanical contrivances which carry his body faster than it was made to go.

Surface Phenomena

The behavior of matter depends to a large extent on its surface area. Large bodies have smaller surface areas in proportion to volume than do small bodies. A mouse has more surface area in respect to its volume than does an elephant. This can easily be computed by simply measuring the surface area of a cube, say 10 inches on a side or 600 square inches, and then its volume, which is 1,000 cubic inches (Fig. 2–2). This is a ratio of 0.6 square inch of surface to every cubic inch of volume. If a cube one inch on a side is cut out of the larger cube, and the surface computed in respect to the volume, it is found that each inch cube will have 6 square inches of surface, a ten-fold increase. Since chemical reactions may take place at

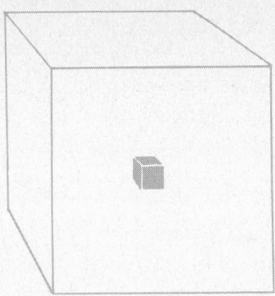

Fig. 2–2. Two cubes, one with dimensions one tenth of the other. In proportion to volume, the smaller has 10 times the surface area of the larger.

surfaces, it is obvious that the activity will be much greater as matter is divided into smaller and smaller particles. This is a very important physical property, as we see when we discuss enzyme action and many other activities that go on within the animal body. This principle is also important in body heat loss in animals. Large warm-blooded animals have little difficulty in securing sufficient food to maintain their constant body temperature; but very small animals, such as the shrews, must eat the equivalent of their own body weight in food each day in order to remain active and keep warm. The relatively large surface area in proportion to volume of these tiny creatures results in an extremely rapid heat loss. The shrew probably has reached the limit in small size for a mammal.

Particles of matter, especially small particles where the effect is more apparent, attract other particles of the same kind and also have an attraction for other particles of a different sort. The first is given the name **cohesion,** and the latter **adhesion.** Particles of matter such as water tend to cling together to the extent that they actually form a film at their surfaces. This gives the effect of a stretched membrane and is spoken of as **surface tension.** Water has a rather high surface tension, but mercury surpasses it by a considerable margin. It is well known that when mercury is dropped it breaks up into hundreds of small perfect spheres. Water gives a similar but lesser response when dropped on a dry, dusty surface. Rain drops are usually

near-spheres. Liquids, when free of external forces, assume the shape of a sphere because the cohesive forces of the particles of which they are made form a surface membrane. This fact makes it possible for certain species of insects to walk on the water (Fig. 2–3). Their bodies are light and the weight is distributed over a large area, so that the cohesive force of the water is sufficiently strong to keep the membrane intact.

Adhesive forces are simultaneously at work with cohesive forces. These have no significant influence on large bodies, but on small particles they become very important, just as important as gravity is to the larger bodies. Certain gaseous molecules, for example, will adhere to carbon particles so tenaciously that it requires high temperatures to remove them; this is the principle of the gas mask and certain purification processes. It is important in biological systems where enzymes, for example, adhere or **adsorb** (the process is called adsorption) to food particles, thus enhancing digestion. This physical property of matter manifests itself in many ways in the bodies of animals.

Composition of Matter

The ancient Greeks conceived of the universe as being composed of four **elements:**

Fig. 2–3. Some insects, like the water strider, can walk on water because their evenly spread weight is not sufficiently great to break the surface tension.

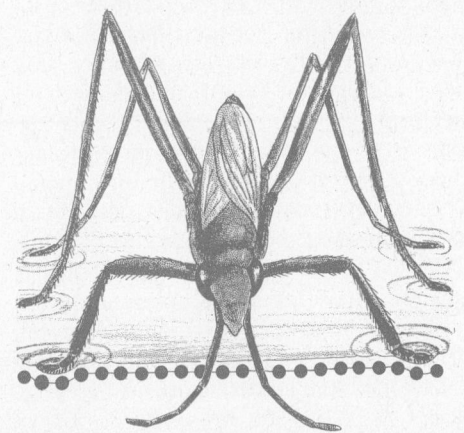

fire, air, water, and earth. This idea persisted until the eighteenth century when it was shown that water could be separated into two components, hydrogen and oxygen, both gases and neither one resembling water. Likewise air was found to be composed of several gases, the two most important being nitrogen and oxygen. From these experiments it was obvious that the meaning of "element" had to change to something less than originally thought. The term still persists today and is defined as the simplest form of matter that possesses specific chemical properties. An element cannot be decomposed into simpler particles with unique "chemical" properties. However, the reduction of elements to lesser particles has been accomplished, but that is discussed later.

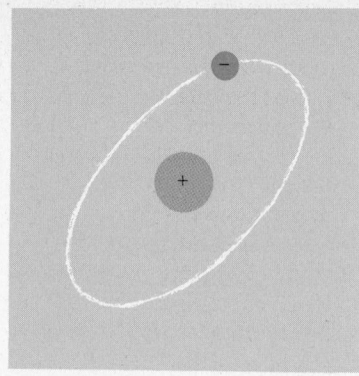

Fig. 2–4. The possible structure of the hydrogen atom. The proton with its positive charge lies at the center and the negatively charged electron revolves about it.

Atoms. The universe is composed of 92 naturally occurring elements, although scientists have produced others in the laboratory. The **atom** is the smallest unit of matter that possesses all of the chemical properties of an element. Physicists have been occupied in recent years studying the structure of atoms; consequently we know a great deal about these units. The atom is composed of still smaller particles, units which have been weighed, counted, checked for speed, and measured for their electrical properties. The significance of atomic structure came to the attention of everyone when physicists were able to manipulate atoms in such a manner as to have them release tremendous amounts of energy in the form of atomic explosions.

One might think of the atom as a miniature solar system with its relatively large heavy **nucleus**, the "sun," and its revolving **electrons**, the "planets" (Fig. 2–4). Reminiscent of the universe, the most striking characteristic of an atom is the vast amount of space between the various particles of which it is composed. If we imagine the atom to be the size of a balloon 100 feet in diameter, it would appear as a hazy, transparent sphere. At the center would appear a nucleus, the only clearly visible part, about the size of a small marble. The electrons would be invis-

ible, but because of their terrific speed they would appear as a dim blur, giving a vague limit to the entire structure. In some atoms there would be other concentric rings within the outer shell and these would appear as hazy as the outer shell; they might be intertwined with one another. Only the speed of the electrons would make these limits apparent, because the electrons themselves seem to have very little, if any, mass. They are "intangible units of energy," whirling at unimaginable speeds, yet maintaining remarkable stability. Atoms can be "smashed," as we all know, if unbelievable amounts of energy are directed at them. As a result of "smashing," the nature of the atom itself changes.

The nucleus of the atom consists of a dense cluster of positively charged particles, **protons,** and uncharged particles, **neutrons** (Fig. 2–5). These are held together by some unexplained intra-nuclear force. The protons and neutrons make up almost all of the mass of the atom. For every positively charged particle (proton) in the nucleus there is an **electron,** which carries a negative charge, in one of the orbits. Thus the total atom is neutral, that is, it carries no apparent charge. It is interesting to note that all of the 92 elements are made up of these units, differing one from another only because of the relative number and arrangements of protons, neutrons, and electrons.

A specific atom behaves the way it does because of the number of protons and neutrons in its nucleus and the number of electrons in its orbits. While there is always the same number of protons in the nucleus as there are electrons in the orbits, there may be a varying number of neutrons present in any specific atom. Since the chemical characteristics of the atom are controlled by the electrons in the outer shell or orbit, and since the number of electrons is controlled by the number of protons at the center, any additional neutrons will be without effect on the chemical properties of the atom. The only difference will be in its weight. Physicists have found that they can add or knock out neutrons as well as electrons, and thus change the physical properties of the atom itself. When only the number of neutrons is changed the resulting atom is called an **isotope;** isotopes have the same **chemical properties** as the naturally occurring atom but they have different weights, and therefore can be identified or "tagged." Tagging atoms has made it possible to trace various chemicals as they pass through the animal body. This has been very helpful in determining what happens to certain substances in normal life processes. If the isotope happens to be radioactive, that is, if it happens to give off radiations that can be detected with a sensitive instrument (Geiger counter) or by some other means, then the problem of tracing the chemical becomes less difficult. We are gaining a great deal of knowledge today from this type of so-called tracer research, and the future holds out much promise in this field of investigation.

It would seem simple, then, to arrange all of the various elements in a series from 1 to 104, according to the number of electrons (or protons) in each individual atom. This has been done and we call such an arrangement the **periodic table.** It is possible to diagram this atomic sequence (Fig. 2–6). The **atomic number** corresponds to the number

of electrons in the orbits, or the number of protons in the nucleus, limited in range, of course, from 1 to 104. The **atomic weights** are arbitrary figures assigned to each atom, and they depend on the number of protons and neutrons in the nucleus. In the past oxygen was used as the base. However, in 1961 physical scientists agreed to use the carbon atom ($_6C^{12}$ isotope) as a reference standard. The atomic weight arbitrarily assigned to carbon is 12.000 which makes most of the other elements essentially whole numbers (H = 1.008). The atomic weight of any atom is thus defined as the relative weight compared with that of carbon (12). Therefore, the weight of a neutron or proton is one-twelfth the mass of carbon. Since a single atom cannot be weighed, these figures must be based on measurements that include a great many individual atoms, and hence are not absolute figures.

There are seven possible concentric shells of electrons among all of the atoms, the one closest to the nucleus containing two electrons, and each of those beyond having varying numbers. Hydrogen has one proton and one electron but contains no neutrons. It has the atomic number 1. Helium, the next in the series, has two electrons and, therefore, the atomic number is 2. However, it has an atomic weight of four, because it also has two neutrons in its nucleus along with the two protons. Lithium starts a new outer orbit containing one electron; thus, with its two inner electrons and the one in the outer

Fig. 2–5. A hypothetical explanation of how helium is formed by the combination of two neutrons with two hydrogen atoms.

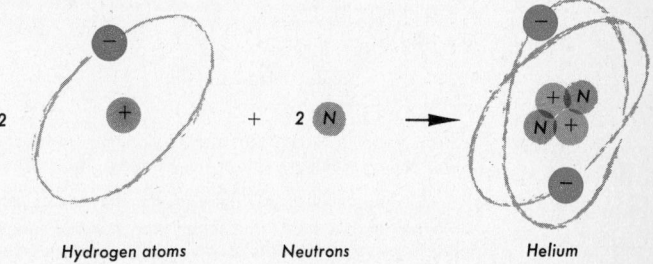

Hydrogen atoms   Neutrons   Helium

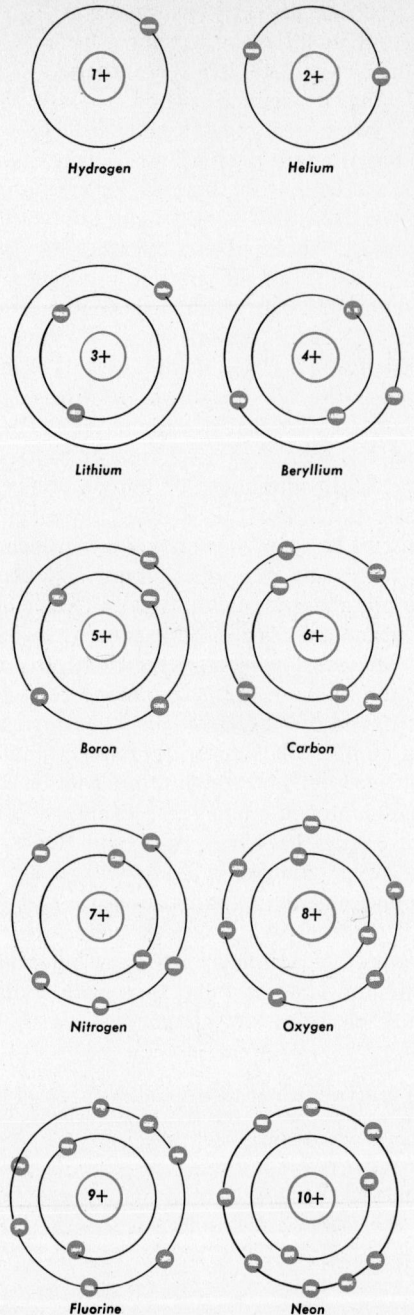

Hydrogen 1+	Helium 2+
Lithium 3+	Beryllium 4+
Boron 5+	Carbon 6+
Nitrogen 7+	Oxygen 8+
Fluorine 9+	Neon 10+

Fig. 2–6. A small portion of the Periodic Table. The number at the center (nucleus) indicates the atomic number. The electrons lie in the orbits, of which only two are shown here. The inner orbit requires only two electrons while the second needs eight, which is satisfied in neon. Of the ten elements indicated, hydrogen, carbon, nitrogen, and oxygen are important constituents of living material.

orbit, it has three altogether, giving it the atomic number 3. Sodium (not shown in Fig. 2–6) is the first atom that has a third ring, with 11 electrons in all. Usually the inner ring must be completed before the next one is formed. Following this principle the numbers of protons, neutrons, and electrons of nearly all 104 elements have been determined.

The concentric orbiting shells of electrons are visualized as an electron cloud. They represent different energy levels which increase progressively from the positively charged nucleus outward. Those in the outermost shell have the greatest energy and are least attracted to the nucleus. The ones between have progressively less energy and are more attracted to the nucleus. This is why the outermost electrons of atoms are most active in forming chemical bonds with other atoms initiating chemical reactions.

The inner shell of electrons is called the K shell; others progressing outward are identified by the letters L, M, N, O, P, and Q. Some heavier elements have all seven shells; others have fewer. The K shell can have only two electrons while all others can have a maximum of eight (there are exceptions).

When the number of electrons in the outer shell is less than half the total number it can hold, it may lose them, or if it has more than half, it may gain others to complete the shell. Any change in these numbers of electrons changes the electrical nature of the atom; if it gains electrons it becomes negative, and if it loses electrons it becomes positive. Whenever the atom is out of balance in respect to its electrons it is an **ion;** if it possesses an excess of electrons it is called an **anion** (because if placed in an electrical field it will move to the positive pole, the anode); if it has lost electrons it is known as a **cation** (because it moves to the negative pole, the cathode). A solution containing ions will conduct an electric current and the compound producing them is called an **electrolyte.**

Chemical Properties of Atoms. Atoms possess their special chemical properties depending on the number of electrons in their outermost shell. There is a special chemical stability of those atoms whose outermost shell contains eight electrons (two exceptions are hydrogen and helium which have only the K shell). It seems that when there are eight electrons in the outermost shell, the atom rarely undergoes a chemical reaction because the electrons do not interact with the outer electrons of other atoms to form chemical bonds. Hydrogen is a notable exception because it is highly reactive—the reason being that it has but one electron in its K shell. It readily gives up its electron (thus becoming a proton or hydrogen ion) or it may gain an electron by sharing one with another atom such as carbon, oxygen, or nitrogen, thus forming a large variety of compounds. Helium, on the other hand, has two electrons in its K shell which renders it extremely stable; it rarely reacts with other atoms. The five **noble gases** (neon, argon, krypton, xenon, and radon) are also inert atoms because they possess a full complement of eight electrons in their outermost shell. All other atoms react with one another.

Atoms have a tendency to react with one another in such a manner as to complete their outermost shell with eight electrons (two in the case of hydrogen). Each atom has a certain number of bonds which is a characteristic of the atom. The number depends on the number of electrons it must gain or lose to fulfill the eight in the outermost shell (two in the case of hydrogen). Hydrogen has one chemical bond, oxygen two, carbon four, and so on.

Of all the atoms in living things, carbon certainly is the most important. This may be due in part to its physical make-up. It possesses just one half the maximum number of electrons in its outer shell, which means that it does not lose or gain electrons; it unites with a large variety of other atoms by simply sharing its electrons. This arrangement permits combinations with other carbon atoms to produce long chains or rings which may then join up with a large variety of other atoms to produce immense molecules. It was undoubtedly this nature of car-

bon that made it the central atom around which life was built. We find it an integral part of all biological systems and playing important roles, not only in the construction of living materials, but also in storing and releasing energy which is essential in life processes.

Molecules. Atoms almost never exist singly in nature; they are usually found in combination with other similar atoms or with different atoms. When such a combination of like or unlike atoms occurs a **molecule** is formed. For example, hydrogen in the gaseous state consists of two identical atoms, and hence is designated as H_2; likewise, the gas produced by living things called carbon dioxide is composed of one atom of carbon and two of oxygen, and hence is designated as CO_2. Molecules that are composed of two or more different atoms are called **compounds.**

It is useful in studying compounds in living things to be able to express them quantitatively. This can be done in very precise terms. As mentioned earlier, the atomic weight is the sum of the masses of the protons and neutrons. When this number is expressed in grams, it is referred to as the **gram atomic weight.** The number of atoms in a gram atomic weight is 6.02×10^{23}, a fantastically large number. The usefulness of this number becomes apparent when we learn that the atomic weights of all elements have exactly the same number of atoms. For instance, one gram of hydrogen has the same number of atoms as does 16 grams of oxygen. This information is useful in computing the composition of compounds. For example, we know that in CO_2 one atom of carbon combines with 2 atoms of oxygen. Both carbon and oxygen have the same number of atoms but since their gram atomic weights are 12 for carbon and 16 for oxygen, it will require 32 grams of oxygen to react with 12 grams of carbon. Similarly, the **molecular weight** is the sum of the atomic weights of the atoms in a molecule. When this is expressed in grams, it is referred to as the **gram molecular weight,** or **mole.** This means that the number of molecules in a mole of any compound is equal to the number of atoms in a gram atomic weight of that compound; for example, 32 grams of oxygen has the same number of molecules as does 44 grams of CO_2 or 18 grams of water. The value in this observation is that when one compound reacts with another in a ratio of one to one, it becomes possible to calculate the weights of the reacting compounds. Consider the reaction that takes place when carbon dioxide (CO_2) is bubbled through water (H_2O), forming carbonic acid (H_2CO_3). Chemists have devised a kind of shorthand to express this reaction:

$$H_2O + CO_2 \longrightarrow H_2CO_3$$

A reaction expressed in this form is called an equation because the compounds on one side of the arrow are equal to those on the other side. This equation can be read to mean that one mole of water reacts with one mole of carbon dioxide to give one mole of carbonic acid. By substituting gram molecular weights the equation reads: 18 grams of water react with 44 grams of carbon dioxide to produce 62 grams of carbonic acid. The usefulness of this information is obvious when certain unknowns are present in the equation because one can compute their exact quantities. Reactions always occur in whole-number proportions of moles; that is 1:1 or 1:2, etc., regardless of the molecular weights of the compounds.

Under certain conditions molecules tend to dissociate into their constituent ions. For example, in water the electrons shared by hydrogen and oxygen are more firmly attached to the oxygen atom than the hydrogen (Fig. 2–7). Because of the high rate of velocity of their spin, there is a slight tendency for the molecule to fall apart. As a result the nucleus (proton) of one hydrogen atom separates leaving the remainder of the molecule which consists of an oxygen atom, a hydrogen atom, and the surplus electron. This is indicated as follows:

$$H_2O \rightleftharpoons H^+ + OH^-$$

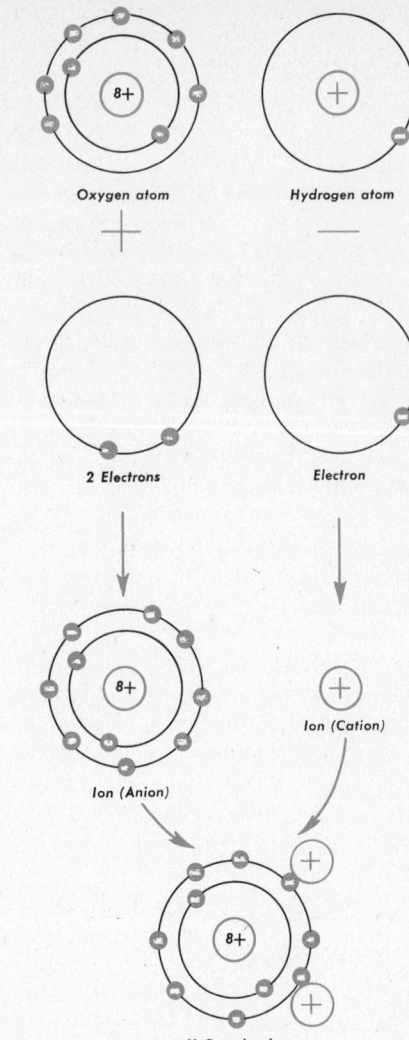

Fig. 2–7. A possible explanation for the formation of a molecule of water from two atoms of hydrogen and one of oxygen. Note that the hydrogen ion is a naked proton, and the oxygen ion is formed by the addition of two electrons. The union of the two kinds of ions produces the molecule of water.

These products are referred to as the hydrogen (H^+) and the **hydroxyl** (OH^-) **ions.** The hydrogen ion is simply a proton. Both ions wander about in the water independent of one another. Water dissociates into its ions only very slightly (at any one time in pure water, one molecule out of 200 million) and since they are always equal in number, any effect they might have goes unnoticed. Since all living things are composed of large quantities of water one might expect that any unequal quantities of these two ions might be detrimental, which is the case.

In some hydrogen-containing compounds there is a much greater tendency for the protons of hydrogen to separate from the remainder of the molecule than is the case with water. Such ionized compounds are called **acids,** and they are classified as strong or weak depending on the percentage of protons produced. Sulfuric (H_2SO_4), nitric (HNO_3), and hydrochloric (HCl) are strong acids because each molecule yields a proton since they dissociate completely in water. In weak acids such as acetic acid (CH_3COOH), only 0.1 per cent of the molecules dissociate.

Hydroxyl ions are likewise produced when compounds which contain them are dissolved in water. Such compounds are called **bases** and they react with acids to produce **salts.** For example, sodium hydroxide (NaOH) when dissolved in water dissociates completely to form sodium (Na^+) and hydroxyl (OH^-) ions. If these are then mixed with hydrochloric acid the hydrogen and hydroxyl ions react to form water and the sodium and chloride ions remain in solution. If the water is removed by evaporation salt crystals (NaCl) will form.

Living things will not tolerate even small excesses of either hydrogen or hydroxyl ions. This is understandable since they are composed primarily of water which contains very few of these ions; those that are present exist in equal numbers. For this reason biologists have found it necessary to find some method of detecting these ions in a very precise manner. This is done by several methods, the most precise being to measure the electrical conductivity of the solution. Such an instrument is called a pH (hydrogen potential) meter, which measures the number of hydrogen ions (protons) present at any specific time in solution. An arbitrary pH scale ex-

tends from 0 to 14, in which a pH of 7.0 means that the numbers of H^+ and OH^- ions are equal, as they are in pure water. When the H^+ ions are in excess the solution is said to be acid and is indicated by (pH) numbers less than 7.0; numbers higher than 7.0 indicate a basic solution, that is, a solution in which the OH^- ions are in excess. An indication of the importance of these figures becomes clear when we note that human blood has a pH of 7.45; if it rises to 7.6 or drops to 7.2, death results.

Chemical Reactions. Recalling the structure of atoms, we know that atoms will unite to form molecules depending on the number of electrons they can deliver or consume in their ionic state. For example, the potassium atom has 19 electrons; when it loses one electron it becomes a potassium ion. The bromine atom lacks one electron in its outer shell; when this is gained it becomes a bromide ion, an anion. When the potassium and bromide ions are present in the same system, they are brought together because "unlike charges attract"; since bromine requires one electron to complete its outer shell, one potassium atom is required to do the job. The result is a molecule of potassium bromide, a stable compound. In general the reaction takes place between different atoms because certain combinations of electrons associated with specific atoms are more stable than others. Among the most common elements found in organisms, stability is accomplished when the number of electrons is two, ten, or eighteen. Those with fewer or greater are the most reactive because in acquiring or losing electrons, they become stable. Reactions always move in the direction to provide stable end products. For example, sodium with 11 electrons, tends to react with chlorine which has 17, the former having one more electron than the stable number (10) and the latter, one less (18). They react and satisfy the deficiency of each, producing a stable compound, sodium chloride (table salt). The sodium chloride is found to weigh exactly the same as the sodium and chlorine from which it was derived; hence, nothing was gained or lost. This is the **law of the conservation of matter.**

Chemical reactions also obey another rule, called the **law of the conservation of energy,** which means that neither mass nor energy can be created or destroyed. This can be observed by measuring the amount of heat that is liberated when hydrogen combines with oxygen to form water; indeed this may be an explosive reaction because both atoms are very unstable. In this reaction energy is released and when this happens it is called an **exergonic reaction.** On the other hand, if electrical energy is applied by means of electrodes (electrolysis) to water, the molecule can be degraded into its component parts, namely hydrogen and oxygen. However, this requires as much energy as was released when the gases combined to form water. This type of reaction is called an **endergonic reaction.** Since there is no loss or gain in energy in these reactions it is obvious that the law of conservation of energy holds.*

Free Energy and Entropy. The total energy possessed by a chemical substance is known as **free energy.** This is the sum of the heat content when the substance is completely oxidized, plus the energy tied up in its configuration, called **entropy.** Entropy is a state of randomness or orderliness of a system. As entropy increases, the system becomes more disordered or random and, conversely, as the system becomes more ordered, entropy decreases. Changes in entropy, there-

fore, relate to changes in orderliness or randomness. If left alone, all systems tend toward an increase in entropy or randomness which is the most probable state. Less probable is a decrease in entropy, that is, a more ordered system because it requires energy to reach this condition. Living things are highly ordered (less entropy); energy has been required to bring them to their present state.

Any system, if left unattended, spontaneously loses energy (does work) and becomes more random (increases in entropy). In other words, all animate and inanimate systems in the universe tend to run down toward the lowest useful energy state (greater entropy). This is the fundamental principle underlying the **second law of thermodynamics.** The only way it can be reversed is by utilizing free energy supplied by other systems. Living systems maintain their orderliness by energy ultimately coming from the sun. If such energy failed, all life would finally become disordered and death would result. The essence of the second law is that all systems tend to approach a state of equilibrium by a loss of useful energy as it is transformed from one kind to another. The energy becomes less useful with each downward step, ultimately being lost in the form of heat. Whereas this principle is utilized primarily by chemists, it is helpful in the understanding of living systems because they obey the same law.

One might ask, where is this energy? It is tied up in the molecule somewhere; that is, it is **potential energy** or **stored energy.** It is thought that the energy is stored or concentrated in the bonds between the atoms which make up the molecule. This is called **bond energy.** Any change in the linkages between atoms results in a loss (exergonic) or gain (endergonic) in energy. Exergonic reactions, when strong, can produce very high temperatures as in the hydrogen-oxygen torch. By mixing ignited hydrogen and oxygen continuously, a very hot flame is produced. Of course in living systems bond energy is utilized at a much lower temperature over long periods of time, as we discuss later.

---

* With the formulation by Einstein in 1906 of the famous equation $E = mc^2$, in which E is energy in ergs, m is mass in grams, and c is the speed of light in centimeters per second, it was agreed that matter and energy were different forms of the same thing and equivalent to each other. As a result, the Laws of Conservation of Energy and Mass were combined into a single general law known as the **First Law of Thermodynamics,** which means that mass–energy can neither be created nor destroyed. The universe consists of so much mass–energy and, whereas the proportions of both mass and energy may change, the total mass–energy cannot.

Whereas the law of conservation of energy pertains to all living systems, scientists in recent years have been able to convert matter to energy by changing the nuclei of atoms, which has been most spectacularly dramatized in atomic and hydrogen bombs. Since these reactions have little to do with living things (except to destroy them) we do not consider them in our discussion.

Activation Energy. The molecules in matter do not all have the same energy; that is, some have more energy than others. The sum total of all of them constitutes the average energy. Molecules react through collision with one another and those with the most energy (moving the fastest) collide more often than those with less energy (moving slower). It is only the energy-rich molecules that take part in a chemical reaction, and this number is usually small compared to the total, but this depends on other circumstances (temperature, etc.). This extra energy possessed by these molecules is called the **activation energy.** The amount of extra energy needed by reacting molecules is dependent on the particular type of reaction

and the nature of the reacting substances. Those molecules with less energy simply bounce off one another. In fast reactions more molecules possess the extra energy, whereas in slow reactions only a few are so endowed. It seems that the activation energy is that energy sufficient to bring molecules close enough together to overcome the repulsion caused by their electron clouds (like charges repel). Once this happens, the appropriate bonding occurs, producing the reaction products. There is an intermediate stage in this process, called the **activation complex,** which is responsible for quickly making the proper electron adjustments to produce the final compounds.

An analogy may be helpful in visualizing activation energy (Fig. 2–8). It requires a certain amount of energy, delivered by the boy, to pull the wagon up the hill, the initial barrier. This would be similar to the activation energy required to initiate a chemical reaction, and the summit of the hill would resemble the activation complex. Once over the barrier, the wagon rolls down the hill. The analogy can now be interpreted in chemical terms. In a hypothetical reaction, molecules a and b react to form molecules c and d, producing an intermediate ab, the activation complex. This can be written this way:

$$a + b \longrightarrow (ab) \longrightarrow c + d$$

The difference between the average energy of molecules a and b and (ab) is the activation energy. If the energy given off as a and b produce c and d, via (ab), is greater than the activation energy, energy will be released and the reaction is **exergonic.** If the difference of the energies is less than the activation energy, then an input of energy is required and the reaction is said to be **endergonic.**

Temperature greatly influences the rate of chemical reactions. As the temperature increases, not only do the molecules move faster, but they collide more frequently and a greater proportion possess the extra energy needed for chemical reactions to occur. More

Fig. 2–8. Activation energy. See text for explanation.

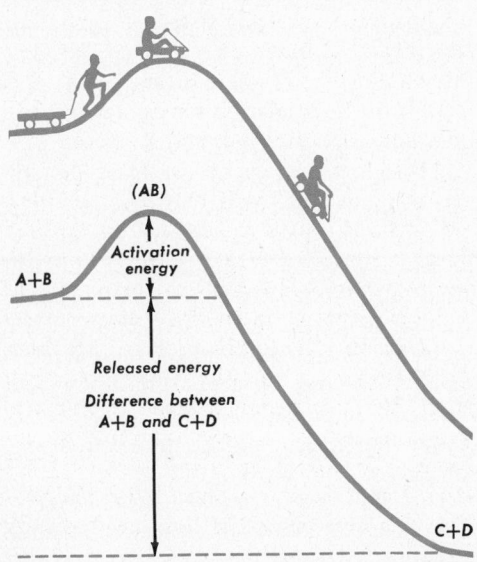

molecules, therefore, participate in the reactions which proceed at an increased rate.

Note that the conversion of water to hydrogen and oxygen, or the opposite reaction, the combination of hydrogen and oxygen to form water, is a so-called **reversible reaction**. This is theoretically true of all reactions. This concept then can be written symbolically:

$$A + B \rightleftharpoons C + D$$

in which A and B are the starting compounds and C and D the final products. The significant aspect of this concept is that energy is consumed or released depending upon which way the reaction goes. If energy is required for it to move to the right, the same amount of energy will be released when it goes to the left. This is an important principle with which we are confronted again and again in living systems.

ENERGY IN BIOLOGICAL SYSTEMS

In living things free energy is stored in the carbon-hydrogen bonds of compounds such as carbohydrates. As these compounds are transformed to lower energy carbon-oxygen compounds, energy is released. This chain of events is called respiration, a topic which we discuss in detail later. These high energy bonds arise in nature during the process of photo-synthesis in green plants. The sun's energy is utilized to convert the lower energy carbon-oxygen bond of carbon dioxide to the high energy carbon-hydrogen bond of molecules such as carbohydrates, fats, and proteins. Whereas plants obtain their energy directly from the sun, animals get theirs by utilizing the energy-rich molecules from plants or other animals. During respiration energy is always released; some of it is utilized in the business of living and some of it is wasted. Less energy is passed on than is received, as food is passed on from one organism to another. Consequently, the original energy stored in plants is gradually dissipated until it is finally lost in the form of useless heat. Thus, entropy increases in living things just as it does in the inanimate world. The pattern in living things is a struggle for free energy to decrease entropy. Once the sun burns itself out, which it will ultimately do, the source of energy will dry up and life will cease shortly thereafter.

Properties of Matter

Matter exists in space in one of the three forms: **solids, liquids,** or **gases.** Water is a convenient example of matter. Below freezing ($-0°C$) it exists as ice, a solid; between freezing and boiling ($0°C$ and $100°C$ at sea level), it exists as water, a liquid; above boiling ($+100°C$ at sea level) it exists as a vapor or steam, a gas. In all of these states the substance is still water, its chemical composition is unchanged. The particular state in which matter exists depends on the speed of the movement of the individual molecules. In the solid state they vibrate, but stay in fixed positions within the solid; in the liquid state they move faster and are free to move about within the liquid, whereas in the gaseous state they attain high speeds, so high that they exceed the intermolecular attractions and separate from one another to become independent free-moving bodies. The rate of molecular movement is reflected in the phenomenon of **temperature.** Molecules move faster in hot bodies than in cold ones. When there is no movement the body is as cold as it is possible to get, a condition called **absolute zero** ($-273°C$).

All specific chemical substances exhibit the same changes in state that water does, although they are not always so easy to observe. There are thousands of different kinds of matter, each composed of a specific kind of molecule, such as oxygen, hydrogen, or water. Since life is composed of matter, it must then also be composed of different kinds of molecules, and the properties of these molecules must be reflected in the properties of living things.

It is the various combinations of elements that make up all the thousands of chemical substances existing naturally or that can be produced in the laboratory. For example, glucose (blood sugar) is a complex molecule. It is written this way: $C_6H_{12}O_6$, which means that it is composed of 6 atoms of carbon, 12 of hydrogen, and 6 of oxygen. It is obvious that with 104 building units millions of combinations of atoms are conceivable. Many of these molecules appear in living things and it is necessary to know something about their individual behavior in order to have some understanding about their combined effect as it occurs in a cell. Chemists have been studying these for a long time, and their knowledge is so significant to the study of animals that today zoologists are dependent to a large extent on this information to aid them in solving some of their complex problems.

The arrangement of atoms in a molecule gives to that molecule its properties. Sugar is sugar because of the arrangement of the atoms in its molecule. If any of the atoms are removed or even changed in their respective positions within the molecule, the substance is no longer the same; the properties are different. For example, if the hydrogen and oxygen in the water molecule are separated, we no longer have water but two gases, neither of which acts like water in any way. When the molecules are all alike, we speak of the aggregate as a **substance**; if, however, there are several different kinds of molecules or substances present, we refer to the combined material as a **mixture**. A lump of sugar is a substance; when it is placed in a cup of coffee, the result is a mixture. The sugar exhibits specific properties which are always the same, whereas a mixture displays variable properties. Living things are composed of mixtures and therefore respond as mixtures. Mixtures are much more difficult to understand than substances, and because life exists in a mixture, a very complex mixture, it, likewise, is difficult to understand.

ORIGIN OF LIFE

Of the countless billions of stars, of which our sun is but one, it is reasonable to assume

that planets encircle many of them and some form of life may exist on a small percentage of these planets. It is certainly possible that the physical conditions that made life possible on planet Earth would be found on planets of other solar systems. It is also possible that planets exhibiting quite different environmental conditions than those on this planet could support some form of life, markedly different from that we know, but a self-reproducing and self-directing system which would qualify as a living being. The proof of the possibility of life on other planets may be forthcoming during the next few decades.

Once the earth reached a relatively stable condition, the stage was set for the beginning of this most remarkable drama, the inception and subsequent unfolding of life in all of its variety and complexity. It was once thought that life might have come to earth through interstellar space, originating from some other planet. This seems unlikely because of the intense heat it would endure en route. Life as we know it certainly would be destroyed by such high temperatures. Once a body enters the atmosphere at high speeds it burns to incandescence, thus destroying any life that might be inside. However, with continued success in the realm of space travel, a new interpretation of the origin of life on earth may be considered.

It is possible that life evolved to a very high state on other planets, resulting in organisms with very high intelligence who travelled freely in space. Such intellectuals could have visited our planet when it was young and inhospitable. Finding it uninhabitable they departed, leaving behind spores of certain forms of life which subsequently gave rise to all life that we see today through the process of evolution.

Whereas the above notion is certainly possible, let us speculate as to how life might have originated on the earth without reference to extraplanetary influence. Geologists tell us that there was a time about 2 billion years ago when conditions were such that life could have started. Those conditions do not exist today, so it is unlikely that life is being generated now. Assuming that life did take its inception a very long time ago on this planet, let us consider the most logical steps that could have taken place. Many scientists, among the first of whom were Oparin (a Russian) and Haldane (an Englishman), have given a great deal of thought to this problem and together have formulated a theory which is backed up by some experimental evidence. As plausible as their theory may seem, it is far from proven, and it is improbable that satisfactory proof will ever be forthcoming.

From our earlier discussion we learned that the first atmosphere contained large quantities of water (H_2O), ammonia (NH_3) and methane (CH_4) and that the young oceans contained the latter two gases as well as dissolved minerals. It was in these bodies of water that reactions among the numerous contained chemicals took place, from which the first living forms emerged.

Carbon Compounds. The versatile carbon atom (Fig. 2–6) was the element around which life was initiated. It has a bonding capacity of 4, as illustrated with methane:

$$
\begin{array}{c}
\text{H} \\
| \\
\text{H--C--H} \\
| \\
\text{H}
\end{array}
$$

Any one or all of the H atoms can be replaced with other atoms, for example:

| Methyl chloride | Chloroform | Carbon tetrachloride |

Obviously these are all radically different compounds. Other atoms may replace the H atoms, producing a large variety of compounds. Methane also may react with itself, forming short or long chains, or the end carbons of a chain may combine forming a ring:

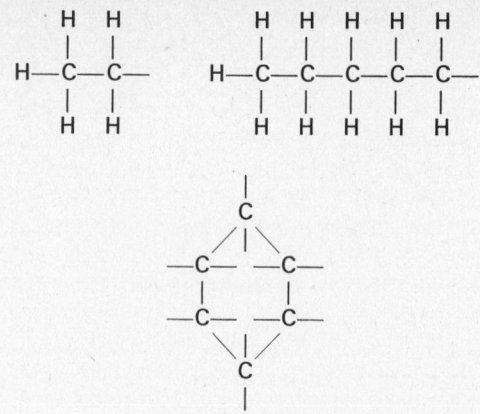

The chains may branch and be combined with other chains or with rings occupying three as well as two dimensions. The free bonds of these carbon skeletons combine with other atoms, thus making innumerable complex compounds each with its own distinctive properties.

The opportunity was present in the young oceans for these reactions to take place and we assume that they did. Methane was in great abundance at this time which made possible the linkage of carbon atoms to one another and to other substances, indeed such combinations still go on in living things. Such carbon-containing compounds occur almost exclusively in matter which either is alive or has been alive at one time. Consequently, such molecules are called **organic** compounds to distinguish them from **inorganic** compounds which contain no linked carbon atoms, such as metals, salts, water, and minerals. The nonliving world is composed of inorganic compounds, although some of them have become a part of living matter and play a very important part in the fabric of all living things.

Pre-Life Compounds. Whereas a great many compounds were undoubtedly formed in the early seas, a few became precursors to life. These can be grouped into four classes of compounds all of which are conspicuous in living things today. They probably were very simple at first but with time they became very complex as they formed the structure of living matter, which itself became complex.

Sugars and Polysaccharides. Perhaps some of the first and simplest reactions were those between methane and water, producing short carbon chains and involving only the atoms C, H, and O. Since the latter two came from water the ratio of 2 H's to 1 O was maintained. These compounds are called **carbohydrates,** two of the most familiar being **glucose** and **fructose.** They are composed of a chain of 6 carbons to which 12 atoms of hydrogen and 6 of oxygen are combined in the following configurations:

Glucose

Fructose

Most sugars (called **monosaccharides**) contain 6 carbons, although some very important ones have only 5 as we shall soon see. Moreover, sugars may combine with one another by the simple process of losing a molecule of water, which is called **condensation.** For example, a molecule of fructose and one of glucose can combine to form one molecule of **sucrose,** a **disaccharide** (double sugar) as follows:

$$C_6H_{12}O_6 + C_6H_{12}O_6 \xrightleftharpoons[+H_2O]{-H_2O} C_{12}H_{22}O_{11}$$

Fructose   Glucose   Sucrose

Incidentally, the sucrose molecule can fall apart by the simple addition of water. This

process is **hydrolysis** which, as we see later is important in digestion. Many glucose molecules, as well as other sugars, can combine in this fashion forming huge molecules, called **polysaccharides.** Some familiar polysaccharides are **starch** (plants) and **glycogen** (animals). These molecules are important as building materials and as sources of chemical energy for living matter.

Glycerol, Fatty Acids, and Fats. Several other compounds were formed from C, H, and O atoms. The simplest of these is **glycerol,** consisting of 3 carbons to which 8 hydrogens and 3 oxygens are attached. Another group of compounds are the **fatty acids** in which the carbon atoms are arranged in chains, from 2 to 18 or even more. There are many fatty acids and in all of them H atoms are attached to all of the carbon atoms except the last one to which 2 O atoms are attached, one of which is combined with another H atom. In water a hydrogen ion is freed which gives the molecule its acidic properties. Glycerol and a general fatty acid have the following configurations:

Glycerol

Fatty acid

Fatty acids and glycerol may have combined in the early oceans to form fats of various kinds. Fats are complex molecules which were used in the architecture of living material and served as excellent sources of energy. They possess less oxygen than the polysaccharides.

Glycerol and fatty acids combine to form **fats** by the loss of water, just as sugars do to form polysaccharides. Likewise the fat molecule can be degraded to fatty acids and glycerol by the addition of water.

$$C_3H_8O_3 + 3C_{18}H_{36}O_2 \underset{+H_2O}{\overset{-H_2O}{\rightleftarrows}}$$
$$C_{57}H_{110}O_6 + 3H_2O$$

Glycerol   Stearic acid   Fat

Amino Acids and Proteins. Another group of compounds formed which was composed of not only C, H, and O atoms but N atoms as well. These are called **amino acids.** They are identified by the presence of at least one NH_2 group, the **amino group,** which probably came from NH_3 (ammonia) by the loss of one H atom. These may be illustrated by the simplest one, glycine:

Glycine

Amino acids also combined in many ways forming the most highly complex of all compounds, the **proteins.** These molecules became the most important of all in the fabrication of living material. Today we find over 20 amino acids which make up the structure of all living things; hence in the early days of the earth at least this many were formed. Amino acids can combine in a large variety of patterns, forming an almost infinite number of proteins. Since they can possess thousands of amino acids, protein molecules reach great sizes, indeed they are the largest and most complex of all molecules. This is important because we see in the proteins a means by which a great variety of living matter can be constructed; hence through the building of protein molecules a system was initiated from the very beginning by which living things could be different from one another, and this they are.

Proteins are formed from amino acids in much the same way that polysaccharides and fats are formed, namely by the loss of water. However, the linking together of the amino acids is done in a special way which we discuss later (see page 44). For simplicity at this point we merely say:

$$\text{Amino acids} + \text{Amino acids} \underset{+H_2O}{\overset{-H_2O}{\rightleftarrows}} \text{Proteins}$$

Some special protein molecules which are an integral part of the process that brought about living matter must have come into existence at this early period. These we now call **enzymes** and we have much more to say about them later. At this point we need only consider the fact that in order for reactions to take place high temperatures are often required, temperatures that far exceed those compatible with life. If the reactions we have been talking about and many others occurred, some conditions other than heat must have been present. In the laboratory we can create necessary conditions by the use of **catalysts** which are substances that speed up reactions. Enzymes are such catalysts which operate in living systems; indeed these special proteins must have appeared as some of the very first proteins because they were essential in bringing about the multitude of reactions that were to follow in the synthesis of the first living things.

Purines, Pyrimidines, and Nucleotides. Other compounds were formed which contained molecular skeletons composed of rings; these were the **pyrimidines** with one ring and the **purines** with a double ring, thus:

Pyrimidine

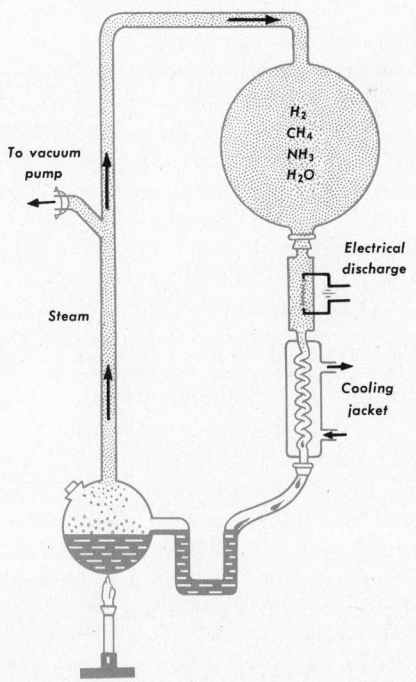

Purine

These compounds combined with two other substances, a sugar with 5 carbons and phosphate (PO_4), forming complexes called **nucleotides**. They may be designated as **pyrimidine-sugar-phosphate** and **purine-sugar-phosphate**. This tendency to combine continued with nucleotides uniting with one another forming highly varied and exceedingly complex macromolecules called **nucleic acids**. Finally, as you might expect, a pinnacle was reached in chemical synthesis when the tremendously complex nucleic acids united with the equally complex protein molecules to form **nucleoproteins**. We may indicate these steps as follows:

Pyrimidines + sugar + phosphate → nucleotides
Purines + sugar + phosphate → nucleotides
Nucleotides + nucleotides → nucleic acids
Nucleic acids + proteins → nucleoproteins

It was with the formation of the nucleoproteins that the probability of life was assured, because as we shall see later, they were sufficiently complex for a living system to be built around them. As plausible as this story so far may seem, it would be helpful if some experimental evidence could be found that would support it.

It would seem feasible to simulate in the laboratory the conditions that existed during early geological time and to attempt to find some of the molecules that were synthesized then. This Professor H. C. Urey and his student, S. L. Miller, did in 1953. They heated the gases ammonia, methane, and hydrogen in the presence of water and an electrical discharge simulating lightning which produces ultraviolet light (Fig. 2–9). After several weeks of this treatment a number of amino acids, as well as other organic compounds were recovered from the contained fluids. This simple experiment offers convincing evidence for the assumption that life could have been synthesized from simple compounds.

The energy source for synthesis was primarily ultraviolet light which could come both from the sun and from lightning which was violent as the earth was cooling (Fig. 2–10). Whereas in the experiment mentioned above, which took place in three weeks, only relatively simple compounds were formed, it is conceivable that if comparable conditions existed for millions of years a great variety of simple as well as very complex compounds would be formed. Indeed, since the Miller–Urey experiment scientists have been able to produce a large number of macromolecules, such as **polysaccharides**, **polypeptides**, and **nucleic acids**,

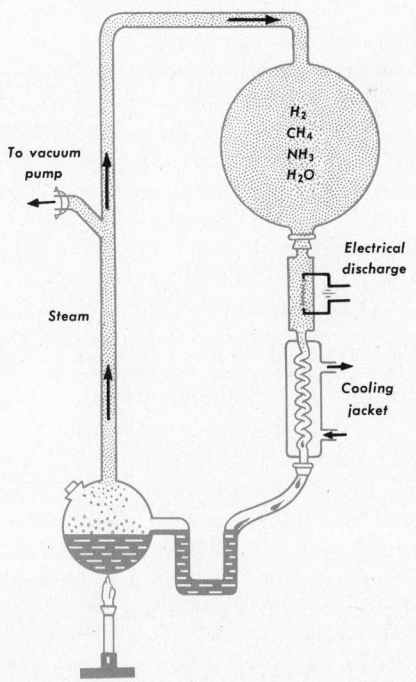

Fig. 2–9. The apparatus built by Urey and Miller to simulate conditions at the surface of the young earth.

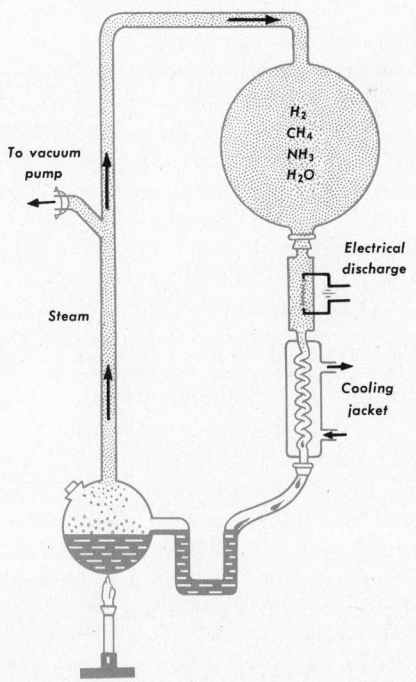

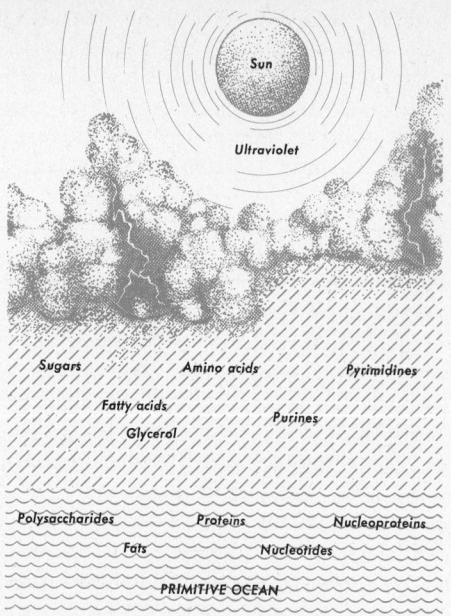

Fig. 2–10. Ultraviolet light from the sun, together with that produced by violent electrical storms, bought about the synthesis of complex molecules which accumulated in the primitive oceans.

starting with simple precursors and subjecting them to conditions similar to those in the Miller–Urey experiment. All of these macromolecules are found in living things today. These results from laboratory experiments offer some evidence for the assumption that life could have originated from simple initial compounds through a long series of chemical syntheses to the tremendously complex macromolecules we observe in living things today.

As time passed the source of energy gradually diminished owing to decreased electrical activity in the clouds and the screening effect of the atmosphere. Consequently, fewer and fewer organic molecules were formed, and a time was reached when there must have been complete cessation of chemical synthesis. This situation once reached might have stayed that way indefinitely had not the **nucleoprotein molecules** begun to reproduce themselves. Just how this is done we discuss in detail in a later chapter. Once this happened the simpler molecules,

such as sugars, amino acids, fatty acids, etc., served as raw materials in the manufacture of the larger nucleoprotein molecules. Today we would think of them as nutrients, the food essential for life. Because no more of these food molecules were being synthesized one would imagine that within a period of time the oceans would contain only nucleoproteins. This, however, was not the case, as we see later.

Nucleic acid and protein molecules are often very complex, consisting of thousands of atoms. We can guess that some of the simpler nucleic acids and proteins could have arisen by random associations, perhaps in combination with lipid-protein complexes which possessed membrane-like properties. Eventually the pattern which is now universal for life on earth was achieved, in which one type of nucleic acid (DNA) served as a template for the synthesis of other types, (mRNA, rRNA, tRNA), which in turn acted to provide for the assembly of amino acids into proteins. Some of these proteins were used as enzymes to synthesize more DNA and RNA; others were used for metabolic and structural functions.

This general scheme of DNA $\longrightarrow$ RNA $\longrightarrow$ Protein (called the **Central Dogma** of molecular biology), once established, proved so much more efficient than any other that it soon outdistanced its "competitors." A key feature was its ability to make copies of itself, and to the extent that a given copy was imperfectly made, it would either be more or less efficient in assembling its molecular food and in turn successfully reproducing itself, thereby either increasing or decreasing in numbers disproportionately. Gradually the more efficient combinations became dominant, photosynthetic forms arose, and then came animals.

What we have been discussing is the essence of **organic evolution**, which biologists believe is the underlying principle that has been responsible for the development of life from the simplest to the most complex forms.

Energy Sources. When the ultraviolet decreased to a point where it could no longer serve as a source of energy for the synthesis of the food molecules, a new energy source had to be utilized if organisms were to survive. We are familiar with the fact that the combination of oxygen with other molecules (oxidation) is an excellent source of energy but at this early period there was probably no free oxygen in the atmosphere. It is thought that oxygen appeared after living things were well established on earth. Previous to the dawn of life all oxygen was tied up in the form of oxides, water, and organic compounds. One source of energy available to the nucleoproteins for reproduction at this early period was that locked up in smaller molecules already in existence. The most likely candidates were the energy-rich sugar molecules because they could be easily broken apart releasing their stored energy.

Fermentation. Splitting the sugar molecule into two fragments we now call **fermentation,** and this process might well have been the first by which energy was released from molecules. The manner in which this is done can be written this way:

$$C_6H_{12}O_6 \longrightarrow 2 \ C_2H_5OH +$$
$$2 \ CO_2 + Energy$$

Sugar	Ethyl alcohol	Carbon dioxide

The energy released in this reaction could be used by the large protein molecules in producing more of their kind. It is interesting to note that organisms on earth today which depend on fermentation exclusively for their energy source have never achieved a high state of development; they are all lowly microorganisms.

Photosynthesis. Since no more molecules of energy-rich compounds (sugars) were being formed, sooner or later all of them would have been used up; and if some new source of energy had not been found the entire ingredients in the "soup" would have gradually disintegrated and thus the end would have come to this "pre-living" material. Fortunately, this catastrophic day did not come

because a method of trapping the sun's energy evolved. This we called **photosynthesis** (Fig. 2–11). It is simply a method of storing the sun's energy in the sugar molecule, in the presence of a green plant pigment, **chlorophyll,** using as raw material carbon dioxide and water, two compounds very abundant in nature. It can be written like this:

$$Energy \ (sun) + 12 \ H_2O +$$
$$6 \ CO_2 + Chlorophyll \longrightarrow$$
$$C_6H_{12}O_6 + 6 \ O_2 \uparrow + 6 \ H_2O$$

Glucose

It is clear that the energy from the sun is locked up in the sugar molecule by this process. Green plants probably initiated this method of storing energy and have retained it. They are the ultimate source of all utilizable energy on earth today, atomic energy and direct solar absorption excepted.

Respiration. When did animals come into this story? Once the photosynthetic process

Fig. 2–11. A schematic representation of how energy is stored in the glucose molecule and subsequently released for use by cells. By the process of photosynthesis, the sun's kinetic energy is stored in the glucose molecule using the simple molecules of CO_2 and H_2O. This molecule possesses high potential energy. As the glucose molecule degrades to CO_2 and H_2O during respiration, the stored energy is released for various cell functions.

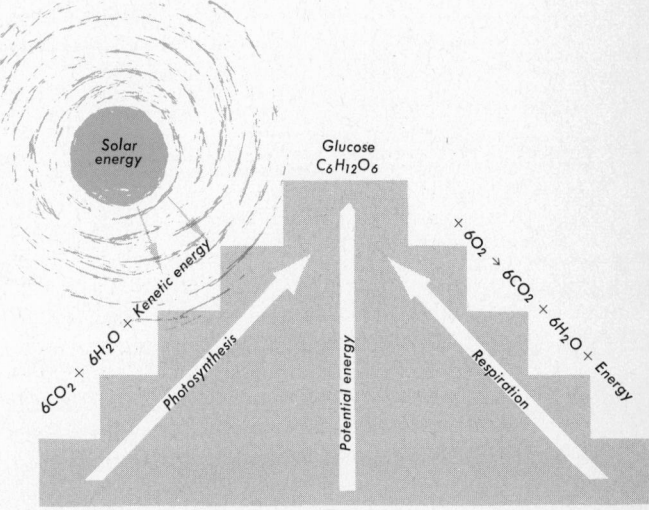

was able to supply all of the energy needed for protein synthesis, it would have been possible to produce enough plants to cover the surface of the earth, or until the CO_2 was used up. However, this never had a chance to occur because of the emergence of animals. The first animal was probably a degenerate plant, a plant that could not make sugar; therefore it became dependent on the green plants for a source of energy. The only way it could get the energy out of the sugar molecule was to burn or oxidize it, that is, permit the union of the now abundant oxygen in the atmosphere (put there as a by-product of photosynthesis) with the sugar molecule, thus degrading it once again to carbon dioxide and water. This process we now call **respiration,** the opposite to photosynthesis (Fig. 2–11), and it goes like this:

$$C_6H_{12}O_6 + 6\,O_2 \longrightarrow$$
$$6\,CO_2 + 6\,H_2O + Energy$$

By this means the animal could derive energy to reproduce itself and do all the other things animals have learned to do. Once this basic principle had evolved, there was a balance between plants and animals. A cycle was established which has operated rather satisfactorily up to the present, but with man's continued pollution of both air and water, it is possible the cycle may be irreparably altered.

Up to this point this story presents a plausible explanation of the chemistry that was necessary during these early days, but what about the morphology of these primitive entities? What did they look like and do they have counterparts in existence today? Let us try to answer these questions.

First Morphology. During the following discussion it will be helpful to refer to Figure 2–12. It is impossible to say what the first "pre-life" looked like but we are quite certain that it contained large nucleoprotein molecules, probably in combination with other molecules. As a result of competition among these molecules some must have been selected out which were sticky and tended to

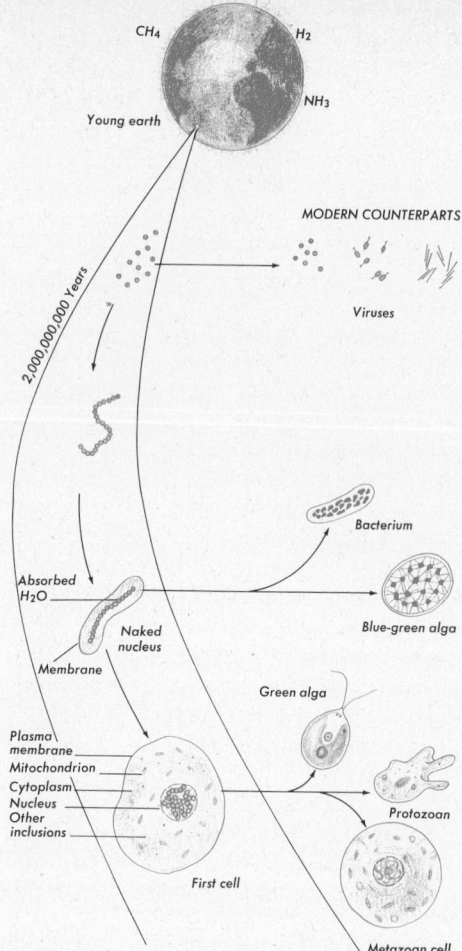

Fig. 2–12. A schematic representation of how the first organisms might have originated on the earth.

aggregate into masses. Modern proteins, and nucleoproteins in particular, are sticky compounds as anyone knows who has observed the white of an egg or certain types of glue. This property of stickiness would make it possible to collect materials in the immediate vicinity of the nucleoprotein aggregate. If these materials should consist of nutrients, molecular aggregates that possessed a high degree of cohesiveness would have an advantage over those with less, since having nutrients in close proximity would certainly be an advantage in the business of reproduction.

Thus such molecular aggregates would gradually dominate the population of molecules in the young seas.

Do we see anything resembling such bodies as we have just described on earth today? The first one that comes to mind is the modern **virus**, the invisible (under light microscopy but visible under the electron microscope) organisms, so minute that they pass through the pores of a porcelain filter (Fig. 2–13). They cause many plant and animal diseases, including such dread maladies as poliomyelitis, yellow fever, and many others. Viruses are composed of nucleoprotein molecular aggregates surrounded by a shell of protein molecules (Fig. 2–13). They vary widely in morphology, some being spheres and perhaps like the first viruses, and others possessing tails which may be an advanced form. One would expect evolution among them as we see in all living things that followed.

Viruses seem to have some of the prerequisites of our postulated preliving material. However, they reproduce only inside of living cells where ribosomes and other components of the host cell are available for help in the construction of viral proteins and nucleic acids. It is argued by some that viruses may be degenerate parts of previous cells now turned into parasites. It seems equally likely that viruses may represent modified versions of one of the earliest forms of life. In this case, one might think of the cells of plants and animals as similar to the primeval oceans, both serving as a source of nutrients for reproduction of the virus. In the early days the primitive viruses could find an adequate supply of food molecules in the oceans but later this source disappeared. In the meantime cells had evolved and if the viruses were to survive they were forced to make their way inside the newly formed cells,

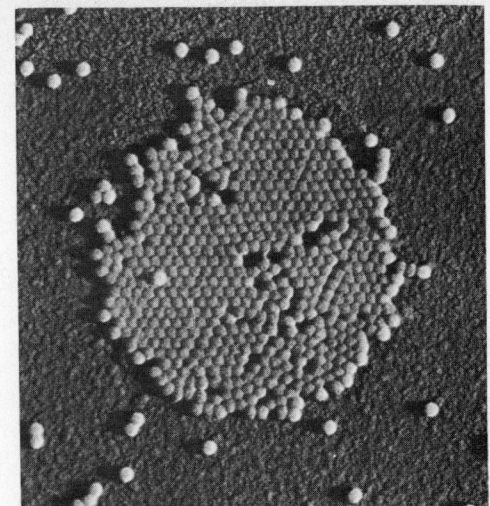

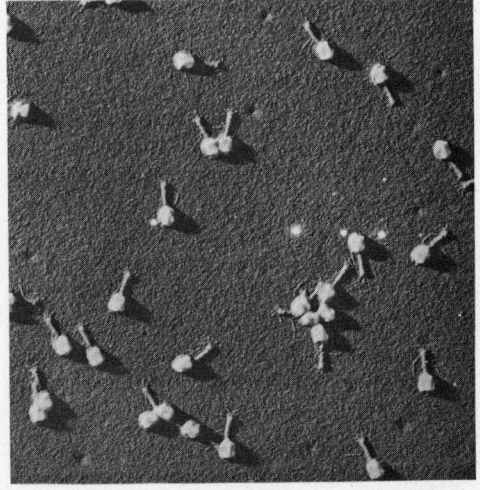

Fig. 2–13. Pictures of viruses taken through an electron microscope. (Top) Poliomyelitis virus Type II [× 75,000]. Note the regular arrangement of the particles. (Middle) Influenza A virus [× 35,000]. (Bottom) T4 bacteriophage [× 50,000], a virus that attacks bacteria. Note the characteristic head and tail.

thus becoming the first parasites. Such viruses may well have been the ancestors of those that we find today.

The question often arises, are viruses alive? As we see later (p. 36) they do not possess all of the characteristics we assign to living things. They also demonstrate characteristics which living things do not possess. For example when properly treated, at least one, the tobacco mosaic virus about which we know a great deal, forms crystals and remains inactive for indefinite periods of time. Living things may hibernate, form resting cysts, or otherwise remain relatively inactive for periods of time, but such vital processes as taking in oxygen and giving off carbon dioxide are still observable although at very low levels. The crystalline virus does not demonstrate these properties. Indeed, for all purposes it seems to belong to the inanimate world. Moreover, it is possible to fragment the virus; that is, break it into its component parts and some time later, reassemble the parts into the original virus. When dissociated, the fragments show none of the properties of the whole virus; but when put back together again the resulting compound differs in no way from the original virus, either chemically or in its ability to infect and grow in tobacco leaves. No truly living thing will withstand such treatment. We can say then that, whereas viruses do not possess all the properties of life, they probably re-semble the borderline forms which were the first to split off from the nonliving world and became the precursors to all life that followed.

The First Cells. The process that brought about the aggregation of nucleoprotein molecules and their protein shells to form the first viruses could have continued in a much more complicated fashion. In addition to proteins, other molecules, both organic (particularly lipids) and inorganic, could have accumulated at the surfaces of the nucleoprotein core. Many of these molecules would function as nutrients at first but later serve other functions. For the sake of efficiency these organized molecules could be walled off from the surrounding environment by means of a thin membrane, thus forming tiny packages of diverse molecules, each with one or more nucleoprotein aggregates. These could have been the first cells, endowed with the properties of life.

These cell prototypes retained all of the properties of the viruses but they added some new ones. Their nucleoproteins endowed them with the capacity of reproduction, mutation, and evolution. They were able to withdraw food from the surrounding sea and to synthesize these raw materials into their own structure. With the newly added diverse molecules they developed the remaining functions which brought them to the level

Fig. 2–14. An electron micrograph of the bacterium, *Bacillus subtilis*, showing the thick cell wall (W) and thin plasma membrane (C) lying beneath it. The fibrillar chromatin material occurs in a region called the nucleoid (N) and contains DNA. The cytoplasm contains many small, dense granules, ribosomes (R) which measure about 180 Å in diameter. Mesosomes (M) are thickened infoldings of the plasma membrane which play a role in cell division and also contain respiratory enzymes. The nucleoids (N) appear to be attached to the mesosomes (arrows). (Courtesy of A. Novikoff, E. Holtzman, and Holt, Rinehart & Winston. Micrograph courtesy of A. Ryter.)

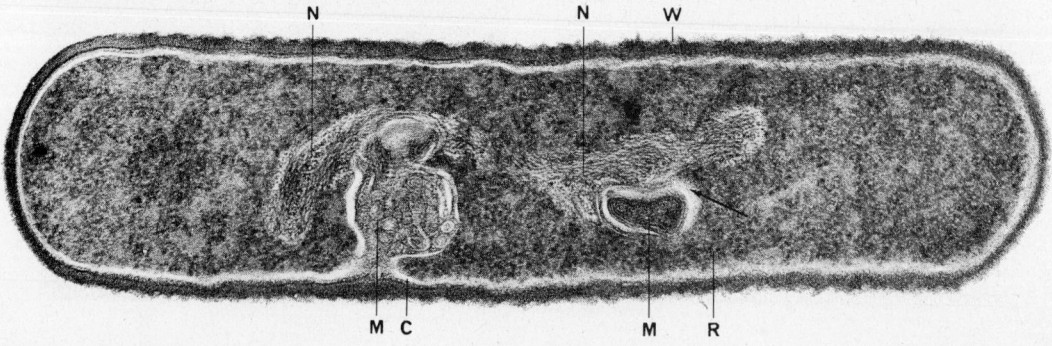

used by the nucleoprotein molecules for reproduction, not only of themselves but of the entire cell. Here we see in bacteria and certain algae counterparts of our postulated first cell. It is possible to select a series of algae which demonstrate succeeding stages from completely dispersed chromosomes to those with the chromosomes centrally located.

The next step in the path toward greater complexity was the accumulation of more materials from the environment in the immediate vicinity of this primitive cell, forming a second shell including a limiting **membrane**. The material trapped in this outer shell constitutes the **cytoplasm** and with this addition the "modern" cell came into being (Fig. 3–10). It consists of a **nucleus** which is not unlike bacteria and algae mentioned earlier, and cytoplasm. We have seen then, with the evolution of these first complete cells requiring at least two billion years, the transition from the nonliving to the living world. Further evolutionary steps involved aggregation of cells, which was another remarkable series of events that resulted in all the organisms we see on earth today.

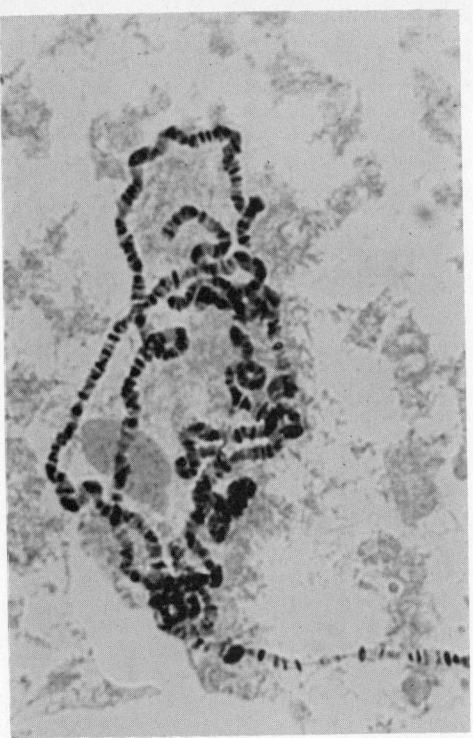

Fig. 2–15. Chromosomes are composed primarily of nucleoprotein molecules which become brightly colored when fruit fly salivary gland cells are stained. The colored bands of nucleoprotein probably correspond to the location of the genes. [× 2,500]

of living single-celled organisms. Do we see such primitive cells on earth today?

Let us first consider the anatomy of various kinds of cells that can be examined in the laboratory. We discuss only those parts that are pertinent to our discussion; later we go into a very detailed study of cells. The simplest kinds of cells are the **bacteria** (Fig. 2–14) and **blue-green algae**. They consist of an outer **membrane** which encloses a watery fluid in which **chromosomes** and a large variety of molecules and molecular aggregates are suspended. The chromosomes are composed of nucleoprotein aggregates (Fig. 2–15), not unlike viruses in chemical composition. The nucleoprotein aggregates that make up the chromosome include the **genes** (p. 477). Many of the surrounding molecules are the raw materials which are

THE NATURE OF LIFE

We have discussed at some length the origin of life but have not attempted to define it. When we attempt a definition we are confronted with an extremely difficult task. All of us at times have experienced the problem of defining something complex. We say we know what it is but are hard-pressed to formulate a satisfactory definition. Biologists have the same difficulty in defining life, hence the definition resolves itself into a description of living things and how they function. This turns out to be the goal of this entire book, the contents of which could hardly be called a definition. Nevertheless, this is what we are forced to do although at the end of this chapter we sum up our thinking in a few sentences.

The basic pattern of life was laid down in the first cells and all forms that followed

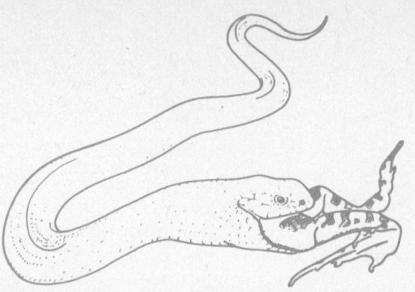

Fig. 2-16. Living things take in food from which they derive energy to carry on their life processes.

merely varied this pattern. Certain characteristics became incorporated in these first cells which distinguish all living things. Whereas most of them are simulated in the inanimate world, only living things possess all of them simultaneously.

Metabolism. In a broad sense metabolism includes **nutrition, respiration, synthesis,** and **excretion.** All living things must have the raw materials (or nourishment) to carry out the processes that are essential for life. The incorporation of raw materials is the process of nutrition. The raw materials come in the form of large insoluble molecules which must be degraded to smaller soluble molecules that can be utilized as a source of energy or as building materials (Fig. 2-16). Animals and some lower plants are equipped with digestive enzymes which bring about this degradation. The molecules that are destined to be used as a source of energy are burned by the process of respiration, releasing the energy which is required in all vital processes. This is often referred to as **catabolism,** or **destructive metabolism.** Some of the energy is utilized to transform other molecules into the physical structure of living matter. This process of synthesis is referred to as **anabolism** or **constructive metabolism.** This constant turnover of energy and synthesis is called **metabolism.** The removal of waste products of metabolism such as carbon dioxide and nitrogenous wastes is called **excretion,** which is an integral part of metabolism.

Metabolism may be simulated by inanimate machines. For example, a machine utilizes fuel (nourishment) and the energy released from the burning fuel could operate the mechanism to manufacture its own parts, and perhaps even put them together in the form of the parent machine. Even the replacement of worn parts might be built into the design. Such a machine has never been built but it seems within the realm of possibility. Therefore metabolism may be simulated in the inanimate world. However, there are other characteristics which definitely separate the animate and the inanimate world. As a result of metabolism, energy is available for other living processes to be carried out.

Movement. Some of the energy resulting from metabolic processes is utilized in movement, a characteristic of all living things, from the imperceptible movement of plants to the flutter of the grouse's wings in flight (Fig. 2-17). To be sure, water moves in a river, stones roll down hill, a car moves along the highway, but all of this movement results from forces acting externally. The water in the river bed flows, the stone rolls down hill, both as a result of the pull of gravity; the car

Fig. 2-17. This type of movement is characteristic of living things.

moves because the exploding fuel within the engine drives the wheels, causing a forward movement of the vehicle. The movement seen in the vibrating wings of the grouse or the slow movements of an opening flower are quite different. The energy for such movement comes from within the organism, as a result of oxidizing food molecules. This type of movement is said to be **autonomous,** that is, independent of all outside control. While this is not absolutely true (all life is dependent to some extent on outside controls), it certainly can be easily distinguished from the type of movement seen in the inanimate world.

Irritability. Responsiveness or irritability was implied in the preceding paragraph. All living things respond to their surroundings, that is, to their environment. If living things are to profit by their association with their environment they must at all times be aware of the nature of their immediate external world. They are, therefore, equipped with the means to sense its characteristics and respond to them. The information comes to the organism in the form of **stimuli** to which **responses** are made that are appropriate for the preservation of that organism. The rapid response an organism may make to an external stimulus is illustrated by the runner waiting for the discharge of the pistol which is the signal for him to start the race (Fig. 2–18). At other times the response may be very slow, as when with the onset of cold weather, the activities of the cold-blooded animals are gradually retarded. Animals move into favorable environments and out of unfavorable ones owing to their ability to respond to certain physical properties of those environments. Such responses play a very important role in the survival and ultimate success of a species. Most animals are equipped with an elaborate set of sense organs that permit them to keep in constant contact with their external world; the functioning of these organs makes the difference between life and death of the species.

This capacity to respond to both internal and external stimuli provides the organism

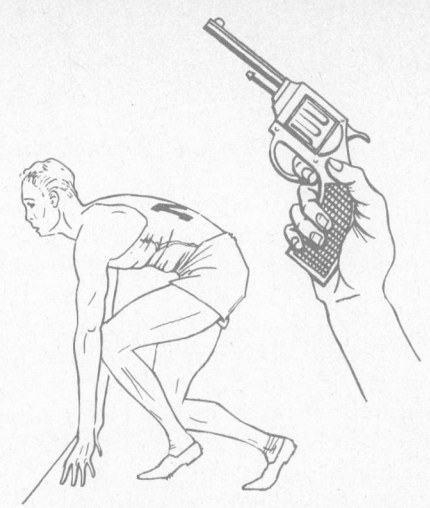

Fig. 2–18. Living things are irritable. They respond to their environment in very definite ways.

with controls to maintain a so-called **steady state** within itself without which it would die. These steady-state controls operate to supply the organism with nutritive materials when it is hungry and to adjust the oxidation and synthetic rates in keeping with the demands of the organism at any particular moment. In addition they operate in self-preservation of the organism by swift movements in offense or defense. When injury occurs steady-state controls initiate a complex series of reactions that bring about the repair of damaged parts. Many parts of the organism are constantly being replaced which also come under these steady-state controls. Man has not yet reached the point in his technological knowledge to be able to build a machine that will operate as a self-repairing and self-healing instrument, although he has constructed some remarkable self-feeding and self-adjusting machines. Organisms became self-preserving millions of years ago and this accomplishment separates them from the inanimate world.

Growth. Steady-state controls also permit another characteristic of all living things to operate. When the synthetic forces of an organism exceed the destructive forces there is a net gain in body mass, which is called **growth** (Fig. 2–19). This is typical of all organisms, particularly during their early life. However, a stage is reached where there is a balance between these two forces, and this is referred to as **maturity;** at the time of maturity the organism merely holds its own, becoming neither larger nor smaller. As life continues the synthetic forces fail to keep pace with the destructive forces, and the entire bulk of the organism loses ground and eventually dies. This involves the processes of aging, the nature of which is not well understood. In its final demise it is likely that only one or two aspects of the steady-state control mechanism break down, but these are of such importance that the entire machinery collapses, a process we call **death.**

No machine has been built, or perhaps ever will be built, that grows. A crystal may grow in size by the addition of other similar crystals to its own bulk, but the pattern and the method of executing it are quite unlike that of a living organism. The crystal merely adds other crystals to its external mass (accretion), as a mason adds bricks to a wall; whereas the living organism takes the raw materials within, and there makes them an integral part of its own structure (**intussusception**).

Reproduction. We mentioned earlier that the failure of certain vital processes results in the death of the organism. This would mean the end of the organism if some method of continuing its life had not interrupted this inevitable course. With the raw

Fig. 2–19. Growth is a characteristic of all life.

Fig. 2–20. All living things reproduce and the offspring resemble the parents in most respects.

materials and energy supplied the organism has grown in size, which in itself permits subdivision and subsequent growth in numbers (Fig. 2–20). This is **reproduction.** The young are composed of materials just as old as the parent; but for some reason not at all understood, the young take on renewed vigor, and are said to be **rejuvenated.** Moreover, the young usually are smaller than the parent and frequently quite unlike it in structure. As the young mature they undergo changes which we speak of as **development,** the details of which we consider in a later chapter.

One of the remarkable outcomes of reproduction is the continuity of pattern from generation to generation. Offspring are endowed with structural and physiological characteristics that are almost exact duplicates of those found in the parents. The transmission of these characteristics from parent to offspring is known as **heredity,** another subject which we consider in some detail in a later chapter.

There is nothing in the inanimate world that reproduces in any manner similar to that seen in living things. No machine has been built that reproduces itself (a worthy goal for inventors) to say nothing of one that undergoes development and demonstrates rejuvenescence.

Adaptation. Even with all of the characteristics so far described, living things must have added one more in order to survive through the rigors of past ages. With time the face of the earth has constantly changed and with it the environment in which organisms must survive. A basic pattern set down in the bodies of the first organisms which satisfied their needs adequately at that time may be quite unsatisfactory several million years later when the environment has drastically changed. If these ancient characteristics were faithfully reproduced in the offspring generation after generation, it is obvious that the time would be reached when the organisms could no longer survive. This indeed happened to many, but some acquired the capacity to change and so to fit more closely their changing environment. This capacity is called **adaptation,** a characteristic of all living things. Here again we see nothing in the inanimate world that possesses this characteristic. The possibility of building a machine with this capacity seems highly unlikely, although we must never underestimate the ingenuity of man.

We have attempted to describe the characteristics of living things and any structure that possesses these characteristics is said to be alive. We still need to learn more about the nature of life, and this can be done by studying its physical state.

PROTOPLASM

Life resides in material given the name **protoplasm** over 100 years ago. The word means the first form (**protos**—first, **plasma** —form). If tiny pieces of any plant or animal are examined under the microscope, they will be found to be composed of **cells,** each of which contains protoplasm. Recalling our earlier discussion of the origin of life, it must be concluded that protoplasm is very complex. It is so delicately adjusted physicochemically that any attempt to find out how it behaves or of what it is made has often meant almost instantaneous loss of the very thing sought for. Protoplasm, by definition, encompasses that "something" called life, and it can only be handled with the utmost care without "killing" it. At the outset it would seem impossible to study protoplasm. This is not quite so because some ingenious

biologists have been able to build delicate pieces of apparatus that will permit a study of this fascinating material. One of these is the phase-contrast microscope (p. 60), with which it is possible to examine the insides of cells, thus observing protoplasm and its constituent parts.

One of the easiest cells to study is the common protozoan, *Amoeba*, since it is large (barely visible with the naked eye) and rugged (Fig. 2–21). Its protoplasm appears grayish in color and is usually moving about in a more or less haphazard manner, even though the general flow is in the direction the amoeba is going.

In order to learn more about protoplasm, a simple experiment is helpful. If we drop a little alcohol or mercuric chloride on the amoeba, all activity abruptly stops and the

Fig. 2–21. A living amoeba, when seen under the light microscope, gives us a distant view of the nature and activity of protoplasm. There is a distinction between the clear protoplasm at the edges of the cell and the granular interior. The protoplasm is in almost continuous movement, which can be followed by observing the tiny particulate matter inside.

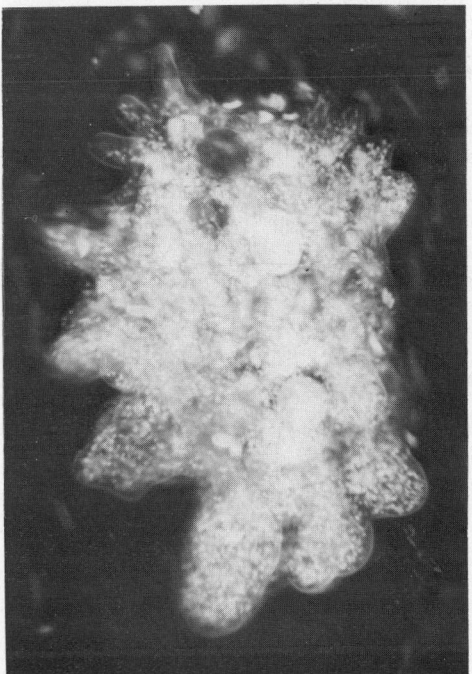

entire cell becomes rigid, much like the white of an egg when cooked. The cell has died and, so far as we know now, this is an irreversible reaction. If we raise the temperature momentarily to the boiling point of water, the same reaction occurs. Both the chemical treatment and high temperature cause irreversible damage particularly to proteins and nucleic acids so that essential functions of molecules like **enzymes** (which are proteins) are destroyed. Protoplasm is a very complex mixture of organic and inorganic molecules.

Let us suppose that we could magnify the protoplasm of an amoeba until its molecular structure became visible. The most striking characteristic would be the violent activity of molecules of many shapes and sizes. By far the most numerous would be the water molecules, which because of their small size, would move faster than the others. The outer limits of the cell would be made up of huge molecules bound together, thus forming a continuous network perforated with holes of different sizes. This would be the plasma membrane, through which the water molecules pass freely in both directions. Many other molecules pass through also, but some are stopped because of their large size, whereas others are stopped because they possess electrical charges that prevent them from getting past the electrical charge on the membrane. Oxygen molecules enter freely, uniting with fragments from the large glucose molecules which suddenly break into smaller water and carbon dioxide molecules. The smallest particles that result from this process are the hydrogen atoms which exist only momentarily before combining with other molecules.

Many other molecules, including proteins, would be moving about performing duties that are essential to the life of the cell. Some of these molecules are large and complicated, and composed of thousands of atoms. In addition, there are membranes which tend to divide the protoplasm and separate classes of molecules into compartments for certain specialized activities. We will examine these relationships in greater detail later.

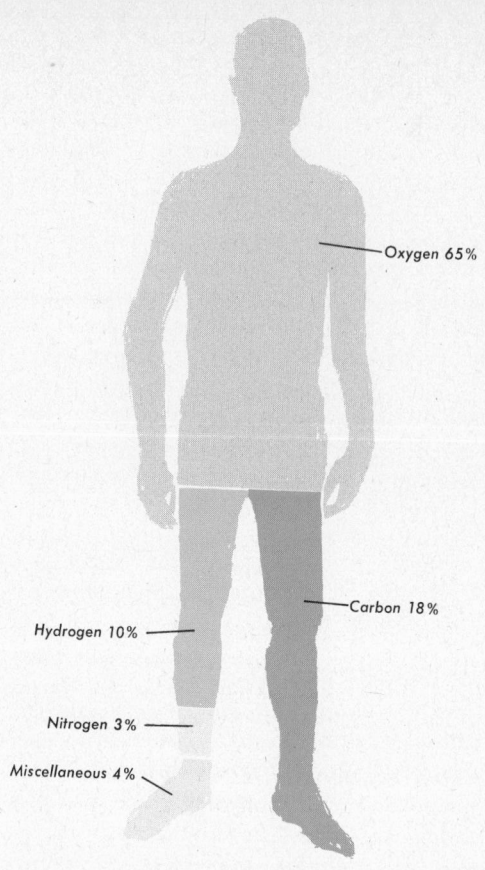

Fig. 2–22. Percentage of various chemical elements in man. Numerous lesser elements are included in miscellaneous elements.

CHEMICAL COMPOSITION OF PROTOPLASM

No matter what the origin of protoplasm is—rat, bee, cow, or plant—we would find, from a careful chemical analysis, that it is composed of molecules and atoms. Moreover, no matter what function of protoplasm we study we always find that it can be explained in the light of the chemical and physical properties of molecules and atoms. It is the combination of atoms and molecules that gives protoplasm its unique characteristics.

If we examine protoplasm derived from a

number of plants and animals, including man (Fig. 2–22), we find a remarkable similarity in their chemical composition. The four principle elements are carbon (C), hydrogen (H), oxygen (O), and nitrogen (N). Carbon is the core element of the complex molecules found in protoplasm, probably because of its physical nature which was discussed earlier. Additional elements regularly present in protoplasm are phosphorus (P), calcium (Ca), iron (Fe), potassium (K), sulfur (S), iodine (I), magnesium (Mg), sodium (Na), and chlorine (Cl). There are many others existing only as traces but nevertheless essential for the normal activities of protoplasm.

The chemicals found in protoplasm may be divided into **inorganic** and **organic,** the characteristics of which were described earlier. We learned that the former make up most of the earth, whereas the latter are always derived from living things. The inorganic compounds that are important in protoplasm are **water, inorganic salts,** and certain dissolved gases, such as **oxygen, nitrogen** and **carbon dioxide.** The important organic compounds are **carbohydrates, fats, proteins** and **nucleic acids.** The last are usually found in combination with proteins and are therefore called **nucleoproteins.**

Inorganic Compounds in Protoplasm

Water. Of all the inorganic compounds in protoplasm the most important and most abundant is water, the substance in which all other materials are suspended and transported. Life without water would be impossible, because there would be no means of mixing and dispersing the energy-yielding and building materials of protoplasm. Water is important in protoplasm because of its many unique properties.

Water **dissolves** more substances than almost any other liquid. This property of water makes possible the mixing of a large variety of substances that would not dissolve in any other liquid. Hence, the great complexity of protoplasm has come about because so many different substances could come together and mix freely in one common medium. Furthermore, interactions occur more readily between substances in a fluid condition where ample freedom of movement of the molecules is permitted.

Water has a high capacity for **holding heat.** It is "reluctant" to take on or lose heat, a physical property that is very important in living things. Anyone living near large bodies of water is aware of their tempering effect on the climate of the surrounding areas. Since our bodies are primarily water (approximately 65 percent), we respond to heat much the same as water does. If this were not the case, it would be almost impossible for us to tolerate changes in temperature.

Water possesses some interesting **chemical properties;** for example, it has more ability than any other substance to dissociate molecules into their **ions.** Salts such as sodium chloride (NaCl) ionize readily in water; others such as sugars and starches do not ionize at all. Water itself ionizes slightly forming hydrogen (H^+) and hydroxyl (OH^-) ions. Both H^+ and OH^- ions are extremely active and for this reason they seem to be tolerated only in very small numbers by protoplasm; that is, any increase of either over the other brings about prompt changes in the activity of protoplasm, and any marked increase or decrease terminates life. Protoplasm is approximately neutral all of the time, the numbers of H^+ and OH^- ions being approximately the same.

Salts. Many inorganic salts of sodium, potassium, calcium, and magnesium exist in protoplasm. Inorganic acids are also found in small concentrations. The role played by these various compounds comes out in later chapters.

Gases. Because of their diffusing qualities, gases tend to enter protoplasm the same as they enter other material. Since the atmosphere contains over 20 percent oxygen, considerable quantities of this gas are found in protoplasm. **Oxygen,** of course, is very important in the release of energy. **Nitrogen,** on the other hand, is even more abundant, but because of its chemical inertness it takes part in no reactions that are important in metabolism. However, certain soil bacteria do convert atmospheric nitrogen into nitrates which are essential for plant growth. **Carbon dioxide** exists in small amounts both in the atmosphere and in protoplasm.

Organic Constituents of Protoplasm

The carbon atom is the starting point for all organic compounds as we have already learned. We have also learned that very complex molecules can be built when carbon atoms bind with other atoms and with themselves, forming chains and rings. These organic compounds not only form the fabric of protoplasm but also hold and release the energy that makes all activity possible. Some of the important organic compounds in protoplasm are carbohydrates, fats, proteins, including enzymes, and nucleoproteins.

Carbohydrates. We have referred earlier to the atomic composition of carbohydrates, from glucose to such complex molecules as starch and glycogen. In the plant leaf, CO_2 and H_2O unite under the influence of the sun's rays and in the presence of a green plant pigment, called **chlorophyll,** to form glucose. By condensation of the 6-carbon sugars (glucose and fructose) with the loss of water, the disaccharide sucrose is formed. Further combining, called **polymerization,** results in the formation of the huge polysaccharide molecules such as starch (plants) and glycogen (animals). Carbohydrate molecules also may play a structural role in cells, for instance, in bacterial and plant cell walls; they also combine with proteins to form **mucopolysaccharides** and **glycoproteins.** Nevertheless, we normally think about carbohydrates in connection with the storage and release of energy originally derived from the sun, without which animals could not exist. Just how this happens we will discuss in Chapter 3.

Fats. These include all of the **fats, oils,** and **fat-like substances** (lipids) identified by a greasy texture. They are relatively insoluble in water, but soluble in hot alcohol, ether, and chloroform. With a few exceptions they all contain C, H, and O like carbohydrates, the chief difference being their low oxygen content. This is easily illustrated by examining the formula of a common fat taken from beef tallow, $C_{57}H_{110}O_6$. Note that there are only 6 oxygen atoms as compared to 57 carbon and 110 hydrogen atoms. This means that when fats burn, they require more oxygen than does glucose. It also means that the molecule can release more energy per gram when burned. Their insolubility makes fats desirable material for the storing of energy.

The fat molecule is composed of **glycerol** $(C_3H_8O_3)$ and fatty acids. A common fatty acid is **stearic acid** with the formula $CH_3 (CH_2)_{16} \cdot COOH$. This is called an acid because the carboxyl group (COOH) can liberate one hydrogen ion when dissolved in water. It can be written as indicated below.

Going back to the plant leaf again, fatty acids are joined with glycerol to form oils (fats) of different kinds. This is done by the loss of water (see top of facing page).

As in the case of starch, the reverse action, hydrolysis, takes place during digestion when water is added, splitting the fat into fatty acids and glycerol.

There are other fats such as **steroids** and **phospholipids,** the latter of which has phosphate and a nitrogen compound in place of the fatty acid. They resemble fats in many ways, but have other characteristics that tend to set them off in a group by themselves. Some of them play their role in the plasma membrane where they are responsible for the selective action of this delicate structure. They are also important in some of the intricate chemistry of the animal body which we touch on later.

Proteins. Proteins are composed primarily of C, H, O, and N; in addition they usually contain sulfur (S) and phosphorus (P). They are huge molecules and, like starch and

$$\text{HO—C}\overset{\displaystyle H}{\underset{\displaystyle O}{|}}\text{—C—C—C—C—C—C—C—C—C—C—C—C—C—C—C—C—C—H}$$

Stearic acid (from beef fat)

$$
\begin{array}{c}
\quad\; H \\
\;\;| \\
H\!-\!C\!-\!O\,|\,H \\
\;\;| \\
H\!-\!C\!-\!O\,|\,H \qquad + \qquad
\begin{array}{l}
HO\,|\,OC\ C_{17}H_{35} \\
HO\,|\,OC\ C_{17}H_{35} \\
HO\,|\,OC\ C_{17}H_{35}
\end{array}
\rightleftharpoons
\begin{array}{l}
\quad\; H \\
\;\;| \\
H\!-\!C\!-\!OOC\cdot C_{17}H_{35} \\
\;\;| \\
H\!-\!C\!-\!OOC\cdot C_{17}H_{35} \\
\;\;| \\
H\!-\!C\!-\!OOC\cdot C_{17}H_{35} \\
\;\;| \\
\quad\; H
\end{array}
\qquad + \qquad 3\ H_2O \\
\;\;| \\
H\!-\!C\!-\!O\,|\,H \\
\;\;| \\
\quad\; H
\end{array}
$$

| 1 molecule | 3 molecules | 1 molecule | 3 molecules |
| Glycerol | Stearic acid | Fat | Water |

fat, can be hydrolyzed to **amino acids.** There may be hundreds or thousands of amino acid molecules in a single protein molecule; but when broken down, it yields about 20 amino acids, all of which are not usually found in any one kind of protein. The proportion of the different amino acids depends on the nature of the original protein molecule.

The formula for a simple amino acid, glycine, was mentioned earlier. Let us now consider the general formula which identifies all amino acids:

$$
\overset{\displaystyle R}{\underset{\displaystyle COOH}{\overset{|}{H\!-\!C\!-\!NH_2}}}
$$

when R represents the main portion of the molecule. The remainder is found in virtually all amino acids. Note that, in addition to the **amino group** (NH_2), there is another group, the **carboxyl group** (COOH). These two groups are responsible for the behavior of amino acids. It is obvious that the presence of the carboxyl group gives the molecule acidic properties, just as is true of any organic acid. Strangely enough, an amino acid can also act like a base owing to the presence of the NH_2 group. It responds like

a base by removing hydrogen ions, not by delivering hydroxyl ions. This is demonstrated by the following equation:

$$R\cdot NH_2 + H^+ + Cl^- \longrightarrow R\cdot NH_3 + Cl^-$$

In the presence of an acid, then, an amino acid acts like a base by absorbing or removing the hydrogen ions. On the other hand, in the presence of a base it acts like an acid by delivering hydrogen ions from its carboxyl group, thus:

$$R\cdot COO^- + H^+ + Na^+ + OH^- \longrightarrow$$
$$R\cdot COO^- + Na^+ + H_2O$$

A substance that responds in this fashion is said to possess **amphoteric** properties. It therefore always has a tendency to bring a solution to neutrality, that is, to balance the number of hydrogen and hydroxyl ions. For this reason, amino acids tend to prevent too much fluctuation in the number of hydrogen and hydroxyl ions in a solution. Such a substance is spoken of as a **buffer**; amino acids are good buffers.

Amino acids unite to form proteins in the plant leaf by the same process that starch and fats were produced, namely, by the loss of water. Likewise, in every cell of the animal body proteins must be assembled from the amino acids that come to it through the bloodstream. Just how this is done puzzled chemists for a long time, but it is now known to occur through the production of a **peptide linkage.** The following illustrates how this occurs:

Glycine + Glycine

Glycyl-glycine + H_2O

In this manner the various amino acids are tied together to form molecules that range in size from several hundred to several thousand atoms. Their size can be estimated by the relative molecular weights. For example, glycine has a molecular weight of 75, whereas that of the oxygen-carrying protein of the crayfish's blood (hemocyanin) is over 5 million. Obviously, by combining the 20 amino acids in various ways, limitless numbers of different proteins can be formed.

The remarkable feature of this formation is that these amino acids are not linked together at random; rather, they are connected together in particular sequences according to instructions received from the genes of the cell. A chain of such amino acids is called a **polypeptide,** and any given protein is made up of a particular set (one or more) of these chains. The characteristics of the protein are determined by the **primary amino acid sequences** of its constituent polypeptides (i.e., the order and kinds of amino acids in the polypeptide chains), and that seemingly simple, only recently determined fact has to be one of the most exciting discoveries ever made about the mysteries of how living systems work. We will discuss this remarkable process in detail in Chapter 23.

Proteins are the most characteristic and most abundant material (exclusive of water) in protoplasm. Besides forming the actual supporting structure of protoplasm, proteins provide the specific catalytic "encouragement" by which the large majority of chemical reactions essential for life take place.

Enzymes. Enzymes are **protein** molecules which are called **organic catalysts** and may be defined as substances that hasten a chemical reaction at living temperatures but are not themselves consumed by the reaction. They arose early in the evolution of life and have played an essential part in all reactions that go on in protoplasm ever since. Without enzymes, protoplasm loses its ability to start and maintain the multitude of activities that go on within it. They are essential in the business of living but the details of their operation are only now beginning to be understood.

A simple experiment demonstrates how an enzyme works. A watery starch solution is placed in two test tubes, to one of which is added saliva, to the other, nothing. Any attack on the starch can be detected by using an appropriate test for **maltose,** one of the products of starch breakdown. Within a few minutes this sugar can be demonstrated in the tube containing saliva, whereas in the tube without it there will be no maltose for many hours. The enzyme **amylase** brings about this reaction. The same breakdown can be accomplished without the enzyme, but it requires drastic treatment with strong acids at high temperatures, conditions that could not be tolerated by protoplasm. It is obvious, then, that enzymes can bring about a difficult chemical change at body temperatures and in a very short period of time. All of this is essential if the many reactions that go on in protoplasm are to occur as quickly and smoothly as they do.

Just how does an enzyme work? It is generally agreed among enzymologists that the enzyme combines temporarily with the reacting molecules, thus placing them in close proximity in order that a reaction can take place at a much faster rate than would be possible by the fortuitous collision of these molecules without the enzyme. The enzyme, like all protein molecules, is large and has a vast array of surface configurations, each specific for each enzyme. It apparently is this surface geometry that is the critical point in enzyme reaction. The surface must be such as to fit precisely the surface of the reacting substances, called **substrates**. For example, in the reaction:

$$A + B \longrightarrow AB$$

A combines with B in an accelerated fashion under the influence of the enzyme, providing the surfaces of both A and B fit the surface of the enzyme in juxtaposition thus bringing their reacting surfaces close together (Fig. 2–23). Any alteration in the surfaces of either enzyme or substrate destroys the effect.

Fig. 2–23. A scheme representing how enzymes work. The substrates A and B are brought into juxtaposition by finding their places on the active site of the enzyme. The enzyme has no effect on substrate C, owing to its wrong configuration. If a molecule (inhibitor) can take the place of the substrate, the enzyme is inactivated.

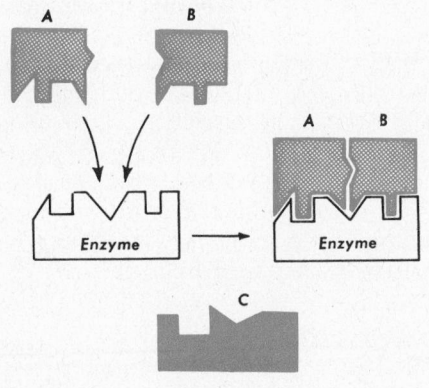

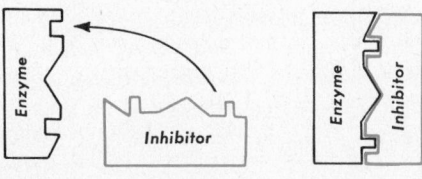

The mystery of how certain poisons work has been cleared up by applying this principle. Any substance that has the configuration of the normal substrates, even if only in part of its structure, may take its place on the surface of the enzyme, thus making it unavailable to the substrates. For all purposes the normal reaction stops, which may well be lethal for the organism. On the other hand if the poison incapacitates the enzymes of a disease-producing organism, and at the same time does not affect the enzymes of the organism itself, it can be beneficial. This explains the beneficial effects of many drugs used in the treatment of disease.

The enzyme may be used over and over again to bring about the reaction because the moment the chemical reaction has taken place the combined complex is released and the enzyme is ready to join two more substances. Hence, we see that the enzyme functions as a very special structure designed for trapping substrate molecules and positioning them so that chemical reactions can take place much more quickly than they would otherwise. Indeed, the breakdown of starch, mentioned earlier, occurs quickly in the presence of amylase, but without the enzyme the reaction hardly takes place at all at body temperatures.

This complexing of enzyme and substrate has been referred to as the "lock-and-key" hypothesis. The enzymes may resemble locks into which only particular keys, substrates, fit. This thought also brings out the precise specificity of enzymes; each reaction is usually carried out by only one enzyme. Since proteins have almost an infinite number of configurations it follows that there can be as many different kinds of enzymes, each specific for a particular reaction; and there are thousands of these going on simultaneously in every living thing. This does not mean that every protein is an enzyme, only those with configurations that fit specific substrates.

Another characteristic of enzymes is that they accelerate reactions in either direction depending on the concentration of the reacting substances. In all reactions the speed with which it takes place is proportional to the concentration of the participating substances. This is called the **law of mass action.** Let us see how this works in the following reaction:

$$\text{Glycerol} + \text{Stearic acid} \underset{}{\overset{\text{lipase}}{\rightleftharpoons}} \text{Fat} + H_2O$$

The direction it proceeds depends on the concentration of glycerol, stearic acid, and fat at any particular moment. If glycerol and stearic acid are more concentrated than fat, the reaction goes to the right; if fat is higher it goes to the left. If the reaction is started with only glycerol and stearic acid, and no fat, it moves rapidly to the right. As fat forms there is an ever-increasing movement to the left. Eventually the reactions are moving in both directions at equal speeds and an **equilibrium** is reached. There is no net increase of fat and for all purposes the reaction stops since it is moving in both directions at equal speeds. This reaction may be influenced by removing fat from the system. In that case it continues to the right until all of the glycerol and stearic acid molecules have combined forming fat. Again the reaction may be pushed to the right by adding a continual supply of glycerol and stearic acid, or it may be forced to go to the left by increasing the amount of fat. The law of mass action operates for all reactions in protoplasm but the reactions are almost never in equilibrium. If the end-product is an escaping gas, such as CO_2 when food is burned, the reaction proceeds in one direction continuously until the food is gone; which, of course, never happens if life is to be maintained.

Enzymes are classified in accordance with the substrate they influence; for example, those accelerating reactions of proteins are called **proteinases.** Likewise those involved with carbohydrates are known as **carbohydrases. Lipases** affect lipids (fats and fat-like substances). Note that the suffix -ase denotes the enzyme, although they are also identified by names which describe their action. For example, we may speak of them as synthesizing (synthetases), splitting (hydrolases), or transferring (transferases) enzymes.

The activity of enzymes is controlled by the movement of molecules just as all chemical reactions are. As the temperature rises from 0 to 40 degrees Centigrade and consequently molecular motion increases, enzyme action increases. With each 10 degree rise in temperature the action doubles, up to a critical temperature of about 40 degrees when all activity ceases. At this temperature, enzymes, like most proteins, apparently change their surface geometry and are no longer able to orient the substrate molecules to affect their union or separation.

It is interesting to observe that cold-blooded animals (called **poikilothermal,** that is, those that cannot maintain a constant body temperature, such as insects, frogs, and lizards) are forced into hibernation because at reduced body temperatures enzymatic activity is slowed to a point where normal response is impossible. Warm-blooded animals (called **homiothermal,** that is, birds and mammals) are not bothered in this respect because their bodies maintain a constant temperature. It is not sheer coincidence that the body temperature of these animals happens to be the point of maximal activity of the body enzymes.

Enzymes are also very sensitive to hydrogen and hydroxyl ions, as well as to certain specific ions such as calcium. Enzymes work most efficiently in solutions with particular pH values, which vary with different enzymes. High pressure as well as certain metallic ions stimulate or inhibit the action of enzymes. Here again, the effect is probably due to the alteration of the surface configuration of the enzyme molecule.

Nucleoproteins. The possible origin of these supermolecules, as well as their composition, was discussed earlier in this chapter. At this point we shall concern ourselves only with certain components of these molecules, namely the nucleic acids. Nucleoproteins hydrolyze to protein and nucleic acids, the latter existing in two forms, **ribose nucleic acid** (RNA) and **deoxyribose nucleic acid** (DNA). DNA is located primarily within the nucleus and constitutes the information-carrying part of the chromosomes (genes);

and RNA exists primarily in the cytoplasm although some is found in a special body, the **nucleolus,** within the nucleus, as well as in and associated with the chromosomes. We have much more to say about both of these compounds in a later section. At this point we merely wish to note their occurrence in protoplasm.

We have considered the chemical composition of protoplasm and found it to be extremely complex. It consists of a great variety of atoms and molecules but the most impressive fact is the constant activity of these components. Highly involved chemical reactions are going on continually and these are integrally related with both internal and external changes. These reactions obey the same physical laws that operate in the inanimate world and we must consider those now.

PHYSICAL NATURE OF PROTOPLASM

Having examined the chemical composition of protoplasm, we have found it composed of particulate matter in a vast array of sizes. These particles obey the same physical laws whether in or out of protoplasm. They behave in particular ways when isolated from others of the same kind, or when in close association with those of similar structure. If each particle follows specific laws of behavior when among its own kind it will behave differently when mixed with others of a different sort. Since protoplasm is made up of many kinds of molecules, it follows that the operating forces become extremely complex. In spite of this almost hopeless confusion, each particle seems to take its part in a definite pattern so that an orderly procession of reaction occurs. Let us examine some of these physical properties of protoplasm.

Size of Protoplasmic Particles

It was implied in an earlier chapter that the size of particles had a profound effect on their behavior. We should, therefore,

have some appreciation of the relative magnitude of the innumerable particles of matter existing in protoplasm.

In order to speak with any degree of accuracy about the size of these tiny particles, it is necessary to apply some unit of measurement to them. Scientists throughout the world employ the **metric system** of measurement almost exclusively. Fractions of the meter, **micrometers** (μm),* are used by the microscopist because these units are convenient for measuring objects that fall within the range of the microscope. For example, red blood cells in man are about 7 μm in diameter. Smaller objects, such as the particulate matter in protoplasm, are measured in **nanometers** or **Angstrom** units. It is possible to measure particulate matter with the same degree of accuracy that can be attained in the macroscopic world.

To demonstrate the relative sizes of objects we may use a cell as the starting point. A large cell of your body, a cell lining your mouth for example, would occupy a space about the size of a needle point (Fig. 2–24). If this were magnified ten times, you would see it quite easily with the naked eye but the parts would not be very well defined. Magnifying it another ten times ($100\times$) brings the nucleus and cytoplasm into full view; even the nuclear structure can be made out. Another tenfold increase ($1,000\times$) shows the chromosomes in outline, but not in detail. A magnification of 100 to 3,000 is the range in which the microscopist works with a light microscope. Any further increase in size must be viewed through the electron microscope, which operates much the same as the light microscope except electrons are used instead of light waves for a source of illumination (p. 58). Of course, these cannot be seen directly with the eye because our eyes are sensitive only to light rays and not to electrons, but such objects can be photographed (Fig. 2–25). Furthermore, the treatment of the material is so drastic that living

* 1 meter (m) = 1,000 millimeters (mm); 1 millimeter = 1,000 micrometers (μm); 1 micrometer = 1,000 nanometers (nm) = 10,000 Angstrom units. (Å).

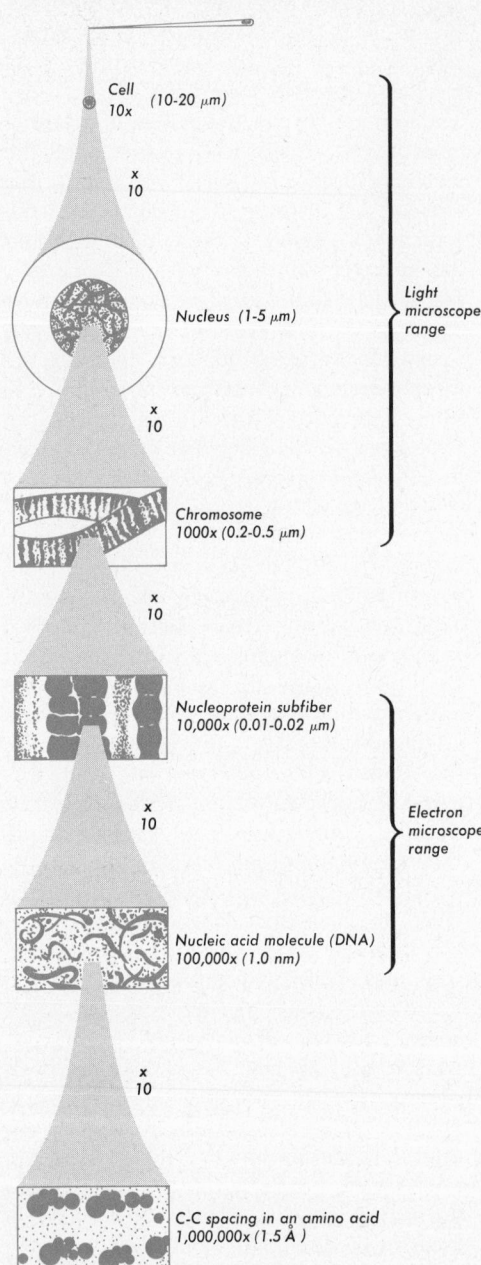

Cell
10x   (10-20 μm)

x
10

Nucleus (1-5 μm)

Light
microscope
range

x
10

Chromosome
1000x (0.2-0.5 μm)

x
10

Nucleoprotein subfiber
10,000x (0.01-0.02 μm)

Electron
microscope
range

x
10

Nucleic acid molecule (DNA)
100,000x (1.0 nm)

x
10

C-C spacing in an amino acid
1,000,000x (1.5 Å)

Fig. 2–24. Relative sizes of things that are of interest to the biologist (approximate average sizes). Magnifications greater than 100,000 are useful to the physicist and chemist.

things cannot be studied under an electron microscope. In order to study any material it must be sliced into extremely thin sections (about $\frac{1}{40}$ μm). With this instrument, magnifications can go up to 1,000,000 diameters, and resolution approaches 1 Å. (Instruments with 2 Å resolution are currently being supplied commercially. It should be noted in passing that *resolution* is a better test for microscope performance than magnification; resolution is a measure of the ability of the objective lens of the instrument to separate two small objects in space, whereas magnification is a measure of the ability to make such an image large enough to be visible to the human eye. If magnification exceeds that required by the eye, but resolution is too low, such magnification is called "empty" and is no help in seeing desired detail in the specimen).

The electron microscope allows us to see molecules above about 10 Å in size, which includes many proteins and nucleoprotein complexes. Beyond this, we must rely on methods familiar to the physicist to demonstrate the size and shape of particulate matter, methods that are beyond the scope of this book. Physicists are able to measure the size, shape, and behavior of molecules and atoms, and are now working on the nature of the components of the atoms. For the present discussion, it is only necessary for us to think in terms of the size of particles at the molecular level, because protoplasm is molecular.

Colloids and Crystalloids

If a solid is ground to particles the size of dust, and these placed in water, they form a murky fluid and after a time settle to the bottom of the container. If the particles are ground still finer they reach a size when they remain in suspension and do not settle out even after a long time. These particles are then in the **colloidal state**. Therefore, whether or not a substance exists as a colloid is merely a matter of size. Physicists have set an arbitrary figure for colloids; they state that particles ranging in size from 0.1 to 0.001

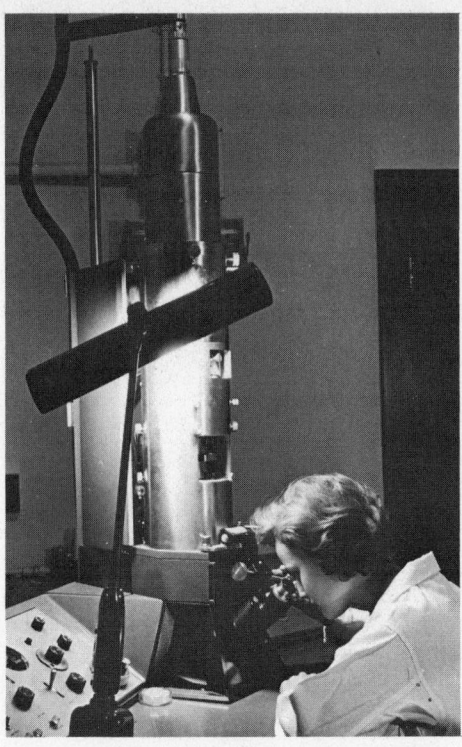

Fig. 2–25. The electron microscope is a precision instrument which makes it possible to observe and photograpgh the fine structure of cells and large macromolecules. Current instruments can resolve particles only 2 Å apart.

prevents them from passing through an animal membrane. If egg albumin is placed in a loop of frog skin and submerged in water, very little, if any, of the albumin is found in the water even after hours have elapsed. Furthermore, colloidal particles move slowly when compared to smaller particles such as atoms or ions. This might be expected from our knowledge of the movement of objects that come within our experience. Physicists tell us that the movement of particles is dependent on the absorption of **heat**; the higher the temperature, the faster the movement and the lower the temperature, the slower the movement. Movement is governed by the size of the particle; when the diameter of the particles is halved, the rate of movement is doubled. Therefore, the huge lumbering molecules in a colloidal system move slowly compared to the tiny molecules of a salt solution.

Particles that are smaller than 0.001 μm in diameter are called **crystalloids**; sugar or table salt dissolved in water forms a crystalloid solution. Such systems appear clear and transparent to the naked eye because light passes directly through without being changed in any way by the tiny molecules of salt and sugar. Furthermore, crystalloids pass readily through some membranes and their individual particles move much faster than those in the colloidal state.

Protoplasm contains numerous crystalloidol and colloidal particles in the form of atoms, molecules, and molecular aggregates. The particles remain evenly dispersed and do not respond to the pull of gravity because of their continuous movement. Each particle is being bombarded by others of its own kind as well as by those of a different sort. This can be verified by observing even larger particles, such as certain pollen grains, under the highest powers of a light microscope. They will be seen to jostle about in a random manner, seeming to get nowhere. The apparently aimless motion has been given the name **Brownian movement**. In an aqueous solution much of the activity is due to the bombardment of water molecules which, of course, are much smaller. It requires millions

μm are in the colloidal state. **A colloidal system** can often be observed with the naked eye. For example, if egg albumin, which is composed of large protein molecules, is placed in water, the solution has an opalescent appearance. Light rays will strike the suspended particles and be scattered, rather than pass directly through as would be the case if the particles were smaller. We do not see the individual particles, only the effect produced by scattered light. However, if the particles are larger than 0.1 μm in diameter, the light is blocked altogether and the system appears opaque, as it does in milk, for example. The larger particulate matter in milk, of course, separates out (cream on the surface) and is therefore not colloidal.

The large size of colloidal particles also

of hits of water molecules to move the huge visible particles, and these must be concentrated more on one side than the other if movement is to occur in any one direction. It is the change in concentration of bombardment that causes the particle to move in the random fashion that is observed. This type of activity is essential in keeping the particulate matter dispersed.

Other factors complicate the behavior of particles in solution. We learned earlier that atoms and molecules may become ionized; that is, they may carry an **electrical charge, positive** when electrons are short, and **negative** when electrons are in excess. We also know that particles of the same charge repel one another while those of unlike charges attract.

The electrical charge on colloidal particles, in addition to Brownian movement, keeps them dispersed. The solid particles of a colloidal system are either positively or negatively charged, hence are kept apart by

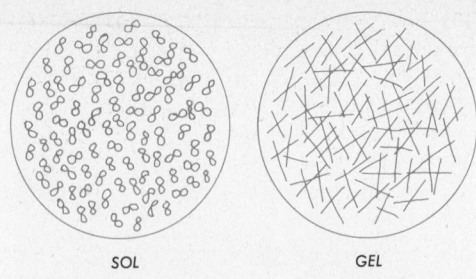

SOL   GEL

Fig. 2–27. A schematic explanation of the formation of sols and gels. In a sol the elongated molecules are folded and flow smoothly past one another in a more or less fluid state. In a gel, a lattice-work effect is produced when the elongated molecules unfold and become linked together to produce a colloidal network. Such a physical arrangement of the particles produces an elastic material which may also be contractile.

the charge. If the charge is neutralized by an opposite electrical charge, the colloidal particles will aggregate and settle out (Fig. 2–26). Since protoplasm is primarily a colloidal system, these facts have a profound effect upon its physical nature.

Colloidal particles possess another important characteristic which accounts for the behavior of protoplasm. This is referred to as **phase reversal** or **sol-gel transformations.** This can be illustrated by adding many colloidal particles to a system, or withdrawing water, so that the particles move close together—finally touching. When this happens the particles become intricately entangled, forming a spongelike network, called a **gel** (Fig. 2–27). The original dispersed colloidal particles become continuous, holding water and other dissolved particles in isolated droplets. As the name implies this has the consistency of gelatin. It is this property of protoplasm that gives to such fragile forms as amoebas and jellyfish their body shape.

If the particles are withdrawn or water added to a colloidal system, the opposite effect results, that is, it becomes fluid and is said to be in the **sol** state. This alternate switching of the sol and gel states is characteristic of protoplasm. Which state it is in at any one time depends on a number of con-

Fig. 2–26. Colloidal particles of like charge repel one another and so remain in suspension as shown in A. If particles of opposite charge are added (B) the particles are neutralized and settle out as shown in C.

COLLOIDAL PARTICLES

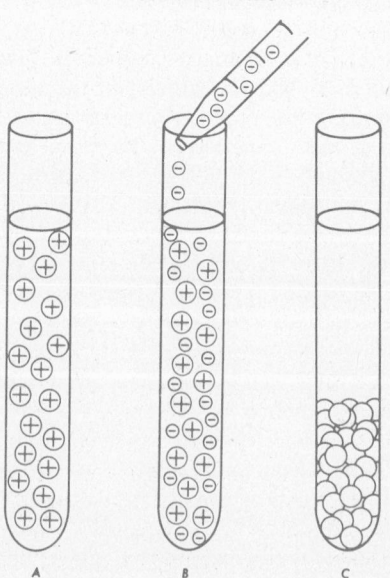

A   B   C

ditions. As the temperature rises the particles move more rapidly and the gel is converted to a sol, a familiar fact to anyone who has worked around the kitchen. Other factors such as extremes in *p*H or pressure affect the sol-gel state of colloidal systems.

Sol-Gel Interactions in Cell Movement. Most animal cells show protoplasmic movement to a greater or lesser extent; at the minimum they round up at division before separation into two cells, and some cells such as white blood cells (**leucocytes**) and amoebae move very actively by means of protoplasmic extensions called **pseudopodia** (false feet). See Chapter 6 for a more extensive description of amoeboid movement. Figure 2–28 illustrates such an amoeba (*Pelomyxa pelustris*) which has only one pseudopod and appears to move by a continual conversion of sol to gel at the anterior end and gel to sol at the posterior end, a conversion apparently accomplished by use of energy to fold up the molecules. Recently it

has been shown that many cells contain small (50–60 Å in diameter) elongate particles called **microfilaments** in areas corresponding to gel regions, and that these filaments are chemically very similar to the muscle protein **actin,** which could thereby be involved in "contracting" the gel. This pressure would be applied primarily at the lateral and posterior outer margins of the amoeba; we thus have the "toothpaste tube" model for amoeboid movement, which combines the ideas of a contraction of the outer gel with gel-sol and sol-gel interconversions. This is a very active area of research at this time, and a much more precise molecular understanding of cell movement seems close at hand.

Limiting Membranes

Protoplasm is always confined within containers which have special properties. The container acts as the boundary between the colloidal system of the protoplasm and the external world which is quite different from the protoplasm itself. It is a physical principle that at the region where a colloidal system contacts the surrounding medium (water, air, or another colloidal system), called the **interface,** a **membrane** forms owing to the physical forces placed upon the molecules in this region. They become tightly packed and arranged in a definite orientation that produces the membrane as illustrated by the surface film on boiled milk. Such a boundary on protoplasm is the **plasma membrane.**

The plasma membrane is composed of two protein layers between which is a layer of lipid molecules (Fig. 3–10). The membrane is normally in the gel state, a condition which is maintained only as long as the proportion of calcium and potassium ions are in the proper levels. Any change in the concentrations of these ions, either within or without, drastically affects the membrane. For example, if an amoeba is placed in a solution high in potassium ions and low in calcium ions a new membrane fails to form if a rent is made in it, whereas in a solution

Fig. 2–28. Conversion of sol to gel by folding and unfolding of molecules in an amoeba. In the center, cytoplasm flows forward, slows down, and changes to gel as it reaches the anterior end of the cell. Eventually, as the amoeba moves forward, a given gel region becomes located posteriorily, and then is converted back to sol. Spacing and size of arrowheads indicate relative amount of plasmasol movement. Dashed arrows at posterior region (uroid) suggest energy input.

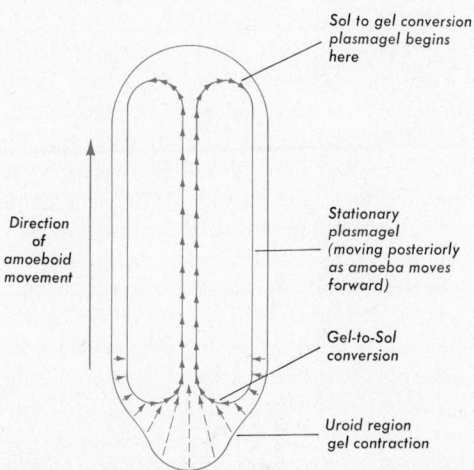

Sol to gel conversion
plasmagel begins
here

Direction
of
amoeboid
movement

Stationary
plasmagel
(moving posteriorly
as amoeba moves
forward)

Gel-to-Sol
conversion

Uroid region
gel contraction

with the proper proportions of these ions a new membrane forms at once. If the concentration is high in calcium ions and low in potassium ions, the entire cell gels, in other words, it loses all of its fluidity and becomes a congealed corpse. Therefore, a careful balance must be maintained between these two ions if a normal membrane is to form around the protoplasmic mass.

Such a sheet of protein and lipid molecules must possess some unique properties if it is to perform the necessary functions of keeping the protoplasm from disintegrating. This is accomplished by the **physical nature** of the membrane itself. The lipid and protein molecules cling together but between them are small openings where they do not fit quite snugly. These openings allow certain molecules to pass through in both directions, whereas others, owing to their large size or chemical nature, are prevented from being transported across. These membranes are thus said to be **selectively permeable or semipermeable.**

Water and other small molecules may pass freely from one side to the other. Fatlike particles can dissolve in the liquid molecules forming the membrane, pass inside, then slowly move out of the other side to the interior of the cell. Larger protein molecules cannot penetrate the membrane because the openings are too small to admit them. Glucose and amino acid molecules pass through the openings readily but some of the ions, even though much smaller, are unable to get through owing to the electrical charge which they bear.

Just how the many complex molecules move in and out of the cell is not completely known, but the movement of smaller ones such as water and certain ions is understood.

Diffusion and Osmosis

Diffusion is a process by which molecules tend to move from an area where they are highly concentrated to an area where few of these molecules exist (Fig. 2–29). This occurs in the presence or absence of a membrane as long as the membrane is permeable

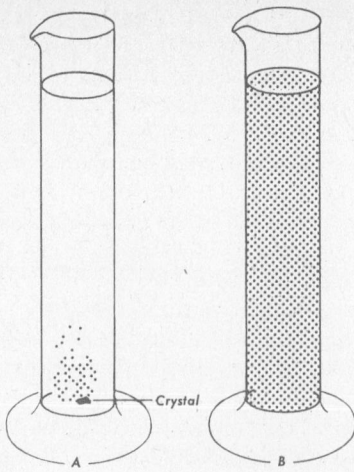

Fig. 2–29. Molecules tend to move from higher to lower concentrations. Thus in A, a crystal of salt is placed at the bottom of a cylinder of water. It immediately begins to dissolve; that is, the molecules mingle with those of the water. An equilibrium is established at some later time as shown in B.

to the molecules. Molecules which easily pass through the cell membrane tend to equalize their numbers on both sides of the membrane. Thus, in Figure 2–30 water molecules will move into the cell from the outside of the membrane, since there is a higher concentration of water molecules on the outside. Once the molecules—water, ions or other particles—pass through the membrane of a cell, they diffuse throughout the protoplasmic mass until they have become evenly distributed, that is, they have reached an equilibrium.

A different situation exists when some of the particles surrounding a membrane cannot penetrate. Under these conditions water molecules will move in order to establish an equilibrium. This unequal movement of water molecules through a semipermeable membrane is called **osmosis.** In a situation where a membrane is permeable to water, but impermeable to large sucrose molecules, the sucrose molecules constantly strive to intermingle with the water molecules on both sides of the membrane, but since they cannot pass through there is an uneven distribution of the water molecules on the two sides. The water and sucrose molecules bombard the membrane with about equal hits on one side; corresponding hits are made by the water molecules on the other side, but since there are only water molecules present, one passes through for every molecule of sucrose on the other side which cannot get through. The water molecules are passing toward the sugar side more rapidly than they pass in the other direction, thus building up a hydrostatic pressure on the sugar side (Fig. 2–30).

From biophysical evidence the mechanism of osmosis may be more complex than simple diffusion which does not account for the rapid flow of water through the membranes of plant cells. Where diffusion may initiate the process, hydrostatic differences within the membrane itself may be responsible for most of the flow of water through the pores in the membrane.

Osmosis can be easily demonstrated by placing a strong sugar solution in a bag of thin skin, such as frog skin, and tying it to a small-bore tube (Fig. 2–30). The water passes into the bag, or toward the sugar, thus building up a pressure, the **osmotic pressure,** within the bag which is registered by the rise of fluid in the small tube. If properly constructed, such an apparatus demonstrates a

Fig. 2–30. An osmometer is made from the skin of a frog's foot into which a strong sugar solution is poured and a small-bore tube attached. The diagrammatic view at left shows why the membrane is permeable to water and not to sugar.

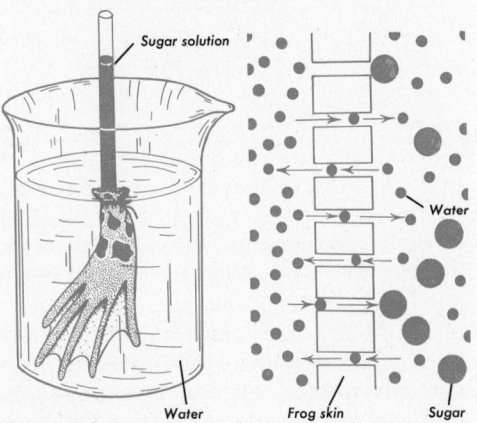

Sugar solution

Water

Water   Frog skin   Sugar

rise of the fluid column to many feet, the height being determined primarily by the effective semipermeability of the bag. Usually such a preparation is not absolutely semipermeable; some sugar goes the other way, that is, after a time sugar molecules will have wedged their way through the membrane to mingle with the uniform water molecules outside.

The practical significance of osmosis can be demonstrated with red blood cells. These tiny disks have a limiting semipermeable membrane which encompasses a large variety of particles that constitute the internal protoplasm of the cell. While these corpuscles are suspended in the fluid of the blood, water passes in and out of the cell with equal speed, so the membrane is uninfluenced by the movement. Other particles pass in and out, but in so doing there is always an even distribution on both sides of the membrane; that is, for every particle going inside another comes out, so that the numbers, not necessarily the kinds, are approximately the same all of the time. Such a surrounding fluid is said to be **isotonic** to the corpuscles (Fig. 2–31). If the corpuscles are now separated from the fluid portion of the blood and placed in distilled water, a very rapid and sudden change occurs. The water moves into the cell because the dispersed particles are greater (less water molecules) inside than out (where there are more water molecules), so the water flows in, causing the membrane to swell and eventually burst (Fig. 2–31). The water in which the cells were placed is said to be **hypotonic** to the blood cells. Hypotonic solutions should not be injected into the blood stream of an animal because the destruction of red cells, as well as others, could prove fatal.

Again, if the cells were placed in a salt solution in which the numbers of dispersed particles were much greater than on the inside of the cell, the direction of flow would be in the opposite direction, namely, out of the cell, causing it to shrink to only a fraction of its normal size (Fig. 2–31). The reason here is the same as in the previous case. Such shrunken cells are called **crenated cells,**

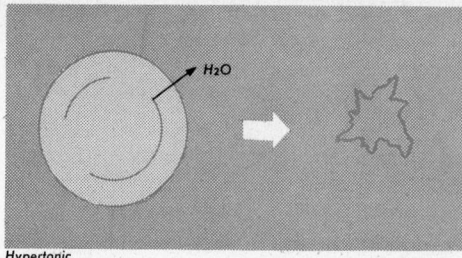

Hypertonic

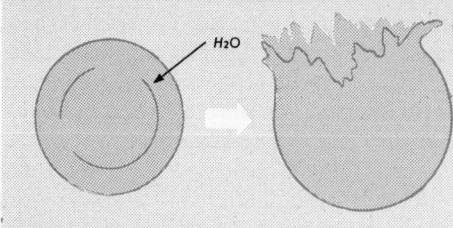

Hypotonic

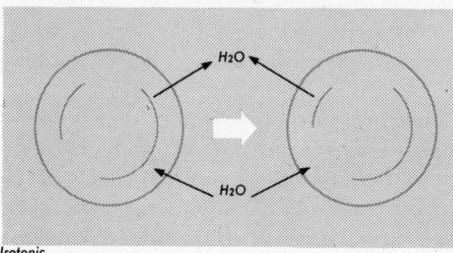

Isotonic

Fig. 2–31. The effect of salt solutions of various concentrations on the red blood cell.

and such a concentrated salt solution is said to be **hypertonic** to the blood cells. If such a hypertonic solution were injected into the blood stream, it might also prove fatal because of the rapid destruction of blood cells. It is important, then, whenever cells are placed in any kind of solution, that the medium have the same number of particles, or, in other words, it must have the same osmotic pressure as the cells themselves, if severe trouble is to be averted.

Another illustration of the effect of osmotic pressure might be cited because of its practical application. Perhaps you have wondered why a person, floating upon vast quantities of water, must die at sea if he has no fresh water to drink. Sea water is heavily laden with salts and its osmotic pressure is considerably above that of the blood and tissues of man and all other land animals. If, then, he should take sea water into his stomach it would extract from his stomach the precious water that is already short, eventually filling his stomach so that he would be forced to throw it up. This would leave his body with less water than it had before he swallowed the sea water. That is why drinking sea water can be fatal.

Passive Transport

The movement of particles across membranes is of two sorts depending upon whether or not energy is required. The movement of water through a semipermeable membrane, as illustrated in osmosis, is spoken of as **passive transport**, because the direction the water migrates depends on the relative concentrations of these and other molecules on each side of the membrane and also because no energy is required. In Figure 2–32 A, water flows into the cell owing to the fewer molecules of water inside the cell (or conversely, the greater the number of particles of the dissolved substance). Alternatively, in B the water flows out of the cell for the same reason (Fig. 2–32 B). In Figure 2–32 C, however, the effect of the electrical charge is responsible for the direction of flow of the different ions. If there is a high concentration of positively charged ions on the inside and negatively charged ones on the outside, an electrical potential gradient is established across the membrane. This brings about a gradual migration of the ions across the membrane. Hence, positively charged particles move out of the cell and negatively charged ones move in, eventually establishing an equilibrium.

Active Transport

For the proper functioning of some cells it is necessary to maintain an unequal distribution of ions on either side of the membrane. In such cases, ions must move from an area of low concentration to one of high concentration or against a concentration gradient for which energy is required to

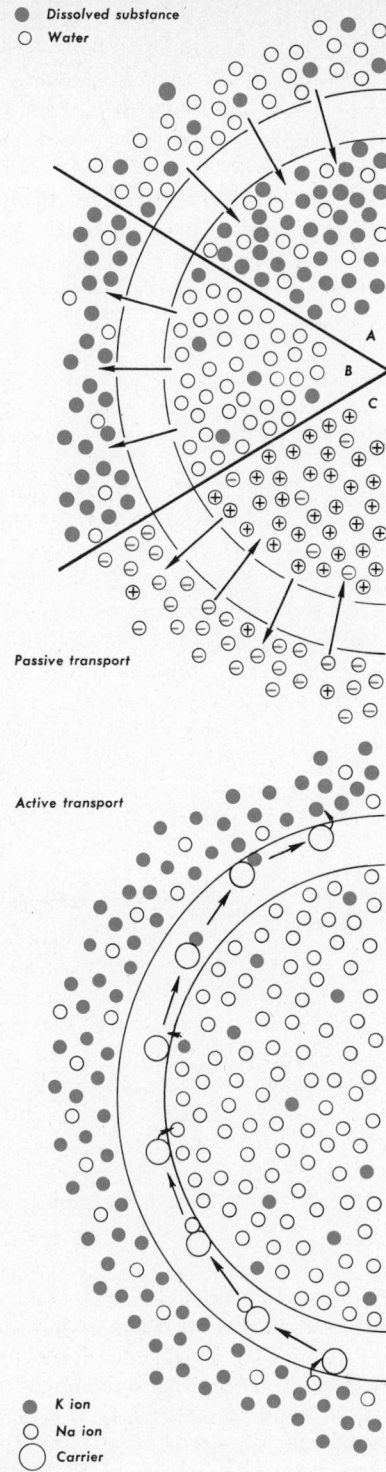

● Dissolved substance
○ Water

A
B
C

Passive transport

Active transport

● K ion
○ Na ion
◯ Carrier

counteract the forces of diffusion which tend to bring about an equilibrium of the particles. This type of movement is called **active transport,** because energy is involved in moving particles. The manner in which this is accomplished is shown in a highly schematic fashion in Figure 2–32.

It has been postulated that a carrier molecule located in the membrane itself may pick up a sodium ion from the outside forming a temporary compound, and carry it inside the cell, where it is released. Alternatively, it may transport potassium ions outside the cell. The carrier molecule requires energy supplied by the cell for its action. This active transport system makes possible the maintenance of different concentrations of specific ions on the two sides of a membrane.

DEFINITION OF LIFE

We have seen the characteristics of living things, but this is merely a list and really does not get at the heart of a definition. We have acquired sufficient understanding of the common basic features and mechanisms which account for these characteristics. Our newer knowledge, accumulated primarily within the past two decades, has resulted in an interpretation of biological structure and function in terms of chemistry and physics. Two hybrid sciences, biochemistry and biophysics, have given us a basic understanding of some areas in biology at the molecular and macromolecular levels; indeed they have merged structure and function to the point where we think today only in terms of both rather than, as in the past, of separate disciplines. As a result, problems in biology are approached from this viewpoint which has yielded some unifying basic patterns at the

Fig. 2–32. Highly schematic representation of how particles migrate through membranes. Carrier molecules may not move circumferentially as far along the membrane as illustrated and in many cases are large enough to extend from one side of the membrane to the other. Given sufficient energy, some carriers apparently can move a particular ion directly through the membrane with little change in position.

molecular level which are common to all living things. For example, all living cells possess DNA, RNA, and proteins made from a nucleic acid code. We are impressed with the similarities of living things in terms of physical and chemical makeups, including the "standard approach" to such things as protein synthesis, which we will discuss later. Cells do not break the laws that are operative in the physical world; rather they seem to be an extension of a once complex nonliving system which acquired, in the course of two or more billion years, the added features of reproduction and growth, thus emerging as a living organism. On this basis we can define life in the broadest terms based on a mechanistic viewpoint. *A living organism consists of a highly organized, complex system of particulates, from atoms to macromolecules, capable of self-direction, growth, and reproduction, utilizing exogenous matter and energy.*

SUGGESTED SUPPLEMENTARY READINGS

Books

*ABERCROMBIE, A., HICKMAN, C. J., and JOHNSON, M. C., A *Dictionary of Biology*. Great Britain: Hunt Barnard, Ltd., 1973.

*BAKER, J. J. W., and ALLEN, G. E., *Matter, Energy and Life*. Reading, Mass.: Addison-Wesley, 1965.

BERNAL, J. D., *The Origin of Life*. New York: World, 1967.

CALVIN, M., *Chemical Evolution*. New York: Oxford University Press, 1969.

*KEOSIAN, J., *The Origin of Life*. New York: Reinhold, 1964.

*McELROY, W. D., *Cell Physiology and Biochemistry*. Englewood Cliffs, N.J.: Prentice-Hall, 1964.

*MILLER, S. L., "The Origin of Life," in W. J. Johnson and W. C. Steere, eds., *This is Life*. New York: Holt, Rinehart & Winston, 1962.

*NOVIKOFF, A., and HOLTZMAN, E., *Cells and Organelles*. New York: Holt, Rinehart & Winston, 1970.

*OPARIN, A. I., *Life: Its Nature, Origin and Development*. New York: Academic Press, 1961.

*STEEN, E. B., *Dictionary of Biology*. New York: Barnes and Noble, 1971.

WHITE, A., HANDLER, P., and SMITH, E., *Principles of Biochemistry*. New York: McGraw-Hill, 1968.

WOLFE, S., *Biology of the Cell*. Belmont, Calif.: Wadsworth, 1972.

* Available in paperback.

Articles

GAMOW, G., "The Origin and Evolution of the Universe." *American Scientist*, February, 1951.

PONNAMPERUMA, C., and MARINER, R., "The Formation of Ribose and Deoxyribose by Ultraviolet Irradiation of Formaldehyde in Water." *Radiation Research*, Vol. 19, p. 183, 1963.

PONNAMPERUMA, C., LEMMON, R. M., MARINER, R., and CALVIN, M., "Formation of Adenine by Electron Irradiation of Methane, Ammonia and Water." *Proceedings of the National Academy of Sciences*, Vol. 49, pp. 737–740, 1963.

STEIN, W. H., and MOORE, S., "The Chemical Structure of Proteins." *Scientific American*, February, 1961.

UREY, H., "The Origin of the Earth." *Scientific American*, April, 1952.

WALD, G., "Origin of Life." *Scientific American*," August, 1954.

3

UNITS OF LIFE- CELLS

Going back to the story of the origin of life, we remember that protoplasm came into being very slowly, and its organization into cells must likewise have required a very long time. All cells, from the free-living one-celled protozoa to the many-celled animals, are extremely complex, and certainly the result of hundreds of millions of years of evolution. All cells perform essentially the same functions whether they are deep in the muscles of an elephant or exist as independent individual units like the amoeba. Before we examine the general cell structure more carefully, a little of the **historical background** of cell studies may help us in our perspective of modern biology.

Long ago, about 1665, an English biologist by the name of Robert **Hooke** cut thin slices of cork and placed them under his newly fashioned microscope. He noted that this material was composed of numerous tiny compartments to which he as-

signed the name "cells," a name that has come down to the present time. He gave them this appellation because they reminded him of the cubicles in monasteries in which monks of his time lived. Today any beginning biology student can repeat Hooke's experiment and be rewarded with a much better visual image of cells, although he would probably lack some of the enthusiasm that compelled this inquiring man of the seventeenth century to make the discovery.

Everyone who was sufficiently curious to follow the exciting hobby of looking through the newly invented microscope of this early period saw that all parts taken from living things were composed of these tiny "bladders," as **Grew** called them. Their first and only interest seemed to lie in the fact that animals and plants were made up of these tiny units, and they paid little or no attention to the internal parts of the cells for over 100 years. By 1824, **Dutrochet** in France had noted that plants grew by an increase in number and size of the cells of which they were composed. By 1831, **Brown** had seen and described the centrally located body within the cell which he called the **nucleus.** This, together with the increased use of the microscope, centered the attention of investigators on the internal parts of the cell. **Dujardin** conceived the idea of a universal "life-substance" common to all cells and gave it the name **protoplasm.** About this time (1839), two Germans, **Schleiden** (a botanist) and **Schwann** (a zoologist), advanced the **cell theory,** which was merely a concise statement of what had been learned by a great many men up to this point, namely, that all living things were composed of cells.

The cell theory gave biologists for the first time a tangible idea upon which to base further studies. It meant to them what the molecular theory did to the chemist and physicist, and as a result, biology became more and more a study of cells rather than a study of the organism as a whole. Because cells occupy the central core of studies in embryology, reproduction, growth, heredity, and behavior, investigators have given them steadily increasing attention through the years. Once a thorough understanding of the cell is acquired, many of our most perplexing problems in biology will be closer to solution.

METHODS OF EXAMINING CELLS

Microscopy. Since Robert Hooke saw cells for the first time, great strides have been made in the production of instruments which have made it possible to explore the microscopic and submicroscopic anatomy of the intact cell. If we examine the drawing of a cell which represents cytologists' knowledge over five decades ago (Fig. 3–1), and compare it to a recent drawing (Fig. 3–10), the difference is striking. The empty regions are now filled with bodies of various dimensions and the structures that were drawn in the earlier sketch now have specific mor-

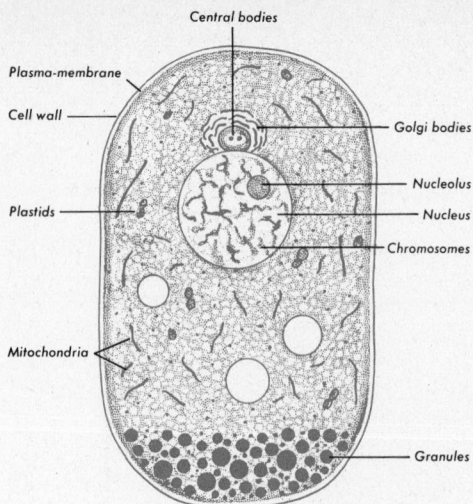

Fig. 3–1. A diagram showing what was known about cell morphology in 1925, according to E. B. Wilson.

phology. These changes have come about as a result of improvements in instruments, particularly in the magnification and resolving power of our modern microscopes.

Essentially all of our information about morphology relies finally on the human eye. In terms of studying cells, however, it has two principal limitations—**magnification power** and **resolving power.** Depending on individual sensitivity, the eye has a resolving power of 0.05 to 0.1 millimeter. Objects with a smaller diameter than this are invisible and two lines closer together than these distances will be seen as one line. The magnification power of the eye is fixed.

Microscopes both **magnify** and **resolve.** The resolution limit ultimately depends on the wavelength of the illuminating source (either light or electrons) and the design of the objective lens. Figure 3–2 illustrates the similarity between light and electron microscopes. The **condenser** concentrates (condenses) the illumination on the object, the **objective** gathers the information from the **object,** and the **projector** projects this information to a photographic plate, retina, or screen.

Resolution depends upon the information

gathered by the objective. Magnification is the function of the projector (eyepiece of light microscope), and must be sufficient to allow the eye to comfortably resolve the finest details gathered by the objective. Once that point is reached, further magnification is harmful because of loss of intensity and reduction in the field of view; such excess magnification, since it does not improve resolution, is called **empty magnification.** The light microscope uses glass lenses, and the electron microscope uses magnetic lenses. Because of the nature of the interaction between matter and light or electrons, the best resolution obtainable is equal to about one-half the wavelengths of the illumination used. For the light microscope, the average wavelength is 5500 Å, which corresponds to the center of the visible region of the electromagnetic spectrum. Pure 5500 Å radiation (monochromatic) would appear apple-green in color, and light microscope optics are usually corrected to perform optically at this wavelength. The eye is also most sensitive to this region. One-half of 5500 Å (0.55 μm) is 0.275 μm, which is the usual resolution limit; by going to the blue end of the spectrum, i.e., 4000 Å, one can achieve the maximum resolution of 0.2 μm. To obtain this resolution the objective lens must be perfectly ground and corrected. The optical limit of the light microscope was reached many years ago, which meant that particles or structures smaller than 0.2 micron could not be seen. As a result little new information was gained concerning cell structure until an instrument of greater resolution was invented.

The solution of this problem came about when physicists explored other sources of illumination. Among these the most rewarding has been a beam of high-speed electrons which resulted in the <u>T</u>ransmission <u>E</u>lectron <u>M</u>icroscope (TEM).

The source of electrons is a tungsten filament heated to incandescence by 50,000 to 100,000 volts of electricity. The electrons are focused by electromagnetic lenses so that they pass through the specimen where they are deflected or absorbed depending on the density of various parts of the specimen (Fig. 3–2). The resulting black-and-white image impinges upon a fluorescent screen which makes it visible to the eye. Since we cannot see electrons we must rely on the screen to visualize the image; moreover, the image can be recorded only with a photographic plate for the same reason.

The wavelength of a beam of electrons produced by 50,000 volts is about 0.05 Å which means that theoretically the electron microscope should resolve objects with a diameter of 0.025 Å, less than the smallest atom. In reality, however, it has been impossible to perfect the lens system so that the resolving power is better than 1–2 Å, even in

Fig. 3–2. Comparison of the light and transmission electron microscope. See text for explanation.

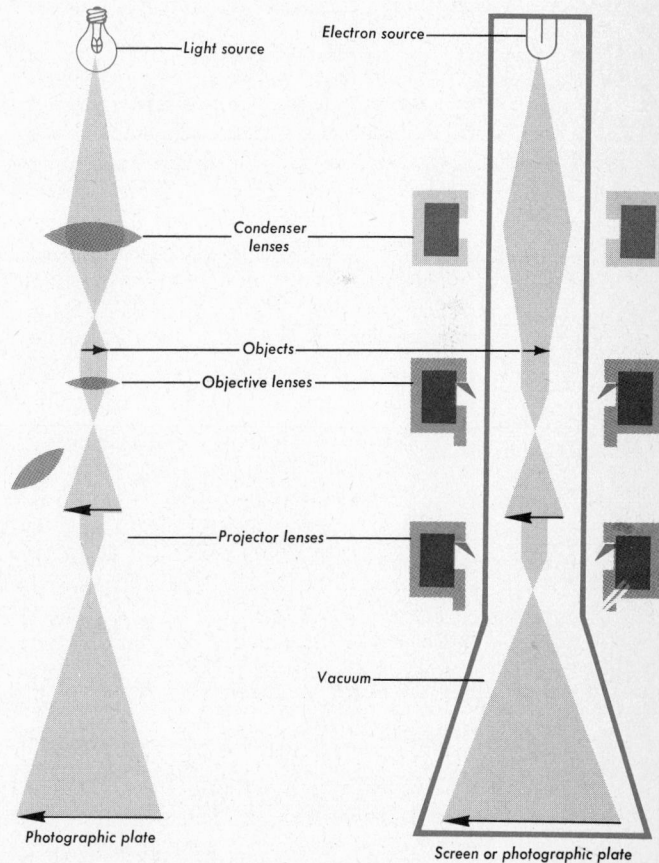

the best microscopes; this is still about one-thousand-fold improvement in resolution over the best light microscopes. A problem with the TEM is that it must pass electrons through the specimen, and since 50–100 KV electrons have very small penetrating power, the specimen must be very thin. Whole cells are almost always too thick and must be cut into thin slices or **sectioned**. Whereas cells and tissues are normally sectioned at 5–10 μm for the light microscope, the TEM requires sections in the range of 0.03–0.1 μm. Seeing one entire cell in three dimensions involves the very painstaking collection of several hundred **serial sections**, as many micrographs, and the difficult task of reconstructing the cell from these many pictures, a task that has been completely accomplished only a few times. In order to improve penetration so that thicker sections can be examined, high voltage transmission electron microscopes have been developed recently which can operate at an accelerating voltage of one million electron volts (1-MEV), are three stories high, and cost a dollar or more per volt. It is hoped that such instruments, by examining sections 1 μm or so in thickness, will allow visualization of internal relationships of such complicated three-dimensional objects as chromosomes. Efforts are also being made to design special ultrathin chambers containing living cells which can be put into the vacuum chamber of 1-MEV microscopes.

Another kind of electron microscope that has just recently come into prominence is the Scanning Electron Microscope (SEM), which is normally used to look at **surface structure** of many kinds of animals and cells. Figure 3–3 shows a diagram of such an instrument, in which electron lenses are used to form a very small spot of electrons which rapidly scan across the specimen in a series of sweeps similar to a television system. The spot, on striking the surface of the tilted specimen, causes the specimen to give off secondary electrons which are collected in a scintillation detector. The current produced in the detector is amplified and used to regulate the intensity of the spot of light on the face of a display tube, which then in effect gives a television picture of the surface of the specimen. The tenfold improvement in resolution of the SEM over the light microscope is coupled with a significant improvement in the depth of field (the amount of the specimen which is in focus). Pictures taken by photographing the display tube image give a composite view of surface relationships unobtainable by other methods.

Electron microscopes have not been very useful in recording internal details of **living** cells. For this purpose, the most effective instrument has been a special form of the light microscope, the **phase contrast microscope** invented by the Dutch scientist **Zernike** in 1932, for which he received the Nobel Prize. By utilizing an optical trick, this instrument makes use of very

Fig. 3–3. A schematic view of a scanning electron microscope (SEM) as described in the text. Only the surface of a specimen can be examined, and the resolution is dependent on the size of the spot falling on this surface, the spread of energy in the specimen, and the speed of the scan. Redrawn from G. Meek, *Practical Electron Microscopes for Biologists* (New York: Wiley Interscience, 1970).

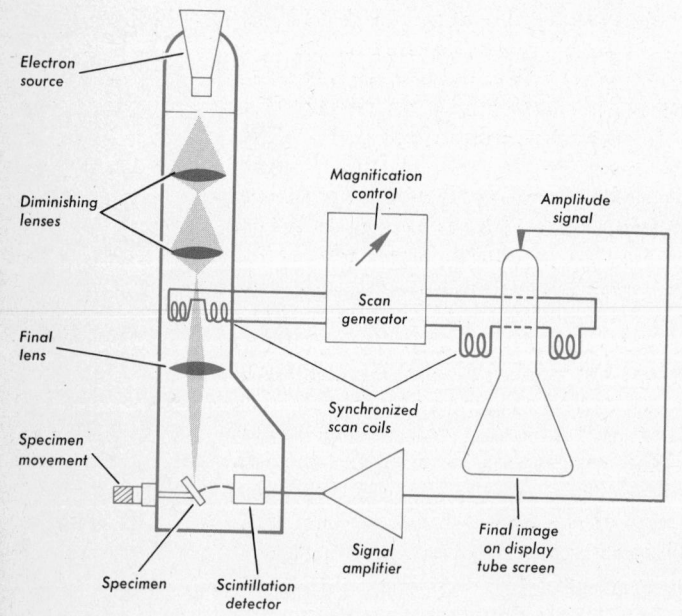

Electron source

Diminishing lenses

Final lens

Specimen movement

Specimen

Scintillation detector

Magnification control

Scan generator

Synchronized scan coils

Signal amplifier

Amplitude signal

Final image on display tube screen

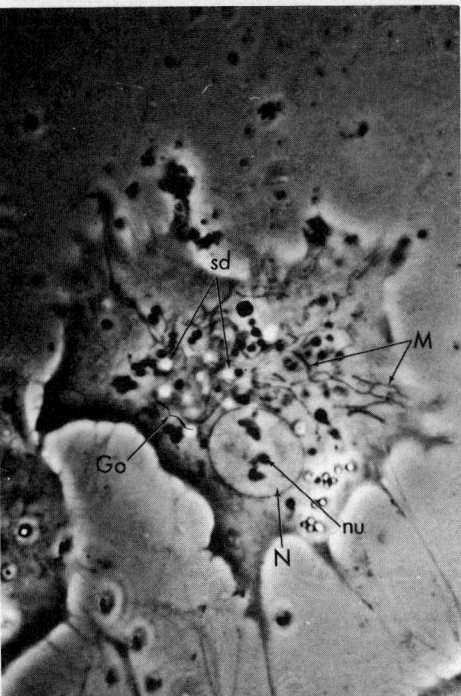

Fig. 3–4. A phase contrast photograph of a living tissue bone cell of a 14-day-old chick cultured for 26 days. The parts of the cell shown are Golgi body (Go), nucleus (N), nucleolus (nu), mitochondria (M), and secretory droplet (sd).

Cell Fractionation. A commonsense method for finding how something works is to take it apart and examine its components. Taking the cell apart is called **cell fractionation**; in principle it is easy to understand, although in practice there are many technical innovations which are required, depending in part on the kinds of cells and tissues to be examined. Many cells are needed and liver tissue has been a favorite experimental material because the liver is such a large organ and is composed of predominately one type of cell, the **hepatocyte**. The tissue is first cut up into very small pieces, put into a sucrose buffer solution, and placed in an ice bath to essentially stop all cellular activity; in all subsequent steps the preparation is maintained at ice-bath temperature (about 4°C). The minced tissue is then placed into a "homogenizer," which is most often a teflon pestle fitting into a glass tube (Fig. 3–6). The pestle is turned by an electric motor, and the device breaks the cells by forcing them in and out of the space between the pestle and tube wall. The resulting homogenate consists of some unbroken cells, fragments of membranes, and intact organelles including nuclei and mitochondria. The tube is lightly centrifuged to pack pieces of tissue and whole cells in the bottom of the tube (pellet), and the supernate is placed in an upper portion of a clean tube, which contains a gradient of increasing concentrations of sucrose (Fig. 3–7), with the most concentrated or dense solution at the bottom. The tube is again centrifuged and the heaviest components, chiefly nuclei, are concentrated at the bottom of the tube. The lighter mitochondrial fraction accumulates in a band above the nuclear fraction, and the microsomal fraction above that. The microsomes consist of broken pieces of membrane mostly from the endoplasmic reticulum. The top layer (supernate) now contains various soluble components including enzymes and other proteins, ribosomes, and lipid granules. Each of these fractions can be biochemically analyzed to determine how much and what kinds of protein, nucleic acid, lipids, etc.,

small differences in the speed at which light passes through the various cellular components, converting these differences, which the eye cannot detect, into differences in amplitude contrast, which the eye can detect. The details of this microscope are beyond the scope of this book; suffice it to say that, within its resolution limits of about 0.275 μm, detailed examination of cells, such as the living bone cell seen in Figure 3–4, have tended to confirm the view that the sizes, shapes, and general relationships obtained from studies of fixed and stained cells are reasonably close to those found in life.

Figure 3–5 compares pictures taken with the TEM, the light microscope, and the SEM.

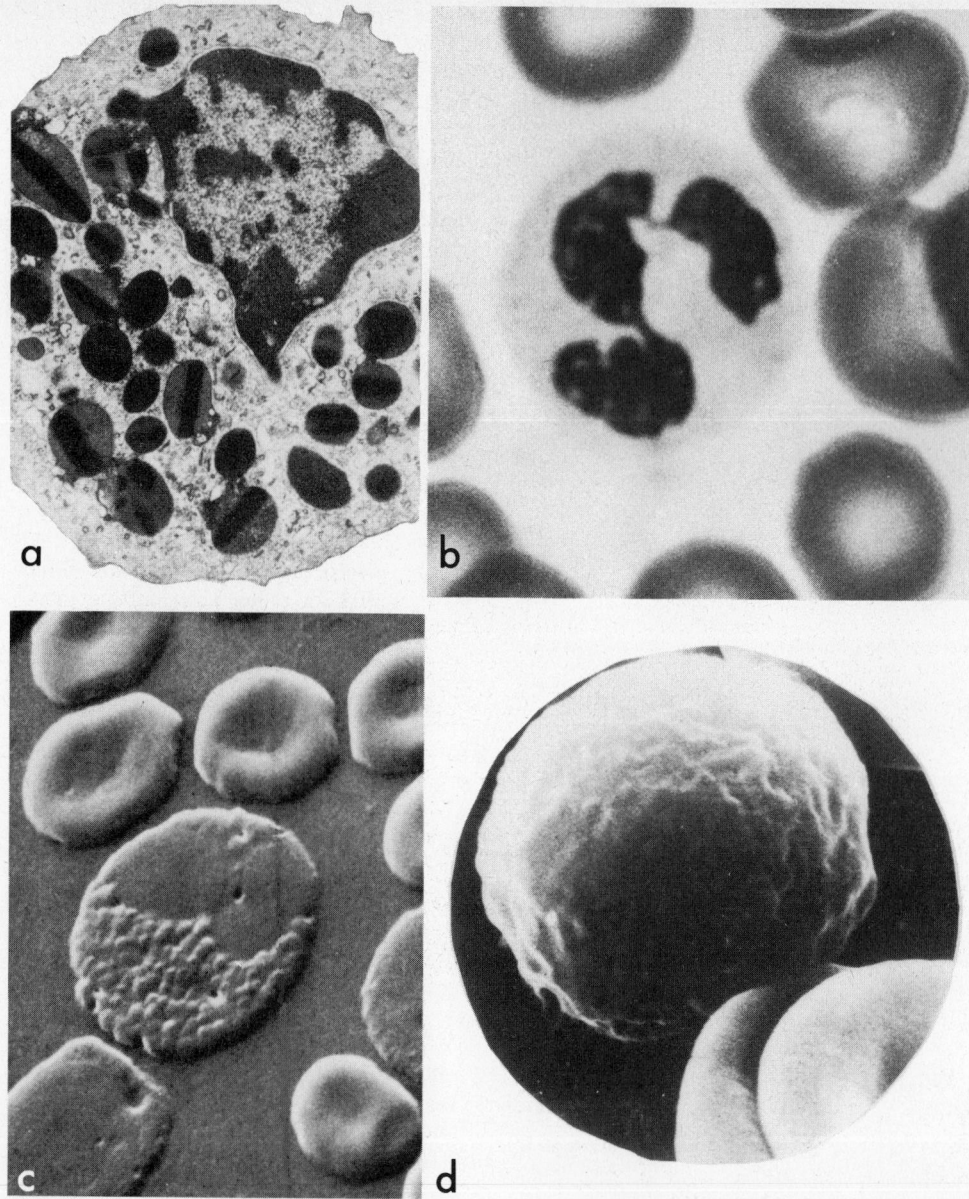

Fig. 3–5. A comparison of pictures of fixed blood cells with the electron (TEM), light, and scanning (SEM) microscopes. In (a) the high resolution capability of the TEM is illustrated; the micrograph is a thin section showing essentially a two-dimensional image of a white blood cell (eosinophilic leukocyte), which contains many specialized cytoplasmic granules. In (b), a light micrograph of a similar leukocyte illustrates the multi-lobed appearance of the nucleus (N) evident in the whole cell but not in the thin section. Some three-dimensional information is given in the light micrograph, but resolution is relatively poor. Part (c) shows a scanning micrograph of another leukocyte; drying has caused a flattening of the cell so that the granules and nucleus are visible. In (d), a lymphocyte was fixed to more accurately preserve its shape before it was examined in the SEM. In both (c) and (d) the biconcave-disc shapes of the red blood cells adjacent to the leukocytes are strikingly illustrated. Note that the resolution of the SEM is intermediate between the light microscope and the TEM, but that more of the cell is in focus; i.e., the depth of field in the SEM is much greater. (Composite arrangement from R. Dyson, 1974.)

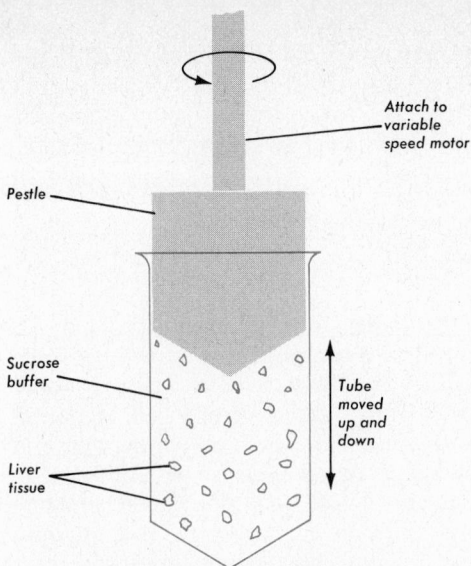

Fig. 3–6. Cell homogenizer. Small pieces of liver are placed in a sucrose solution and the cells gently broken open between the walls of the tube and the moving pestle. The tube is then centrifuged to remove unbroken tissue and debris and the supernate is placed on top of another sucrose solution illustrated in Figure 3–7.

are present, and further purified as desired. These methods were used, for example, to establish that most of the energy-producing, enzyme-catalyzed processes occur within the mitochondria.

Cell Culture. While it is convenient to make use of a large organ like the liver as a source for cell fractionation, there are many other kinds of cells present in the liver besides hepatocytes, for example, red and white blood cells, cells of the blood vessels themselves, secretory cells, and duct cells. The discovery of **tissue culture** by **Harrison** (1907) and its subsequent development by **Carrel** allows one to excise small bits of tissue, preferably embryonic tissue, from a living organism and place it in a medium which contains all of the nutrients required for growth and development of the tissue cells. For a long time this medium consisted of such complex substances as clotted blood plasma or embryonic extracts, but there has

been an intense search for more simplified media and biologists have found a chemically defined medium which supports growth of some kinds of cells.

This tissue culture technique has recently been refined one step further, a step which holds great promise for a better understanding of how cells function. It is now possible to isolate single cells from certain tissues and obtain thousands of cells from this parent cell. This technique is a very old one among microbiologists, but has only recently been successfully employed with tissue cells of higher animals. Such cell-lines are called **clones** and they have the advantage of being offspring of a single type of cell (e.g., a hepatocyte). As we shall discuss in greater detail in Chapter 23, the daughter cells arising from division of one hepatocyte are differentiated, tending to assume the characteristics of the parental type **in vivo** (i.e., in its normal site in the liver of a living animal); whereas when isolated in tissue culture, such clones tend to maintain hepatocytelike characteristics for a time, but eventually become less and less similar to their parental type; i.e., they become dedifferentiated **in vitro.**

This is the same kind of behavior exhibited by cancer cells in vitro, and the causes have been a matter of great interest. Normally these changes are not a result of failure attributable to division; mitosis generally results in a faithful distribution of exact copies of parental genes to the daughter cells. Rather, the evidence suggests that the activation or repression of different sets of genes is a principal cause of these changes. This is closely related to developmental processes, in which the embryo is able to **differentiate** its cell types to form different tissues and organs. Thus the liver cell, although genetically identical to the kidney cell, has some of the same genes turned on as the kidney cell, it has others turned on that are shut off in the kidney cell, and it has some turned off that are actively functioning in the kidney cell. The understanding of how this is controlled is probably closely related to understanding cancer mechanisms, and in that sense alone is worthy of intensive study.

Tissue culture clones, nevertheless, allow one to obtain a much more homogeneous population of cells in large numbers and thus to investigate, for example, how and why these cells are different from cells from another part of the body. An extension of these methods are the cell fusion techniques, wherein two cells of different animals can be fused into a single cell and development followed. We are still evaluating some of the potentials of these techniques.

Another method of investigating the nature of living cells is by **microsurgery,** which consists of introducing fine needles, electrodes, and pipettes into cells for the purpose of injecting or removing parts or measuring electrical potentials of various parts under a variety of conditions. The needles and pipettes are operated by an instrument of high precision; indeed, it is so accurate that even chromosomes or nucleoli can be removed from individual cells without destroying them. This instrument has added much information to the knowledge of the essentiality of parts of the cells, the physical nature of the protoplasm during different stages of growth, and cell differentiation.

CELLULAR ARCHITECTURE

In order to understand how the cell works, it is essential that we study the living cell in the intact state as well as isolated particulates. The study of killed, sectioned, and stained cells has also been invaluable in adding to our knowledge of cellular morphology. There is no doubt that structure and function are intimately related; thus, the structure of an enzyme is highly relevant to its function, and similarly energy production will not proceed in the mitochondrion if its structure is too severely damaged. In the following discussion of cellular architecture it is important to realize that any given structure probably has had a long evolutionary history, and that its size and shape have thereby been selected for optimum performance.

Living Cells. Groups of isolated cells, such as cells from blood, or protozoa, or tissue culture cells, or other free cells taken directly from an animal, are easiest to study. When one examines such a living cell, say a bone cell like the one in Figure 3–4, with an ordinary light microscope one sees a translucent jellylike mass of material whose outer margins are defined by differences in the refractive index of the cell and the medium surrounding it. We know that this cell is held together by a plasma membrane whose thickness ($0.01 \ \mu m$) is below the limits of resolving power of the light microscope. By carefully adjusting the condenser diaphragm on the microscope, one may see within the cell the **nucleus,** which is about $2 \ \mu m$ in diameter and is distinctly larger than any other intracellular structure. The remainder of the cell, the **cytoplasm,** contains bodies of various shapes which seem to be in constant motion. This is about all one sees under these conditions. However, if we now place the cell under the phase contrast microscope, we observe the same structures, but much better, and, in addition, we see other things (Fig. 3–4). The nucleus contains many small particles of uniform size and one or two large spherical bodies, the **nucleoli.** In the cytoplasm numerous tiny rodlike bodies, the **mitochondria,** can be made out. The cell membrane can be seen as a more distinct limiting boundary although one still cannot see that it has thickness. Other bodies may be present in the cytoplasm and again movement of the particles is characteristic. If we are going to learn more about the structure of the intact cell we must prepare it in a variety of ways, but in so doing we must kill it.

Killed or Fixed Cells. For over 100 years the detailed structure of cells has been studied from material that has been killed and stained in a very precise manner. There are hundreds of methods for doing this and the methods vary with the type of cell studied. In general, they involve certain common steps. The cells, whether they are single isolated cells or those incorporated in tissues, are killed suddenly with a **fixative,** which is a chemical compound that preserves the protoplasm with as little distortion as possible. The cells are then treated in a variety of ways, all very exacting, designed to set off the various parts of the cells in a contrasting manner so that they may be distinguished. This often involves cutting the material into thin slices which are then stained with a variety of dyes or treated with other compounds which are more or less specific, each attacking only a certain part of the cell, thus making that part stand out from other parts. These techniques have been very useful in identifying the internal morphology of cells and are still employed today (Fig. 3–8).

With the invention of the electron microscope various techniques were employed to prepare cells for study. The most useful killing agents turned out to be gluteraldehyde and osmium tetroxide which imposed the least amount of distortion on the cell parts. The most difficult problem to overcome was that of cutting the cells into sufficiently thin sections so that the electron beam would pass through and give a picture. This requires a cutting instrument (microtome) of high precision. Staining involves the immersion of the sections in a solution which deposits heavy metals on the cellular parts. This causes them to form heavier shadows thereby producing a picture with more contrast. By examining cells with the light, phase contrast, and electron microscopes we can describe rather accurately the morphology of cells.

Fig. 3–8. A young egg cell (Oocyte) of the fish, *Fundulus heteroclitus.* Large spherical nucleus contains several dark, round nucleoli and threadlike chromosomes. Clear inclusions in the cytoplasm are yolk vesicles.

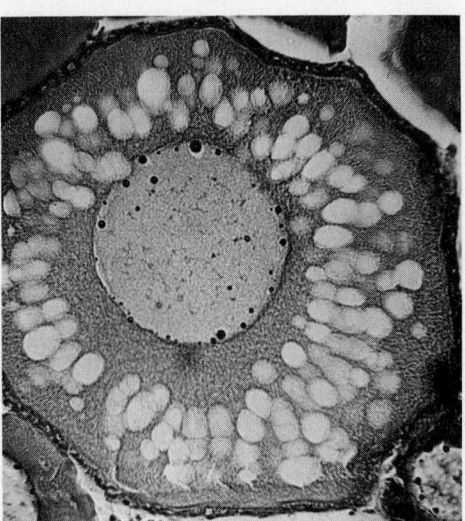

KINDS OF CELLS

Cells can be divided into two major groups, those lacking a nucleus (**prokaryotes**) and those possessing such a nucleus (**eukaryotes**). This classification has turned out to be a fundamental one, reflecting many important differences, although

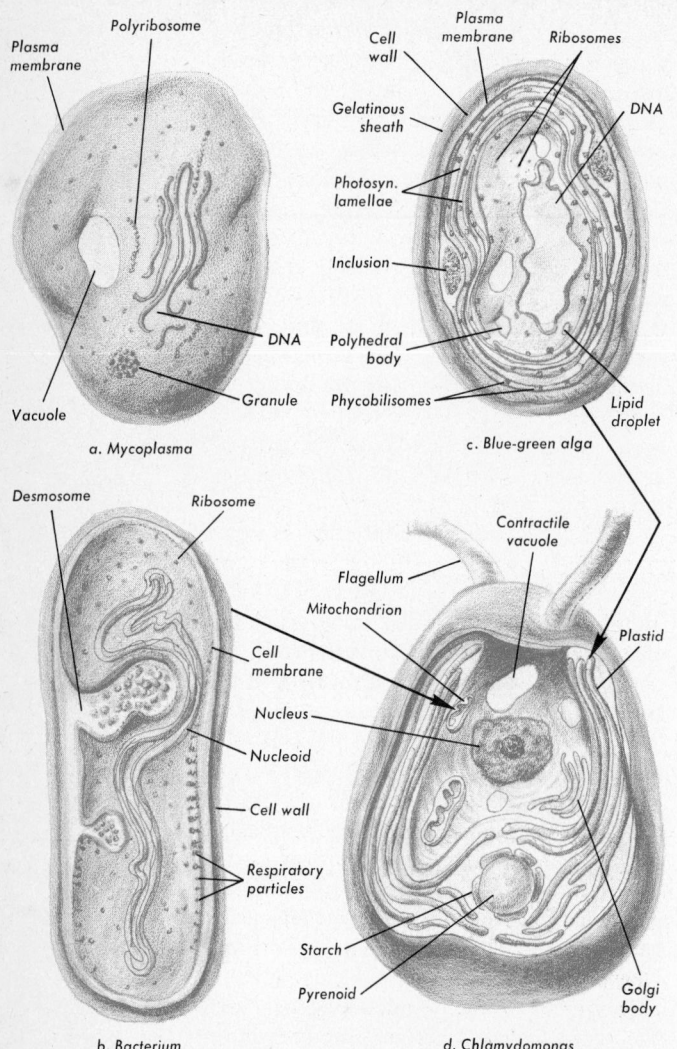

Plasma membrane
Polyribosome
Vacuole
DNA
Granule

a. Mycoplasma

Cell wall
Plasma membrane
Ribosomes
Gelatinous sheath
DNA
Photosyn. lamellae
Inclusion
Polyhedral body
Phycobilisomes
Lipid droplet

c. Blue-green alga

Desmosome
Ribosome
Cell membrane
Nucleoid
Cell wall
Respiratory particles

b. Bacterium

Contractile vacuole
Flagellum
Mitochondrion
Nucleus
Plastid
Starch
Pyrenoid
Golgi body

d. Chlamydomonas

Fig. 3–9. Diagrams of major cell types. (a) *Mycoplasma*, a simple and perhaps degenerate prokaryote. (b) A bacterium, showing relationship of DNA and mesosomes. (c) A blue-green algal cell showing photosynthetic lamellae and special pigment granules called phycobilisomes. (d) *Chlamydomonas*, a eukaryotic protozoan and green algal cell. Arrows indicate the current view that the plastid and mitochondrion were probably derived from the establishment of a symbiotic relationship with a blue-green alga and a bacterium in primitive eukaryotes ancestral to *Chlamydomonas* and other plant and animal cells. Cell sizes: (a) 0.3–1.0 μm; (b) ~1.0 μm; (c) ~1.0 μm; (d) 7–10 μm.

the scheme of the **Central Dogma** (DNA → RNA → protein) and other basic life processes appear very similar in both groups. Stated another way, we now conclude that there is only one principal pattern of life on earth, with a major division into two groups, prokaryotes and eukaryotes.

Prokaryotes

Mycoplasma. Prokaryotes are the simpler of the two groups even though they are still exceedingly complicated cells. The simplest of the prokaryotes and the smallest cells known are a group called **Mycoplasma** (formerly called PPLO for **pleuropneumonialike organisms**). Their cell sizes range from 0.2–1.0 μm, overlapping with the larger viruses and smaller bacteria. Figure 3–9(a) illustrates one of the smaller forms, which contains about $\frac{1}{10}$ the amount of DNA and $\frac{1}{100}$ the number of ribosomes of a typical bacterium. These organisms probably represent the minimum size and complexity which can exist on earth today. Prolonged antibiotic treatment of some kinds of bacteria can produce clones without any cell walls (protoplasts) which then seem identical to some mycoplasmas. Such protoplasts appear to be equivalent in every way to similar organisms isolated from nature. Once the cell wall is completely removed, the organisms seem unable to synthesize a new wall again even though removed from the antibiotic treatment, suggesting that a portion of the old cell wall is a necessary part of making a new one, perhaps serving as some sort of template.

Bacteria. The biggest, most successful group of prokaryotes, bacteria, are characterized by rigid cell walls outside of the cell membrane. Figure 3–9(b) is a line drawing (similar to the electron micrograph seen in Figure 2–14) illustrating the bacterium, *Bacillus subtilis*, showing the DNA-containing regions called **nucleoids**, which have no membrane around them and therefore are different from eukaryotic nuclei, two mesosomes, ribosomes in the cytoplasm, and the cell wall. The nucleoid region appears to be attached to the mesosomes, which are infoldings of the plasma membrane and are thought to play a role in division of the cell and of the nucleoid. Probably one DNA molecule containing the genetic code is attached to each mesosome; these DNA molecules are circular, measuring about 0.003 μm wide and 1000 μm around the circle. The inner or cytoplasmic side of the cell membrane and mesosome contain particles consisting of enzymes involved with respiration.

Blue-Green Algae. These organisms, Figure 3–9(c), can be most easily distinguished from bacteria by the fact that they give off oxygen as a by-product of photosynthesis. Blue-green algae have a prokaryotic type of cell wall quite similar in structure and chemical composition to the bacterial cell wall, and an enveloping sheath. Their photosynthetic pattern is similar to that seen in higher plants and in fact the entire cell somewhat resembles a chloroplast of higher plant cells, although the organization is less complex. It is now thought that a primitive blue-green algal cell "infected" a primitive eukaryotic cell, and that such a relationship evolved into the present photosynthetic eukaryotes.

Eukaryotes

Chlamydomonas. This green unicellular organism is probably the best example of a much-studied living cell which typifies the characteristics of the ancient eukaryotic stem cells which we believe gave rise to most of the plants and animals on earth today. The cell has two anterior flagella, a cellulose wall which lies outside of the plasma membrane, and a typical eukaryotic nucleus. Most of the cytoplasmic space of the cell is taken up by a single, cup-shaped chloroplast with its specialized starch-accumulating region, the pyrenoid. This chloroplast is reasonably typical of chloro-

plasts in general, and thus has extraordinary implications for life on earth, since these organelles are the site of CO_2 fixation from the atmosphere and trap the sun's energy inside more complex carbon compounds such as starch. About 90 percent of the weight of a *Chlamydomonas* or a tree comes from the atmosphere either directly or as CO_2 dissolved in water. The evolutionary idea now shared by many biologists that chloroplasts were derived from an ancient symbiotic association between a blue-green alga and a eukaryotic cell, resulting in a *Chlamydomonas*-like ancestor for plants and animals, is both fundamental and exciting.

Chlamydomonas possesses mitochondria which it uses for the complete breakdown of stored carbohydrates and other molecules especially when light is not available. These resemble a mycoplasma or bacterial type of endosymbiote as previously indicated. Since all animals rely ultimately on "plants" for food, it seems reasonable to suppose that animals evolved after plants became plentiful; this view is supported by the belief that the earth's early atmosphere contained little free oxygen, a condition which would not be detrimental to plants, since they can make their own, but which would be much more difficult for animals to cope with. Plant activity is presumably responsible for most of the earth's oxygen, which is now about 20 percent of the atmosphere. The question of when plants and animals acquired mitochondria is open. There are anaerobic (not requiring air, generally meaning oxygen) prokaryotes and anaerobic eukaryotes, and particularly for the former the anaerobic state could have been the primitive condition. Also, there are some very primitive amitochondrial ciliates still living in anaerobic muds or sands at the bottom of fresh and marine waters, which may represent primitive forms and which have been suggested as ancestral to metazoa. Even so, for several reasons including the fact that the mitochondria of *Chlamydomonas* and other plants appear to be essentially like those of

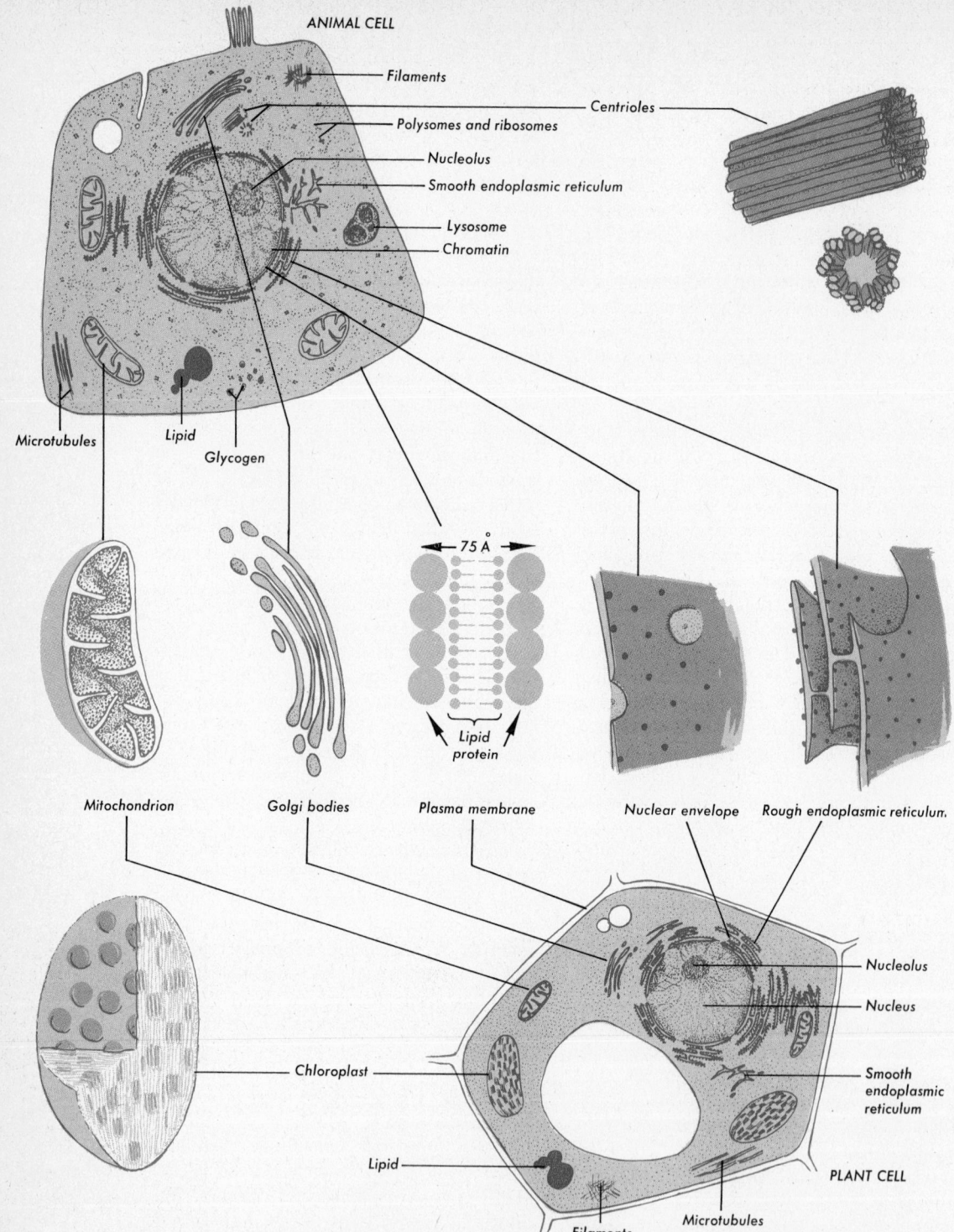

ANIMAL CELL

Filaments

Centrioles

Polysomes and ribosomes

Nucleolus

Smooth endoplasmic reticulum

Lysosome

Chromatin

Microtubules

Lipid

Glycogen

75 Å

Lipid
protein

Mitochondrion

Golgi bodies

Plasma membrane

Nuclear envelope

Rough endoplasmic reticulum.

Chloroplast

Nucleolus

Nucleus

Smooth
endoplasmic
reticulum

Lipid

Filaments

Microtubules

PLANT CELL

Fig. 3–10. A generalized cell based on electron microscope observations. See explanation in text.

animal cells, most biologsts favor a primitive phytoflagellate such as *Chlamydomonas* as the ancestral prototype of all eukaryotes, including animals.

Chlamydomonas has other standard organelles. The nucleus has a nuclear envelope and contains genes (DNA) necessary for cellular, chloroplast, and mitochondrial function. There is a typical Golgi apparatus, some vacuoles of undetermined content and function, a contractile vacuole, and two flagella.

THE GENERALIZED CELL

It is impossible to designate any one plant or animal cell as representative of all cells, because each cell type has certain unique features; but since all cells have certain parts in common it is possible to construct a generalized animal cell which will serve for discussing common characteristics. Figure 3–10 diagrams the typical parts of an animal cell. The plant cell has **chloroplasts** in the cytoplasm and a **cell wall** outside of the plasma membrane; otherwise it is identical to the animal cell.

The Plasma Membrane

We may think of the cell as a factory consisting of a number of subunits, each of which performs specific tasks essential to the existence of the entire cell. The outer fence around the factory is the plasma membrane. This membrane is so thin that it cannot be seen in the light miscroscope, but in the electron microscope it appears as two thin lines separated by a space, a sort of railroad track. From indirect evidence the two dense lines are thought to consist of protein macromolecules and the space to be filled with lipid molecules arranged in a specific pattern (Fig. 3–10). They are said to be polarized so that the hydrophilic (water-loving) ends lie in or near the protein layers and the hydrophobic (not water-loving) ends point toward one another near the middle of the space. The overall thick-

ness of the membrane is 75–100 Å and the entire structure is called a **unit membrane**. The protein layers contribute to the elasticity of the membrane and the lipid probably functions in transport of ions and molecules through the membrane. It has been suggested that most, if not all, cell membranes are of this type. It is difficult to imagine that a single structure of this design could perform all of the functions known to occur in the diverse membranes of cells and in fact, biochemical studies of isolated membranes have shown that they contain different enzymes and other molecules. It is becoming clear that they are probably different in chemical composition, in function, and therefore in structure.

Our current picture is one of a membrane consisting of a protein-lipid bilayer as diagrammed in Figure 3–10, but superimposed on this are various protein and polysaccharide moieties (singly or in complexes) some extending all the way through

Fig. 3–11. These are cells growing in tissue culture, originally from the ovary of a hamster. The spherical cell is about ready to divide into two cells; the surface of the cell is extended into long, thin, filamentous projections some of which attach it to the substrate, presumably for the purpose of helping the cell pull itself into two daughter cells. The neighboring flat cells are still in interphase, but will soon also begin to round up for division. This remarkable scanning electron micrograph was made by Keith R. Porter, David M. Prescott, and Jearl F. Frye of the University of Colorado. Magnification is 6,000 diameters.

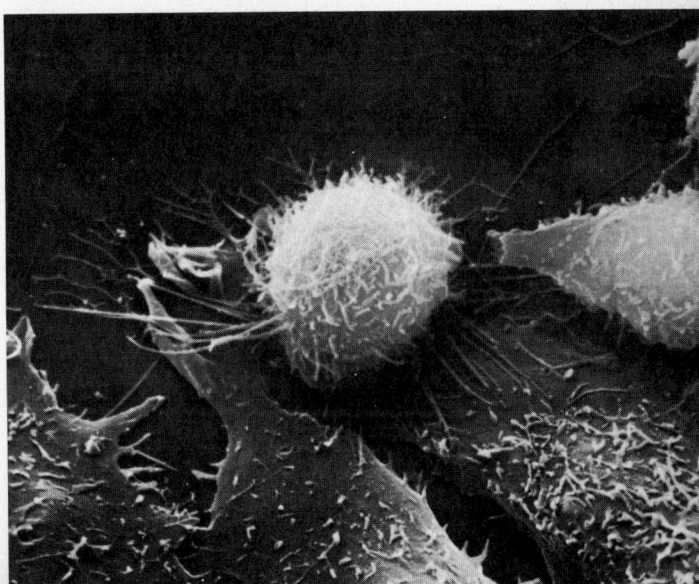

the membrane, some localized on either side, and each possessing greater or lesser mobility laterally along, or possibly back and forth through the membrane bilayer. We know that the membrane itself is not static. The dynamic state of the cell membrane is particularly well illustrated in dividing vs. nondividing cells as evident in the SEM micrograph of hamster cells in tissue culture (Fig. 3–11).

The cell membrane is the barrier which isolates the interior of the cell from its external world, and it is across this barrier that all vital materials must enter and all waste materials must exit from the cell. We know it performs these functions most efficiently. Materials pass through the membrane by passive and active transport as mentioned earlier. The combination of these forces maintains the concentration of essential substances at the proper levels for normal functioning of the cell.

Many cells, however, possess an additional method of taking in substances. Cells such as white blood cells, tumor cells, intestinal lining cells, and amoebae readily take in materials by **phagocytosis** (cell eating) and **pinocytosis** (cell drinking). At the beginning of pinocytosis in the amoeba,

where it is most easily observed, the cell membrane extends outward to form a ruffled **pseudopodium**. A tubelike channel pushes inward and pinches off small vesicles at the inner tip (Fig. 3–12). These channels are short-lived but new ones are constantly forming so that the cell, when "drinking," may have 50 to 100 such channels. Small particles (macromolecules) may be included in the droplet of fluid and thus are carried inside the cell. A great deal of fluid enters the cell by this means, particularly water. Salts and proteins readily induce the process, whereas carbohydrates are inert in this regard. An interesting experiment with amoeba is to place it in a solution of radioactive glucose and by autoradiography note whether or not glucose enters the cell by pinocytosis. Almost no glucose penetrates the cell. Add a protein, like egg albumin, and large quantities of the glucose enter the cell along with the protein. Larger particles up to and including bacteria and small protozoa are taken in a similar manner by white blood cells and other cells such as the amoeba; this process is called **phagocytosis** (Fig. 3–12).

Cytoplasmic Structures

The Nucleus. Inside the cell membrane are particles of different sizes; the largest is the nucleus. We know that this is the most important structure in the cell, because removal of the nucleus by micromanipulation will inevitably lead to the death of the cell. The nuclear envelope consists of two **unit membranes** in close apposition, forming a space of quite uniform dimensions (about 100–150 Å) called the **perinuclear space.** These two membranes coalesce in various places to form **pores** which appear octagonal in surface view and which generally seem to be plugged with fibrous material that resembles chromatin. Presumably the pores are differentiated regions designed to permit the selective passage of materials (Figs. 3–10 and 3–17) including RNA; they may be coincident with, or close to, attachment sites for the chromosomes. The outer mem-

Fig. 3–12. Amoeba, as well as many other cells, feeds two ways. Along with water, it takes in soluble particles by pinocytosis; and it engulfs whole organisms by phagocytosis. Vacuoles containing the food are formed in the cytoplasm; these acquire hydrolytic enzymes which digest the food.

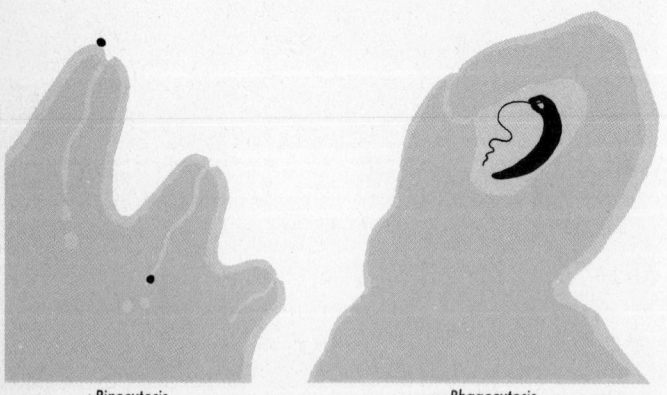

Pinocytosis   *Phagocytosis*

brane of the nuclear envelope is studded with ribosomes and in some regions is continuous with rough endoplasmic reticulum (RER). This architecture would appear to allow materials in the spaces of RER to enter the nucleus or nuclear material to leave the nucleus. Just how the nuclear membrane carries out its complex functions is not at all clear.

It has been known for a long time that the nucleus contains **chromosomes** in a largely invisible state; they "condense" and become visible at the time of cell division. Between divisions (interphase) much of the chromosome material is uncoiled; in this form the DNA can make copies of itself and make RNA. Some parts of the chromosomes are visible as **chromatin granules** (granules which stain strongly with certain dyes) and these probably represent condensed, inactive gene regions.

The nucleus contains one or more conspicuous bodies, the **nucleoli** (singular, **nucleolus**) which are composed of a fibrillar portion (Fig. 3–16, F) and a granular portion (Fig. 3–16, G). The latter consists of many small granules resembling the ribosomes in the cytoplasm. They are rich in ribosomal RNA (rRNA) in precursor form. This rRNA is coded for by only about 2 percent of the DNA of the cell, but many rRNA copies are made and somehow transported to the nucleolus and then to the cytoplasm, probably via the pores in the nuclear envelope.

All eukaryotic cells at some time have a nucleus. The nucleus directs the synthesis of proteins and other constituents of the cell as a whole and contains its hereditary material. With the exception of the mitochondria (and chloroplasts in plant cells) all DNA is localized inside the nucleus, packaged in the **chromosomes**.

In some cells, such as developing amphibian eggs and dipteran salivary gland cells, the extension of the surface is greatly increased by loops extending laterally from the axis of the chromosome which gives it a fuzzy appearance, and for that reason these are termed "lampbrush" chromosomes (Fig.

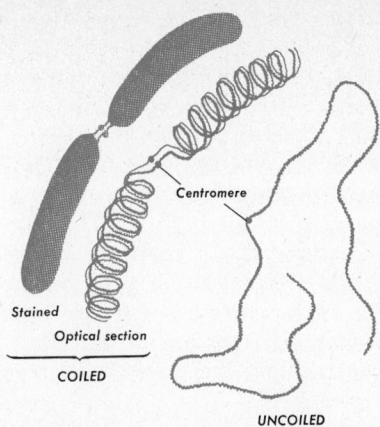

Fig. 3–13. A single chromosome is shown here in the coiled and uncoiled state. Stained preparations observed under the light microscope appear solid black. The chromosome appears to be composed of a tightly coiled pair of threads (chromonemata). The uncoiled chromosome loses its major coils but retains the minor ones.

24–24) or chromosome "puffs." These chromosomes, like all chromosomes, consist of a thin **axial filament**, sometimes called a **chromonema**, to which thick dense swellings, the **chromomeres**, are attached (Fig. 3–13). These bodies contain DNA and protein (DNP) and probably represent different stages of coiling and supercoiling. There is no membrane bounding the chromosome. We say more about the chromosome when we discuss division.

Mitochondria. There are several particles, mitochondria, in the cytoplasm which are primary sources of energy. Mitochondria were seen and clearly described at the turn of the century. They were identified in living cells by their affinity for the stain, Janus green. Their function and intricate morphology have since become known largely as a result of biochemical studies and electron microscopy. They are oval or rod-shaped, usually, although they can vary in morphology under different metabolic conditions of the cell. They range in size from 0.5μ to 3μ as spheres or ovals; others may be very elongated, such as those in skeletal muscle cells. The change in shape and size depends on the type of cell, the age of the cell, or their own chemical activity. They are more numerous in metabolically active cells, such as secretory and muscle cells. In the living cell they are often in motion as a result of protoplasmic streaming of the cytoplasm. They are absent in some parasitic protozoa and mature red blood cells (although they are present in the earlier stages), but nearly all other eukaryotic cells contain them. They are semi-autonomous organelles which contain their own DNA.

Mitochondria have a highly organized fine structure which is designed to provide the maximum surface area for enzyme action. One of the great surprises electron microscopists encountered was the elaborate architecture of mitochondria (Figs. 3–10 and 3–14). It is bounded by a double membrane; the outside membrane is smooth whereas the inner one is thrown into folds (**cristae**) or tubules which extend deep into the **matrix** (contained

fluid). Both membranes are designed in such a way that they can enlarge (swell) under certain metabolic conditions of the cell. The outer membrane is selectively permeable so that some, but not all, substances can pass through it. Small dense granules appear in the matrix. They consist of cations of calcium and other ions. Their function is not known.

Mitochondria can be isolated from cells and concentrated in large numbers so that their biochemistry can be studied in great detail. Fortunately they seem to function **in vitro** (outside the cell in a test tube) much the same as they do when in the intact cell (**in vivo**). It has been shown that mitochondria supply most of the energy and for that reason they are often spoken of as the "power houses" of the cell. The bulk of the energy from foodstuffs is retrieved by mitochondrial enzymes. This is called **aerobic respiration** (p. 87). Other chemical reactions occur in the mitochondria, such as the breakdown of fats. They are rich in protein and fat as well as RNA and small quantities of DNA.

With higher resolutions of the electron microscope combined with special staining techniques, it has recently been possible to see more structure in both the outer and inner membranes of the mitochondrion. Both membranes are composed of repeating units. Small spheres rest on the outer membrane and other particles make up the inner membrane (Fig. 3–14). By using very refined techniques it has been possible to isolate the outer membrane, the enzymes of the outer space (Fig. 3–14), the inner membrane, and matrix materials. The isolated inner membrane units resemble lollipops with a headpiece, a stalk, and a base piece; they can be further fractionated into separate subunits and chemically characterized. The bases will, under appropriate conditions, reassemble themselves into vesicular membranes, while the headpieces apparently contain an enzyme (an ATPase) that can break down ATP. The outer membrane (Fig. 3–14) strongly resembles other cytoplasmic membranes, such

Fig. 3–14. High resolution micrographs and biochemical fractionation studies of the mitochondrion demonstrate that enzymes are associated with both membranes. The inner membrane contains so-called "elementary particles," which contain enzymes involved in the production of ATP.

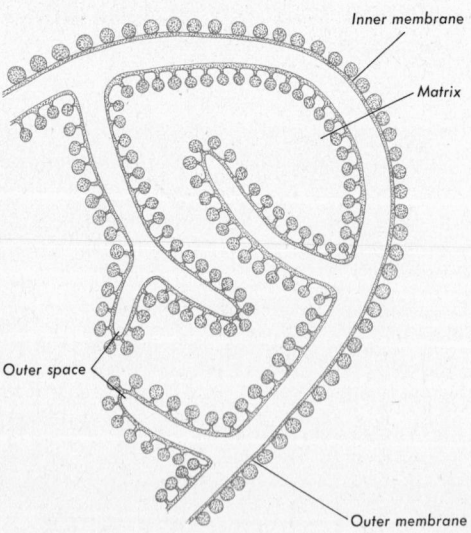

Inner membrane

Matrix

Outer space

Outer membrane

as Golgi and ER, in that it has similar enzymes associated with the membrane and also appears to have particles in or on the membrane which are not so regularly oriented as the particles on the cristae. The outer space contains only a few enzymes, while the matrix is so concentrated it is almost a protein gel and contains most of the soluble enzymes of the Krebs cycle and fatty acid degradation cycle, to be discussed later. The matrix also contains DNA and ribosomes.

Thus we see the mitochondrion as one of the most fascinating of cellular organelles, intimately involved with the efficient use of foodstuffs for energy production by the slow-moving amoeba as well as by the 50-mile-per-hour antelope. Mitochondria can divide and multiply, usually keeping pace with the growth of the cell.

They seem to be able to drastically alter the shape of their cristae. This shape change appears to be related to the energy state within the particular mitochondrion at a certain time. Figure 3–15 illustrates three examples of different shapes that mitochondria can assume. Why they exhibit these conformations is not known. The membranes of mitochondrial cristae contain, or have adsorbed to them, respiratory enzymes that carry out the final steps of making ATP. Since ATP is the primary driving force of all biological reactions and hence life itself, one can hardly over emphasize the importance of mitochondria and the oxidative phosphorylation process.

Membrane Relationships in the Cytoplasm

At this point we need to examine some of the other membranous systems in the cell as diagramed in Figure 3–10 and illustrated in the hepatocytes shown in thin sections by TEM in Figures 3–16, 3–17, and 3–18. First it is necessary to get a feeling for the magnifications involved.

There are two conventional ways in which magnifications are indicated. The best method is to place a bar in one corner of the micrograph; the length of the bar

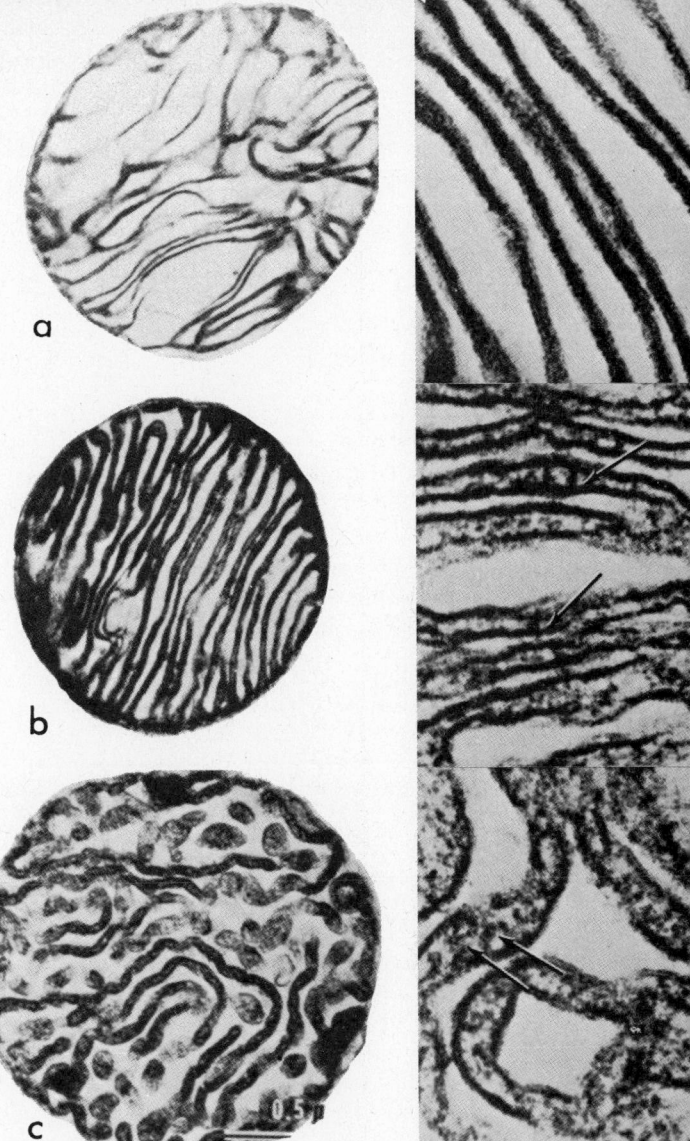

Fig. 3–15. Changes in the shape of mitochondrial cristae in isolated beef heart mitochondria. Low magnification on the left, high magnification on the right. (a) Inactive state, in which cristae are so swollen that the elementary particles are tightly pressed together by the adjacent cristae. (b) Intermediate stage, in which the elementary particles are partly energized such that they push the adjacent cristae apart. (c) The "energized twist" configuration of mitochondria in which the elementary particles have pushed the cristae fully apart and the stalks and head-pieces can be clearly recognized (arrows)

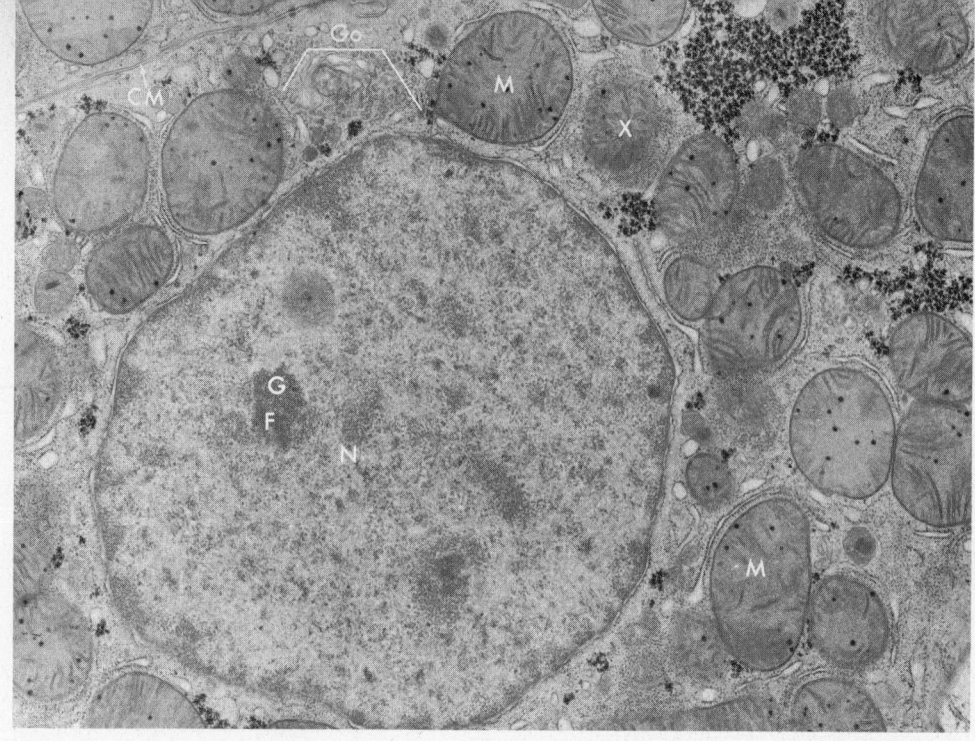

Fig. 3–16. An electron micrograph of the rat liver showing the cell membrane (CM), glycogen granules (Gl), Golgi body (Go), mitochondria (M), nucleus (N), and a small piece of the nucleolus, consisting of fibrillar (F) and granular (G) portions. Mitochondrion sectioned at one edge (X). [× 13,000] Bar = 1 μm. (Courtesy of Keith R. Porter.)

Fig. 3–17. An electron micrograph (top of facing page) of a rat liver cell showing mitochondria (M), rough endoplasmic reticulum (ER) with dark granules on the surface (ribosomes), smooth endoplasmic reticulum (SER), nucleus (N) and the nuclear membrane showing distinct pores. [× 25,000] (Courtesy of Keith R. Porter.)

Fig. 3–18. An electron micrograph (bottom of facing page) of two contiguous cells from rat liver, showing lysosomes (L) and peroxisomes (P), in addition to the structures previously seen (Figs. 3–16 and 3–17), namely: rough endoplasmic reticulum (ER), cell membrane (EM), mitochondria (M), and the nucleus (N). Glycogen and smooth ER are also present. [× 9,000] (Courtesy of Keith R. Porter.)

may be indicated directly on the micrograph or the caption. A less satisfactory alternative is to place a magnification value in brackets; for example, for Figure 3–16 the magnification is indicated by [×13,000]. With experience one learns to "read" a micrograph in terms of relative sizes of objects, based on certain relatively standard dimensions such as:

0.1 μm	1,000 Å
Unit membranes	75–100 Å
Ribosomes	200 Å
Small subunits of glycogen (beta particles)	150–300 Å
Mitochondrial width	~0.5 μm
Nuclear membrane pores	~400 Å
Microtubules	250 Å

In the following discussion of various cytoplasmic organelles and membrane systems it will be found helpful to compare structural features as seen in Figure 3–10 as well as in sections of hepatocytes in Figures 3–16, 3–17, and 3–18.

Lysosomes. These include a heterogeneous group of vacuoles enclosed by a unit membrane and containing digestive enzymes (hydrolases). Together, they make up what

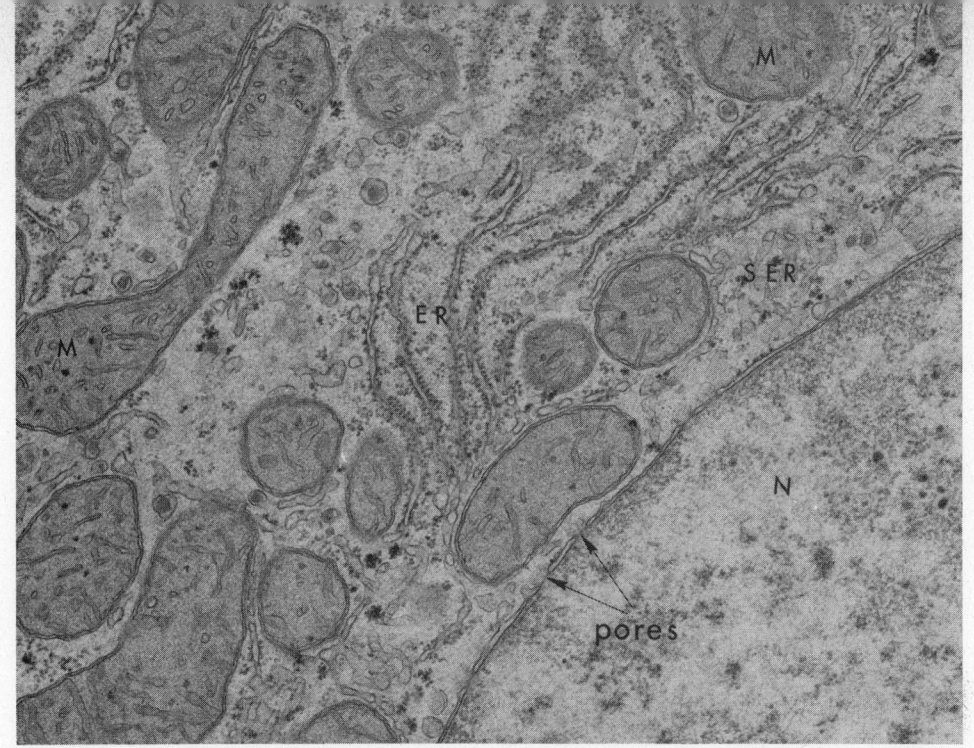

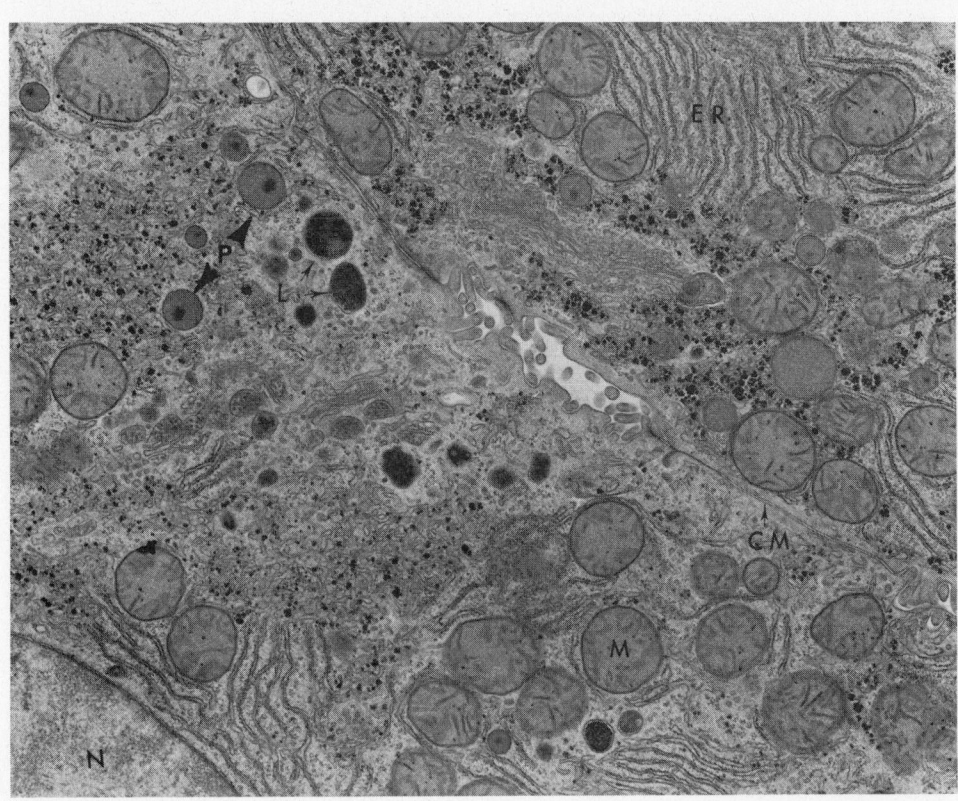

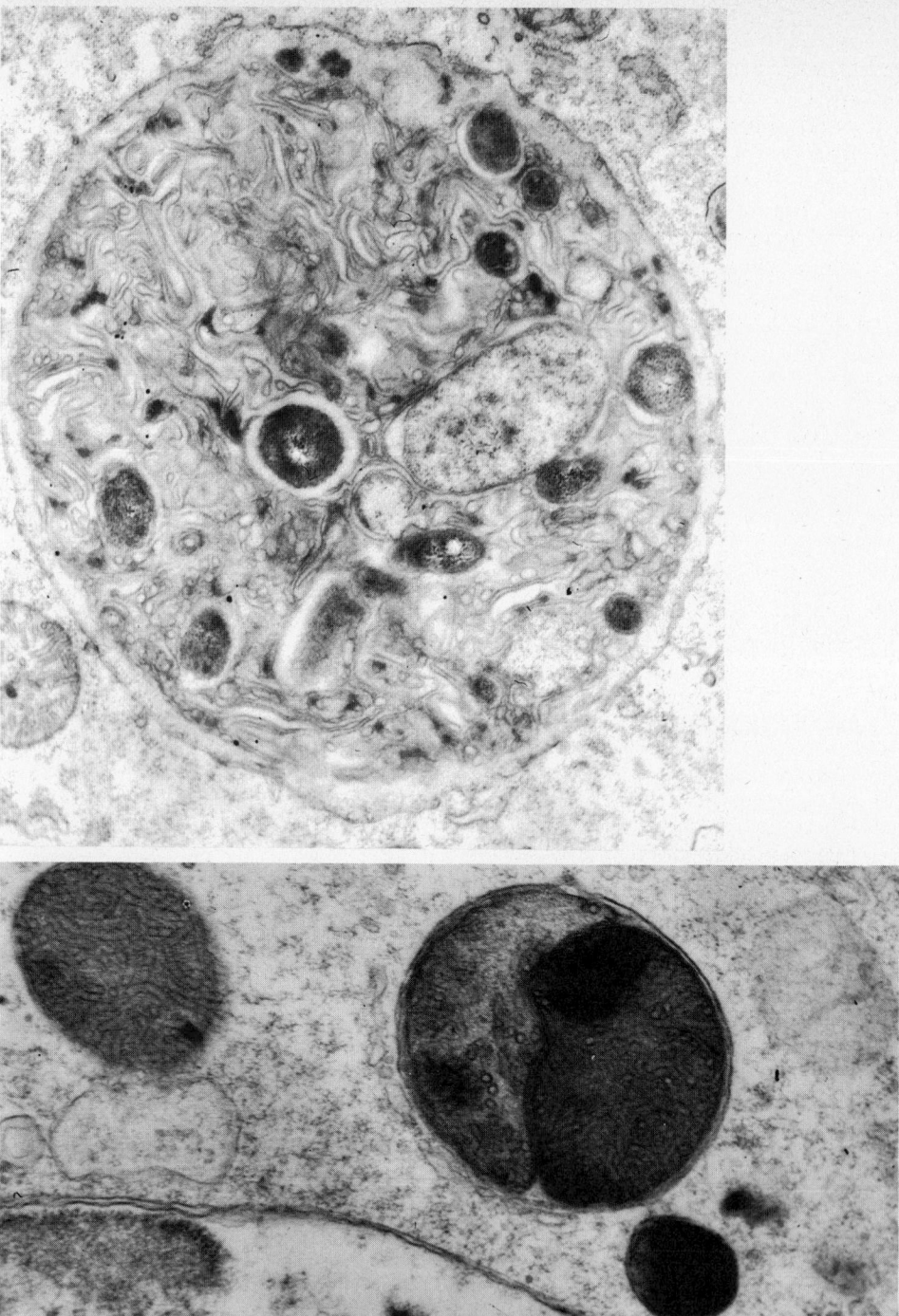

Fig. 3–19. These are electron micrographs of secondary lysosomes in the protozoan, *Tetrahymena pyriformis*. The food vacuole contains bacteria undergoing digestion (top). Only membranes are left of some bacteria while others are in early stages of digestion. The autophagic vacuole (bottom) contains two mitochondria, one of which has started to digest. This cell has the capacity to utilize its own substance as a source of energy when stressed. [Top, × 25,000; bottom, × 40,000.]

amounts to an intracellular digestive system comparable to that found in multicellular animals. According to Professor C. de Duve of Rockefeller University, who first identified lysosomes biochemically, there are two general types of lysosomes. One, called **primary lysosomes**, contains newly synthesized enzymes which have not yet engaged in a digestive event. They are usually smaller than mitochondria and are uniformly dense with little or no internal structure (Figs. 3–18, 3–19). It is thought that they contribute hydrolases to the second type of lysosome, the **secondary lysosomes**. These are vacuoles of varying size and morphology depending on their contents. It is in these vacuoles that the digestion of food takes place, in other words, they are food vacuoles. Many protozoa, as well as various types of metazoan cells (leucocytes, intestinal lining cells, and others) engulf food, forming food vacuoles. These acquire their digestive enzymes by fusing with primary lysosomes. Once digestion is completed, the end-products (amino acids, glucose, etc.) pass out of the vacuole into the cytoplasm where they are metabolized. Undigested material remaining in the vacuole (now called a **residual vacuole**) is extruded outside the cell.

Another type of secondary lysosome is the **autophagic vacuole**. These resemble food vacuoles in morphology but contain cell particulates such as mitochondria, bits of endoplasmic reticulum and other cellular debris (Fig. 3–19). They probably acquire hydrolases much the same as food vacuoles and the subsequent digestive events are the same. Autophagic vacuoles appear in cells under certain pathological conditions and during starving and aging. However, they also occur in normal cells during embryogenesis when cells are destroyed, such as during the loss of the larval tail in amphibian metamorphosis. It is the cell's way of utilizing its own fabric as an economy measure for survival.

Golgi Apparatus. Near the end of the last century the Italian cytologist, Camillo

Golgi, identified structures in the cytoplasm which were later named after him. They are now known as the **Golgi apparatus,** or simply **Golgi bodies.** Under the electron microscope the Golgi apparatus appears as stacks of smooth membraned, flattened saccules (little sacs) which pinch off tiny vesicles at their ends that pass into the cytoplasm (Figs. 3–16 and 3–20). By employing specific stains, it has been demonstrated that the saccules and vesicles in some cells contain hydrolases. Some investigators believe that the Golgi apparatus stores and secretes these enzymes in vesicles (primary lysosomes) which are responsible for intracellular digestion as mentioned above. It is reasonably certain that some enzymes are synthesized on the ribosomes clinging to the endoplasmic reticulum (ER). The enzymes find their way into the channels of the ER and then to the Golgi apparatus (Fig. 3–21). The ER seems to be contiguous with the saccules of the Golgi apparatus. As reasonable as this sequence of events seems, it has been demonstrated in too few cells (with much contrary evidence) to be considered a gen-

Fig. 3–20. This electron micrograph shows the Golgi apparatus, in this case a set of 5–7 flattened membranous sacs called Golgi stacks in which material is processed either for export outside of the cell or for use elsewhere inside of the cell. The cell is *Euglena gracilis*, a protozoan [× 10,000].

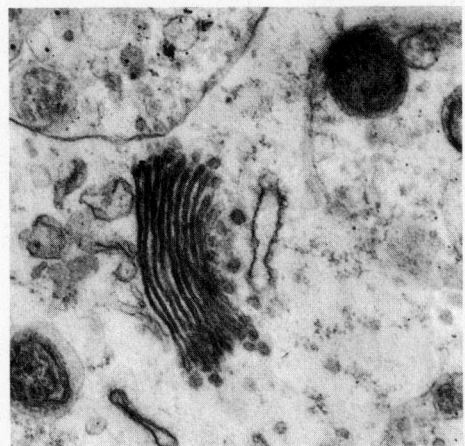

eral concept for all cells. In some protozoan flagellates, the Golgi saccules appear to be the site for the synthesis of cortical structures which are then transferred to the outside of the cell. There may be other functions of the Golgi apparatus. There is much left to be learned before the true function or functions of this structure can be stated.

Peroxisomes. Another class of particles has been observed in the cytoplasm of some cells, but until recently no function was assigned to them. These have been called **microbodies** in the past, but now that certain enzymes that have to do with hydrogen peroxide metabolism have been found in them, they are renamed **peroxisomes** by Professor de Duve of Rockefeller University. They are finely granular and in some a crystalline body or crystalloid appears in the matrix (Fig. 3–18), as is often seen in liver cells. They contain the enzymes **uricase, catalase,** and **d-amino acid oxidase.** At present just how these enzymes fit into the overall metabolism of the cell is not clear.

Endoplasmic Reticulum. Professor E. B. Wilson drew a generalized cell in 1925 which summed up what was known at that time. In this cell much of the cytoplasm seemed to be empty, although a wispy or vacuolar type of ground substance was indicated in the drawing (Fig. 3–1). With the aid of the electron microscope we now see that this apparent empty space is composed of a complicated system of membranes called the **endoplasmic reticulum** by Professor K. R. Porter of the

Fig. 3–21. This sketch represents a hypothetical pathway of hydrolytic enzymes from their origin to their sites of action .The enzymes arise as a result of the interaction between mRNA, sRNA, and the ribosome which contains rRNA. The information from the messenger is translated by the ribosome into polypeptides which are somehow directly placed, probably by the ribosome, into the cisternae of the endoplasmic reticulum. These polypeptides spontaneously undergo changes in shape and sometimes several may associate together to form particular enzymes. They then pass through the Golgi apparatus, finally emerging as primary lysosomes which then fuse with the food and autophagic vacuoles, contributing their contents (hydrolytic enzymes) to them. Digestion occurs within these vacuoles and the end-products pass into the cytoplasm of the cell where they are metabolized. Undigested material in the vacuoles leaves the cell by exocytosis. Vacuoles containing enzymes for export leave the cell the same way. Such enzymes are utilized elsewhere in the organism for digestion of foods.

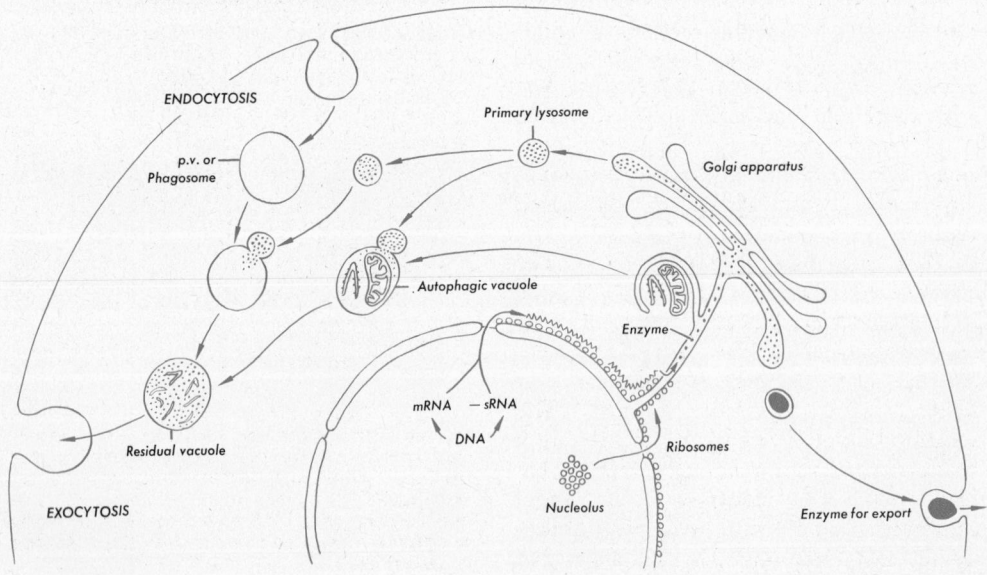

University of Colorado (Fig. 3–17). These membranes extend from the outer nuclear membrane into the cytoplasm, forming an elaborate system of channels that seem to swell and deflate in the living cell. Some of the membranes are "smooth" (SER), others are "rough" (RER), the difference being that the latter possess tiny granules regularly arranged along their outer surfaces whereas the former have no such granules. These granules are called **ribosomes** and contain a high percentage of **ribonucleic acid** (RNA). Other ribosomes lie free in the cytoplasm. They are the sites of protein synthesis as we see later. The amount of endoplasmic reticulum varies among cells; some, such as protozoa, have little, whereas cells specialized for the production of proteins (enzymes) for export, such as liver and pancreas cells, possess a great deal, particularly of the rough variety. Some of these membrane relationships are summarized in Figure 3–21.

Centrioles and Kinetosomes. Most animal cells, some protozoa excepted, contain a pair of tiny bodies called **centrioles.** These are barely visible under the light microscope; they are best observed during cell division although they are present during all stages in the life cycle of the cell. Their remarkable structure was revealed by the electron microscope. They are composed of a cylinder of nine triplet microtubles or fibers (Fig. 3–10 and Fig. 3–22). They always lie at right angles to one another during interphase, are located near the nucleus, and, surprisingly, have essentially the same structure as **kinetosomes** which give rise to cilia and flagella (Fig. 3–22). The centriole and kinetosome are not only alike in their structure, but they also seem to be self-generating; i.e., the new "daughter" centriole or kinetosome typically arises at right angles to the older "parent." The building blocks are molecules of tubulin, a common protein in many cells. Tubulin has the capacity to assemble itself, under appropriate conditions, into long, hollow tubules called **microtubules,** each about 250

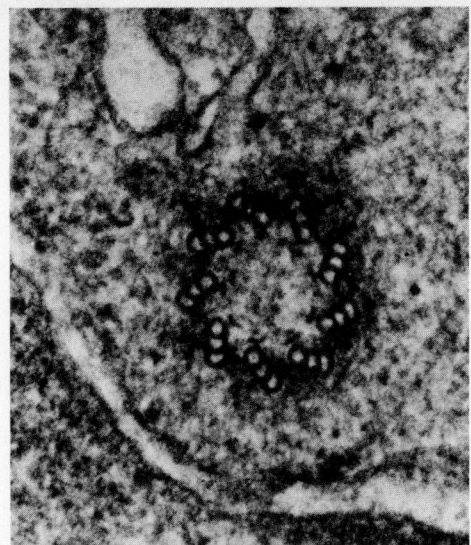

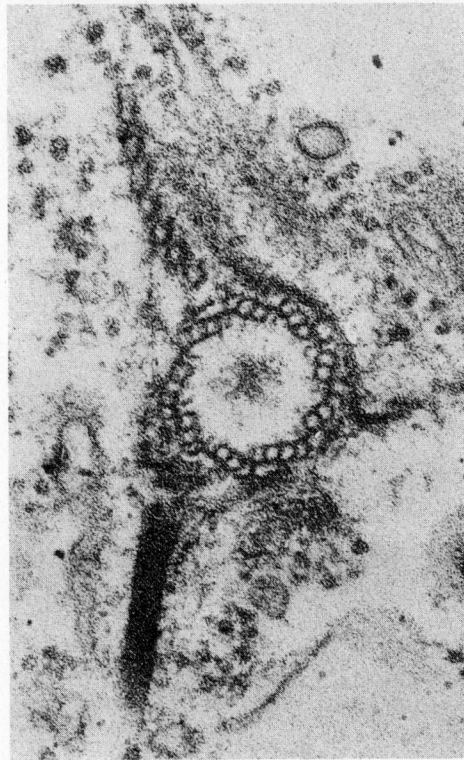

Fig. 3–22. (Top) An electron micrograph of an embryonic chick pancreas showing a centriole in cross-section. (Bottom) Compare this cross section of a protozoan kinetosome with the centriole above. In both cases the diameter measures about 0.2 μm, and consists of a cylinder of nine sets of triplet microtubules arranged in a pinwheel. (Top, courtesy of Jean André.)

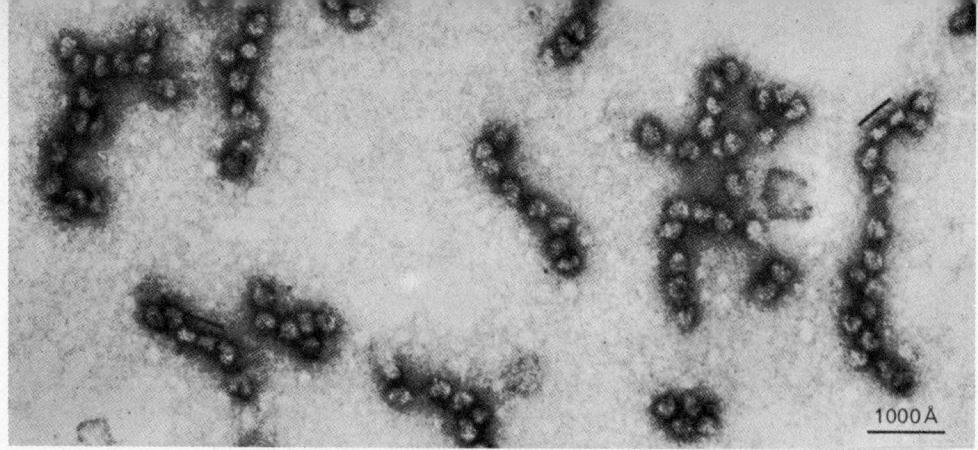

Fig. 3–23. Isolated rat liver polysomes, consisting of several ribosomes attached to a single mRNA. The entire ribosome is called an 80S particle because it sediments in the ultracentrifuge at that rate (in Svedberg units). The bars indicate places where only the small subunit, called 37S, is attached. The other subunit (60S) presumably attaches later and translation can then proceed. (From Y. Nonomure, G. Blabel, and D. Sabatini, *J. Mol. Biol.*, **60**, 303, 1971.)

A in diameter. In centrioles and kineto-somes, these microtubules are uniquely arranged into nine sets of three, called triplets. The triplets are skewed with respect to the circumference of the centriole-kinetosome in a standard way. The proximal region of these organelles contains a hub and spoke arrangement called a cartwheel, and the clockwise, inwardly directed skewing (commalike) of the triplets in Figure 3–22 indicate that the observer is looking from the proximal towards the distal end of the centriole-kinetosome. Centrioles appear to be involved with the proper arrangement of the microtubules that make up mitotic spindles and asters, while kinetosomes have a similar function in relationship to cytoplasmic microtubules in protozoa. The inner two microtubules of each kinetosomal triplet are continuous with the doublet tubules of cilia and flagella. This standard organization of centrioles and kinetosomes is most remarkable and geometrically precise in eukaryotes, and its ubiquity strongly suggests common ancestry. Kinetosomes and centrioles are thus probably homologous structures which had a common origin but through evolution have taken on other functions in different cells.

Other Cytoplasmic Structures. Glycogen accumulates in many cells, such as those of the liver, which provides a carbohydrate storage pool for energy. Glucose enters the cells and the surplus is polymerized into huge glycogen molecules—first as beta particles (150–300 Å in diameter) and then into aggregates of beta particles, called alpha particles (Fig. 3–16). Likewise, fatty acids and glycerol combine to form fats which are visualized at the ultrastructural level as dense membraneless lipid droplets (Fig. 3–10). These also serve as energy reserves. Ribosomes are randomly distributed throughout the cytoplasm; these are not attached to membranes. They exist as singles or sometimes they form clusters, called **polysomes** (Fig. 3–23). This arrangement of ribosomes probably is related to their function in protein synthesis.

Single **microtubules** are often seen, usually in small groups in the cytoplasm, not only during division but frequently during interphase as well. They are particularly abundant in cells which have special shape; e.g., they serve in nerve cells as an aid in maintaining their extreme length; in some animals like the giraffe with its long legs and vertebral column, some nerve cells are over six feet long! In addition to microtubules, **microfilaments** occur abundantly in many cells, often just below the cell membrane. These microfilaments measure 40–60 Å in diameter, appear to be essentially identical to the muscle protein actin, and are probably involved with cell movement and division. An additional filament type about 100–120 Å is also seen frequently, and it is tempting to equate these 100–120 Å filaments with myosin, since the diameter is identical with that of thick filaments in striated muscle, but the correlative biochemical and ultrastructural characterizations are less certain.

Nevertheless, the list of cell types from which both actinlike and myosinlike proteins have been isolated and identified is impressive, and includes the following: striated (skeletal and cardiac) muscle, smooth muscle, leukocytes, blood platelets, amoebae, epithelial cells, fibroblasts, chondrocytes, slime molds, and brain. These proteins seem to be generally involved with cell movements.

Differences between Cells

We have considered the generalized cell in this section and in so doing have emphasized the common features of cells which we believe reflect a common origin. However, cells vary tremendously in size, shape, and special functions, and we would delude ourselves if we oversimplified the complexity of the biological world with its many adaptations and specializations, acquired after many thousands of years.

Size. There is as much difference in the size of cells as there is between most animals, an elephant and a mouse, for example, but there is no correlation between the size of the animal and the size of its cells. The larger animals simply have more cells. Cells must not be confused with viruses which live in and at the expense of other cells and are essentially metabolically inert when outside the host cell. The smallest cells belong to the genus **Mycoplasma** and measure only 0.3–1.0 mm, as described earlier. Bacteria can only be seen with the best high resolution light microscopes; their structure has been more successfully elucidated since they have been

examined with the TEM and SEM. Most cells that compose the bodies of animals are considerably larger than the largest bacteria; on the average they are about 7 microns in diameter. Nerve cells are very long and thin, particularly those reaching from the tip of the toe to the spinal cord in the back. The largest animal cells are found among bird's eggs, in which what is popularly called the "yolk' is a single cell. The largest living cell known is the yolk of the ostrich egg which reaches a diameter or 2 or more inches. The large quantity of stored food (yolk) in eggs is responsible for their great mass.

Shape. The shape of cells depends a great deal on their function (Fig. 3–24). If they perform a tensile function, such as cells found in tendons, they are long and thin because of the constant stretching action placed upon them. If they are conductive, as in the case of nerve cells, they must also be long and thin. On the other hand, red corpuscles are tiny discs, fitted by shape to float in the blood stream and exchange oxygen with tissue cells. Smooth muscle cells are spindle-shaped, which is the ideal shape for a cell that must shorten or contract. Cells that line the respiratory tract in our bodies are small cylinders with minute, vibratile, hair-like structures (**cilia**) on one end, which function in carrying the mucus along the tract by their whip-like action. Other cells, such as the amorphous white blood cells, resemble tiny bits of jelly that move by rolling along, taking in and destroying bacteria and other foreign particles in the blood and tissues.

Number. A glance through a microscope at a thin slice taken from any animal will demonstrate the fact that there are a great many cells in even a small animal like the mouse or spider. The larger the animal, the more the cells, although swift-moving and very active animals such as insects and birds usually have fewer cells per unit volume than do sluggish, slow-moving creatures such as the salamander. Every

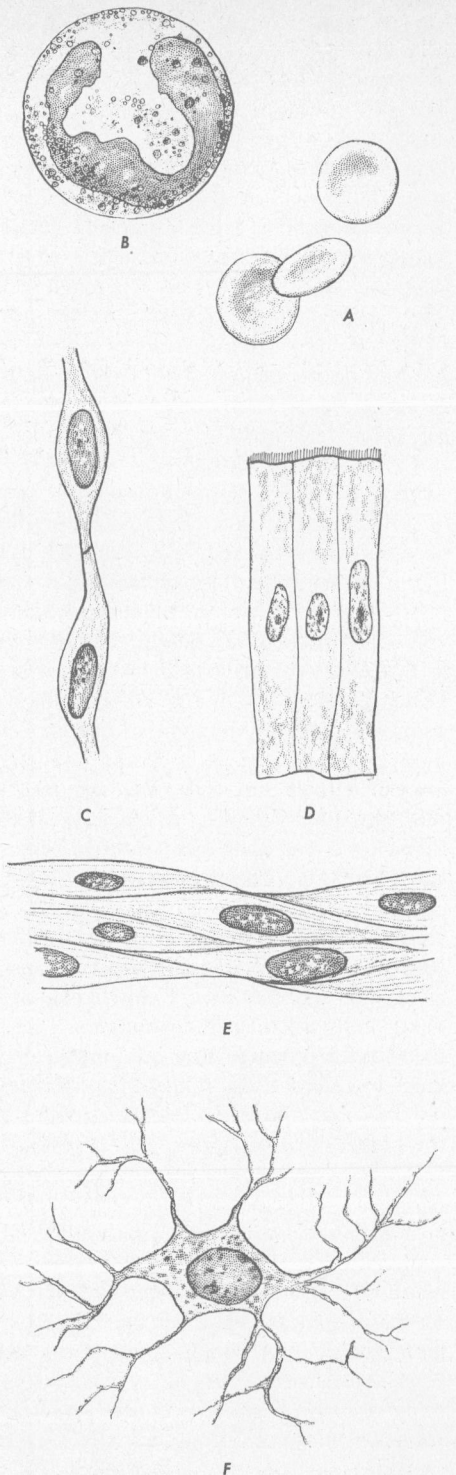

cubic millimeter of human blood contains about 5 million red blood cells, and the total number in the entire blood stream approximates 30 trillions. The human brain alone has billions of cells; the number in the whole body thus takes on astronomical figures.

THE FUNCTION OF CELLS

As we have already noted, cells resemble one another anatomically, therefore one would expect that they would also function in much the same manner. This is essentially true; all cells have certain requirements which are satisfied in much the same way. Whereas there may be differences in the way they acquire their food, the processes of utilizing the food are quite the same in most cells. As pointed out earlier, these vital processes are included under the term metabolism. Therefore let us consider **cellular metabolism,** remembering that the overall operation of the entire organism, be it a flea or dog, is ultimately at the cellular level. An understanding of metabolism at the cellular level is fundamental to understanding the metabolism of communities of cells as they are found in complex animals such as man.

Nutrition

All of the activities of an organism that have to do with securing, consuming, digesting, and absorbing food may be considered **nutrition** in the broad sense. At the higher organism level this involves a complex of organ systems which need not concern us at this time. We are only concerned with the ways in which a cell may receive nourishment.

There are two methods by which cells

Fig. 3–24. Various kinds of cells. (A) Red corpuscles. (B) White corpuscle. (C) Flat (squamous) epithelial cells. (D) Columnar ciliated epithelial cells. (E) Smooth muscle cells. (F) Nerve cell (neuron).

acquire food. Those cells that are able to synthesize their own food, that is, manufacture all of the complex molecules that are used for building materials and for energy, are said to prossess **autotrophic** (self-nourishing) nutrition and they are called **autotrophs.** Those cells that must depend on outside sources for their food are said to have **heterotropic** nutrition and are identified as **heterotrophs.**

The autotrophs include green plants and certain types of bacteria which obtain their energy for building food molecules from the sun. Some bacteria are able to gain energy for food synthesis by oxidizing certain inorganic compounds such as hydrogen sulfide (chemosynthesis). Others oxidize organic compounds under the influence of sunlight to obtain energy for synthesis of food molecules. In the whole gamut of living things these organisms are relatively insignificant. The most dominant type of photosynthesis, of course, occurs in green plants. They utilize, as raw materials, carbon dioxide, water, and a few inorganic ions. Green plants are superb chemists in that they combine these simple compounds into all of the complex molecules that are eventually found in the mature plant, something that the human chemist has found no economical way of doing as yet.

Heterotrophs vary somewhat in their method of securing food. Some protozoa, the fungi, and most bacteria absorb their food in the form of small soluble molecules directly from their surrounding environment. They are referred to as **saprophytes** (saprophytic nutrition). They possess powerful hydrolytic enzymes which are excreted into the immediate environment. The large fat, carbohydrate, and protein molecules are broken down (digested) to soluble molecules of fatty acids and glycerol, sugars, and amino acids which are then taken inside the cell (absorbed) and utilized.

Other cells such as certain protozoa (amoeba) and those found in simple invertebrates (hydra) ingest the large molecules and then digest them while they are enveloped in tiny sacs (food vacuoles) with-

in the cell itself. Cells that receive nourishment in this manner are called **phagotrophs** and the method **phagotrophy,** or **phagocytosis. Pinocytosis,** referred to earlier, is another method by which cells take in fluids and dissolved materials. Phagotrophy in cells should not be confused with eating in multicellular animals such as a cat eating a mouse, although the end result is the same. In this case the large molecules which compose the mouse are taken into a tube (digestive tract) where enzymes, produced by special cells, break them down to soluble molecules which are then taken up by the cells lining the wall of the tube. This is called **holozoic nutrition** and usually refers to the method of food taking of most animals and is discussed later.

It may be useful at this point to compare the requirements of a typical autotroph cell (green plant) and a typical heterotroph cell (animal), remembering that there are intermediate types which do not follow the rule. The self-nourishing plant cell requires only water, minerals (including nitrate NO_3), organic carbon in the form of carbon dioxide, and the presence of chlorophyll for photosynthesis. The heterotrophic animal cell also requires water and minerals, but in addition, it must have organic carbon in the form of a complex compound, such as glucose, and amino (NH_2) nitrogen which it usually obtains in the form of amino acids. It cannot make some amino acids, hence requires them in their intact form; these are called **essential amino acids.** Moreover, the animal cell requires **vitamins,** some of which it also cannot make. The plant can manufacture the vitamins and all of the amino acids; therefore, the animal cell must depend directly or indirectly upon the autotroph to supply these nutritional requirements. The animal cell is only required to provide the hydrolytic enzymes that are necessary to break the plant (or other) molecules to such form that they can be absorbed and utilized. The plant cell, on the other hand, needs nothing from the animal cell (except CO_2) and is therefore self-sustaining.

The basic pattern of nutritional requirements of cells must be very old because we see, even among the single-celled forms (algae and many protozoa), essentially the same requirements that are found in the multicellular plants and animals. The tree is an autotroph and the dog is heterotroph. One wonders how the animal cell became so dependent upon the plant cell. Perhaps at one time the animal cell may have manufactured most of its requirements through an enzyme system as elaborate as that of the plant cell. With time, mutations, perhaps, blocked certain metabolic pathways by destroying the effectiveness of certain enzymes. As long as the animal cell could obtain the essential nutrient which it could no longer make from plants or perhaps other animal

cells, it could survive. Any defect caused by mutation would be inherited so that the progeny of the altered cells might continue to exist and even prosper. A continual loss of enzyme effectiveness might persist until further reduction in synthetic capacity could not be tolerated. The state of heterotrophy reached may have been while cells were still in the one-celled stage; and for that reason all higher animals have essentially the same basic nutritional requirements. Hence man, like a protozoan, requires about ten amino acids and certain vitamins in his diet. There are minor variations, but many of the same nutrients are required by both men and protozoa.

Regardless of how the cell came by its

Fig. 3–25. A schematic representation of the sites of physiological reactions in a typical cell. There remains some doubt about the exact location of some of the reactions.

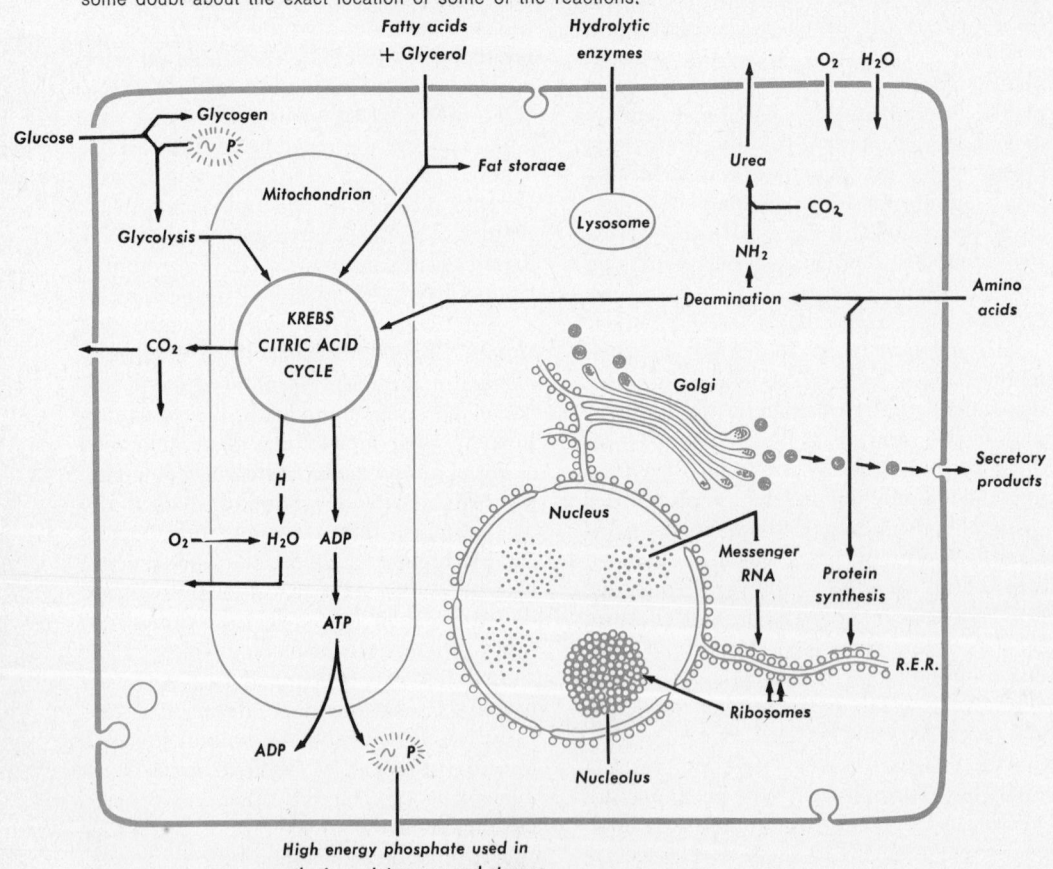

raw food molecules, once they are inside the cell, the processes by which energy is released from these molecules and the way products are synthesized are remarkably similar in all cells. Let us consider these now.

Intermediate Metabolism. **Metabolism** includes the processes by which energy is released from the energy-rich compounds and building materials are incorporated into the structure of the cell. The business of providing the cells with food and removing the waste products is merely accessory to the real job of extracting energy from the food and using it to build new protoplasm. This process involves a great many enzymes and the chemical reactions are tremendously complicated. The remarkable similarity of these numerous enzymes and their corresponding chemical reactions among all living organisms is strong evidence for common descent of all cells. It seems quite clear that primitive living things evolved a system of obtaining energy and constructing proteins that has remained more or less unchanged up to the present. Undoubtedly thousands of chemical pathways have been "experimented" with through past ages, but long ago the situation, as we see it today, emerged as the most compatible with the environment on this planet and has become firmly established in all cells.

The history of food molecules from the time they enter the cell until they leave as waste products involves some rather complicated chemistry, most of which has become understood only in recent years. We speak of the entire process as **intermediate metabolism.** It has been known for over a century that cells take in glucose and convert it to carbon dioxide and water, yielding energy during the process. But understanding of the steps involved in this degradation of glucose, the intermediate steps, has come to light primarily through the aid of traceable atoms such as radioactive carbon (C^{14}), radioactive phosphorus (P^{32}), and others which have been used to *tag* the large molecules involved in this process. This **tracer** technique is very useful and has been respon-

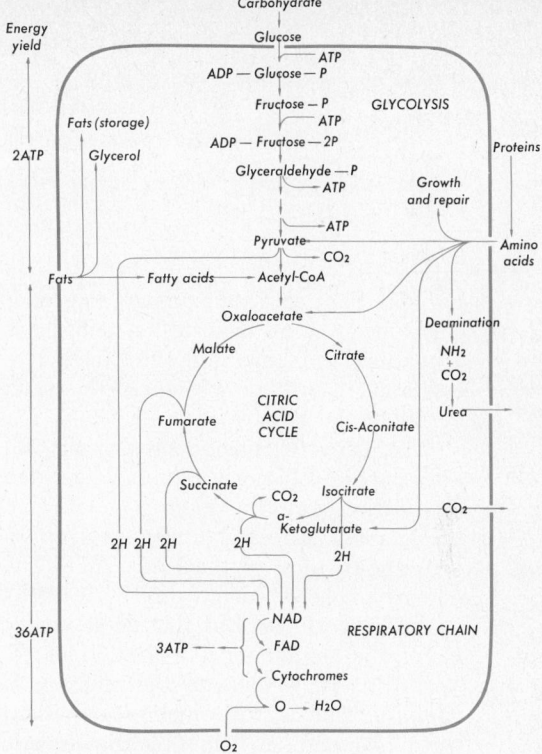

Fig. 3–26. A simplified schematic representation of cellular intermediary metabolism. See text for explanation.

sible for much of our newer knowledge in biology.

While studying the following description it will be useful to refer to the accompanying diagrams. Figure 3–25 indicates graphically where the chemical reactions probably occur in the cell and Figure 3–26 shows a skeleton outline of the reactions themselves. Glucose arrives at the cell membrane, following the hydrolysis of starch or glycogen, and passes into the cell where it must first combine with phosphate to form **glucose-6-phosphate.** This is accomplished by the enzyme **hexokinase** and the process is called **phosphorylation.** Since this is an endergonic reaction, energy must be supplied for it to take place. This energy comes from the breakdown of the high energy phosphate compound, **adenosine triphosphate,** called ATP

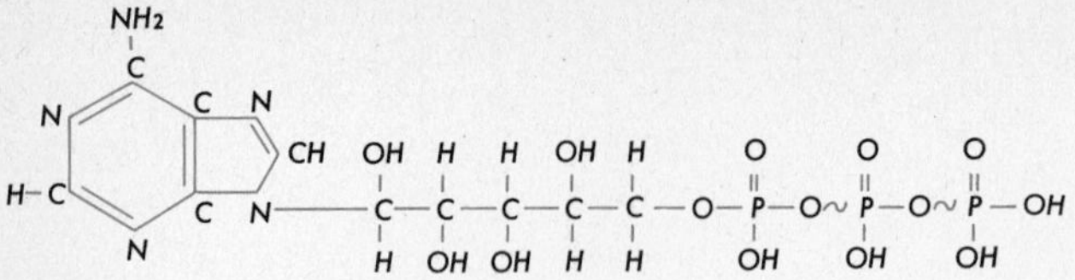

$$NH_2$$

Fig. 3–27. The structural formula for adenosine triphosphate.

for short. Since this is the primary source for quick energy in all biological systems, it is fitting that we discuss it at this time before continuing with the discussion of the metabolism of the cell.

This unique compound supplies the cell with a source of energy that can be called into service instantly. When a muscle contracts, a cell divides, a firefly turns on its light, the energy is supplied from ATP. The primary source of energy goes back to the oxidation of glucose but the usable energy is stored in the form of ATP as we shall see. The structural formula of the molecule is given in Figure 3–27 in which the important part for our purposes here is the triphosphate portion. The two chemical bonds that link the phosphate with the adenosine are called the **high energy phosphate bonds.** The uniqueness of this molecule is demonstrated when these bonds are broken. Two or three times as much energy is released when this happens in ATP as compared to when most

other kinds of chemical bonds are ruptured. This bond energy in ATP directly or indirectly drives essentially all of the energy-requiring processes of living things.

If you examine the structure of ATP, you will note that it is composed of the purine, **adenine,** plus the sugar **ribose,** to which is attached three phosphate groups. The skeleton of the molecule with one phosphate group is a nucleotide (adenylic acid), a compound with which we are familiar since it appeared very early in the evolution of life. Undoubtedly ATP also evolved during this early period and supplied the energy that was essential for further reactions to take place.

Energy is released from ATP when the terminal high-energy phosphate bond is ruptured forming **adenosine diphosphate** (ADP) (Fig. 3–28). As indicated, the reaction runs in a cycle. Obviously it requires the same amount of energy to form ATP as was released. The source of this energy is the oxidation of glucose; it is not a simple oxidation process and there are many steps involved. It is necessary to degrade the glucose molecule in a slow orderly fashion so that the energy can be stored in ATP at body temperatures. If it were oxidized rapidly the energy would be lost in the form of heat and thus could not be made available to the cell for doing all the necessary jobs that it must do.

This degradation of glucose proceeds in a series of step-wise reactions, each resulting in some loss of the energy present in glucose; such reactions are called **catabolic** ones. If a given step involves a large enough energy loss, it may be coupled to the regeneration of ATP from ADP; in this event it is called a **coupled reaction.** This coupled reaction is reversible; i.e., ATP can be used to add energy to make a particular step go; such reactions are called **anabolic.** An example of such a reaction would be written as follows:

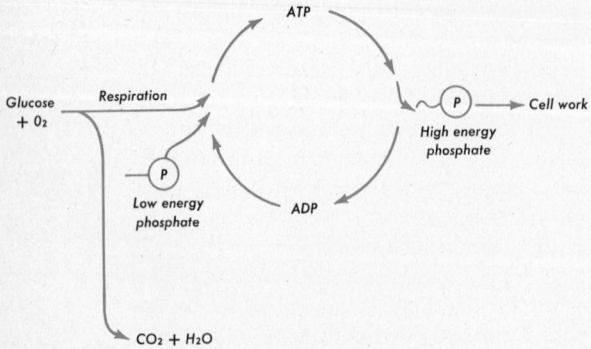

Fig. 3–28. A schematic representation of how energy is released from glucose by the ATP–ADP cycle to do cell work.

ATP ADP

Glucose $\xrightarrow[\substack{hexokinase \\ Mg^{++}}]{}$ glucose-6-phosphate

$$+ \; H^+$$

86

The enzyme hexokinase and the magnesium ion (Mg^{++}) serve to catalyze the reaction. By contributing one ATP at this point and another in converting fructose 6-phosphate to fructose-1, 6-diphosphate (Fig. 3–26), the cell is providing **activation energy** so that the step-wise process may proceed. Each hexose molecule (6 carbons) is later cleaved into 2 glyceraldehyde (3 carbons) molecules from each of which 2 molecules of ATP can be regenerated, yielding a net gain of 2 ATP's and two molecules of pyruvate. Thus, the cell, through use of a series of small step-wise reactions, each catalyzed by specific enzymes, has captured energy from the original glucose molecule and stored it in the form of ATP's.

The series of steps taken so far in degrading the glucose molecule has yielded a little energy and you will note that no oxygen has been used. For that reason we say this is an **anaerobic** process and the overall reaction to this point is referred to as **glycolysis.** Microorganisms, such as some bacteria, can obtain energy anaerobically only through the process of **fermentation** which involves the utilization of glucose and the evolution of gas. They cannot carry the process any further, owing to the lack of specific enzymes to do the job. It is obviously a very wasteful method of obtaining energy because most of the energy still remains locked up in the waste products, alcohol, lactic acid, or acetic acid. Most cells, however, evolved enzymes to carry the degradation all the way to carbon dioxide and water, thus removing almost the last trace of energy from the glucose molecule.

Pyruvic acid, which is the starting point for the next series of reactions, is converted to acetic acid, and then enters a series of reactions which move sequentially in a cycle, yielding energy as acetic acid is converted to carbon dioxide and water. This circular pattern of reactions has been called the **Krebs citric acid cycle,** in honor of the man who first postulated it. The sequence of events starts when acetic acid (in combination with coenzyme A) combines with oxaloacetic acid (a 4-carbon compound) to form citric acid (a 6-carbon compound), which is then converted into seven other compounds, coming back finally to oxaloacetic acid. Each conversion is under the influence of a specific enzyme. If any one of these enzymes is destroyed, the cycle stops at that point, and the compound formed at that time accumulates. By employing specific enzyme poisons, it is possible to identify each compound in the cycle; indeed, this is one of the methods that was first employed in working out the cycle.

As the Krebs cycle turns, carbon dioxide and hydrogen fall away, and the former is ultimately lost from the cell as a waste product. The process by which hydrogen is released is called **dehydrogenation** and the enzymes involved, **dehydrogenases,** of which there are a number. The released hydrogen ultimately unites with oxygen to form water, but many steps are encountered before this final stage is reached. A number of compounds, collectively called the **respiratory chain,** combine temporarily with hydrogen before it unites with oxygen to form water (Fig. 3–29). Some of these contain the vitamins niacin and riboflavin, which is the main reason why they are essential in the diet. After hydrogen is passed along from one vitamin-containing compound to another, it is directly passed to a series of **cytochromes,** protein compounds related to hemoglobin (p. 401) which change color when they accept or lose hydrogen. The cytochromes, of which there are usually three (a,b,c), finally donate the hydrogen to oxygen to form water. It is interesting to note that the oxygen in the waste CO_2 comes from some substrate molecule in the glycolytic and Krebs cycles, whereas the oxygen in released water comes from the oxygen that is taken in when breathing.

The Krebs citric acid cycle has been referred to as the "intracellular energy wheel" because, as the wheel turns, a large amount of fuel is made available for the respiratory chain where it is captured and stored in ATP molecules. As illustrated in Figure 3–29, this fuel is initially in the form of hydrogens or protons, designated simply as H or H^+.

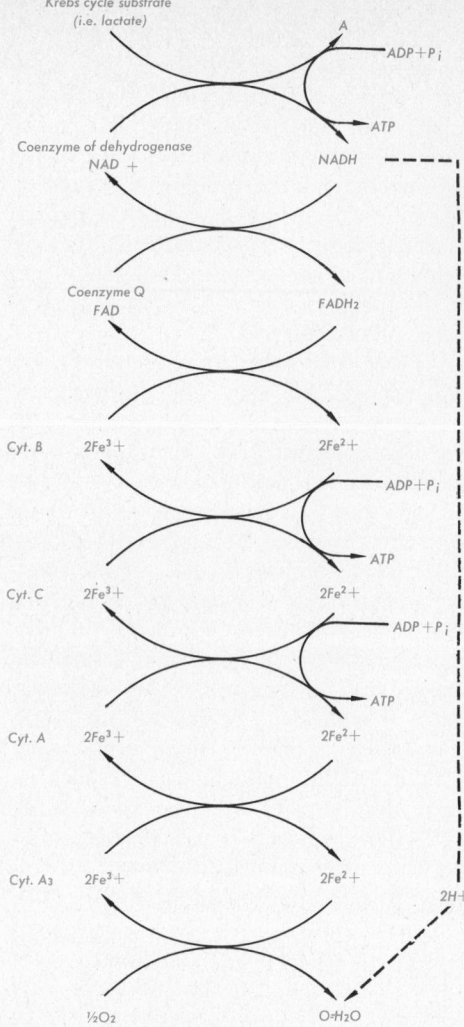

Krebs cycle substrate
(i.e. lactate)

A

ADP+P_i

ATP

Coenzyme of dehydrogenase
NAD +

NADH

Coenzyme Q
FAD

FADH$_2$

Cyt. B   2Fe^{3+} 2Fe^{2+}

ADP+P_i

ATP

Cyt. C   2Fe^{3+} 2Fe^{2+}

ADP+P_i

ATP

Cyt. A   2Fe^{3+} 2Fe^{2+}

Cyt. A$_3$ 2Fe^{3+} 2Fe^{2+}

2H+

½O$_2$ O+H$_2$O

Fig. 3–29. The electron transport chain.

These hydrogens are produced by enzymes called dehydrogenases and since they will be eventually combined with oxygen to make water, the fuel is in fact being oxidized. In chemical terms, an **oxidation** can also be defined as the removal of an electron from a molecule or atom; this can only be done by the simultaneous **reduction** of some other atom or molecule, and such reduction then is the addition of an electron. The result is a **redox** reaction, i.e., a combined reduction

and oxidation, and it is characterized by the presence of an electron donor and an electron receptor. If a hydrogen donates its electron, the hydrogen is oxidized, and therefore energy can move during pyruvate oxidation either as hydrogens or as electrons, symbolized as H$^+$ or e$^-$, respectively. In the **electron transport chain** (Fig. 3–29), hydrogens from the Krebs cycle are transferred to an important coenzyme, the hydrogen acceptor molecule NAD$^+$ (nicotinamide adenine dinucleotide). The resulting compound, NADH, is now in its reduced from and is said to possess "reducing power," which means that it will be used in the next step in which flavoprotein FAD will be reduced to FADH$_2$ and NAD$^+$ regenerated. FADH$_2$ will be reoxidized by the next compound (Coenzyme Q) which will in turn be oxidized by the new component, etc. Thus energy and electrons flow "downhill" like a series of small waterfalls, each one representing a lower state of energy or "redox potential."

If, in a given step along the electron transport chain, sufficient energy is available, the oxidation-reduction can be coupled with the production of an ATP molecule from ADP. This process is called **oxidative phosphorylation** to emphasize its dependence on oxygen in the final step. In all, 36 ATP molecules are generated from one molecule of glucose, as compared with only 2 net ATP molecules from glycolysis. The key to the capturing of energy lies in the numerous, small reactions; the reason for the evolution of such a method is probably related to the analogy of two coin systems, one using only quarters and one using nickels. Since the only capturing device is ATP, if the cell needs only a nickel's worth of energy to convert ADP to ATP, then a coin system using only quarters is wasteful.

The mitochondrion is the site of the activities of the Krebs cycle and the electron transport chain, whereas glycolysis occurs in the cytoplasm. If a primitive cell really did acquire in the ancient past a prokaryotic symbiote which became a mitochondrion, it is clear that such a system could extract the energy remaining in the molecular fragments left over from glycolysis and fermentation much more efficiently, thereby having a distinct advantage for survival and reproduction.

The entire process of converting glucose to CO_2 and H_2O in the cell is respiration; and while we can write it simply:

$$C_6H_{12}O_6 + 6O_2 \longrightarrow 6CO_2 + 6H_2O + Energy$$

which quite amply describes the net result, the intermediate reactions are far from simple. One might ask, why bother with all of these steps when an end result is all that we need to know? This knowledge of the intermediate steps has been necessary in understanding such questions as why certain substances are poisons, why antibiotics work, and many others. Heavy metals, such as arsenic, mercury, and lead, are poisons because they block the activity of many enzymes in the Krebs cycle. They accomplish this typically by combining directly with sulfhydryl (SH) groups on enzymes, thereby inactivating them. This reaction tends to be nonspecific, affecting several enzymes at the same time. In the case of arsenic, Erlich (1913) discovered that greater specificity for the intracellular parasite *Trypanosoma* could be obtained by making an organic derivative of arsenic, and thus began the first scientific approach towards finding antibiotics which selectively attack an invader. Penicillin represents the outstanding example of an antibiotic that inhibits a special metabolic pathway occurring in bacterial cell wall synthesis. Such cell walls contain **peptidoglycans** which must be linked together by an enzyme (transpeptidase), and it is believed that this enzyme is specifically inactivated by penicillin. Since animals do not have cell walls, penicillin, even in very large doses, is harmless to man unless he develops an **allergic** reaction to the drug. Allergies are **immune responses** of the organism to foreign substances as a defensive mechanism; such responses to penicillin can be very serious but have nothing to do

with cell wall synthesis. We discuss immune responses later.

Antibiotics have been extremely effective against prokaryotic "invaders" of the human and other vertebrate systems, and massive screening programs have been instituted to find new antibiotics. Penicillin of course is a natural product originally "discovered" by a mold (*Penicillium notatum*) to improve its own growth and survival opportunities in competition with bacteria. Some bacteria respond by producing resistant **mutants**; these mutants can make an enzyme which quickly inactivates the penicillin. Unfortunately the mechanism of action of most antibiotics is still not fully understood, and the problem is compounded by the fact that certain antibiotics like chloromycetin which are very effective against bacteria also apparently effect mitochondrial functions with extended treatment. Streptomycin can be used to "bleach" the protozoan *Euglena*; i.e., it destroys chloroplast function, including eventually the chloroplast itself, and there is some suggestion that in very large doses and extended treatment, streptomycin may be harmful to mitochondria. Of course, many treatments (e.g., arsenic compounds) are only slightly more "poisonous" to the pathogen than to the human. Until we know more about the biochemistry of the action of a given antibiotic or antimetabolite, it is essential that we recognize the dangers inherent in treating prokaryotic invaders with substances which could be potentially dangerous to our own prokaryotic-like mitochondria.

The action of heavy metals and penicillin then is to inactivate or poison certain enzymes, but the exact mechanism is not completely understood. We normally distinguish two methods of poisoning cells. The first involves the application of substances or treatments which result in the formation of nonfunctional enzymes or cofactors, i.e., an attack on the catalytic agents themselves. This can be done at several levels; one method is use various **analogs** (closely related compounds) of substances required by the cell, such as vitamins, amino acids, or nucleosides. The active site or binding site of an enzyme depends upon the proper amino-acid sequences in the polypeptide chain which allows the correct three-dimensional folding; this can be influenced by the binding of a cofactor at another binding site on the same enzyme. Substitution of one or a few of the incorrect amino acids by alteration of the DNA or misreading of the messenger RNA (chapter 23) could affect the active site to the extent of destroying all activity or reducing the efficiency of the enzyme.

The second method is to offer the enzyme a substrate which looks like the natural substrate, but which in fact is not. The enzyme combines with the substrate analog, but cannot make the required chemical change. The analog competes with the normal substrate, and the process is called **competitive inhibition.** Its effectiveness depends on the concentrations of the substances involved. An excellent example is sulfanilamide, which was the first "wonder drug," and has very high activity against bacteria. Sulfanilamide is a competitive inhibitor of p-aminobenzoic acid (PAB).

NH$_2$ NH$_2$

COOH   SO$_2$NH$_2$

p-Aminobenzoic acid   Sulfanilamide

The effect of sulfanilamide can be reversed by adding excess PAB; sulfanilamide is called an **antimetabolite** which competes with PAB for the same active site on the enzyme. It is easy to see the similarity in structure between the two compounds. In the bacterium PAB is used in a series of reactions leading to the synthesis of folic acid and eventually translation of mRNA on the ribosome. Man requires folic acid in his diet but uses a different set of enzymes leading to translation in the cytoplasm. A distinction should be made between

bacteriocidal (bacterial death) and bacteriostatic (cessation of growth) conditions. It is generally only necessary to stop the tremendous growth rate of the bacteria in animals and man, and the white blood cell phagocytes will soon remove the bacteria.

From what we have said so far, it would appear that the only source materials that keep the Krebs cycle going come from glycolysis. If this were true, cells could not utilize amino acids and fats as sources of energy. We know that these foods are used and that animals can live very well on protein and fat, with very little carbohydrate in their diet. This is possible because some portion of the amino acids that enter the cell are **deaminated,** that is, lose their amino group (NH$_2$), and the remaining fragment of the molecule, which contains only C, O, and H atoms, is fed into the Krebs cycle via pyruvic acid, thus furnishing material to keep the cycle operating. The amino groups may leave the cell in the form of ammonia, or they may combine with carbon dioxide to form urea (Fig. 3–25) and then leave the cell. The amino acids that are not utilized as an energy source are utilized by the cell for growth and repair of worn-out protoplasm, i.e., in new protein synthesis.

The fats that arrive at the cell membrane pass through it and are hydrolyzed to fatty acids and glycerol. The fatty acids then are converted to acetic acid, which moves into the Krebs cycle and is oxidized. Unoxidized fats are stored in the cell for future use. Thus we see that the three foods—fats, carbohydrates, and proteins—are utilized as sources of energy by cells.

Just where do all of these reactions occur? Let us now take another look at the cell with this thought in mind. The site where some of the chemical reactions occur has been ascertained by studying cell fragments and particulates. As was pointed out earlier, it is possible to homogenize cells and by centrifugation separate fractions which contain the different parts of the cell. By testing these fractions for the various reactions we have been describing, it is possible to determine where they occur in some instances. However, it is not possible as yet to say where they all occur because not all are completely understood, to say nothing of their position in the cell.

There seems to be no doubt about where the Krebs cycle enzymes are located because oxidative metabolism occurs in mitochondrial fractions. The respiratory chain enzymes are localized on the mitochondrial membranes. Glycolysis apparently occurs in the cytoplasm and is not associated with particles. ATP is found throughout the cell which is understandable because energy is needed for nearly every activity of the cell. Fats are stored in vacuoles and the hydrolytic enzymes are confined to membraned sacs, the lysosomes. The secretory products of the Golgi bodies pass from the cytoplasm out of the cell and perform their function extracellularly for the most part, although they may have intracellular functions as well. Deamination and urea formation probably occur in the cytoplasm with no relationship to particulates. DNA and RNA are manufactured on the chromosomes inside the nucleus, with rRNA accumulating in the nucleolus. The smaller-

molecular RNA, called transfer RNA (tRNA), passes through the nuclear membrane and migrates to the ribosomes where it functions with the messenger RNA in the synthesis of proteins, including enzymes. Just how this occurs is discussed in a later section when cell duplication is considered.

We have seen that cells, while highly variable in size and morphology, all have the same fundamental needs which must be cared for whether they are a part of a many-celled organism or live as isolated individuals. Some cells, however, have additional unique functions, such as the production of compounds like enzymes and hormones, or contraction as in muscle fibers. One very interesting property found in special cells of many organisms is the capacity to produce light.

Bioluminescence

Light can be emitted by an atom or molecule when an electron changes its energy level, and this can be accomplished by light (photoluminescence), by electrical energy (electroluminescence) and by a chemical reaction (chemiluminescence). Our partic-

Fig. 3–30. This is a comb jelly (*Mnemiopsis leidyi*) found along Cape Cod. The combs are luminescent and this photo was taken using their own light.

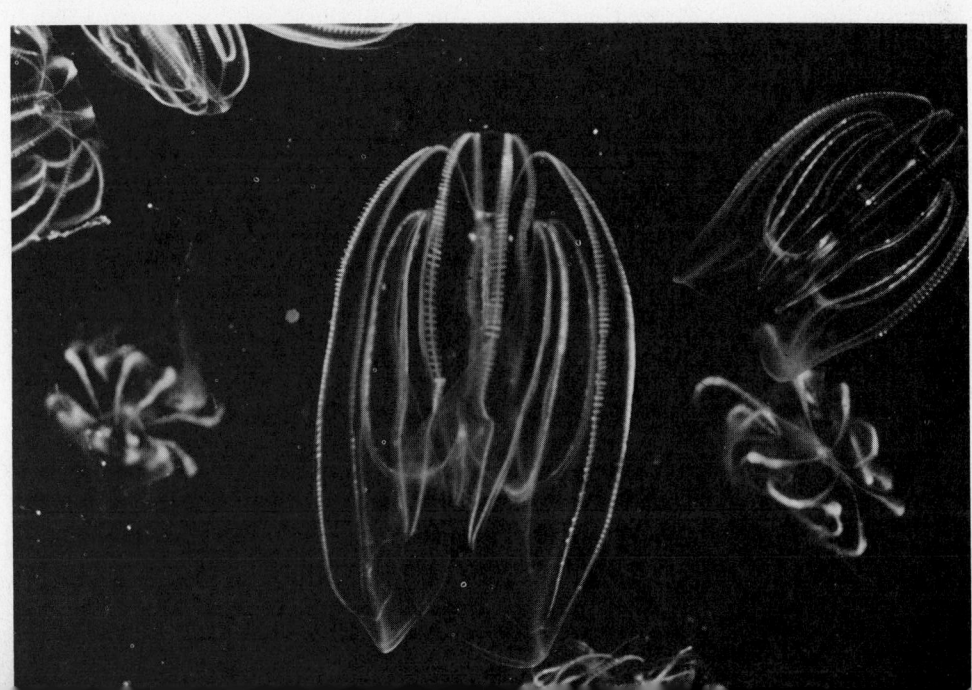

ular concern is with one type of chemiluminescence occurring in living organisms
which is called **bioluminescence.** The capacity of producing light is widespread;
indeed, luminous species occur in all the
major phyla, except the Platyhelminthes
and the Nemathelminthes. The sea is
teeming with luminescent bacteria, protozoa, and a large variety of metazoa (Fig.
3–30). These microorganisms, as well as
some of the smaller metazoa, cause a brilliant display of light in the wake of a ship
on the ocean on dark nights at certain
times of the year. Some deep sea forms, particularly fishes, are endowed with **luminescent organs** which provide light in an
otherwise eternally dark world. These
organs are rather remarkable in that they
themselves do not emit light but are really
cavities in which luminescent bacteria live.
The organ is well supplied with blood
vessels to permit adequate oxygen for the
bacteria. The bacteria glow continuously
but the fish can turn the light off by means
of a flap of skin which functions much like
an eyelid.

The best known terrestrial luminescent
animals are the fireflies and the glow
worms, although many others have the ability such as certain beetles, flies, millipedes,
centipedes, snails, and earthworms. A very
interesting fly (*Arachnocampa luminosa*),
living in caves in New Zealand and elsewhere, supplies sufficient light to dimly illu-
minate these dark caverns. It is actually the
larval stage that clings to the roof of the
cave, sending down threads of mucus as
"fishing nets" for trapping midges (insects) living in the river below, their sole
food supply. The glowing light attracts
the midges which fly up from the stream
and become enmeshed in the mucus
threads and are then devoured by the larval
glow worms. To visit one of these caves is
a fascinating experience.

The mechanism that brings about
bioluminescence is not understood although a great deal of work has been done
with the problem. Its solution could have
far-reaching effects in our economy because
these organisms produce "cold light" in
which very little energy is lost in the form
of heat; this energy loss is the big problem
in our present methods of producing light.
In 1887 it was suggested by a Frenchman,
Dubois, that the light in the luminous clam,
Pholas dactylus, was produced when a
substance, **luciferin,** was oxidized. We now
know that luciferin is destroyed in the
presence of the oxidative enzyme, **luciferase,**
and oxygen. This reaction causes the light
to be emitted. If the two substances are
placed in a test tube in the presence of
oxygen and ATP, luminescence occurs. It
turns out, however, that luciferin is not a
single substance present in all light producing organisms, but occurs in many different forms which suggests that bioluminescence evolved independently many
times during the long course of organic
evolution. It has been suggested, because
of the close chemical relationship between
bioluminescence, photosynthesis and vision,
that the former is the reverse of the latter
two. That is, in vision and photosynthesis
the reaction is excited by light and
chemical compounds result, whereas in
bioluminescence the reverse happens; as a
result of a chemical reaction light is produced. Moreover, oxygen is a by-product of
photosynthesis and an absolute requirement
for bioluminescence. As more is learned
about the energy transformations in these
reactions, the evolutionary story will perhaps be better understood.

Light emission may have occurred in
early times in most organisms as a by-
product of oxidative processes, and has
gradually been lost owing to a shifting of
the end product, to ATP for example. Do
modern organisms today make any use of
this illumination? It seems quite clear that
some do whereas in others its value is
questionable. For example, bacteria and
fungi make little, if any, use of their light;
likewise, what use does *Chaetopterus*, living
on the ocean floor in a dark tube from
which it never escapes, have for a light? On

the other hand it has been shown that the intermittent flash of the firefly is intimately tied up with the mating process. The light emitted by deep sea animals could be helpful in recognition of one another or as warning signals for predators. It would be interesting to know what use the remarkable railroad worm of Uruguay has for its row of green lights along its two sides and the two red ones on its head.

SUGGESTED SUPPLEMENTARY READINGS

Books

DE ROBERTIS, E. D. P., NOWINSKI, W. W., and SAEZ, F. A., *Cell Biology*, Philadelphia: Saunders, 1970.

DYSON, R., *Cell Biology, A Molecular Approach*. Boston: Allyn & Bacon, Inc., 1974.

GIBBS, M., ed., *Structure and Function of Chloroplasts*. New York: Springer-Verlag, 1972.

GIESE, A. C., *Cell Physiology*. Philadelphia: Saunders, 1974.

LEHNINGER, A. L., *Biochemistry: The Molecular Basis of Cell Structure and Function*. New York: Worth Publishing Co., 1970.

*LOEWY, A. G., and SIEKEVITZ, P., *Cell Structure and Function*. New York: Holt, Rinehart & Winston, 1963.

*NOVIKOFF, A., and HOLTZMANN, E., *Cells and Organelles*. New York: Holt, Rinehart & Winston, 1970.

PORTER, K. R., and BONNEVILLE, M. A., *An Introduction to the Fine Structure of Cells and Tissues*. Philadelphia: Lea and Febiger, 1964.

SAGER, R., *Cytoplasmic Genes and Organelles*. New York: Academic Press, 1972.

*TONDLER, B., and HOPPEL, C., *Mitochondria*. New York: Academic Press, 1972.

WHITE, A., HANDLER, P., and SMITH, E., *Principles of Biochemistry*, 5th ed. New York: McGraw-Hill, 1973.

WOLFE, S., *Biology of the Cell*. Belmont, Calif.: Wadsworth Publishing Co., 1972.

* Available in paperback.

Articles

BEERMAN, W., and CLEVER, U., "Chromosome Puffs." *Scientific American*, April, 1964.

BRACHET, J., "The Living Cell." *Scientific American*, September, 1961.

CAPALDI, R. A., "A Dynamic Model of Cell Membranes." *Scientific American*, March, 1974.

CRICK, F. H. C., "The Genetic Code: III." *Scientific American*, October, 1966.

CRICK, F. H. C., "Nucleic Acids." *Scientific American*, September, 1957.

DE DUVE, C., "The Lysosome." *Scientific American*, May, 1963.

DE DUVE, C., "Tissue Fractionation, Past and Present." *Journal of Cell Biology*, 50,20D, 1971.

GREEN, D. E., "The Mitochondrion." *Scientific American*, January, 1964.

HAYASHI, T., "How Cells Move." *Scientific American*, September, 1961.

HURWITZ, J., and FURTH, J. J., "Messenger RNA." *Scientific American*, February, 1962.

KOSHLAND, D. E., JR., "Protein Shape and Biological Control." *Scientific American*, October, 1973.

94

MAZIA, D., "The Cell Cycle." *Scientific American*, January, 1974.

NIRENBERG, M. W., "The Genetic Code: II." *Scientific American*, March, 1963.

PUCK, T. H., "Single Human Cells in Vitro." *Scientific American*, August, 1957.

ROBERTSON, J. D., "The Membrane of the Living Cell," *Scientific American*, April, 1962.

III

THE

ORGANIZED

ANIMAL

4

FROM SINGLE CELLS TO MANY CELLS

Once the fully organized cell had become well established on earth, it probably "explored" many possibilities in structural and physiological patterns in order to improve its situation, which usually means better adaptation to its environment. Undoubtedly, as numbers increased, cells moved into many different environments. Some found niches in which they were "satisfied," where they have remained more or less unchanged, through the succeeding millions of years. We find them still occupying these same, or similar, places today. The demands of their environment remained essentially static, hence they were not forced to change in order to survive. Others, however, were forced into adverse environments which compelled them to change or perish. It was from this group that we might expect to find not only new varieties of single cells, but also some that banded together in small groups for apparently enhanced survival. We shall never know why cells aggregated into groups since there are no fossil remains to guide our thinking, but we can speculate a little about some of the advantages that accrue from such assemblages of cells.

First of all, there is the matter of size. Small single-celled forms, moving about with tiny whiplike flagella, have very little control over their movement. Whereas they can orient themselves to a limited extent, they are predominantly under the influence of water currents and must go wherever these currents take them. With increased bulk this influence is diminished; indeed as size increases the organism has more command over where it shall go. Hence, it is likely that those cells which aggregated, increasing their overall bulk, coped better with certain environmental hazards, and therefore were able to survive where single cells would have perished.

An alternative method of increasing size

was apparently attempted by some single cells, and to a point it was successful. The cell itself increased in size by adding more protoplasm, resulting in huge cells. We see these today among the ciliated protozoa where some reach a length of a millimeter. This increased protoplasmic mass required additional nuclear material to maintain normal metabolism, hence, in addition to an ordinary nucleus they acquired extra nuclear material in the form of the large macronucleus. Even with this bit of accessory equipment, size limits were soon reached. These cells are all comparatively small when compared to most animals. The solution to the problem of size limitation had to be solved in some other way. The most obvious possibility was for cells to aggregate into masses and this apparently happened. Just how this came about has inspired considerable speculation on the part of many biologists for a long time.

Let us assume that during the millennia required for the unfolding of higher animal life, some animals found satisfactory niches and have remained essentially unchanged up to the present. If this is so it should be possible to find such representative animals and to arrange them in the order of their complexity, forming a continuous series from single cells to the most complex animals alive today. We would expect to find gaps between groups, and we would also expect the animals we do find not to be exact duplicates of the originals. They, too, would undoubtedly have undergone changes during this long period of time, even though they might have remained in a relatively unchanging environment. By forming such a series, the story of evolving life can, possibly, become a little clearer.

Biologists agree that the many-celled living things, both plants and animals, took their origin from single-celled forms. The only part of this story we wish to consider in this discussion is that dealing with the transition from the single- to the multi-celled organisms. Of the several theories that have been formulated by biologists during the past hundred years, we consider the two which

are currently most popular, although neither is completely satisfactory. Consult Figure 4–1 during the following discussion.

The Colonial Theory. Surveying the thousands of unicellular organisms (protozoa and protophyta), we find the most likely candi-

Fig. 4–1. A schematic representation of the colonial and syncytial theories for the origin of the metazoa.

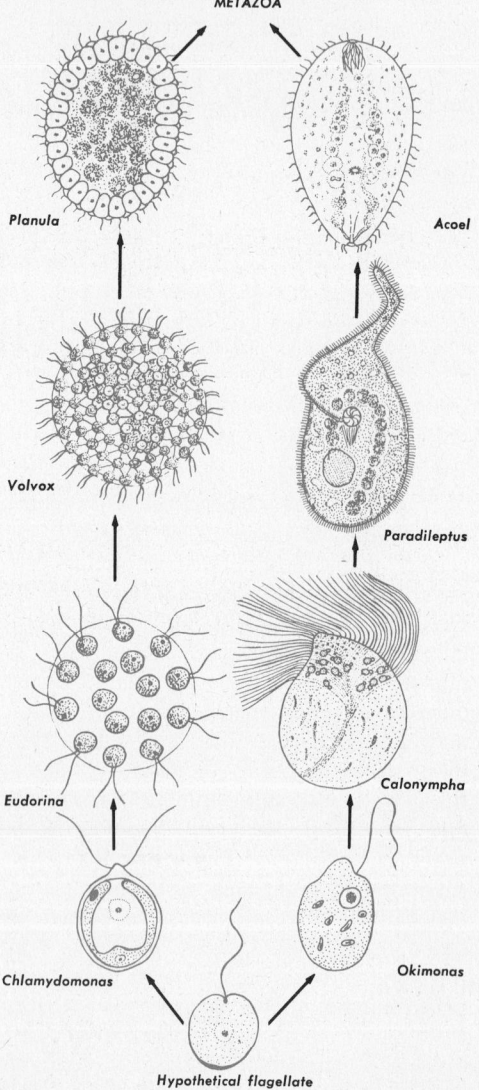

METAZOA

Planula

Acoel

Volvox

Paradileptus

Eudorina

Calonympha

Chlamydomonas

Okimonas

Hypothetical flagellate

COLONIAL THEORY

SYNCYTIAL THEORY

dates for the most primitive cells are those which bear **flagella.** Among these, the flagellates which contain **chlorophyll,** a green plant pigment, are usually selected as the most primitive and are therefore thought of as the starting point for all subsequent evolution. This group of flagellates contains a large assemblage of those that exist singly to those that exist in groups called colonies. Starting with the single-celled *Chlamydomonas,* one can arrange a graded series where individuals differ only in number of cells that cling together. *Eudorina* is composed of a ball of sixteen cells embedded in a matrix of jellylike material. Each of the cells is not greatly different from *Chlamydomonas.* The combined beating of their flagella causes the entire colony to roll along in a graceful manner. Another form, *Pleodorina,* is composed of many more cells, clustered in the shape of a hollow sphere; aside from the increased number of cells there is little difference between this one and *Eudorina.* We do note one rather interesting dissimilarity. Not all of the cells are the same size; some are smaller than others, and during reproduction the smaller ones are unable to produce new colonies, in other words, they are **sterile.** Therefore, in this form there are two kinds of cells, **reproductive cells** and sterile or **soma cells.**

A much larger aggregate is illustrated by *Volvox,* a beautiful hollow spherical colony consisting of several thousand cells. Again these cells resemble *Chlamydomonas* in most respects, although there are tiny bridges between individuals which tend to lock them together more securely than the loose jelly of other forms. Most of the cells are alike, although there are some here and there that are larger and have a different appearance. These are the reproductive cells; all others are soma cells. A more careful observation will reveal that the reproductive cells are of several kinds. Some are bundles of tiny bodies, the **sperm** or male cells, while others are large ovoid **egg** cells. These special **sex cells** reproduce the colony by a **sexual process;** that is, the sperm are released into the water where they swim to and unite with the egg. This subsequently becomes a **zygote,**

which hibernates in a heavy-walled zygospore. Other reproductive cells merely divide and move into the hollow of the sphere where they become small colonies, known as **daughter colonies.** These eventually burst out, destroying the mother and becoming adult colonies themselves. The larval stage of most coelenterates, truly metazoan animals, is the planula which is much like *Volvox* except that the internal space is filled with cells and it is oval in shape and spirals on its long axis when it swims. After swimming about for a time it settles down and grows into a mature coelenterate with many kinds of soma cells.

Two striking events occurred in this gradual association of cells. First, similar cells aggregated into a mass which apparently succeeded better; that is, there was strength in union probably because of increased size. Second, **division of labor** was initiated among the cells, some becoming sterile and functioning only in locomotion and food-getting, whereas others retained the primitive condition of colony-reproducing. Some of the reproductive cells became highly modified into eggs and sperm while others merely retained their primitive characteristics of reproducing by simple fission. In other words, a differentiation of function took place among the cells of the aggregate, definitely marking it off from the isolated single cells and at the same time creating the first step in the organization of a complex animal through the loss of the power of reproduction by some of the cells. Once this step was taken, differentiation of the soma cells continued in various directions toward greater and greater complexity, and thus to higher animals.

This gradual advance in complexity might be compared to the evolution of our own society. The protozoa may be compared to primitive man who lived alone and was compelled to obtain all of his own food, make his own clothing, and provide his own shelter. Existence by this crude means made chances for survival poor, and mortality high. Later, man associated himself with others in the common interest of survival and of making the drudgery of life less grueling. The first groups were made up of the immediate family; they lived together, hunted together, and made shelters together. In other words, they performed all the duties acting as a group rather than singly as heretofore. Food was easier to secure because they could surround and kill larger animals, their shelters could be more elaborate, and the burdens which fell upon each individual were not as great as when each lived alone. Such aggregation grew to include larger groups until small villages were formed; with the increasing numbers of individuals participating in mass efforts, less responsibility fell to each one, and what was more important, each shared in the results of the mass efforts. They all lived better and longer. This method of living has finally developed into our modern civilization.

Following this analogy, we can think of primitive society as resembling the single-celled animal and modern society as the complex metazoan, such as the mammal. As the cells began to aggregate into groups, individual cells specialized in particular jobs, and the group as a whole became more complex. There are animals all along the evolutionary scale which represent steps in increasing complexity.

If we look at some of the simplest metazoan animals alive today we may be able to guess which ones were the most likely candidates for the next step toward greater complexity. In order to fulfill the requirements the animal should be some modification of the spherical ball seen in *Volvox*. If such a sphere of cells pushed in on one side, it would become a two-layered sac. The coelenterate, *Hydra*, is such an example (Fig. 7–4). We study this fascinating creature later in detail; it is only necessary at this point to examine its anatomy briefly to carry further the idea of increasing complexity.

Hydra is made up of many cells, mostly soma cells, which are arranged in two layers. Some of the cells in the outside layer (**ectoderm**) have differentiated into "nettle cells" for stinging purposes in defense or offense. Others are able to lengthen and shorten during locomotion and to convey impulses (**neuromuscular cells**). Still others have the ability to give rise to sperm and eggs, and to new individuals by bud formation. Here, then, we see that the soma cells have differentiated into several kinds, while the sex cells remain much like they were in *Volvox*. Division of labor has started among the soma cells which is the next step in the development of more complex animals.

The Syncytial Theory. Whereas the Colonial Theory may seem to be a reasonable explanation for the origin of the metazoa, and most biologists adhere to the hypothesis, still it is not the only one. Let us consider the Syncytial Theory which stems from the fact that there exists a group of microscopic marine metazoan animals, called **acoel turbellarians** (p. 169) which possess primitive characteristics (Fig. 4–1). Their bodies are composed of a protoplasmic mass in which nuclei are present with no cell membranes. Such an arrangement is called a **syncytium** and it is not infrequently found among both simple and complex animals. For example, skeletal muscle of vertebrates is a syncytium. Among the protozoa there are some multinucleated ciliates, such as *Paradileptus*, which appear to be quite similar to the acoels. They approximate them in size and are covered with short cilia. Going back a step farther we find among the flagellated protozoa such forms as *Calonympha* which likewise has many nuclei and numerous long flagella. Most protozoologists believe that the ciliates arose from the flagellates and that the ancestral type from which these complex flagellates were derived were the simple colorless flagellates such as *Oikomonas*. The starting point for both theories is a chlamydomonas-like flagellate.

Regardless of which theory is correct, there is no doubt among biologists that the metazoa arose from protozoan ancestors. Once cells aggregated into masses and began to differentiate, each group of cells performing specific tasks, animals became larger and were able to invade more niches (p. 626), and hence became more varied in form and function. Through this process, the tremendous variety of animals that live on the earth today have emerged.

From the simplest metazoans to the most complex, division of labor has occurred among the soma cells, resulting in specialized cells. The major types of such specialized cells are not great in number but each type has several kinds. Let us examine these cells and how they are organized in higher animals.

ORGANIZATION OF CELLS INTO TISSUES

Division of labor among the soma cells spread throughout the metazoa until a wide variety of cells was produced, each doing a specific job. Cells of the same kind grouped together in a continuous mass form a **tissue**. A particular kind of tissue is not necessarily limited to one region of an animal body, but usually is found in several different places, where it may or may not perform the same function. There are five major kinds of tissues, **epithelial, supporting, blood, muscular,** and **nervous**. All of the tissues are discussed in this section except blood which is considered in Chapter 19. Although tissues are found in all groups of animals except the protozoa, for simplicity we consider only those found in the vertebrate such as man.

Epithelial Tissues. These are the surface tissues which cover and line not only the outside of the body but the cavities within as well. They are composed of closely fitting cells forming continuous membranes and with very little intercellular material binding them together. Since the jobs performed by epithelial tissues vary greatly, these tissues exhibit a wide variety of form (Fig. 4–2). The tissue is usually named according to the shape of its constituent cells, for example, **squamous** (flat), **cuboidal** (cubes), and **columnar** (columns or pillars). They may also be described in terms of accessory structures such as **flagella, collars,** or **cilia**. Finally, the tissue may be referred to as **stratified** if

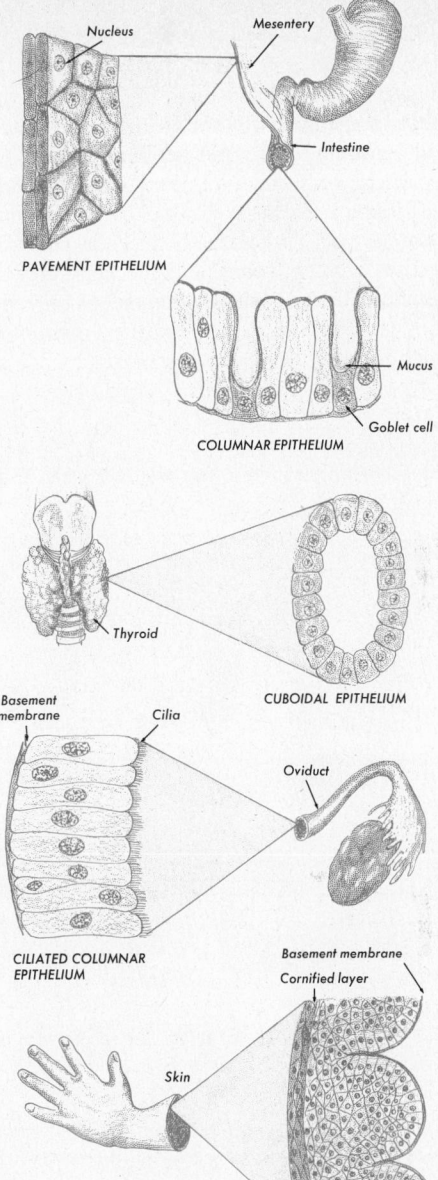

Fig. 4–2. Epithelial tissue.

the cells have different forms and lie several cells in thickness.

In addition to protection, epithelial cells that line cavities usually have the function of **secretion,** which is the production of special substances used by the organism in various ways. For example, the cells lining the digestive tract are mostly secretory in function. These cells form the secreting portion of **glands,** whether the gland is single- or many-celled. In multicellular glands the secreting cells may lie beneath the surrounding surface forming simple tubes (**tubular glands**) or flask-shaped pockets (**alveolar glands**). Such tubes and pockets may be single structures (**simple glands**) or they may be grouped into aggregates (**compound glands**) (Fig. 4–3).

Fig. 4–3. Various kinds of glands that arise from epithelial tissue. (A) A simple flask-shaped gland (alveolar) such as the mucus-secreting gland found in the skin of the frog. (B) A compound alveolar gland such as that found in the salivary glands. (C) A simple tubular gland such as that in the lining of the intestine. (D) A compound tubular gland such as that found in the stomach lining.

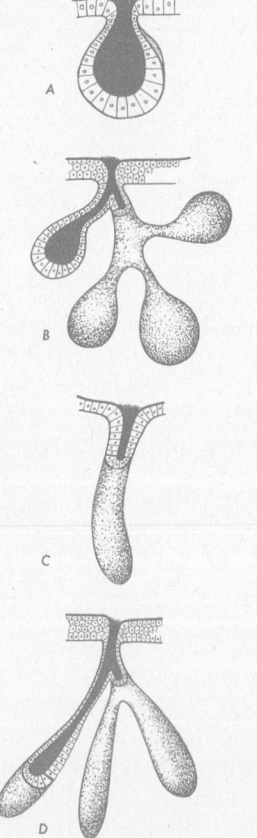

Supporting Tissues. These are the tissues that give the body form and support. They are composed of cells embedded in a **matrix,** secreted by the cells and usually occupying more space than the cells themselves. The matrix may be composed of fluid, gelatinous material, long tough fibers, or hard mineralized material. There are several different kinds of supporting tissue, all dependent on the type of matrix (Fig. 4–4).

The tough **ligaments** that fasten the bones together and the **tendons** that connect the muscles to the bones are composed mostly of tough fibers forming a matrix about the cells which produce them. Also, many of the internal organs of the body are laced together by sheets of similar tissue called **mesenteries.** In this type of tissue the fibers lie at random with no particular arrangement; this results in a thin layer of tissue that is soft and pliable, yet tough. A similar type makes up most of the deeper portions of the skin lying below the superficial epithelium. It is this supporting tissue that gives the skin the qualities essential for an adequate body covering.

Animals, particularly land forms, require a very rigid skeleton to support their massive weights. This is provided by **bone** and **cartilage,** types of supporting tissue that are composed of large quantities of matrix secreted by isolated cells. In the case of cartilage, the matrix is a spongy semisolid mass in which cells are embedded in tiny cavities (**lacunae**). The cells are usually single, although as they divide there may be as many as four in one cavity before they finally separate. Cartilage is excellent material to resist shock; therefore, it is found between bones such as the vertebrae. It also provides ideal support for the tip of the nose and the external ear where retention of shape and pliability are essential. Bone, on the other hand, consists of a mineralized matrix (calcium carbonate and phosphate) which is very rigid, imparting an element of solidarity to the entire structure. In this case also, the matrix is formed from cells embedded in tiny spaces (lacunae) within the matrix itself. These usually take on definite patterns

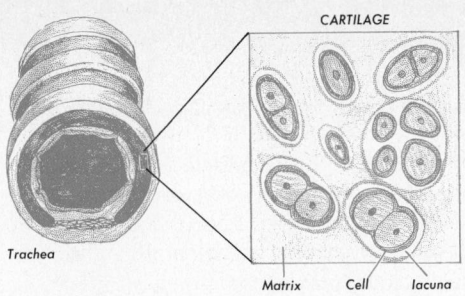

CARTILAGE

Trachea

Matrix Cell lacuna

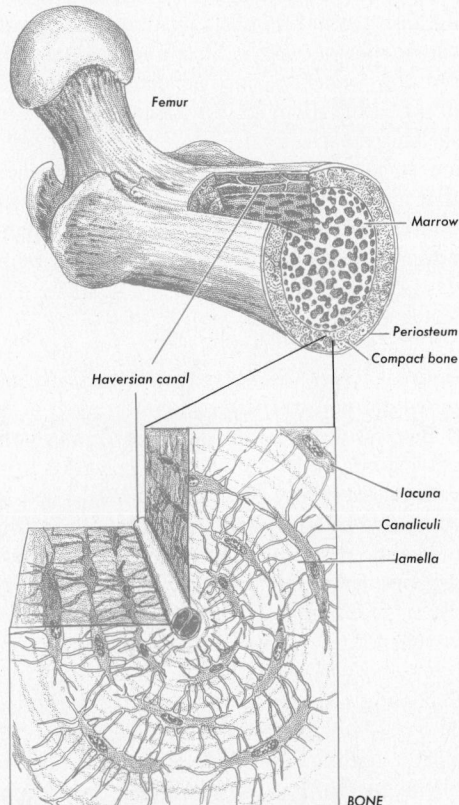

Femur

Marrow

Periosteum

Compact bone

Haversian canal

lacuna

Canaliculi

lamella

BONE

Fig. 4–4. Various kinds of supporting tissues.

around blood vessels and nerves, called **Haversian systems.** All of the cells have access to a food supply from the blood system by means of tiny canals (**canaliculi**), for these cells are alive and must be nourished like any other cells (Fig. 4–4).

Tissue in which fat is stored is often classified as supporting tissue, primarily be-cause there seems to be no other category for it. It performs no mechanical function other than to occupy space. The fat is stored within the cell itself and these cells are located under the skin and in the abdominal region as well as other well-known areas of the human body. During periods of starva-tion it is very scanty, but during good times it may be stored in quantities far beyond any usefulness to its owner, as attested by many overweight people.

Muscular Tissue. This tissue, called muscle tissue, has the ability to shorten, that is, to pull its two ends closer together. This apparently very simple action is responsible for all of the movements of most organisms. Muscle tissue consists of elongated cells or fibers whose internal parts consist of **myo-fibrillae,** tiny contractile fibrils, lying in a fluid protoplasm called **sarcoplasm.** There are three well-defined kinds of muscle tissue, **visceral, skeletal,** and **cardiac,** each of which differs in its appearance under the miscro-scope (Fig. 4–5).

Visceral or **smooth** muscle is found in the walls of the digestive tract, and other places in the body which are not under voluntary nervous control; its activity is primarily auto-matic. The cells are spindle-shaped with centrally located, flattened nuclei, and with myofibrillae running lengthwise in them. It is the shortening of the myofibrillae that pulls the two ends of the muscle cell closer together. Visceral muscle contracts and re-laxes slowly, a behavior which is quite satis-factory for the kind of job it has to do.

The **skeletal** muscles are usually attached to bones and they constitute the large muscles of the body. It is this muscle-bone combination that is responsible for the move-ment of the body as a whole. These muscle cells are peculiar in that they are not marked off by definite cell membranes, and a single skeletal muscle fiber is composed of many cells whose nuclei lie at regular intervals along the periphery of the fiber just under the surrounding membrane (**sarcolemma**). The fiber is called a **syncytium,** referred to earlier (p. 100). Another marked difference

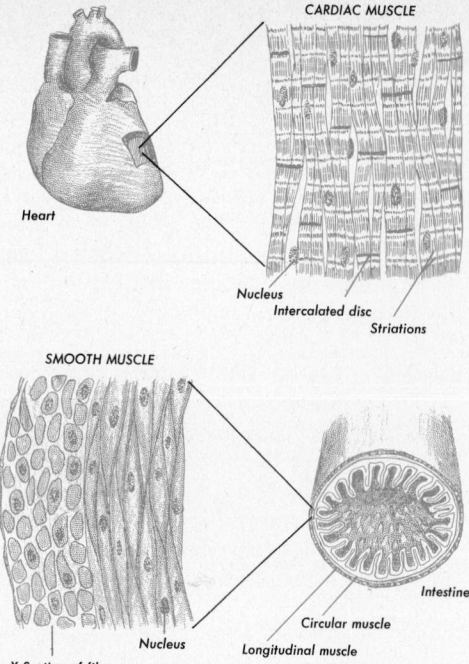

CARDIAC MUSCLE

Heart

Nucleus
Intercalated disc
Striations

SMOOTH MUSCLE

Intestine

Circular muscle
Nucleus   Longitudinal muscle

X-Section of fiber

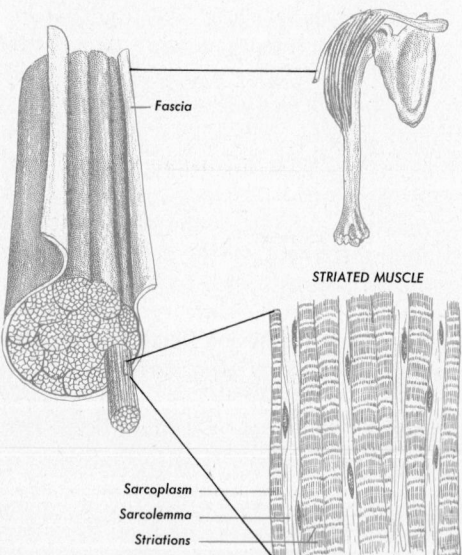

Fascia

STRIATED MUSCLE

Sarcoplasm
Sarcolemma
Striations

Fig. 4–5. Various kinds of muscle tissue.

between this muscle and the preceding is that there are dark and light transverse bands extending throughout the fiber. These **striations** identify the tissue as striated muscle. The details of these fibers are discussed in Chapter 15. The skeletal fibers contract suddenly with considerable force, a feature essential for moving the body. They can contract rapidly again and again with only momentary rest periods.

Cardiac or **heart muscle** is characteristic of vertebrates and is not found in the heart of any of the lower forms. It differs from skeletal muscle in that all of the fibers are connected with one another so that the entire organ functions as a unit, that is, as a **syncytium**. This is an apparent advantage because the nature of its job requires almost continuous operation. Striations are present, but the nuclei are located deep within the fibers rather than at the surface as in the skeletal muscle fibers. Cardiac muscle cells are so closely connected with one another that a single nerve impulse sets them all contracting at once, thus executing a single powerful contraction. This is obviously very desirable in a pump such as the heart.

Nerve Tissue. This type of tissue is composed of **neurons**, special conducting cells that are found throughout the brains and nerve cords of all animals that possess a nervous system (Fig. 16–10). The nerve cell is composed of a **cell body**, which contains the nucleus and surrounding cytoplasm. Extending out from the cell body are threadlike fibers, consisting of numerous **dendrites** which normally convey impulses to the cell body and a single **axon** which usually conducts impulses away from the cell body. The cell body maintains the nutrition of the entire neuron, and if it is destroyed, the fibers die. However, nerve fibers severed from their cell body usually are replaced by new fibers growing out from the cell body.

Cell bodies are concentrated into masses, the most conspicuous of which are in the **brain** and in the **spinal cord**; other masses called **ganglia** have special locations in the body. The nerves that we see on dissection are made up entirely of fibers, each of which is insulated by a fatty sheath, the **myelinated sheath**. These units go to make up the complex nervous system which we study in more detail in Chapter 16.

ORGANIZATION OF TISSUES INTO ORGANS AND ORGAN SYSTEMS

By definition any structure which performs a given function is an **organ**. Obviously, a single contractile cell could be an organ under this general definition. However, in the usual, restricted sense, an organ is a group of tissues assembled for the purpose of performing a specific function. The small intestine, for example, is an organ whose function is the digestion and absorption of food. It is composed of layers of different tissues—an outer layer of epithelial tissue covers the gut throughout its length; immediately inside this are two layers of muscle tissue, then a layer of connective tissue, and finally a thin layer, one cell thick, of lining epithelium. Nerves and blood vessels are interspersed among the other tissues. All of these tissues perform specific jobs in bringing about the greater function of digestion and absorption of food. Even so, the small intestine is not adequate to complete the job of ingestion, digestion, absorption, and egestion as a single organ. This greater function involves a series of organs, the mouth, teeth, esophagus, stomach, small intestine, liver, pancreas, colon, and anus. In other words, the entire job is done by a **system of organs**. Likewise, circulation, breathing, excretion, and indeed all bodily functions are performed by different organ systems. The combined organ systems in integrated action constitute an **organism,** or an **individual.** This can be relegated to the cellular level, as in the case of an amoeba in which all of the activities take place within a single cell. On the multicellular level, tissues, organs, and organ systems have been assembled to make up an organism which functions as a unit, just as the single cell functions as a unit.

The Consequences of Organization

As animals grew in bulk and complexity the problems of transport and coordination had to be solved. Such activities as nutrition, respiration, and excretion, which were performed in a simple manner when the cell was in constant contact with its fluid world became difficult or impossible when it was separated from this environment by even a few covering layers of cells. Such inner cells would have to depend on diffusion to carry oxygen and food to them and to remove wastes from them. At best this is a slow process and certainly not rapid enough to allow an animal to grow very big or become very active. Therefore, specific organ systems had to be evolved if animals were to grow in bulk and activity.

In the following discussion we compare some of the activities of organisms at the cellular and multicellular levels, pointing out the problems involved in becoming complex and indicating how the metazoan animal has solved them. We can use paramecium for the single-cell level, hydra for a simple metazoan, and a mammal for the multicellular level.

Locomotion and Coordination. With increase in size of cellular aggregates, the problem of getting around in the environment required a new approach. Flagella were quite inadequate to move a really large animal, although they were quite satisfactory in single-celled forms, even quite large ones, as well as in small metazoans. Moreover, some means had to be established whereby the individual cells could be connected to each other by some communicating system, so that they would do whatever was necessary in a coordinated fashion. Chaos would have resulted had this not been accomplished.

The Metazoa solved the problem of movement through the agency of contractile cells which are called **muscles.** These are very simple fibers distributed in certain strategically located cells of the primitive metazoan; in more advanced forms these cells became highly specialized. Their job was to move the organism in all of its parts in an efficient manner, if it was to survive. Indeed, its very life depended on the efficiency of these contracting cells, both in securing food and defending itself against injury. Hence, with increasing complexity, a more and more elaborate system of muscles evolved reaching the pinnacle of perfection in such animals as insects and mammals.

Along with the evolution of muscles, other cells specialized into conducting units, called **nerves**, which developed to a high degree the capacity of conveying signals from one cell to another, frequently over long distances. This made possible a closely knit communicating system which was essential if animals of any great size were to succeed. How elaborate this system has become in metazoans is demonstrated by a careful analysis of the communicating system of large mammals such as the whale or man.

Requirements of Metabolism. At the one-celled level the problems of securing food, digesting it, absorbing it, metabolizing it, and removing the waste products, are a relatively simple procedure. The *paramecium* (Fig. 4–6) draws its bacterial food supply into its gullet, forming food vacuoles which might be thought of as miniature intracellular "stomachs." Digestion takes place in these tiny sacs and the end products (amino acids, glucose, fatty acids, and glycerol) are absorbed into the surrounding cytoplasm. The food is oxidized and the released energy is utilized by the protozoan in the many ways that are essential for its life.

Fig. 4–6. The problem of consuming food, digesting it, burning it, and excreting the waste products is essentially the same at all levels of animal organization. Here it is compared in the single-celled animal, *paramecium*, the simple metazoan, *hydra*, and in a highly complex form such as a sheep. The problem is always reduced to the level of the cell and must be solved at that level in all animals. The letters EP refer to end products and E refers to enzymes.

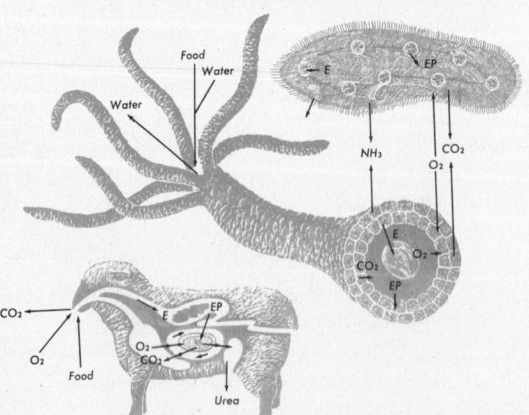

The simple metazoan, such as *Hydra*, captures and digests its food by the cooperative effort of many cells, some of which are contractile and conducting. The food is captured by the tentacles and conveyed to the body cavity where it is retained until the lining cells secrete enzymes to digest it. The end products are then absorbed directly by the lining cells where some are used for metabolism and the remaining portion sent along by diffusion to the external layer of cells.

In a complex metazoan, such as a sheep (Fig. 4–6), the simple process of diffusion will not suffice to transport the food molecules to the millions of cells that lie great distances from the digestive tract. A **circulatory system** is required in any metazoan of great size. This is provided by a system of tubes in which a fluid (blood) is in constant movement in a one-way path. Some portion of the system has thin-walled tubules which come very close to every cell in the organism and which are so thin that they are no more than one or two cells thick. These tubules are called **capillaries** (p. 405). With such a system it is possible to nourish all the cells in the metazoan.

In *Hydra*, all that is needed is a place where the food can be held for sufficiently long a time that digestive enzymes pouring into it have time to break the complex food molecules into their end products. Then proper facilities are needed for absorption, that is, large surface areas are required.

These conditions are met in the digestive tract of all higher animals very satisfactorily. They possess a portal of entry or mouth, which may or may not be armed with teeth for macerating food, and a long tube into which digestive glands empty their food-splitting enzymes. Undigested food leaves through the end of the tube, the anus. Once the food has reached the soluble stage, it is absorbed into the blood and transported to each cell, which picks and chooses the particular energy-giving and constructive materials it requires.

Once the food molecules are delivered to the cells, they must be oxidized and this requires oxygen. In *Paramecium* this is cared for very efficiently and simply by diffusion through the cell membrane (Fig. 4–6). Once in the cell, the oxygen diffuses to the food molecules with which it combines, forming carbon dioxide and water as end products. Carbon dioxide likewise diffuses throughout the protoplasm, ultimately passing through the cell membrane into the external fluid world. The only essential need is sufficient oxygen in the immediate environment.

This gaseous exchange is essentially the same in the simple metazoan where each cell is in contact with its external world. Since water circulates within the cavity, the lining cells are adequately supplied with oxygen. Diffusion, therefore, is quite adequate to take care of the gaseous exchange of this simple animal.

In the complex metazoan the same problems arise as were encountered in supplying nourishment to each cell. Since most of the cells lie deep within the organism, there is no possible chance for gaseous exchange by diffusion with the external world, especially since the organism is usually covered with skin which is impervious to such gases. The complex metazoan solves the problem by developing a **breathing system.** This, in combination with the transportation system, would make it possible to deliver and remove the gases from each cell. In aquatic forms such as fish, the breathing organs are **gills;** in land forms such as the sheep, they are **lungs.** Both must meet certain requirements if they are to function as breathing organs. They must be constructed so that oxygen in the surrounding medium (water or air) can pass readily into the transporting medium (blood) which will convey the gas to each body cell. This means that the blood must flow in capillaries very close to the external environment in order that the exchange can be readily accomplished. Microscopic examination of a gill filament or lung sac reveals that this condition is met. Blood flowing through the capillaries passes within two cells of the outside world (Fig. 4–6), which makes the gaseous exchange possible by simple diffusion.

The transporting system delivers oxygen to the cells of the body and collects the carbon dioxide from them. The latter is conducted to the breathing organs where it is lost to the outside environment. Thus, by the combined operation of the breathing and transporting systems, each cell of a complex body consisting of millions of cells can be as readily supplied with oxygen and relieved of its carbon dioxide as can a single cell. This accomplishment removed one of the size limitations on metazoan animals.

Other end products of metabolism must be removed from the cell constantly, if it is to survive. These are those resulting from the metabolism of nitrogen compounds, such as amino acids. They are nitrogenous compounds such as ammonia and urea which are toxic if retained within the cell or its immediate vicinity. They are normally eliminated through the cell membrane, and simple diffusion is quite adequate for the single cell. At the multicellular level, however, diffusion is quite inadequate to do the job and without some means of removing these toxic substances the cells cannot survive. Hence, an effective **excretory system** was required in any animal where the cells were located any distance from the surface.

The simple metazoan, like the single-celled animal, gets rid of its nitrogenous wastes by simple diffusion (Fig. 4–6), where there are but two layers of cells, each in contact with the external world, the problem of excretion is readily solved.

In most complex metazoans, an elaborate system of selectively absorbing tubules, in combination with the transportation system, evolved to take care of these waste products (Chapter 20). By a process involving filtration, selective reabsorption, and secretion, wastes are removed from the blood and conveyed out of the body. Most metazoans above the two-layered animals possess such a system of excretory tubules. In the most primitive animals there are many units scattered among the cells so that fluid bathing these cells can find its way to one of these tubules and be relieved of its load of nitrogenous wastes. In more advanced forms the many excretory units become compactly arranged in a single organ, the **kidney.**

Reproduction. Certain other complications also arose when cells aggregated into huge masses. The single-celled forms depend primarily on increasing their numbers by the process of simple *fission*, that is, the cell merely splits into two offspring, each retaining the exact characteristics of the parent cell. Whenever *Paramecium*, for example, reaches a certain size, it divides and continues growing. The rate at which it can increase its numbers under favorable conditions is limited only by its food supply and its capacity to build protoplasm. The progeny are thus all alike and, barring accidental death, live forever.

Most multicellular animals do not employ simple fission as a means of increasing their numbers, but rely on special cells, **eggs** and **sperm**, which, with a few exceptions, must unite to produce a new individual. This process has the advantage of uniting two lines of protoplasm which results in variation in the offspring. The progeny always possess a combination of the parental characteristics and are therefore different from either. Variation seems to have some advantage in survival of the species and probably has been important in the gradual evolution of complex forms.

These specialized reproductive cells of higher animals are generated in special organs called **gonads.** Eggs are produced in the **ovary** and sperm in the **testis.** To insure the union of the sex cells, special tubules are usually necessary to conduct these cells out of the body, and ultimate union is still more effectively assured in some higher animals by the development of copulatory organs. To insure greater survival, the offspring of many higher animals are either retained within large egg shells or the body of the mother for varying periods of their early development. All of this machinery apparently has survival value because it is found in complex metazoans.

The Penalty of Organization. What are the sacrifices, if any, that animals have made for becoming complex? The organization of cells into masses and the subsequent specialization of different kinds have resulted in organisms that are able to penetrate many different environments because of their greater motility and intricately adjusted bodies. This means an increased number of species as well as individuals; and we think of this as biological success. Along with all of the benefits derived from specialization, however, has come at least one rather severe penalty, and that is natural death.

Single cells reproducing by simple fission, barring accidental death, live on forever. You will recall that the *Volvox* mother colony and the soma cells of the slime mold die when reproductive cells are produced. Obviously, death put in its appearance when cells became specialized to function in capacities other than reproduction. Such specialized cells which had lost the power of reproduction were sacrificed and those which retained reproductive capacity were able to continue.

It is difficult to understand why cells that are separate, free from other cells, may continue living forever, whereas others that are bound together into a mass must eventually die even though, apparently, all of their basic needs are satisfied. Perhaps during the process of evolution the organization was not quite perfect; that is, the individual cells were not completely cared for, or perhaps the whole organization slowed down after a certain period of time and could not keep pace with the demands of all the cells. This point has long intrigued biologists and has resulted in some very fruitful research.

If it were possible to grow tissues away from the animal of which they are a part, it would be possible to determine whether or not cells once released from their intended environment could survive like single isolated cells. In 1907 Ross G. Harrison succeeded in growing embryonic tissues in flasks by feeding them special nutrients. Alexis Carrel, employing similar methods, kept embryonic chick heart tissues alive for over 30 years. At the end of this period of time, about three times the normal life span of a chicken heart remaining with its owner, the cells were active, although some changes had occurred. These cells apparently have the capacity to live forever, just as single-celled animals do.

Many tissues have since been kept in culture for years and, whereas slight modifications do occur in their morphology and perhaps in their physiology, they do remain alive. It seems that metazoan cells have retained their immortality, but the subtle changes that occur when they are removed from the animal may be the same ones that bring about senescence when they are in the intact animal. Recent studies have shown that cells both in the intact animal and in the test tube gradually change with time. Some protozoa demonstrate a definite life cycle, showing youth, maturity, and old age, each stage demonstrating physiological and morphological change. So we cannot say that all single cells are immortal, although some seem to be. As in most problems in biology, there is no simple answer to aging. The process may vary in different cells in both unicellular and multicellular animals, but until we know a great deal more than at present, the problem remains unsolved.

SUGGESTED SUPPLEMENTARY READINGS

Books

BLOOM, W., and FAWCETT, D. W., *A Textbook of Histology*, 9th ed. Philadelphia: Saunders, 1968.

ROMER, A. S., *The Vertebrate Body*, short version. Philadelphia: Saunders, 1971.

*THOMPSON, D. W., *On Growth and Form*. New York: Macmillan, 1942.

* Available in paperback.

5

ORDERLINESS AMONG ANIMALS

To the casual observer the plants and animals on the earth may seem to exist without order, being so varied in shape and size that it would be useless to try to arrange them in any kind of order. Early scholars were confronted with this problem and many attempted to place organisms into groups, that is, classify them. It seems to be man's innate tendency to catalog everything about him when it exists in sufficient numbers. The first recorded effort to classify plants and animals was made by the Greek scholars Aristotle and Theophrastus. According to their scheme, they divided the plants into herbs, shrubs, and trees; the animals into land-dwellers, water-dwellers, and air-dwellers. This system served the single purpose of arranging organisms so that one might find the name of any one of them, so long as the number of known organisms was small, but with increasing knowledge it became cumbersome and almost useless. It was not until the sixteenth and seventeenth centuries that significant steps were taken to find a better system of classification. John Ray in England in the seventeenth century and Carolus Linnaeus in Sweden in the eighteenth century established **taxonomy** (the science of classification) on a firm footing. We may remember Ray for defining the category **species**, which previously had meant a rather unclear group of organisms. He described a species as a closely related group of organisms whose parents were similar and who passed their characteristics on to their offspring. Once the species concept was established, an entirely new approach was taken in classifying plants and animals.

LINNAEUS (1707–1778) is generally recognized as the father of taxonomy because he gave us the system of classification that is in current use today (Fig. 5–1). He was a Swedish physician who developed an interest in natural history that continued from childhood throughout his lifetime. In his early youth he recognized the need for a better system of classification and soon set down the basic principles on which he later built a satisfactory method of cataloging plants and animals. Linnaeus had the insight to select important fundamental characteristics as bases for his classification. This was fortunate because not only did he give us a system which was workable and sound for an infinite number of additions, but it also closely corresponded to the theory of evolution which became widely accepted after 1859. Linnaeus' system is a branching type of system, just as evolution is, so the two are compatible, although not actually intended to be so by Linnaeus.

Linnaeus used such fundamental structures as the skeleton, scales, hair, feathers, and so forth in classifying the larger animals; for the soft-bodied invertebrate types he used characters like the foot of the mollusk, the body segments and exoskeleton of the arthropods. Likewise with plants he placed those bearing flowers in a group distinct from those without this structure.

All animals and plants in this system of classification were given two names, a **generic** (a noun) and a **specific** (an adjective) name. This is now known as the **binomial system of nomenclature.** The generic name is comparable to our own family name, whereas the specific name is like our given name. Linnaeus decided that these names should be written in a universally familiar language that would cause the least amount of international jealousy and therefore selected Latin.

Plants and animals that are most alike were placed in one species, such as *sapiens,* the specific name for all men alive today—there have been other species of men but they are all extinct. Likewise, man also belongs to the genus *Homo;* there have been other Homos but they, too, have been extinct many thousands of years. Under the Linnaean classification, therefore, man is known as *Homo sapiens.*

Linnaeus grouped all the various **genera** (plural of genus) into larger groups which he called **orders;** while these organisms re-

sembled one another in certain respects, they differed much more than did the various species in the separate genera. He further grouped the orders into six **classes,** his largest category. Since his time, of course, a great many biologists have unearthed information about more and more organisms so that it became necessary to enlarge his classification extensively. This was done by adding two more groups, namely, **families** (between genera and orders) and **phyla** (the largest group of all). Furthermore, each group has been subdivided again and again, so that we now have the following general categories: phylum, subphylum, class, subclass, order, suborder, family, subfamily, genus, subgenus, species, and subspecies. However, not all of these divisions are necessary in the classification of every organism.

The differences are less and less as the

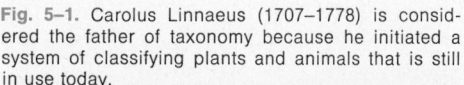

Fig. 5–1. Carolus Linnaeus (1707–1778) is considered the father of taxonomy because he initiated a system of classifying plants and animals that is still in use today.

selection moves from the phylum to the species. For example, the differences between the horse and the earthworm are various and striking; each belongs to a separate phylum. The differences between the horse and the alligator, while many, are not nearly as numerous as those between the horse and the earthworm; they belong to the same phylum but not the same class. They show more differences than are observed between the horse and the dog, both mammals belonging to the same class. The differences between the dog and the horse are sufficient, however, to place them in different orders, families, genera, and species.

In many cases, however, the problem of separating species is much more difficult. In fact, a point is reached where biologists are frequently in doubt as to whether or not a given organism is actually a separate species at all. This is what one would expect if the theory of evolution holds, namely, that plants and animals have originated and are indeed still originating from a common ancestral stock. Undoubtedly, new species are forming at the present time and will continue to do so as long as there is life on earth.

The following example illustrates the use of the Linnaean system in classifying man according to his distinguishing characteristics and in outlining the basis of his relationship to other animals:

Phylum Chordata—notochord, gill slits, nerve cord
Subphylum Vertebrata—backbone
Class Mammalia—mammary glands, hair
Subclass Theria—marsupium and placenta
Infraclass Eutheria—true placenta
Order Primates—superior nervous system.
Suborder Anthropoidea—flattened or cupped nails
Family Hominidae—no tail or cheek pouches
Genus Homo—manlike
Species sapiens—present-day man

Thus *Homo sapiens* includes all living men today. To distinguish between different colors and other characteristics of men, the subspecies is given. The scientific name then becomes trinominal, such as *Homo sapiens africanus*, which identifies a particular race of living men, the African black.

MODERN TAXONOMY

As you would guess, a great deal has been learned about animals since Linnaeus' time; consequently much new information is available for the taxonomist to work with. Moreover, our idea of species has changed. Whereas Linnaeus conceived of species as immutable, today we are certain that species are constantly evolving, forever changing. This idea, though very old, was crystallized by Charles Darwin when he supplied abundant evidence that species are constantly evolving. We discuss his theory of evolution in some detail later (Chapter 26). All we need say now is that all modern taxonomists have completely abandoned the idea that a species is static, and have developed an entirely different interpretation of the Linnaean Binomial System.

Today, the systematist (taxonomist) conceives of the species as a population strongly influenced by ecological and genetic factors. Instead of one animal or plant being the taxonomic unit, he now considers the entire population when he speaks of a species. This requires a great deal more knowledge on the part of the systematist because he must be familiar with statistical methods and must understand the role genetics play in an ever-changing population. He must also possess a thorough knowledge of morphology in order to recognize subtle differences in organisms, morphology still being very important in describing species.

Perhaps the best definition of a species is based on genetics, namely, a species is a population of interbreeding organisms. This is a straightforward definition that has the possibility of proof. The difficulty is that it is not practical to carry out crossbreeding experiments with every species that is or has been described. Unfortunately only one in ten thousand species has been so tested, hence this definition will probably never be completely satisfactory. Structural characteristics as well as behavior, physiology, and distribution are now relied upon for describing species.

Using interbreeding populations as a criterion for species has certain pitfalls. For example, among the ciliated protozoa there are some morphological species that have among them a number of interbreeding populations. This is true of *Paramecium aurelia* and *Tetrahymena pyriformis*. In the latter there are at least ten interbreeding populations, called **varieties,** or **syngens,** which are genetically isolated from one another, even though two or more may inhabit the same environment. Morphologists agree that they are very similar in structure. It has been suggested that such populations be called **syngens,** which identifies the genetic species, leaving the term species for the morphologist's species. Since this problem is still in a state of controversy, it is best that we use the term species where it includes both genetic and morphological characteristics.

The modern systematist studies the structure of organisms from the point of view of relationships, that is, he searches for **homologies** between structures when he is describing a taxonomic category (**taxon**). Homologous structures are those parts that are similar in form and number in different organisms and can be arranged in a continuous series showing only slight differences from one to the next. This is usually illustrated by comparing the anterior appendage of vertebrates, for example, the frog, man and the horse (p. 561). A species, then, is a group of organisms that share the greatest number of homologous structures. The number of homologous structures becomes fewer and fewer in each succeeding higher taxon. The dividing line between taxa, of course, becomes a matter of the taxonomist's judgment and must be arbitrary; this cannot be avoided. It has been said that a system of classification is no better than the taxonomist who designed it.

Because of the confusion that has existed in categorizing plants and animals precisely, a new school of taxonomists, called **biosystematists,** has come into prominence. They have included many additional characteristics, such as blood proteins, chromosomes, behavior, and ecology, in their definition of relationships. Sometimes 100 or more measurements are made, which has compelled the use of the computer to more precisely determine the numerical degree of affinities among individuals. However, not all taxonomists accept this method and only time will tell whether it will receive universal acceptance.

SOME BASIC CHARACTERISTICS OF ANIMALS

The whole scheme of animals has been studied by biologists for centuries in an effort to set down certain characteristics that can be used to separate large groups into still smaller categories. Is it possible to determine certain characters which can be found in many phyla and which will establish relationships among them? As more animals have been studied, it has been possible to establish a few fundamental categories which are agreed to be sufficiently basic to include all animals living today. These are: (1) **grade of organization,** (2) **types of symmetry,** and (3) **kinds of body cavities,** if present.

Grade of Organization. This merely indicates the way cells are arranged in an animal,

Fig. 5–2. Two kinds of symmetry.

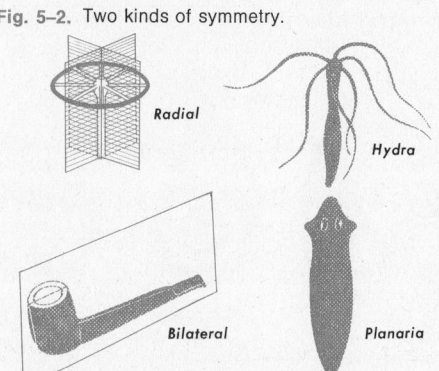

Radial

Hydra

Bilateral

Planaria

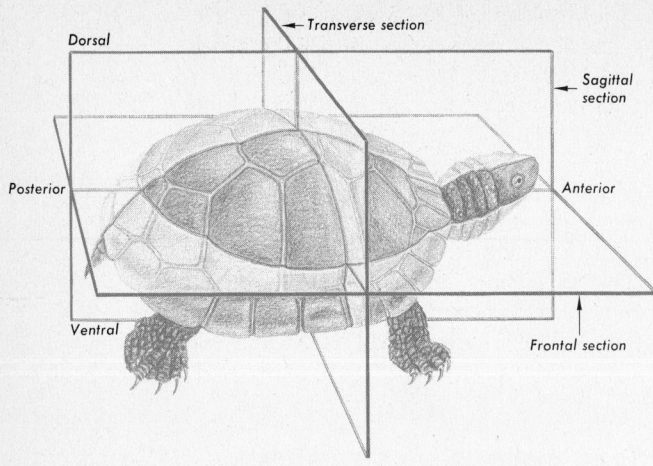

Fig. 5–3. Orientation of an animal possessing bilateral symmetry.

whether the animal is single-celled or multi-cellular. The unicellular Protozoa are distinct entities with the exception of the colonial protozoa, which give indications of some sort of organization among individuals in the colony. The Porifera, distinctly metazoan, demonstrate the **cellular** grade of organization, whereas all higher metazoa represent the **tissue** grade of organization. All animals can thus be separated into these large categories.

Types of Symmetry. As cells became organized into groups, they formed definite

relations to one another, imposing upon the resulting animals a particular shape that can be described in terms of symmetry (Fig. 5–2). Symmetry refers to the arrangement of parts in relation to points, planes, and straight lines. The metazoa can usually be conveniently divided into two types of symmetry. Probably the first to appear in evolution was **radial symmetry,** which is characteristic of the Coelenterates and the Ctenophores. These animals are distinguished by a principal axis around which the parts are arranged in a radiating fashion. All higher forms possess **bilateral symmetry.** In these one can divide the animal by a plane which results in two approximate mirror halves. Such animals are usually oriented into an **anterior** end, or head, a **posterior** end, or tail, a **dorsal** side, or back, a **ventral** side or belly, and a **right** and **left** side (Fig. 5–3). These terms are handy in locating organs and parts of animals.

In general, there is a correlation between types of symmetry and modes of life of the organisms. Primitive and attached animals possess radial symmetry, whereas higher and active animals are bilaterally symmetrical. Sessile forms require a body form that permits them to explore their food-laden world in all directions, whereas an animal that pursues and apprehends its prey can succeed in this task much better with a body that permits one-way movement, a head equipped with numerous sense organs to locate the prey, and some means of propelling itself, such as appendages. These latter features we see in nearly all bilaterally symmetrical animals.

Kinds of Body Cavities. Whereas several body cavities developed in the most primitive metazoa, such as gastrovascular cavity of the coelenterates, we wish to consider here only those animals with bilateral symmetry which either possess or do not possess a cavity surrounding the digestive system (Fig. 5–4). Those that have no such cavity are called **acoelomate** animals (nemertines and flatworms), and those with such a cavity are termed **eucoelomate**

Fig. 5–4. Various types of animal cavities.

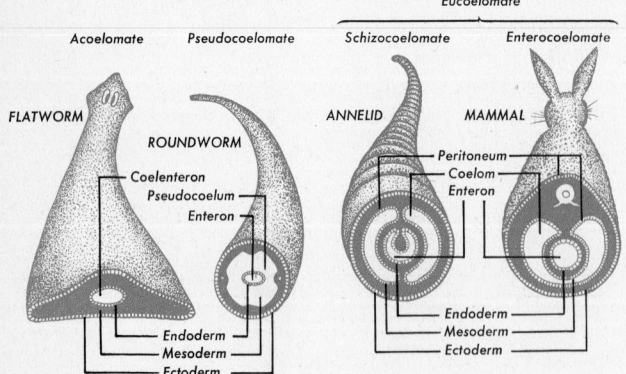

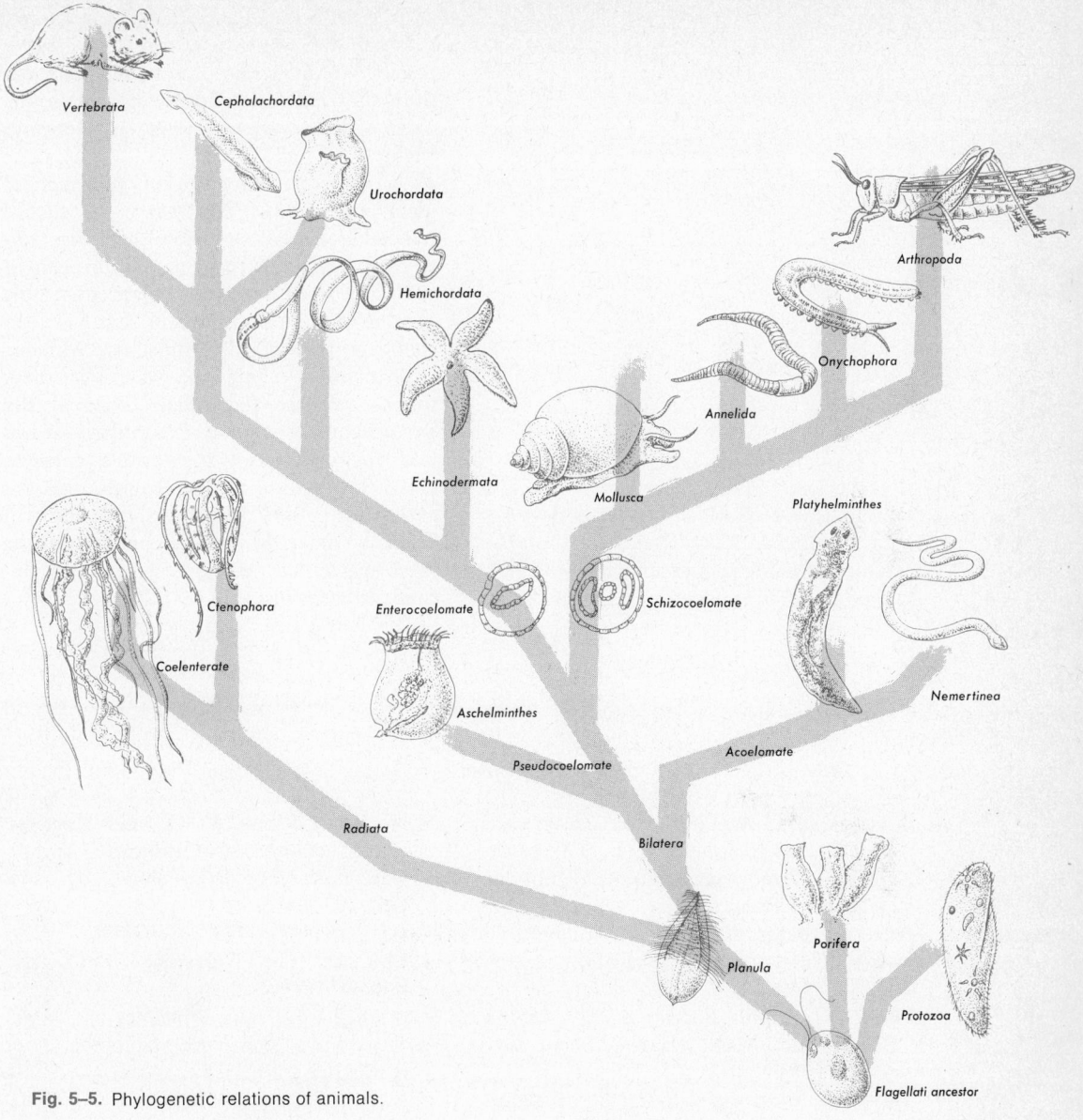

Fig. 5–5. Phylogenetic relations of animals.

animals. The cavity of this latter group is usually spoken of as a true **coelom,** which is a cavity lined with a thin layer of mesodermal cells called the **peritoneum.** A few animals such as the nematodes and the rotifers have a cavity, but it is not lined with mesoderm, hence is not a true coelom. These are called **pseudocoelomate** animals.

The eucoelomate animals may be divided into two large groups: (1) those in which the coelom is formed from solid mesoderm (**schizocoelom**) and (2) those in which it forms from mesodermal pouches from the gut (**enterocoelom**) (Fig. 5–5). The manner in which the coelom forms is evident only during early embryology. In the

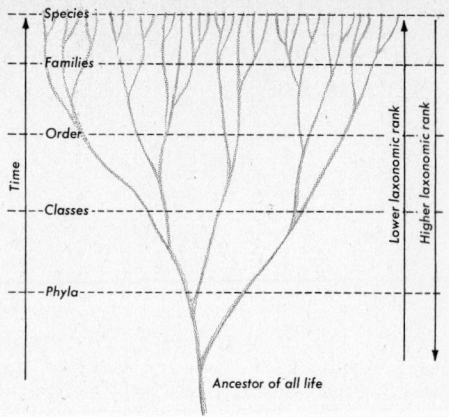

Fig. 5–6. The bush of life.

schizocoelomates, the mesoderm arises as a solid mass of cells which very early develops a cavity, the coelom. The coelom of the enterocoelomates is present when the out-pocketing of the endoderm, or primitive gut,

first appears to form the mesoderm. The end result in both cases is the same.

With these three general categories it is possible to show the evolutionary relationships of the entire animal kingdom in the form of a branching system, as shown in Figures 5–5 and 5–6.

As we gaze upon the great profusion of plant and animal life today, we should remember that each living thing has survived following a struggle by its forebears to provide a plant or animal organization able to cope with an environment that has been changing through the millennia. Millions have perished along the way, but these species you see today have survived the struggle and, even now, are fighting to maintain their place on the earth's crowded crust. The nature of the struggle and the path which each species has followed will probably never be known, but as one of these species, we can appreciate the magnitude of the problem.

SUGGESTED SUPPLEMENTARY READINGS

Books

BORRADAILE, L. A., POTTS, F. A., EASTMAN, L.E.S., and SAUNDERS, J. T., *The Invertebrata*. Cambridge, England: Cambridge University Press, 1958.

HYMAN, LIBBIE H., *The Invertebrates*. 6 vols. New York: McGraw-Hill, 1940–1967.

MAYR, E., *Animal Species and Evolution*. Cambridge, Mass.: Harvard University Press, 1963.

MAYR, E., *Principles of Systematic Zoology*. New York: McGraw-Hill, 1969.

MAYR, E., LINSLEY, E. G., and USINGER, R. C., *Methods and Principles of Systematic Zoology*. New York: McGraw-Hill, 1953.

ROMER, A. S., *The Vertebrate Story*. Chicago: The University of Chicago Press, 1959.

SIMPSON, G. G., *The Principles of Animal Taxonomy*. New York: Columbia University Press, 1961.

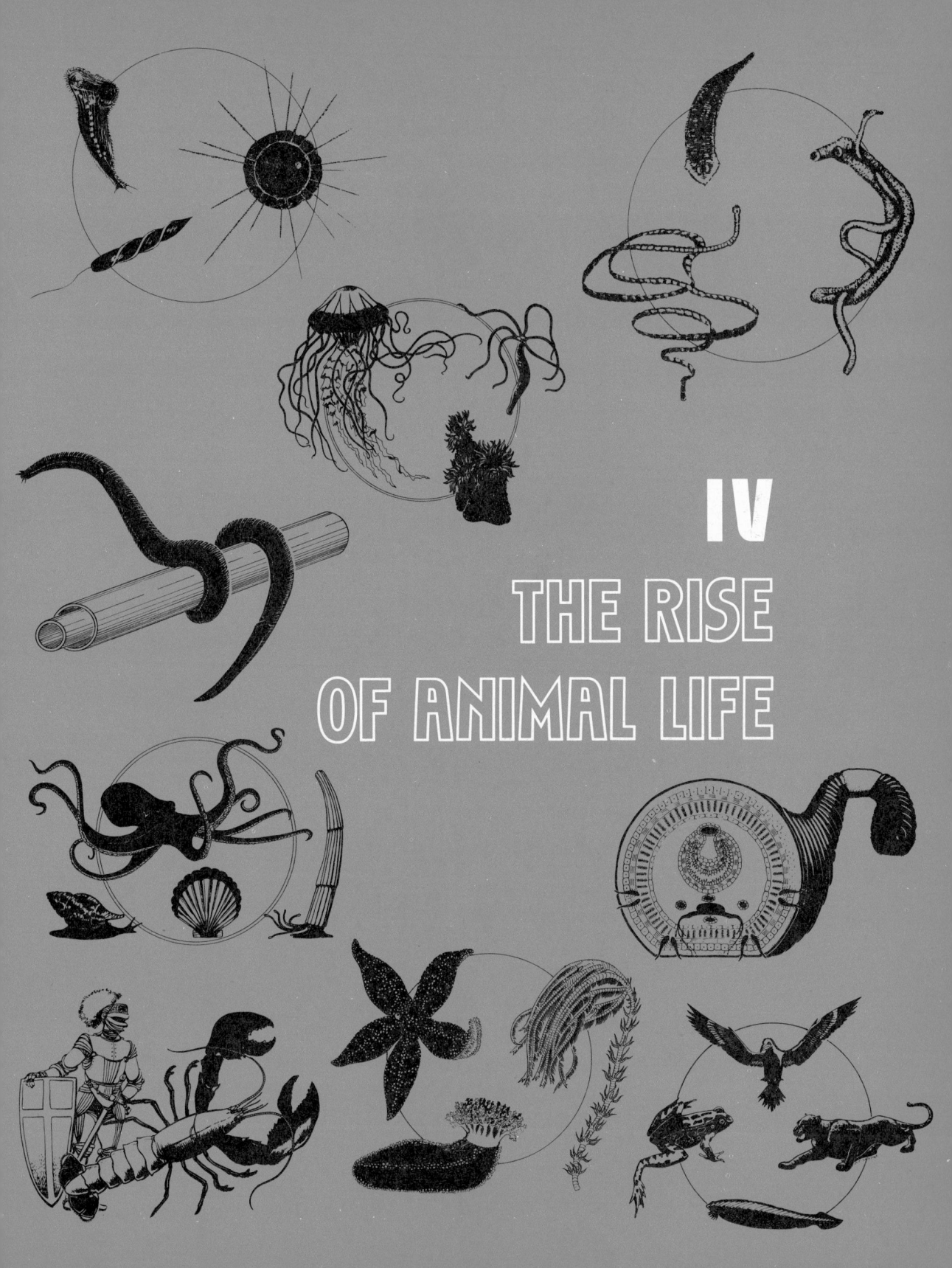

IV
THE RISE
OF ANIMAL LIFE

6

THE FIRST ANIMALS- PROTOZOA

There is little doubt that animals started from simple one-celled forms and gradually increased in complexity of structure and function. Of the 1,500,000 different species described, we consider only one or two representatives of the major groups with the hope that some understanding will be gained of the process of evolution. It is important to keep in mind the story that is being told, rather than the details concerning individual animals. Specific facts are required about representative animals, but the panoramic view is more important than isolated details, no matter how fascinating they may be.

MOVEMENT IN THE SINGLE CELL

At one stage in the evolution of living things, the predominant type was in the form of single cells. These explored all the available niches where they might survive and increase their numbers. With time, competition for favorable niches undoubtedly occurred resulting in diversity of form and function within limits of one cell. One of the first problems that must have faced these first cells was that of getting from place to place to secure food or to secure protection. This apparently was accomplished in two ways (Fig. 6–1); which came first is a matter of conjecture.

Flagella and Cilia

One way to move was by means of a vibrating, flexible, thread-like structure, which we call a **flagellum** (plural, flagella) or a **cilium** (plural, cilia). The simplest type of flagellar movement is illustrated in Figure 6–2. As a result of active bending and straightening of the flagellum, a wave-like motion passes down the flagellum which exerts a force on the surrounding fluid

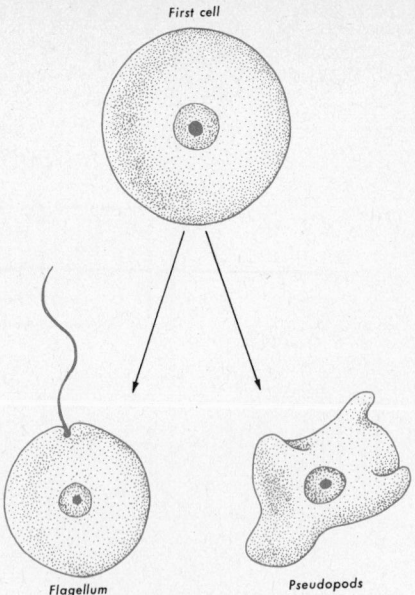

First cell

Flagellum   Pseudopods

Fig. 6–1. Two methods of movement evolved among the first cells, one by means of a vibrating flagellum and the other by pseudopods.

nine pairs located around the inside of the membrane and two at the center often spoken of as the "9 + 2" arrangement. The filaments arise from the kinetosome which has 9 triplet filaments and no central filaments. The kinetosome, which as we mentioned in Chapter 3 appears identical to the centriole, interacts with the cell cortex to produce a cilium and to act as an anchor for it. The exquisite fine structure of both the kinetosome and cilium is intriguing because we know it must relate to its function. Our particular interest at this point is how the cilium moves; apparently its structure cannot be materially altered and still function, because cilia and flagella show essentially identical fine structure throughout the animal and plant kingdoms. This is an ancient organelle of motility, present essentially unchanged in *Chlamydomonas*, *Paramecium*, and the human trachea. Figure

Fig. 6–2. Flagellar movement requires that the flexible "thread" (flagellum) undulate, thus creating a force against the surrounding fluid which results in pushing or pulling the cell. Cilia move the cell by paddlelike action.

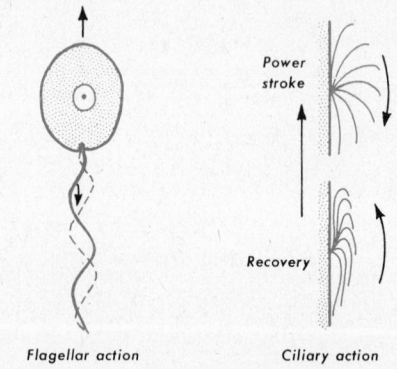

Power stroke

Recovery

Flagellar action   Ciliary action

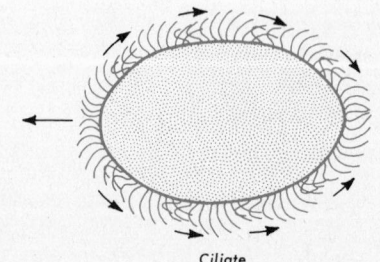

Ciliate

causing the cell to move forward. This type of flagellar movement was gradually improved upon as indicated by the fact that only the simplest cells observed today employ it.

Another type of movement, employing the same thread-like structure, is ciliary movement. In this case the numerous cilia are short and beat much like the back and forth movement of oars in propelling a boat. The power stroke exerts a backward force on the surrounding fluid which causes the cell to move forward as shown in Figure 6–2. This type of mechanism is employed by many single cells and is also found among the metazoa.

How can a structure so small as a flagellum or cilium bend on itself, performing the complex movements that we observe under the light microscope? Considerable light was shown on this problem when electron micrographs became available. They show that these structures consist of tiny filaments (microtubules) and a covering membrane (Figs. 6–3, 6–4). There are

6–4 shows the fibrillar part of the cilium which is called the **axoneme.** It consists of doublet **microtubules** on the outside of the axoneme, and two single microtubules called the central pair of microtubules. These microtubules serve as skeletal elements, capable of bending but not of shortening or lengthening. There are two sets of connecting elements: (1) the **arms of the doublets,** each of which connects to or adjoins with the next clockwise doublet— these have an ATPase and slide along the adjoining doublet creating both bending and rotational forces, and (2) the **radials,** each consisting of a link and link head, which apparently slide up and down along the sheath with its transitional junctions. Ciliary movement is believed to be the result of **microtubules sliding past each other,** such sliding being powered by the arms on the ciliary doublets and possibly the links between these doublets and the central sheath (Figs. 6–3, 6–4).

Figure 6–5 shows cross sections of cilia in the oral region of *Tetrahymena pyriformis,* a ciliate closely related to *Paramecium.* Note that the cilia are all oriented in the same direction, reflecting the fact that they have a preferred plane of beat. This preferred plane is defined by the central microtubules such that both of the microtubules lie within the plane and thus the power and recovery strokes (Fig. 6–2) would be at right angles to this plane. The oral cilia in *Tetrahymena* function to direct appropriate particles (usually bacteria) into the mouth. It is also the rule in ciliated epithelial systems like trachea and oviduct that all of the cilia are oriented in the same way; thus they can transport particles in the appropriate direction. The cilia beat in coordinated waves resembling waves of water or grain. Such coordinated activity is called **metachronal beating.**

Amoeboid Movement

Quite a different method of locomotion must also have appeared in the very earliest of eukaryotic cells. This is by means of

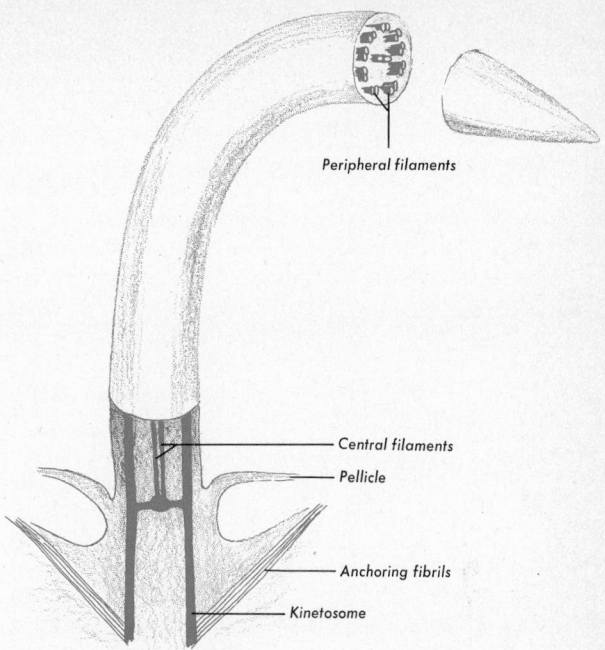

Fig. 6–3. A schematic representation of the cilium. Note that the cilium arises from the kinetosome. The peripheral filaments originate from its walls whereas the central filaments arise from a dense body at the level of the pellicle. The filaments are enclosed in a covering which is a continuation of the outer membrane. The coordinating fibrils may function in bringing about unified action of all the cilia.

protoplasmic flow, which includes such things as **streaming** in plant cells and so-called **amoeboid movement** which involves **pseudopodia.** The observation of a cell such as an amoeba "flowing" along has enthralled observers ever since the microscope was invented—it appears to be the simplest type of movement and many ingenious theories have been advanced to account for it. For example, a drop of oil in water will "follow" a toothpick dipped in soap solution, and a drop of mercury will "follow" an acid source; thus someone proposed a surface-tension theory, in which a localized reduction in surface tension at the tip of a pseudopod would cause the amoeba to move in that direction. The sol-gel phase reversal observations (chapter 3) and the widespread occurrence of actin-like and myosin-like proteins, suggest some sort of universal sliding microfilament model for amoeboid movement. Yet the biochemistry of all this still eludes us. We know much more about muscle contraction, as we shall see later, than we do about what probably was a predominant form of motion when the earliest eukaryote cells were crawling around in the ancient primordial ooze. Let us consider the descriptive basis for amoeboid movement, giving one "explanation" with a brief reference to another, realizing that we must know the complete molecular story before we can really understand amoeboid movement.

It is most interesting to look at an amoeba like *Amoeba proteus* from the side rather than from above. Figure 6–6 illustrates how such an amoeba moves. There is a general cytoplasmic flow in a central channel, continuing essentially unabated into usually one pseudopod, although sometimes there is a temporary struggle between two different pseudopods as to which will finally advance and determine direction. Simultaneously, the temporary posterior end gives up its position, and the protoplasm in this region moves forward, filling the region left by the protoplasm that is actively forming the new pseudopod. Thus the entire cell slowly moves in the general direction of

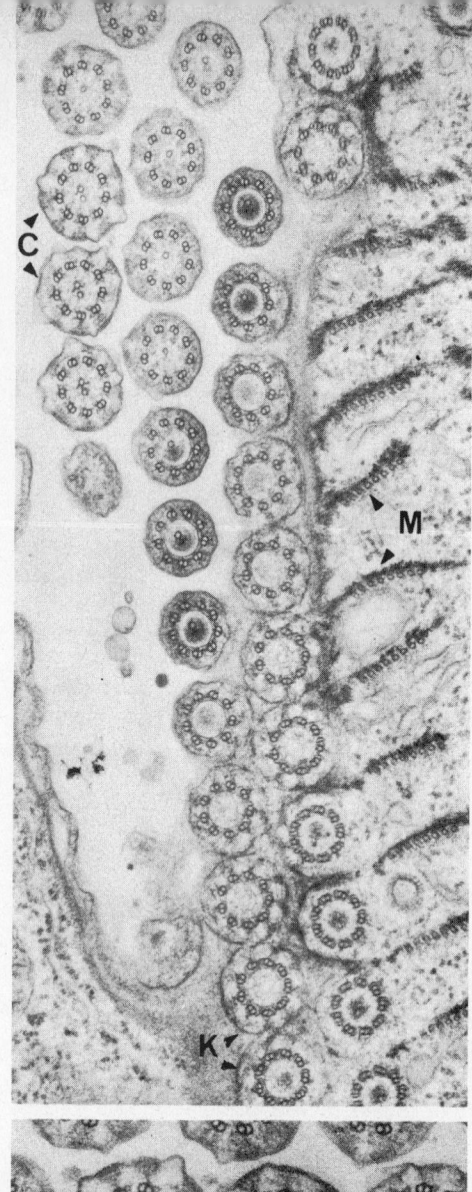

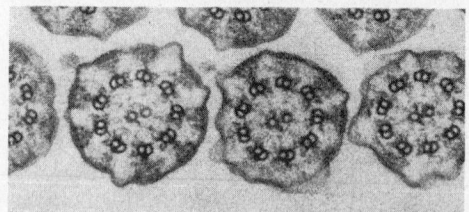

Fig. 6–4. Schematic representation of a typical ciliary or flagellar "9 + 2" axoneme as viewed from the base of the organelle to its tip. Probable mechanochemical interaction to produce movement occurs between the dynein arms and neighboring doublet subfiber b (sliding) and between the radial spoke head and central sheath (bending).

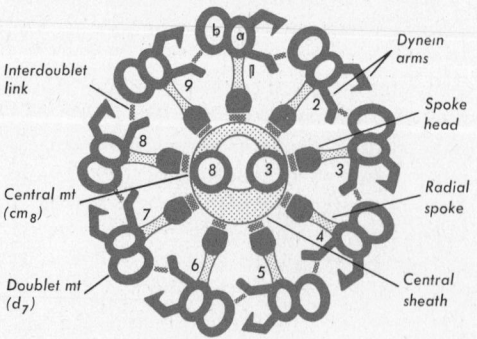

Fig. 6–5. The upper electron micrograph shows the cilia and kinetosomes in cross section taken in the mouth region of the protozoan, *Tetrahymena pyriformis*. Note the nine peripheral doublets and central pair of filaments in the cilia; also the enveloping membrane (C). There is no membrane around the kinetosomes (K) and the peripheral filaments are triplets; also there are no central filaments, only a dense body. Microtubules (M) probably function as supporting structures. The lower insert shows the cilia at higher magnification.

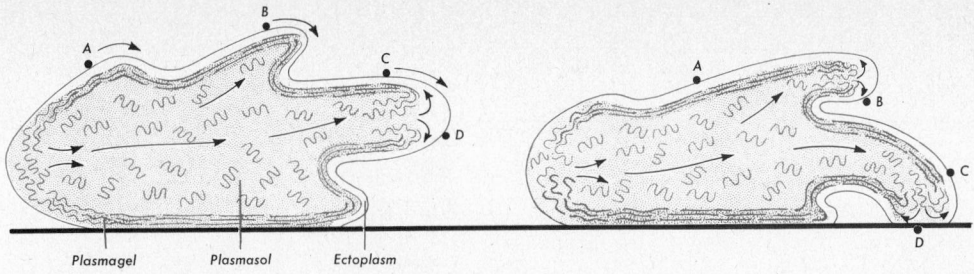

Fig. 6–6. A schematic representation of amoeboid movement as seen in side view. Note that the particles (A, B, C, D) move forward as the amoeba advances. Plasmasol changes to plasmagel at the anterior end and the reverse action occurs at the posterior end.

the advancing pseudopod. In order to understand this entire process we must examine the different regions of the protoplasm more carefully.

The ectoplasm is clear and seems to have little to do with movement. The **endoplasm** on the other hand is made up of two parts, the outer **plasmagel** and the inner **plasmasol**. These two may quickly change from one phase to the other, a change which is responsible for the movement. Precisely, the plasmasol flows in the direction in which the pseudopod is to form, and changes to plasmagel as it spreads out at the tip. This describes what happens, but why does the plasmasol flow in the direction the pseudopod forms? There are two popular theories offered to answer this question.

An explanation that has been accepted for some time now is that the plasmagel in the rear and around the sides contracts, forcing the sol forward in the manner of squeezing tooth paste from a tube. A more recent view is that the contraction of the plasmagel occurs near the anterior portion of the pseudopod thus pulling the plasmasol forward. (There is the added complication that the amoeba, *Amoeba proteus*, shown in Figure 6–6, sort of walks on the "toes" of its pseudopodia, which enables it to surround its prey.)

Neither of these theories really explains amoeboid movement, although they probably help localize the most interesting regions of molecular interactions, and it is most interesting to be able to see almost directly and vividly biochemical changes taking place. The need to understand this basic process is essentially pragmatic; it relates to our desire to be able to control or intercede in fundamental cellular events, in order to encourage good health, reduce disease, slow down the aging process, prevent cancer, etc. In this chapter we will consider some basic responses of cells to simple stimuli, where each response involves cell movement. The implicit question is always dichotomous as follows: (a) Does the stimulus affect the motile system directly, or (b) does the stimulus affect some other sensor which then directs the motile system? In either case, we must know how the motile system works before we can hope to predictably influence results.

Flagellar and amoeboid movement probably originated in cells when they were isolated individuals. Once established, however, these "patents" were retained throughout all higher groups of animals that were to follow. For example, we see amoeboid movement in the white blood cells of man as well as ciliary movement in the cells lining his trachea. We would expect to observe these types of locomotion in modern single cells, and we do.

Before going on to study these first animals, perhaps we should discuss their place among microorganisms. Microscopic cells exist in a large variety of forms, illustrated by bacteria, yeasts, molds, algae, and protozoa. Some are marked off by rather well defined characteristics, but others show close affinities; indeed it is often difficult to be certain in which category they belong. Some have plant-like characteristics, that is, they resemble higher plants, whereas others are definitely animal-like and still others seem to possess both animal and plant characteristics. One has no difficulty in distinguishing between a tree and a cat, but at the unicellular level frequently it is difficult to clearly separate these microscopic organisms into distinct categories. To overcome this problem, and for convenience, some biologists have created one large taxon to include all of these unicellular organisms. They call this group **Protista** and the individuals, **protists.** Since we are dealing in this book primarily with animals, we have little need to refer to this term and will confine our attention to the animal-like protists, the protozoa.

PHYLUM PROTOZOA

The protozoa display the full potentialities of protoplasm within the confines of a single cell. With few exceptions they are unicellular, yet in spite of this limitation they have explored almost every possible design in form and function so that we see them doing nearly all the things multicellular animals do. They exist in a vast array of sizes, shapes, and habitats. It has been estimated that there is at least one protozoan associated with every metazoan, either as a commensal or a parasite, and since commensals and parasites are usually host-specific, each host harboring a different species, it would appear that there must be several million species of protozoa, although only about 50,000 have been described. They range in size from 3 to 15,000 μm and live in almost any environment, from soils to the red blood cells of vertebrates.

Biologists have argued for a long time over the question of whether the protozoa fit the definition of a cell, that is, whether they are equivalent to the metazoan cell. Quite obviously a metazoan cell is only a part of a greater unit and each cell plays

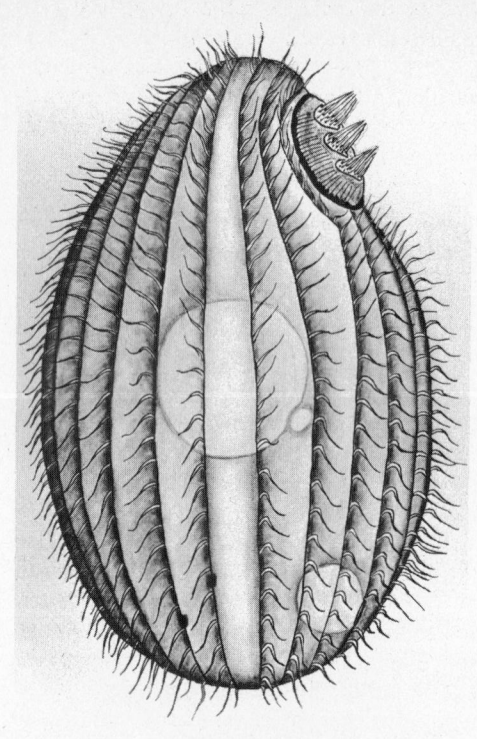

Fig. 6–7. This ciliated protozoan, *Tetrahymena pyriformis,* has many of the nutritional requirements of higher animals, including mammals, and for that reason has become valuable in research. The study of microorganisms has become an important field of research where the fundamental workings of protoplasm are under investigation.

a particular role in the operation of the whole organism, whereas in protozoa, the cell is the organism. Some biologists prefer to call protozoa **acellular,** leaving the term cell for the units making up many-celled organisms. Those who prefer to think of them as **unicellular** argue that most of them contain a nucleus and other structures found in all cells and this is sufficient evidence to call them one-celled animals. Since all functions are performed within the confines of the single cell, numerous structures have evolved within protozoa, some far more elaborate than those found in any metazoan cell. In other words, differentiation has developed at the one-cell level, just as it developed later among the cells of the multicellular organisms.

The question of what the protozoa should be called can be reduced to a matter of semantics and need not concern us here. We refer to them as unicellular.

Since the protozoa are the most primitive animals one might expect that they possess many, if not most, of the characteristics of animals. Cells possess common biochemical and morphological characteristics whether they reside in the body of a man or are single individuals such as a paramecium or an amoeba. It is frequently difficult or even impossible to study individual cells of a higher organism in the manner that is required for understanding its basic mechanisms, whereas isolated protozoa in a test tube might yield the desired information which then can be extended to metazoan cells. For example, the ciliate *Tetrahymena pyriformis* (Fig. 6–7) has proven to be an excellent cell for research in nutrition, metabolism, cytology and genetics. Several

species of paramecium have been utilized as experimental animals for studies in genetics, and amoeba has had a long history as an ideal cell for the study of behavior, cytology, movement, biochemistry, and many other problems in biology. Such protozoa as *Stentor* and *Tokophyra* (Fig. 6–8) are ideally suited for studies of embryology. The latter organism is unique among one-celled organisms in producing internal buds which become "embryos," mature, and are "born," a process reminiscent of higher animals. Such organisms provide excellent research material for studying the way new cells are formed from parent cells. Many other protozoa are currently being used in a great variety of experiments which are designed to solve problems in basic biology. As more is learned about handling these tiny animals in the laboratory, they will become more and more useful in research.

When one attempts to classify this heterogeneous group of organisms into any sort of logical system, he is at once faced with some difficult decisions. In the past, biologists utilized the very obvious character of methods of locomotion as the single criterion for assigning groups. At first it appears to be rather easy to separate protozoa into groups which possess flagella, cilia, pseudopods, and no obvious means of locomotion. However, it has become increasingly clear as more knowledge accumulates that certain fallacies exist in this method of classification. For example, it is now known, as a result of electron microscope studies, that cilia and flagella are structurally the same (p. 119). The fact that cilia are usually short and numerous is not a distinguishing characteristic, because there are flagellates which possess numerous flagella and ciliates without cilia. Distinctions between amoeba and flagellates-ciliates are not always clear either; for example, there is a group called amoebo-flagellates, perhaps best illustrated by the protozoan *Tetramitus rostratus*, which exists as an amoeboid form under certain cultural conditions and when these are changed it transforms into a four-flagellated organism (Fig. 6–9). Moreover, many protozoa which

Fig. 6–8. This protozoan, *Tokophrya*, remains attached by means of a stalk throughout its adult life and feeds through sucking tentacles. It is unique in that it gives rise to a free-swimming offspring which forms as an internal bud, is "born" through a birth pore, swims about for a time, then becomes attached to the substratum by its caudal cilia, grows a stalk and tentacles, and loses its cilia as it metamorphoses into the adult.

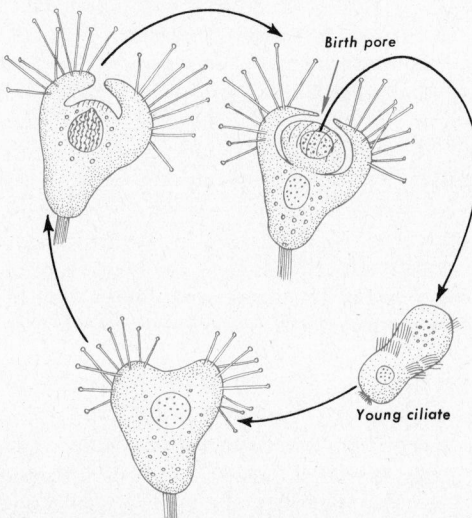

Birth pore

Young ciliate

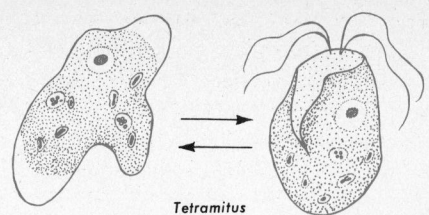

Tetramitus

Fig. 6–9. The life cycle of *Tetramitus* involves both amoeboid and flagellated stages indicating the close affinities of these large groups, the sarcodinids and mastigophorans.

appear to be without means of locomotion, such as the malarial parasite, do have stages in their life cycles which are motile.

The problems inherent in classifying protozoa were recently attacked by a group of specialists who, over a period of years, agreed on a system which is generally accepted by protozoologists today.* We adhere to this system.

Owing to their great diversity it is impossible to obtain more than a glimpse of the phylum protozoa as a whole. We can, however, study two representatives in some detail in order to have some understanding of their morphology and how they function. Later we can briefly describe representatives of the various groups to gain some notion of their diversity. We have selected **amoeba** as a simple form and **paramecium** as one of the most complicated protozoans.

AMOEBA

The common amoeba, *Amoeba proteus* (Fig. 6–10), spends its life today, as it probably has for many millions of years, in an aquatic environment from which it receives all of its nourishment and into which it deposits all of its wastes. Its continual search for food keeps it on the move, crawling over vegetation on the bottom of ponds and streams. It resembles an amor-

* This system of classification appears in *The Journal of Protozoology*, Vol. II (1) (1964), pp. 7–20.

phous blob of nearly transparent jelly and its chief outward manifestation of animal behavior is its movement, which appears mysterious because there seems to be no propelling mechanism.

Locomotion

The amoeba moves by means of lobe-shaped projections of its protoplasm, called **pseudopods** (false feet). We have already discussed the mechanism involved in their formation (p. 121). The **ectoplasm** is separated from the surrounding fluid environment by a thin membrane, the **plasmalemma** (Fig. 6–10). Electron micrographs of this membrane demonstrate that it is not a simple unit membrane but possesses tiny finger-like projections which extend outwardly. These probably function in pinocytosis and phagocytosis. They become incorporated inside the vacuoles as they form. The **plasmasol** and **plasmagel** show numerous solid fibers in the electron microscope, the function of which is not clear. The most obvious indication of movement is the flowing of the numerous particles suspended in the plasmasol. The large oval **nucleus**, as well as **food vacuoles**, also flows freely during movement. The **contractile vacuole** is conspicuous by its rhythmic pulsation.

Pseudopods form vertically and anteriorly, that is, in the direction of the cell's general progress. The amoeba is able to "step over" particles which are not food and to engulf those which are a part of its diet. Most descriptions of this process point to the fact that the pseudopods form in several directions at one time. There follows a "tug-of-war" until finally the pseudopods on one side or the other accumulate the bulk of the protoplasm, causing the others to retract and follow the cell body in a specific direction.

Ingestion of Food

The activity of the amoeba is about the same during locomotion and during food-getting, with one or two minor exceptions. When it approaches a motionless particle of food, such as an immotile alga, the pseudopods spread around and over the plant cell in close proximity until they meet, forming a **food vacuole.**

Once the food vacuole is within the cell body, digestion begins. It is probable that primary lysosomes (p. 77) fuse with the food vacuole, contributing their contents (digestive enzymes) to it. Digestion is then initiated and when it is complete, the end products (amino acids, monosaccharides, fatty acids, and glycerol) pass through the membrane into the cytoplasm where they are metabolized by amoeba. Undigested material remaining in the food vacuole (now called a **residual vacuole**) is discharged to the outside as the amoeba moves on its way.

Gas Exchange and Excretion

The exchange of oxygen and carbon dioxide is simple in amoeba. Oxygen moves into the protoplasm of the cell by diffusion whenever its concentration is greater outside than inside. Likewise carbon dioxide diffuses out of the cell into the surrounding water when it forms as a result of metabolism within the organism.

As a result of the metabolism of nitrogen-containing compounds (amino acids), poisonous wastes tend to accumulate in the cell. These are removed by diffusion to the outside fluid world also. They are not allowed to accumulate because of their toxic effects.

There is a prominent organelle (little organ), located variously in the cytoplasm, called the **contractile vacuole.** This is a small, clear, spherical area which forms in the cytoplasm, soon grows to maximum size, then suddenly disappears. It forms and disappears at regular intervals and probably functions only as a device for getting rid of excess water that accumulates inside the cell. Because of the hypertonicity of the amoeba, water is constantly flowing into its protoplasm. If it were not removed the little

animal would soon become waterlogged. Furthermore, if the concentration of dissolved substances in the surrounding water is increased (hypertonic), as it is in sea water or in a solution with high salt content, the contractile vacuole disappears. If the marine amoeba, which has no contractile vacuole, is placed in less hypertonic fluid, vacuoles form very soon. All evidence points, therefore, to the fact that the contractile vacuole functions as a hydrostatic organelle.

Behavior

Another quality of amoeba which makes possible its survival, is its responsiveness to changes in its external world. It responds usually to protect itself from harm or to lead itself to a rich food area. This is spoken of as its **behavior.** Although gradual changes in intensity of light elicit very little or no response, intense, sudden light causes it to move away from the light source (Fig. 6–11). Likewise, if a concentrated salt solution is placed near the amoeba, a definite response is noted. If it is floating free in the water and one of its pseudopods contacts a

Fig. 6–10. Amoeba with internal anatomy shown in detail.

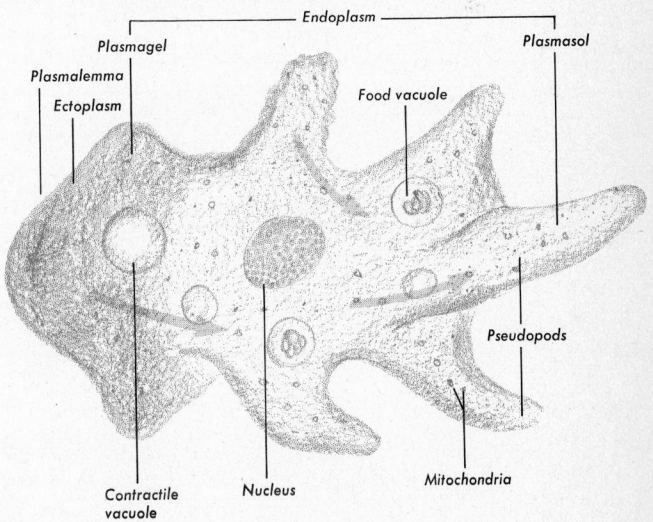

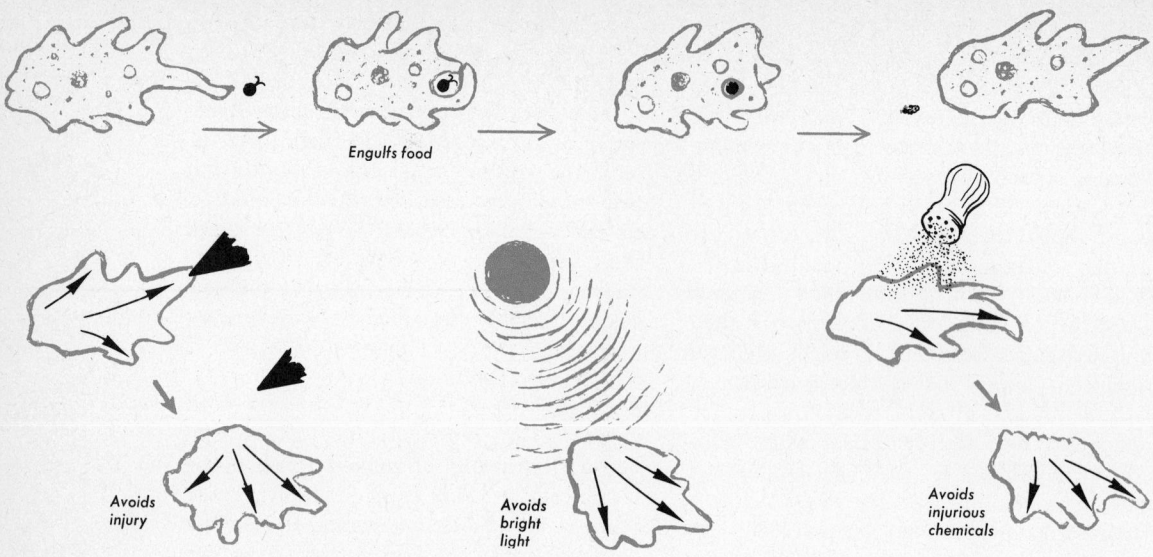

Engulfs food

Avoids injury

Avoids bright light

Avoids injurious chemicals

Fig. 6–11. Amoeba responding to various conditions in its environment.

surface, it at once adheres to the substratum. However, if a sharp point is pressed into its protoplasm at the surface, an immediate avoiding reaction follows. Amoeba thrives best at room temperature. If it is subjected to lower temperatures, all of its activities slow down and cease altogether as the freezing point is reached. Activity also ceases if the temperature is raised to approximately 30°C.

Reproduction

Amoeba, like all other living things, reproduces. It does this in the simplest way, namely, by dividing into two equal parts, a process known as **binary fission** (Fig. 6–12). After the cell has grown to a certain size, it rounds up into a ball. The nucleus divides first, then the entire cell cleaves into two parts which are usually spoken of as "daughter cells" (they are called daughter cells perhaps because they give rise to other cells; "son cells" would be unable to do this). The amoeba seems to have to reach a certain size before dividing, for if a small piece of cytoplasm is cut off periodically as

the amoeba grows, never allowing the animal to reach the proper size, it will not divide.

Amoeba is "immortal," as was pointed out in an earlier section. If death occurs, it comes only through accident. Occasionally its watery environment may dry up, leaving it to dessicate and die. When such times come, the amoeba (there seems to be some question about A. *proteus* forming a cyst although other amoebas do) secretes an impervious outer covering called a **cyst**, which allows life to continue at a very low ebb until it is once more submerged in water. The cyst then splits open and the amoeba resumes its active life.

PARAMECIUM

Not all protozoa are as simple as amoeba. A study may now be made of a protozoan which is perhaps one of the most complicated of all single cells—paramecium, the "slipper animalcule" (Fig. 6–13). This form has been experimented upon and studied as much as, if not more than, the amoeba, and a great deal of fundamental biological information has been derived from it.

The animal is pointed at the posterior end and blunt at the anterior end. A groove extends throughout most of its length and the **mouth** or **cytosome** (cell mouth) is formed in the groove on the oral side, about two-thirds back from the anterior end. Paramecium has an outer covering, the **pellicle**, which is sufficiently rigid to maintain a constant shape. The covering is made of minute hexagonal plates, and the middle of each plate is perforated by a central opening through which a **cilium** passes. The animal moves by the combined rhythmic beating of these cilia. At the junction of the plates are other tiny holes through which threads are thrust when the animal is disturbed. The threads originate from small bodies lying just beneath the pellicle, called **trichocysts**. They are apparently used in defense and perhaps also in attaching the animal to detritus in the water. Adding a

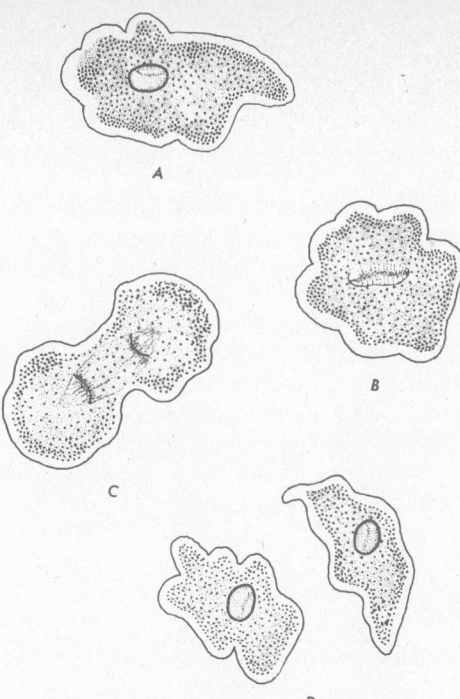

Fig. 6–12. Amoeba undergoing binary fission.

Fig. 6–13. Paramecium with internal parts shown on the left. The detail of cortical structures is shown on the right.

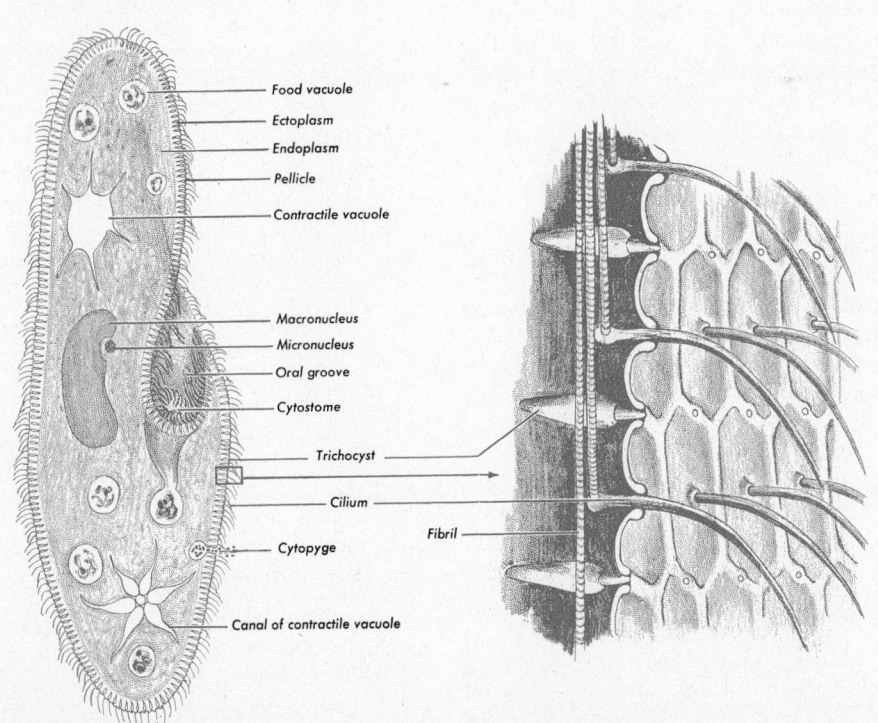

Food vacuole
Ectoplasm
Endoplasm
Pellicle
Contractile vacuole

Macronucleus
Micronucleus
Oral groove
Cytostome

Trichocyst

Cilium

Fibril

Cytopyge

Canal of contractile vacuole

small amount of acetic acid to the water near the paracecium discharges them.

Locomotion

The power stroke of the cilia is diagonal, so that the animal turns on its long axis. Since the cilia in the oral groove are more numerous, the anterior end describes a circle and causes the animal to swim in a spiral manner (Fig. 6–14). When the posterior end is stationary, the long axis of the body describes a cone. When the animal is confronted with an obstacle, the cilia reverse their effective beat so that the cell moves backward a short distance, turns slightly, then moves forward again. If it meets the obstacle again, the process is repeated until the animal passes around and goes on its way (Fig. 6–15). This is known as an **avoiding reaction.**

The cilia move in a coordinated manner producing waves, called **metachronal waves** (Fig. 6–2) which are responsible for movement of the cell. There are structural connections between the kinetosomes as well as an elaborate system of fibers lying beneath the pellicle. Experiments show that cutting these fibers interrupts the rhythm, which implies some sort of conducting mechanism similar to nerves in higher animals. These experiments led Kofoid and his students in the 1920s and 1930s to the idea of a "motorium," a sort of "brain" with fibrillar connections just underneath the pellicle leading to all parts of the cell including the cilia. Contemporary ultrastructural studies fail to confirm the presence of a single motorium, but instead emphasize the repetitive nature of ciliary units, consisting of one or two cilia, a striated fibril at the base of the kinetosome

(Fig. 6–13), and accessory microtubules arising near the kinetosome and running along underneath the pellicle in fixed directions and patterns both laterally and anteroposteriorly. These studies suggest that the subpellicular fibrils provide cytoskeletal support. They do not rule out the possibility of a conductive function, but make the cutting experiments extremely difficult to interpret, since the membranous as well as the fibrillar parts of the cortex were damaged. Since impulse conduction apparently universally involves cell membrane depolarization (discussed later) in metazoa, there seems no reason to except protozoa from similar conductive membrane phenomena at this time.

Ingestion of Food

A simple experiment can be performed to demonstrate how paramecium feeds. Some yeast cells that have been heavily stained with Congo red (a dye) are placed on a glass slide containing paramecia. By studying the region of the gullet, one can see the cilia beat in such a way as to pass the yeast particles along its oral groove and down into the gullet. There they are rounded up into a mass which finally pinches off into the cytoplasm as a food vacuole. Once in the cytoplasm the yeast cells remain deep red for a time, but gradually begin to turn blue as they approach the anterior end. This means that the contents of the food vacuole are alkaline at first (Congo red is red in alkali, blue in acid), just as the mouth of man is alkaline. As digestion proceeds, the vacuoles become acid as indicated by the blue color; the human stomach is acidic as well. The digested material passes through the wall of the vacuole and out into the protoplasm where it is metabolized, the same process that was noted for amoeba. Finally the undigested portions left in the vacuole pass through a tiny opening to the outside, the **cytopyge,** which is equivalent to an anus in higher animals.

Some species of ciliates are able to receive nourishment from dissolved organic

Fig. 6–14. Path taken by paramecium when it moves freely through the water.

matter in the medium. In fact, *Tetrahymena pyriformis* (Fig. 6–7), which feeds on bacteria in its natural habitat, grows in a culture medium consisting only of dissolved organic compounds. The exact composition of this medium is known and it is about the same as that of higher animals, including man. It consists of vitamins, amino acids, nucleic acids, glucose, and salts. Because its diet is so precisely known, it has become a useful tool in nutritional research, such as assaying (testing) for qualitative and quantitative value of certain foods. It has also been useful in determining the amounts of certain nutrients in the body fluids of people suffering from deficiency diseases. It is very likely that more protozoa will be used in the future for such studies.

Gas Exchange and Excretion

Gas exchange and excretion of nitrogenous waste take place in paramecium much the same as in amoeba. The **contractile** or **pulsating vacuoles** which lie at either end of the cell contract alternately at about 15-second intervals. Several **radiating canals** empty into each contractile vacuole, becoming most obvious when the vacuole is nearly empty. Each vacuole discharges its contents to the outside through a minute **pore** in the pellicle. The rate of contraction varies with temperature, activity of the animal, and the concentration of salts in the surrounding medium. As in the amoeba, the contractile vacuole functions normally as a bailer to rid the animal of excess water constantly entering the cell.

Behavior

Paramecium is very sensitive to the relative acidity and alkalinity of its environment (Fig. 6–15). Paramecium reacts positively to a weakly acid medium, such as acetic acid or an environment containing CO_2 bubbles which render the medium slightly acid. Once in the acidic medium, it swims slowly and tends to stay there by utilizing the avoiding reaction when it approaches a less acidic region of the en-

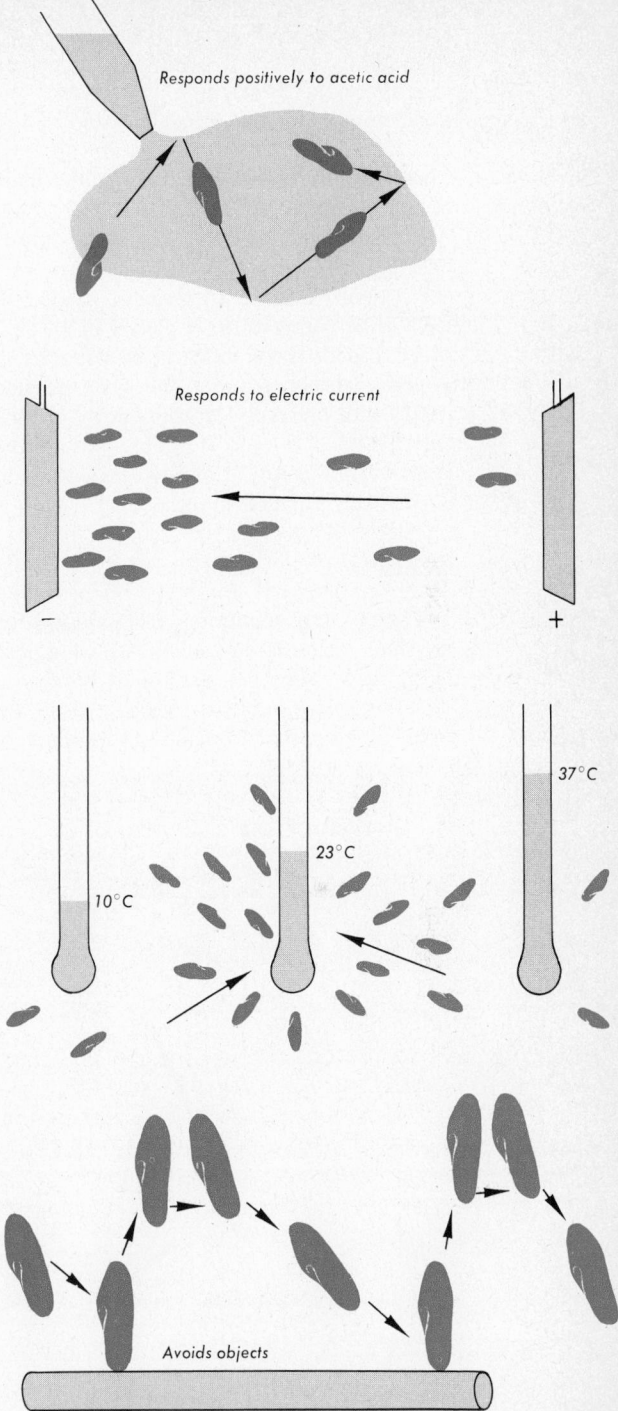

Fig. 6–15. Paramecium responding to various conditions in its environment.

vironment. It also avoids the strongly acidic medium very close to a bubble of CO_2. It thus moves in such a fashion as to remain in a slightly acid medium. Why does it do this? Apparently the adaptive value of this reaction is related to its food, bacteria, which usually reduce the pH (more acid) of the medium.

Paramecium selects a temperature optimal for its activities, usually around 23°C (Fig. 6–15). If given a choice, it seeks this temperature. If placed in an electrical field of direct current, it responds in a very definite manner, always orienting itself with respect to the flow of the current (Fig. 6–15). It moves toward the negative pole, indicating that externally it is positively charged.

Reproduction

Paramecium maintains its numbers by dividing, transversely across its long axis (Fig. 6–16). The first sign of division is in the tiny **micronucleus** which moves out from the large **macronucleus** and undergoes

a change in shape from a sphere to a spindle-shaped body. The micronucleus functions as a carrier of hereditary traits and gives rise to new micronuclei and macronuclei during conjugation. The macronucleus functions in providing RNA for protein synthesis during the vegetative stages of the life cycle. Only the macronucleus is indispensable for the life of the paramecium as well as some other ciliates; this is borne out by the fact that some ciliates have been kept alive, dividing normally, for many years without micronuclei. Once the micronucleus has spun out into two equal parts and these deposited at the two ends of the cell, the macronucleus divides by pinching into two parts. A second gullet and two more contractile vacuoles then appear, followed by indentations in the mid-region where ultimate separation is to occur. Other structures also duplicate before the two daughter cells finally separate. After a growth period they are ready to divide again. Under optimal conditions, division occurs about every six to twelve hours. Under natural conditions they very soon cease

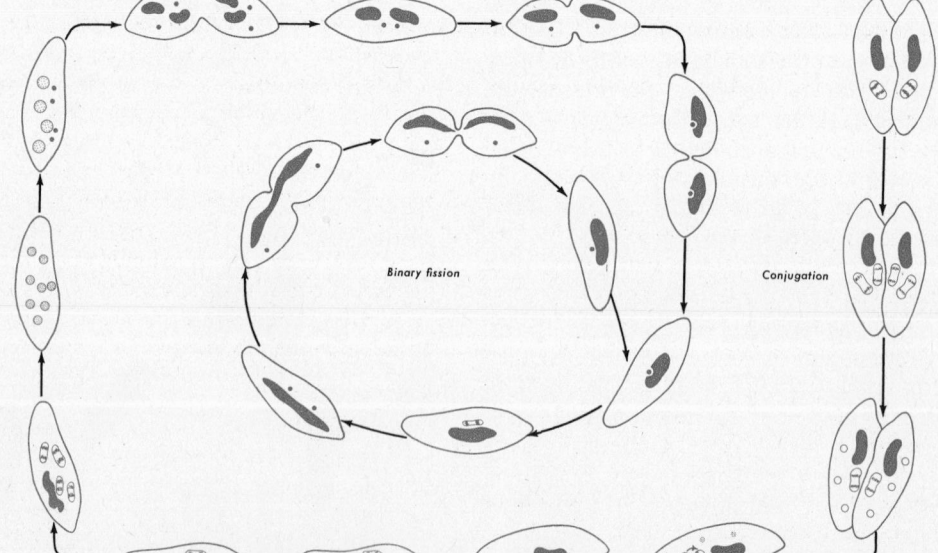

Fig. 6–16. Conjugation and asexual reproduction by binary fission in paramecium.

Binary fission

Conjugation

dividing, owing to the accumulation of wastes, lack of food, or other adverse environmental factors.

Under partially starved conditions, paramecia periodically undergo a sexual process called **conjugation** (Fig. 6–16). During the process two individuals interchange micronuclear material, which means that genes are involved. This is similar to sexual reproduction in that there is a fusion of genetic material, although there is no immediate increase in numbers as the term *reproduction* usually implies. Paramecium undergoes a type of self-fertilization, called **autogamy**, at regular intervals. Moreover, it seems that unless autogamy or conjugation occurs periodically, certain species of paramecia cannot continue dividing and finally die. Another ciliate, *Tetrahymena pyriformis*, has been maintained in continued culture for over 35 years with neither conjugation nor autogamy taking place. Therefore, nuclear reorganization seems to be essential in some ciliates but not in others.

The process of conjugation is well known for a large number of ciliates and, in a few, **mating types** have been found. Mating types were first discovered by T. M. Sonneborn in *Paramecium aurelia* in 1937. Mating types are distinguished from one another by their conjugation reactions and have been identified for convenience by Roman numerals. Thus, an animal designated as mating type I will mate with ciliates of mating type II, but not with others like itself, that is, mating type I. This might suggest sex in these protozoa, male and female. The difficulty arises when other information is considered. Jennings and his coworkers found as many as eight different mating types in *Paramecium bursaria*, and more recently ten have been found in *Tetrahymena pyriformis*. Since two sexes are usual in both plants and animals, it is perhaps better that we retain the designation *mating types* for this unusual condition found only among the ciliates and a few lower plants.

Conjugation in ciliates involves a series

of rather precise steps. In *Paramecium caudatum*, for example, when animals of opposite mating type are mixed under proper conditions, the surface membranes of the cells seem to become sticky which causes them to adhere in pairs, rather securely attached at their oral grooves where a narrow protoplasmic bridge extends from one to the other (Fig. 6–16). Very shortly thereafter, the micronuclei in both animals divide twice forming four micronuclei; during one of these divisions the chromosome number is reduced to one-half, which is equivalent to meiosis in metazoa (see p. 490). At this time the macronucleus and three of the newly formed micronuclei disintegrate and disappear, leaving one micronucleus which then divides once more giving rise to two micronuclei each containing half the number of chromosomes possessed by the original micronucleus. A significant event now takes place—one of these micronuclei from each animal migrates across the protoplasmic bridge into the opposite cell and fuses with the micronucleus remaining there. The migratory micronucleus is sometimes thought of as a *sperm cell* and the stationary one as an *egg*, although they are not comparable to their implied counterparts in higher forms. The **fusion nucleus** now possesses the full number of chromosomes, one-half from the migratory micronucleus and one-half from the stationary micronucleus. The paired cells now separate and the fusion nucleus undergoes three successive divisions producing eight micronuclei, four of which remain micronuclei and four become new macronuclei. Both paramecia then divide twice without nuclear division, producing eight individuals in all, each one of which resembles the original parents. These then undergo ordinary asexual fission until conjugation is once more induced.

OTHER PROTOZOA

This brief introduction to two different protozoa gives us a basis for considering the importance of the group as a whole in

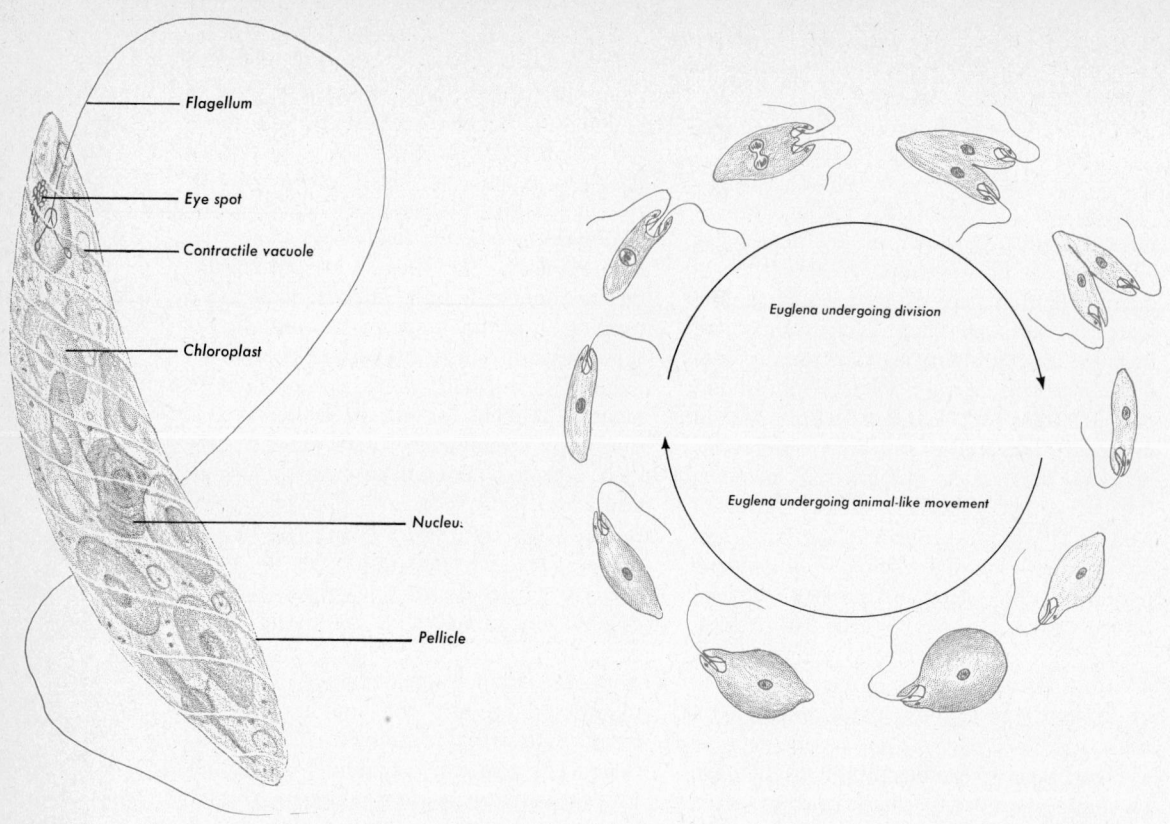

Flagellum

Eye spot

Contractile vacuole

Chloroplast

Nucleus

Pellicle

Euglena undergoing division

Euglena undergoing animal-like movement

Fig. 6–17. Euglena in detail and undergoing some of its life processes.

respect to numbers, variety, and classification. Over 50,000 protozoa have been described as distinct species, and in numbers of individuals they exceed all other animals. They live in water of all kinds, in soil and dust, in and on bodies of plants and animals. Some of them cause the most destructive diseases known to man, malaria, for example. Most protozoa are free-swimming, although some are sessile; most live singly, some form colonies. Many form a source of food for aquatic animals such as fish, but they are of little value to man, except a few which are useful in sewage treatment. They are such a large and varied group of animals that some biologists have considered placing them in a group larger than the phylum, that is, a subkingdom.

The protozoa are divided into four subphyla, based on a number of characteristics. The subphylum **Sarcomastigophora** includes those that move by means of flagella and pseudopods. Those that have no means of locomotion in the adult stages and are all parasites belong to the subphylum **Sporozoa.** Protozoa with cilia and two types of nuclei (macro- and micronucleus) belong to the subphylum **Ciliophora.** Each of the subphyla is divided into several subgroups. There is some evidence to show that the protozoa may have evolved in the order indicated by the classification, although there undoubtedly was some overlapping and there is no precise indication as to which types actually preceded other types. For our purposes, the order above is followed; that is, we consider the Sarco-mastigophora as the most primitive and the Ciliophora as the most complex.

SUBPHYLUM SARCOMASTIGOPHORA

Superclass Mastigophora

This is a widely diverse group of protozoa in which some members are colored and live independently like plants, whereas others are colorless and require food from the outside, such as the parasites living in the intestinal tract of termites.

Class Phytamastigophorea

Colored Flagellates. Although there are wide divergences in structure and habitat of flagellates, a brief description of **Euglena** (Fig. 6–17) will be of help in understanding the group as a whole.

Anatomically, *Euglena* is quite different from amoeba. It has a rather definite general shape, which is something like a spindle, although it is sufficiently elastic to be able to undergo animal-like movements when confined to a small space (Fig. 6–17). It moves by means of a single, hair-like, vibratile **flagellum** which, when active, pulls the organism through the water in a spiral path. In its cytoplasm euglena bears bodies known as **chloroplasts**, which contain the green plant pigment, **chlorophyll**. Because it utilizes photosynthesis like higher plants, and at the same time possesses certain animal characteristics, *Euglena* is thought to be intermediate between the plant and the animal world, and is frequently referred to as a **plant–animal** type.

There is a **reservoir** at the anterior end, the walls of which give rise to two flagella, one of which emerges, the other terminates within the reservoir. Lying near the reservoir is the **contractile vacuole**. In the immediate region of the reservoir is the **eye-spot**, a conspicuous red dot which apparently functions in aiding the cell to find the proper light intensity for photosynthesis. In order to manufacture its own food, *Euglena* must receive the proper amount of light, hence the significance of the eye-spot. *Euglena* can, however, live in the dark providing nutrient materials are present in the surrounding medium, in which case it absorbs its food directly.

Euglena divides longitudinally, that is, in an anterior-posterior direction, splitting the cell into two equal parts (Fig. 6–17). In a rapidly growing culture, cells can be observed in all stages of division. This protozoan, like many others, can be grown in sterile medium which is not true of some other microorganisms. The exact nutritional requirements are known for *Euglena gracilis*. Of significance is that it requires B_{12}, a vitamin which, when deficient in man, causes a fatal type of anemia. Since *Euglena* requires the vitamin, it has become a valuable organism for determining the levels of B_{12} in the fluids of patients suffering from the disease.

Other Colored Flagellates. All kinds of fresh water as well as the oceans are teeming with free-living flagellates. The **dinoflagellates**, for example, live in the oceans for the most part and constitute a large portion of the diet of small crustacea and other animals. Some members, such as *Noctiluca* (Fig. 6–18), possess luminescent properties which cause flashes of light when the surface of the water is disturbed on a dark night. This is a particularly attractive sight in the wake of a boat. Another interesting dinoflagellate, *Gymnodinium brevis* (Fig. 6–18), has appeared several times during the past hundred years along the Florida coast in extremely large numbers (50,000,000 per liter—a normal count is about 100,000 for all kinds of protozoans). Furthermore, this protozoan secretes a by-product which is lethal to all other kinds of animal life in the vicinity. Each year billions of fish are destroyed along our coast.

Fresh water, particularly that containing a considerable amount of decomposing

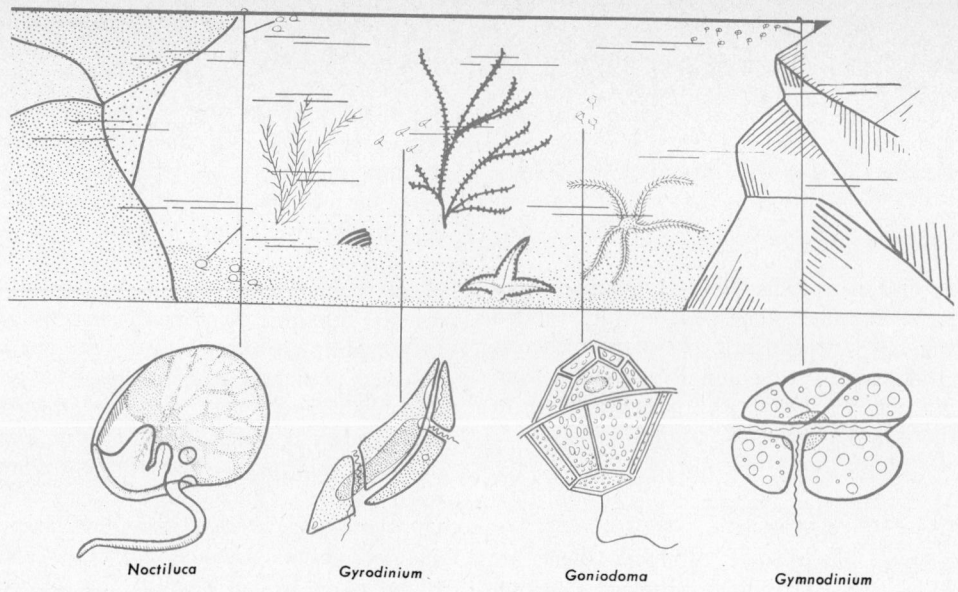

Fig. 6–18. Various types of marine flagellates.

Noctiluca Gyrodinium Goniodoma Gymnodinium

organic matter, supports a large variety of flagellates (Fig. 6–19). Many of them play a very important function in providing food for aquatic animals, especially during their early life when their mouths are so small that no other food but a protozoan could be ingested. Without these tiny animals there would be no fish in many of our lakes and streams. Some, like *Haematococcus pluvialis* (Fig. 6–19), are bright green in color and can reach unbelievable numbers in small pools. They produce haematochrome (a red pigment) at certain times of the year, imparting a reddish color to the water. In some Alpine passes they have been responsible for the so-called bloody snow, a familiar sight to mountain climbers.

Many protozoa form colonies. None is more beautiful than the large *Volvox* which contains up to 50,000 cells (Fig. 6–20). It tumbles over in a graceful fashion as it moves through the water. In spite of the fact that it seems to roll in a random fashion, careful studies show that it always turns in such a way that certain cells lead the colony. In other words it actually possesses an orientation along one axis so that it demonstrates radial symmetry.

Class Zoomastigophorea

Colorless Flagellates. Some colorless forms, such as *Peranema* and *Chilomonas* (Fig. 6–19), live in stagnant water where they feed upon bacteria or other smaller protozoa. Although in its normal environment *Chilomonas* seems to live on a complex diet, it can be grown in a test tube on a diet consisting of ammonia (as a source of nitrogen) and carbon dioxide (as a carbon source). Here apparently is an organism that in nature lives much like an animal but in the laboratory can be forced to live like a plant or even more simply, since it does not require nitrogen in the form of nitrates.

Most colorless flagellates live as single cells, although some live in colonies, such as the branched and stalked form, *Codosiga* (Fig. 6–19). This colony possesses a collar at the anterior end of each cell from which the flagellum emerges. With this exception, collared cells such as these are found only among the sponges. This group of protozoa may well have given rise to the sponges.

Parasitic flagellates. The flagellates have been very successful in making their way

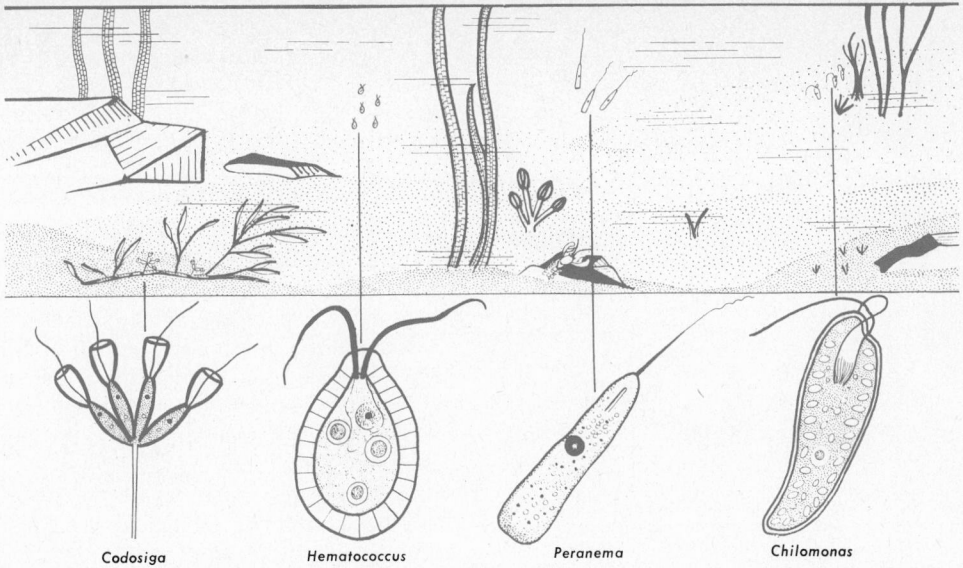

Codosiga Hematococcus Peranema Chilomonas

Fig. 6–19. Various types of freshwater flagellates.

Fig. 6–20. Photomicrograph of Volvox colonies, all containing daughter colonies, each of which will eventually become a mature colony.

into the tissues, blood, and body cavities of almost every group of animals. They are particularly common in the digestive tract and the blood of vertebrates, including man. Some have become serious parasites and today cause untold misery in many parts of the world, not only because they infect man directly but also because they destroy many of his domestic animals. Among the most common offenders is a closely related group of tiny elongated flagellates, called **trypanosomes**. They are polymorphic, that is, they

Fig. 6–21. Closely related parasitic flagellates which cause serious diseases of man and other animals.

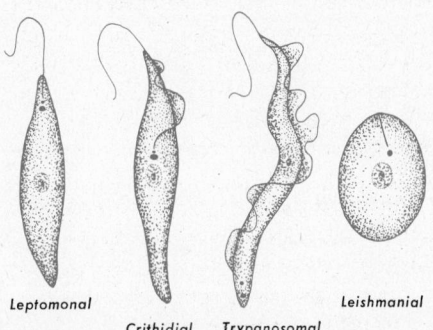

Leptomonal Crithidial Trypanosomal Leishmanial

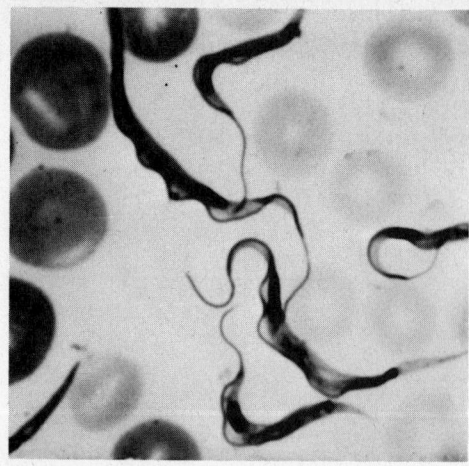

Fig. 6–22. These tiny leaf-like trypanosomes live in the blood of vertebrates and other animals. This is a blood smear showing the parasites among the red blood cells.

change body structure during their life cycle (Fig. 6–21). The four body forms are designated as **leptomonal, crithidial, trypanosomal,** and **leishmanial.** The parasite may exist in one or more of these forms, passing readily from one to the other depending on which host it occupies. Most of these parasites are transmitted from one vertebrate to another by means of an intermediate host, by either a blood-sucking

Fig. 6–23. Life cycle of a trypanosome that causes African sleeping sickness.

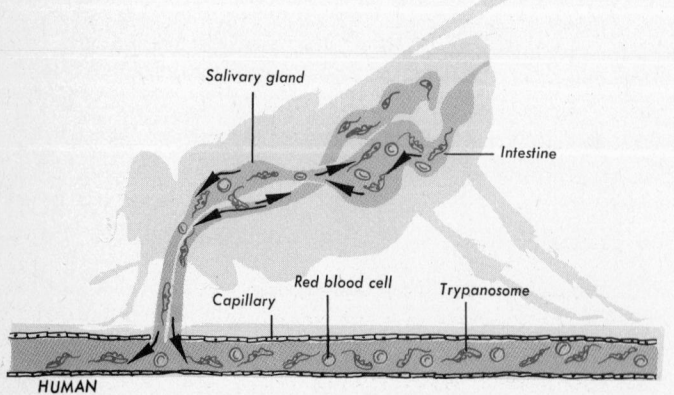

TSETSE FLY

Salivary gland

Intestine

Red blood cell

Capillary

Trypanosome

HUMAN

insect or some other arthropod. Moreover, the leptomonal and crithidial stages are usually found in the invertebrate host, whereas the trypanosomal and leishmanial stages infect the vertebrate host. The disease is usually recognized and named after the organism found in the blood of the vertebrate. We consider several of these parasites because of their medical importance in the tropical parts of the world where many nations are presently emerging.

One of the most common and well-known diseases of equatorial Africa is sleeping sickness caused by two species of trypanosomes (*T. gambiense* and *T. rhodesiense*). These tiny parasites possess a single flagellum which lies in a fold of the outer membrane throughout its length, forming an *undulating membrane* (Fig. 6–22), an effective propelling apparatus in blood and other body fluids. This parasite normally lives in the large wild animals, being transmitted from one to the other by means of the *tsetse fly*; man becomes infected when the fly bites him. Unless treated in its early stages the disease is usually fatal after several months or years.

The trypanosome is sucked up into the gut of the fly during its blood meal (Fig. 6–23) where it transforms to crithidial and leptomonal stages. It finally makes its way into the salivary glands, a common procedure among parasites of blood-sucking insects. Some time later, if the fly bites another person, the parasite is injected along with the saliva. It remains free in the blood for a time but eventually reaches the cerebral-spinal fluid. In this stage the metabolic products of the parasite cause paralysis resulting in the characteristic drowsiness of the victim. He finally falls into a fatal coma.

Vast regions of Africa are denied to man because of the ravages of this disease. The parasite is harmless, however, to the local wild animals, which act as reservoirs, always keeping it circulating in substantial numbers. For this reason the wholesale destruction of the tsetse fly is the only satisfactory control measure. The use of massive amounts of DDT for this purpose is not without environmental hazards, including the accumulation of potentially dangerous levels in the oceans. Figure 6–24 shows a close-up view of the head region of the tsetse fly taken with a scanning electron microscope. The picture illustrates one of several newer research approaches which focus on various special features of insects which may prove amenable to attack. In this case, the sense organs of these "problem" insects are being examined in an effort to understand how these devices allow the insect to locate its hosts. Then an attempt will be made to develop ways of inactivating or bypassing this sensory equipment. Another type of approach with other flies has involved growing and releasing of large numbers of sterile males which copulate with normal females but produce no viable offspring. Man has not yet won his battle with certain diseases and insect vectors.

Another tragic disease that incapacitates millions of people is *kala azar* caused by

Fig. 6–24. This is a SEM micrograph of the tsetse fly showing the proboscis, pointing towards the viewer, by which the fly punctures the skin of a warm-blooded host. The mosaic eyes are the two large domes on either side of the head. It has a pair of antennae and various bristles and hairs. (From C. P. Gilmore, *The Scanning Electron Microscope World of the Infinitely Small.* Greenwich, Conn.: New York Graphic Society, 1972.)

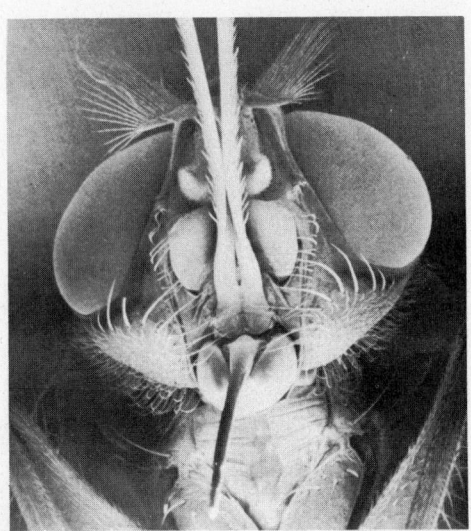

Leishmania donovani, a member of this group of flagellates. It occurs in large sections of North China, various parts of India, the Sudan and South America. In the past it has ravaged the inhabitants, but with the advent of knowledge concerning epidemiology, diagnosis, and treatment, many of the evil effects of the disease have been greatly reduced.

Leishmania are tiny (2–4 microns) ovoid parasites which attack the cells of tissues associated with the circulatory system, including the spleen and liver. Upon entering the cells, they multiply (Fig. 6–25), until eventually the host cell bursts and the released parasites attack other cells. Some enter the blood stream where they are picked up by the intermediate host, the sand fly (*Phlebotomus*). In the gut of this insect they become flagellated and change considerably in shape. When the insect bites another person, some time after it has received the parasite, the flagellated forms are injected directly into the blood where they attack the lining cells of the blood vessels and the cycle is complete. Like most blood-sucking insects, the sand fly introduces a small amount of saliva, which has an anticoagulating effect on the blood. If this were not so, a blood clot would shortly interrupt its meal. The parasite is introduced along with the saliva.

The disease runs its course in a matter of months or several years, frequently ending in death. It has been shown that the flies also bite dogs, which in turn act as reservoirs for the disease. So the problem of eradication consists not only of preventative measures and treatment of infected persons, but also control of the dog population of any affected community. The most common method of prevention is the destruction of the breeding places of the fly as well as the fly itself.

Superclass Sarcodina

Although members of this superclass resemble the amoeba to some extent, there is wide variation in form and structure in the group as a whole. Among the fresh-

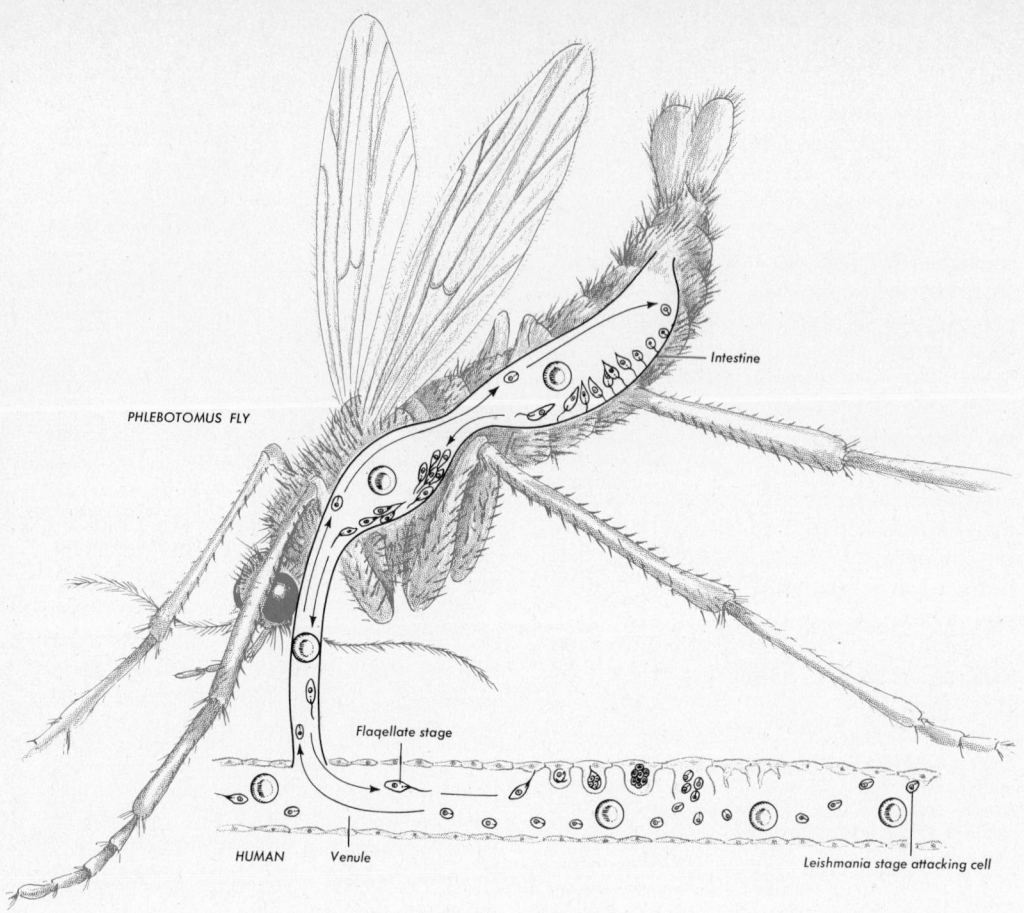

PHLEBOTOMUS FLY

Intestine

Flagellate stage

HUMAN   Venule

Leishmania stage attacking cell

Fig. 6–25. Life cycle of a *Leishmania* that causes kala azar.

water forms there are those, such as *Dif-flugia* (Fig. 6–26), that build houses for themselves. This tiny animal gathers grains of sand and cements them together to form a pear-shaped outer covering into which it may retreat when in danger. *Plagiophrys* and *Arcella* (Fig. 6–26), likewise, build houses for themselves, but in these cases they are secreted by the animals. When observed through the microscope, the shell of *Arcella* resembles a doughnut. Corresponding to the hole in the doughnut is the opening through which the amoeboid form passes as it retracts or extends itself from the shell.

Another freshwater sarcodinid of interest is the "sun animalcule," *Actinophrys sol*

(Fig. 6–26), which resembles a miniature sun when it is floating in the water. The radiating, ray-like pseudopods have a paralyzing effect upon other protozoa, such as *Euglena*, which serve as a food source. There are many relatives of this spectacular protozoan. Together they constitute the subclass **Heliozoa**, a group that is commonly found in the oceans of the world. One, *Oxnerella* (Fig. 6–27), is a particularly beautiful heliozoan.

A large group of forms resembling Heliozoa form the subclass **Radiolaria**. These are also distributed throughout the oceans of the world and float near the surface of the water. Most of them possess a siliceous skeleton (Fig. 6–28) which sinks

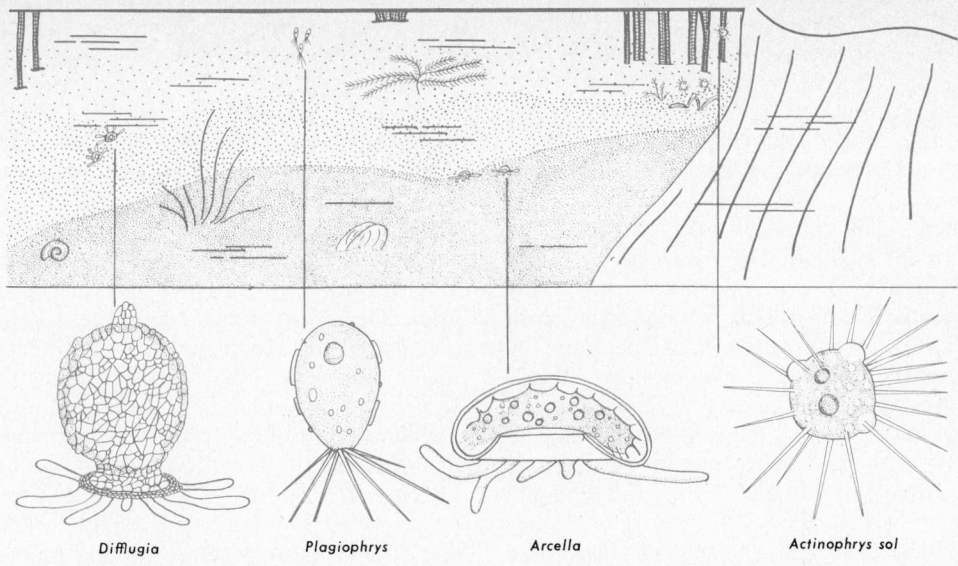

Difflugia   Plagiophrys   Arcella   Actinophrys sol

Fig. 6–26. Various types of freshwater sarcodinids.

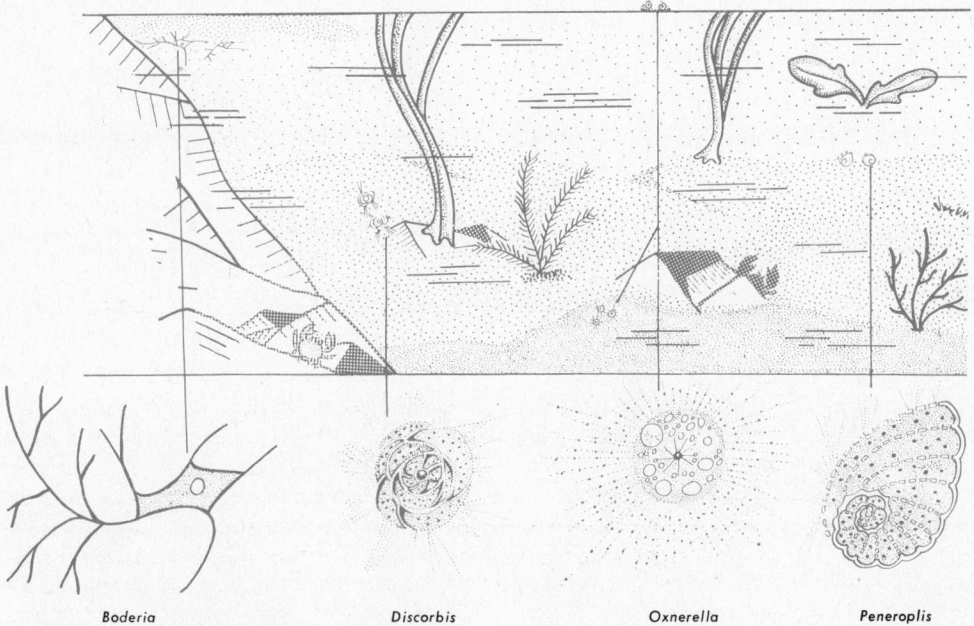

Boderia   Discorbis   Oxnerella   Peneroplis

Fig. 6–27. Various types of marine sarcodinids.

to the ocean floor when the animal dies, forming a thick, mucky layer called "radiolarian ooze." This is particularly extensive in the Pacific and Indian Oceans. Skeletons of these animals are also found in rocks and

have been used by geologists in learning about the history of the earth.

Marine sarcodinids that have even greater significance to geologists are found in the order *Foraminifera*, which secrete shells of almost pure calcium carbonate. They have a rather complicated life history as illustrated by the pelagic (floating at the surface) foraminifer, *Tretomphalus sp.* (Fig. 6–29). The cycle includes an orderly succession of sexual and asexual generations in which two distinct types of individuals and three kinds of shells are involved. The asexual generation (microspheric stage) consists of a large individual with many chambers in its shell. It produces many small individuals asexually which become the sexual generation (megalospneric stage). These grow and produce a gas-filled chamber which permits them to float at the surface of the sea. When mature they become associated with another of the opposite mating type and remain united until they produce prodigious numbers of biflagellated gametes which fuse to form zygotes. These settle and become attached to vegetation where they develop into tiny amoeba-like forms which then grow into the microspheric stage. Thus the cycle is completed and starts over again.

Most of the marine foraminifers form shells, and these are highly varied, although some, such as *Boderia* (Fig. 6–27), possess extremely thin, delicate outer coverings. The variety of shell forms is illustrated by *Discorbis* and *Peneroplis* (Fig. 6–27). These many-chambered, snail-like shells are produced as the animals grow larger, and are perforated with tiny holes through which fine pseudopods project. When they die their skeletons, like those of the Radiolarians, form a "globigerina ooze" (named after the most dominant form, *Globigerina*). This eventually becomes chalk many hundreds of feet thick. The Cliffs of Dover are an outstanding example of this phenomenon. Wherever this chalk appears on land, one can be certain that it was once at the bottom of the sea. Globigerina deposits appear in certain strata of the earth's surface and have a definite relation to the formation of petroleum. Therefore, knowledge of foraminiferans has been useful to geologists in predicting the location of oil deposits.

Parasitic Amoebae. Some amoebae, like some flagellates, have become adapted to life in the body cavities of many different

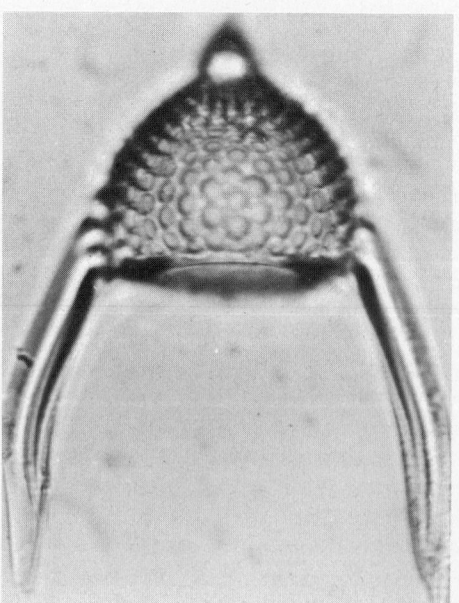

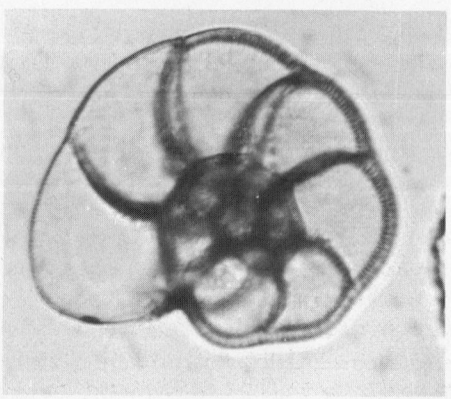

Fig. 6–28. The oceans of the world are teaming with radiolarians and foraminiferans which upon death leave beautiful skeletons, as shown by these two photographs. (Left) the siliceous skeleton of a radiolarian. (Below) The calcium carbonate skeleton of a foraminiferan. Note the snail-like nature of the shell.

animals, including man. Of the half-dozen or so amoebae that inhabit the various cavities of man, only one, *Entamoeba histolytica*, causes any great harm. This amoeba is responsible for **amoebic dysentery**. Though not common in the population as a whole, it is a debilitating disease in the tropics. Native villages are often infected 100 percent, providing a rich source of parasites for spreading the infection to newcomers.

The parasites are transmitted from person to person by contaminated food and water (Fig. 6–30). There may also be an indirect transfer by way of flies and other insects that pick up the infective stages on their feet and proboscis, carrying them directly to food and water. Hence, the obvious method of control is to instigate sanitation in respect to human excreta and to destroy the flies.

Some notorious outbreaks of amoebic dysentery have occurred in the United States which have been traced directly to faulty plumbing. Studies among groups in underdeveloped areas of this country show an infection rate as high as 23 percent, although for the nation at large it is around 5–10 percent. Varying degrees of success have been achieved with a great many substances in attempts to combat the disease. Of the antibiotics, **aureomycin** and **terramycin** have proven the most effective.

SUBPHYLUM SPOROZOA

This class includes a large and heterogenous group of organisms, all parasites, which have puzzled taxonomists for a long time. They show relationships to such diverse organisms as the fungi and some of them are so bizarre that they are included among the sporozoa for want of a better place to put them. Sporozoa are usually immotile, although many have flagellated stages, and they usually lack contractile vacuoles. They reproduce asexually by a process called **schizogony.** At some stage in their complex life cycle they produce sex cells, **macro-** and

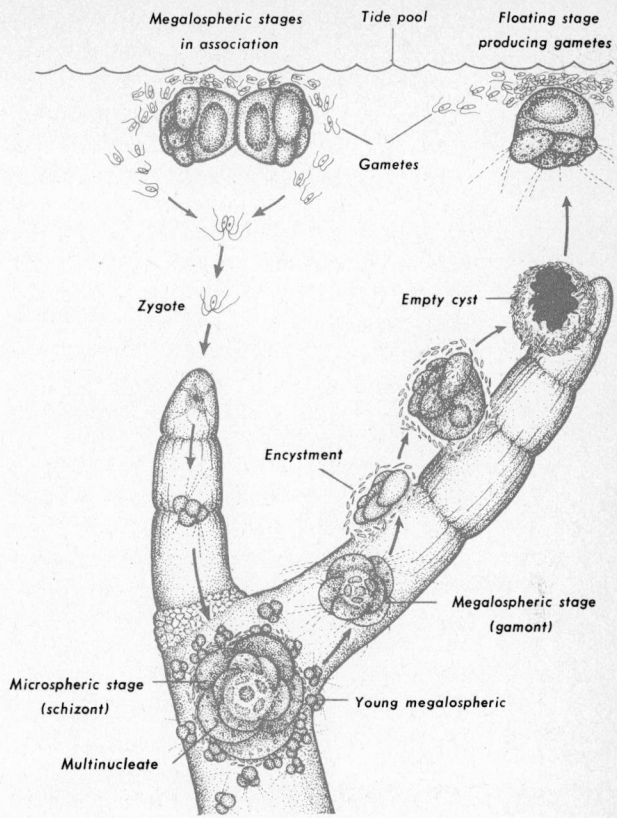

Fig. 6–29. The life history of the foraminifer, *Tretomphalus.*

microgametes, which fuse to form **zygotes**, and then by a process of **sporogony** form resistant **oocysts.** The asexual stage gives rise to the sexual stage in a cyclic manner, resulting in alternation of generations spoken of as **metagenesis.** Some sporozoa complete their life cycle in one host, others require two hosts, the parasite being transferred from one to the other in cycles. They parasitize animals from protozoa to man, invading cells, fluids and cavities. Some seem to cause little damage to their hosts whereas others are devastating; for example, millions of fowl are destroyed each year by sporozoa and malaria is the most important single disease of man.

The subphylum is divided into three classes, the most important of which is

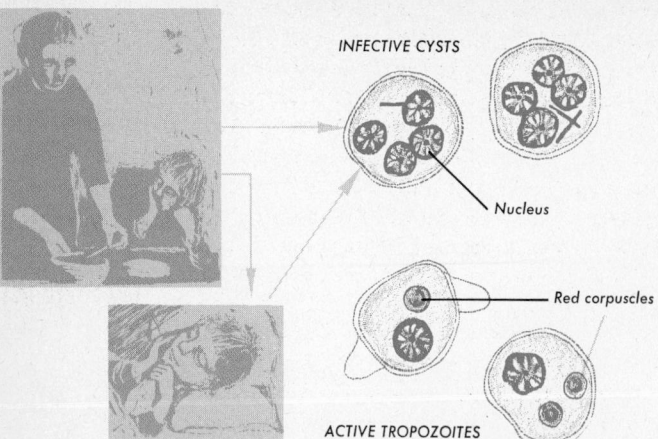

INFECTIVE CYSTS

Nucleus

Red corpuscles

ACTIVE TROPOZOITES

Fig. 6–30. Life cycle of the dysentery amoeba (*Entamoeba histolytica*). Infective cysts are carried on the hands of food handlers and transmitted directly to uninfected people on uncooked food. The trophozoites emerge from the cysts in the intestine, where they multiply and feed on the tissues and blood of the host, thus producing serious illness.

Teleosporea. This class includes two subclasses, **Gregarinida** and **Coccidia.**

Gregarinida. Members of this order either invade cells or live in body fluids. A common example is *Monocystis lumbricis*, which parasitizes the seminal vesicles of earthworms (Fig. 6–31). The worm becomes infected by eating spores which are enclosed in a capsule, the oocyst. These pass down the digestive tract where they "hatch" and become active **sporozoites** which move through the digestive wall into the bloodstream. They are carried to the seminal vesicles where they invade the "sperm mother" cells, growing at their expense to become the **trophozoites.** Some of the potential sperm-producing cells of the earthworm are thus destroyed, but not enough to interfere effectively with its reproductive powers. At maturity the trophozoites are freed as **sporonts** which become associated in pairs (syzygy) as **gametocytes** inside a **gametocyst.** Nuclear division (sporulation) then occurs followed by a partitioning off of the cytoplasm by cell walls thus producing numerous micro-

gametes and macrogametes. These then unite forming the zygotes which are soon encased in a hard shell, the **oocyst.** Three nuclear divisions occur producing eight nuclei, each of which acquires cytoplasm and cell membranes to become the sporozoites. The oocysts containing the sporozoites are left behind in the burrows in some way that is not clear, where they are picked up by another worm and the cycle continues.

Coccidia. This subclass includes several subgroups, the most important of which are the **Eucoccida** and the **Haemosporidia** to which parasites of great significance to man belong.

Eucoccida. Members of this order inhabit epithelial cells of many vertebrates and a few invertebrates. Their chief infection site is the lining cells of the intestine, although they are also found in other organs. Multiplication of the organism takes place by schizogony and sporogony. Infection with these parasites is called **coccidiosis** which is frequently fatal. Domestic animals such as poultry and barnyard mammals, as well as their relatives in the wild, may be heavily infected. Rabbits are frequently infected by the coccidian, *Eimeria stiedae,* whose life cycle is typical (Fig. 6–32).

Rabbits become infected by eating vegetation contaminated with oocysts. The action of the digestive juices releases the sporozoites which then enter the epithelial cells where they undergo rapid multiplication by schizogony, producing **merozoites.** These break out into the lumen and repeat the cycle by entering other cells. This permits the production of millions of sporozoites, and with each cycle more and more of the epithelial cells are destroyed thus causing the disease. After several cycles some of the merozoites become gametocytes which then develop into macro- and microgametes. These unite to form zygotes which shortly secrete an oocyst. The oocysts pass out of the body of the rabbit with the feces and undergo further development. Two divisions occur producing four spores, each containing two sporozoites. This is the infective stage and when taken in with food by another rabbit it continues the cycle.

Haemosporina. Members of this order infect the blood cells and tissues of birds and mammals. Unlike members of the previous two groups, the haemosporidians do not produce resistant spores but are transferred directly by blood-sucking arthropods in which a part of the life cycle takes place. The best known example of this group is *Plasmodium*, the causative agent of malaria.

The malarial parasites (*Plasmodium vivax* is the most common) infect large numbers of warm-blooded vertebrates besides man. In fact, the life cycle was worked out originally in birds by Ronald Ross in 1898. The widespread occurrence of the disease in human populations is indicated by the fact that prior to 1945 over 300 million people were infected all the time. It was estimated that 3 million died of malaria each year—over half of all the deaths in the world. This certainly places it in the number one position among deadly diseases. These figures come as a surprise to most Americans because we now have the disease under control, although a hundred years ago it was responsible for a great many deaths in the South. During World War II it once again became a very important health problem for men in the tropics, and the large number of men continuing to suffer from the disease attests to the fact that we were not altogether successful in our preventive program.

Two factors are necessary for the propagation of malaria, a large population of the appropriate species of *Anopheles* mosquito and infected humans (Fig. 6–33). Only female mosquitoes bite. In order to become infective the female anopheles must bite a person suffering from the disease and withdraw blood that contains gametocytes. There are two kinds of gametocytes, male and female, both of which must be taken into the stomach of the mosquito, where each type of cell undergoes certain modifications. One remains substantially unchanged, producing a single large macrogamete, whereas the other produces 6 or 8 smaller motile, threadlike microgametes. The macrogametes and microgametes unite in pairs to form zygotes which are able to bore through the stomach wall under their own power. In the outer part of the stomach wall each zygote multiplies many times, producing a great many tiny infective sporozoites. These swollen zygotes protrude from the outside walls of the mosquito's gut like tiny beads. They puzzled Ross when he first saw them, and one can imagine his surprise when he squeezed them and saw thousands of spindle-shaped bodies emerge. Normally they burst into the blood which fills the

Fig. 6–31. Life history of *Monocystis lumbricis.*

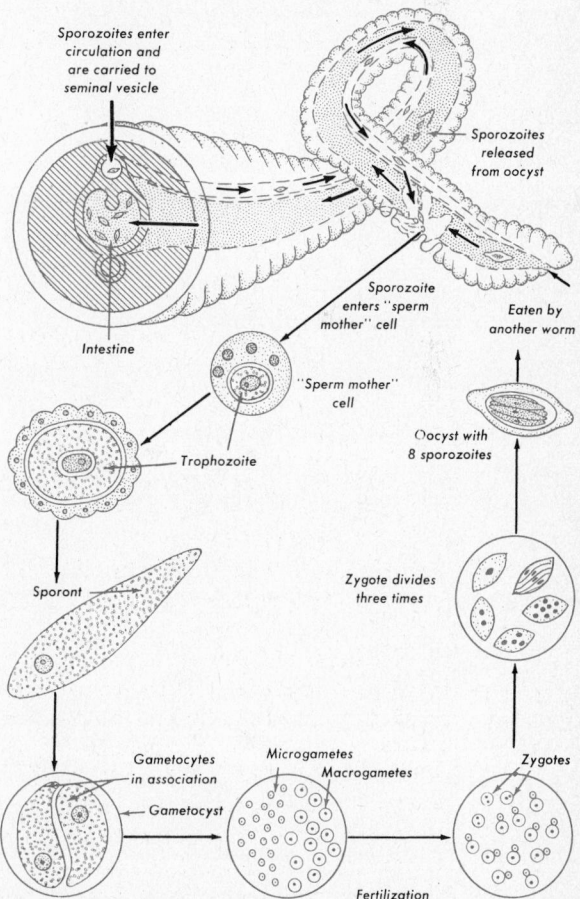

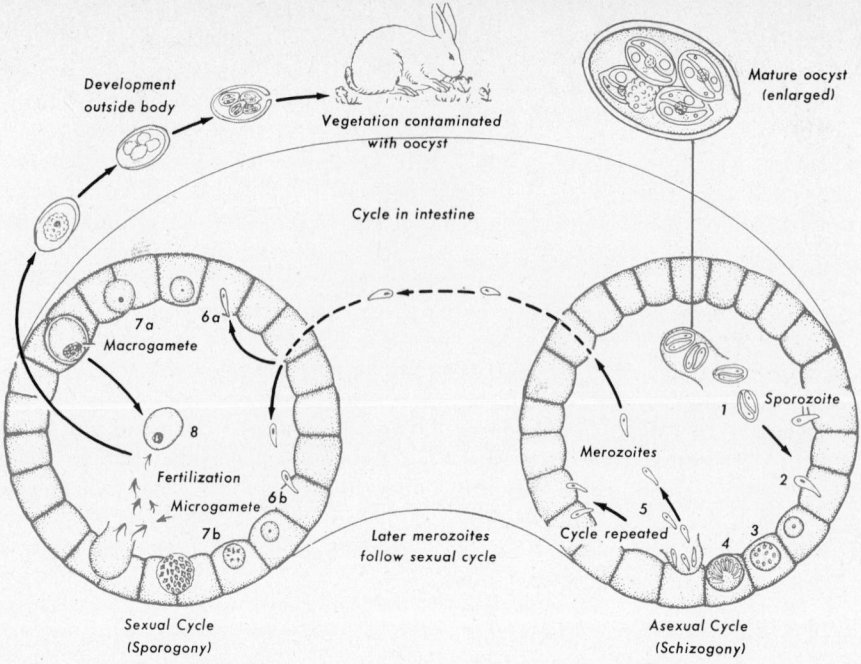

Development
outside body

Vegetation contaminated
with oocyst

Mature oocyst
(enlarged)

Cycle in intestine

7a   6a
Macrogamete

8

Fertilization

Microgamete   6b

7b

Sexual Cycle
(Sporogony)

Merozoites

Later merozoites
follow sexual cycle

Sporozoite

1

2

5

3

Cycle repeated   4

Asexual Cycle
(Schizogony)

Fig. 6–32. Life history of *Elmeria stiedae.*

space between the gut and the body wall, and via the blood, the sporozoites make their way into the salivary glands. With each bite of the mosquito from this time on, sporozoites are injected into the blood of the next host.

After entering man's blood stream the sporozoites seem to disappear for a few days. This fact puzzled biologists for many years until it was discovered that they undergo their early multiplication stages in various tissues of the body, notably certain cells of the liver. At any rate, within ten days some of the parasites are in the blood stream, each entering a red blood cell where it grows and multiplies by schizogony. After a remarkably regular period of time— namely, 48 hours in *P. vivax*, the most common form of malaria—the infected red cells burst, each releasing 10–20 merozoites. These immediately enter other cells, and so the infection keeps increasing in intensity. When there is a sufficient number of infected cells, the person suffers chills before and a fever just following the bursting of the red

cells which occurs at 48-hour intervals. The symptoms become more intense for the next two weeks when either the person is unable to combat the infection and dies, or he is able to and lives, although intermittent chills and fever continue for a long time, sometimes for several years. During this time some of the merozoites become modified into gametocytes. If some of these are taken up by the mosquito with its blood meal, they pass to the stomach and thus complete the cycle.

The treatment for malaria has an interesting history. About 1640 a countess visiting Peru became ill with malaria and when given extracts from the bark of a tree, since named cinchona in her honor, she recovered. She was so impressed with the drug that she brought some back to Europe where it was shown to be extremely effective in the treatment of malaria. From that time to the present, **quinine**, the effective drug in cinchona bark, has been used as a specific treatment for malaria. It has probably saved more lives and relieved more suffering than any other drug ever discovered by man, the recent antibiotics included. During World War II, our supplies of quinine were cut off, so we had to rely on substitutes such as **atabrine** and **plasmochin.** The attempt to synthesize quinine continued during the war years with unrelenting vigor until finally the synthesis was successfully accomplished

Fig. 6–33. Life cycle of the malarial organism, *Plasmodium vivax.*

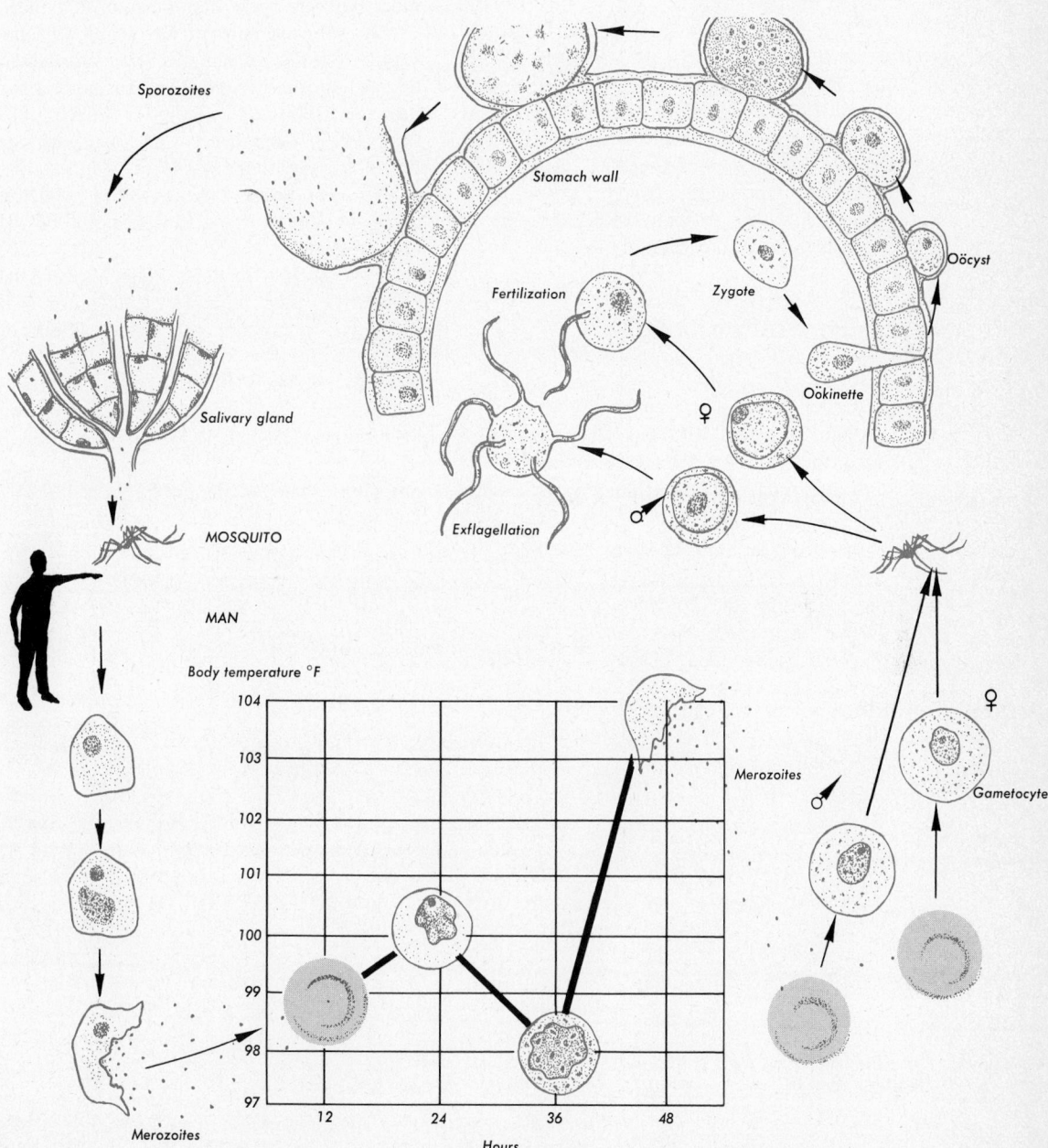

at the end of the war. A constant search is made for a new specific for malaria and some success is reported from time to time. The drugs of choice today are **proguanil** and **pyrimethamine**. Unfortunately, the parasites are becoming resistant to these drugs, so that they are no longer effective. This means that there must be a continuing search for new drugs.

The best and most successful methods have been to destroy the mosquito. This has been so effective that today there are about 100 million cases of malaria as compared to 300 million thirty years ago. World health groups predict that the disease may well be eradicated from Europe, the Americas, North Africa, and a large part of Asia at some future time.

SUBPHYLUM CILIOPHORA

Class Ciliata

The members of this class are distinguished by the possession of **cilia** and two or more nuclei. These characteristics were observed in paramecium, which is a representative of this group. The class has been subdivided according to the arrangement of the cilia into four orders. The more interesting members of three of these are described briefly.

Those belonging to the subclass **Holotricha** possess evenly distributed cilia over most of their body. Paramecium is typical.

The subclass **Spirotricha** includes a large variety of diverse ciliates, one of which, *Spirostomum*, is a veritable giant among the protozoa. This cell reaches a length of 3 millimeters and can easily be seen with the naked eye. Another large form, *Stentor* (Fig. 6–34), vase-shaped and colored a beautiful greenish blue, is a spectacular sight for the microscopist.

This subclass includes several species in which the long cilia are fused into stiff bristle-like organelles called **cirri**. Cells of this type are flattened dorso-ventrally and seem to use their cirri in "walking" along the substratum. One of these is *Stylonichia* (Fig. 6–34), which is about the size of paramecium but whose actions are quite different. It moves along in a jerky fashion,

Fig. 6–34. Various types of freshwater ciliates.

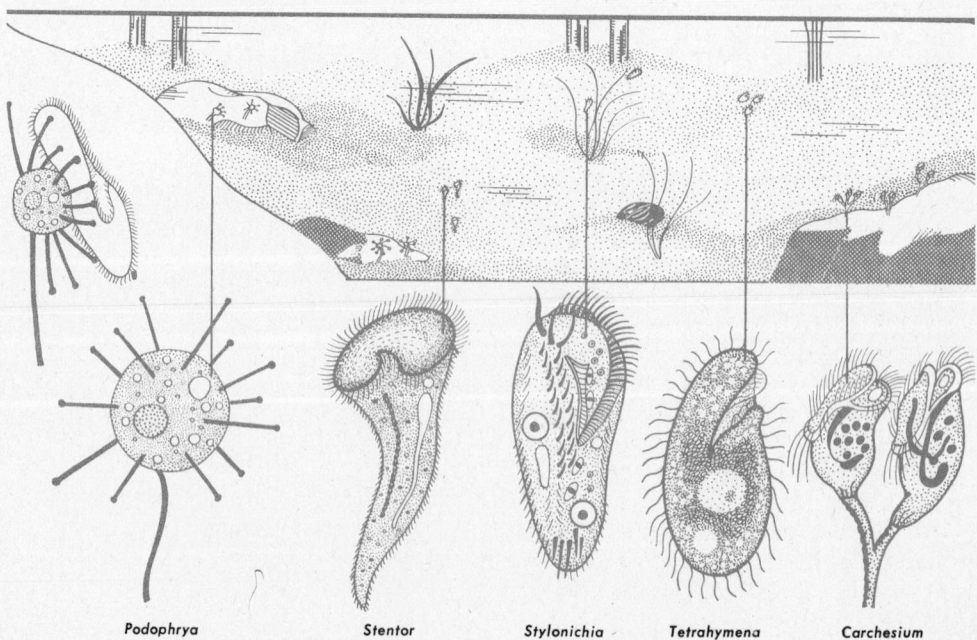

Podophrya   Stentor   Stylonichia   Tetrahymena   Carchesium

This class includes a very peculiar group of stalked protozoa which possess long "tentacles" instead of cilia in the adult stage. *Tokophrya* and *Podophrya* (Fig. 6–34) are examples. The young stages, however, resemble other ciliates in that they possess cilia and are free-swimming. It is because of this evidence that the suctoria are placed with the ciliates. The adults have no mouth but acquire their food by means of the knob-tipped tentacles. When other ciliates which serve as food touch the tentacles they are immobilized and their internal parts drawn through the tentacles into the suctorian, much like fluids sucked through a straw.

Parasitic Ciliates. Very few ciliates have developed the parasitic habit. Only one, *Balantidium coli*, infects man. However, there are a fantastic number of ciliates living in the two parts of the complex stomach of ruminants and in the colon of other herbi-

Fig. 6–35. Anton van Leeuwenhoek (1632–1723) was the first man to see and describe many protozoa as well as other microorganisms. He was not trained in science, but his devotion to the disciplines of the field places him among the foremost scientists of his day.

darting forward and backward, and sometimes crawling along on the bottom.

Ciliates belonging to the subclass **Peritrichia** have their cilia conspicuously arranged in the anterior region. Most of these forms are vase-shaped and many are stalked. Common examples are *Carchesium* (Fig. 6–34) and *Vorticella*. Of the two, the latter is more common and is familiar to anyone who has examined stagnant water under the microscope. It was seen and described for the first time by Anton van Leeuwenhoek (Fig. 6–35) in Holland during the seventeenth century. Vorticella is usually attached to the substratum by means of its contractile stalk. When disturbed or sometimes for no apparent reason, it suddenly contracts and at the same time the cilia around the funnel-shaped mouth disappear and the entire cell rounds up into a ball. Shortly, it emerges again and starts its oral (mouth) cilia beating so that food particles floating by are wafted into its mouth.

Fig. 6–36. This protozoan, *Epidinium ecaudatum*, lives in the digestive tract of ruminants. The external anatomy is shown in the left drawing and the internal structure in the right.

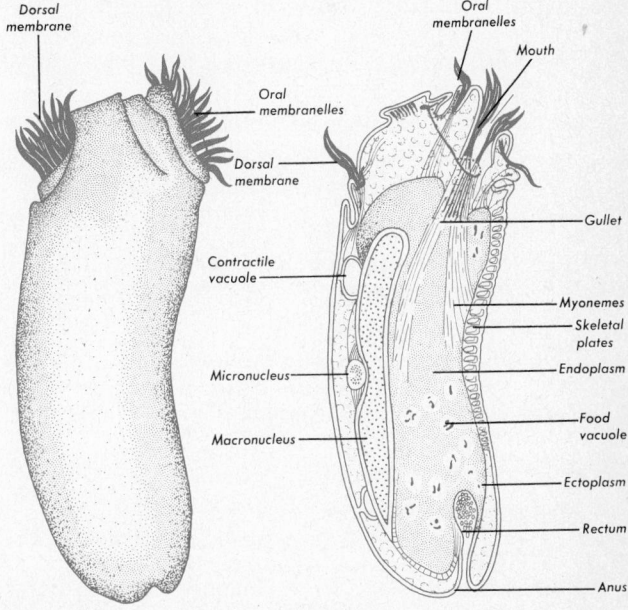

vorous animals. There may be as many as 10 billion in the stomach of a cow.

These ciliates feed on plant fragments and bacteria and when they die they serve as nourishment for the host. Even though they seem to be consistently present in ruminants, the protozoa are not essential for the health of the host because they can be experimentally removed (copper sulfate) without detectable injury to the cow. The ciliates are highly variable in structure and size; one, *Epidinium ecaudatum* (Fig. 6–36), will illustrate their general anatomy. It is one of the most complex of ciliates, even exceeding paramecium in complexity. It has a mouth surrounded by membranelles which are attached to a collar that can be protruded or retracted. This is accomplished by the action of **myonemes** which function like muscles. It has a terminal opening, the **anus,** through which waste products are removed from the body. In addition, the most complete coordinating system of any protozoan, consisting of a **nerve center** with connecting fibers, is found in *Epidinium.* This protozoan maintains its form by means of **skeletal plates,** and most of its body is devoid of cilia.

The *Epidinium* illustrates how far differentiation can go within a single cell. It maintains organelles for locomotion, feeding, swallowing, digestion of food, egestion of wastes, excretion, and continuity of its germ plasm by means of its nuclear structures. This exhausts all the accomplishments metazoan cells have achieved.

SUGGESTED SUPPLEMENTARY READINGS

Books

CORLISS, J. O., *The Ciliated Protozoa.* New York: Pergamon Press, 1961.

DOBELL, C., *Antony van Leeuwenhoek and his "Little Animals."* London: John Bale Medical Pub., 1932.

ELLIOTT, A. M., ed., *The Biology of Tetrahymena.* Stroudsburgh, Pennsylvania: Dowden, Hutchinson, and Ross, 1973.

GILMORE, C. P., *The Scanning Electron Microscope World of the Infinitely Small.* Greenwich, Conn.: New York Graphic Society, 1972.

HALL, R. P., *Protozoology.* Englewood Cliffs, N.J.: Prentice-Hall, 1953.

*HALL, R. P., *Protozoa—The Simplest of All Animals.* New York: Rinehart & Winston, 1964.

HYMAN, L., *The Invertebrates, Vol. I, Protozoa through Ctenophora.* New York: McGraw-Hill, 1940.

*JAHN, T. L., and JAHN, F. F., *How to Know the Protozoa.* Dubuque, Iowa: Wm. C. Brown Co., 1949.

KUDO, R. R., *Protozoology.* Springfield, Ill.: Charles C. Thomas, 1966.

MACKINNON, D. L., and HAWES, R. S. I., *An Introduction to the Study of Protozoa.* Oxford, England: The Clarendon Press, 1961.

MANWELL, R. D., *Introduction to Protozology.* New York: St. Martin's Press, 1961.

PITELKA, D. R., *Electron-Microscopic Structure of Protozoa.* New York: Pergamon Press, 1963.

TARTAR, V., *The Biology of Stentor.* New York: Pergamon Press, 1961.

WARNER, F., "Macromolecular Organization of Eukaryotic Cilia and Flagella." In E. J. DuPraw, ed., *Advances in Cell and Molecular Biology,* Vol. 2. New York: Academic Pess, 1972.

WICHTERMAN, R., *The Biology of Paramecium.* New York: Blakiston, 1953.

WOLKEN, J. J., *Euglena.* New York: Appleton-Century-Crofts, 1967.

* Available in paperback.

ALLEN, R. D., "Ameoboid Movement." *Scientific American*, February, 1962.

ALVARADO, C. A., and BRUCE-SCHWATT, L. J., "Malaria." *Scientific American*, May, 1962.

MARGULIS, LYNN, "Symbiosis & Evolution." Scientific American, August, 1971.

SATIR, P., "How Cilia Move." *Scientific American*, October, 1974.

7

SPONGES AND THE RADIAL ANIMALS-COELENTERATES

The origin of the metazoa having been discussed in Chapter 4, we describe, in this chapter, the most primitive metazoan forms, which are predominantly simple sacs consisting of two well-defined body layers, **ectoderm** and **endoderm**, with the beginnings of a third layer, the **mesoderm.** Even though simple, these animals are very successful, abounding in the waters of the world (Fig. 7–1).

PHYLUM PORIFERA

This unique group of animals is not in the direct line of ascent to higher forms, and therefore are thought of as "blind alley" animals—that is, they apparently reached their present stage of development millions of years ago and there is no evidence that they gave rise to higher forms. The word **porifera** means "pore bearer," which is a characteristic of the group. The flagellated collared cells, **choanocytes,** that line their cavities resemble one group of flagellated protozoa (Chapter 6), indicating their close affiliation with these one-celled animals. Digestion is wholly intracellular and, except for simple epithelia, there are no tissues. For this reason sponges are considered to have gone no farther than the **cellular grade of organization.** Unlike the protozoa where the single cell carries on all activities, the sponge demonstrates division of labor where the cells are specialized to perform different functions; some support the body while others gather food and still others function in reproduction.

Living sponges have a gelatinous texture, quite different from the familiar bath sponge sold on the market. Being sessile, they are easily mistaken for plants. Sponges assume many shapes, vary in height from 1 millimeter to more than a meter, and are usually drab in color. Some forms, however, take on shades of red, yellow, blue, black, or green, the last being caused by the green alga, *Chlorella*, living in the body cells. Their most common habitat is a number of feet below the tidal zone of the sea (upper and lower limits of the tide) (Fig. 7–1), although some, like the glass sponge (Fig. 7–2), live as much as three and a half miles below the surface of the ocean. Freshwater forms are also known, some of these attaching themselves to twigs and rocks in the streams.

Morphology of a Simple Sponge

In order to have some knowledge of the morphology of the sponge, it is best to discuss a simple sponge, *Leucosolenia*, as an example. It lives beneath the low tide level in the sea and consists of many slender upright tubes which are joined at their bases in a many-branched common tube (Fig.

7–3). The upright portions are thin-walled sacs perforated with hundreds of microscopic holes (**incurrent pores**) and one large opening, the **osculum (excurrent canal)**, at the upper tip. The cavity of the sac is called the **spongocoel**. Its wall is made up of an outer **epidermis** or skinlike layer of flat cells, and an inner continuous layer of the flagellated collared cells (**choanocytes**). A third jellylike layer, the **mesenchyme**, lies between these two, and in this layer several kinds of amoebalike cells, called **amoebocytes**, are present. This layer also contains the skeleton formed by **spicules**, which resemble crystals. In the glass sponge the spicules consist of silicious material and in horny sponges, of fibers and spongin. In most forms, however, they are composed of calcium carbonate. A combination of spicules and fibers also occurs.

Scattered among the ordinary epidermal cells are the tubular **pore cells**, each with a central canal or pore. A pore cell, together with "helpers" surrounding it, controls the flow of water into the sponge. The vigorous beating of flagella on the collared cells lining the spongocoel causes water to move in through the pores and out through the osculum. This movement causes a constant stream of water, heavily laden with microscopic organisms, to pass within reach of the choanocytes. The manner of beating of the

Fig. 7–1. A few sponges and coelenterates that live in the oceans.

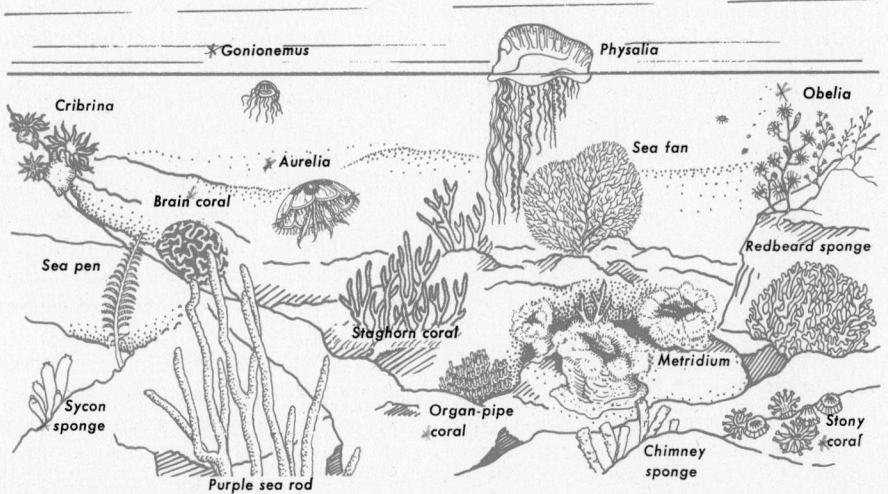

flagella in the collared cells propels tiny food particles to the cell body which engulfs it, forming food vacuoles much like amoeba. Any food not needed by the choanocytes is passed to the amoebocytes for further distribution. Waste products are borne out through the osculum in this same current of water.

Sponges in General

Sponges reproduce asexually by means of **internal** and **external buds,** as well as sexually by means of eggs and sperm. Buds form on the outside of the sponge and sometimes move away, but as often remain a part of the parent sponge. During unfavorable conditions, as in drought or cold winters, sponges develop internal buds, called **gemmules,** which are merely masses of cells with a hard outer covering. They drop to the bottom of the stream or sea during these adverse conditions and grow into sponges the next season when circumstances are again favorable.

Some sponges are **monoecious** (of one household), that is, both sexes are present in one animal; others are **dioecious** (of two households), the sexes being found in separate animals. There are no special sex organs, and the sperm and eggs simply develop from certain of the amoebocytes in the middle layer or mesenchyme. Fertilization occurs *in situ* (in place) and is followed by rapid division until a **blastula** is formed with many flagellated cells (future choanocytes). The flagella are directed inward toward the central cavity, the **blastocoel.** A hole appears later and the embryo turns inside out, bringing the internal flagellated cells to the outer surface. In this stage, called the **amphiblastula,** it leaves the mother sponge through the osculum and swims around for a few days, finally settling and attaching itself to a rock or some other solid object where it grows into a sponge (Fig. 7–3). The larger cells of the larva overgrow the flagellated cells and completely surround them. The cell layers of the sponge seem to be the reverse of those found in higher forms and for that reason are not comparable to them.

Sponges have canal systems which vary in complexity. The simplest, the **Ascon** type found in *Leucosolenia,* is merely a thin-walled sac. The first step in increasing complexity occurs in the **Sycon** type, which has two kinds of canals, **incurrent** and **radial,** the latter being lined with choanocytes. The most complicated of all is the **Leucon** type, which possesses a vast array of multiple tubes and chambers, with choanocytes lining only certain restricted chambers. Leucon sponges may reach more than a meter in diameter. Since the sponges are not on the direct line of ascent to higher forms, we leave them in

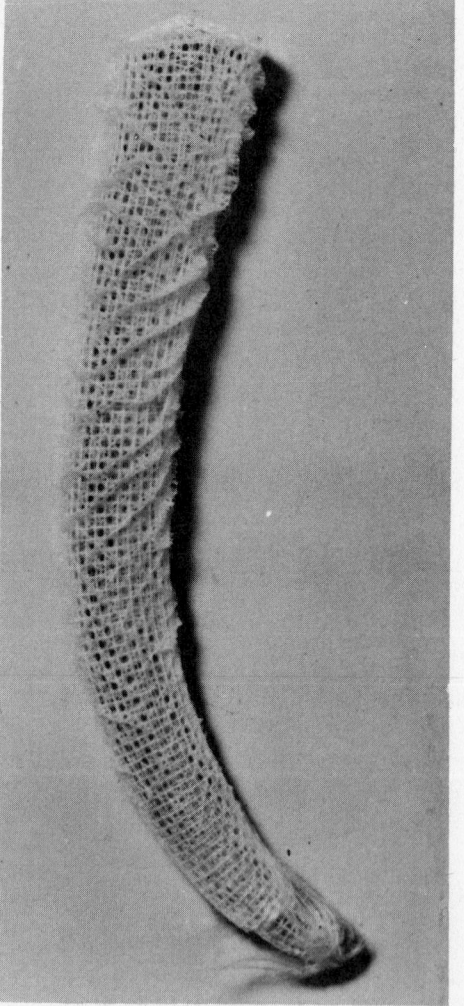

Fig. 7–2. The skeleton of the glass sponge, *Euplectella.* Note the fine bridging of the fibers to form the skeleton.

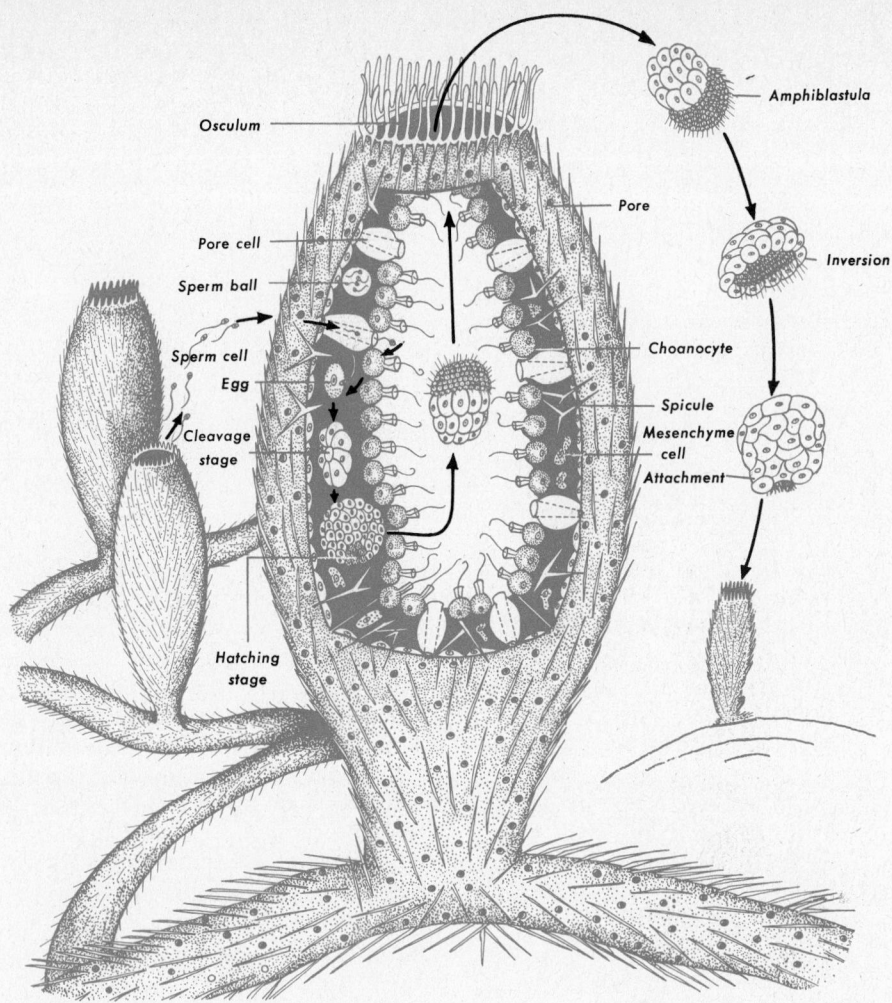

Fig. 7–3. Anatomy and life history of a simple sponge.

their isolated position without further reference and pass on to the next group which has much more significance in our story of the rise of animal life.

PHYLUM COELENTERATA

These sac-like, radially symmetrical, diploblastic animals are the first metazoans to have their somatic cells organized into definite **tissues**, hence have reached the **tissue grade of organization**. Actually they have about the same number of tissues as higher

animals and they function in much the same way. Special epithelial cells, called **cnidoblasts**, produce the **nematocysts** used for offense and defense. In many coelenterates there are two types of individuals, the **polyp** representing the asexual phase of the life cycle, and the **medusa**, the sexual phase. In their life histories each generation successively gives rise to the alternate type, a phenomenon called **metagenesis**. The asexual polyp is tubular in shape, with a mouth at one end, surrounded by tentacles richly supplied with nematocysts. The other end of the tube is closed and forms an attachment organ, the foot. These animals are usually sessile or nearly so. The free-swimming sexual medusa is a delicate transparent animal, shaped like an umbrella. Around the periphery or edge of the umbrella are tentacles, which are also heavily fortified with nematocysts. Both the polyp and the medusa have primitive muscle fibers which make movement possible.

The coelenterates are water-inhabiting animals, most of them marine. One form, **hydra**, has invaded fresh water and is very successful. The large marine **jellyfishes** and the **sea anemones** are members of this group (Fig. 7–1). In both the sea anemones and their close relatives, the corals, the medusae have been completely lost.

With the exception of some jellyfish, which may annoy bathers by the stings of

Fig. 7–4. Hydra shown in detail, A portion of the body wall is magnified here in order to distinguish its cellular structure. Note the variety of cells in both the epidermis and gastrodermis.

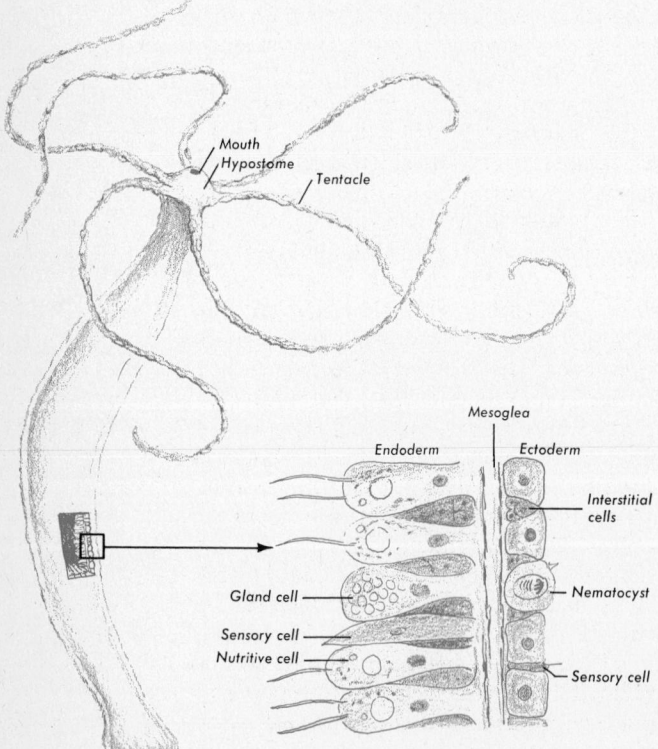

their nematocysts, and the coral animals, which are used for jewelry, the group as a whole has little direct significance to man.

There are three classes of coelenterates (**Hydrozoa**, **Scyphomedusae**, **Anthozoa**) whose characteristics can best be understood by studying two representatives, *Hydra* and *Obelia*, of the class Hydrozoa. Members of the other two classes are discussed briefly later in the chapter.

Hydra

This tiny animal (Fig. 7–4) may be found clinging to underwater vegetation in nearly any freshwater pond, lake, or stream. It moves about very little, hence seeks its food by means of its long tentacles which may reach a length of 10–12 inches. The entire body extends and contracts remarkably so that it may be very long and thin or so short that it resembles a ball with the contracted tentacles as tiny stumps. The hollow tentacles surround a raised conical **hypostome**, in the center of which is the **mouth**. The mouth opens into the digestive cavity, the **coelenteron** or **gastrovascular cavity**, which continues into the tentacles. The outer layer, the **ectoderm**, sometimes called the **epidermis**, is made up of cells which form a protective covering. Some of these cells have an inner contractile portion which enables them to serve as muscle cells. Finally, scattered among the epidermal cells are specialized sensory cells. The inner layer of cells, the **endoderm** or **gastrodermis**, provides all other cells with nourishment. These tall, glandular cells secrete the **enzymes** which digest food that is brought into the coelenteron (Fig. 7–4). These cells also possess muscle fibers at their bases which run at right angles to those in the epidermis. It is by the combined action of these muscle cells with those of the ectoderm that hydra is able to contract and extend itself. Lying between these two layers, the ectoderm and the endoderm, is a thin, noncellular layer, the **mesogloea**, which lends support and holds all the cells together. Lying embedded in both the ectoderm and endoderm are the nerve cells, which are con-

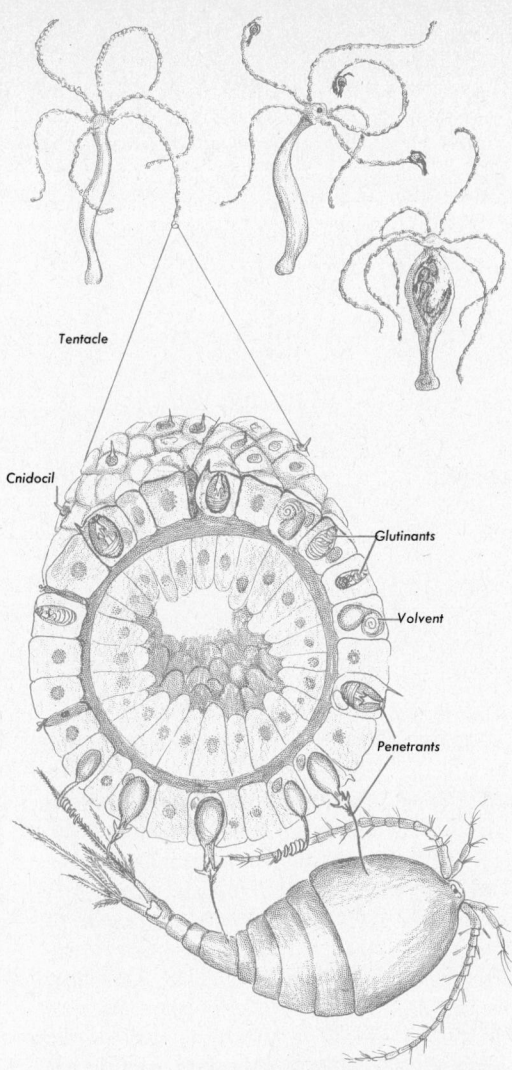

Fig. 7–5. One tentacle of this feeding hydra is enlarged to show both discharged and undischarged nematocysts in detail. The largest type (penetrant) discharges its thread with such force as to penetrate the body of small animals such as the crustacean shown here. The volvent type coils about the appendages of the prey. The glutinants (two types) secrete a sticky substance used in locomotion as well as in feeding.

there is no centrally located mass which could in any way be compared to the brain of higher forms. External **stimuli** initiate impulses in the **sensory cells** which are conveyed to the nerve net and through it to the **contractile fibrils.** This is the simplest form of a metazoan nervous system, but it contains the fundamental elements of which all higher nervous systems are built.

An interesting and intricate part of hydra's response mechanism is the **cnido-blast,** with its contained **nematocyst.** Cnidoblasts are located in nests or "batteries" along the tentacles, for the most part, although they may occur over the entire body with the exception of the foot, or basal disc, on which the hydra "walks." Nematocysts are derived from the interstitial cells and are usually arranged with one large and several small ones in each battery. There are several different kinds of nematocysts, each having a different use. The largest and most conspicuous type is the penetrating or stinging nematocyst, which upon discharge pierces the body of small crustacea or other aquatic animals that happen to touch the tentacles (Fig. 7–5). This nematocyst contains a hollow, coiled thread which everts through the trigger-like action of the **cnidocil,** a slight projecting bristle, when it is touched or stimulated in some other way. It is ejected with such force that it penetrates the soft and even some hard parts of the prey, injecting a small amount of poison which has a paralyzing effect on the victim. Once paralysis sets in, the tentacles move in a manner that draws the prey into the mouth and thence into the coelenteron. Other types of nematocysts function in a mechanical rather than chemical manner. When discharged, some wrap their threads about a portion of the attacked animal and hold it securely. Others fasten themselves to a portion of the substratum and by contractions of the tentacles make possible a slow, somersaulting type of locomotion (Fig. 7–6).

The presence of food in the coelenteron stimulates the gland cells to secrete digestive enzymes into the cavity where the soft parts of the ingested animal are partially digested

nected by minute fibrils to the sensory and muscular cells. They function in coordinating the activity of all the cells. The nerve cells form a network, called the **nerve net,** which connects all parts of the animal, although

Fig. 7–6. Hydra somersaulting.

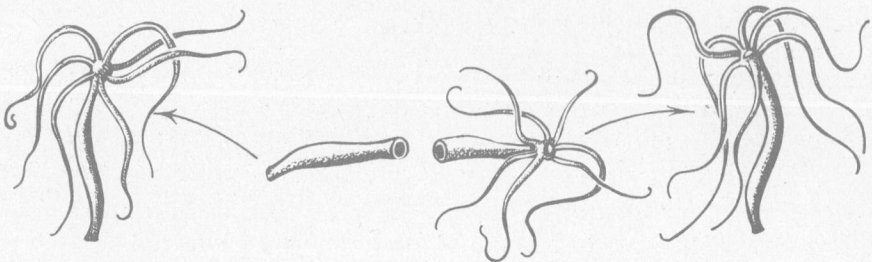

Fig. 7–7. Regeneration in hydra.

(extracellular digestion). The hard outer coverings are indigestible and are eventually regurgitated through the mouth. Thus a single opening functions both as a mouth for the entrance of food and as an anus for the exit of undigested food.

In a suitable environment, hydra reproduces asexually by forming one or more buds. Since buds form so readily from almost any part of its body wall, it should follow that hydra could perhaps be made to reproduce itself experimentally by simply cutting off small pieces of the body (Fig. 7–7). This idea apparently occurred to a Swiss naturalist, Abraham Trembley, around the middle of the eighteenth century (1744). He did just that and gave us our first experiments on **regeneration** in animals.

Hydra also reproduces sexually by the production of eggs and sperm (Fig. 7–8). Usually both ovaries and testes are formed on the same animal, although as a rule not at the same time. In formation of the sperm, interstitial cells (undifferentiated cells) lying in the ectoderm undergo a series of divisions, forming a protuberance which gradually grows large as the sperm matures. A sexually mature hydra may have several "ripe" testes along its walls. An opening appears in the end of the gland and the mature sperm cells swim out into the surrounding water in search of a mature egg on another individual. The eggs develop from interstitial cells also, the difference being that only one egg is formed in an ovary, whereas thousands of sperm cells are produced in a single testis. The eggs at first resemble large amoeboid bodies. As they mature they become spherical in shape, resting on the outer wall of the hydra attached to a cuplike depression in the epidermis.

After fertilization, cleavage starts and continues until a hollow **blastula** forms. An outer resistant shell then develops and simultaneously the cavity (**blastocoel**) is filled with cells from the lining. The young embryo then drops off and lies quietly until favorable conditions eventually arise, when it emerges as a very small hydra with blunt tentacles. Sexual reproduction usually occurs in the fall of the year and seems to be a safeguard for passing the winter months because the young embryo resides in a capsule which resists adverse temperatures.

It has been claimed by Loomis that the differentiation (Fig. 7–8) of sexual organs is controlled by the pressure of CO_2 in the surrounding water. He showed that as hydra became crowded in a small amount of water the sex organs appeared, whereas if a single hydra was placed in a large volume of water it never developed ovaries or testes. Water normally contains about 0.03 percent CO_2 and the sex organs appeared when it reached 1.2 percent. Moreover, he was able to completely suppress the development of buds and tentacles by increasing the CO_2 pressure to 10 percent. Obviously the level of CO_2 in the vicinity of cells has a pronounced effect on differentiation in hydra and probably does in other animals. Such information gained from the study of simple animals often has far-reaching implications which direct studies on higher animals, including man.

Hydra possesses no medusa stage and hence does not exhibit metagenesis, which is common among most Hydrozoa. The next representative is a more typical example of this class and is discussed primarily for that reason.

Obelia

This tiny colonial coelenterate may be found attached to seaweed and other objects lying in the clear water of tide pools and below low tide level to a depth of several

Fig. 7–8. Life cycle of hydra. One hydra is cut transversely through a bud and testis.

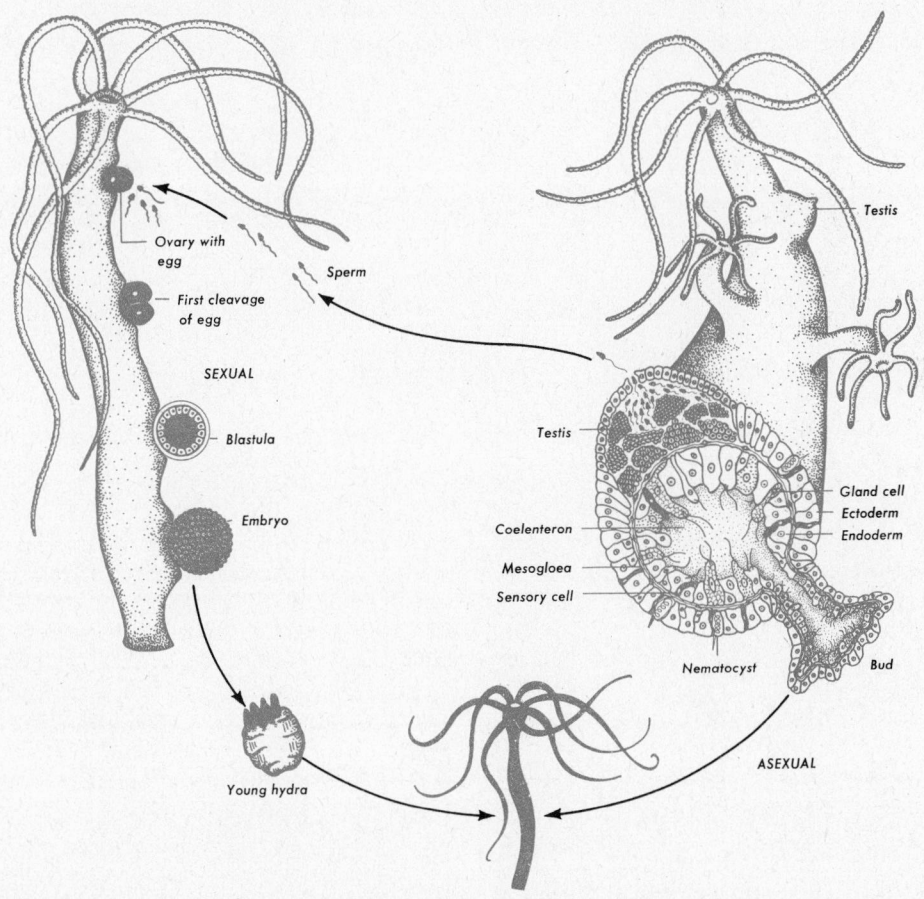

Ovary with egg

Sperm

First cleavage of egg

SEXUAL

Blastula

Embryo

Young hydra

Testis

Testis

Gland cell
Ectoderm
Endoderm

Coelenteron

Mesogloea

Sensory cell

Nematocyst

Bud

ASEXUAL

fathoms. It is attached to the substratum by means of a horizontal branching basal portion.

An obelia colony begins as a **single polyp** which by budding and the subsequent clinging together of the buds forms a colony (Fig. 7–9). The process is reminiscent of the bud-upon-a-bud condition observed in hydra. The tissues and gastrovascular cavity are thus continuous throughout the colony. There are two types of polyps in an obelia colony, the feeding polyp, or **hydranth**, and the less common reproductive polyp, or **gonangium**. The hydranth is not greatly different from hydra except that it possesses solid instead of hollow tentacles and it is surrounded by a tough, horny outer covering called the **perisarc**, which invests the entire colony. The

Fig. 7–9. Life cycle of *Obelia*.

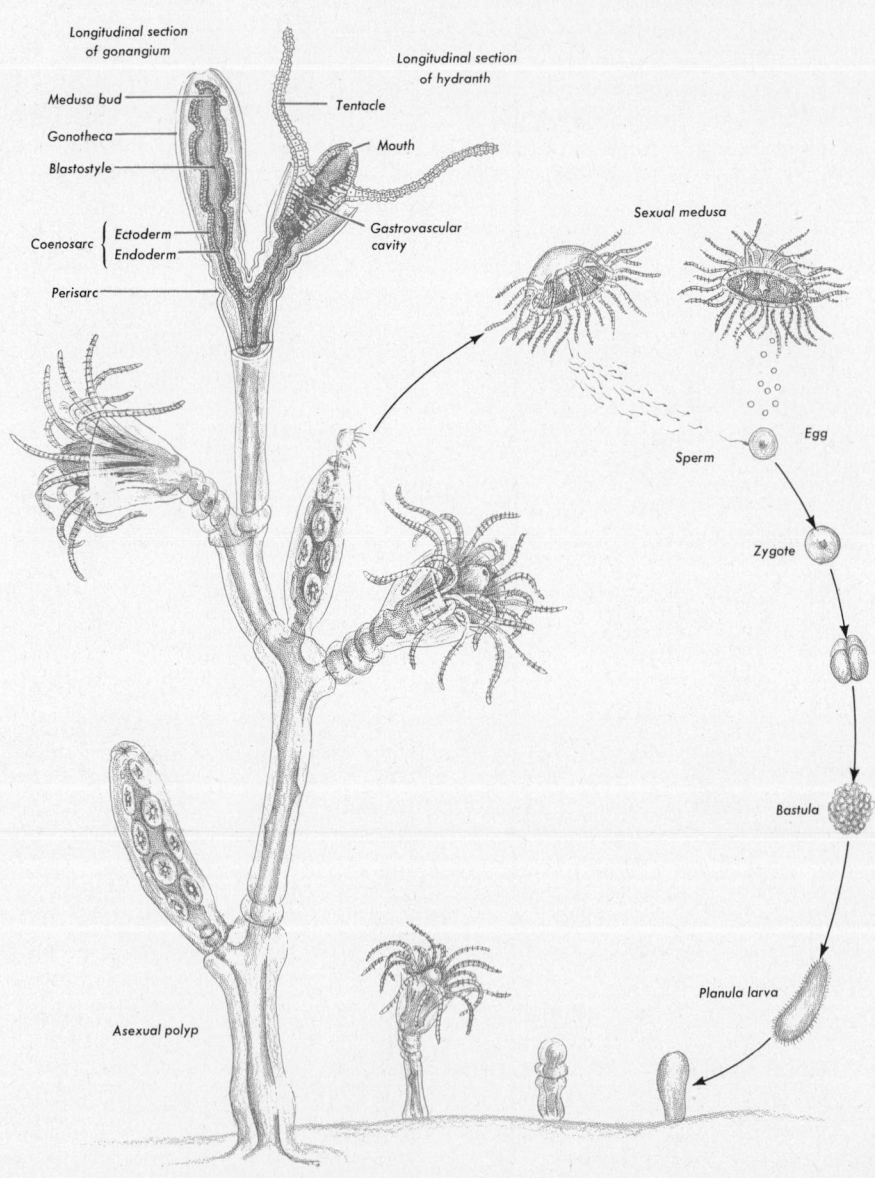

transparent vaselike portion of the perisarc surrounding a hydranth is called the **hydrotheca.** The cellular portion just beneath the perisarc is known as the **coenosarc.** After food has been captured and partly digested by an individual hydranth, it is carried through the gastrovascular cavity by the beating flagella which line the cavity. Thus all polyps share in the good fortune of any one. Digestion is finally completed intracellularly in the lining cells.

Obelia reproduces asexually by forming buds either on the horizontal parent stalk or on the upright stalk. Asexual reproduction also occurs in a second type of individual, the **gonangia** (singular, **gonangium**), which have no tentacles and no mouth. This cylindrical polyp is covered by the transparent **gonotheca,** a continuation of the perisarc. Each gonangium contains a central stalk, the **blastostyle,** upon which are borne small **medusa buds.** As the buds mature they approach the distal end of the blastostyle, finally leaving the gonotheca through a terminal opening and swimming away. These free-living medusae are the familiar jellyfish, which, since they are able to move about, spread the species to new areas.

Medusae develop gonads which produce eggs and sperm that unite in the sea water. The resulting embryo, the **planula,** swims about for a time but eventually settles to the bottom, becomes attached to the substratum, and grows into an asexual colony, the polyp, thus completing the cycle. This species is an excellent illustration of **metagenesis.**

The medusa of obelia appears to be quite different from the polyp, but basically they resemble each other closely. The former has a central hanging **manubrium,** located on the concave side. At the center of the manubrium is the mouth, which opens into four **radial canals,** continuing into the circular **ring canal** in the margin of the bell. This constitutes the coelenteron and is equivalent to the same organ in hydra or the polyp of obelia. The space between the epidermis and the coelenteron is filled with the rather extensive gelatinous mesogloea. The medusa of obelia is microscopic and these structures can best be seen in larger jellyfish.

Obelia illustrates the beginning of division of labor among the polyps, but this is carried to a greater degree of efficiency among some of the other Hydrozoa. **Polymorphism** (many shapes), the name applied to this type of division of labor, is striking in *Physalia,* the **Portuguese man-of-war** (Pl. 1), which comprises at least four types of individuals. Some gather food, some protect the colony, while others reproduce. Physalia also has a type that forms a gas-filled float which supports the colony on the surface of the sea as it is borne by the wind and currents in a never-ending search for food. The tentacles bear large nematocysts which are occasionally a menace to bathers if they happen to become entangled in them. This can easily happen in some species of Physalia, since the tentacles trail as much as 50 feet beneath and behind the float. Physalia has little difficulty in paralyzing a fish several inches in length and eventually consuming it. The many types of individuals in this colony illustrate polymorphism in its most advanced form among the coelenterates.

Other Coelenterates

The second class of coelenterates, the **Scyphomedusae,** includes most of the larger jellyfishes (Pl. 2B) which either have a reduced polyp stage or lack it altogether. *Aurelia* (Fig. 7–10), one of the most common examples, is found in great numbers up and down the coasts of the United States. Its rather flattened umbrella is fringed with small tentacles that are interrupted at eight equally spaced spots where a **sense organ** is located. The rhythmic contractions of the circular muscle in the **bell** are responsible for the graceful movement of this beautiful creature. Four long **oral lobes,** which are located on the short manubrium, are heavily armed with nematocysts and function like tentacles in capturing food and directing it into the mouth. The coelenteron is divided into four large **gastric pouches** from which radiate smaller **canals** that connect with the **ring canal** at the periphery of the bell. The gonads, lying in the gastric pouches, form four horsehoe-shaped bodies when viewed

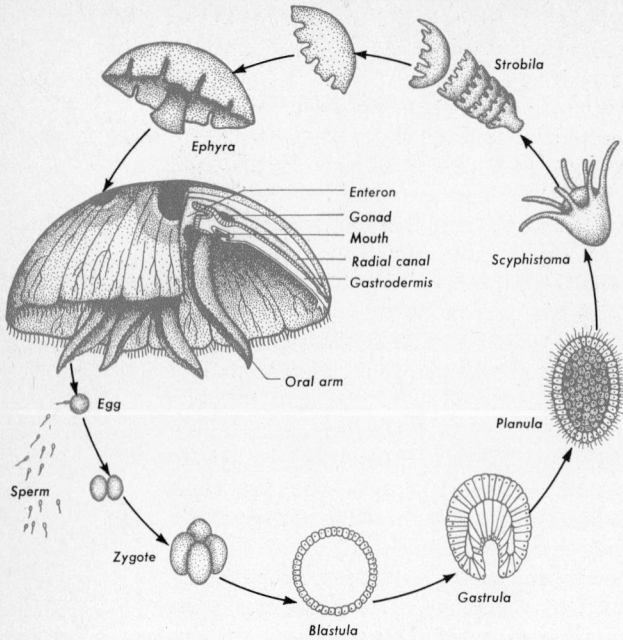

Fig. 7–10. Life cycle of *Aurelia*.

lenteron of the female where the eggs are fertilized. The resulting zygotes move to the oral lobes where they develop into small ciliated larvae, the **planulae**. These soon swim free of the parent and settle to the bottom where they become attached. They then develop into polyps (**scyphistoma**) which are able to feed. These grow and may produce other polyps from lateral buds. However, in the fall the polyps transform into **strobila** which consist of a series of transverse sections that resemble miniature stacked plates. This process is called **strobilization**. These flat eight-lobed medusae (**ephyrae**) separate, invert, and swim free in the sea and eventually grow into the adult jellyfishes.

Representatives of the next class, **Anthozoa**, are characterized by a heavy body, supported by numerous **septa**, transverse sheets of tough tissue. Members of this group possess no medusa stage, eggs and sperm being produced within the body and discharged into the surrounding sea. The most common representatives are the **sea anemones** and **corals**; others, less common, are the **horny, black,** and **soft coral**, the **sea pens**, and **sea pansies**. They compare favorably with the jellyfish in respect to beauty and number. Both the Atlantic and Pacific Coasts, extending from Alaska and Maine to Southern California and Florida, abound with anthozoans (Pl. 2A). Some are found in the polar regions, some at great depths, but they are most numerous in shallow waters in the warmer seas.

The sea anemone usually remains in one place for a long period of time; some have been observed to live for years in a small depression in rock just below low tide level. It can move slowly on its pedal disk and some of the smaller ones are able to swim by

from the aboral (opposite the mouth) side, and constitute a ready mark of identification.

The sense organs, consisting of **eyespots** (sensitive to light) and **statocysts** (sensitive to gravity), are located at the edge of the bell where the nerve net is somewhat centralized. The statocysts are hollow spheres containing small calcareous granules which, as they tumble about in the cyst, stimulate nerve endings and indicate to the animal its relation to the rest of the world. In other words, they function as an organ of equilibrium, much the same as the semicircular canals in our ears. This is the first appearance of such an organ among the metazoa.

The sexes are separate, but resemble one another closely. Sperm, released by the male, pass into the mouth and then into the coe-

Fig. 7–11. Various types of coral reefs.

beating their tentacles. It feeds on any un-suspecting crustacean, mollusk, or even fish that comes within reach of its tentacles. Once the prey is paralyzed by the nemato-cysts it is taken into the coelenteron, and digestion goes on much the same as in other coelenterates. The sea anemone, in spite of its tough outer covering, is preyed upon by a variety of animals such as fish, starfish, and crustacea. When in danger it can retract its tentacles, fold them inside the body, and contract its entire body until it is nearly flat against the substratum. In this condition it is very difficult to remove; in fact, the body is often torn apart before its grasp is released. Although it usually reproduces sexually, occasionally an anemone is found undergo-ing fission, either longitudinally or trans-versely.

Other interesting anthozoans, resembling the sea anemones in many respects, are the corals (Pl. 3A, B). These are usually very small and live in stony cups, of specific design, made by the limy secretions from the base. The colonies lie in close proximity

and after thousands of generations produce the massive corals, bits of which are fre-quently seen reposing as mantelpiece orna-ments in many homes. Corals live in warm waters for the most part, although there is one species, *Astrangia*, living as far north as Maine. They abound in many tropical seas of the world, particularly in the Coral Sea.

Perhaps the most interesting thing about coral is its ability to form reefs, some of which reach many miles in length, like the Great Barrier Reef of Australia, for example, which extends over 1,200 miles in length and has an area of over 80,000 square miles. There are three kinds of coral reefs, depend-ing on how they were formed (Fig. 7–11). The **fringing reef** lies along the shoreline of an island or mainland and usually extends for a quarter of a mile into the sea. Boats approaching such shores are in great danger, particularly in rough weather. Sometimes, as the result of a shifting shoreline, a lagoon appears between the main reef and the shore; this type is called a **barrier reef** (Fig. 7–12). The **atoll** is perhaps the most unique of the

Fig. 7–12. An aerial view of a typical coral island — Green Island, located in the Great Barrier Reef. Note the vast expanse of the reef compared to the land that makes up the island itself.

three kinds of reefs. It is a rim of coral taking on varying shapes, usually a completely enclosed circle. These have always fascinated biologists. One of the greatest among them was Charles Darwin who gave an explanation of how they formed, a theory which is still considered fundamentally sound. He thought the peculiar formations started out as a fringing reef around an island, but due to shifts in the earth's surface the island gradually began to sink. The rate at which it submerged was about as fast as the corals were able to secrete lime and keep themselves in the tidal zone. By keeping pace with the sinking island, the corals built the fringing reefs sufficiently high to catch vegetation and support growth of plants while the central portion gradually became submerged below the water's surface. This then produced a rim of coral, inscribing the outline of the old island and producing the strangely shaped atolls seen in tropical seas today.

PHYLUM CTENOPHORA

The coelenterates were a widely diverse group of animals that explored many possibilities of form, structure, and habitat, remaining all the while within the limitations of two body layers, ectoderm and endoderm, and a simple third layer, or mesogloea. Animals could have gone no further in complexity had they remained within these limitations. The ctenophora surmounted this difficulty by elaboration of the third body layer already present in the coelenterates, though primitive, which resulted in a modification in the entire body plan of the group. Commonly known as comb jellies or sea walnuts, these animals are found floating near the surface of the sea. They resemble the coelenterates in having radial symmetry, an epidermis, and a gastrodermis; but in addition they have a true **mesoderm** which gives rise to definite muscular elements. They are also characterized by the possession of eight rows of swimming **combs** or **plates** which are formed from fused cilia. The ctenophora occur in great abundance in the oceans and some are luminescent, adding beauty to the ocean at night by the rhythmic beating of their combs which send off a feeble blue-green light. This light is sufficiently intense to expose a photographic plate (Fig. 3–30).

SUGGESTED SUPPLEMENTARY READINGS

Books

GILLETT, K., and McNEILL, F., *The Great Barrier Reef and Adjacent Isles.* Sydney, Australia: The Coral Press Pty. Ltd., 1959.

LENHOFF, H. H., and LOOMIS, W. F., eds., *The Biology of Hydra and Some Other Coelenterates.* Coral Gables, Florida: University of Miami Press, 1961.

*RAMSAY, J. A., *Physiological Approach to the Lower Animals.* Cambridge, England: Cambridge University Press, 1962.

WIENS, H. J., *Atoll Environment and Ecology.* New Haven, Conn.: Yale University Press, 1962.

YONGE, C. A., *The Seashore.* London: Wm. Collins & Sons, 1949.

* Available in paperback.

Articles

LANE, C. E., "Man-of-War, The Deadly Fisher." *National Geographic,* March, 1963.

LOOMIS, W. F., "The Sex Gas of Hydra." *Scientific American,* April, 1959.

PRIMITIVE BILATERAL ANIMALS- FLATWORMS

Quite different from the symmetrical beauty of the jellyfishes and sea anemones are the representatives of this group of animals, the flatworms. Their flattened, elongated, and bilaterally symmetrical bodies account for the name of the phylum, **Platyhelminthes** (from the Greek, **platy**—flat). The primitive mesoderm seen in the ctenophores assumed much more importance in this phylum, giving rise to rather elaborate organs; hence we consider these animals as having reached the **organ-system grade of organization** which we see throughout all subsequent higher animal groups. It is also noteworthy that they possess no body cavity,

hence are called **acoelomate** animals. The mesoderm has given rise to a good muscular system making possible other modifications. The worm can move on a flat surface, a feat which was probably instrumental in bringing about a change from radial to bilateral symmetry. This means that it acquired head and tail ends, dorsal and ventral sides, and left and right sides. Localization of the sensory system in the head region was initiated in these forms, signifying a definite step toward centralization of the nervous system. Moreover, the animal now moved in one direction to seek food rather than acquiring its meal in a passive manner. All of these

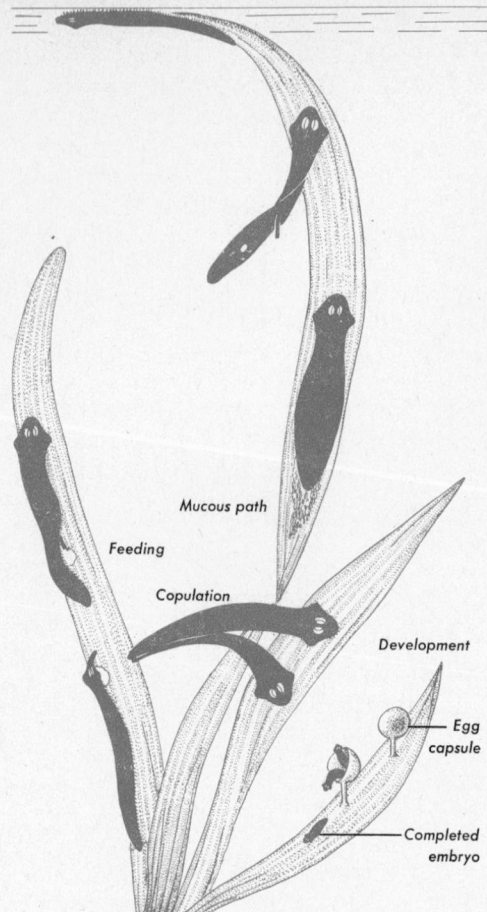

Fig. 8–1. Life history of planaria.

Mucous path

Feeding

Copulation

Development

Egg capsule

Completed embryo

the **Trematoda,** or flukes, and the **Cestoda,** or tapeworms. All members of these two classes are parasitic.

PLANARIA

The study of **triploblastic,** or three-layered, animals begins with planaria, a common inhabitant of fresh water streams. It seeks shelter in darkened, secluded spots and comes out at night to move around in the cool waters in search of food. Planaria is flattened dorsalventrally and is darkly pigmented. Its ventral side is covered with **cilia,** which enable it to glide along the substratum over a mucus path (Fig. 8–1) secreted by glands on the ventral surface of the body. By use of a muscle layer developed in the **mesoderm,** planaria can crawl in true worm-like fashion. Although it may appear that planaria can see, actually its two large **eye-spots** form no image and are only sensitive to varying intensities of light.

Unlike most heads, that of planaria has no mouth, for the mouth is located on the ventral side of the body near the middle. It opens into a muscular **pharynx,** which lies in a sheath extending anteriorly, and when planaria is hungry and in search of food, it often protrudes its pharynx and moves about with it thus extended. When a small piece of meat is tossed into the water, hungry planarians attach themselves to it. A digestive fluid pours out from the pharynx to aid in the disintegration of the meat. The partially digested food is then taken into the digestive tract where digestion is completed. Planaria's chief food consists of small crustacea. When these are its prey, the epidermal slime glands secrete a sticky substance which is sprayed over the victim, rendering it helpless. It is usual for planaria to grip its prey with the head region first and then attach its muscular pharynx to the food, bits of which are drawn into the mouth by a sucking action.

The digestive system of planaria is sac-like, similar to hydra, with but a single opening for the entrance of food and the exit of

changes resulted in a much more complex animal, one that was well on its way toward higher forms.

Of the three classes of the phylum Platyhelminthes, only the class **Turbellaria** includes free-living animals. Members of this group may be found among the rocks in cool streams or ponds, or upon the shady side of submerged plants. Turbellarians are carnivorous, feeding on small animals, either living or dead. There are also marine forms in this group some of which live in the intestines of sea urchins and other forms of ocean life.

The other two classes of the phylum are

waste materials (Fig. 8–2). In some of the platyhelminthes, particularly the parasitic members, the sac is merely a straight and unbranched tube, but in planaria it branches into three distinct parts to form a **tri-clad intestine.** Each part, in turn, ramifies into many smaller branches which supply food directly to the various regions of the body (Fig. 8–2). Large thin-walled, unciliated cells line the gut and secrete digestive juices which carry on extracellular digestion. In addition, the cells lining the intestine are able to ingest solid food by means of pseudopods and digest it intracellularly, as in the case of hydra.

Between the ectoderm and the endoderm is a mass of large star-shaped mesodermal cells, the **parenchyma.** It is possible for food substances to pass not only from the gut into the lining cells, but also from the parenchyma into the lining cells. Thus, when planaria cannot find food, it may consume certain organs in the parenchyma by passing them into the intestinal cells where they are digested. This enables planaria to go without food for quite a long time, through gradual reduction in size.

Special excretory organs appear for the first time in a metazoan among the flatworms. In planaria the excretory system consists of a pair of branching tubes running down each side of the body. The main tubes

Fig. 8–2. The various systems of planaria. See text for explanation.

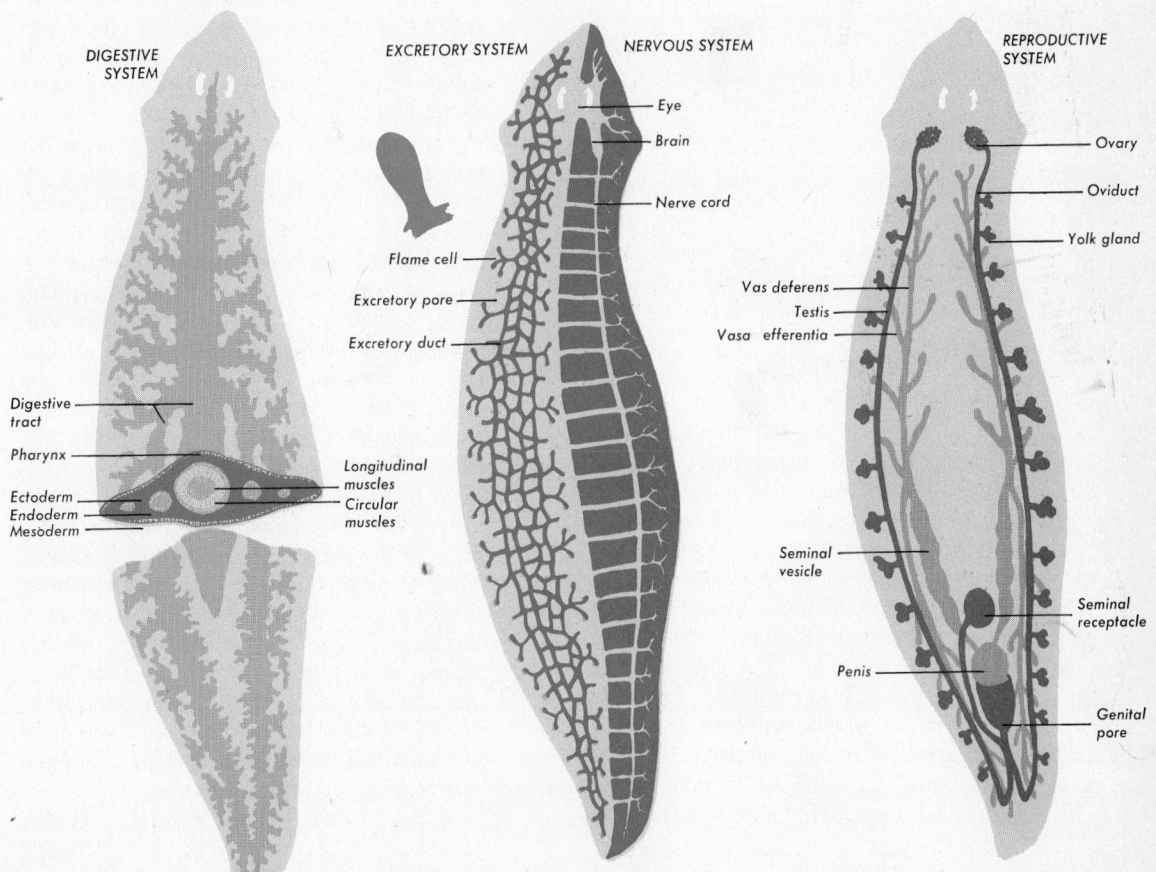

DIGESTIVE SYSTEM

EXCRETORY SYSTEM

NERVOUS SYSTEM

REPRODUCTIVE SYSTEM

Eye
Brain
Nerve cord

Flame cell
Excretory pore
Excretory duct

Digestive tract
Pharynx
Ectoderm
Endoderm
Mesoderm

Longitudinal muscles
Circular muscles

Ovary
Oviduct
Yolk gland

Vas deferens
Testis
Vasa efferentia

Seminal vesicle

Penis

Seminal receptacle

Genital pore

or canals divide into small branches, each of which finally ends blindly in a single **flame cell** (Fig. 8–2). These cells have long cilia which extend into the lumen of the canal, and it is their flickering motion that suggests the name. The movement of the beating cilia carries some of the nitrogenous wastes into the tube and to the exterior through a number of **excretory pores.** Probably the chief function of this system is regulating the water balance, because most of the nitrogenous. wastes are lost through the endodermal cells. Because the tubes of this system are a primitive type of **nephridium** (kidney), they are called **protonephridia.**

The **nervous system** of the planarian is simple (Fig. 8–2), yet it is strikingly more advanced than the primitive nerve net of the coelenterates. Perhaps most unique is the concentration of the nervous tissue in the head region below the eyes. The nerve cell bodies are contained in two masses of nervous tissue, the **cerebral ganglion,** commonly referred to as the **brain.** From this concentrated point two longitudinal **nerve cords** pass posteriorly and two short nerves extend anteriorly to connect with the eyes. Along the two longitudinal cords are many transverse nerves, which are distributed to the various internal structures of the planarian body.

The eyes of planaria are found on the dorsal surface where they appear as two dark spots (Fig. 8–3). There is no lens, as such, although the ectoderm over the eye is without pigment so that light can pass through to reach the **sensory** cells below, which connect with the brain. Without a lens no image is possible, but the eye is sensitive to varying intensities of light and the animal characteristically withdraws from bright light and seeks out moderate illumination. Other sensory cells protrude from the surface of the body and act as receptors for registering changes in the flow of the water or other variations in the surroundings.

Although it is evident that the nervous system of the Turbellaria is still very simple, the increase of special sensory cells, their groupings into such an organ as the simple

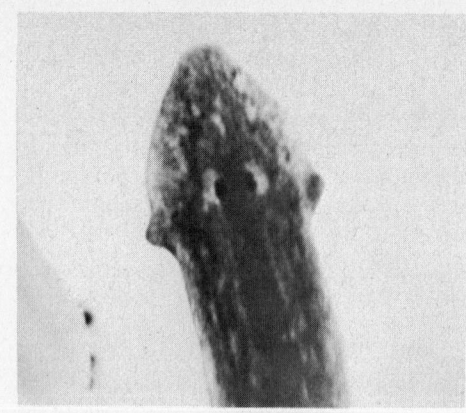

Fig. 8–3. Head of planaria (*Euplanaria* or *Dugesia*) highly magnified. Note the "crossed" eyes and the two earlike extensions of the head region in which tactile sense organs are located.

eye-spot, and the aggregation of nerve cells into the cerebral ganglion are the beginnings of a definite central nervous system which is sufficiently complex to permit a wide range of activities.

Of the various systems, the planarian reproductive systems shows the greatest advancement over that of the coelenterates (Fig. 8–2). In the sexually mature worm, male and female reproductive systems are present in each individual, hence it is said to be **monoecious** (hermaphroditic). Both ovaries and testes develop from the cells of the parenchyma. The numerous testes are rounded bodies which lie along both sides of the body. They give rise to the **spermatozoa,** or **sperm cells,** which are conveyed through small ducts, the **vasa efferentia,** to a larger tube, the **vas deferens,** or sperm duct, running the length of the body on each side. The two seminal vesicles terminate in a pear-shaped **copulatory organ** or **penis.** At rest the copulatory organ opens into the **genital atrium,** which leads into the **genital pore,** the external opening through which the penis is thrust during **copulation.**

The two **ovaries** of the female reproductive system are found near the anterior end of the body; these produce the **ova.** The **yolk glands,** which give rise to the yolk and shell of the egg, are found along the **oviducts.**

The two oviducts lie parallel to the nerve cords and join before entering the atrium. The **seminal receptacle,** a sac for storing sperm, also opens into the atrium, very near the external opening. The genital atrium, therefore, receives the openings of both the female and the male organs.

At the start of copulation the ventral surfaces of the two animals come together, so that the openings of the genital atria are opposite one another. The penis of each is extended into the genital opening of the other and the sperm cells are exchanged. At the time of copulation the ova are also ripe. To prevent self-fertilization, the genital area has been elaborated. The penis, when extruded and dilated, completely fills the atrium and thus blocks the openings into the oviducts, so that neither can the ova escape nor sperm cells enter the oviduct, but sperm can be deposited into the seminal receptacles. At the completion of copulation the penes are withdrawn and the sperm cells then are able to enter the oviduct. The ova are fertilized in the oviduct and, as they move down toward the atrium, the products of the yolk glands are discharged. The mature fertilized egg is released from the atrium through the genital pore, and the capsule-like shell becomes attached by a stalk to submerged objects. The egg cases may undergo a rest period before growing into young planaria.

Planaria also reproduces asexually by means of transverse fission. Indeed, this is much the more common method of reproduction. The worm constricts itself in two and the parts which are missing after the fission has taken place are then regenerated. Planaria is an excellent animal for regeneration experiments. For example, if the head of planaria is split and if the parts are kept separated for a short period of time, a double-headed monster is formed. Should the animal be cut transversely into two separate pieces, two little planaria will result (Fig. 8–4). An animal can be cut into as many as six pieces and each will give rise to a miniature worm one-sixth the size of the original. Members of the class Turbellaria are, for the most part, free-living, some live on the exterior of other animals, and others are true parasites. Some marine forms are beautifully colored (Pls. 4 and 5).

CLASS TREMATODA

The trematodes, commonly called **flukes,** are characteristically flat like all platyhelminthes, but their gut is reduced in complexity. Because of their parasitic life they have lost their cilia and have developed holdfast organs called **suckers.** They have no apparent sense organs, but the reproductive and excretory systems are well developed. In general, the entire anatomy is adapted to a parasitic mode of life.

All grades of parasitism are found in this group, from those that live on the outside of their hosts to those that inhabit the internal cavities. The life history of parasites that cling to the outside of animals such as fast-moving fish is relatively simple, its main characteristic being the development of holdfast organs or suckers. The case of the internal parasite is an entirely different one. There is no great problem of staying put, hence the suckers are small, but in order to complete its life cycle an enormously prolific reproduc-

Fig. 8–4. Regeneration in planaria.

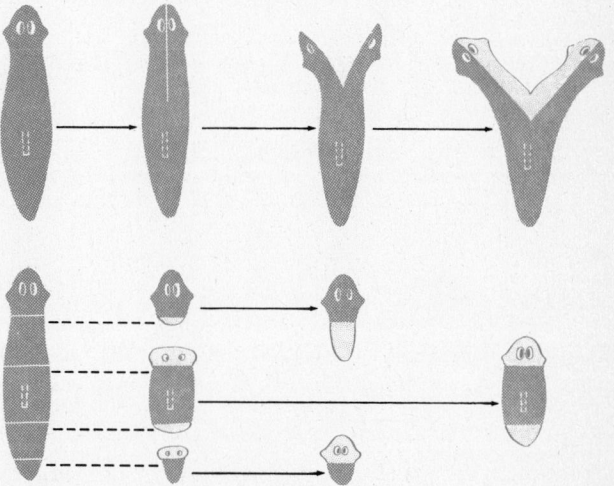

tive system has been developed. Not only is the production of large numbers of potential offspring necessary, but also various kinds of larval stages that are able to pass through several hosts, all of which are instrumental in spreading the parasite far and wide. Let us consider two examples of flukes that infect man, one that lives in the liver (*Opisthorchis sinensis*) and another that lives in the blood (*Schistosoma japonicum*).

The life cycle of the human liver fluke can serve as a typical example of most related flukes that are so prevalent in wild and domestic animals (Fig. 8–5). It involves two **intermediate hosts** which harbor the larval stages of the parasite, and, of course, one **final host** in which the adult lives. The human liver fluke infects 75–100 percent of the people in certain parts of China, Japan, and Korea, constituting a real health prob-

Fig. 8–5. Life cycle of *Opisthorchis sinensis*, the human liver fluke.

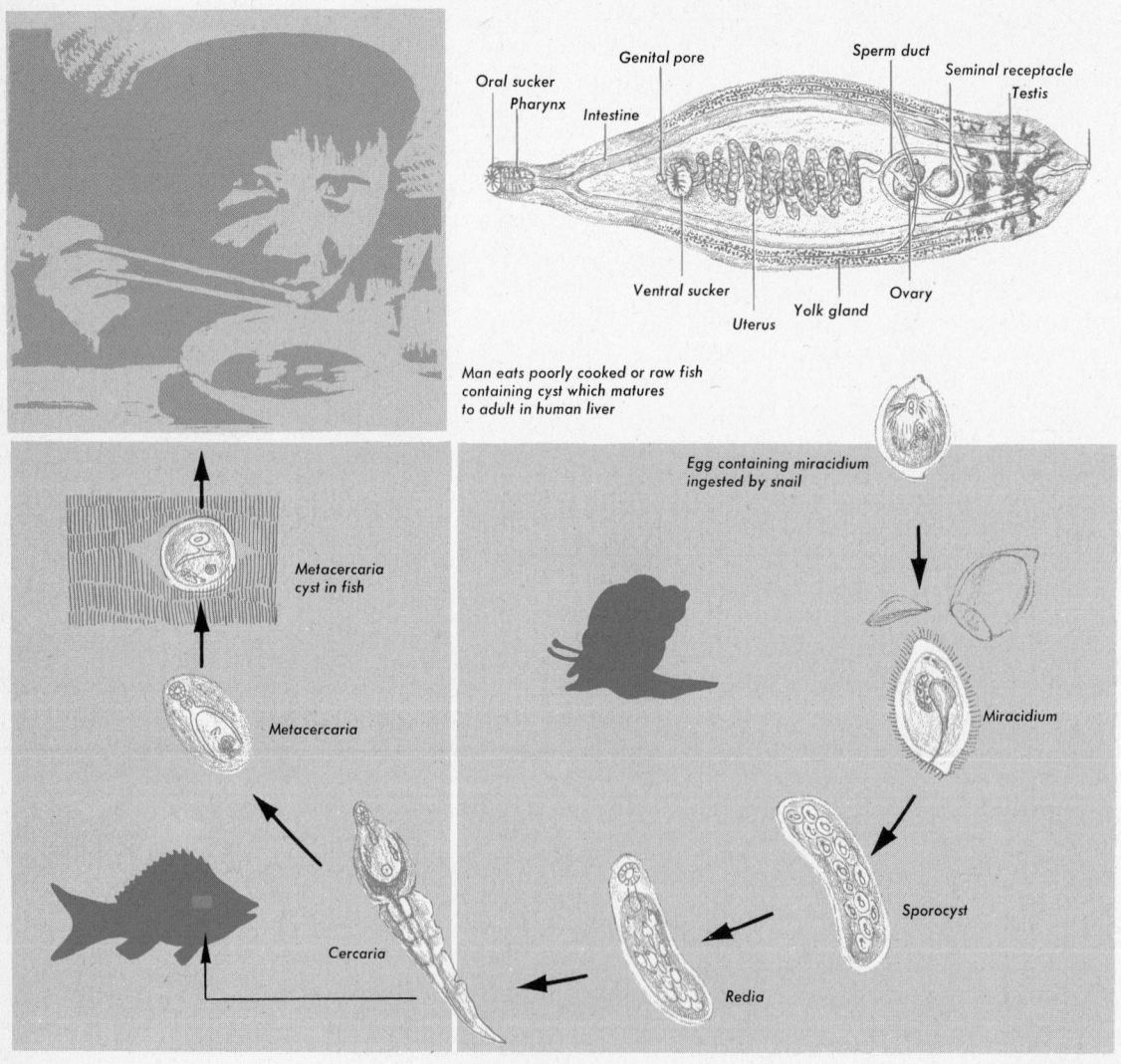

Oral sucker
Pharynx
Intestine
Genital pore
Sperm duct
Seminal receptacle
Testis
Ventral sucker
Uterus
Yolk gland
Ovary

Man eats poorly cooked or raw fish containing cyst which matures to adult in human liver

Metacercaria cyst in fish

Metacercaria

Cercaria

Egg containing miracidium ingested by snail

Miracidium

Sporocyst

Redia

lem in these regions. This situation should be alleviated with the advent of improved sanitation and a better educational program.

The adult fluke (Fig. 8–5) lives in the small bile ducts of the liver, where toxic products excreted by the flukes and the subsequent mechanical occlusions of the ducts may cause serious damage. For a heavily infected individual, this condition may eventually cause cirrhosis, which together with complicating infectious disease usually terminates in death. The adult fluke is about three-fourths of an inch in length and has two suckers, one at the anterior end, another about one-third of the way from the anterior end. It feeds on blood which is drawn in through the anterior mouth.

Eggs laid by the adult pass through the bile duct into the gut and eventually pass out of the body in the feces. Because of the oriental habit of using human excrement as fertilizer in the rice paddies, the eggs usually get into water. Unlike most flukes, this fluke does not hatch into the larval stage known as the **miracidium** until it is eaten by a certain species of snail of the genus *Bythinia*. Inside the snail the egg hatches, releasing the miracidium which makes its way into the tissues of the snail and develops into a **sporocyst**. The next stage, the **redia**, develops inside the sporocyst. Both the sporocyst and the redia stages make possible a tremendous increase in numbers by asexual reproduction. The redia finally develops **cercariae** within its walls, which make their way out of the snail into the surrounding water where they swim about by means of their large vibratile tails. The cercaria then becomes attached to the next host, one of several different fish and, after losing its tail, bores its way into the flesh of the fish. It rounds itself into a ball and produces a cyst wall; in response to the parasite, the fish secretes another wall around the invader. Here it lies until the raw fish is eaten by man in whose gut the cyst wall is digested away, releasing the young worm which makes its way up the bile duct and finally into the smaller tubes of the liver where it grows to maturity.

The control of the disease is obviously very simple; destroy the snails or cook the fish, either of which interrupts the cycle and kills the parasite.

Some of the most important flukes are the blood-inhabiting **schistosomes** such as *Schistosoma japonicum* (Fig. 8–6). Unlike *Opisthorchis*, these worms are **dioecious,** that is, there are two separate sexes. They are long slender worms, beautifully adapted for living in the small blood vessels. A strange relationship exists between the males and females; the male holds the extremely slender female in a groove on his ventral side, from which she ventures forth during the business of laying eggs. Her slender threadlike body is ideally adapted to fit in the tiny blood vessels of the intestinal wall or over the bladder where she lays her eggs. The eggs have a single sharp spine by which passively they work their way through the wall into the cavity of the intestine or bladder where they are voided with the urine or feces.

Again through the use of human excrement for fertilizer, the eggs usually find their way into water. They hatch into **miracidia** which penetrate the tissues of snails, and follow stages of development similar to those of *Opisthorchis* with minor variations. Instead of encysting on a fish, the cercaria burrows through the skin of a person who is unfortunate enough to be in the vicinity and makes its way into the vascular system. It passes through the heart, lungs, and liver, eventually maturing in the blood vessels (veins) that drain the intestines and bladder. Here it grows rapidly, feeding on blood, and when sexually mature lays its eggs, thus completing the cycle.

The several species of blood flukes infect people of tropical America, many parts of Africa, and the Orient, particularly China. In some irrigated regions the infection runs as high as 90 percent among the adult males who are constantly in contact with the water, hence exposed to the cercariae. In spite of all efforts at control of schistosomiasis, the disease is more widespread than ever, particularly in Africa where irrigation has greatly expanded in recent years. Treatment consists of giving large doses of antimony com-

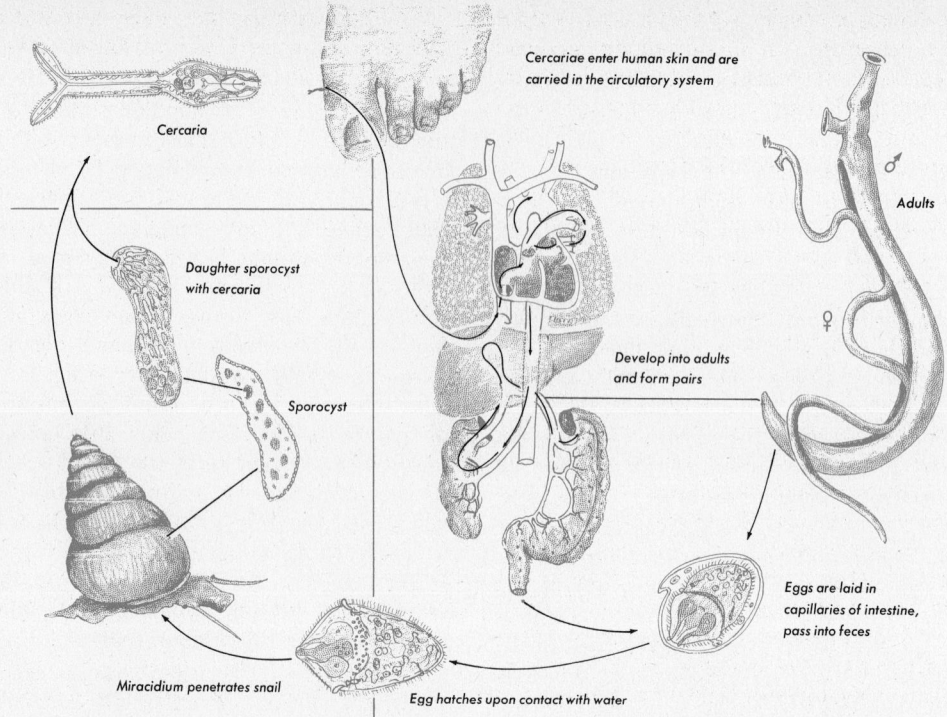

Cercaria

Daughter sporocyst
with cercaria

Sporocyst

Miracidium penetrates snail

Cercariae enter human skin and are
carried in the circulatory system

Adults

Develop into adults
and form pairs

Eggs are laid in
capillaries of intestine,
pass into feces

Egg hatches upon contact with water

Fig. 8–6. Life history of a typical human blood fluke.

pounds and, if the patient can stand the treatment, he can be cured. Like all of the complex parasites, blood flukes can be controlled by removing the intermediate host.

Water birds such as ducks, terns, and gulls, have their own varieties of blood flukes which apparently cause them no particular harm. However, if this type of cercaria cannot find its proper final host, it does penetrate the skin of any person who happens to be near, causing a severe itching which has been called **schistosome dermatitis**, or "swimmer's itch." The cercariae apparently are not able to penetrate the tough thick mammalian skin, but in their attempts to do so enter it and cause intense irritation. There are several different species of cercariae that follow this pattern. Some are found on the sandy bathing beaches in the lake regions of the North Central states, especially Michigan and Minnesota, where they sometimes become such a nuisance that bathing is actually prevented. Elaborate methods of treating the beaches with copper sulfate in order to destroy the snails have been developed and have had reasonable success.

CLASS CESTODA

These are the **tapeworms**, a group of parasites that the layman has long erroneously associated with lean hungry adolescents. The worm gets its name from its long ribbonlike appearance, a feature that is common to this large and varied group, members of which infect almost all, if not all, vertebrate animals.

The tapeworm is, perhaps, the most degenerate of all animals, a condition indicating that the association with its host is one of long standing. At the same time it is perfectly adapted to its specialized environment, the vertebrate gut. It is indeed the supreme parasite among parasites. It is provided with excellent hold-fast organs to keep it in place in the gut of the host (Fig. 8–7). All nourishment is received from the contents of the gut or from the gut wall directly, so the animal has not bothered to retain even a semblance of a digestive tract. Its nervous and excretory systems are very rudimentary, and its ability to move has been reduced to very feeble contractions. However, it has evolved an extremely elaborate and prolific reproductive system, an essential feature in its survival since the possibility for any individual egg to reach maturity is very small. Although it has degenerated in other respects, it has gone all out in this one phase of its life, and measured in terms of biological success, the shift in emphasis has apparently been satisfactory.

The common beef tapeworm (*Taenia saginata*) of man is a typical example of this group (Fig. 8–8). It consists of two principal parts, the head or **scolex,** and the **proglottids,** sectional pieces attached to one another, and growing larger as they proceed posteriorly. The scolex possesses hold-fast organs which

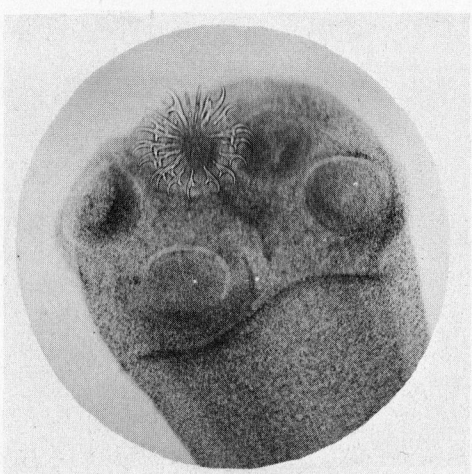

Fig. 8–7. The scolex of the dog tapeworm (*Taenia pisiformis*). Note the sharp hooks and sucking discs used as attachment organs.

make it possible for the worm to maintain its position in the gut. The proglottids, which are budded off from the region just back of the head, called the **neck,** mature as they move progressively posteriorly. The younger proglottids are therefore anterior to the older. The mature proglottids, gorged with eggs containing young embryo worms, break away from the worm and pass out of the body in the feces. Because of the close association of cattle and their keepers in certain parts of the country, it is not unusual for the eggs to be picked up by grazing cattle. Once in the gut of this host the egg membranes and shell are digested away and the young **six-hooked embryo (hexacanth)** emerges. It soon bores its way through the gut wall into a blood vessel where it floats to the muscles, particularly heart and jaw muscles. Here it develops into a **bladder** and a tiny inverted tapeworm scolex grows from the wall of the bladder. Beef so infected is said to contain "bladder worms" and is usually unmarketable. If such beef is poorly cooked and then eaten by humans, the bladder worms "hatch." The tiny scolices evert and become attached to the soft intestinal wall where they immediately begin budding off proglottids.

An examination of the proglottid demonstrates the fact that it is almost a complete individual itself. Indeed, the tapeworm is sometimes considered a colony in which each proglottid is an individual, much like the buds in hydra or the polyps in obelia. Besides rudimentary nervous and excretory systems it possesses male and female sex organs, which are capable of producing prodigious numbers of sperm and eggs. The **testes,** numerous and scattered throughout the proglottid, are connected through fine tubules to the **sperm duct** which opens to the outside through the **genital pore.** The paired **ovaries** produce eggs which pass through a small duct (oviduct) where they receive sperm from another proglottid, or another worm, via the **vagina.** Here they also obtain the food material called **yolk** from the **yolk gland,** while the **shell gland** secretes material for forming the shell. The fertile eggs then are deposited into the **uterus** which eventually becomes greatly distended as the eggs begin to develop crowding all other structures out of place. Such a **gravid** proglottid (full of developing embryos) breaks off and follows the cycle indicated above.

Meat inspections in this country have greatly reduced the incidence of this parasite. Control of tapeworms is very simple:

merely cook the suspected meat. There is a similar tapeworm in pork (*Taenia solium*) which is also becoming rare.

In review, we have seen that with the advent of the mesoderm and with it several important organ systems, the flatworms far outstripped the coelenterates in complexity. They are, however, still small creatures and relatively simple when compared to a mammal; in other words, still further important

Fig. 8–8. Life history of the beef tapeworm (*Taenia saginata*) of man.

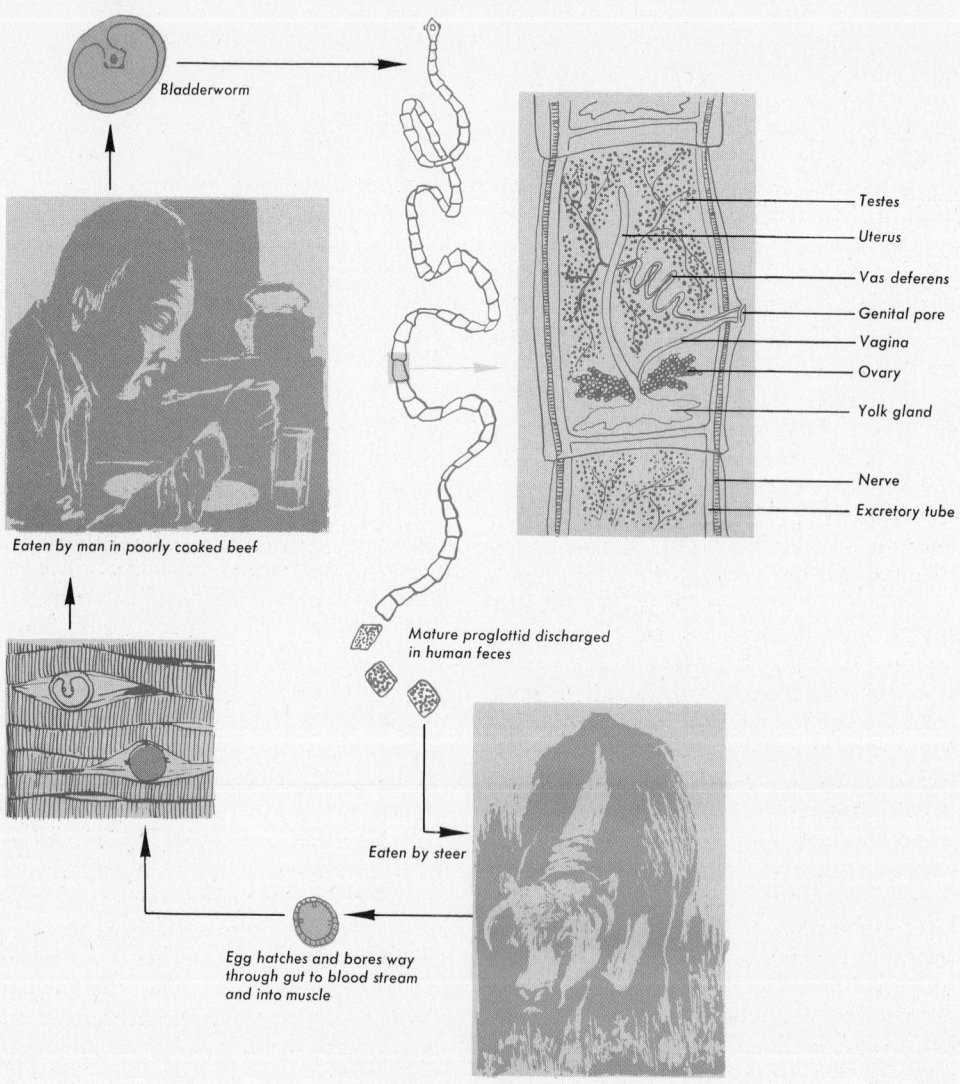

Bladderworm

Testes

Uterus

Vas deferens

Genital pore

Vagina

Ovary

Yolk gland

Nerve

Excretory tube

Eaten by man in poorly cooked beef

Mature proglottid discharged in human feces

Eaten by steer

Egg hatches and bores way through gut to blood stream and into muscle

changes must have taken place in subsequent groups of animals.

PHYLUM NEMERTINEA

The Nemertines

The development of a circulatory system first appears in this phylum, representatives of which are sometimes called "band worms" because of their long ciliated flat bodies. The nemertines live primarily in the ocean where they crawl among the rocks. They are highly colored and greatly elongated, sometimes measuring as much as 80 feet in length. If captured, their bodies stretch so that they often break into two parts under their own weight, but regeneration occurs as readily as it does in planaria. They are also able to break their bodies into many parts, a process called **autotomy** (self-cutting), which is not an uncommon characteristic among invertebrates.

In addition to a complete digestive tract, the nemertine possesses a very primitive circulatory system (Fig. 8–9). It consists of three blood vessels that extend throughout the length of the body, but which do not break up into tiny capillaries as in higher animals. In these forms, oxygen and food still diffuse some distance through fluid before arriving at the cells. Although still inefficient, this method is a considerable improvement over that found in lower forms, where diffusion of digested material and oxygen from sources of intake to the cells must pass a greater distance. The circulating fluid (blood) contains red cells in some species. The red color is due to **hemoglobin**, which makes up the major portion of the material found in these cells.

Another deficiency of the nemertine circulatory system is the lack of a pumping station. The only propulsive force for the blood is furnished by the contractions of the animal as it swims.

With the nemertines we have seen the introduction of a transport system which was not present in the metazoa studied earlier. Although very primitive it served the pur-

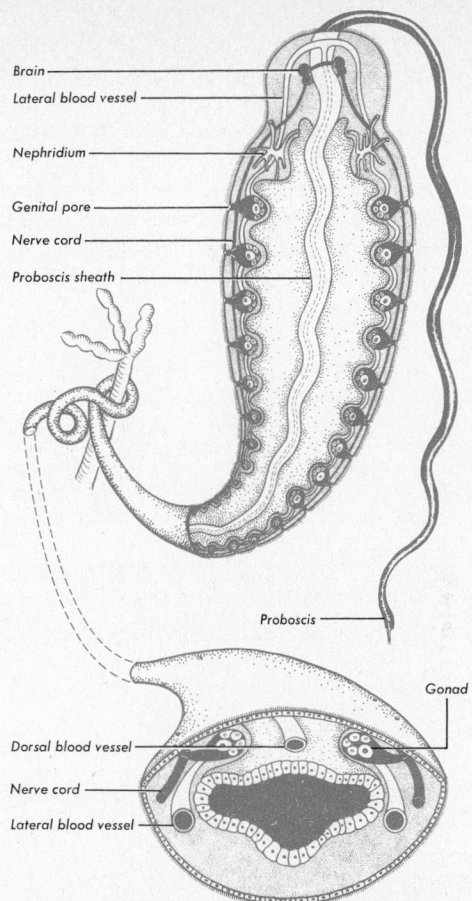

Fig. 8–9. A longitudinal and cross section of a nemertine showing the tube-within-a-tube body plan and also the beginnings of a circulatory system.

pose for these simple animals. For more complex animals it would need to be greatly improved; which it was.

The nemertines possess the pocketed gut of planaria but with the addition of an opening, the **anus**, at the posterior end of the animal. Thus the undigested food leaves the body through this opening rather than being regurgitated as in planaria. The excretory system is also more specialized yet the reproductive system is less complicated than in planaria. The sexes are usually separate and each gonad opens to the exterior by a separate pore. Ova and sperm are discharged into the sea where fertilization occurs in most species; in some it takes place within the body cavity of the female. The zygote develops into a ciliated larva, the **pilidium**, which later metamorphoses into the adult.

It is interesting to note that not all improvements appeared in any one group of animals. For example, the nemertines possess the complete digestive tract and a circulatory system, yet retain the primitive excretory system of the flatworms. The Aschelminthes, where we see the "tube-within-a-tube" body plan, possess more elaborate digestive, nervous, and muscular systems than the nemertines but have no circulatory system. All of these advances are consolidated in the mollusks where we observe animals far more complicated than any studied so far.

SUGGESTED SUPPLEMENTARY READINGS

Books

CHANDLER, A. C., and READ, C. P., *Introduction to Parasitology*. New York: Wiley, 1961.

NOBLE, E. R., and NOBLE, G. A., *Parasitology, The Biology of Animal Parasites*. Philadelphia: Lea and Febiger, 1964.

PENNAK, R. W., *Fresh-Water Invertebrates of the United States*. New York: Ronald Press, 1953.

WARD, H. B., WHIPPLE, G. C., and Edmondson, W. T., eds., *Fresh-Water Biology*. New York: Wiley, 1959.

Articles

BEST, J. B., "Protopsychology." *Scientific American*, February, 1963.

9

THE TUBE-WITHIN-A-TUBE BODY PLAN

A mixture of widely diverse animals, of which the most important are the nematodes (roundworms) and the rotifers, are placed in the phylum **Aschelminthes.** All have two characteristics in common: they have a complete digestive tract, mouth to anus, and they are **pseudocoelomates** (p. 115). Since there is much confusion among taxonomists about the lesser members of the phylum, we consider only the classes **Rotifera** and **Nematoda.**

Class Rotifera

These tiny animals are often called "wheelbearers" because their two tufts of cilia on the anterior end, when active, move in such a manner as to appear as two wheels turning in opposite directions (Fig. 9–1). Actually the cilia beat in this manner to propel microorganisms into the mouth during feeding. Rotifers are highly variable in body form, all microscopic in size, and they are abundant both in salt and fresh water. They are, of course, bilaterally symmetrical, and possess a complete digestive tract, although in some forms the anus apparently does not function. A characteristic organ in the gut is the **mastax,** a grinding organ which can be easily observed when in action through the body wall. They also possess a pseudocoelom but no circulatory system.

Class Nematoda

These are the "roundworms," a name which well describes their anatomy (Fig. 9–2). In numbers they exceed all others with the exception of the arthropods and the protozoa. The most notorious members are parasites but the greatest numbers exist as free-living forms in soil and water. Their characteristic whipping movement readily identifies them. Most of them are very small,

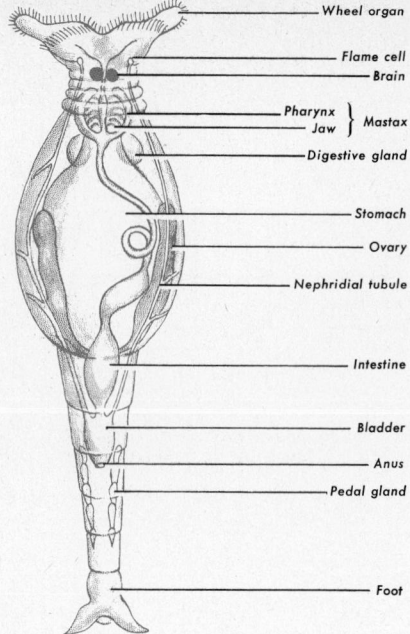

Fig. 9–1. A rotifer.

Labels (top to bottom): Wheel organ, Flame cell, Brain, Pharynx, Jaw, Mastax, Digestive gland, Stomach, Ovary, Nephridial tubule, Intestine, Bladder, Anus, Pedal gland, Foot

Fig. 9–2. The body plan of the roundworms includes two tubes: one, the gut, within the other, the body wall. This figure shows the anterior end of the worm with its three teeth surrounding the mouth, and a cross section taken anterior to the gonads.

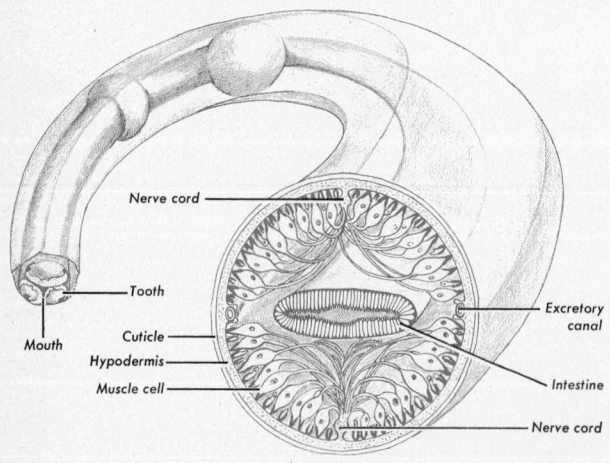

Labels: Nerve cord, Tooth, Mouth, Cuticle, Hypodermis, Muscle cell, Excretory canal, Intestine, Nerve cord

almost microscopic, although there are a few —the "horsehair worms," for example—that may reach a length of 1 yard. Some of the ascarid parasites of horses may reach a length of 10 to 12 inches.

Their complete digestive tract consists of a slender tube, without pockets, extending from one end of the body to the other. Digested food is absorbed directly through the gut wall and diffuses into the fluid which fills the pseudocoelom, thence to the body cells. This is an inefficient system since it relies to a large extent on diffusion, although the fluid-filled pseudocoelom does permit a much more rapid dispersion of the food and oxygen than was possible without this cavity.

A pair of tubes extend internally along each side of the body, forming **excretory canals,** but they lack any cells comparable to the flame cells found in planaria. The two tubes unite into a single one, which opens ventrally to the outside through a minute **excretory pore.** Another feature of the nematodes is a complex **nervous system** which consists of several nerves extending the length of the body and terminating anteriorly in numerous ganglia.

The body wall contains a thick muscular component, separated into four bands of **muscle cells** extending lengthwise, and so attached that the animal can flex its body only in a dorsal-ventral manner, a rather ineffective method of locomotion in water. In fluids of high viscosity or in soil, however, it is more effective.

The rather elaborate reproductive system lies free in the pseudocoel. Since the sexes are separate, only one set of organs is found in each animal. Females, which are usually larger than the males, possess two ovaries in the shape of long coiled tubes. The two ovaries continue into two **oviducts,** which enlarge to form two **uteri** (singular, **uterus**). These join to form a single short **vagina,** which opens externally on the ventral side in the anterior portion of the body. The mature eggs are stored in the uteri. In the male, sperm are produced in a long coiled tube, the single **testis,** which joins the **vas**

deferens and then becomes the **seminal vesicle,** the storage place for the sperm. A pair of bristles at the posterior end aid in conducting the sperm from the male to the female during copulation. The opening of the male reproductive system is close to the posterior end of the animal near the base of the bristles.

Parasitic Nematodes. Because of the economic significance of the parasitic nematodes we discuss representative forms, particularly those that attack man. Although over 50 different species are human parasites, a still greater number affect man indirectly by their ruthless destruction of his domestic plants and animals. They invade almost every organ of the body, their damage depending on the kind and number of individuals. Like most parasites, the nematodes tend to remain with a specific host, although they are a little more careless in this regard than some. Occasionally they attack a variety of hosts and may produce a serious disease when they enter a new one. For example, a hog may be riddled with *Trichinella* with no apparent damage, whereas a man with a similar infection is apt to die because he is the "accidental host" while the hog is the normal host. The hog has had trichinella in its tissues so long that it has built up some resistance to the parasite. Since man gets the parasite only occasionally, he has developed little or no resistance.

Ascaris lumbricoides (Fig. 9–3) is a common intestinal parasite of many domestic animals as well as man himself. It is an excellent example of the usual life cycle of parasitic roundworms, although there are wide modifications. It is not infrequently found in the digestive tract of children, since they are apt to get ascaris eggs on their hands from the soil and transfer them to the mouth. The embryonated eggs pass through the stomach to the intestine where they hatch into tiny worms (0.2–0.3 mm. long). These bore through the intestinal wall into the lymph, then the capillaries, and finally the general circulation. They are carried

through the heart to the lungs where they grow somewhat in size. Eventually the larvae break through into the air sacs, migrate up the trachea, and are swallowed, arriving in the intestine for the second time. Here they mature, copulate, and lay eggs which pass out of the host with the feces. The eggs may be

Fig. 9–3. A schematic representation of the life cycle of *Ascaris lumbricoides* as it occurs in humans.

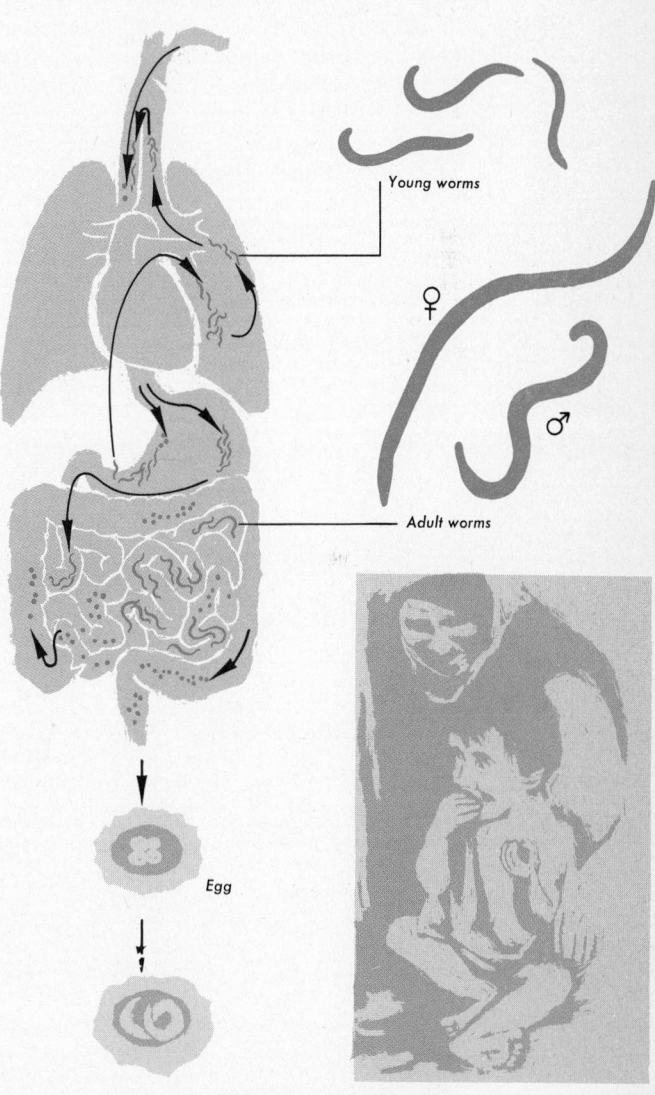

Young worms

♀

♂

Adult worms

Egg

Embryo worm within shell—*EATEN*

picked up directly by another host, or they may become desiccated and blow around in the dust to be engulfed at some later time.

The adult ascaris probably maintains its place in the intestine by active movements, since it does not possess an attachment organ such as the flukes and tapeworms do. It feeds on the food in the gut by a pumping action of its bulb-like pharynx. To keep from being digested by the enzymes secreted by the host, ascaris, like all intestinal parasites, is protected by a tough **cuticle**, through which probably are secreted substances that counteract the hydrolyzing power of the enzymes. Metabolism is primarily anaerobic (without oxygen), since there is very little oxygen in the gut. The only hope for survival is to produce a great many eggs, which it does most effectively. A large female has

been known to contain 27,000,000 eggs, 200,000 of which she lays every day. As in all parasites, the chance for any one egg to produce a mature worm is extremely remote, but by this colossal effort to bring forth potential offspring ascaris has been very successful in the world, as attested by its universal occurrence.

A notorious relative of ascaris is the hookworm (*Necator americanus*), which is directly responsible for untold misery and indirectly for the death of millions of people throughout tropical and subtropical regions of the world. People in certain parts of the world are heavily infested. It is little wonder that they are demoralized and debilitated when their intestinal walls are teeming with hookworms sapping their strength.

The adult hookworm differs from ascaris in that the mouth is provided with teeth so it can cling to the soft mucosal lining of the intestine from which it withdraws its food, blood (Fig. 9–4). Fertile eggs pass out with the feces of the host and are deposited on the ground where they hatch into larval worms. After a brief growth period the larvae are ready for their next host. They gain entrance by holding on to the foot or any other part of the body of the host, boring in, and finally getting into the bloodstream. They then follow the same path described for ascaris, eventually reaching the intestine.

Preventive measures are so simple that one wonders why there are any cases of hookworm at all. Wearing shoes prevents the worms from getting into the host; proper methods of disposing of human excreta would also stop the infection very swiftly. Both of these methods have been tried with reasonable success. Worldwide measures could stamp out the disease altogether, but such suggestions are only wishful thinking at the present time.

Another roundworm parasite that is of grave importance to man is **trichina** (*Trichinella spiralis*), a worm whose normal hosts are the pig and rat, although it has been found in other vertebrates as well. Man is an **accidental host** and is therefore perhaps even more severely affected by the parasite.

Fig. 9–4. Life cycle of the common hookworm (*Necator americanus*). The adult worms are shown attached to the lining of the intestine. The fertilized eggs begin development while still in the digestive tract. They pass out with the feces and eventually hatch in the soil where they lie in wait for their next host.

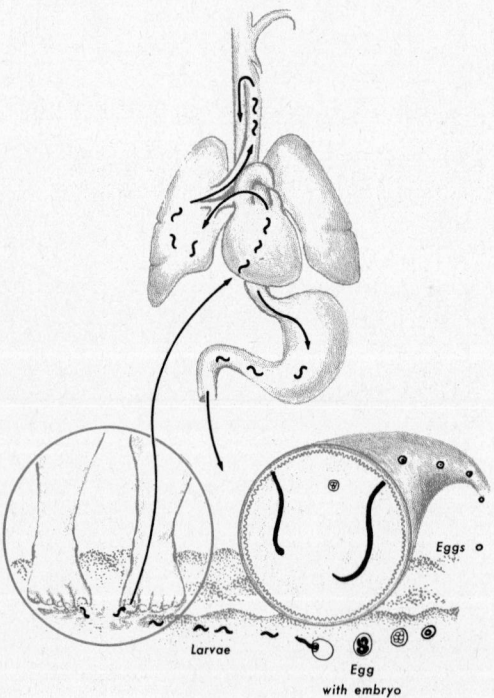

Eggs

Larvae

Egg
with embryo

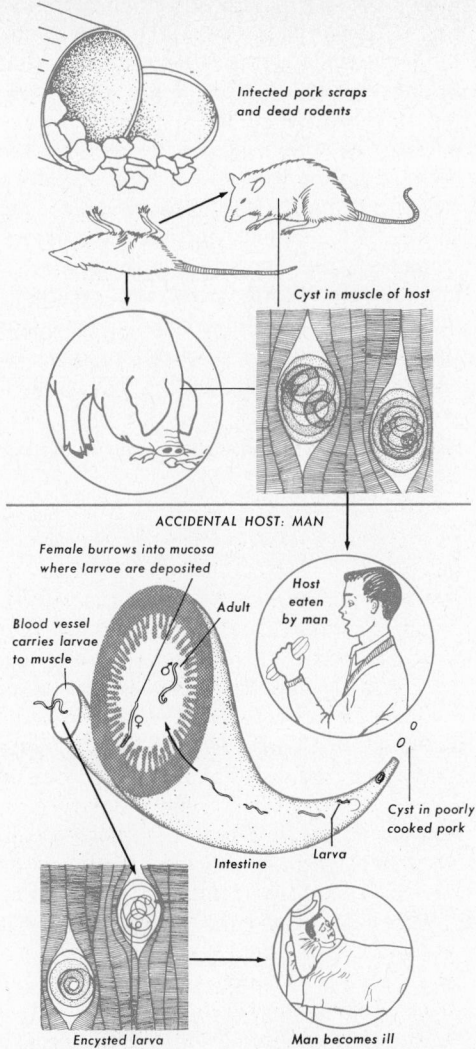

and copulate. The very tiny worms deposited by the female bore through the intestinal wall into the blood stream and distribute themselves through the muscles of the body, attacking particularly those of the tongue, eyes, diaphragm, and ribs. It is this migratory period that is dangerous because, in addition to the mechanical injury that millions of worms can inflict, there is also the likelihood of bacterial infections. The disease at this stage is characterized by high fever, intense muscular pains, and frequently partial paralysis. The intake of a sufficient amount of infected meat can cause death at this time, but if the infection has been light enough not to cause permanent damage to nervous and muscular tissue, the symptoms will subside and the person recover. Infections are far more common than records indicate. For example, it is estimated that 2.2 percent of Americans were infected in 1973.

Preventive measures are even simpler than for hookworm: merely cook pork. There is no treatment for the disease once an infection has gotten under way. It is also well to

Fig. 9–6. Larval *Trichinella* cysts in the muscles of a rat.

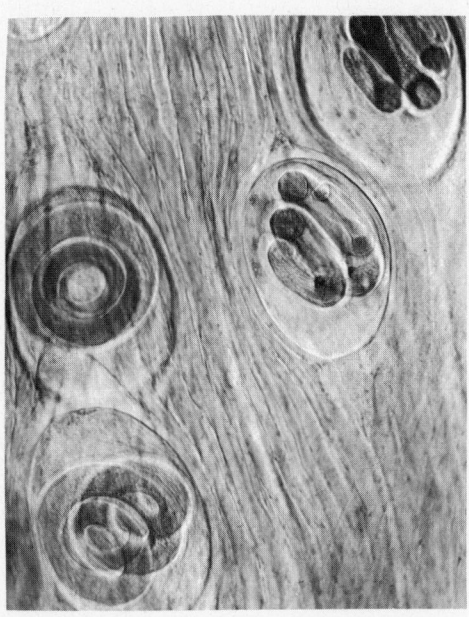

Fig. 9–5. Life cycle of trichina (*Trichinella spiralis*).

The life cycle of trichinella varies somewhat from the two preceding examples of roundworm parasites (Fig. 9–5). The common source of human infection is through the muscle of the pig, which harbors trichina in its infective stage, small **cysts** containing **larvae** (Fig. 9–6). If these are eaten, uncooked, the tiny worms (1 mm. long) emerge in the intestine where they mature

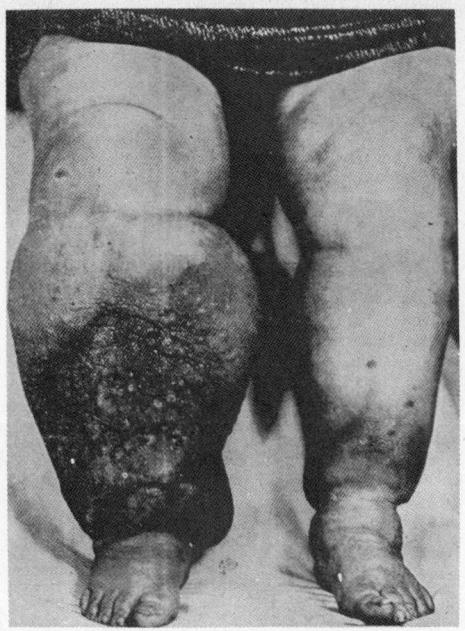

Fig. 9–7. A case of elephantiasis. (From Smith and Gault, *Essentials of Pathology*, 1938, D. Appleton-Century Company.)

important (Fig. 9–7). The life cycle of this worm differs from that of other nematodes in that it requires two hosts. The larvae, called **microfilariae,** circulate in the blood of the infected person (Fig. 9–8). An interesting adaptation is that these tiny worms come out in the peripheral blood vessels in the evening, a time when the mosquitoes which are the carriers (intermediate host) are active. The mosquito picks up the microfilaria with the blood as it feeds; inside the mosquito the worm grows and eventually makes it way out through the proboscis of the host. During the biting process the

Fig. 9–8. The microfilariae that cause the disease elephantiasis live in the blood of man and can be seen in blood smears taken at certain times of the day. The tiny worm is clearly visible in this picture and its relative size can be determined by comparing it to the small irregular objects which are the white corpuscles.

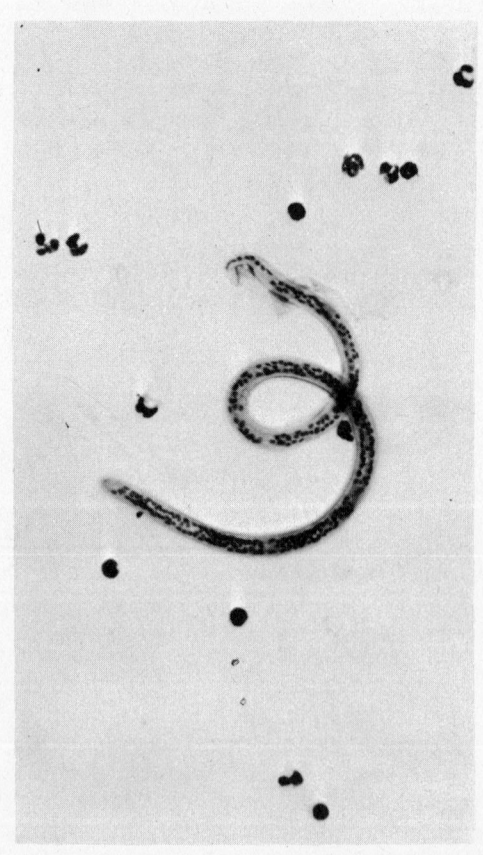

remember that meat sold on the market is not inspected for trichina, primarily because it is too difficult to detect light infections. One poorly cooked "pink" pork chop can contain thousands of worms, which are adequate to kill a person. It is true that there are fewer and fewer cases of trichinosis reported, probably because the practice of feeding meat scraps to hogs is less prevalent and also because a general war on rats has cut down the rat-hog cycle which normally keeps the worms going.

There are numerous parasitic nematodes that cause bizarre diseases in the tropics, diseases which are normally known only to parasitologists and medical men who are experts in the field of tropical medicine. However, during World War II the tropics became the battleground for many American men and consequently tropical diseases suddenly loomed as a significant health problem. Among the numerous roundworm parasites the one that causes **elephantiasis** (*Wüchereria bancrofti*) is perhaps the most

microfilaria slips off the proboscis onto the skin of the next host and immediately bores its way in. Once in the blood stream of the final host it moves into the lymph glands where it becomes mature. The worms become so numerous that they can effectively clog the lymph passages, which results in huge swellings, usually in the extremities. A leg may grow to weigh 100 pounds, hence the name *elephantiasis*.

Preventive measures consist of preventing oneself from being bitten by infected mosquitoes. Light infections are not serious because eventually the body forms new lymph channels so the swelling is reduced to normal. The danger lies in continual infection where the same individual is bitten again and again by infected mosquitoes.

Members of this phylum have acquired the rudiments of organ systems which we see greatly elaborated in subsequent phyla.

SUGGESTED SUPPLEMENTARY READINGS

Books

BELDING, D. L., *Textbook of Parasitology.* New York: Appleton-Century-Crofts, 1965.

CHANDLER, A. C., and READ, C. P., *Introduction to Parasitology.* New York: Wiley, 1961.

CHITWOOD, B. G., "Nematoda," in *McGraw-Hill Encyclopedia of Science and Technology*, Vol. 9. New York: McGraw-Hill, 1966.

LEE, D. L., *The Physiology of Nematodes.* San Francisco: W. H. Freeman, 1965.

LEVINE, N. D., *Nematode Parasites of Domestic Animals and Man.* Minneapolis: Burgess Publishing Co., 1968.

NOBLE, E. R., and NOBLE, G. A., *Parasitology, The Biology of Animal Parasites.* Philadelphia: Lea and Febiger, 1964.

PENNAK, R. W., *Fresh-Water Invertebrates of the United States.* New York: Ronald Press, 1953.

10

SHELLED ANIMALS-MOLLUSKS

From here on we deal with **eucoelomate** animals which apparently split off into two divergent stocks, the difference between the two being the manner in which the coelom formed (p. 115). The **Schizocoela** stock gave rise to the mollusks, annelids, and the arthropods, whereas the **Enterocoela** stock gave rise to the echinoderms and the chordates. In both, large, swift-moving animals evolved. Each gave rise to great numbers of species and individuals which dominate the living world today; hence each stock was successful.

The phylum **Mollusca** includes a wide variety of well-known animals. Most of them possess shells which greatly affect their mode of existence and which have been very important in giving us a more or less continuous story of their evolution, because shells produce excellent **fossils.** Fossils are imprints or traces of ancient animals or plants that have been preserved in the earth's crust. Ancestral mollusks appear in Cambrian rocks which

were laid down 600 million years ago; throughout the ensuing ages they have prospered, giving rise to millions of species of which at least 80,000 exist on earth today. The phylum includes animals of wide diversity of form, such as the common slugs and snails, oysters and clams, slow moving chitons, swift darting squids, slithering octopuses, and the chambered nautilus of poetic fame. They have invaded most habitats, sea, fresh water, land, and even deserts. A few forms that live in the seas of the world are shown in Figure 10–1. They constitute a significant part of the world fauna today and are important in our lives also, as we observe later in this chapter.

The fossil records give us little evidence as to the origin of mollusks. It is generally believed that they were derived from "worm-like" ancestors, although these were not the familiar annelids (see Chapter 11) of today. Larval studies indicate a close relationship with the annelids; but that relationship extends far back in time, since the larval stage of the mollusk, the *trochophore*, shows no segmentation in the mesoderm (Fig. 10–2). It seems rather strange that this group of animals failed to acquire a body arrangement that is so successful among the annelids, arthropods, and chordates. Since the trochophore larva is common to both the mollusks and annelids, the lack of segmentation in this larva may mean that these two great groups of animals split off and went their separate ways before segmentation was introduced.

The molluscan body plan has certain characteristic features that appear consistently in all of the species in the group. One of these is a muscular organ, the **foot**, an organ which serves for several types of locomotion. The snail and chiton crawl on it, the clam digs a wedge-shaped path with it and also walks on it, while the squid uses it to capture prey as well as to crawl over the ocean floor (Fig. 10–1). Another new character is the **mantle,** which is an envelope of tissue covering the entire animal. The mantle gives rise to the **shell,** common in so many members of this group. The original shell appears in the larva as a product of the mantle epithelium and gradually expands as the animal grows.

Those mollusks that possess shells use them as an abode which is readily available for a retreat in case of danger and, more important for land forms, to prevent desiccation. Although a clam is usually safe in its tightly closed shell, it is nevertheless preyed upon by the starfish, which has the ability to open the shell and devour the soft body parts (Fig. 13–5). Some members of the phylum have no shells, such as the slug and octopus. They are protected only by their coloration, their habits, or the ability of some to discharge a cloud of inky material into the water which totally obscures them. Still others secrete a mucus that is particularly distasteful to enemies of the mollusk.

Fig. 10–1. A few marine mollusks and where they live.

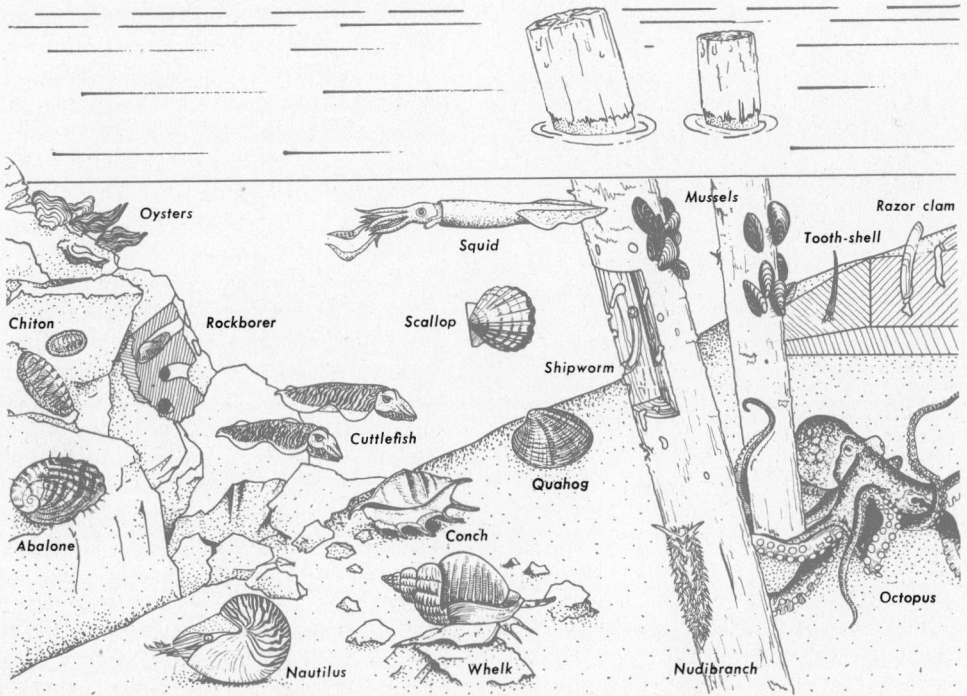

Oysters

Squid

Mussels

Razor clam

Tooth-shell

Chiton

Rockborer

Scallop

Shipworm

Cuttlefish

Quahog

Abalone

Conch

Octopus

Nautilus

Whelk

Nudibranch

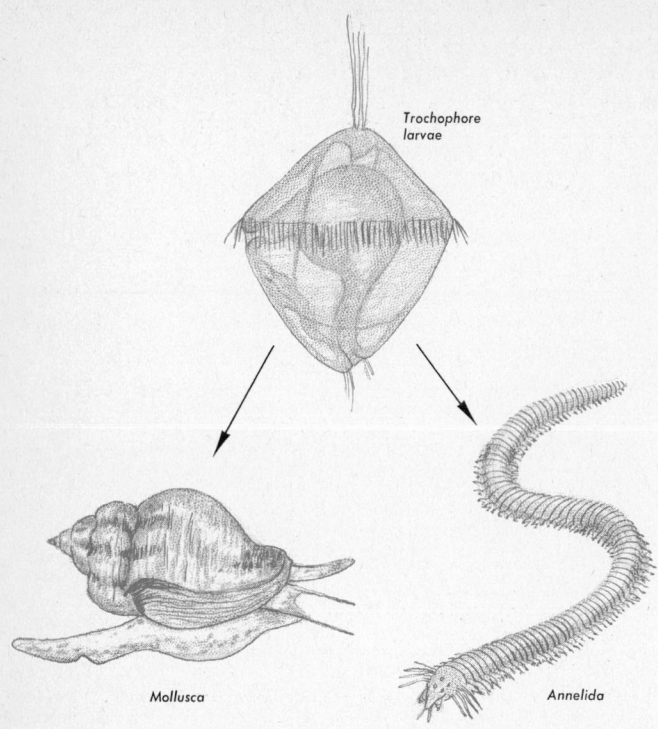

Trochophore larvae

Mollusca

Annelida

Fig. 10–2. The ancestral trochophore larva from which both the mollusca and annelida are thought to have been derived.

The digestive tract is tubular and is coiled in various ways in the different groups of the mollusks (Fig. 10–3). Many of these animals are provided with a peculiar rasping tongue, the **radula,** which is found nowhere else in the animal kingdom. It is used in loosening algae from surfaces, and in tearing bits of plants loose as the animal feeds. Some are carnivorous with radulas constructed somewhat differently than for the herbivores but

Fig. 10–3. Modifications in the body plan of various kinds of mollusks.

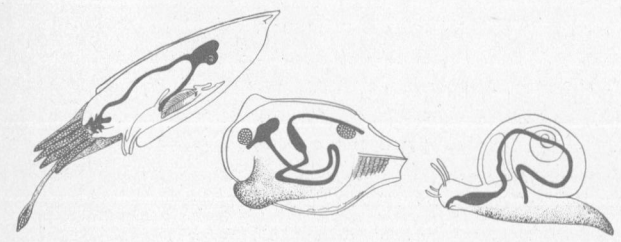

just as effective in tearing bits of animal tissue for food. The radula is a long ribbon of tough tissue, to which many sharp teeth are attached. Muscles are arranged so as to pull the radula back and forth over a projection which is thrust out through the mouth while feeding. It is an interesting and clever device to facilitate feeding.

The Clam

The freshwater clam, although differing in some respects from other molluscan forms, is a familiar representative of the entire phylum. It is a bilaterally symmetrical, "headless" animal, enclosed in a double shell, usually found partly buried in the sand of lakes or streams. By means of its hatchet-shaped, muscular foot, which protrudes from the shell, it is able to plow slowly along, feeding on microscopic forms of life.

When the clam "walks" the foot is thrust forward between the two valves of the shell. This permits blood to flow into the many **sinuses** of the foot, causing it to swell and thus form an anchor. As the retractor muscles contract, the clam is drawn forward an inch or so. The blood then is forced out of the foot so that it thins down again and can be withdrawn from the sand. The process is repeated with each step (Fig. 10–4) and a wedge-shaped path is left behind.

If a clam is molested, its foot is hastily withdrawn into the shell by the **anterior** and **posterior retractor muscles,** and the valves are slowly and tightly shut by two powerful muscles, the **anterior** and **posterior adductors.** This is the only means the clam has of barring its door to intruders. To attempt to pull the valves of the shell open is a nearly hopeless task, unless a thin-bladed knife is first inserted through the edge of the shell to sever the large adductor muscles. The starfish, however, has a novel way of opening the valves. It circumvents the clam, attaches its tube feet to the two valves of the shell, and exerts a steady pull. The pull is resisted by the clam for some time, but finally the muscles are exhausted and begin to relax (Fig. 13–5).

Fig. 10–4. Locomotion of a clam.

The two valves of the clam are hinged dorsally by a ligament, which can be observed when the adductor muscles are cut. The shell itself is usually oval in shape, with a blunt anterior end. Along the dorsal surface is the **umbo,** a bulbous structure which is the oldest part of the shell. From it appear the **concentric lines of growth,** which can be counted to determine the age of the clam.

The outer layer of the shell, the **periostracum,** is produced first, then the **prismatic** layer, and finally the innermost part, the **nacreous** layer. The periostracum is rough and can resist the weak acids produced by the dissolved carbon dioxide in the water. The prismatic layer, which gives strength to the shell, is produced from crystals of calcium carbonate, lying perpendicular to the outer layer. The pearly layer, the portion that interests the shell collector, is also com-

posed of calcium carbonate crystals that are arranged parallel with the shell, resulting in an extremely smooth iridescent layer. The mantle deposits this layer over any irregularities that occur, either in the shell or over loose particles that may lodge in the mantle itself. This is the origin of pearls. Foreign bodies, such as grains of sand or the larvae of certain parasitic worms, sometimes become attached to the mantle or lodged between the mantle and the shell. In such a case, layer after layer of calcium carbonate (pearl) is secreted over the particle, eventually resulting in a pearl. The Japanese produce pearls artificially by inserting glass beads into the mantles of clams or oysters. After several years the pearls can be removed and sold on the market. These are true pearls artificially produced.

Once the valves of the clam are opened, a soft body enveloped in a mantle is exposed. The mantle simply consists of two thin sheets of tissue, or **lobes.** The posterior free ends are muscular, and come together to form the **ventral incurrent** and the **dorsal excurrent** siphons, which permit water to move in and out by ciliary action of the inner mantle cavity (Fig. 10–5). Each side of the mantle adheres to the inner nacreous surface of the two valves. At these points of adhesion, the **pallial line** is formed on the shell. The heavy muscular foot lies directly beneath the mantle and extends anteriorly from the mid-portion of the body. Just posterior to the foot and beneath the mantle are two pairs of **gills.** The dorsal portion of the body, directly above the muscular foot, contains the internal organs of the animal and is called the **visceral mass.**

The four plate-like gills are attached from a point between the siphons to the region just opposite the umbo (Fig. 10–5). They hang freely between the mantle and the visceral mass. Each gill is made up of two plates, the **lamellae,** which are held together by bridges of tissue. The cavity between the lamellae is divided into separate **water tubes.** The lamellae are thrown into vertical folds called **gill bars** and are reinforced by chitinous rods. In addition horizontal rows of ciliated pores, or **ostia**, perforate the lamellae through which water enters the gill. The water tubes lead to a dorsally situated **supra-branchial chamber** that continues to the posterior portion of the gill and opens into the excurrent siphon. Blood from the veins circulates through tiny vessels within the gill to be aerated before returning to the heart. In this manner the constant stream of water flowing through the gills supplies the animal with oxygen.

The beating cilia of the gills and the mantle draw water and food into the mantle cavity through the incurrent siphon. The siphon opening serves to strain out all but very minute food particles such as algae, protozoa, and bits of debris. Mucus secreted by the gills catches these particles which are borne anteriorly by cilia to two pairs of triangular, ciliated **labial palps**. The surfaces of the upper part of the mantle cavity and the gills and those of the labial palps are covered with rapidly beating cilia which

function in carrying food to the mouth (Fig. 10–5). The lower part of the mantle cavity, also lined with beating cilia, collects the heavier particles of debris that cannot be used as food. Material that has been rejected at the mouth as undesirable as food and the heavier fragments that have accumulated in the lower mantle chamber are massed together in clots of mucus. When these have reached considerable size, they are forcibly ejected from the animal by a rapid current of water set in motion by a quick contraction of the adductor muscles. In the buccal cavity more mucus is secreted and the food is carried through the short **esophagus** to the dorsally located, sac-like **stomach**. A pair of **digestive glands** joins the stomach through ducts. Digestion occurs both in the stomach and in the glands themselves. In some species of clams the **crystalline style**, a gelatinous rod resembling a pouch or caecum of the stomach, secretes a cellulose- and starch-digesting enzyme. The intestine, leading from the

Fig. 10–5. The internal anatomy of the clam, side view.

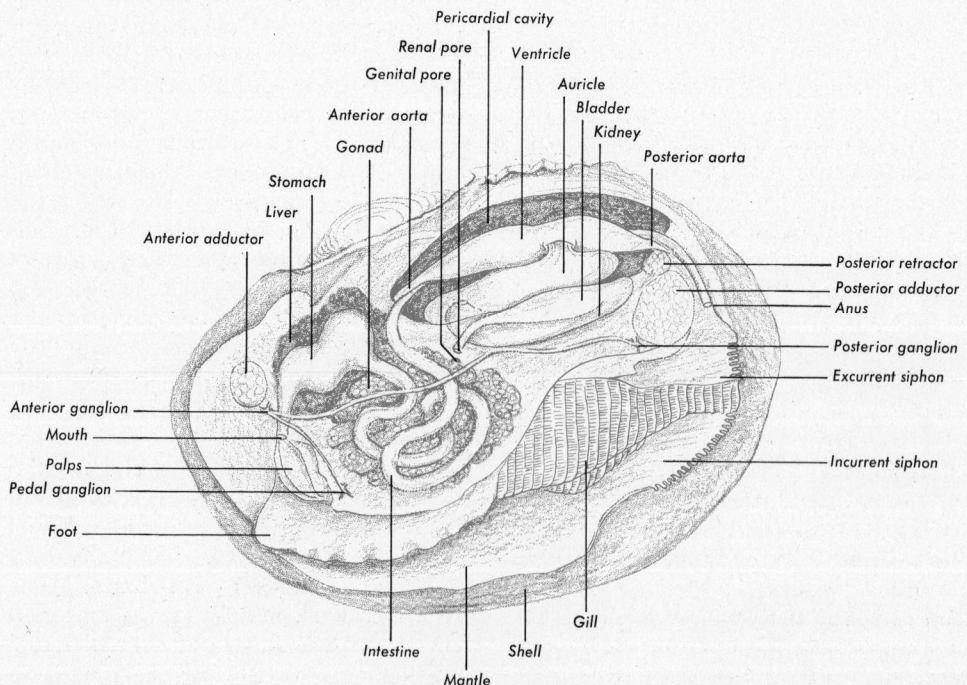

stomach, coil several times through the visceral mass, much of which is the yellow-colored, branched **gonad**, before it turns dorsally to pass through the **pericardial cavity** and the heart. Absorption takes place throughout the length of the intestine, particularly in the portion of the rectum which passes through the **ventricle** of the heart. The **typhlosole** is a longitudinal fold in the rectum. Posteriorly, the intestine opens through the **anus**, located within the excurrent siphon, where the feces are carried away with the outgoing current of water (Fig. 10–5).

The **heart**, lying in the dorsal pericardial cavity, forces the blood through the circulatory system of the clam. The **ventricle**, which is joined by two laterally situated **auricles**, pumps the blood forward into an **anterior aorta**, supplying the muscular foot and viscera, and posteriorly, through the **posterior aorta**, supplying the rectum and the mantle. Blood from those parts of the body supplied by the aortas, with the exception of the mantle, is returned through a vein to the **nephridia**, or **kidneys**, for the elimination of waste products. It then moves to the gills to pick up oxygen and eliminate carbon dioxide. Oxygenated blood is returned from the gills on each side of the clam to the corresponding auricle. The mantle also returns oxygenated blood to the auricles. Unlike the circulatory system of some other animals, some of the arteries and veins of the clam are not joined by capillaries, but end in **sinuses**, without cellular lining. Food and oxygen carried directly to these sinuses can pass into intercellular spaces and are not restricted to absorption through diffusion. This is particularly true in the region of the foot where blood sinuses are numerous.

Two U-shaped kidneys lie ventral to the pericardial cavity. These function in the removal of wastes from the blood and other fluid of the pericardial cavity. Each is a tubular organ, folded upon itself and divided into **glandular** and **bladder**like parts. A ciliated opening from the pericardial chamber into the glandular portion drains this region, while the bladder region opens into the path of the excurrent water, thus carrying wastes out of the body.

Three pairs of ganglia and their connecting nerve cords constitute the nervous system of the freshwater clam (Fig. 10–5). Each pair of ganglia controls the body region in which it is located: the **anterior** or **cerebropleural ganglia** on either side of the mouth; the **pedal ganglia** in the foot; and the **posterior** or **visceral ganglia** ventral to the posterior adductor muscle. Each pair of ganglia is connected by a **commissure** and by two **connectives**, the **cerebrovisceral** and **cerebropedal**, to the other ganglionic pairs. Small nerves extend from the ganglia to the body surface and the muscles. Although the clam is a highly specialized animal in many respects, its nervous system is comparatively primitive. First of all, there is little evidence of a brain. The sensory apparatus is limited to sensory cells on the siphon margins, tactile organs on the mantle, and some areas which are believed to be sensitive to chemicals. Most of these structures resemble similar parts of lower type animals. Clams are slow, sluggish animals and the sensory system required is usually relatively simple.

Freshwater clams, for the most part, are dioecious, but some are hermaphroditic. The **ovaries** and **testes**, yellow in color and surrounding the intestine, constitute much of the visceral mass (Fig. 10–5). The sperm of the male are liberated from the testes through the **genital pore**, just ventral to the aperture of the bladder portion of the kidney. From here they are carried through the body to be discharged through the excurrent siphon. As water is carried into the incurrent siphon of the female it may, purely by chance, carry sperm cells with it. These then enter the **suprabranchial chamber** of the gills where the **ova** discharged from the ovary await fertilization. After fertilization, the eggs are drawn into the water tubes of the gills and attached to them by mucus. While the tubes are carrying eggs they become enlarged and are called **brood chambers**. After a period of development, the zygotes become small larval **glochidia** (singular, **glochidium**), complete with two valves and a larval thread

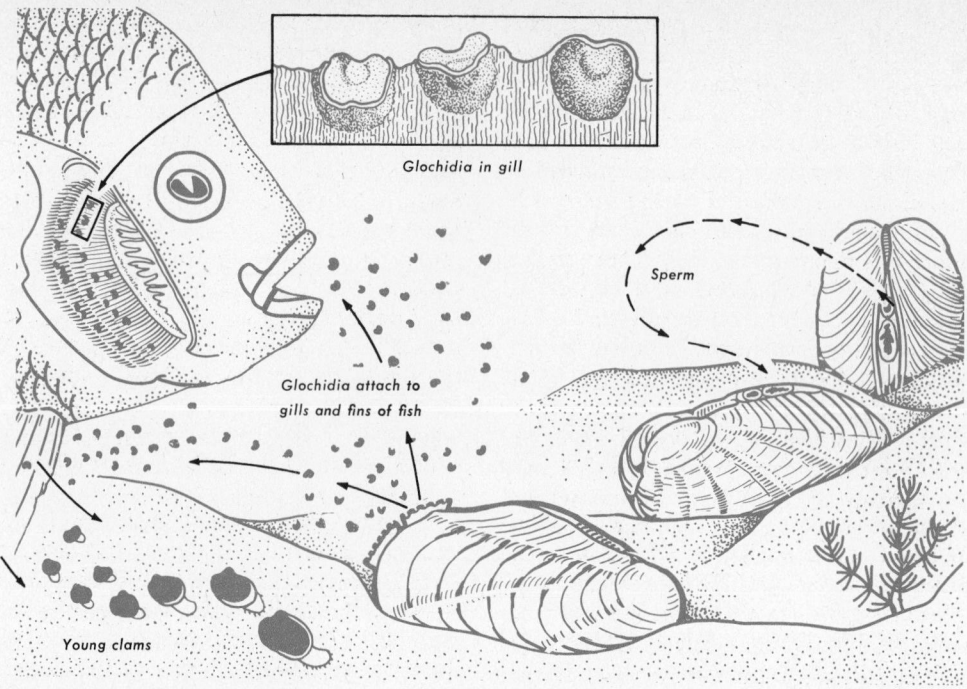

Glochidia in gill

Sperm

Glochidia attach to
gills and fins of fish

Young clams

Fig. 10–6. Life cycle of the clam.

(Fig. 10–6). Many species develop a hooked valve. At this stage they are discharged into the water through the excurrent siphon of the female where they may float in the currents or sink to the bottom. There is no free movement of the glochidium at this time, other than the opening and closing of its valves. Hooked forms try to attach themselves to any fish with which they may come in contact, whereas the hookless forms grasp the gills of fishes by means of their valves. In time, the epithelium of the fish encases the glochidium in a cystlike case. During this period the young clam is entirely parasitic, receiving nourishment from its host through absorption. After the adult organs have developed, the glochidium bursts out of the cyst and sinks to the bottom as an independent free-living animal.

Most pelecypods (from the class name—Pelecypoda) are bottom dwellers. One species *Tridacna*, lives on coral reefs where it lies buried in the coral rock (Pl. 5A). Its exposed mantle, brilliantly colored by sym-

biotic dinoflagellates, varies in pattern from one to another. There seems to be no two alike. Some species of clams even burrow far down into the sand and push their siphons up into the water. Other forms, such as the **oyster,** are permanently attached to rocks or similar objects beneath the water.

There are many species of **shipworms** of which *Teredo navalis* is a common one (Fig. 10–1). The reproductive powers of the shipworms are dramatic, and consequently even though their life span is no more than one year, they manage to infect practically all wood that remains in sea water for any length of time, thus doing untold damage to docks and wharfs. Millions have been spent in research for protecting wood against the invasion of these creatures.

Other Mollusks

Members of the class **Amphineura** are the most primitive forms among the mollusks. In this group are the chitons, which most nearly resemble the probable wormlike ancestors of the phylum (Fig. 10–1). Their dorsal covering of eight calcareous plates has led some biologists to believe that these plates may be the remnant of segmentation. The ocean-inhabiting chitons attach themselves so securely to rocks that it is almost impossible to pry them loose. If they are dislodged they promptly curl into a ball. They are principally "vegetarians," feeding on various kinds of marine algae.

The class **Scaphopoda** includes several "headless" species, the best known being *Dentalium*, the "elephant's tooth" (Fig. 10–1). It has a muscular foot which is modified for burrowing into the sand and is therefore quite sharply pointed. Its elongated body is encased in a tapering shell. Unlike most mollusks, this animal bears no gills and the mantle alone takes care of respiration.

Some of the most interesting and varied forms of the mollusks belong to the class **Gastropoda**. They range in size from microscopic forms to the large whelks and abalones (Fig. 10–7 and Pl. 5B). Although most members of this class possess some kind of shell, forms that are entirely without a shell also appear. Some have adapted themselves to terrestrial life as well as the usual aquatic habitat. The most common form is the snail, an animal familiar to almost everyone. It moves by gliding over a secreted mucus path, using its flattened muscular foot which forms the ventral surface of its body. In some species the foot is actually ciliated to aid the gliding motion; in others movement occurs by rhythmic muscular contraction of the foot. The land snail has a definite head which bears two pairs of tentacles, one short pair, supposedly the center of the sense of smell, and a longer pair with a simple eye at the tip of each. In water forms the eyes are situated at the base of the tentacles.

Judging from its coiled shell, one would expect the snail to have an asymmetrical body. This is only partly true, however, for the head and the elongated flattened foot are bilaterally symmetrical, whereas the remainder of the body, which composes the visceral mass, is asymmetrical, parts of the digestive and circulatory systems being coiled. The shell of the gastropods is univalved, that

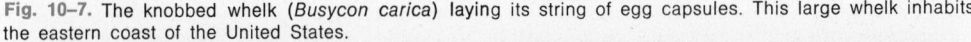

Fig. 10–7. The knobbed whelk (*Busycon carica*) laying its string of egg capsules. This large whelk inhabits the eastern coast of the United States.

is, one piece, but the single valve may vary in shape from tiny flattened spirals to long spindle shapes or even turban or slipper-like forms.

Snails occur in nearly all parts of the world (Pl. 6A, B), even in the Arctic and high snow-covered Alps. Some of the fresh-water snails, such as species of *Lymnaea* and *Heliosoma,* are able to survive for several weeks in cakes of ice, provided they are frozen gradually. Movements of these snails can be observed through the ice. Many snails pass the winter in an inactive state, although some remain active throughout the year. Some aquatic snails are able to hibernate when the water dries up, and with the return of moisture they become active again. Land snails are active only during the warmer parts of the year in temperate zones and show greatest activity at night or immediately following a light rain. As cold weather approaches they seek a protected place for hibernation. During this quiescent period, a membrane is formed over the aperture of the shell, protecting the animal from desiccation.

With the exception of a very few snails, the West Coast abalone (Pl. 5B) being one of them, snails are harmful to man. The herbivorous land snails, especially those that have been accidentally introduced into new

regions, such as the giant African snail, cause considerable damage to vegetation. Some forms serve as intermediate hosts for parasitic flatworms, such as the human blood flukes causing schistosomiasis. It is difficult to calculate the damage done in this fashion by snails, but it must be tremendous. In former times snails were an important source of food for man, but this is negligible today.

The **slugs** are gastropods that do not bear shells. Some are large and drab such as the sea hare (Fig. 10–8). The nudibranch, sometimes called the sea slug, does have a shell during the larval stage but it is lacking in the adult. In other respects it resembles the snail. The name **slug** was apparently given to the lifeless preserved laboratory specimen which takes on a drab, collapsed appearance after it has been exposed to light and preservative. When observed in its natural habitat on rocky coasts, however, these colorful animals are found to be most inappropriately named. Much of their beauty is due to their highly colored translucent gill-like filaments that extend some distance above their backs and are constantly waving in the ocean currents. Their exquisite beauty can only be appreciated when observed under natural conditions (Pl. 7).

Other very interesting relatives of the snail are the **pteropods,** or sea butterflies. Their fleshy foot is modified into a pair of wings which make it possible for these tiny creatures to skip and sail over the surface of the sea. They travel in vast numbers, resembling butterflies in their flight, and some of them swim among the icebergs around Greenland where they serve as the chief food for the whalebone whales.

While the gastropods were of economic significance to ancient man, the pelecypods are most important to modern man. Clam chowder, sautéed scallops, and oyster cocktails have become favorite forms of sea food all over the world. The so-called American oyster, *Crassostrea virginica,* is so commercially profitable that it is cultured and harvested in a very efficient manner. Millions of dollars are involved in this industry. The shells of bivalves were used by primitive man

Fig. 10–8. A sea hare (*Tethys californica*) from the California coast. These large slugs secrete a purplish fluid when disturbed. This one had just completed laying its eggs when taken.

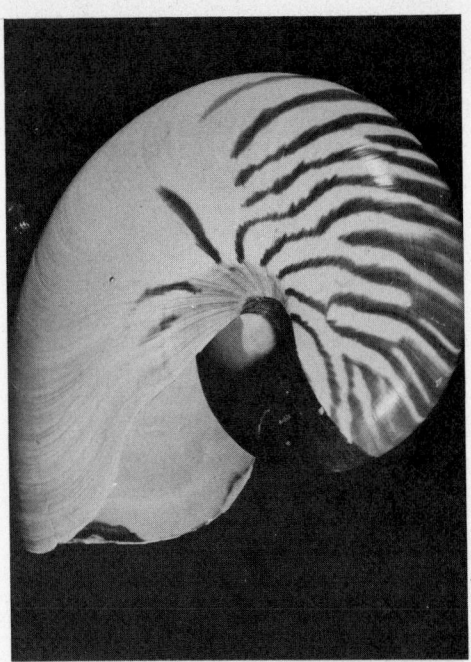

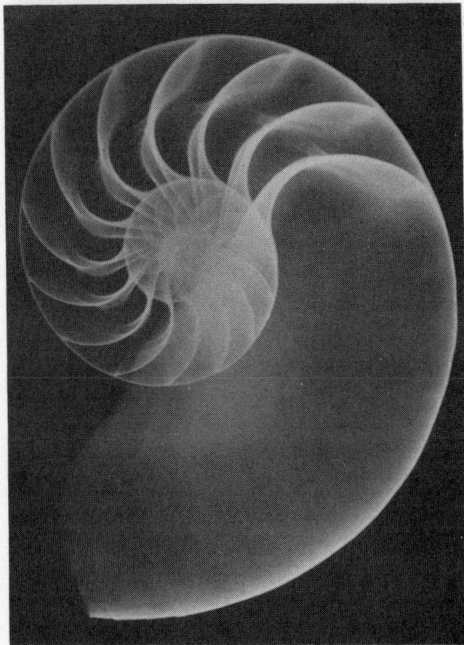

Fig. 10–9. The chambered nautilus, a cephalopod mollusk, lives in the deep waters (1,800 ft.) of the South Pacific. Its shell shown here as it appears from the outside (above) and inside (below) is composed of many chambers, the last and most spacious being occupied by the animal. The bottom picture is an X-ray photograph.

for household utensils and even today find their way into our lives. Bits of shell that are cut and polished into buttons are products of fresh-water clams, and in this day of "substitutes," men still fasten themselves in their shirts by means of these pearly gadgets. The most cherished products of bivalves are the rare jewels secreted by the mantle of fresh-water clams and pearl oysters.

Members of the class **Cephalopoda** are the most highly organized of the mollusks and include the largest species of the invertebrate animals. The head region, as the name implies, is large and well developed, unlike most of the preceding groups. Most forms of this class bear two large complex eyes, resembling the eyes of vertebrates. Some have continuous shells, such as the shell of the *Nautilus* (Fig. 10–9), a member of the group immortalized by Oliver Wendell Holmes in his poem, "The Chambered Nautilus." In others, such as the **cuttlefish** (*Sepia*) and the **squid,** the shell is located internally. In addition to a generally large, fleshy body, the cephalopods usually have long muscular **arms,** or **tentacles,** which are modified portions of the foot.

A very common cephalopod and one that has become an important animal in nerve physiological research is the squid, particularly *Loligo,* which inhabits our eastern seaboard (Fig. 10–10). In the oceans of the world other species abound, varying in size from 1 inch long to giant forms 20 or more feet with tentacles 30 feet long. Fossil remains indicate that squids were one of the most numerous animals during prehistoric times.

The tapering body of the squid, which suggests an arrowhead or rocket, enables it to dart through water with lightning-like speed. This is accomplished by jet propulsion. In emergencies the water within the mantle cavity is forcibly ejected through the siphon which propels the squid in the opposite direction with considerable speed (Figs. 10–11 and 10–12). For ordinary swimming the squid employs its lateral muscular fins which are sufficiently strong to permit forward and backward movement at slow speeds. In spite

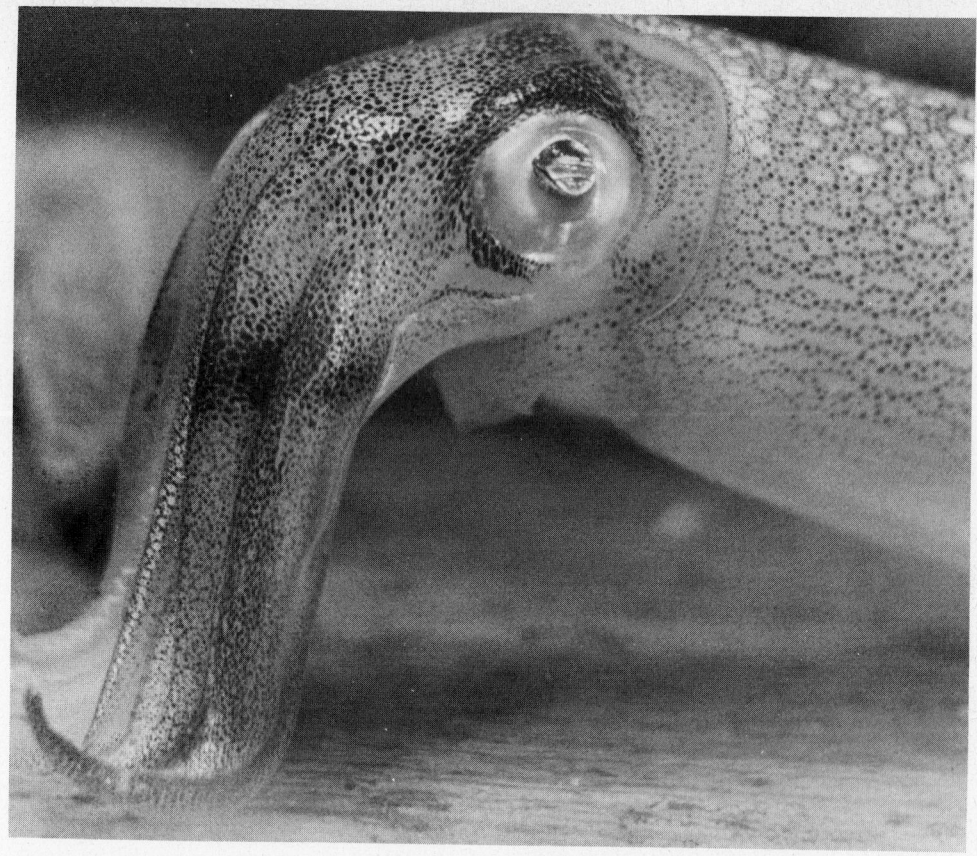

Fig. 10–10. Members of the class Cephalopoda are conspicuous by their many tentacles or arms and their large complex eyes. The squid (*Loligo*) is a common member of the class. This one is resting on the bottom of an aquarium. Note the siphon just below the eye which controls the squid's direction of movement. Also note the contracted chromatophores which cause the animal to appear light in color.

of the fact that the squid is a rapid swimmer, it is often caught by large fish and some whales. When it is hard pressed by its enemy, it will discharge an inky fluid into the siphon, thus spreading a cloud of murky water which obscures the vision, and perhaps numbs the olfactory organs of the enemy. Thus the squid possesses a defensive device which man has only recently used in warfare.

Squids have still another protective device which, though present in many other animals, is here developed to a high degree. This is its capacity to change color. Below the thin epidermis are complex **chromatophores**, color-bearing structures which can change

from pale white to a deep purple in a matter of seconds.

The squid is supported internally by two kinds of skeletons. One consists of the so-called **pen** which, in *Loligo*, is composed of chitin and lies in the mantle. The other is composed of cartilage and makes up a skull enclosing the brain. The cartilage is remarkably similar to that found in vertebrates, which is another case of convergent evolution (p. 103).

Another highly modified structure in the squid is the foot which is composed of ten muscular **arms** each of which possesses sucking disks. These are modified for different functions. One pair (sometimes specified as the tentacles) is longer than the others and is used in capturing prey (Fig. 10–12). The victim is held by the shorter arms while it is being crushed by the **mandibles** and torn into fine bits by the **radula**. The lowest left arm in males is modified for the transfer of sperm capsules to the female during the breeding season. During swimming the arms are clustered tightly lending a fusiform shape to the entire animal, the shape which resists movement in the water least.

Internally the organs resemble those of other mollusks. The digestive system begins with the long esophagus which terminates in the sac-like stomach. A large blind sac, the **caecum,** extends posteriorly (actually dorsally) to the end of the animal. The digestive tract continues as the intestine terminating in the mantle cavity, a handy arrangement for getting rid of wastes, much like in the clam. The large liver lies near the esophagus. The paired kidneys also have their excretory exit in the region of the siphon.

The circulatory system is quite unique in that it has two types of hearts. One (arterial heart) pumps arterial blood throughout the body. The second type consists of two hearts (branchial hearts), one located at the base of each of the two gills; these hearts receive the venous blood from all over the body and pump it through the gills to the arterial heart.

The nervous system is composed of several large ganglia (brain) arranged in a ring around the esophagus. Under the mantle near the head there is a pair of star-shaped ganglia from which radiate a number of giant nerve fibers. These innervate the mantle and are responsible for the rapid contraction of the muscle in this organ when jet propulsion is required. These huge single nerve fibers have become popular with nerve physiologists because they are several hundred times larger than vertebrate nerve fibers. Because of their size they are much easier to use when it is necessary to insert tiny electrodes in determining the characteristics of the nerve impulse.

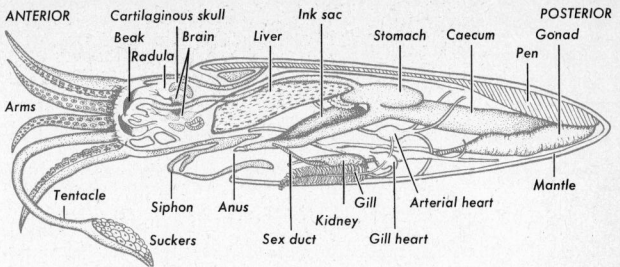

Fig. 10–11. The internal anatomy of the squid.

The squid is aware of its external world through the usual sense organs. It has a pair of **statocysts** located in cavities in the skull which serve as organs of equilibrium. Light stimuli are received by a rather remarkable eye, one that resembles in many respects that of vertebrates. The eye is made up of a transparent cornea, a pigmented iris, a more-or-less spherical lens located in a vitreous chamber, and a retina. The arrangement of the rods is the opposite of that in the vertebrate eye. Its evolutionary significance is discussed in Chapter 26.

Fig. 10–12. During locomotion the squid forces water through its siphon which results in its rapid forward and backward movements. The direction it moves is controlled by the siphon. When laying eggs the female assumes the position shown at the lower left. The egg cases are shown attached to a rock.

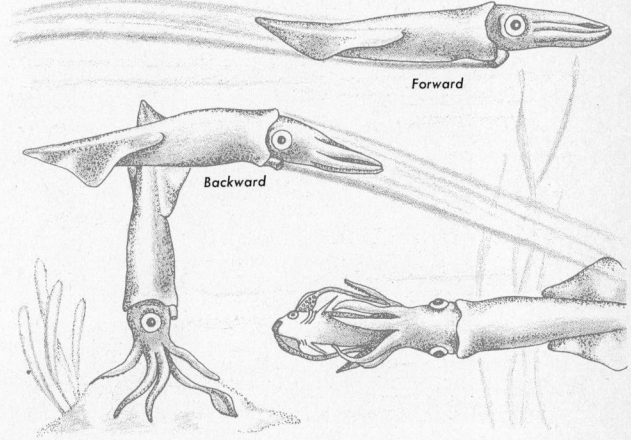

During the breeding season the mature single gonad practically fills the posterior end of the squid. In the male a complicated system of glands package the sperm cells into **spermatophores.** These are "handed over" to the female during courtship, and the sperm are released from them and then fertilize the eggs. The females possess **nidomental glands** which produce the jelly in which the eggs are embedded. Fertilized eggs develop directly into tiny squids which resemble the adults. They swim about in schools for some time, feeding on minute marine organisms.

The octopus (Fig. 10–13), famed in legend as a killer, is actually a very shy animal. Its food does not usually include humans although some octopi, reaching a length of 28 feet can cause trouble if pro-

voked. The bulbous and flexible body of the octopus possesses muscular arms that are fortified with powerful suckers. It usually lurks in shady underwater caverns awaiting its prey, which it seizes with quick movements of the arms. A siphon, similar to that of the squid, enables the octopus to swim, but it more commonly crawls over rocks. Since it is so ambulatory underwater one wonders why it has not moved onto land during its long evolution.

In the mollusks we have seen a very successful group of animals with highly diverse body structure and a high degree of complexity. The coelom is small and rudimentary, although the organ systems have evolved to a point to permit the development of a large body, indeed one of the largest. Their circulatory system is sufficiently good to per-

Fig. 10–13. Cephalopoda. The octopus or devilfish (*Octopus*) is completely naked and has eight very long powerful arms, well provided with suckers, which aid the animal in crawling over rocks in search of crabs which are its main source of food. Its heavy horny jaws are used in cracking the shells of crabs and the radula tears the flesh in tiny bits that are eaten. Most octopuses are small and relatively harmless, although they continue to maintain their diabolical reputation.

mit rapid movement. However, their nervous system is primitive in that it has no well-defined brain, the brain being clearly present in all subsequent higher forms. In the next phylum we see this system more centralized, the coelom expanded, and the introduction of a new type of body organization, segmentation.

SUGGESTED SUPPLEMENTARY READINGS

Books

GILLETTE, K., and McNEILL, F., *The Great Barrier Reef and Adjacent Isles*. Sydney, Australia: The Coral Press Pty. Ltd., 1959.

MacGINITIE, G. E., *Natural History of Marine Animals*. New York: McGraw-Hill, 1949.

MORTON, J. E., *Mollusca*, 4th ed. London: Hutchinson University Library, 1967.

PURCHON, R. D., *The Biology of the Mollusca*. New York: Pergamon Press, 1968.

RICKETTS, E. F., and CALVIN, J., *Between Pacific Tides*, 9th ed. Stanford, Calif.: Stanford University Press, 1968.

WILBUR, K. M., and YONGE, C. M., eds., *Physiology of Mollusca*, Vols. I and II. New York: Academic Press, 1964, 1966.

WILSON, E. P., *They Live in the Sea*. London: Wm. Collins & Sons, 1947.

YONGE, C. M., *The Seashore*. London: Wm. Collins & Sons, 1949.

Articles

KEYNES, R. D., "The Nerve Impulses and the Squid." *Scientific American*, December, 1958.

LANE, C. E., "The Teredo." *Scientific American*, February, 1961.

11

SEGMENTED ANIMALS- ANNELIDS

The features initiated in earlier phyla are further developed in the phylum **Annelida**, and a new one is added. This is **segmentation,** or **metamerism,** which is one of the most obvious differences noted in these animals from the ones previously studied. Segments are serially similar parts, conspicuous on both the outside and the inside of the body. Internal organs are repeated in every segment, each part resembling all others in most respects. There are modifications, however, in certain body regions, as we shall see. There is an intricate connection between the segments, so that the animal is a coordinated whole. Segmentation is retained in higher forms such as the arthropods and the chordates, and is therefore a feature that has contributed toward the greater complexity of animal bodies. This characteristic has made it possible for animals to become larger, swifter, and more complex.

PHYLUM ANNELIDA

The members of the phylum Annelida are mostly marine worms (Fig. 11–1). They abound in the oceans and live near the shore and hide in the sand in burrows. Since they form the basis for further development, it is necessary to examine this group of animals carefully, which we shall do by studying two representatives, the sandworm (*Nereis*), and the common earthworm (*Lumbricus*). Of these, *Nereis* is more typical because it is aquatic and possesses more of the characteristics of the phylum. The earthworm, on the other hand, is terrestrial and in many respects is quite unlike most annelids.

Nereis

Nereis, the common "sandworm" or "clamworm" (Pl. 8A), lives in shallow water on the sandy shores of most oceans of the world, where it is found in burrows with its head and tentacles protruding slightly. When small animals venture too close, the worm suddenly everts its heavily armed proboscis, seizes its prey, and drags it into the burrow to be devoured. The worm leaves its burrow during the breeding season, but only rarely does it leave otherwise.

The segments of *Neresis* are conspicuous externally, especially because each one possesses a pair of laterally placed paddles, called **parapods**, which function like oars on a boat to propel the animal through the water. Paired bunches of bristles (**setae**), located in the parapodia, hold the animal in its burrow, should an outside force attempt to remove it. The head is a distinct structure well provided with sense organs in the form of four eyes, and two pairs of tentacles which appear to be tactile in function (Fig. 11–2). There is a protrusible pharynx which terminates in a pair of fierce looking jaws. The sturdy muscular body is covered with a cuticle which takes on an iridescent sheen in the sunlight.

The tube-within-a-tube body plan is conspicuously evident when one studies *Nereis* internally (Fig. 11–2). The organs are serially repeated in each segment with the exception of the first and last. The gut, a straight tube passing from mouth to anus, is surrounded by a sheet of tissue, the **peritoneum**, which forms the walls of the **coelom**, a very important cavity. The coelom functions in nutrition. A large portion of the food and waste products diffuse into the coelomic fluid and through it reach their ultimate destination.

The circulatory system consists of tubes to which blood is confined throughout its circuit, quite different from the **open** system of the mollusks. However, contraction of the large blood vessels serves the function of the heart. Tiny capillaries allow rapid exchange of oxygen and carbon dioxide, as well as food and waste products. This type of **closed** system is present in most higher forms. The blood contains hemoglobin which is free in the plasma.

Nereis receives its oxygen and eliminates its carbon dioxide through the thin walls of

Fig. 11–1. A few marine annelids in their natural habitats.

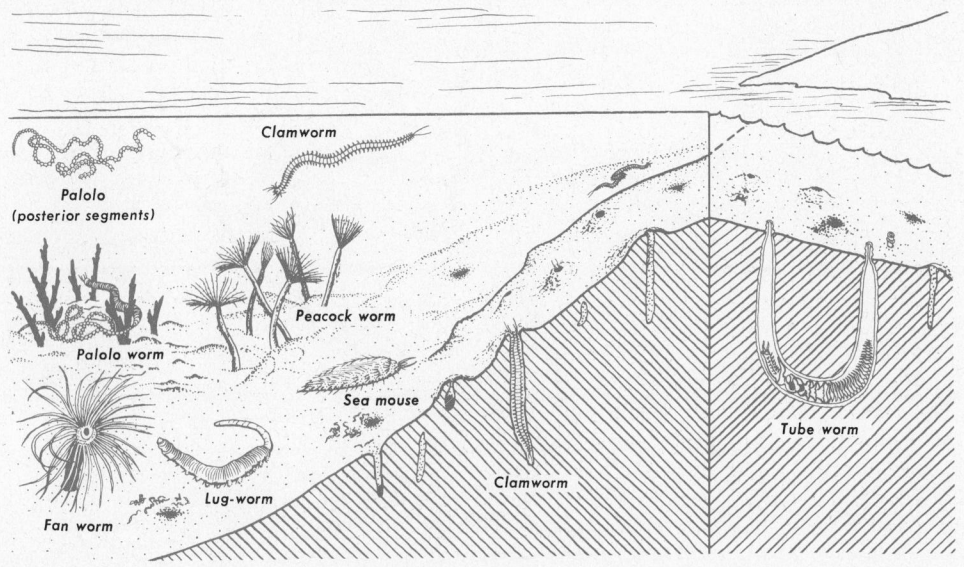

the parapods. The constant waving movement of these organs facilitates rapid gas exchange. The excretory system consists of tiny tubular "kidneys" called **nephridia**, a pair in each segment. The internal opening, the **nephrostome**, is a ciliated funnel into which coelomic fluid bearing waste products enters. As it passes through the long tubule, the wastes are concentrated and excreted through the outside opening, the **nephridiopore**. The glandular portion of the tubule is a special type of tissue that has the unique properties of accumulating and excreting nitrogenous wastes.

Nereis shows more varied and specific responses to the external world than are found in lower forms. Its four eyes and sensitive tentacles aid the worm in getting around in dim light. The centralization of the nervous system, initiated in planaria, is carried much further in *Nereis*. Not only has

the nervous tissue concentrated into two large masses, but also each segment has a similar enlarged **ganglion**. With this organization, the animal has a well-developed means of coordination, a far cry from the nerve net of hydra.

During the nights of late July and August, *Nereis* gather in great numbers along the eastern coast of the United States for the purpose of shedding their eggs and sperm. When a small area of the ocean surface is illuminated, thousands of worms may be seen at this time churning the water by swimming at a rapid, erratic rate, and swirling in apparent frenzy. Suddenly the females discharge their eggs into the sea water like a white cloud. The males, which are smaller than the females, shed their sperm in a similar manner. By dipping up some of this water it is possible to observe the early stages in the embryology of this

Fig. 11–2. *Nereis* in cross section to show the internal body plan of an annelid. A "head on" view demonstrates how well this animal is equipped with sense organs.

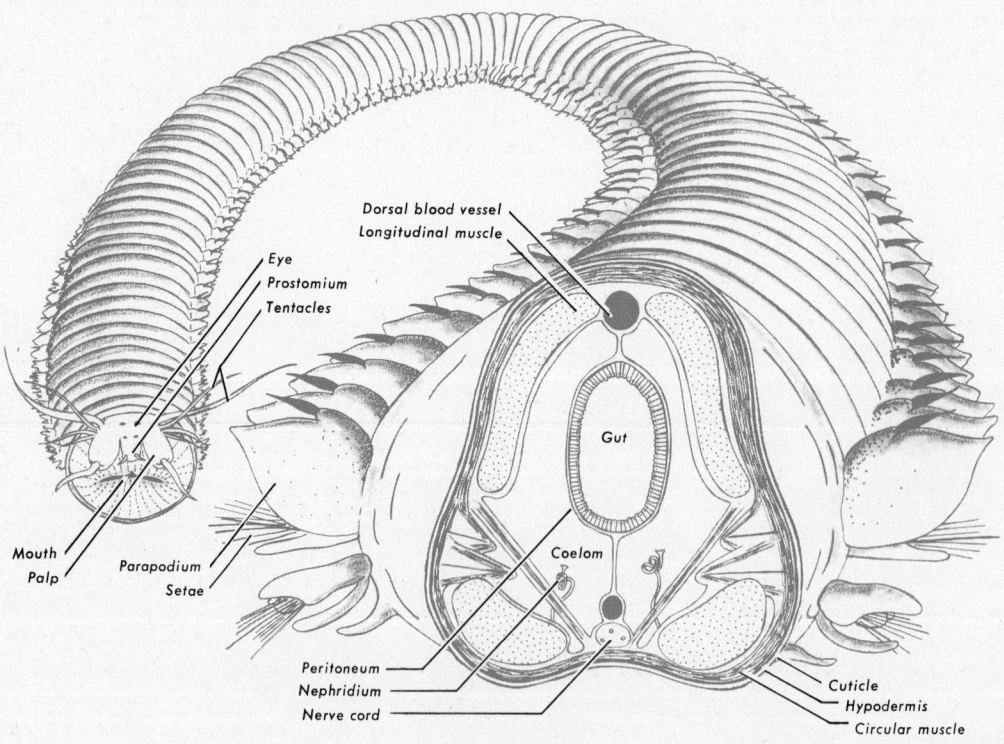

Eye
Prostomium
Tentacles

Dorsal blood vessel
Longitudinal muscle

Gut

Mouth
Palp

Parapodium

Setae

Coelom

Peritoneum
Nephridium
Nerve cord

Cuticle
Hypodermis
Circular muscle

animal. The eggs undergo segmentation and develop into free-swimming ciliated larvae, called **trochophores,** which promptly settle to the bottom and metamorphose into young *Nereis.*

Relatives of Nereis

There are many close relatives of *Nereis* living along the ocean shores, although some are found at great depths. They range from microscopic forms to those that reach 10 feet in length. Some are active swimmers and live in the open ocean catching their prey in flight, whereas others such as the sea mouse (Pl. 8B) crawl over the ocean floor. Many construct burrows out of mucus, such as *Chaetopterus* (Pl. 9A); others bore into rocks to provide a home for themselves. Some are highly colored, such as the peacock worm (Pl. 9B), which could easily be mistaken for a flower.

Many of these worms have spectacular breeding habits. One, the **palolo worm** (*Eunice viridis*) of Samoa and Fiji, spawns in a most remarkably regular and peculiar manner (Fig. 11–1). On the first day of the last quarter of the October–November moon, the posterior portion of the worm, heavily laden with eggs or sperm, breaks off from the parent worm and swims to the surface where the gonadal products are discharged. The surface of the sea is milky white with the great numbers of these cells. Natives, who are familiar with the exact time of spawning, collect these worms when they are about to spawn and feast on them.

Many animals that have external fertilization must depend on rhythms in their environment to be able to synchronize the discharge of eggs and sperm in order to assure fertilization. In the oceans such rhythms as variation in temperature, length of day, and food control the sexual behavior of many marine animals. Also the tides, which produce currents and changes in depth of coastal waters, and lunar and solar light are pronounced environmental changes to which organisms are sensitive. They seem to possess "clocks" which are set to synchronize with these environmental rhythms. These are discussed in Chapter 27.

The Earthworm

No discussion of the phylum Annelida is complete without a study of the lowly earthworm, spurned by the squeamish and cherished by the fisherman and robin. It seems striking that this creature, which is so helpless when removed from its burrow, has been able to spread its kind over nearly the entire surface of the earth. This is even more surprising when it is known that most of its relatives are aquatic forms. It apparently deserted its ancestral watery environment and moved into a terrestrial habitat of semisolid soils. In this transition it lost certain of its ancestral parts and acquired others. There are relatively few species of earthworms compared to many species of aquatic forms.

Earthworms are highly beneficial to man by constantly tunneling the soil, thus permitting a greater circulation of water and air. Charles **Darwin** noted that their castings on fertile soil amounted to as much as 18 tons per acre per year. This constant elevation of subsoils to the surface tends to cover rocks and gravel, thus making the topsoil more tillable. In this sense, too, the earthworms benefit man.

External Anatomy. The large species, *Lumbricus terrestris* (Fig. 11–3), is a good example of an earthworm. Its most conspicuous external characteristic is its segmentation. Mere vestiges of the parapods remain in the form of four pairs of very short **setae** located on each segment. These are used for traction in locomotion. There are no conspicuous sense organs on the head end, like the eyes or tentacles of *Nereis.* Indeed, the animal seems to lack a head, though it does have a protruding "lip," the **prostomium,** which covers the mouth.

The saddle-shaped **clitellum,** which functions in reproduction, rests on the dorsal side about one-third of the way from the anterior end (Fig. 11–4). There are

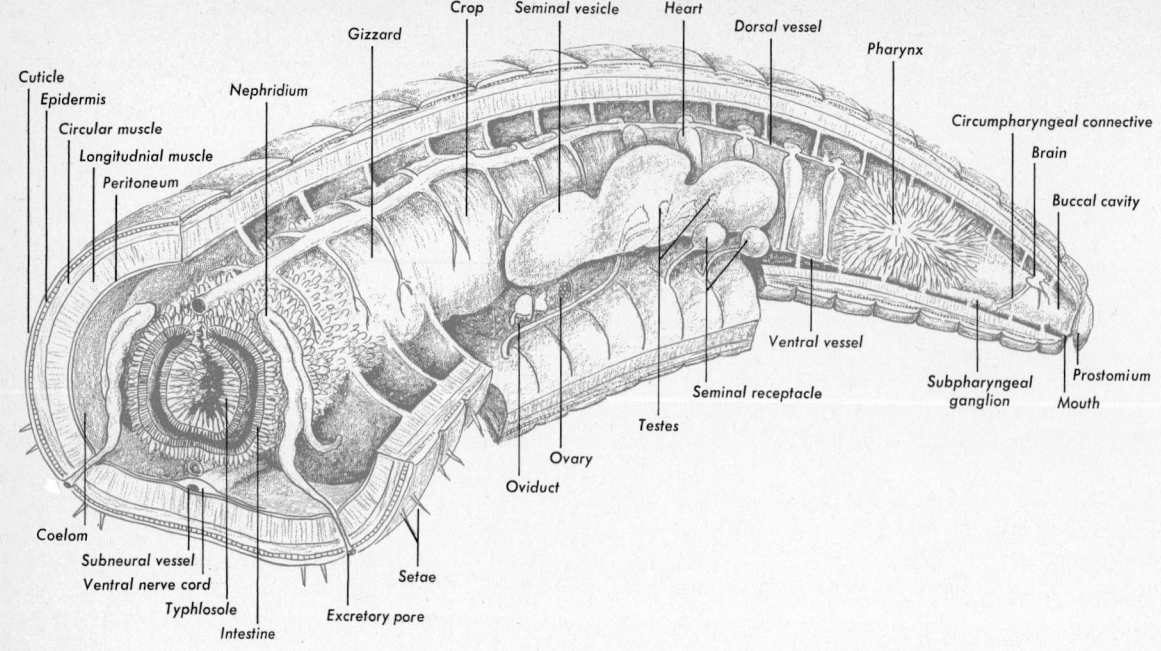

Cuticle
Epidermis
Circular muscle
Longitudnial muscle
Peritoneum
Nephridium
Gizzard
Crop
Seminal vesicle
Heart
Dorsal vessel
Pharynx
Circumpharyngeal connective
Brain
Buccal cavity
Ventral vessel
Seminal receptacle
Subpharyngeal ganglion
Prostomium
Mouth
Testes
Ovary
Oviduct
Coelom
Subneural vessel
Ventral nerve cord
Typhlosole
Intestine
Setae
Excretory pore

Fig. 11–3. Longitudinal and cross section of the earthworm.

also several openings, the most noticeable of which are those of **sperm ducts,** which open on the fifteenth segment. The fourteenth segment bears the smaller openings of the **oviducts,** and each segment except the first three and the last bears a pair of **nephridiopores.** Finally, the four openings which lead into the **seminal receptacles** are located in the grooves between segments 9 and 10, and 10 and 11.

Internal Structures. A section of the body wall (Fig. 11–3) shows the outer thin, tough **cuticle,** which serves as a protective layer for the tall **columnar epithelial** cells composing the bulk of the epidermis. Among these are scattered sensory cells, sensitive to light, touch, and chemical stimulation. Other cells dispersed among the epithelial cells are the mucus-secreting cells responsible for the slimy condition of the skin, which is essential both for respiration and locomotion. Beneath the epidermis lie two layers of muscle, the outer **circular**

and the inner **longitudinal.** Lying beneath the muscle layers and lining the coelom is the **peritoneum.** The digestive tract is a tube, making the tube-within-a-tube plan conspicuous. The segmentation which is so striking externally is just as conspicuous from the inside. Membranous partitions, **septa,** which wall off each segment, are perforated by the **gut, nerve cord, blood vessels,** and **nephridia.**

The **digestive system** starts with the **mouth** which opens into the large muscular **pharynx;** this latter organ functions as a pump to draw food into the mouth. Following the pharynx is the **esophagus,** which opens into the **crop,** a storage sac. This in turn leads into the **gizzard,** which functions in the grinding of food. The remainder of the gut is a long tube, the **intestine.** This organ bears a fold along its dorsal side, the **typhlosole,** which increases the surface of the gut without increasing the volume of the animal.

Various gland cells are located throughout the digestive epithelium. Some produce digestive enzymes, while the secretion of others also lubricates and thus facilitates the movement of food. Lateral to the esophagus and attached to it are three pairs of **calciferous glands,** which function in secreting calcium carbonate for neuralizing any acid soil that may be taken in with the food.

Fig. 11–4. Copulating earthworms, showing the various steps in the fertilization of the eggs and subsequent cocoon formation.

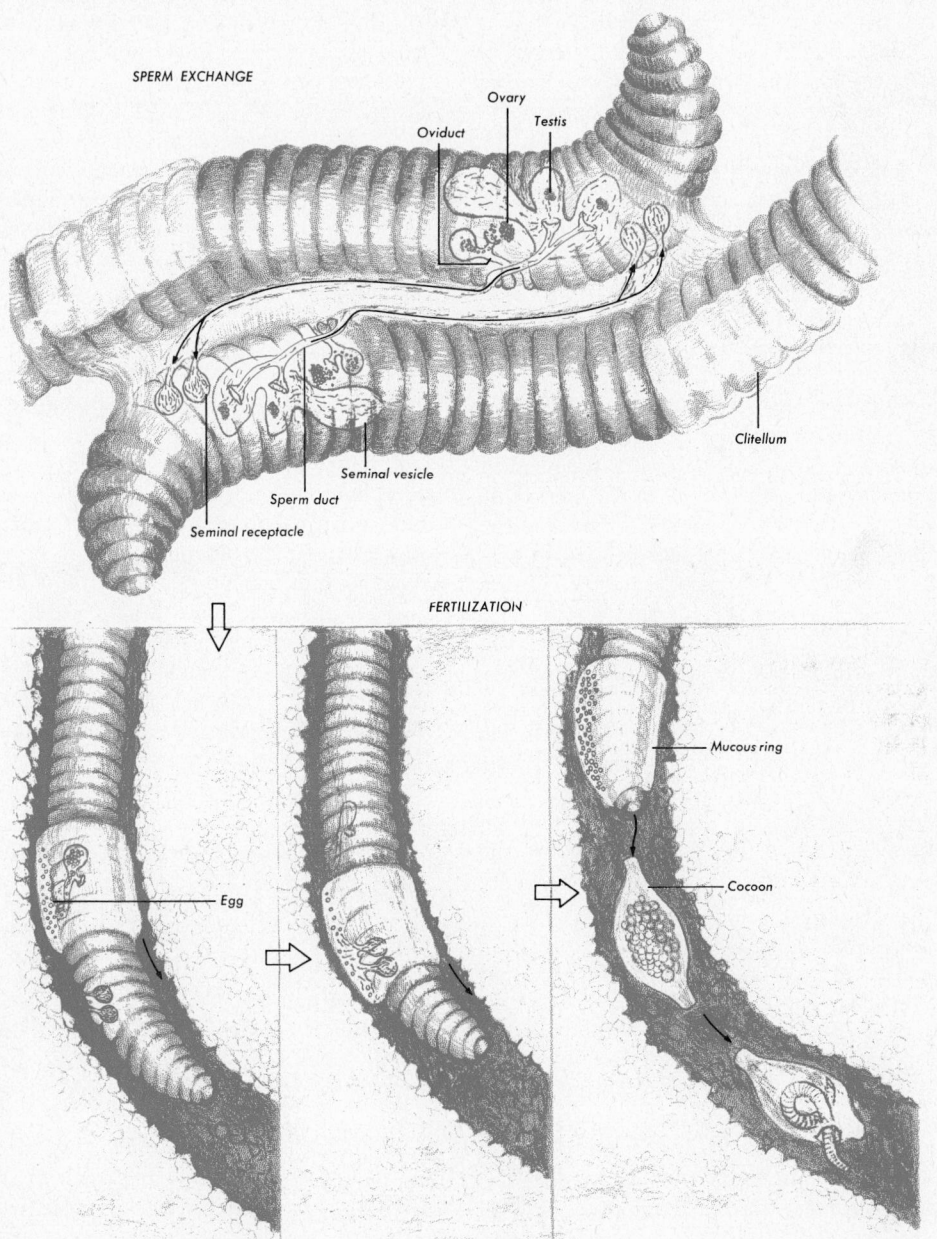

The epithelial lining of the gut secretes fat-splitting, carbohydrate-splitting, and protein-splitting enzymes. The gut is surrounded by **chloragogen cells** which are derived from the peritoneum and probably function in the elimination of wastes from the blood.

Food for the earthworm consists of leaves and any other available organic matter, even bits of meat. Much soil is taken in with the food and used later in the gizzard for grinding the food in preparation for digestion. Food is temporarily stored in the crop before it passes into the gizzard, where it is ground to a fine mass. It then passes on to the intestine where digestion and absorption are carried on. Undigested material passes out through the anus.

The circulatory system (Fig. 11–3) of the earthworm is similar to that of *Nereis*. However, there is an improvement in the pumping system in the form of five pairs of "hearts" which surround the esophagus and connect the **dorsal** with the **ventral blood vessel.** In addition to the peristaltic waves that move the blood forward in the dorsal blood vessel, the "hearts" send it by a contraction of their walls to the ventral blood vessel with considerable force.

The blood of the earthworm contains many white blood cells (**leucocytes**) suspended in the fluid **plasma.** The respiratory pigment is hemoglobin, which is carried free in the plasma, not in corpuscles. The gas exchange takes place where the capillaries come close to the surface of the epidermis. For this reason the surface of the animal must be moist at all times, for a dry membrane will not allow the gaseous exchange to take place.

The arrangement of the nephridia is similar to that in *Nereis*. The nephridium consists of a small ciliated funnel, the **nephrostome,** which opens into a tiny coiled tubule. This penetrates the septum of the next segment, where it coils, gradually becoming larger and finally expanding into a bladder-like sac before opening to the outside through the **nephridiopore.** Waste materials in the blood are picked up by the glandular portion of the tubules and excreted directly.

The center of the **nervous system** is a bilobed "brain," located in the anterior region dorsal to the digestive tract (Fig. 11–3). The **circumpharyngeal connectives** connect to the **ventral nerve cord** which consists of a series of **ganglia** much the same as in *Nereis*. There are a few nerve fibers extending into the prostomium, suggesting that this organ is probably sensitive to touch. The nervous system of the earthworm possesses the components of reflex arcs very similar to those of man. Thus, in a form as low as the earthworm, there is an intricate nervous mechanism which enables the animal to carry out complex operations.

Reproduction. In order to survive in its terrestrial habitat, the earthworm has been forced to undergo some drastic adaptations in its reproductive system. It will be recalled that most annelids discharge their sex products into water where union of the eggs and sperm is purely fortuitous. On land, obviously, some other means must be provided to bring about this union and to insure adequate conditions for the developing embryo. In the first place, the sexes, which are separate in other annelids, are united in the earthworm, that is, it is **monoecious,** or **hermaphroditic.** This has the advantage of making it unnecessary for worms of different sexes to unite; any two worms can exchange sperm.

The **ovaries** and **testes** are located in the anterior end of the worm where ducts provide the proper exit for eggs and sperm (Fig. 11–4). There are two pairs of tiny testes located in the tenth and eleventh segments, surrounded by large sac-like bodies, the **seminal vesicles,** which are storehouses for the sperm cells. Funnels direct the sperm into the sperm ducts which open to the outside on the fifteenth segment. A pair of ovaries cling to the posterior wall of the septum in segment 13, and small funnels catch the eggs and direct them into a sac where they are temporarily stored. Eventually the eggs pass to the outside through the oviduct in segment 14.

The process of exchanging sperm occurs at night, usually following a rain. Two worms become attached along their ventral sides, as shown in Fig. 11–4, 11–5. A slimy material is then secreted mutually by the worms which, as it hardens, encases both animals together in a temporary sheath. Small tubular passageways form between the sheath and the body walls, thus providing a pathway for the sperm, which are forced from each worm along these channels until they reach the **seminal receptacles** of the other worm. After this exchange of sperm cells the animals separate. Later, when the eggs are mature, the clitellum secretes a mucous sheath which slips forward over its "head." In the fourteenth segment eggs are forced into the mucus ring, and as it slips over the ninth and tenth segments, sperm pass out of the seminal receptacles and unite with the eggs. After the mucus ring slips over the anterior end, the ends contract, forming a closed capsule, or **cocoon,** full of fertilized eggs, which is left behind in the burrow. There is no larval stage and the eggs hatch into small earthworms which penetrate the wall of the mucus capsule and begin to shift for themselves.

Relatives of the Earthworm

Earthworms have many relatives which range in size from microscopic forms to species which may reach 10 feet in length. There are well over 2,000 species, and most of them are smaller in size than the common earthworm. Although nearly all live in damp soil, some are found in fresh or polluted waters. One form, *Tubifex*, is encouraged to grow in filter beds of sewage disposal plants in order to keep the filter open. The common blood worm, which is a species of *Tubifex*, is found in tubes at the bottom of freshwater ponds where it feeds on the muck and perhaps aids in the purification

of such waters when they are polluted. Another small white, almost microscopic, form, called *Enchytraeus*, is frequently sold in pet shops as a source of food for small fish.

Leeches

After a swim in the old swimming hole, boys often find small black leeches that cling tenaciously to the skin. When removed, they leave a stream of blood flowing from the wound. It is also common for the fisherman to see a large (12 inches long) leech, *Haemopis grandis*, swimming in beautiful undulating movements near the surface of the water. Another leech, *Hirudo medicinalis*, was once cultured in Europe for the specific purpose of bloodletting when that practice was in vogue. It is interesting to note that during the last century, and before, it was considered beneficial to remove blood in certain illnesses.

The leech has many of the annelid characteristics, but it lacks setae, and it possesses copulatory organs, which other annelids lack. The body has remarkable powers of extensibility and contractibility, enabling it to move like the measuring worm. The holdfast organs are two suckers, one on each end, which are used both in locomotion and in feeding. In the center of the anterior

Fig. 11–5. Earthworms usually copulate during damp periods. These were photographed in the early morning during a drizzling rain.

Fig. 11–6. This onychophoran, *Peripatoides novae-zealandiae*, lives in New Zealand. It can be found under stones and rotting logs where the humidity is high.

sucker is the mouth, which is usually provided with three small cutting teeth that inflict the wound when the leech is feeding upon its victim. The anus is located in the center of the posterior sucker. The digestive tract is sacculated so that it can retain a large meal of blood. Apparently this is provided because meals are usually few and far between, some leeches being able to live a year between feedings.

When securing a blood meal the leech becomes attached to the skin, which it pierces with its teeth. An enzyme (hirudin) is secreted by the salivary glands, which prevents the blood from clotting. Blood is sucked by the pumping action of the pharynx and stored in the large sacculated crop to be passed on into the intestine for digestion a little at a time.

We have seen, so far, the advent of characteristics of key importance to the development of more complex animal bodies. The acquisition of a well-developed coelom, segmentation, and a centralized nervous system have laid the foundation for higher forms. Perhaps the features that have appeared so far in the animal kingdom are more important than any that follow, because later groups, while adding some new characteristics, usually become complex by elaborating features that are already established in these lower animals.

PHYLUM ONYCHOPHORA

We now consider a small group of animals that have puzzled taxonomists for a long time. They are represented by only a dozen genera, of which *Peripatus* is the best known. A close relative is *Peripatoides* shown in Figure 11–6. They are often called "walking worms" and their anatomy would lead one to think of them as closely related to the annelids. However, on close examination one finds that they possess characteristics which resemble both the arthropods and the annelids. Like annelids, they possess a pair of fleshy appendages for nearly every segment. These have clawed feet but they lack joints, which is one of the principal characteristics of the arthropods, the next important phylum. The exoskeleton resembles the cuticle of annelids; it is nothing like the covering seen in all arthropods. The excretory system is annelidlike, as are the tubules of the reproductive system, which are ciliated, a characteristic found among the annelids, but not the arthropods. Only remnants of the coelom remain in the adult, and the main body cavity is the hemocoel. It appears to be segmented, but this is not actually so. The tracheal system resembles that of arthropods but is sufficiently different to be thought of as having arisen independently.

Obviously, *Peripatus* possesses both arthropod and annelid characteristics, which fact has led many zoologists to consider it a connecting link between these two phyla. Rather, the onychophores probably are very ancient animals that evolved from the annelid-arthropod stem and have no close affinities with either modern annelids or arthropods.

Peripatus is a small animal found only in widely separated places—Africa, Australia, Central and South America. Because they are limited to widely scattered, isolated places, it is believed that they were very common all over at one time, but eventually died out in the areas where they do not presently exist. Today they live in the dense, tropical forests, in damp places under logs and other objects found on the forest floor.

SUGGESTED SUPPLEMENTARY READINGS

Books

BORRODAILE, L. A., POTTS, F. A., EASTMAN, L. E. S., and SAUNDERS, J. T., *The Invertebrata*. Cambridge, England: Cambridge University Press, 1958.

DALES, R. P., *Annelids*. London: Hutchinson University Library, 1963.

LAVERACK, M. S., *The Physiology of Earthworms*. New York: Pergamon Press, 1963.

WARD, H. B., WHIPPLE, G. C., and EDMONDSON, W. T., eds., *Fresh-Water Biology*. New York: Wiley, 1959.

Article

GALLOWAY, T. W., "The Common Fresh-Water Oligochaeta of the United States." *Transactions of the American Microbiology Society*, 1911.

12

ANIMALS WITH JOINTED FEET- ARTHROPODS

Although the arthropods resemble the annelids in many respects, they show important innovations. One of these is the hard outer skeleton, the **exoskeleton,** which functions as a rigid coat of armor. This means that the muscles can be attached to the inside of this material and function more efficiently in moving the body. The exoskeleton supports the entire animal, much like the framework of an airplane or automobile. It is composed of a nitrogenous polysaccharide, **chitin.** Some arthropods, such as the lobster, have a high percentage of lime in the exoskeleton, whereas the insects have smaller amounts. The greater the lime content, the

harder the skeleton. For the land dweller, exposed to rapid desiccation, a waterproof outer covering becomes essential. Since the exoskeleton of these air breathers is practically waterproof, one may assume that this condition made it possible for the animal to invade land.

In order to increase in size, arthropods must periodically shed their exoskeletons by a complex process called **molting.** A new skeleton forms under the old one, and the actual process of shedding the old one and hardening of the new one requires several hours in the larger arthropods, during which time they are vulnerable to predators. They

208

seem to be aware of their precarious situation and take special precautions to hide during this period. A single layer of epidermal cells gives rise to the new skeleton, and just prior to shedding, a molting fluid is secreted which digests away parts of the old skeleton, the products of which are absorbed by the body. The animal then swells its body by swallowing air or water which splits the old skeleton open. The arthropod then crawls out, leaving behind the old skeleton which includes not only the covering of the body and legs, but also the lining of the foregut, hindgut, and sometimes the tracheal system. The animal then grows, again by swallowing air or water, thus stretching its new skeleton which hardens by oxidation of the products in the outer layer. In the hard-shelled forms, calcium carbonate is deposited which greatly stiffens the exoskeleton.

Another important difference between the annelids and the arthropods is the presence in the latter of appendages with **joints.** This conspicuous character gives the phylum its name. The feet and legs of animals with hard skeletons need joints for the purpose of movement. In the arthropods many appendages have lost their original locomotor function and have become radically modified into organs of defense, offense, and even sense organs. The exoskeleton and the jointed appendages thus have distinct advantages and have contributed much to the success of this group.

The nervous system, while similar in plan to that of annelids, is more centralized. There are fewer ganglia, and less independence of all parts of the body. This increased integration has resulted in an animal that is swifter, more agile, and better able to cope with its environment.

The coelom is much reduced in size, being replaced to a large extent by a system of blood spaces called the **hemocoel.** There are certain modifications in other systems also, but these are discussed in the various groups, as they occur.

In recent years an elaborate system of hormones has been discovered in arthropods; indeed, they seem to be as numerous and as complex as in vertebrates, although they probably arose independently. Such processes as molting, color change, reproduction, and metabolism are under endocrine control. As in vertebrates, the hormones are related to the central nervous system, and they exist in pairs, each one being antagonistic to the other. One of the most interesting and important discoveries in hormone research was made with arthropod nerves. It was found that the nerve cell body produces the hormone, which passes along the nerve fiber to be secreted at its tip. For example, the so-called **x-organ** in crustacea, which lies one on each side of the optic ganglia, produces a variety of hormones affecting reproduction, metabolism, molting, and body color. The hormones travel through individual nerve fibers to the **sinus gland,** where they are stored until used. All of the hormones were once thought to be produced by the sinus gland itself. A similar situation exists in the vertebrate pituitary gland.

Among insects a hormone is produced in the **intercerebral gland,** located in the brain, which initiates molting. The hormone passes through axons to a pair of small bodies posterior to the brain called the **corpus cardiacum.** When released, this hormone stimulates another hormone to be released from the **prothoracic glands,** located in the thoracic region, which actually sets the machinery going that brings about the molt. This hormone is also responsible for metamorphosis. Another pair of small bodies, located posteriorly to the corpora cardiaca, called the **corpus allatum,** found only in larval insects, is responsible for the juvenile state, and thus is antagonistic to the hormone that brings about metamorphosis. Here we see duplicate hormones that function in an antagonistic manner, much the same as we find in vertebrates.

The arthropods include more species than are found in all other phyla put together; in fact, the species of all other phyla number about 130,000, whereas the arthropods alone add up to 870,000, of which over 800,000 are insects. They literally encompass the earth; they invade the soil, the water, the air, the frigid polar zones, and the torrid equatorial latitudes. Since they feed on the same foods as man, they have become his most serious competitor; indeed, it has been said that the main struggle for survival today is between the arthropods, particularly the insects, and man. Insects are carriers of some of the most serious diseases, and although some species are beneficial, it is doubtful that the benefit to man of some outweighs the damage done by others.

The phylum arthropoda is divided into

Fig. 12–1. A scheme to show relationships among the various groups of arthropods.

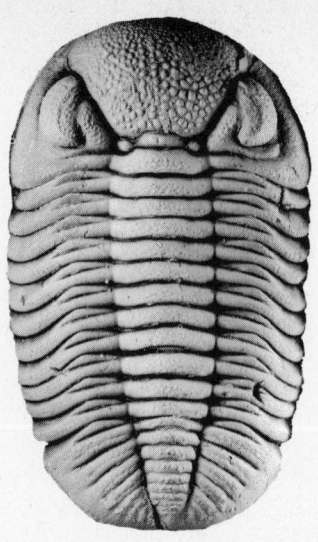

Fig. 12–2. Trilobites were abundant in Cambrian times but later died out. They exist today only as fossils. This one (*Phacops rana crassituberculatus*) shows their arthropod characteristics. Note the broad head bearing compound eyes and the segmented trunk with attached appendages.

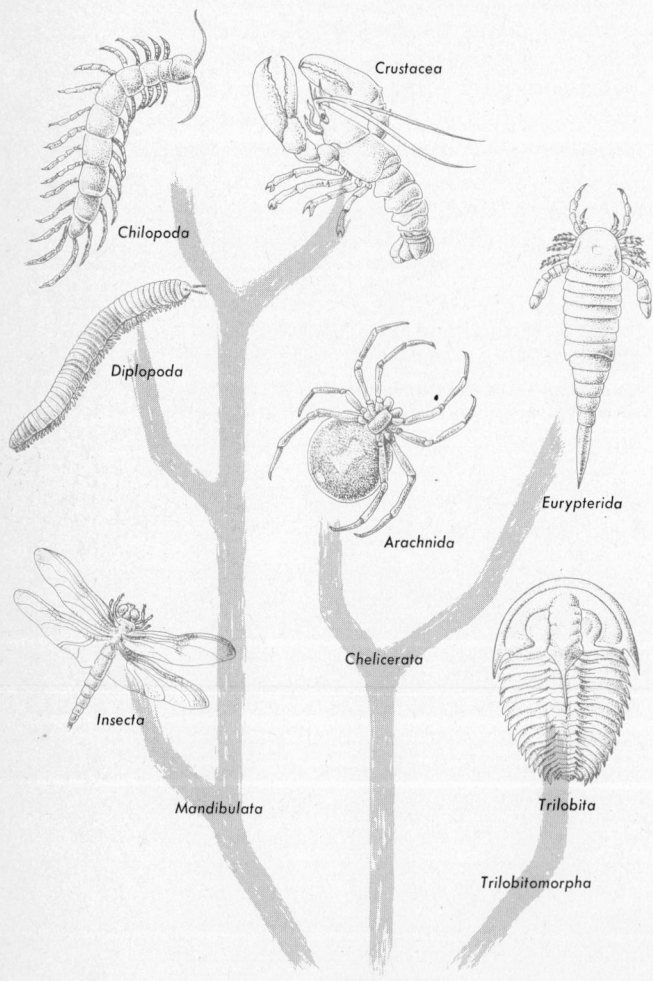

Crustacea

Chilopoda

Diplopoda

Arachnida

Eurypterida

Insecta

Chelicerata

Mandibulata

Trilobita

Trilobitomorpha

ANCESTRAL TYPE

three subphyla, based on the structure of the appendages of the first six segments. They are **Trilobitomorpha, Chelicerata,** and **Mandibulata** (Fig. 12–1). The first includes only extinct arthropods, a common representative being the trilobites (Fig. 12–2). The second subphylum includes those arthropods without antennae and with the first pair of ventral appendages modified into **chelicerae** which are pincerlike structures at the tip of the appendage. This subphylum is divided into four classes: **Xiphosurida,** which includes the famous "living fossil," *Limulus* (Fig. 12–29); **Eurypterida,** extinct sea scorpions; **Pycnogonida,** the sea spiders; and **Arachnida,** the spiders and scorpions. The subphylum Mandibulata includes those arthropods which possess a pair of **jaws,** or **mandibles,** from which the group gets its name. This subphylum includes the most important and numerous of all arthropods. There are a number of classes, of which we shall mention only four. They are **Crustacea** (lobsters, crayfish, and crabs), **Chilopoda** (centipedes), **Diplopoda** (millipedes), and

Insecta (insects). Of this large and complex phylum, we spend most of our time with representatives of only two classes, the Crustacea and the Insecta, because they are by far the most important. Less important groups are discussed at the end of the chapter.

CLASS CRUSTACEA

The Crustacea live on land, in fresh water (Fig. 12–3), and in the ocean (Pls. 10, 11 and Fig. 12–4). They include such common forms as freshwater fairy shrimps and crayfish, seashore crabs (Fig. 12–5), lobsters, and barnacles. The group is of considerable economic significance to man. It is a valuable source of food, as demonstrated by the fact that around $20,000,000 worth of shrimp, crab, and lobster are marketed annually in this country alone. In addition, the Crustacea play an important role in the food chain of fishes (see p. 624). There is a stage in the development of all fish when they must feed on some form of Crustacea; this may be the larval stage of some larger form such as the crayfish, or it may be a minute adult crustacean, such as **cyclops** or **daphnia** (Fig. 12–3). Some Crustacea, on the other hand, act as intermediate hosts for certain dangerous parasites of man, such as the human lung fluke, *Paragonimus westermani*. On the whole, the group is important both economically and biologically.

The Lobster and the Crayfish

These two closely related animals live in different environments, the lobster in the sea and the crayfish in fresh water (Fig. 12–6), yet their bodies have much the same appearance. In fact, a description of one also fits the other fairly well. This discussion is confined largely to the lobster, although frequent references are made to the crayfish.

The lobster is a bottom-dweller, spending its life seeking food, which consists of other Crustacea, small fish, and mollusks. It hides by day and sallies forth at night, crawling

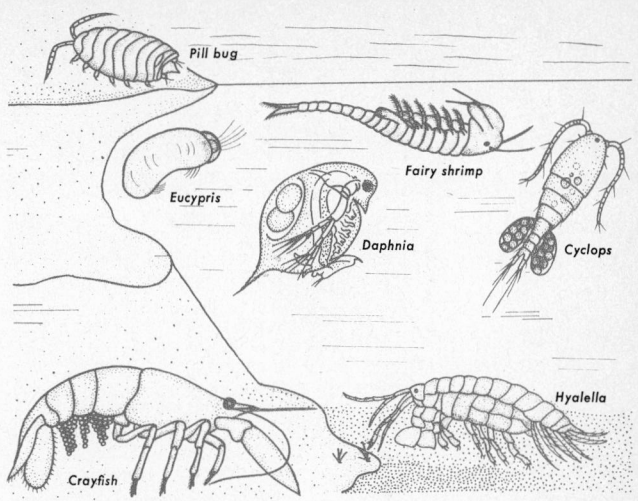

Fig. 12–3. Freshwater and terrestrial crustacea. Although primarily aquatic, the pillbug lives on land.

forward along the sea bottom. If in danger of attack by other predators, it can swim backward with darting speed by powerful strokes of its abdomen. The crayfish employs the same methods in defending itself, and it seeks the same type of food in much the same manner.

Structure. The animal is enclosed in a **chitinous exoskeleton**, containing consider-

Fig. 12–4. Marine crustacea. These vary widely from the sessile barnacle to the burrowing fiddler crab.

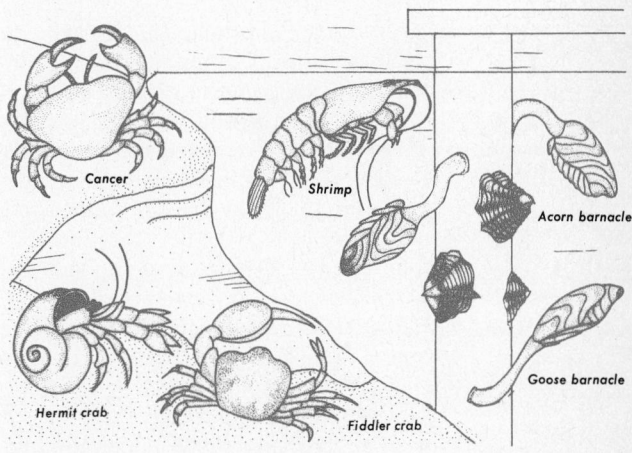

Fig. 12–5. This is the Ghost Crab, *Ocypode cera-tophthalma*, which inhabits the Pacific coral islands. It is a swift marauder that is normally a scavenger but during the hatching season of the large sea turtle eggs, it feeds on the emerging young.

tacea, is said to be **biramous,** because it branches into two parts. It persists in its primitive condition in the **swimmerets,** located on the underside of the abdomen. The single basal portion, the **protopod,** is attached to the body and two branches extend from it, the **endopod** toward the median line and the **exopod** away from it. The original function of such an appendage was locomotion (swimming), and all of the primitive Arthropods, such as the fossil trilobites, possessed just this type and nothing more. Through the ages it became modified in a most versatile manner in all of the appendages except the swimmerets. For example, the antennules and antennae are receptors for tactile and chemical stimuli and resemble the swimmerets very little. One of the most radical departures is the mandible, or jaw, where there is almost no hint of its progenitor. Among the walking legs, as well as other appendages, the exopod has been greatly reduced or lost entirely.

Structures that have a common origin, such as in the case of these appendages, are said to be **homologous,** and when they are

able quantities of lime and sclerotin which make the skeleton heavy and bulky, but a very excellent armor plate (Fig. 12–7). The chitin thins out at the joints, allowing maximum flexibility. The anterior portion of the body is covered by the **carapace,** and each posterior abdominal segment by an arched dorsal **tergum,** two lateral **pleura,** and a ventral **sternum.** Tiny holes perforate the entire skeleton, being particularly numerous in the appendages and tail region. Set into these are bristles, which make the animal extremely sensitive to its surrounding world through tactile stimulation.

The appendages of the lobster or crayfish demonstrate a very interesting series of adaptations and modifications for a particular mode of life (Figs. 12–7, 12–8). There are nineteen pairs of appendages in all, one pair on each segment. The **antennules** and **antennae** are modified for tactile and chemical stimulation; the **mandibles,** or **jaws,** for chewing; the next five, **maxillae** and **maxillipeds,** chiefly for food manipulation; the next pair, the enormous **chelipeds,** for grasping food and for defense; the next four for walking; and the last six for swimming and various other functions. All of these appendages, with their variety of form and function, come originally from a simple appendage with a single function (locomotion) (Fig. 12–8).

The primitive appendage, one which appears in the early embryology of all Crus-

Fig. 12–6. The common crayfish (*Cambarus*) found over most of the United States. They live in ponds and streams where they feed on small fish or any other animal, dead or alive. Note the great area of the expanded tail (uropods and telson). This is very effective in locomotion. When disturbed, the animal contracts its tail powerfully, thus sending it darting backwards. The crayfish is much like its marine cousin, the lobster.

on the same animal they are said to be **serially homologous.** This introduces the principle of **homology** which is illustrated throughout the animal kingdom and is very important in determining animal relationships. Homologous structures have a common embryological origin, therefore when two animals show such structures, even though they may not have the same function in the adult form, the animals are known to be closely related. The more nearly the structures are alike, the closer the relationship,

Fig. 12–7. Crayfish appendages. Most of the right appendages have been removed for comparative purposes. Each retains parts of the original larval appendage, but here it is drastically modified to perform specific functions. This is an illustration of serial homologies among the invertebrates.

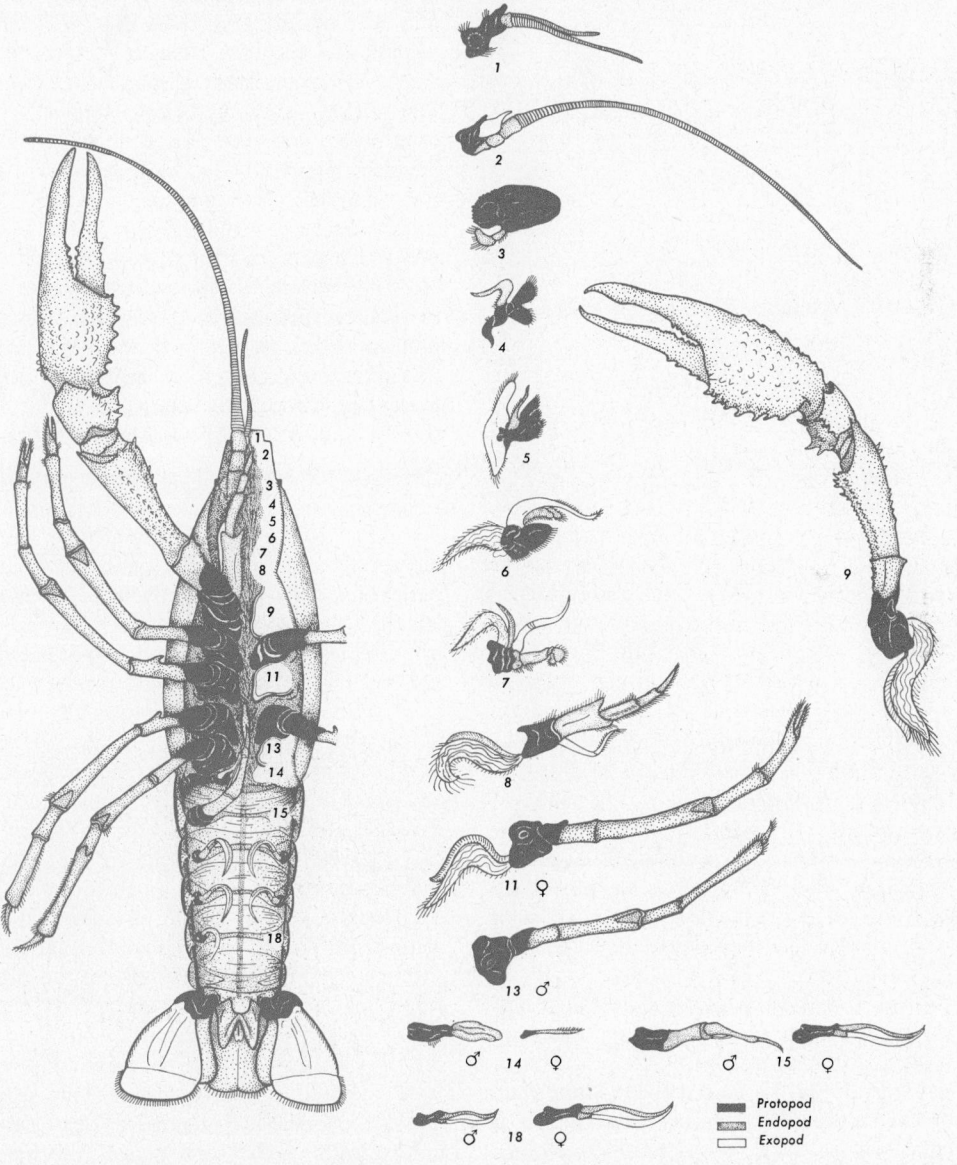

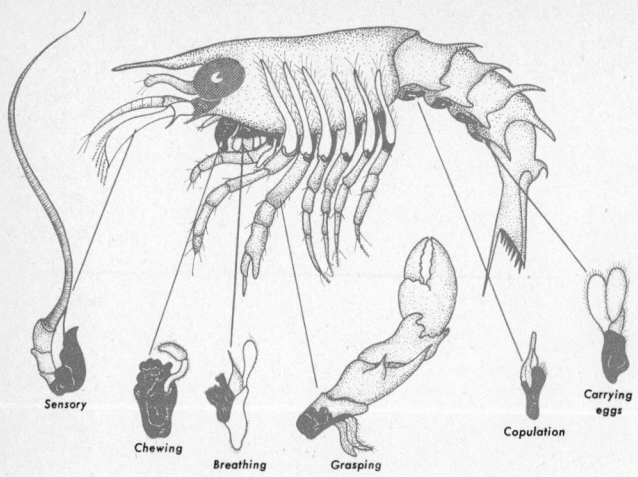

Sensory

Chewing

Breathing

Grasping

Copulation

Carrying eggs

Fig. 12–8. Young lobster, showing the undifferentiated appendages, together with the appendages as they appear in the adult animal. Note how they are modified for specific functions. In the young form the appendages are designed primarily for locomotion, in the adult they take on a variety of functions.

which means that they came from a common ancestor. We discuss this topic again in Chapter 26.

Upon cutting through the body wall, among the first and most obvious structures we note are the muscles, particularly in the abdominal region. The muscles are attached to the internal side of the exoskeleton, where they are always in pairs, and oppose one another in action. One muscle pulls an appendage or any other part of the body in one direction, whereas the other pulls it back again. The principle is followed by all higher animals with skeletons, including man.

Digestive System. The **chelipeds** catch and crush the food, then pass it to the **mouth** with the aid of the **maxillae** and **maxillipeds** (Fig. 12–9). The **mandibles** break the food up still further, before it is swallowed through the short **esophagus** and taken into the large **cardiac stomach,** which functions more as a storehouse for food than as a digestive organ (Fig. 12–9). At the posterior end of this organ is the **gastric mill,** which is composed of three small toothlike bodies

called **ossicles,** two in a lateral position and one in the mid-dorsal line. These fit tightly together and the grinding movement is brought about by muscles attached to the outside of the stomach. Before food can pass on into the second stomach, called the **pyloric stomach,** it must pass through the gastric mill, which grinds all the food finely and renders it digestible. In addition, the two stomachs are separated by filtering "hairs" which act as strainers, allowing only fine particles to pass through. Foods that cannot pass through, such as parts of skeletons, are regurgitated through the mouth. Once the food is in the pyloric stomach some of it passes into the two large multi-lobed **digestive glands.** The enzymes necessary for the complete digestion of the food are secreted by these glands. Digestion occurs for the most part in the upper end of the **intestine,** although some apparently goes on in the digestive glands themselves. Absorption also takes place here, and in the intestine. All undigested food in the intestine passes out through the **anus** as feces.

Circulatory System. The lobster possesses a single pulsating vesicle that lies on the dorsal part of the body, surrounded by a thin membrane and cavity, the **pericardium** and **pericardial cavity,** respectively. The **heart** has three pairs of tiny valved openings, called **ostia,** which allow the blood to enter from the pericardial cavity. Seven arteries lead away from the heart and convey the blood to all parts of the body.

The blood leaves the tiny arterioles and passes out into spaces, called **sinuses,** where it bathes the tissues. This type of system is called an **open blood system,** in contrast to the closed system of the earthworm. Once the blood leaves the tissues, it seeps into the ventral portion of the **cephalothorax,** then through a set of **afferent vessels** (to the gills) where gas exchange takes place. It then makes its way to the pericardial chamber of the heart, through **efferent vessels** (away from the gills), thence through the ostia, and out into the body again. Valves located in the walls of the ostia permit the blood to pass through the ostia in one direction only, namely, into the heart.

The blood contains colorless leucocytes as well as a respiratory pigment, **hemocyanin,** which has a slight bluish color when oxygenated and serves the same oxygen-carrying function as the hemoglobin does in other forms. The blood has remarkable clotting properties. If an appendage is removed forcibly, there is hardly any noticeable loss of blood, the clot forming almost at once and filling the large opening.

Breathing System. When the lateral walls of the carapace are cut away, large feather-like delicate gills are exposed. These are the breathing organs of the animal. While the crayfish lies quietly in running water, the water moves over the gills without any help from the animal. However, when the demand for oxygen is greater or when the oxygen content of the water is low, a special modification of the second maxilla, the **gill bailer** (scaphognathite), waves up and down, causing the water to flow over the gills in a posterior-anterior direction. Since the openings of the excretory organs, which lie at the base of the antennae, are in the path of this outgoing water current, the waste products from these organs are also carried away.

Excretory System. Nitrogenous wastes are withdrawn from the blood and body fluids by a pair of **kidneys,** called the **green glands** because of their color. They are located beneath the antennae in the body and consist of a **glandular** portion and a **bladder** (Fig. 12–9). The bladder stores waste material until it is released through the opening at the base of the antennae, already referred to.

Reproductive System. Sperm cells are produced in a tubular **testis** and pass through the **vas deferens** to an opening at the base of the fifth (considering the chelipeds as the first walking legs) walking legs. The first pair of swimmerets is modified in the male to form a **copulatory organ** for transfer of the sperm to the **seminal receptacle** on the ventral side of the female. The ovaries of the female produce eggs which pass through a straight oviduct to the opening at the base of the third walking legs (Fig. 12–9).

At the start of copulation the male lobster

Fig. 12–9. Internal anatomy of a female lobster.

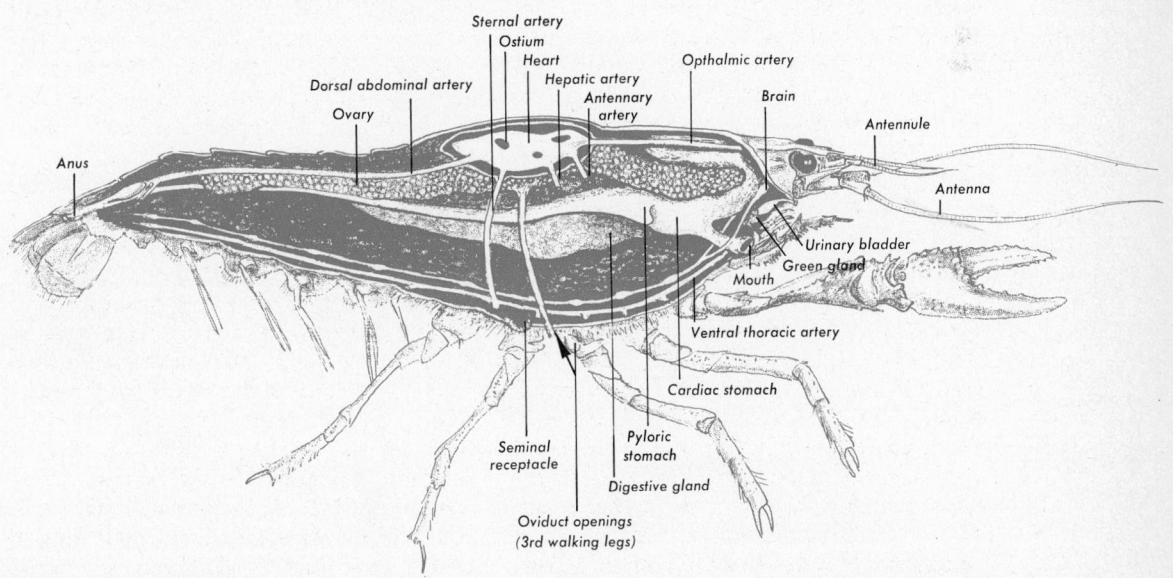

grasps the chelipeds of the female with his pincerlike appendages and turns her over on her dorsal side (Fig. 12–10). The two ventral sides then come together. The first pair of swimmerets of the male is placed near the opening of the seminal receptacle of the female and the sperm are discharged. The female receives the sperm in packets or **spermatophores.**

The female lobster lays eggs only once in two years, usually at the time when the water has reached its summer temperatures (from about mid-June to September 1). Eggs are fertilized after their ejection from the oviducts. As the eggs are extruded, a mucus material is produced by glands in the swimmerets which glues the eggs to the many bristles on these appendages. The eggs are thus carried by the female from ten to eleven months.

Growth.　Molting is necessary in order to provide for increase in the animal's size. Once the old skeleton is cast off, the animal grows rapidly for a period of time as a result of taking in a great deal of water.

Another interesting characteristic of many crustaceans, associated with growth and reproduction, is **autotomy,** the power to sever appendages which are reacquired later by **regeneration.** The value of autotomy is obvious; it enables the animal to escape with the loss of a single appendage, whereas without this ability it might not escape at all.

Nervous System.　The lobster and crayfish are well supplied with sense organs. If a small bit of liver is dropped in one end of an aquarium containing a crayfish, the animal moves directly toward the food. The soluble parts of meat touch tiny hairs on the chelipeds, antennules, antennae, and mouth parts, causing impulses to pass to the central nervous system. This sensitivity to chemicals is equivalent to man's senses of taste and smell.

The animal's ability to move about satisfactorily in total darkness indicates that it must have senses that guide it under such conditions. The sensory structures involved here are the tactile hairs which are stimulated whenever any part of the body comes in contact with an object. They may also be sensitive to vibrations in the water, thus providing the animal with a "hearing" mechanism which functions somewhat like the ears of vertebrates.

The animal is continually responsive to gravity, which means that it must have

Fig. 12–10. Life history of the lobster.

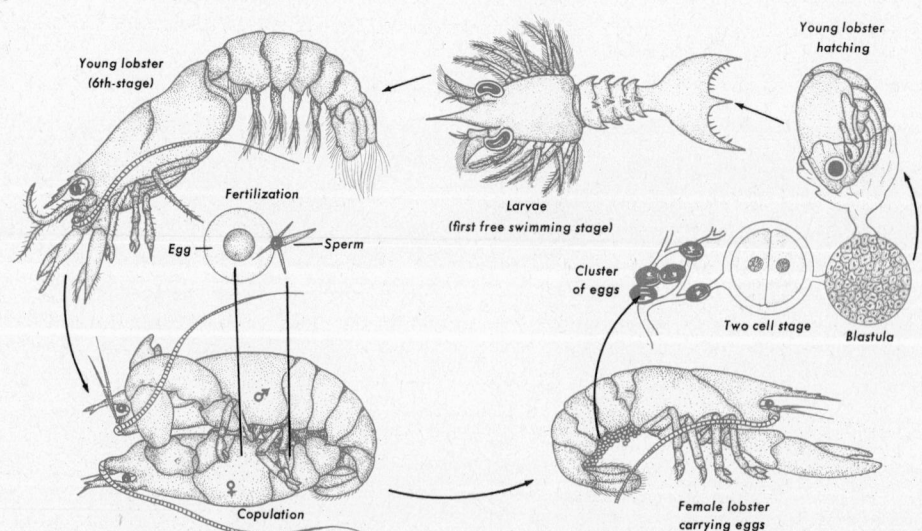

Young lobster
(6th-stage)

Larvae
(first free swimming stage)

Young lobster
hatching

Fertilization

Egg — Sperm

Cluster
of eggs

Two cell stage

Blastula

♂

♀

Copulation

Female lobster
carrying eggs

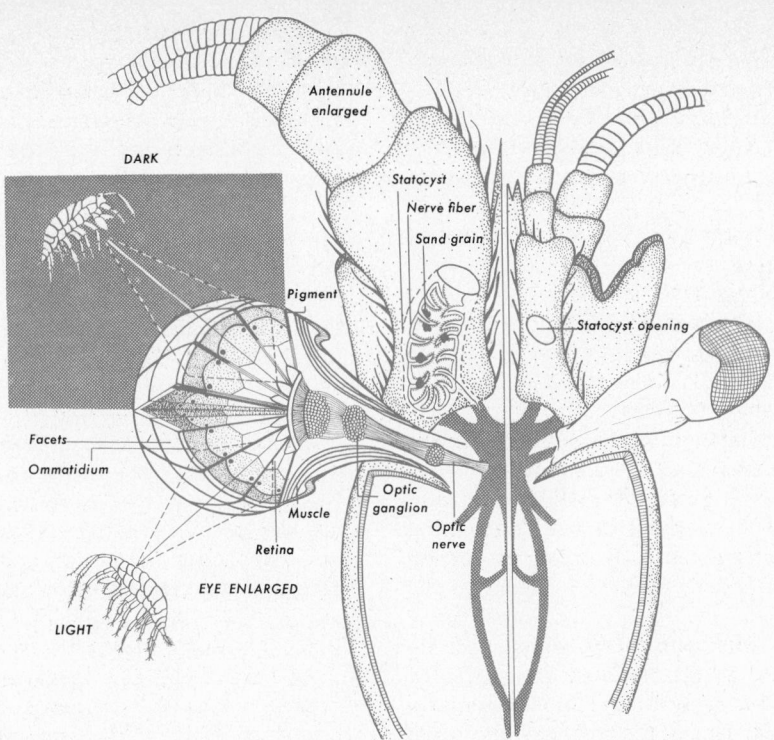

Fig. 12–11. Nervous system of the lobster, dorsal view. The eye and statocyst have been greatly enlarged and simplified in order to show their manner of function. The eye is further divided into two parts, the upper portion shows how it functions in the dark, the lower portion, how it functions in the light.

organs of equilibrium. These organs are the **statocysts,** located at the base of the antennules. They consist of small cuticular sacs, containing tiny grains of sand, which are glued to small sensory hairs (Fig. 12–11). The nerve fibers from these sensory hairs join to form a large nerve leading to the brain.

The visual mechanism is very complex in the arthropods. While arthropod eyes vary considerably, a description for the crayfish eye will convey a general idea for the entire phylum. The eyes are located on long movable stalks and can be protruded or retracted in order to improve the vantage point of the observer. The eye itself differs markedly from any vertebrate eye (Fig. 12–11). It has an outer rounded transparent surface called the **cornea,** which is divided into at least 2,000 tiny sections, or **facets.** The facets merely indicate the outer limits of the units, or

ommatidia, which make up the eye of the crayfish. Each ommatidium is composed of the outer facet, which functions as a **lens,** a series of cells below, which make up a sensitive portion, the **retinula,** and heavily pigmented regions at the outer and inner ends of this cylinder (Fig. 12–11). The sensory cells of each retinula terminate in nerve fibers, which lead directly to the brain, and the fibers from all the retinulae form a nerve equivalent to the **optic nerve** in vertebrates.

The eye functions as a very efficient organ for photoreception. Light falling upon the lens of a given ommatidium at the proper angle is focused on the sensitive region below and stimulates the sensory cells. Since the ommatidia are long cylinders, the animal sees tiny points (one for each ommatidium stimulated) which are parts of objects that come within its vision. The image is thus a **mosaic** in which the slightest movement is readily detected. This is true, however, only in bright light when the pigmented walls of each ommatidium are spread over the entire cylinder. In dim light the pigment recedes toward the two ends of the cylinders, so that the rays of light pass readily from one ommatidium to another. The animal then sees the image superimposed, so that movement is not easily discerned. The image may be more distinct in reduced light, however, than when seen as a mosaic pattern (Fig. 12–11). Thus the animal has a means of seeing under varying light conditions.

The crayfish and lobster possess a bilobed **brain,** relatively large when compared to that found in lower forms. The brain lies between the eyes and is connected with the **ventral nerve cord** by means of a pair of **circumesophageal connectives,** very similar to the arrangement found in the earthworm. There is a large **subesophageal ganglion,** made up of six fused ganglia, followed by a series of ganglia, one for each segment. Large nerves extend out from the ganglia to the appendages and to other parts of the segment. Nervous coordination is supplemented by hormones which control color change, molting, and perhaps other activities (p. 209).

CLASS INSECTA

Insects are so widespread and numerous and touch upon man's life in so many ways that no one is entirely free from their influence. Perhaps the most benefit derived from insects is from their work as pollinators of flowers. Many trees would bear no fruit if it were not for insects, and the same is true of such crops as clover and figs. Shellac is made from a secretion produced by certain lac insects in India; others produce a dye called cochineal. The cocoon of the silkworm is unwound and spun into silk thread used to make fine silk cloth.

On the debit side are those insects which carry disease, such as the mosquito (malaria, yellow fever, filariasis), the body louse (typhus), and the flea (bubonic plague).

Even today these diseases are the cause of a vast amount of human misery. Bubonic plague wiped out from one-half to three-fourths of the population in vast areas of the world several centuries ago. Today approximately one-ninth of the population of the world is made wretched by malaria.

In addition to transmitting human diseases, insects attack man's food and either destroy or actually consume it. Cereal grains both in the storage bins and in the fields are injured or destroyed by various types of insects. Not only furniture, but a house itself can be tunneled and destroyed by termites.

Domestic animals are also harassed by numerous insects. The botfly causes most serious damage to the stomachs of horses, while the ox warble fly larvae bore holes in the skins of cattle causing them distress, and making the hide valueless for leather. Finally, there are the millions of gnats, flies, mosquitoes, and bugs that have a high-grade nuisance value, but do no special harm.

How is it that these small animals have so outstripped all other forms of life? First of all, their rigid exoskeleton has enabled them to invade the air and support themselves outside of water. This wax-coated cover prevents desiccation, an essential feature for an animal that divorces itself completely from an aquatic existence. They are the only invertebrates that have taken to the air. The wings of the insects have made it possible for them to travel long distances, thus not only increasing their ability to find food but also to spread themselves to new areas where they might thrive more successfully. Fortunately, present-day insects have never attained any great size. They range from microscopic dimensions to several inches in length.

Entomologists have grouped insects into over 20 orders based on the following characteristics: (1) type of metamorphosis, (2) presence or absence of wings, (3) character of the mouth parts, (4) the structure of the external genitalia. Many other characteristics are used to identify species, but these are the large general ones that separate the orders. Some of the orders are shown in Figure 12–12. The best way to obtain an idea of

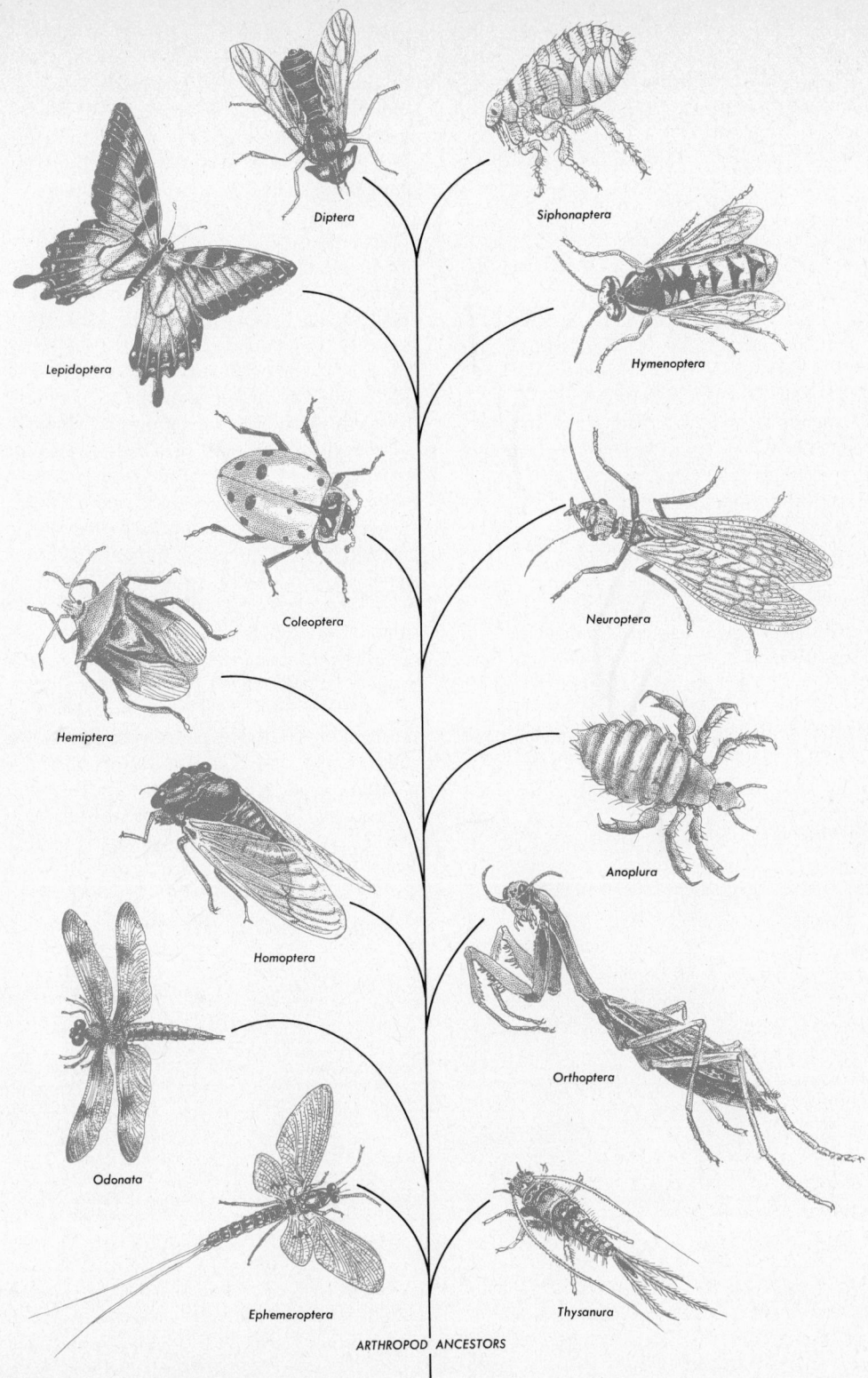

Diptera

Siphonaptera

Lepidoptera

Hymenoptera

Coleoptera

Neuroptera

Hemiptera

Anoplura

Homoptera

Odonata

Orthoptera

Ephemeroptera

Thysanura

ARTHROPOD ANCESTORS

Fig. 12–12. The common orders of insects.

the insect body plan is to study in some detail a generalized form with gradual metamorphosis, the grasshopper.

The Grasshopper

The grasshopper is selected because it shows certain primitive insect characteristics that make it easier to understand than other members of the group.

Structure. Externally, the grasshopper is divided into three parts, the movable **head,** the **thorax,** and the **abdomen** (Fig. 12–13). There is a considerable amount of fusion of segments when compared to the crayfish. For example, the head appears as a single structure, but it is made up of six segments. Likewise, the thorax is composed of three segments, and there is a variable number of segments in the abdominal region, usually eleven. A pair of legs is attached to each of the three thoracic segments, and a pair of wings to each of the last two. The legs have several parts which, named from the body outward, are the **coxa, trochanter, femur, tibia,** and **tarsus.** The hind legs are long and well developed for jumping, whereas the other four are used in walking. The outer

wings are leathery and rigid, serving as protective covers for the more membranous underwings. When the insect is at rest, the underwings, which are the propelling wings during flight, are neatly folded under the outer ones.

The **compound** eyes of the grasshopper are securely integrated into the head skeleton, but in other respects resemble those of the crayfish (Fig. 12–11). In addition, three small simple eyes, or **ocelli,** are located between the compound eyes (Fig. 12–14). The function of the ocelli is not understood. The single pair of antennae varies in length in different species of grasshoppers and functions as tactile as well as olfactory organs. Although most of the head is encased in a solid **epicranium,** the several mouth parts can be traced back to modifications in the crayfish appendage plan. There is a broad upper lip, the **labrum,** which is attached beneath the **clypeus.** A pair of lateral, dark-colored **mandibles** oppose one another in chewing in such a way as to make it convenient for the animal to bite the edge of a leaf without turning its head. Lying outside the mandibles are the paired **maxillae** (each composed of the **galea,** the **lacinia,** and the **palpus**), which are used in manipulating the

Fig. 12–13. External view of the grasshopper.

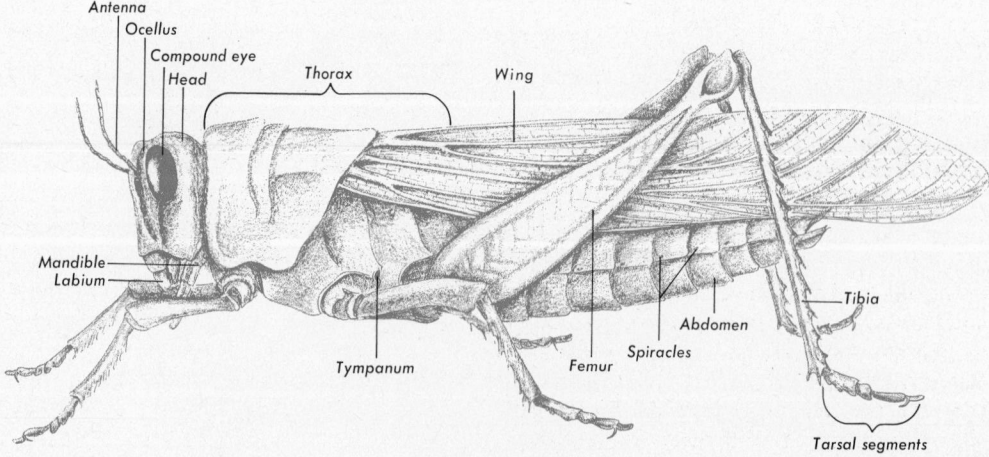

food as it enters the mouth. The lower lip, the **labium,** possesses two small **palpi,** resembling the larger ones attached to the maxillae. Lying in the center of all these parts is the **tongue,** or **hypopharynx.** Together, these make an efficient chewing mechanism.

Digestive System. As food is taken in, it is copiously mixed with colorless saliva, secreted by several **salivary glands.** The food moves through the **esophagus** to the **crop** where it is stored, until it passes into and through the **gizzard** where it is ground to a fine consistency. The food then enters the **stomach** (Fig. 12–15) where digestion occurs by the action of **enzymes,** which are secreted by eight double **digestive glands** or **caeca.** Finally the digested food passes into a large and then a small **intestine.** Small excretory tubules, the **Malpighian tubules,** empty into the anterior end of the large intestine. The gut opens into the **rectum,** and then to the outside through the **anus.**

Circulatory System. The body cavity of the grasshopper is the **hemocoel,** or blood cavity, not the coelom, as was the case in the earthworm. The cavity is filled with a colorless blood which contains only leucocytes. Since the blood does not function as a conveyor of oxygen, it has no oxygen-carrying pigment as the blood of most other animals has. The blood is kept in motion by the tube-like **heart,** composed of a number of chambers into which small **ostia** open (Fig. 12–15). The heart is surrounded by a **pericardial sinus,** which holds the blood before it enters the heart. The heart has no occasion to be so active an organ as it is in many other animals, since respiration is carried on in another fashion.

Breathing System. The breathing system of insects and arachnids is unique in the animal world. Air, with its oxygen, is carried directly to the cells through a system of tubules called **tracheae** (Fig. 12–16). This very complex system consists of tiny tubes which must remain distended so that the

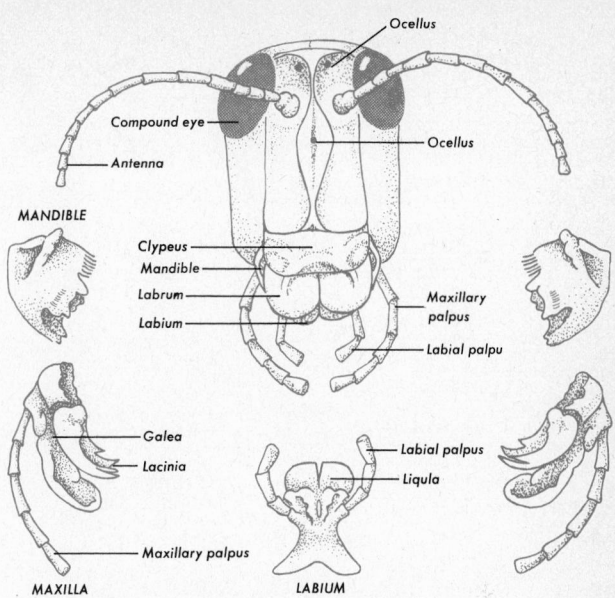

Fig. 12–14. Grasshopper head and mouth parts.

Fig. 12–15. Internal anatomy of the female grasshopper, dorsal view.

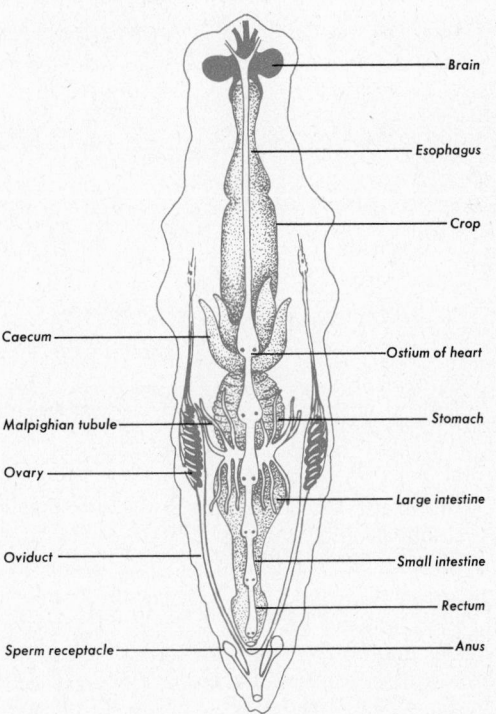

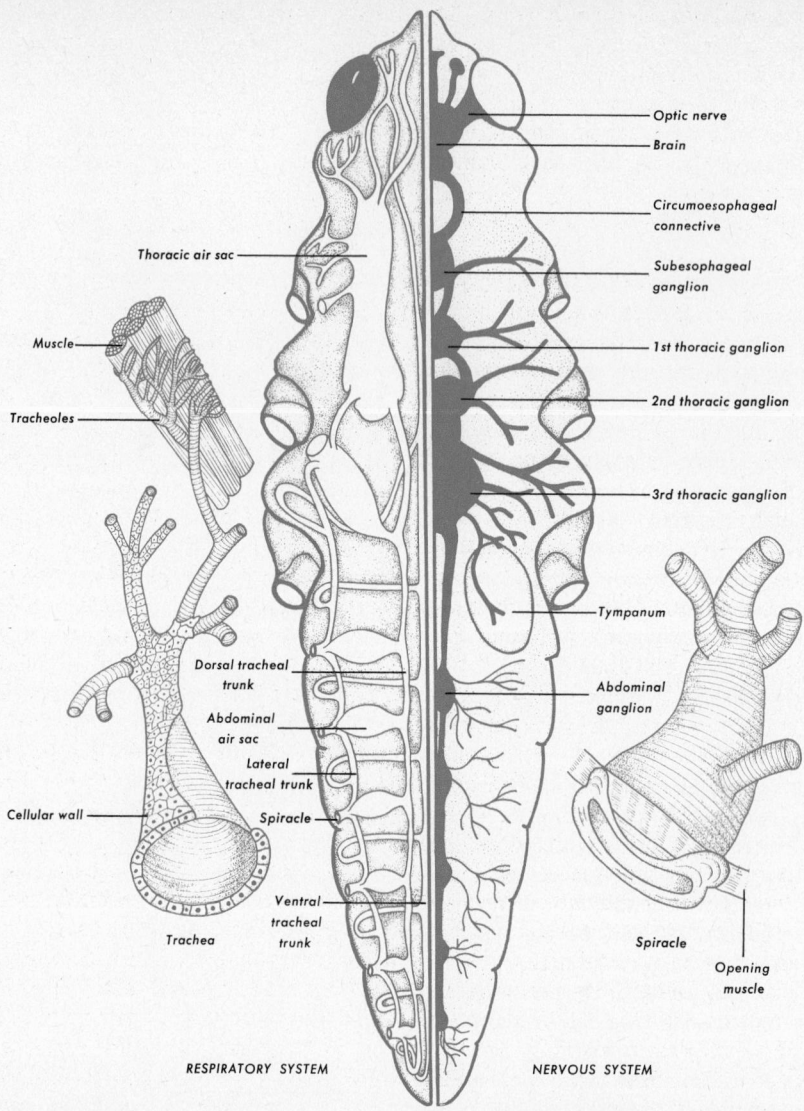

Thoracic air sac

Muscle

Tracheoles

Dorsal tracheal trunk

Abdominal air sac

Lateral tracheal trunk

Cellular wall

Spiracle

Trachea

Ventral tracheal trunk

Optic nerve

Brain

Circumoesophageal connective

Subesophageal ganglion

1st thoracic ganglion

2nd thoracic ganglion

3rd thoracic ganglion

Tympanum

Abdominal ganglion

Spiracle

Opening muscle

RESPIRATORY SYSTEM

NERVOUS SYSTEM

Fig. 12–16. Dorsal view of the breathing and nervous systems of the grasshopper. Trachea and a few muscle fibers are drawn in detail at the left, the spiracle and end of the trachea to the right.

air can pass freely in and out of them. Small chitinous **spiral threads** give support to the larger tubules and control their diameter. The smaller ones are filled with fluid in which gaseous exchange actually takes place (by diffusion). There are several openings into the tracheae along the thoracic and abdominal walls; these are called **spiracles.**

A valve covers the opening so that the spiracle can be opened and closed during the breathing process. Leading in from the spiracles, the tubules become smaller and smaller until they are as small as capillaries and lie directly against the cells, supplying them with oxygen and carrying away the excreted carbon dioxide. The grasshopper also possesses several large air sacs which may be contracted to aid the movement of air through the many small tubules. The grasshopper contracts and enlarges its body, particularly the abdominal region, to facilitate the air flow; and the anterior spiracles open and close alternately with the posterior spiracles so that the air flows only one way.

Nervous System. The eyes and the tactile hairs which cover the various parts of the grasshopper resemble those of the crayfish, and it has organs of chemical sense on its antennae and mouth parts. However, the grasshopper and many other insects have the means of making and receiving sound vibrations. In the grasshopper the organ for hearing is on the first segment of the abdomen (Fig. 12–16). In several respects this resembles the ears of higher vertebrates, in that it is composed of a stretched membrane, the **tympanum,** to which is attached a slender process that is connected to a nerve. The animal makes its characteristic clacking sound by rubbing its roughened hind femur against a wing vein.

As a result of the fusion of three head segments, the grasshopper possesses a rather large brain consisting of three fused double ganglia. Nerves extend directly from the brain to the eyes, the antennae, and to a ganglionated cord running through the body. The first ganglion in the cord, the **subesophageal ganglion,** formed by two great nerves or connectives proceeding from the brain around the esophagus, is followed by three large ganglia in the thorax and five in the abdomen, where some fusion has gone on. The nervous system is somewhat better developed than that of the crayfish, as is indicated by both its structure and function.

Excretory System. The excretory organs, the **Malpighian tubules,** have already been mentioned in connection with the digestive tract. Insects are among the very few animals whose **excretory glands** (kidneys) open directly into the intestine (Fig. 12–15). The long, coiled tubules lying in the hemocoel are bathed in blood so that suspended nitrog-

enous wastes are easily removed. Uric acid has been found to be the end product of nitrogen metabolism in insects.

Reproductive System. On late August days it is common to see grasshoppers copulating. Some days after fertilization the female grasshopper lays her eggs. She uses her powerful pointed **ovipositor** for digging a hole in the soil where the eggs are deposited, together with a mucus substance which cements them into a packet, known as a **pod.** She lays about 20 eggs at a time, and may deposit as many as 10 pods. The eggs begin to develop as soon as they are laid and become well-formed embryos before cold weather sets in. The embryos then undergo a rest period, the **diapause,** in which they pass the winter. In the warm spring days, development is resumed and the embryos hatch in early summer as **nymphs,** which resemble the adult grasshopper without wings. Nymphs undergo several molts during their subsequent rapid growth, the wings appearing and becoming longer at each shedding period. It is during this period of rapid growth, when their bodies demand so much food, that the animals are so destructive to crops.

The sex organs of the male and female grasshopper consist of a pair of **testes** and **ovaries,** respectively. The testes are joined to the **seminal vesicle** by means of the **vas deferens;** the seminal vesicle then joins the ejaculatory duct and copulatory organ, the **penis.** A pair of accessory glands secrete a fluid in which the sperm are suspended. The large ovaries are composed of several egg tubes in which the eggs develop; two oviducts extend from the ovaries and join to form the **vagina** (Fig. 12–15).

Modifications in Form and Function

Although all insects, with few exceptions, have body parts similar to those of the grasshopper, there are wide modifications in these parts in different species. Starting at the anterior end of the insect and working posteriorly, some of the modifications are as

Fig. 12–17. The larval dragonfly is equipped with mouth parts that are adapted for catching other insects in the water, where it lives until it becomes an adult. The upper picture shows the parts thrust out in the striking position; in the lower picture they are retracted where they are held except when in use.

follows: the antennae may be very short, as in the dragonfly, or they may be very long, as in the longhorned grasshoppers, in each case performing a specific function that requires the particular type of antennae in question. The eyes may be extremely large, as in the dragonfly, where they detect the flying mosquitoes which the airplane-like insect pursues. Or they may be absent, as in some termites, which work in the dark.

The mouth parts vary widely in form and function. The larval dragonfly has a formidable weapon for catching its prey (Fig. 12–17). The butterfly has a long tube which is carried in a coil under its "chin" when not in use, but, when stretched out during the process of taking nectar from a flower, it may be as long as the animal itself. The cicada possesses a stiff beak which is used in penetrating plant tissues to obtain the juices on which it feeds (Fig. 12–18). There are also the thin dartlike mouth parts of the mosquito which can pierce the skin very delicately and withdraw its meal of blood, at the same time injecting a small amount of saliva to prevent the blood from clotting. The mouth parts of all insects are homologous; yet, witness the variety of functions they perform.

The thorax of most insects bears two pairs of wings and three pairs of legs, all of which are variously modified in different insects. The appendages of the honeybee are an excellent example. They are modified for pollen gathering and other functions associated with its mode of life (Fig. 12–19). During a visit to a flower, the bee gathers some pollen with its mandibles and moistens it with honey. Pollen is also obtained from the action of the **pollen brushes** on the front two pairs of legs which clean the anterior portion

Fig. 12–18. The mouth parts of the cicada (*Tibicen*) are modified to form a stiff beak, which it uses in piercing twigs in order to obtain the sap, its chief source of food.

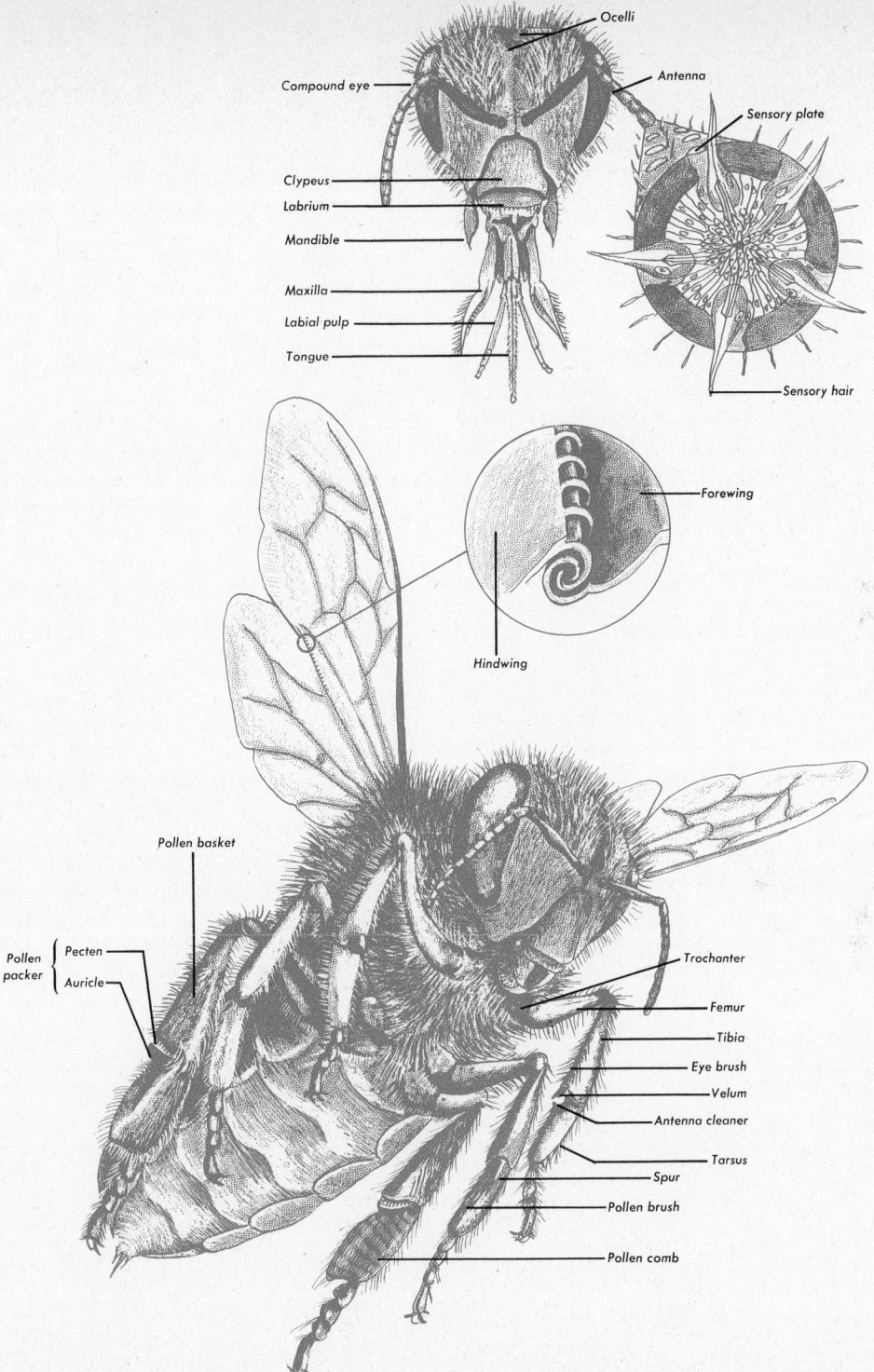

Ocelli

Antenna

Compound eye

Sensory plate

Clypeus

Labrium

Mandible

Maxilla

Labial pulp

Tongue

Sensory hair

Forewing

Hindwing

Pollen basket

Trochanter

Femur

Tibia

Pollen
packer

Pecten

Eye brush

Auricle

Velum

Antenna cleaner

Tarsus

Spur

Pollen brush

Pollen comb

Fig. 12–19. Ventral view of the worker honeybee to show modification of the legs for carrying pollen. In the upper lefthand corner a small portion of the fore and hind wings are drawn to show how the wings may be hooked together during flight. In the upper righthand corner the head is shown (face view) with mouth parts extended. An antenna has been enlarged in cross section to show the details of the sensory end-organs.

of the body. The pollen is then carried to the **pollen packer** (on the third pair of legs) which is composed of two parts, the **auricle** and the **pecten.** Once the pollen has reached this position, the tarsus is flexed on the tibia, packing the pollen from the bottom into the **pollen basket.** Large quantities of pollen may be collected in this manner.

The anterior pair of legs has two cleaning mechanisms, an **eye brush** and an **antenna cleaner,** which the bee uses to remove pollen from these organs. The antenna cleaner is composed of a **velum,** a small flexible projection from the tibia, and a crescent-shaped depression on the proximal end of the tarsus, lined with short bristles. The antenna is brought into this depression and pulled through several times to clean it. In addition to the pollen brush on the middle pair of legs, there is also a **spur,** which is used in picking and transferring wax in the process of comb-building.

Insect legs may all be the same size and used for walking or running, or they may be modified for jumping, as in the case of the grasshopper. Other modifications include the

Fig. 12–20. Cross section of the bee showing how the wings function during flight.

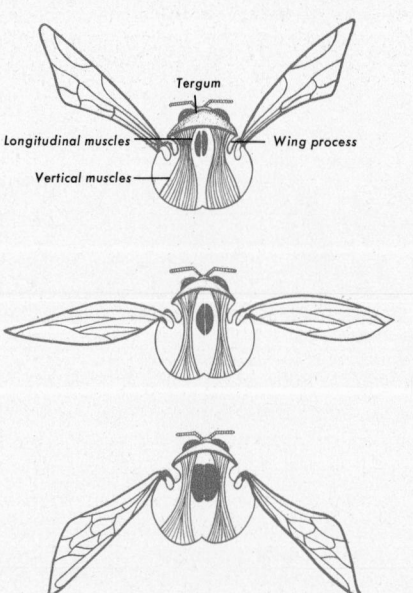

Tergum

Longitudinal muscles

Wing process

Vertical muscles

paddle-like feet for swimming, the digging legs for excavating, and the pincerlike legs for grasping prey, as in the praying mantis (Plate 12A).

Wings of insects are formed as thin sacs, by evaginations from the thoracic wall, through which tracheae make their way. Eventually the sacs collapse and the walls unite and harden, the "veins" being formed by the tracheae. Some insects, such as the wasp, have membranous wings, while others, like the beetle, have hard anterior wings (**elytra**) that fold over and protect the soft posterior membranous wings. Some wings are modified for making a sound, as in the case of the cricket in which the wings are rubbed together to make the characteristic chirp of this insect.

The bee has two pairs of delicate membranous wings which can operate either separately or locked together by means of a row of hooks that fasten into a groove in the posterior margin of the forewing (Fig. 12–19). During a straight flight, where speed is essential, the wings are locked together and the bee flies as if it had only two wings. The question of flight in insects has been a puzzling one, not only to biologists but to engineers as well. Flight in insects is brought about in a peculiar manner, namely, not by wing muscles, as in the case of birds, but by powerful muscles which cause the thorax to vibrate and this in turn forces the wings to flap up and down. In Figure 12–20 this is illustrated. As the anterior-posterior thoracic muscles contract, the dorsal wall of the thorax (**tergum**) is forced upward. This pulls the dorsal basal edge of the wing upward, causing the wing as a whole to be forced downward with considerable force; the body wall acts as a fulcrum. The upstroke is accomplished by the sudden contraction of the dorsal-ventral muscles in a similar manner. Thus the wings move up and down by the throbbing of the thorax and not by any effort on the part of the wing itself. Its pitch can be altered so that the bee can hover, fly forward, or fly backward with ease. Bees are capable of long flights, sometimes as long as 10 miles.

The chief modification in the abdominal region of insects is the ovipositor of the female. This has been described for the grasshopper. In the honeybee it has been modified into a **sting**. This organ has become a complicated apparatus, retaining the muscles which made it possible for the grasshopper to deposit its eggs in hard soil. In the sting these muscles enable the bee to force sharp-grooved **darts** into the tough skin of an intruder. Lying between the upper ends of the darts is the **poison sac** which is in contact by means of ducts with an **alkaline** and an **acid gland.** During the stinging procedure the poison sac is squeezed, and its product is forced into the wound made by the darts. Once a worker stings, its darts become firmly fixed in the skin of the recipient so that the entire apparatus and sometimes other internal organs are torn out when the bee leaves. A day or two later this results in the death of the bee.

In some insects, however, the ovipositor is developed in a most extraordinary fashion. Thus in the Ichneumon fly, it is several times as long as the body and can drill a hole in wood an inch or more deep. Since this insect lays its eggs in the body of the larval wood beetle, such an apparatus is essential.

The over-all color of insects varies as much as it does in birds. The colors are either in the exoskeleton or they are produced by differential interference of light impinging on regular minute depressions and elevations in the cuticula. Some insects resemble other insects or parts of their environment (Pl. 12C). One species of fly, for example, resembles and even acts like a bee, thereby taking advantage of the protection of the bee's sting, even though it has none itself. This is called **mimicry.** Some even mimic sticks or twigs (Fig. 12–21).

There are numerous modifications in the breathing systems. Although most of the insects breathe air, some, such as caddis fly larvae, receive their oxygen by means of thin gills and can get along under water. It is clear that the insects became air-breathing arthropods and that only a few have gone back into the water during their larval life.

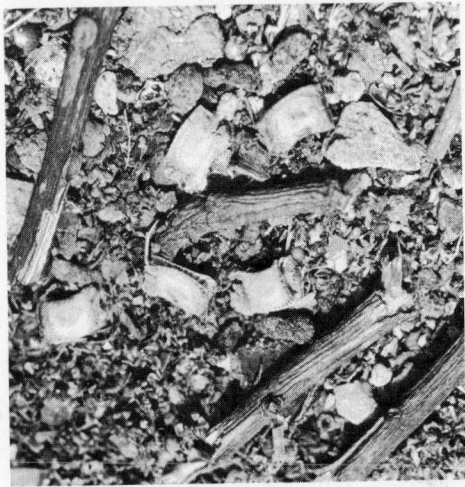

Fig. 12–21. A Panamanian orthopteran (*Pseudoceroys sp.*) which resembles the twigs of its environment. When disturbed, it becomes rigid and takes a pose that resembles a twig. Even its eggs (bottom) resemble the debris on the forest floor.

The digestive systems of insects vary considerably, depending on their type of diet. Furthermore, the diet may differ widely during the larval and adult stages: the butterfly, for instance, feeds on leafy vegetation as a larva, and on nectar as an adult. Feeding may be confined to certain stages of the life cycle and absent in others. Such insects as the May flies and fishflies feed only as larval forms, the adults living but a few days during which time food is unnecessary. The adult

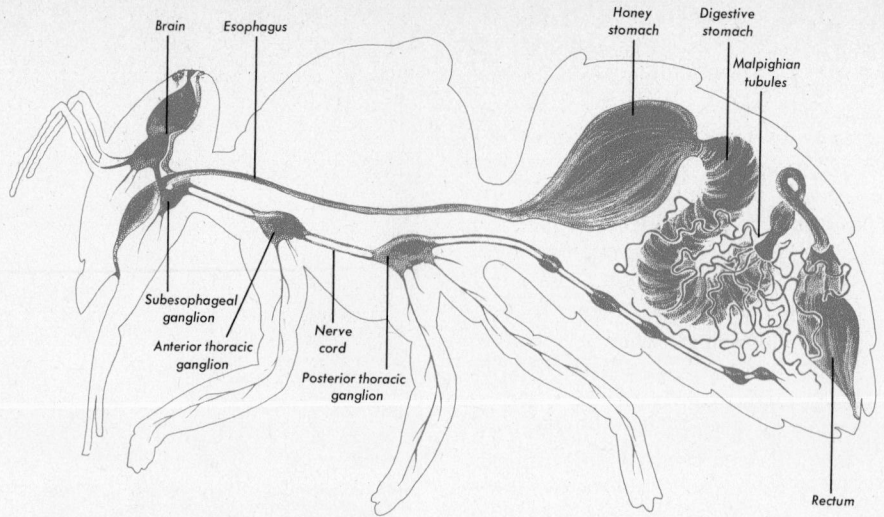

Fig. 12–22. Side view showing the digestive and nervous systems of the bee.

stage is devoted to mating and egg-laying. Among mosquitoes, the male mates and dies, never feeding at any time on blood, whereas the female of some species must obtain a blood meal before her eggs mature. She has a voracious appetite and is able to take a meal of blood equal to several times her body weight. Some insects, like the praying mantis (Pl. 12A), are carnivores (meat eaters), others, like the grasshopper, are herbivores (vegetable eaters), and still others, like the cockroach, are omnivores (both meat and vegetable eaters). However, almost any one of them can be forced to change its usual dietary habits when it is confronted with starvation.

The sense organs and nervous systems have become greatly modified among the insects. The central nervous system of some of the lower insects does not differ greatly from that found in the earthworm, while in others, like the honeybee, there has been a great deal of fusion of ganglia and an apparently higher or more closely knit co-ordination of parts (Fig. 12–22). They are equipped with elaborate sense organs which are responsible for their complicated behavior. Their compound eyes are generally similar to those of the crayfish. Since a great

many experiments have been performed with honeybees, a considerable amount is known about their powers of discrimination. **Von Frisch,** a brilliant Austrian zoologist, discovered that whereas they could not discriminate among various-shaped solid objects, they could detect open figures such as circles, rectangles, squares, or two closely placed lines (Fig. 12–23). He discovered this by placing a food source over cards with various figures

Fig. 12–23. Bees cannot discriminate between the various shaped objects indicated in this sketch but they can distinguish between any pair of objects, open and solid.

and designs on them and observing the association made by the bees between the kind of figures and the food. He concluded that the bee is able to discern only discontinuity in a pattern and not its shape. Its eye records interruptions which the bee counts as it moves past the pattern in flight. Von Frisch also demonstrated that bees see color, but not as we do. They recognize four colors, one of which is ultraviolet, a color we cannot see. The others are violet, blue-green, and yellow-orange. Bees have an excellent sense of smell which is made possibly by sense organs located on the antennae. They, also, have a remarkable communication system which was first investigated by von Frisch. We have more to say about this in Chapter 27.

Fig. 12–24. Life history of the honeybee.

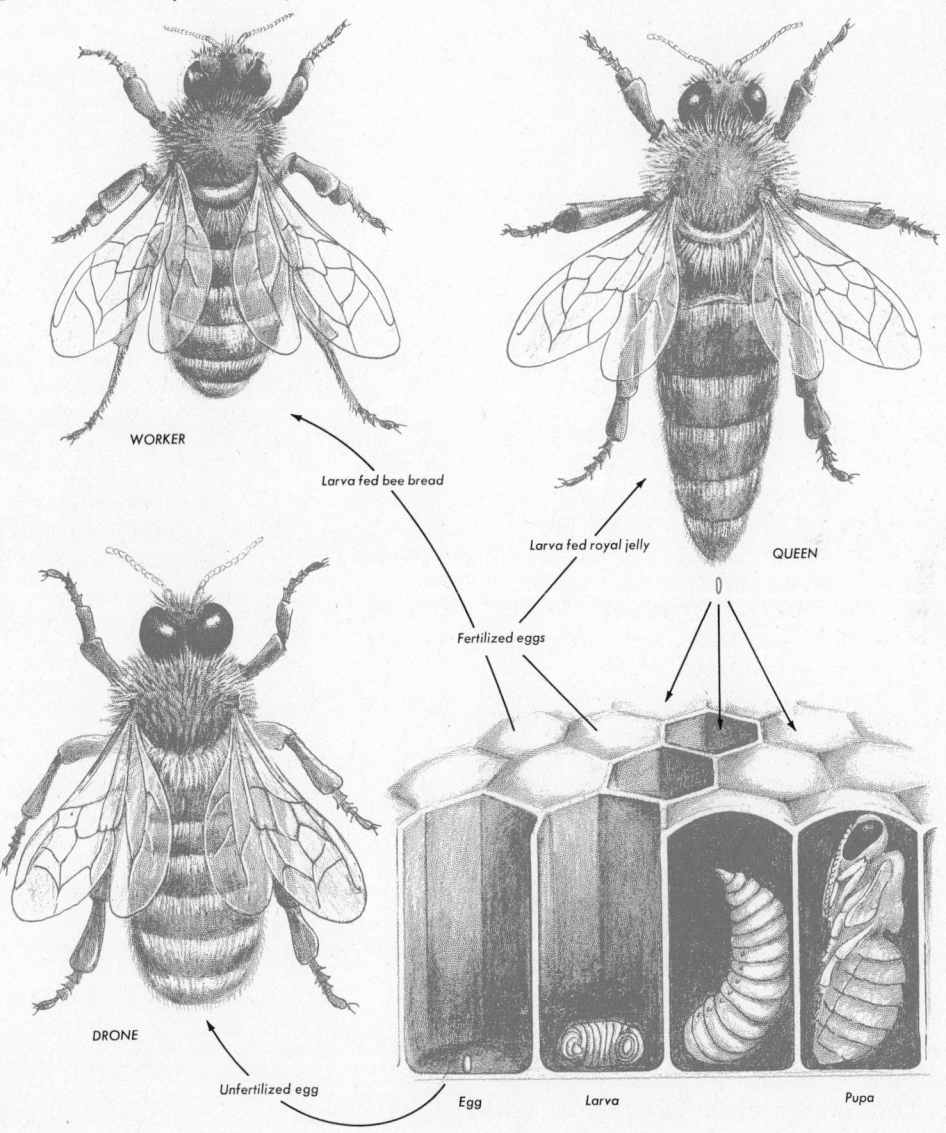

WORKER

Larva fed bee bread

Larva fed royal jelly

QUEEN

Fertilized eggs

DRONE

Unfertilized egg

Egg   Larva   Pupa

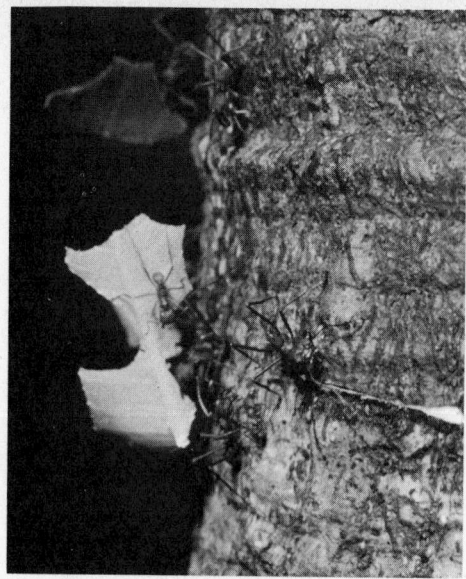

Fig. 12-25. The leaf-cutting ants (*Atto sp.*) possess complex patterns of social behavior. They cut uniformly sized pieces of a leaf, usually from tall trees, and carry them to their nest. The leaves serve as food for fungi, which serve as food for the ant. These ants inhabit Barro Colorado, an island in Gatun Lake, Canal Zone. Note the "hitchhiker" on one of the leaves.

The social insects represent a very high development of the nervous system, perhaps the highest in the invertebrates. There is a long series of gradations from the solitary insects to those which aggregate during hibernation or migration, like some grasshoppers and the monarch butterfly. From such species have come the true social forms which live together in large numbers and have developed various castes, as in the honeybee (Fig. 12-24). The change in the various castes has been fundamental because it involves hormonal changes which in turn alter the anatomy of the caste, as, for example, the development of the large mandibles of certain of the soldiers among some species of termites. In this group, in addition to the soldiers which protect the colony, there are the workers, the males, and the queen. In some ant colonies there are as many as 27 different types of individuals, not all present at once but occurring at some time

during the history of the colony. Each type performs certain duties and fits into the harmonious operation of this complex society.

Ants seem to have followed the food habits of man, at least in their methods of providing a food supply. The more primitive ants merely feed as carnivores, but pastoral

Fig. 12-26. Some species of termites build tremendous homes, called "ant hills," as indicated in the top photograph. These reach as much as 15 feet in height. Others, such as the "magnetic ants," orient their hill north and south, hence their name. Both species occur in northern Australia.

ants take aphids into their nests and feed them, gathering the honeydew (an excretory product) from them in repayment for their efforts. The harvester ants show further development by carrying cereal grains into their burrow to supply them with food during the winter. Finally, the most advanced are those ants which gather a certain species of fungus and plant it in underground gardens, tending it even to the point of adding humus as fertilizer (Fig. 12–25).

The termites also have reached a high level of colonial organization. One of the most striking aspects is the size of the colonies as indicated by the tremendous homes or "hills" that they build (Fig. 12–26). The most spectacular are found in Africa and Australia.

In the reproduction of insects, fertilization is internal (Pl. 12B). Most insects lay eggs (oviparous) and deposit them in a place where development of the larvae is most apt to succeed. In some species as in the housefly, the eggs hatch very soon, while in others many weeks or even months are required. Certain flies and all of the aphids bring forth active young (ovoviviparous). The eggs of some insects develop without fertilization by a sperm, that is, by **parthenogenesis**. The drone bee develops from an unfertilized egg (Fig. 12–24). Parthenogenesis is commonly found among the aphids, where the females lay eggs all through the summer months which hatch only into females. As fall approaches, the eggs produce both males and females. Fertilization then occurs and the resulting eggs remain over winter and hatch into females again in the spring. The number of offspring produced by one individual varies widely among different insect species. Some of the viviparous flies, for example, produce only a few offspring, whereas the queen bee may lay a million eggs in her lifetime. Under optimum conditions the housefly, if unchecked, could increase in one summer to such proportions as to cover the earth completely, for it goes through its entire life cycle in eight days if the temperature is high enough (80–90°F).

The manner of development from egg to

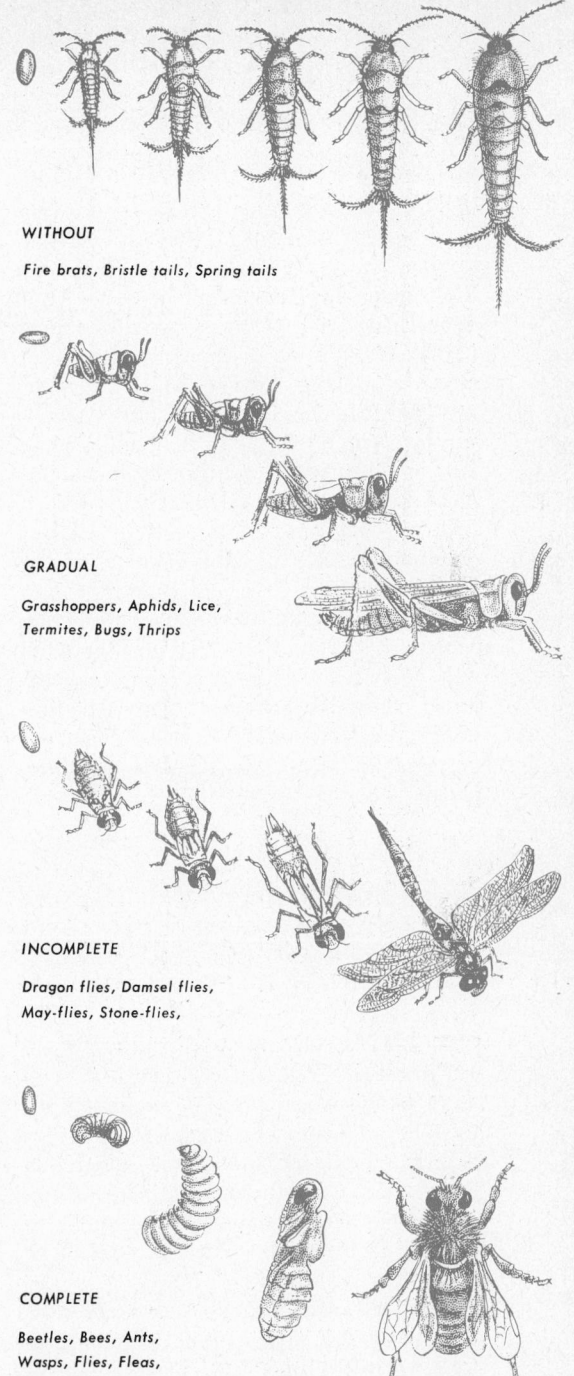

WITHOUT

Fire brats, Bristle tails, Spring tails

GRADUAL

*Grasshoppers, Aphids, Lice,
Termites, Bugs, Thrips*

INCOMPLETE

*Dragon flies, Damsel flies,
May-flies, Stone-flies,*

COMPLETE

*Beetles, Bees, Ants,
Wasps, Flies, Fleas,
Butterflies, Moths*

Fig. 12–27. Various types of metamorphosis, and some of the insects that undergo each type.

adult varies among insects. The term **meta-morphosis**, which means change in form, is applied to any animal that undergoes more or less marked changes of form between the time of hatching and of reaching the adult state (Fig. 12–27). A few primitive insects merely increase in size after hatching, showing no metamorphosis. Others, such as the grasshopper, hatch into a **nymph**, which resembles the adult fairly closely except for the wings which are acquired much later. Such change is known as **gradual metamorphosis**. Some, such as the dragonfly, hatch into a larva (Fig. 12–17), which resembles the adult to some extent, but not as much as the nymph resembles the adult grasshopper. This type of change is called **incomplete metamorphosis**. In the case of the bee, housefly, and June beetle, the larval stage does not resemble the adult in any way (Fig. 12–27); the larva is wormlike and usually its diet varies radically from that of the adult. Moreover, between the larval and the adult stage there is a "resting" stage, known as the **pupa**, during which time the larval body is transformed into the adult body. This type of change is known as **complete metamorphosis** (Fig. 12–28).

OTHER ARTHROPODS

There are several other classes of less important arthropods that need to be mentioned. The horseshoe or king crab, *Limulus* (Fig. 12–29), belongs to the class **Xiphosurida** and is of interest because it is a member of a very ancient group of animals. Since it is the only member left of a once very successful group of animals, it is called a "living fossil." It was dominant in the Cambrian period and has somehow been able to continue to the present, whereas millions of its relatives have become extinct.

The class **Arachnida** includes the scorpions, ticks, and spiders. The scorpions are elongated with large, fierce-looking pincers held out in front when they are on the move (Fig. 12–30). The long, thin abdomen terminates in a sharp-pointed sting which inflicts an irritating, and sometimes fatal,

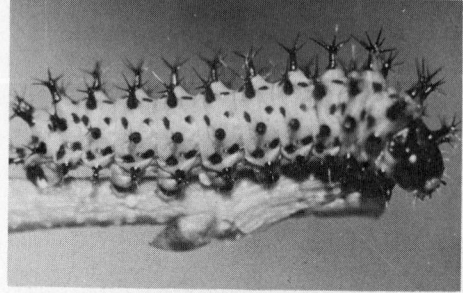

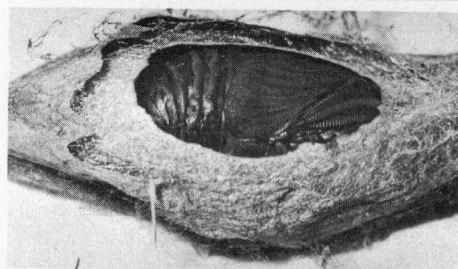

Fig. 12–28. Butterflies and moths undergo complete metamorphosis from egg to adult. These are photographs of the stages of development of the Cecropia moth. (Top) Recently hatched eggs. (Second) Larva a few days old. (Third) Pupa lying within its cocoon. Note the impressions made by the developing antenna and wings. (Bottom) The adult.

wound on man as well as on insects and spiders which make up its diet. Of much greater importance are the ticks and mites, many of which parasitize man and his

domestic animals (Fig. 12–31). Some of these are intermediate hosts for certain microbes which cause serious diseases such as spotted fever, undulant fever, and tularemia.

The most common members of this class are the spiders, which are feared and destroyed by man, yet almost all of them are beneficial because they feed upon insects, many of which are pests. Spiders build delicate webs in which flying insects are

Fig. 12–31. The anatomical characteristics of the tick are well illustrated by this photograph of the Gulf Coast tick (*Amblyomma maculatum*) which bites cattle. Note the eight legs terminating with sharp hooks for clinging to the host and the mouth parts well suited for piercing the skin.

Fig. 12–29. The king crab (*Limulus*), or horseshoe crab, is a distant relative of the spiders and scorpions and is not a crustacean at all. It is a "living fossil" whose close relatives have all become extinct many millions of years ago.

Note the leaflike flaps just back of the legs. These are the book gills, so called because when in use the flaps wave in the water like the pages of a book. The snails (*Crepidula*) find it advantageous to "hitchhike" on the crab, thus affording them a much more extensive feeding area than they could ever attain under their own power. Note the track it makes in the sand when out of water. While clumsy on land, *Limulus* moves effectively along the ocean floor where it shovels in the mud searching for worms of various kinds that make up its diet.

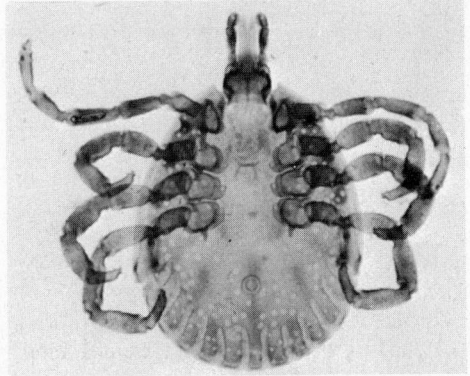

trapped and sucked dry by the owner of the web. The thread of the spider web is made by forcing a fluid containing long protein molecules through a tiny orifice which causes the molecules to align themselves in one direction. The fact that they are so arranged causes them to become semisolid, hence a thread. Other spiders chase and catch their prey; still others leap upon it and kill it by piercing the body with their sharp fangs (Fig. 12–32) and injecting a small quantity of poison that simply paralyzes the insect until it is consumed by the spider. Although most spiders live only a year or so, tarantulas have been kept in captivity for as long as 28 years.

The bite of most spiders is harmless, even the bite of the large tarantula being no more serious than a bee sting. There is one species,

Fig. 12–33. A female black widow spider (*Lactrodectus mactans*) with her egg case.

Fig. 12–32. American tarantula (*Euryplema*), one of the largest of all spiders. They are harmless if properly handled and live to a ripe old age in captivity. This specimen was 28 years old (estimated) when it died.

The dorsal view is shown. Note the four pairs of legs and the "hairy" body. The food of this spider is other arthropods, usually insects, although it will kill a small bird and feed on it. Its bite is relatively harmless to man.

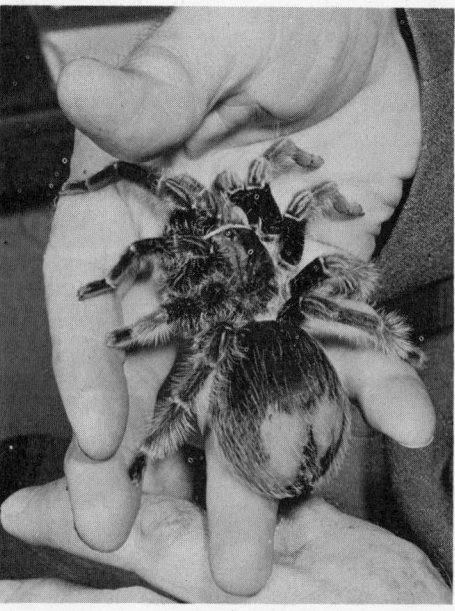

however, which is very common in the United States, which can cause serious illness and death. This is the black widow, *Lactrodectus mactans* (Fig. 12–33), the female of which is three-fourths of an inch long, and glistening black. On the ventral side is a bright red hour-glass-shaped figure, which is a positive means of identification. It lives normally in piles of rocks, lumber, and more recently around buildings, particularly garages. It has been known to cling to the underside of automobiles, thus being transported to all parts of the country.

The life history of the spider is rather unique in some respects. The male is always smaller than the female, and in some cases, such as the black widow, he is hardly recognizable because of his proportionately minute size. He spins a web upon which he deposits his sperm in a mass, which is then picked up by specially formed front appendages and carried while searching for a female. Once he has found a mate, he usually performs an unusual dance and then deposits the sperm bundle into her genital pore. Sometimes the female proceeds to devour her unsuspecting mate. She then spins a tiny ball in which the eggs are laid and often carries it round with her until the young hatch (Fig. 12–33).

The class **Crustacea (Mandibulata)** includes a variety of forms which are divided

Fig. 12–34. Centipedes have numerous jointed legs. Some tropical forms are nearly a foot long and can inflict a painful wound to a man. This is a smaller form common in the United States.

into two subclasses. The **Entomostraca** are small simple forms to which the "fairy shrimps" and barnacles belong. The fairy shrimps are commonly found, for a brief period, in temporary fresh water ponds in the early spring. They swim on their backs by a rhythmic beating of their paddlelike appendages. Another interesting relative is the "brine shrimp" which lives in water of high salinity, such as the Great Salt Lake. The dried "eggs" are collected in great quantities and sold as food for tropical fish. Very unusual members of this subclass are the barnacles (Cirripedia) which are highly modified crustaceans. The larval stages are free-living, but the adults are attached to rocks and shells in the ocean. The rock barnacle, *Balanus*, encrusts rocks between the high and low tides. The free-swimming larval barnacle becomes attached by the anterior end and metamorphoses into the adult, whose appendages have become modified for sweeping small organisms into the mouth. The second subclass, the **Malacostraca**, includes the large crustacea such as shrimps, crayfishes, lobsters, and crabs.

Two other small classes of the Mandibulata should be mentioned. The many-legged centipedes (Fig. 12–34), class **Chilopoda**, are commonly found under stones and logs where they remain inactive during the day. At night, however, they move swiftly about in search of their favorite food, earthworms and insects. They possess a pair of poison claws on the first segment which are effective instruments in securing prey. The millipedes (Fig. 12–35), class **Diplopoda**, occupy similar habitats and resemble the centipedes in some respects. They may have several hundred appendages which move in a wave-like fashion as the animal crawls. They are slow-moving creatures, not at all like the centipedes. They feed on plants and decaying organic matter. When in danger some roll into a ball, whereas others secrete an offensive fluid which serves them well as protection against enemies.

As has been described, the arthropods have acquired a body plan so well designed that it has permitted the group to penetrate

Fig. 12–35. The millipede (*Spirobolus*) possesses two pairs of jointed appendages on most of its segments. These numerous legs move in a rhythmic manner as can be seen from this photograph. Millipedes live in decaying vegetation upon which they feed.

almost every type of land and aquatic environment. They are the pinnacle of the schizocoelomate line and the most diversified of all animals. Now we trace the enterocoelomate line which terminates with the chordates, another very successful group of animals more important to us because we are one of its members.

SUGGESTED SUPPLEMENTARY READINGS

Books

BORROR, D. J., and DeLONG, D. M., *An Introduction to the Study of Insects.* New York: Holt, Rinehart & Winston, 1964.

CHAPMAN, R. F., *The Insects: Structure and Function.* New York: American Elsevier Publishing Co., 1969.

COMSTOCK, J. H., *Introduction to Entomology.* Ithaca, N.Y.: Comstock, 1940.

HERMS, W. B., and JAMES, M. T., *Medical Entomology.* New York: Macmillan, 1969.

LOCKWOOD, A. P. M., *Aspects of the Physiology of Crustacea.* San Francisco: W. H. Freeman, 1967.

METCALF, C. L., and FLINT, W. P., *Destructive and Useful Insects.* New York: McGraw-Hill, 1962.

NEALE, J., ed., *The Taxonomy, Morphology, and Ecology of Recent Ostracada.* Edinburgh: Oliver and Boyd Ltd., 1969.

SWAN, L. A., and PAPP, C. S., *The Common Insects of North America.* New York: Harper & Row, 1972.

VON FRISCH, K., *The Dance, Language, and Orientation of Bees.* Cambridge, Mass.: The Belknap Press, Harvard University Press, 1967.

WHEELER, W. M., *The Social Insects: Their Origin and Evolution.* New York: Harcourt, Brace & World, 1928.

WILSON, E. O., *The Insect Societies.* New York: Academic Press, 1971.

ZINSSER, H., *Rats, Lice, and History.* Boston: Little, Brown, 1935.

Article

VON FRISCH, K., "Dialects in the Language of the Bees." *Scientific American,* **207**, 78, August, 1962.

13

SPINY ANIMALS- ECHINODERMS

This successful group of animals possesses remarkable and unique characteristics. The phylum is named *Echinodermata* (Gr. "spiny-skinned") because its members usually possess external spines. Their close affinities to the next phylum, **Chordata**, are indicated by their skeletons of mesodermal origin and by their similar larval forms (see Fig. 14–3). They have acquired uniqueness in structure which sets them apart from all other animals. They possess radial symmetry in the adult stage, although as larvae they are bilaterally symmetrical. In addition they have evolved a unique method of locomotion made possible by the **water vascular system.** The coelom is well developed whereas the blood-vascular system is so rudimentary as to be nonfunctional and nephridial excretory organs are nonexistent. This puzzling situation in the adult is understood when one considers the evolution of the echinoderms. It seems quite clear that

ancestors of the echinoderms were bilaterally symmetrical, free-swimming animals with coelomic pouches arranged in segments. At some later time they became attached forms which resulted in radial symmetry with five parts, reminiscent of coelenterates. The bilateral symmetry was completely obscured and can be seen today only in the larva. Echinoderms are all marine and usually slow-moving. Some are permanently attached forms living at the bottom of the sea whereas others are found along the seashore among the rocks or in the sand (Fig. 13–1).

PHYLUM ECHINODERMATA

The phylum is divided into two subphyla, the **Pelmatozoa** which includes mostly fossil forms, and the **Eleutherozoa** containing modern echinoderms. The only living Pel-

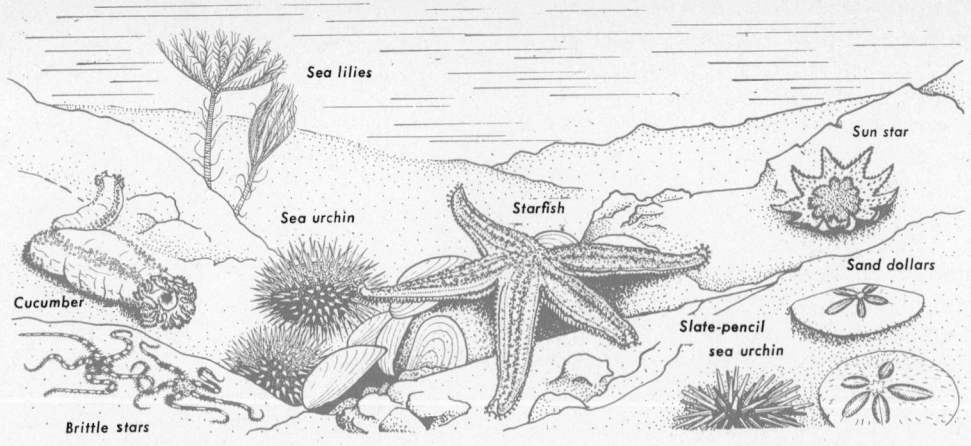

Fig. 13–1. All echinoderms are marine forms. A few of them are shown here in their habitat.

matozoa are in the class **Crinoidea** (feather stars and sea lilies). The four classes of the subphylum Eleutherozoa are: **Asteroidea** (starfish and sea stars), **Ophiuroidea** (brittle stars, basket stars, serpent stars), **Echinoidea** (sea urchins and sand dollars), **Holothuroidea** (sea cucumbers). The starfish is the typical representative which we detail.

The Starfish

Starfish, found in abundance along most seacoasts, vary greatly in size from tiny species about one-half inch in diameter to the giant starfish, which measures about 18 inches. The common starfish, *Asterias vulgaris* (Fig. 13–2), is found chiefly upon rock seashores and bottoms where mollusks, its main food, are also most abundant. Starfish resemble the conventional five-pointed star pattern, the five radiating arms rising from a central disc (Fig. 13–3). It usually moves with the two arms adjacent to the **madreporite** (a sievelike structure through which water enters) pointing forward. The upper portion of the body, the **aboral** surface, is covered with **spines,** along the bases of which are small pincerlike structures, the **pedicellariae,** which serve to keep the body free from foreign material. Because the animal is so well armored with various types

of sharp projections, it is little wonder that it is not chosen as food by other animals. The **oral** side, or undersurface, on which the

Fig. 13–2. The common starfish (*Asterias*), like most starfish, is found along rocky shores crawling over the hard surfaces in search of mollusks, particularly clams, which constitute its main food. This one is crawling over the shells of clams, many of which are empty because their soft bodies were sacrificed to satisfy the hunger of this and several other starfish.

mouth is centrally located, serves two main purposes, locomotion and food collection.

An outstanding feature of the echinoderms is the appearance of a unique device, the **water vascular system**, consisting of two rows of tube feet which extend from the mouth down the oral side of each of the five rays (Fig. 13–3). The tube feet enable the animal to move slowly over rocks or along the ocean floor, to twist and turn its body, and to capture food. If a single foot is examined, the portion that protrudes externally from the oral surface is found to be an elongated tube. Internally, at the opposite end, a bulbous structure, the **ampulla**, joins a central tube, the **radial canal**, which extends up the ray of the animal to join a central circular tube, the **ring canal**. A short tube, the **stone canal**, is joined to the ring canal and runs up to the

dorsal surface of the disk, opening externally into the madreporite plate.

Water enters through the madreporite plate, passes into the stone canal, then the ring canal, and to the ampullae. At the margin of the ring canal are located the **Tiedemann bodies** which produce the **amoebocytes** found in the fluid of the system. When the ampulla contracts, the fluid is forced into the tube foot, which is thus elongated. If the suckerlike tip of the foot touches and attaches to an object, the muscular wall of the foot contracts, forcing the fluid back into the ampulla, thus causing the foot to be shortened. Since the foot adheres to the object it has touched, the shortening of the foot draws the body forward. In this manner the starfish is able to move.

The endoskeleton, which is produced by

Fig. 13–3. Starfish cut in such a manner as to show the internal anatomy.

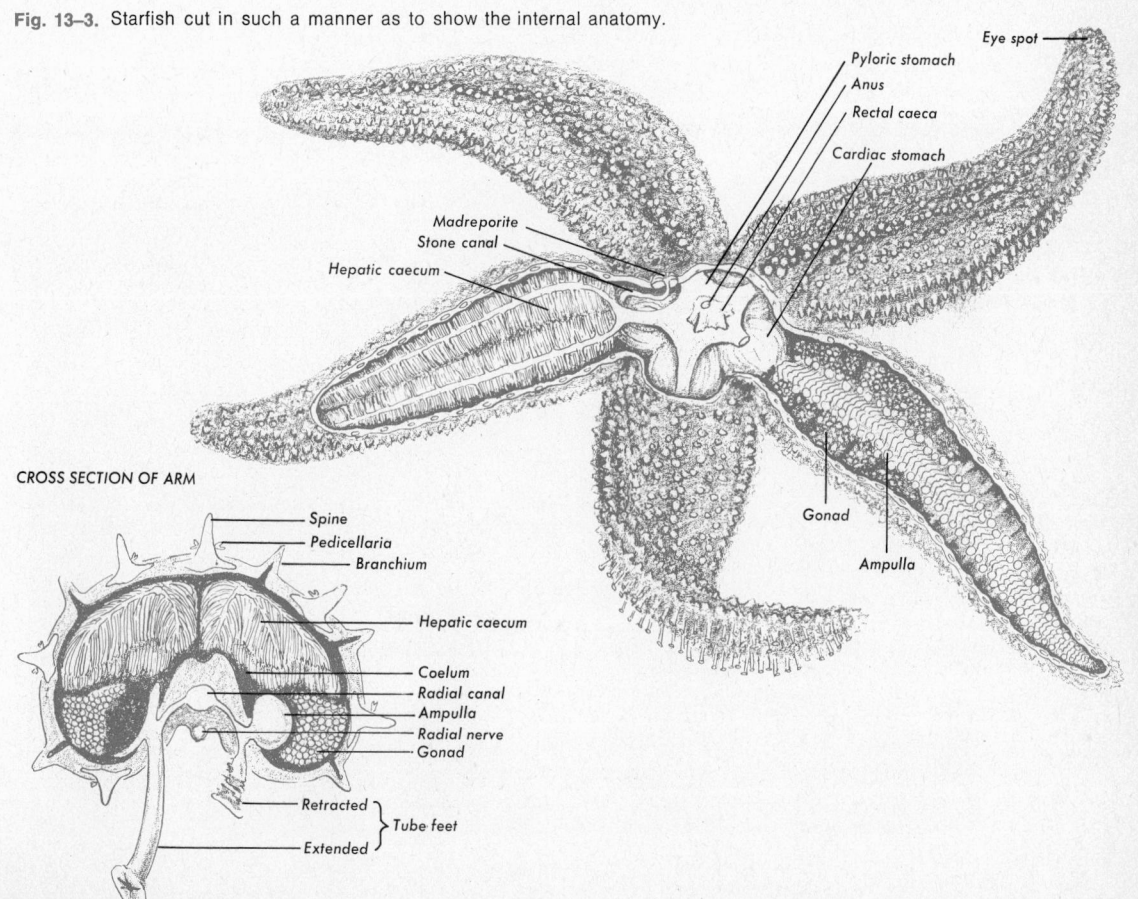

CROSS SECTION OF ARM

the mesoderm, is a calcareous framework composed of many **ossicles,** most of which are arranged in a definite pattern. Even though the starfish has this strong endoskeleton, it is capable of autotomy. Thus it can break off an arm and readily regenerate the lost part in a very short time (Fig. 13–4).

The **ossicles** of the endoskeleton are joined together by a network of connective tissue and muscle fibers. Lying within the skeleton and extending through all portions of the body is the **coelom.** It contains the internal organs and a lymphlike fluid which carries free amoebocytes. In certain regions the coelom comes close to the external epidermis which forms a tiny fingerlike extension, and in these structures, called **branchiae,** the respiratory exchange of gases takes place. The amoebocytes gather waste materials and escape from the body through these same structures.

When feeding, the starfish holds the victim with its arms and secures its grasp by attaching the tube feet (Fig. 13–5). The saclike **stomach,** which consists of a large lower portion, the **cardiac stomach,** and the smaller upper region, the **pyloric stomach,** is then everted through the mouth. If the captured animal is small enough, the stomach may completely surround it. Retractor muscles in the arms, just below the digestive glands, draw the everted stomach back into the body to complete digestion. If the animal is large it is digested in portions while the stomach remains everted. The digestive juices flow from the pyloric region of the stomach and the **hepatic caeca** (paired digestive glands of each ray) until the remaining food is small enough to be withdrawn into the pyloric stomach. The digested food passes into the coelomic fluid where it is distributed. Attached to the dorsal portion of the pyloric stomach is a short intestine with rudimentary **rectal caeca** and a small **anal opening** on the aboral surface of the disk.

The circulatory system is reduced to

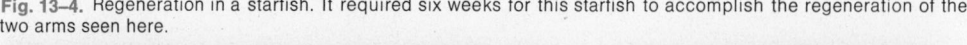

Fig. 13–4. Regeneration in a starfish. It required six weeks for this starfish to accomplish the regeneration of the two arms seen here.

Fig. 13–5. Starfish frequently feed on mussels. In the top picture the starfish is huddled over the mussel. In the bottom picture the starfish has been turned up on one side. Note the long tube feet attached to the two valves of the mussel. Their continued pull eventually weakens the mussel so that it gapes open, allowing the starfish to consume its soft body.

pendently. The starfish has only a few sense organs. An **eye** and a **tentacle** are located at the tip of each ray, and the **pedicellariae** function as dermal sense organs.

Sexes of the starfish are separate. The reproductive system consists of five paired gonads, lying close to the hepatic caeca and attached in the angles between the arms where their external openings are located (Fig. 13–3). Ova and sperm cells are discharged into the sea water, where fertilization occurs (Fig. 13–6). The larva, or **bipinnaria**, is at first bilaterally symmetrical; later, as the pentagonal shape of the adult form appears, radial symmetry becomes evident. Larval forms are partially ciliated and free-swimming. After a period of swimming near the surface of the water, sometimes for several weeks, the larva finally drops to the bottom of the sea, where it undergoes metamorphosis into the adult starfish.

Other Echinoderms

While the starfish gives us a good idea of the echinoderm body plan, there are many modifications of this plan within the phylum (Fig. 13–7). The most primitive forms are the stalked **crinoids**, commonly called **sea lilies** and **feather stars**. They have five greatly branched arms and are attached by means of a stalk to the bottom of deep seas. Food is carried down to the mouth by cilia in the ambulacral grooves. The tube feet resemble tentacles and are probably sensory in function. Their general appearance resembles a flower, hence the name.

The class **Ophiuroidea** includes the **brittle stars** (Fig. 13–7) and the **basket stars** (Pl. 13A). Both forms possess small disks and long, slender, motile arms, the arms of the basket star being branched to form a kind of basket. The brittle stars, characterized by five long, serpentine arms, can move more rapidly than any other echinoderm. Mild stimulation can cause an arm to be snapped off immediately; the rapid regeneration of a new member makes this a valuable means of escape for these animals.

The sea urchin (Pl. 13B) and sand

such an extent that it can scarcely be called a circulatory system at all. There are vessels encircling the mouth and extending down into each ray, but they are too inadequate to transfer the digested material to all parts of the body. Instead, the fluid of the coelomic cavity transports the food to various parts of the body.

The nervous system of the starfish shows the same radial symmetry seen in the other parts of the body and is, in general, simple. It consists of a nerve ring surrounding the mouth, giving off five banches, one to each arm, called **radial nerves** (Fig. 13–3). Two other systems of nerves lie internally, one on the oral side and another near the aboral side. Each part of the nervous system seems to function inde-

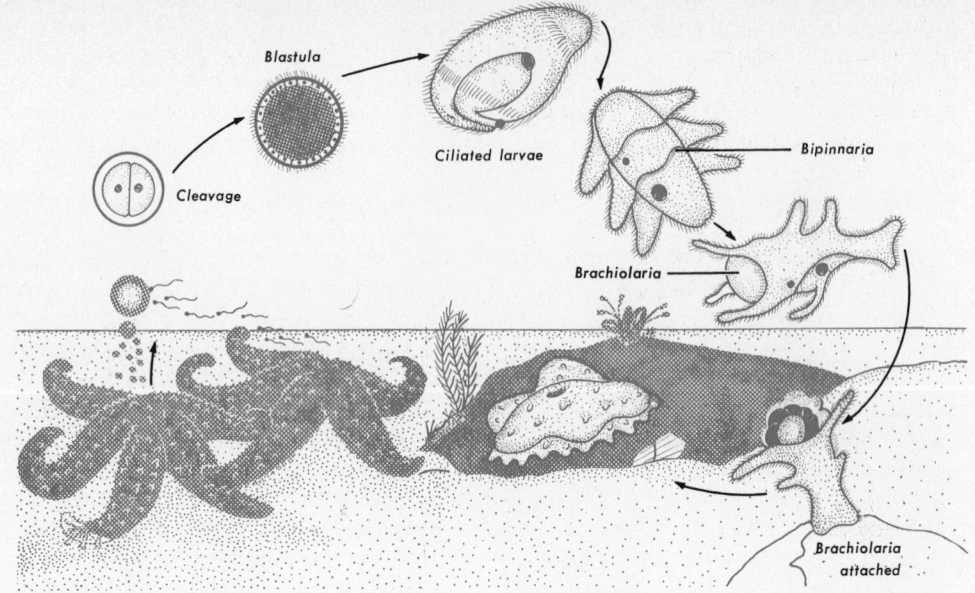

Fig. 13–6.

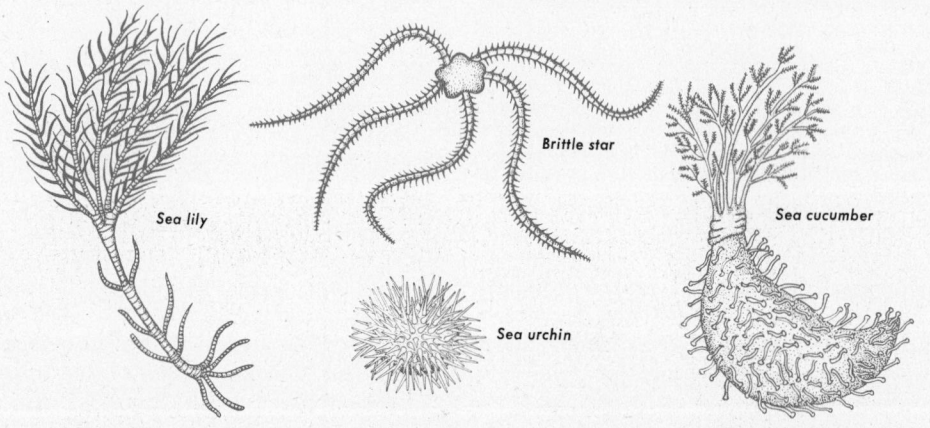

Fig. 13–7. Various kinds of echinoderms.

dollar are also seashore oddities and belong to the class **Echinoidea.** A common species of sea urchin is usually purple in color and is often found lying in the small pockets of rocks, homes that may be occupied for many years. It has remarkably long spines that are used in locomotion as well as in securing prey. The tube feet which cover the rounded body are also used in locomotion. An interesting organ, characteristic of this animal, is **Aristotle's lantern,** a complicated arrangement of teeth in the mouth that is used for picking food apart. In most other details the sea urchin resembles the starfish. The sand dollar is extremely flattened dorsalventrally, so that it resembles a flat disk.

Otherwise, it possesses the organs common to other forms already described.

The **sea cucumber** (Fig. 13–7), a bizarre type, belongs to the class **Holothuroidea**. It is quite unlike the starfish in its general appearance, although its fundamental structure is similar. These animals resemble cucumbers with a fringe of tentacles on one end. The surface of the body seems to be devoid of calcareous plates, but microscopic examination reveals tiny plates embedded in the soft tissue of the body wall. The branched tentacles are located at the anterior end, surrounding the mouth; the anus is at the opposite end of the animal. The animal is able to crawl in wormlike fashion. It feeds by allowing the tentacles to become covered with detritus from the muddy ocean bottom and then pushing them, one at a time, into the mouth.

Breathing is carried on by means of a pair of **respiratory trees** which extend from the lower end of the digestive tract anteriorly in the body cavity. Water is taken into the cloaca through the anus and circulated through these ramifying tubes. It is likely that excretory wastes find their way to the outside through these organs as well.

This animal possesses remarkable powers of autonomy and regeneration of lost parts and, in fact, even employs this behavior as a mechanism of defense. If disturbed, the sea cucumber suddenly contracts its muscular walls until considerable pressure is built up within. Then it splits open, almost explosively, near the anus, everting the respiratory trees which secrete a mucus fluid that becomes stringy and tough when it contacts sea water. The unfortunate enemy, usually a lobster, thus becomes hopelessly enmeshed in this mass of threads so that it is no longer concerned with the sea cucumber as a prospective meal. The sea cucumber is able to break the trees loose at their base and regenerate a complete set within a short time.

PHYLUM HEMICHORDATA

Before proceeding to the phylum Chordata, mention must be made of a small controversial group of animals which have

Fig. 13–8. Acorn worm *(Saccoglossus kowalevkyi)* from the sand flats of Cape Cod. Note the long proboscis for burrowing in the sand and mud.

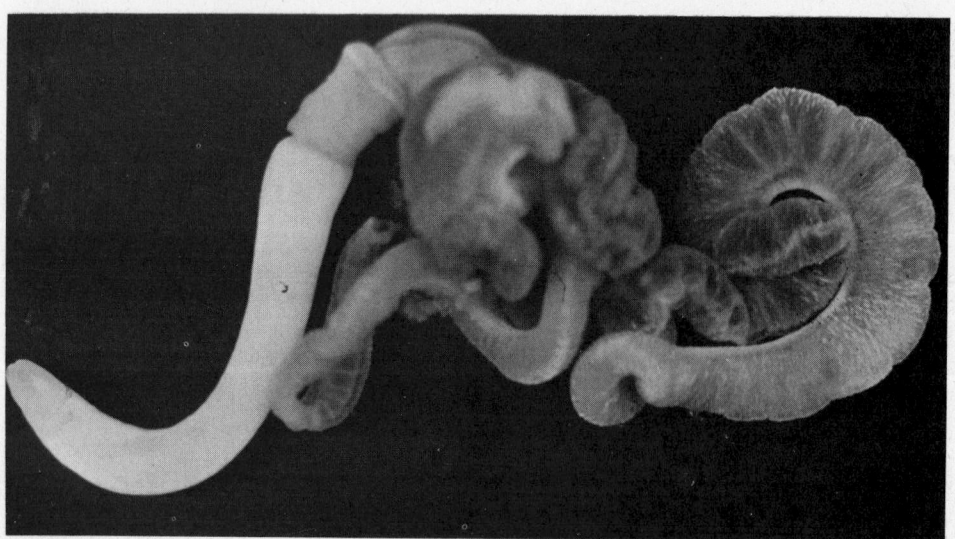

characteristics that closely approach those of the chordates and, hence, are often regarded as primitive members. Since they do not measure up in all respects, we discuss them following the echinoderms because they also have affinities with this phylum.

The phylum Hemichordata are small, sedentary or burrowing, wormlike animals that live exclusively in the sea. Their body is divided into the **proboscis, collar,** and **trunk.** A "half chord" supporting structure, from which the phylum gets its name, is present in the proboscis and the collar contains a dorsal **nerve** component. The pharynx wall is perforated by numerous **gill slits** which permit water entering the mouth to pass out of the body. These characteristics have been used to place them with the chordates but only the gill slit is a true chordate characteristic, the others are hardly comparable to similar structures in chordates. A representative example is the **acorn worm** (*Saccoglossus*) (Fig. 13–8).

SUGGESTED SUPPLEMENTARY READINGS

Books

BOOLOOTIAN, R. A., ed., *Physiology of Echinodermata.* New York: Wiley-Interscience, 1966.

CLARK, A. M., *Starfishes and Their Relations.* London: British Museum, 1962.

GILLETTE, K., and McNEILL, F., *The Great Barrier Reef and Adjacent Isles.* Sydney, Australia: The Coral Press Pty. Ltd.,

MacGINITIE, G. E., *Natural History of Marine Animals.* New York: McGraw-Hill, 1949.

MILLOTT, N., ed., *Echinoderm Biology.* New York: Academic Press, 1967.

NICHOLS, D., *Echinoderms.* London: Hutchinson University Library, 1966.

RICKETTS, E. F., and CALVIN, J., *Between Pacific Tides*, 9th ed. Stanford, Calif.: Stanford University Press, 1968.

GENERAL REFERENCES—INVERTEBRATES

The general reference books listed below will be useful to those students who wish to examine the invertebrates more carefully. Many minor groups, not included in this book, are discussed in some detail.

BARNES, R. D., *Invertebrate Zoology*, 2nd ed. Philadelphia: Saunders, 1968.

BARRINGTON, E. J. W., *Invertebrate Structure and Function.* Boston: Houghton Mifflin, 1967.

BAYER, F. M., and OWRE, H. B., *Free-living Lower Invertebrates.* New York: Macmillan, 1968.

BROWN, F. A., *Selected Invertebrate Types.* New York: Wiley, 1950.

HEGNER, R. W., and ENGEMANN, J. G., *Invertebrate Zoology.* New York: Macmillan, 1968.

HYMAN, L. H., *The Invertebrates*, Six vols. New York: McGraw-Hill, 1940–1967.

KAESTNER, A., *Invertebrate Zoology.* Three vols. New York: Wiley-Interscience, 1967–1969.

MEGLITSCH, P. A., *Invertebrate Zoology*, 2nd ed. London: Oxford University Press, 1972.

PENNACK, R. W., *Fresh-Water Invertebrates of the United States.* New York: Ronald Press, 1953.

14

THE ANIMAL CLIMAX-CHORDATES

The last and most diversified group of animals is the phylum Chordata, to which man himself belongs (Fig. 14–1). Not only are the chordates the largest animals in existence today, but they have adapted themselves to more modes of existence than any other group, including the arthropods. They are found in the sea, in fresh water, in the air, and on all parts of the land from the poles to the equator. They range in size from the tiniest fish to the great whales, which reach a length of nearly 100 feet and a weight of 150 tons and more (Fig. 14–38). Birds and mammals have been able to penetrate cold climates because they have a constant body temperature, something no other animals have.

All chordates possess at some time in their life cycle three characteristics which are not found among the animals studied so far (Fig. 14–2). The first is a **dorsal tubular nerve cord,** which varies from a more or less undifferentiated tube extending throughout the entire length of the body of the lower chordates, to a shorter, highly differentiated tube, with a greatly enlarged anterior portion, the **brain,** in the higher forms. In lower forms the nerve cord is solid, but in all chordates it is tubular or hollow.

A second characteristic of the chordates is the presence of an internal supporting rod, or skeleton, the **notochord.** This may be thought of as a precursor to the vertebral column in vertebrates, but it must not be considered identical. Although the notochord is found in the embryos of all vertebrates, it persists only in the adults of the most primitive. The notochord is made up of a gelatinous matrix, surrounded by a tough, outer sheath, which is inadequate to support a large animal in water, much less on land. In all higher forms, therefore, it is replaced by the more rigid **vertebral column.**

The third characteristic is the presence of pharyngeal **gill slits.** It is obvious that adult land animals have no gill slits, but

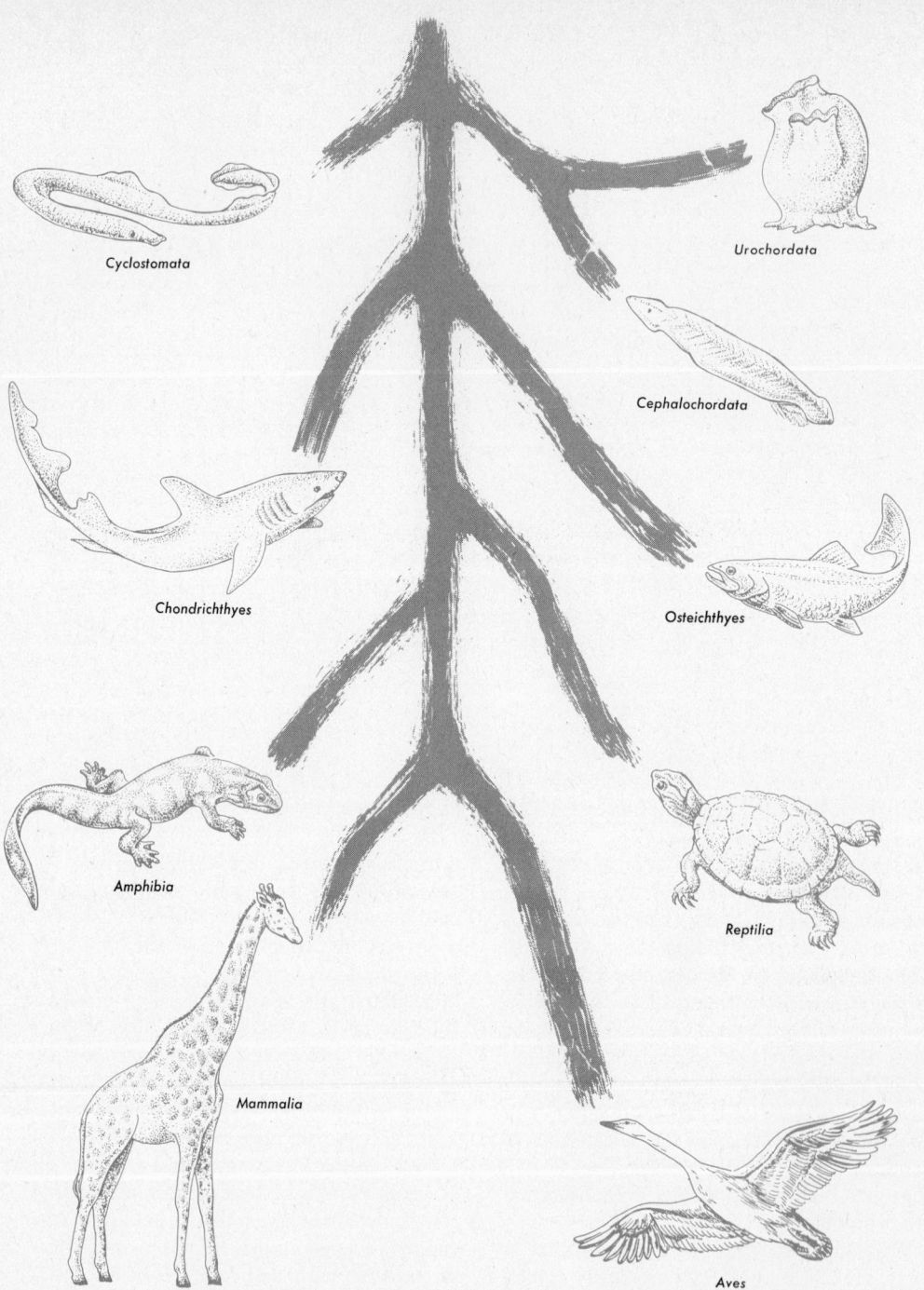

Fig. 14–1. The phylum Chordata is composed of widely diverse animals as indicated by representatives from each of the major groups.

Cyclostomata

Urochordata

Cephalochordata

Chondrichthyes

Osteichthyes

Amphibia

Reptilia

Mammalia

Aves

during embryological development gill slits do appear at some stage. The structures which originally produced functional gill arches in fish produced other structures in higher forms, such as the sound-making apparatus (middle ear bones). These fitted the animal better for a terrestrial existence and gave it a greater chance for success.

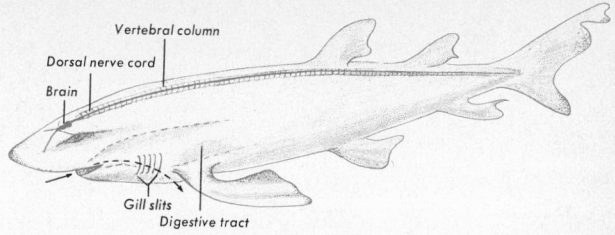

Fig. 14–2. The vertebrate body plan.

CHORDATE BEGINNINGS

Scientists have been perplexed about the origin of the chordates and have been unable to determine which lower forms gave rise to this last and perhaps most specialized group. Fossils have provided us with a great deal of information about other animals but have failed thus far to reveal any substantial remnants of the early chordates. The reason for this is that these soft-bodied animals did not remain intact sufficiently long to become fossilized. In spite of the lack of evidence concerning the early progenitors of the chordates, there has been a great deal of speculation as to their origin.

The phyla Echinodermata, Hemichordata, and Chordata show common ancestry by the embryonic origin of the coelom: all are enterocoelomates (p. 115). In all, the anus forms from the blastophore, and for that reason they are sometimes spoken of as **deuterostomes** in contrast to the schizocoelomates in which the mouth is clearly a part of the blastopore and hence are called **protostomes.** Serological studies demonstrate that the proteins of these three phyla are more closely related to one another than to those of any other phyla. Moreover, the remarkable similarities between the echinoderm (bipinnaria) and hemichordate (tornaria) larvae (Fig. 14–3) is taken as good evidence for common ancestry.

PHYLUM CHORDATA

We usually think of the chordates as synonymous with vertebrates and while certainly the vertebrates do make up the vast majority of the animals in the phylum, there are some inconspicuous and primitive chordates (prochordates) that share the three characteristics which define the phylum. The phylum is divided into three subphyla: **Urochordata, Cephalochordata,** and **Vertebrata.** We consider the first two briefly, but spend some time with the last.

The **tunicates,** or **sea squirts** (subphylum Urochordata) (Fig. 14–4), are commonly attached to rocks along the seashores where they live by forcing water in and out of their saclike bodies through **siphons,** resembling the clam in this respect. The water passes into a large perforated pharynx which strains out the tiny food particles that are carried into the digestive tract. Gills line the many openings in the pharynx wall, but

Fig. 14–3. A study of the larval stages of the acorn worm and echinoderms has lent support to the idea that both came from a common ancestor.

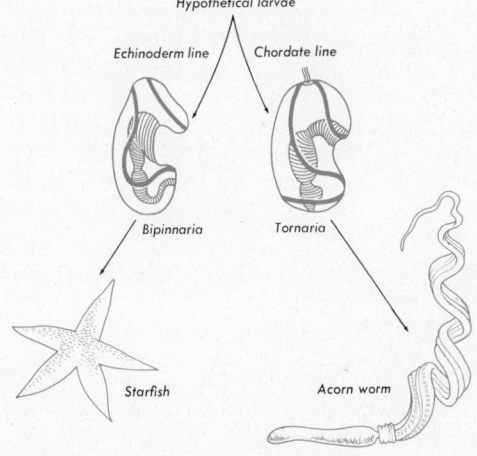

aside from this one chordate characteristic, the adult appears to have no claim to membership among the chordates. However, the larval form demonstrates its true chordate relationships, for the larva possesses a notochord and dorsal tubular nerve cord, in addition to the gill slits. As an embryo, the animal is tadpole-shaped and swims actively in the sea water (Fig. 14–4). Late in embryonic life, however, it settles on a rock and metamorphoses into the sessile adult. Tunicates are very numerous in the oceans of the world and range from microscopic size to more than 12 inches in diameter. They may live in shallow or deep water. The group as a whole has no economic significance.

There is another tiny animal (2 inches long) that cannot be mistaken for anything

Fig. 14–4. Life cycle of a tunicate. Note the definite nerve cord and notochord in the larva, both of which are reduced or absent in the adult. Three stages in the metamorphosis of the larva into the adult are shown in the upper right. A colonial prochordate is also seen, attached to the piling.

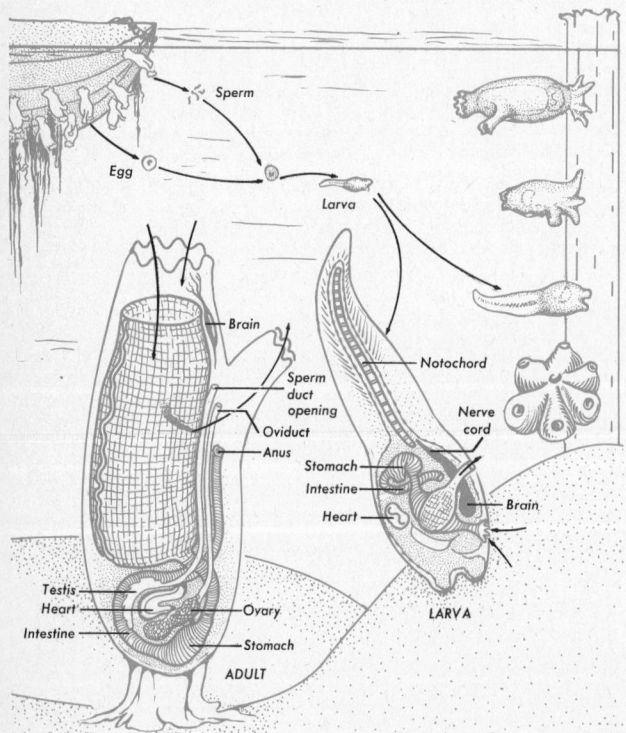

Sperm

Egg

Larva

Brain

Notochord

Sperm duct opening

Oviduct

Anus

Nerve cord

Stomach

Intestine

Heart

Brain

Testis
Heart
Intestine

Ovary

LARVA

Stomach

ADULT

but a primitive chordate and, moreover, it possesses body structure that forces us to believe that some such form might have given rise to the vertebrates. This is **lancelet** or **amphioxus** (subphylum Cephalochordata (Fig. 14–5). It is an ocean dweller, found in relatively few, widely separated regions, and reaches such numbers along a part of the shores of China that it is utilized as a source of food. Not only does it possess the three chordate characteristics exhibited by the preceding group, but it also has a body plan that closely resembles that of the vertebrates.

Amphioxus has a general shape not unlike that of a slender fish, with two longitudinal folds of skin extending throughout most of its length, which may be forerunners of appendages. Its notochord functions as a semirigid supporting internal skeleton, extending from one end of the animal to the other. The muscles are segmentally arranged and by their rhythmic contractions make possible lateral undulations of the body used in swimming. Above the notochord is the hollow nerve cord which terminates anteriorly in the eye-spot. Numerous gills function in breathing. Its digestive and circulatory systems are relatively simple. In fundamental plan, however, these organ systems show great similarity to the vertebrates and thus point to the possibility that the vertebrates may have come from a form not greatly unlike it.

THE FIRST ANIMALS WITH BACKBONES—VERTEBRATES

We now come to the large group of animals known as **vertebrates** (subphylum Vertebrata) because they possess backbones. All animals studied up to this point are called **invertebrates**. The characteristics of the vertebrates, in addition to the three which identify the chordates, are an internal skeleton of bone and cartilage; a vertebral column or backbone which replaces the notochord of the prochordates; usually two pairs of jointed appendages which may be

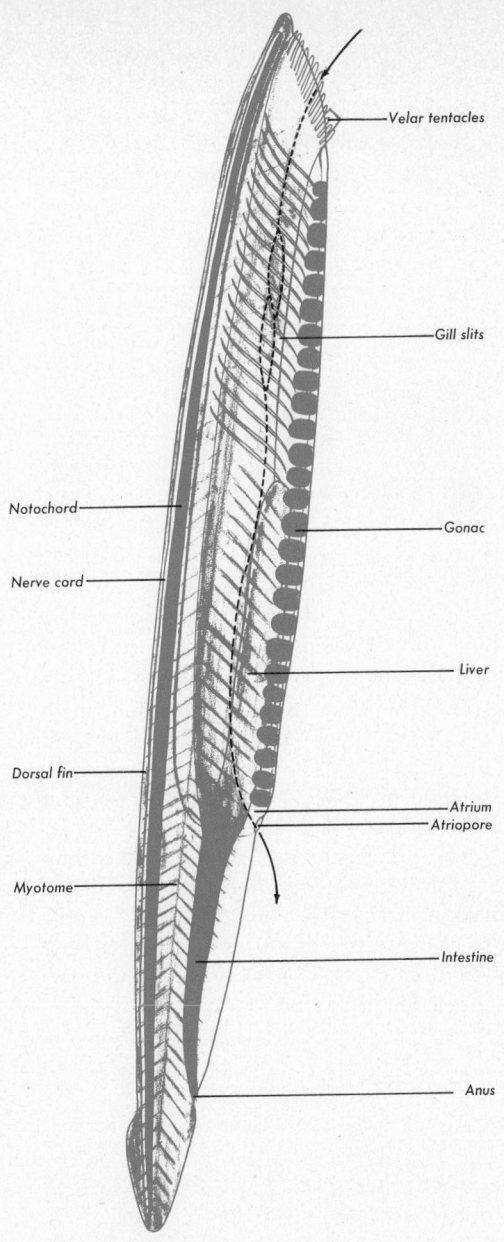

Fig. 14–5. Amphioxus in longitudinal section.

Labels: Velar tentacles, Gill slits, Notochord, Nerve cord, Gonac, Liver, Dorsal fin, Atrium, Atriopore, Myotome, Intestine, Anus

quite certain that without this structure the group could not have become so diversified in body form. To invade the land and air required a substantial internal support upon which a body could be built that would succeed in these environments. This was accomplished by the development of the backbone to which appendages and other structures were attached. Let us examine representatives of the eight classes of vertebrates.

The class **Agnatha** includes only one group of living representatives, the **cyclostomes** or "round-mouthed" lampreys. They have no appendages and no jaws, only a circular mouth lined with denticles, small toothlike structures that aid in clinging to prey. When the lamprey seizes a bony fish, its usual prey, it first attaches itself with the sucking mouth and then removes small bits of tissue with its rasping tongue. If the point of attachment happens to be in the abdominal region, a perforation is made through the body wall and the internal organs are injured so severely that the fish usually dies. However, if the injury occurs on the dorsal side over the large muscles, the effect is usually not fatal. The common sea lamprey, *Petromyzon*, has invaded rivers and streams where it has become a formidable foe of fish populations (Fig. 14–6). These ravages have been particularly severe in the Great Lakes region where in many areas commercial fishing has all but ceased.

Internally as well as externally, the lampreys shows its lowly origin. It retains a notochord similar to amphioxus, but also has the beginnings of a spinal column and other internal skeletal parts which are composed of cartilage. There is a well-developed brain, together with an olfactory organ, a pair of poorly developed eyes, and a pair of simple semicircular canals on each side of the head, used in balancing. It has seven pairs of gills—fewer than amphioxus, but more than the bony fish. Internally, also, the complexity of its body structures far exceeds that of amphioxus.

A group of animals called **ostracoderms** that lived about 400 million years ago

fins or limbs; a closed circulatory system with a two-or-more chambered heart; and a large coelom containing vital organs. Of these, perhaps the most important is the possession of a rigid internal skeleton. It is

Fig. 14–6. Sea lampreys (*Petromyzon marinus*), attacking a freshwater fish. Note the scar from a previous injury on the dorsal side just above the fore-fins.

(Silurian Period) resembled the present-day cyclostomes in many respects. They developed heavy armor plates on their external surfaces, possessed a ventral mouth, and were without appendages. Their heavy exoskeletons were essential in order to survive the onslaughts of their invertebrate enemies, the water scorpions (eurypterids). It is agreed now that later descendants of the ostracoderms lost their plates as these enemies disappeared. With the development of jaws, it became possible for them to pursue and capture their prey more effectively.

THE FIRST APPENDAGES AND JAWS

The lower forms such as the cyclostomes do very little free swimming in the water and get along satisfactorily with a broad fin in the tail region and a dorsal fin to aid in locomotion. In order to be more maneuverable, vertebrates gradually developed more elaborate appendages. Sharklike fossil remains of forms possessing many paired fins (Fig. 14–7) seem to indicate that they originally had many and, according to Romer, only later settled down to the

orthodox two pairs, the pectoral (chest region and pelvic (hip region). From these two pairs of fins, which became so prominent in the early sharks and all later fishes, evolved the appendages of land forms (Fig. 14–8).

The prehensile jaws of the early primitive fish were another important acquisition, making it possible for them to become predators, searching out and capturing their prey. A clue to the development of the jaws can be found from a study of the shark's gill arches. These differ but little from the

Fig. 14–7. Primitive sharks had many paired appendages extending throughout the length of the body as shown in the above figure. The two conventional pairs (pelvic and pectoral) appeared in later forms and in all present-day vertebrates.

jaw itself and, in fact, they are so much alike in this animal, as well as in many fossil forms, that it is generally agreed that the jaws have developed from the first gill arch.

The teeth found in the shark's jaw occur in rows and show a remarkable similarity to its scales. It is thus clear that the scales in the region of the mouth opening merely enlarged and became the teeth of the shark. These teeth simply grow over the edge of the mouth and are continuously shed as they wear out. Later it is shown that all teeth are modified scales, including those of man.

The sharks and their close relatives, the rays, show diversification in body form (Fig. 14–9). The latter is greatly flattened, with enormously developed pectoral fins that look more like wings (hence the name, sea bat) as they undulate in the sea. The tail is drawn out to a long whiplike structure which, in the sting rays, bears a spine at the tip. When annoyed, the ray can inflict a painful wound. Some rays produce electricity in considerable quantity and employ it as a mechanism of defense. Shocking devices are also found among a few species of eels (p. 349). The mechanism of this device is not thoroughly understood.

THE FISHES

The cartilaginous fishes, the sharks and rays (class **Chondrichthyes**), developed appendages and true jaws as important adjuncts to success in the water. Once these became established as a permanent part of the anatomy of aquatic vertebrates, a great

Fig. 14–8. Primitive vertebrates, according to some zoologists, had fin folds much like amphioxus (upper figure). Portions of these folds were retained in the pectoral and pelvic regions to become the paired appendages; other parts became the other fins (middle and lower fish figures). Coming out on land necessitated greater development of the appendages, as is seen among the amphibians, reptiles, and mammals. At first they dragged their bodies over the ground, later the appendages supported the body above the ground, which made it possible for the animal to move more swiftly. This reaches its peak in such cursorial animals as the horse.

Fig. 14–9. Cow-nosed rays (*Rhinoptera bonasus*). Note their huge pectoral fins that give them the appearance of flying when swimming.

deal of "experimentation" apparently ensued. The result was the modern bony fishes (class **Osteichthyes**) which have gone "all out" in exploring possible body shapes, sizes, and colors that best suit them for their particular aquatic niches. They range from ordinary fish such as the common perch to the vicious garpike (Fig. 14–10) and the bizarre sea horse (Fig. 14–11). They are found in the oceans of the world—from the surface to great depths, where they have developed extraordinary luminescent organs. In fresh water they are found in swift-moving streams as well as stagnant pools. Some, such as the salmon and eel, can survive satisfactorily in either fresh or salt water and migrate seasonally from one

environment to the other, movements which are synchronized with their breeding cycle.

Structurally, the bony fish are similar to the sharks with a few minor exceptions. For example, the gills are reduced to four pairs and are covered with a thin bony structure, the **operculum.** Their bodies are covered with large overlapping scales. The fins are highly variable both in position and in size. There is nothing strikingly different about their internal anatomy with the possible exception of the **swim bladder,** which functions in regulating the buoyancy of the body. As the fish moves to different depths, the gases (CO_2, N_2, and O_2) increase or decrease in the swim bladder automatically,

Fig. 14–10. The long-nosed gar (*Lepidosteus osseus*) is a vicious carnivore which feeds on other fishes. This fish inhabits the Great Lakes and most of the streams of the Mississippi Valley.

adjusting the specific gravity of the fish to the corresponding depth, but if a fish is suddenly pulled from great depths to the surface the expanding bladder may force the stomach out of the mouth.

Most present-day fish possess bony skeletons which made it possible for them to migrate onto land. Although this movement began with the fish, it was not completely accomplished until the advent of the land egg, many millions of years later. The hard bones made it possible for appendages to become sufficiently strong to support a body in the air.

Among the ancient fish there were some that had a fleshy portion, or "lobe," which extended some distance out into the fin. This contained certain skeletal elements that have been found to correspond directly with similar bony elements in the appendages of true land forms. Descendants of these fish undoubtedly were able to migrate onto land and at a later time to give rise to the great array of land vertebrates. "Lobe-finned" fish were long thought to be extinct until a fisherman off the coast of South Africa caught one (*Latimeria*) in 1938. Others that have been found since have been studied carefully and described in detail (Fig. 14–12). Relatives of the "lobe-finned" fish are not uncommon today. The

Fig. 14–11. The sea horse (*Hippocampus kuda*) swims in a vertical position by means of its dorsal fin. Note its prehensile tail, used to cling to vegetation. The male has a pouch under the tail where the eggs are brooded until they hatch.

Fig. 14–12. Ancient lobe-finned fish (*Latimeria*) found off the coast of Africa in 1938, supposedly extinct for millions of years.

lungfish, for example, inhabits certain tropical parts of the earth where frequent droughts occur (Fig. 28–2). Since this form followed a different path of evolution, it does not possess well-developed appendages and, in spite of the fact that it has lungs, probably did not give rise to the land forms.

One other absolute essential for migration from water to land was some means of utilizing the oxygen of the air. The "lobe-finned" fish of the past, as well as the lungfish of today, accomplished this rather satisfactorily. Apparently these fish lived in regions where nearly all the water dried up for extended periods during the year, and in order to survive they found it necessary to come to the surface of drying pools and take in air, since there was little oxygen in the water. These animals developed a pair of saclike lungs from the ventral side of the pharynx which allowed them to gulp air during periods when their gills were useless. This, it must be remembered, is a primitive condition. Its counterpart is found in present-day fishes in the form of a swim bladder which functions as a hydrostatic organ rather than a lung. It therefore appears that once this lung-like structure

developed among the "lobe-finned" fish, it was utilized on land where it eventually became the complex organ that is found in such animals as birds and mammals of today (Fig. 14–13).

Thus two features, the bony appendage and the lung, made it possible for animals to attempt the greatest of all transitions—from the water onto land.

INVASION OF THE LAND

Of all the changes that have occurred in vertebrates during their long evolution to present forms, the most striking was when fishes left their aquatic life for life on land. According to Romer, this was the "result of a happy accident." They would hardly have left the water in search of food, since during these times most animals were aquatic, and fish would hardly leave a food-laden world for one almost devoid of food. They had already supplied themselves with a means of breathing air, so this could not have been the cause. Romer reasons that if drought periods were too extensive, those fish which could breathe air and walk about on the land were able to move to other ponds and survive. Thus the appendages and lungs aided them in finding water rather than leaving it. However, during these excursions some may have found abundant food near the water's edge, whereas others which could not stand drought may have found it more profitable to wander from pond to pond in search of food. Again, some may have found members of their own group to feed upon, while others may have changed to a herbivorous diet, since vegetation was abundant. From such beginnings the great variety of life among land vertebrates appears to have developed.

THE AMPHIBIANS

There were undoubtedly numerous un-successful attempts by many groups of fishes to make the transition onto land. The ancestors of the "lobe-finned" fish were

apparently successful and gave rise to the amphibians which include our present-day frogs, toads, and salamanders. As the name **amphibian** implies, these animals live both in and out of the water. Their larval stages are always spent in water or some moist place, whereas the adult of most species are able to live on land, although they do not venture far from moist places. The degree to which amphibians are linked to water varies with different species. Some have divorced themselves completely from an aquatic environment, as for example, the South American toad, whose eggs brood in fluid-filled sacs upon its back. Other species such as the mud-puppy, spend their entire life in the water and cannot leave it. A curious intermediate is a variety of tiger salamander (*Ambystoma tigrinum*) which normally spends its entire life in the larval form, but which if fed thyroid extract or high levels of iodine, can be made to lose its gills, develop lungs, and come out on land just as its relatives do (Fig.

Fig. 14–13. History of lungs and swim bladder.

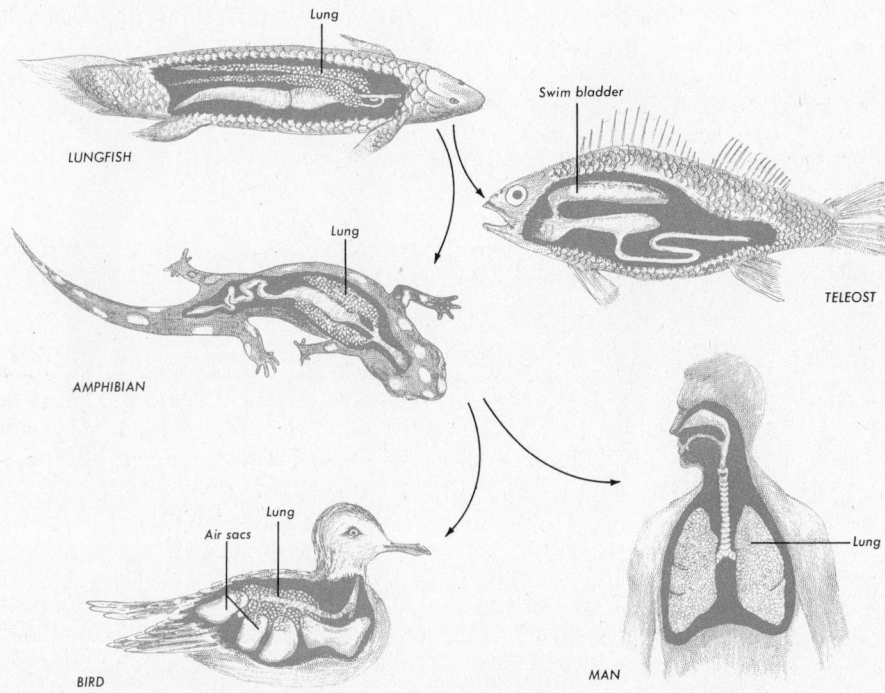

14–14). This tiger salamander larva was thought to be a different species from the usual adult and was called the **axolotl.** Among the amphibians, then, there are those which attempt to leave the water altogether and those which tend never to leave it.

Other characteristics of the amphibia are moist glandular skin without scales; two pairs of appendages used in locomotion (except the legless caecilians); a pair of nostrils connecting with the mouth cavity; breathing by means of lungs, gills, or skin; a heart possessing two auricles and one ventricle; and external fertilization of the eggs in toads and frogs, internal in salamanders.

The class is divided into three orders; the **Apoda** (limbless)—caecilians; the **Caudata** (tailed)—salamanders; and the **Salienta** (tailless)—frogs and toads. Of these we consider one—the frog—in some detail.

The Frog: The Halfway Vertebrate

If numbers are taken as a criterion for success, the amphibians were much more successful at an earlier time than now. At that time they did give rise to very successful groups of animals, the reptiles, birds, and mammals. The amphibians seem to have reached the halfway mark between the aquatic and the land forms and for that reason they show some very interesting in-

Fig. 14–14. The common tiger salamander (*Ambystoma tigrinum*) normally undergoes a typical amphibian metamorphosis resulting in the adult. One variety of this species living in western North America, particularly in the Southwest, becomes sexually mature while still a larva and never reaches the adult stage. The specimen on the extreme left is a young larva, while the one next to it is a sexually mature larva. This is as far as development proceeds in nature. Several specimens similar to this one were placed in water and a high level of iodine. During the next few weeks the "adult larvae" metamorposed to typical adults as the next two specimens show.

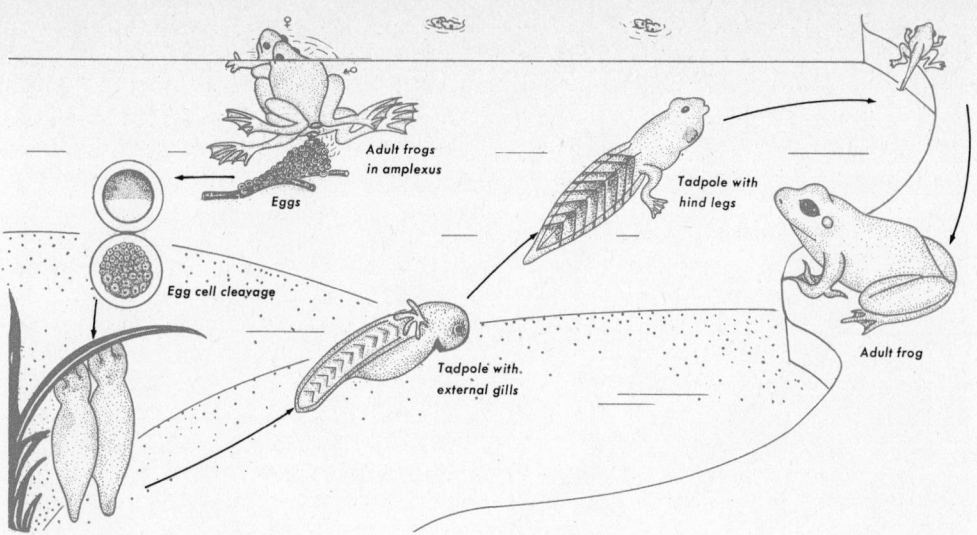

Fig. 14–15. Life history of the frog.

termediate structures. To study a bird or a mammal without reference to the frog would be like studying the present government of the United States without recourse to the struggle for independence. Understanding of a mammal can only come from a historical approach, which means that it is essential to examine an intermediate type. There is no better form to use for such a study than a representative amphibian, and the frog lends itself especially well for several reasons. First, aside from its jumping legs, it possesses most of the typical ancestral amphibian characteristics. Second, it occurs universally. Lastly, it is of such a size that it is easily handled in the laboratory. A thorough knowledge of the "halfway" animal at this point provides the background for a better understanding of the mammal.

Life History. (Fig. 14–15). One of the first harbingers of spring is the familiar croaking of the frogs (Fig. 14–16). While these sounds are pleasant to hear, they have more important meaning to the frogs. In temperate climates the male frogs usually emerge from hibernation first and begin croaking; the females follow some days later. At this time of the year, the eggs in the body of the female are fully mature. In most species the male mounts and clasps the female with his front legs, grasping her just back of her front legs and pressing the small swollen parts of each thumb (**nuptial pads**) against her chest. This process is called **amplexus** (Fig. 14–15), and is a kind of copulation. As the female lays her eggs, the male discharges sperm, thus fertilizing them. Very shortly thereafter, the gelatinous matrix surrounding the eggs swells, causing them to adhere to twigs or any other underwater debris. Under the microscope, one can see the egg divide, and finally develop a small embryo. Some time later a tiny tadpole emerges from the jelly mass. Presently, it begins to swim about and can be seen to feed by a scraping movement of its mouth as it moves along a leaf of a water plant. In this stage it breathes by means of gills. It is a vegetarian, feeding exclusively on algae and other plant life.

After some months or years, depending on the species of frog, the tadpole rather suddenly begins to develop miniature legs while its tail becomes shorter. At the same time its mouth grows larger and wider, and its digestive tract shortens. It gradually seeks shallower water and occasionally comes

Fig. 14–16. A tree toad (*Hyla versicolor*) croaking. Note the greatly extended vocal sacs which aid in producing its shrill sound.

to the surface for air as its lungs develop. These trips for air become more frequent until finally the frog hops away from the water, sans tail and strictly carnivorous. Thus, in a brief period it has reenacted the entire history of this long and arduous migration out of the water onto land, a most remarkable feat!

The Frog Body Plan. Frogs range in size from the tiny chorus frog (*Pseudacris*), about an inch long, to the bullfrog which may be a foot overall. Some are very colorful (Pl. 14). In general, their features are so similar that a description for one fits them all. During the following study refer to Figure 14–17 for general anatomical features.

When a study is made of a living frog, the moist, slippery skin is at once con-spicuous even if the frog is kept away from water for some time. This is due to tiny **mucus glands** in the skin which constantly pour out their fluids to keep it wet. The frog receives oxygen through its skin and must therefore have a moist skin.

The protruding **eyes** permit the frog to come to the surface of the water and see without exposing the rest of its body. The large **eardrums** are a part of the hearing mechanism. The **nostrils**, which have valves that can be opened and closed, function in breathing. The **mouth** is very large and is kept shut all of the time except when the frog feeds. At the posterior end is the **anus**, which is the terminal opening of the **cloaca**.

Outer Covering. Like all vertebrates, the skin of the frog consists of an outer thin **epidermis** and a thicker under layer, the

dermis. The outer layer, which is shed periodically, is made up of flat cells. The dermis contains many glands which provide the mucus for keeping the skin moist. The dermis is also heavily vascularized for its function in breathing.

The dermal scales of the fish are noticeably absent among the modern amphibians, although fossil remains indicate that their ancestors were well covered with scales. Scales offer excellent protection from attack and it seems strange that the amphibians have given up this apparently valuable aid in self-preservation. It must be remembered, however, that in present-day species the skin is an "accessory" lung and very important in respiration.

Locomotion and Support. The most conspicuous difference noted between the frog and fish is the remarkable development of the muscles that operate the appendages. Although many of the muscles approximate the position and seem to function much the same as similar muscles in man, only a very few of them are identical. Some of the muscles of the frog are named in Figure 14–18 and should be studied in terms of their function in locomotion and not from a comparative point of view.

Although the internal skeleton of the frog is made of bone and in many respects resembles that of man, in other respects it must be considered as the skeleton of a "specialized" vertebrate rather than a "generalized" form because it differs so markedly from primitive vertebrates (Fig. 14–18). The appendages are attached to the vertebral column by means of girdles, the **pectoral** in front, the **pelvic** behind. The pectoral girdle consists of three principal pairs of bones attached to a series of mid-ventral bones called the **sternum.** The **scapulae** are located on the dorsal side of the trunk (the flat extension is called the suprascapula), and this structure is similar to the human shoulder blade. It joins ventrally with the **clavicle** and **coracoid** which in turn fuse to the sternum. The clavicle (collar bone) is well developed in

man but the coracoid is only a small "bump," fused to the scapula. The pelvic girdle is composed of three pairs of bones, which in the adult are fused into a single structure. The long, flat, anteriorly directed **ilium** joins posteriorly with the **ischium** and ventrally with the **pubis** (the latter remains as cartilage) to form each half of this girdle.

The front and hind legs of the frog are homologous, that is, they are very similar, possessing approximately the same bones although of somewhat different proportions. A single bone, the **humerus,** which fits into a cavity (**glenoid fossa**) of the pectoral girdle, forms the top of the front leg; this is followed by a pair of bones, the **radius** and **ulna,** which are fused together in the frog but separate in most other vertebrates. The wrist is composed of several bones called **carpals;** these are followed by the **metacarpals** and **phalanges** of the **digits.**

The posterior appendage likewise has a single bone, the **femur,** which fits into a socket (**acetabulum**) in the pelvic girdle; this bone is followed by a pair of bones, the **tibia** and **fibula,** which again are fused in the frog. The **tarsals** are next, and two of

Fig. 14–17. Longitudinal section of the frog, showing internal organs.

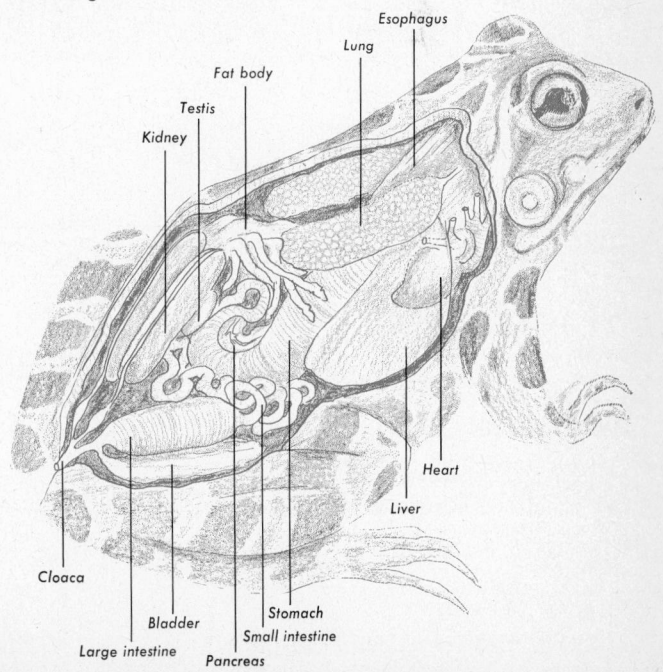

Esophagus

Lung

Fat body

Testis

Kidney

Heart

Liver

Cloaca

Bladder

Large intestine

Stomach

Small intestine

Pancreas

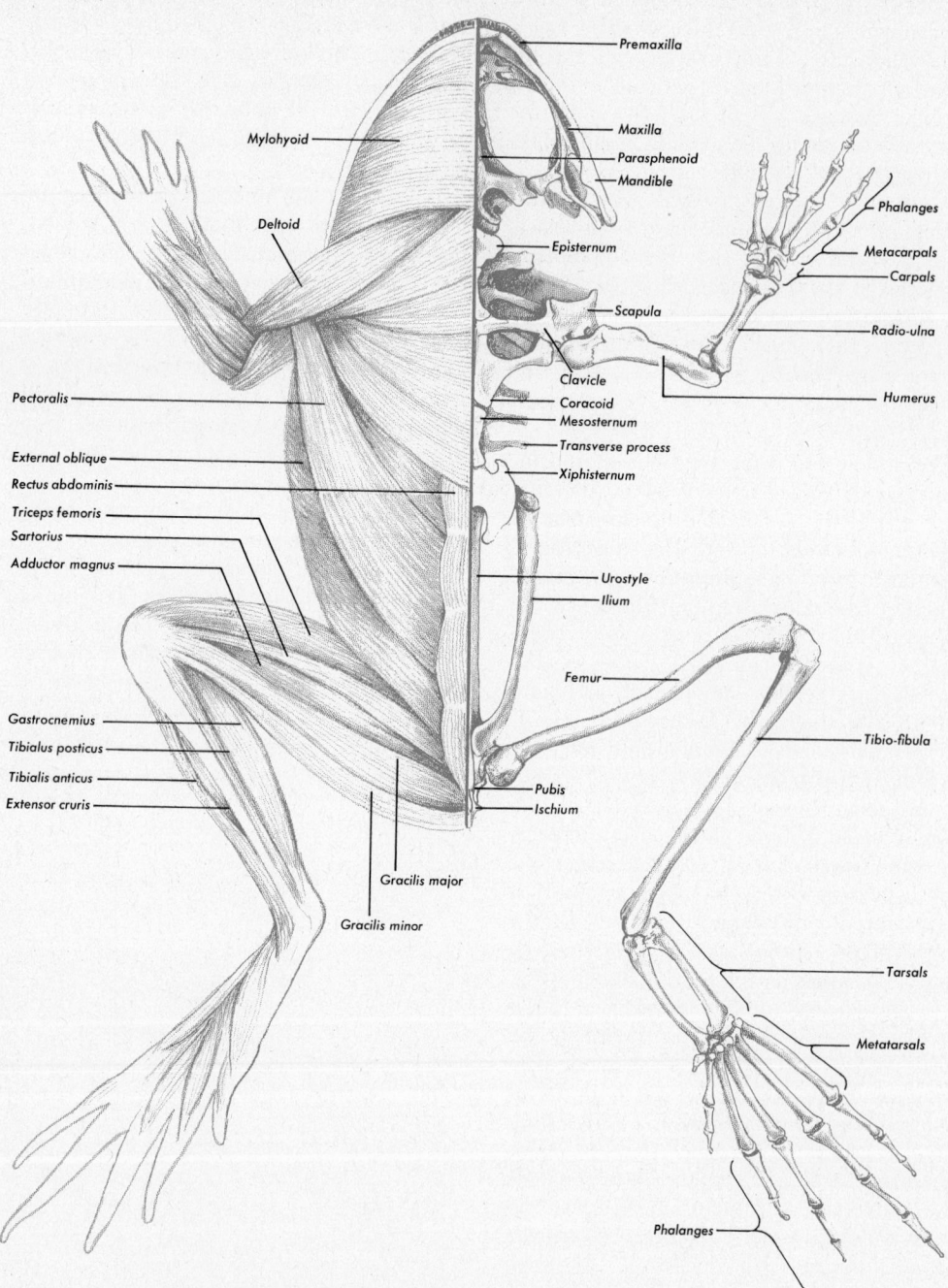

Fig. 14–18. The muscular and skeletal system of the frog.

Premaxilla

Maxilla

Parasphenoid

Mandible

Phalanges

Metacarpals

Carpals

Radio-ulna

Humerus

Mylohyoid

Deltoid

Episternum

Scapula

Clavicle

Coracoid

Mesosternum

Transverse process

Xiphisternum

Pectoralis

External oblique

Rectus abdominis

Triceps femoris

Sartorius

Adductor magnus

Urostyle

Ilium

Femur

Tibio-fibula

Gastrocnemius

Tibialus posticus

Tibialis anticus

Extensor cruris

Pubis

Ischium

Gracilis major

Gracilis minor

Tarsals

Metatarsals

Phalanges

these are enlarged to add a joint in the hind legs, thus facilitating jumping. Following the tarsals are the **metatarsals** and finally the **phalanges.** The bones of these appendages have remarkably similar counterparts in the human skeleton (Fig. 15–7).

The anterior end of the spinal column articulates with the base of the skull which houses the brain. Other parts of the skull make up the food-getting and breathing mechanism. The jaws of the frog are made up of three bones and resemble the gill arches of the shark from which they were derived. Behind the jaws and lying between them is the **tongue,** which possesses a support, the **hyoid apparatus.** The hyoid apparatus, like the jaw, has been derived from the gill arches. Other gill arches have been modified into supporting structures for the **larynx,** the sound-making apparatus. Here is seen a phenomenon constantly found in evolution, that is, the formation of new structures from old ones which no longer function in their original manner.

Nervous System. The nervous system of any animal begins with the sense organs which receive stimulations from the outside world. The nostrils open into nasal passages which are lined with sensory cells that join the olfactory nerve (Fig. 14–19). The frog has a rather good sense of smell.

The frog eye differs from the human eye in minor details only. For example, the lids will not cover the eye completely, and the lens is fixed in place so that the focus cannot be changed as in man. Therefore, the frog sees clearly only at one distance, and it is near-sighted in air and far-sighted under water. The **rods** and **cones** of the **retina,** which are the parts of the eye sensitive to light, are scattered, rather than concentrated in one spot as in man (p. 325). Consequently, the frog probably does not see as distinctly as higher forms do.

The conspicuous **eardrum** of the frog is exposed to the outside world, whereas in higher forms it is buried deep inside the head. Lying beneath the drum is a cavity

in which a single bone, the **columella,** extends from the thin eardrum to a tiny bit of sensory tissue which is stimulated by the vibrations as they are passed to it. Because of the rather primitive nature of the auditory organ, the frog probably cannot discriminate between sounds that vary slightly in pitch. The organs of equilibrium (**semicircular canals**) are similar to those of both lower (shark) and higher (man) forms.

With the exception of a few major modifications, there is remarkable similarity between the brain of the frog and that of man. Starting at the base of the brain and progressing forward, the five parts of the brain can be seen.

The first enlarged portion is the **medulla**

Fig. 14–19. Frog nervous system, dorsal view.

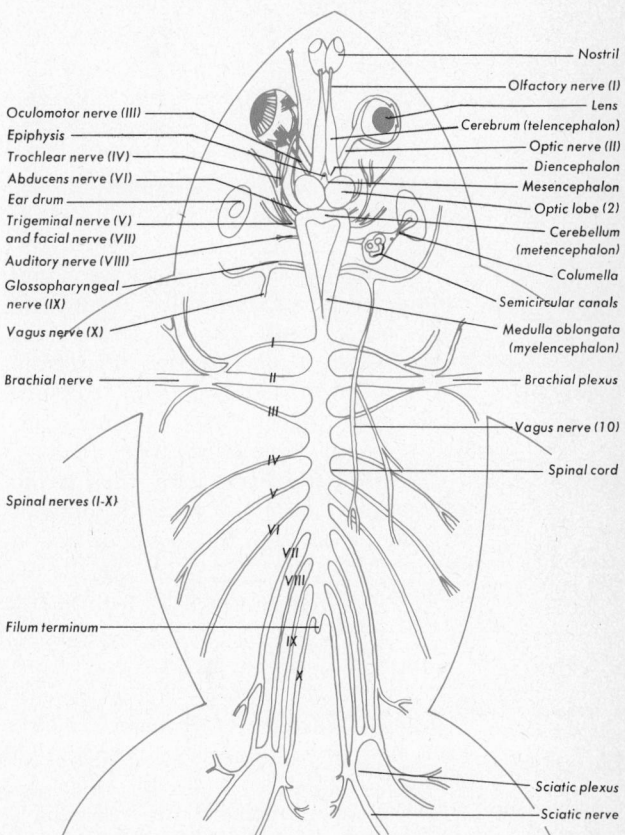

Oculomotor nerve (III)
Epiphysis
Trochlear nerve (IV)
Abducens nerve (VI)
Ear drum
Trigeminal nerve (V) and facial nerve (VII)
Auditory nerve (VIII)
Glossopharyngeal nerve (IX)
Vagus nerve (X)
Brachial nerve
Spinal nerves (I-X)
Filum terminum

Nostril
Olfactory nerve (I)
Lens
Cerebrum (telencephalon)
Optic nerve (II)
Diencephalon
Mesencephalon
Optic lobe (2)
Cerebellum (metencephalon)
Columella
Semicircular canals
Medulla oblongata (myelencephalon)
Brachial plexus
Vagus nerve (10)
Spinal cord
Sciatic plexus
Sciatic nerve

oblongata (myelencephalon—5) which gives rise to most of the cranial nerves (Fig. 14–19). These have to do with most of the automatic functions of the body, just as they do in man. A slight projection which runs transversely across the medulla is the **cerebellum** (metencephalon—4). The most conspicuous objects of the entire brain are the **optic lobes** (outgrowths of the mesencephalon—3) in which nerves from the eyes terminate. These lobes seem to function in inhibition of spinal cord reflexes, rather than as the centers for sight. The small projection just anterior to the optic lobes (diencephalon—2) is the **epiphysis,** an organ of doubtful function. On the ventral side of this same region is a tube-like stalk which terminates in an enlargement, the **pituitary** (hypophysis), a very important gland of internal secretion.

The anterior part of the brain (telencephalon—1) is poorly developed in the frog. It is composed of a pair of lobes which are partly divided transversely. The two anterior parts are the **olfactory lobes** to which the **olfactory nerves** are attached. The posterior parts of these lobes make up the **cerebral hemispheres,** the function of which is not clear.

The frog has only ten cranial nerves, but reptiles, birds, and mammals possess well-developed eleventh and twelfth cranial nerves. There is some evidence that primitive amphibians, too, had these additional nerves, but strangely enough, the frog seems to have lost them somewhere along the way. The spinal nerves that pass to the legs are grouped together in two plexi, the anterior **brachial plexus** and the posterior **sciatic plexus.** From these regions the nerves spread out again and pass to all parts of the appendages.

Digestive System. This system starts with a disproportionately large mouth which when fully open can enclose a body one-fourth the size of the frog itself (Fig. 14–17). The jaws are armed with teeth which function only in holding prey and are incapable of crushing or chewing. The large protrusible tongue lies on the floor of the mouth with the two pointed tips directed down the throat. When in use, the mouth is opened wide and the tongue is flipped out with lightninglike speed, so fast, in fact, that it has no difficulty in capturing agile insects. There are no digestive enzymes in the saliva and hence no digestion occurs in the mouth.

Openings in the mouth cavity are those of the **eustachian tubes** which connect with the cavity under the eardrums, the **internal nares,** connecting with the nostrils, and the **esophagus** which leads abruptly into the large U-shaped **stomach.** The back part of the mouth, the **pharynx,** is lined with cilia which beat continuously and help carry food down to the stomach. The stomach is merely a portion of the digestive tract which is enlarged for storage of food, and the frog has occasion, indeed, to use a sac of such ample proportions. Some digestion takes place in the stomach. The lower extremity is marked by a constriction, the **pyloric sphincter** (a band of circular muscles which, when contracted, closes the opening). The stomach is followed in turn by the **small intestine** which receives the pancreatic juice and bile from a single duct. The **pancreas** is a long, light-colored, ribbonlike organ lying between the stomach and the first part of the intestine. The **liver** is composed of three lobes, two large lateral lobes and one smaller median lobe. The **gall-bladder** usually lies dorsal to the smaller lobe, and the **bile duct** passes from it through the substance of the pancreas on its way to the intestine, picking up the **pancreatic duct** along the way. The gall-bladder is green in color due to the **bile** which it contains.

The small intestine of the herbivorous tadpole is very much longer than that of the carnivorous adult frog, a distinction that generally separates animals that feed on vegetation from those that feed on meat. The small intestine of the frog opens directly into the short expanded **large intestine** which soon constricts down to the **rectum** and then opens into the **cloaca** (sewer). Here the **genital** and **urinary** ducts also empty. Undigested food deposited in this region is soon voided to the outside through the **anus.** The cloaca is found among reptiles, birds, and low mammals, but among all higher mammals the urogenital and digestive tubes have separate openings to the outside of the body.

Circulatory System. In fish, the heart is a simple pump in which all of the blood is carried through a single circuit. In the air-breathing vertebrates, however, a change occurs. There are two blood circuits. One circuit carries blood rich in oxygen to the tissues, then brings the "used" blood, poor in oxygen, back to the heart. The other circuit carries this depleted blood to the lungs and brings it back, after oxygenation, to the heart. Because the only source of oxygen is from the lungs, these two circuits must be kept separate in order to do an efficient job.

A study of the blood pathway through the frog heart will help us to understand the evolution of heart circulation in land vertebrates. Because both **auricles** (atria, singular, atrium) pour their contents into the single **ventricle,** it has long been a controversial question as to whether or not the blood was completely mixed when it left the heart. The position of the entrances of the blood into the ventricle (Fig. 14–20), together with the existence of the longitudinal valve, was thought to be sufficient evidence to prove that blood returning from the body to the right auricle and that coming from the lungs to the left auricle remained separated when it left the ventricle upon its contraction. Recent studies of the oxygen content of the blood in the heart and major blood vessels indicate that the carotid arteries receive highly oxygenated blood from the left auricle, whereas the pulmocutaneous arteries receive blood low in oxygen from the right auricle. The aortas receive mixed blood. Apparently the amphibian heart separates the oxygenated and reduced blood much as is done in the bird and mammalian heart, although not as efficiently.

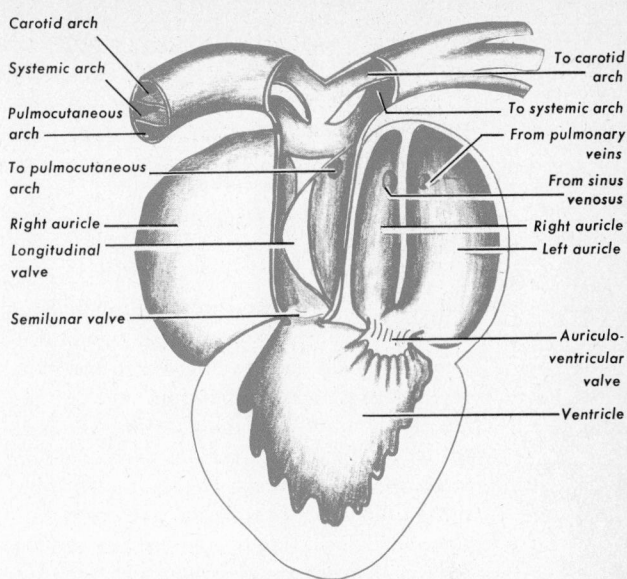

Fig. 14–20. Ventral view of the frog's heart, sectioned to show chambers and valves.

The extensive cutaneous circulation in the frog explains why lung breathing is only supplementary to the gaseous exchange that takes place in the skin. The frog spends much of its time underwater where only skin breathing is possible, and even on land the frog receives much of its oxygen through the buccal membranes. The lungs are brought into use only when the animal is hard-pressed, the skin and buccal membranes being quite adequate for most activities.

Blood coming from all parts of the body first enters the saclike **sinus venosus** and then the **right auricle;** simultaneously blood from the lungs enters the **left auricle.** Blood in the right auricle contains considerable oxygen, because part of it has just returned from the skin and buccal region where it picks up oxygen. That in the left auricle, of course, is rich in oxygen. The auricles then contract, the right slightly before the left, forcing the blood into the single **ventricle** where a certain amount of mixing occurs. The ventricle then contracts, forcing the blood into the **truncus arteriosus** and to all parts of the body through the arteries. The **longitudinal valve** probably serves as a mechanical support for the truncus, making it possible to produce a higher blood pressure when the truncus contracts. The modern amphibian heart, while not as efficient as the bird or mammalian heart, certainly is quite adequate for amphibious life.

Arteries. There are three pairs of large arteries leaving the heart: the **pulmocutaneous** which goes to the lungs and skin, the **systemic arches** which join and become the **dorsal aorta,** and the **carotids** which go to the head and neck regions (Fig. 14–21). Each of these vessels divides many times until a network of capillaries is formed, and these networks supply all portions of the body with oxygen and food.

Capillaries. The arteries terminate when their walls become one cell layer in thickness. Through vessels of this diameter blood cells can only pass single file, and these tiny, thin-walled tubes are the **capillaries.** These are the most important tubes of the entire vascular system because it is through the capillary walls that food and oxygen can get to the cells. The larger vessels in such a preparation are the **arterioles** and **venules,** which can be distinguished from each other by the fact that the blood flows in spurts in the arterioles and only gently in the venules.

Veins. After leaving the capillaries, the blood flows through **veins** which carry it back to the heart. These veins grow larger and fewer in number as they approach the heart (Fig. 14–21). The blood from the hind legs has an alternate course in getting back to the heart. It may pass via the kidneys through the **renal portal system,** or via the liver through the **hepatic portal system.** A portal system is a system of veins which starts and ends in capillaries. The frog, like other lower vertebrates, has two such systems, whereas man and the higher vertebrates have only the hepatic portal system. It can be seen that this system is most important in carrying the blood, heavily laden with food, to the liver where it can be stored and otherwise processed. The two **precavas** and the single **postcava** veins enter the saclike sinus venosus through three openings before proceeding on to the **right auricle.** Blood coming from the lungs in the **pulmonary veins** empties into the **left auricle,** thence into the ventricle where it joins the blood from the sinus venosus.

Breathing System. The tadpole breathes by means of gills. As it metamorphoses into the adult frog, it gradually develops a pair of lungs. The **larynx** is located at the point of junction with the mouth cavity, and contains the vocal cords which, when vibrated, produce the characteristic sounds of the frog. The air passes into the mouth cavity through the nostrils and then is forced into the trachea and lungs through the **glottis,** a slitlike opening in the rigid circular larynx (Fig. 14–17). The trachea is short and immediately branches into the two saclike **lungs.**

Excretory System. The excretory system of the frog is essentially the same as in invertebrates such as the earthworm, as far as the individual units are concerned. It is made up of a great many nephridia massed together into a pair of organs, the **kidneys.** The blood coming forward from the posterior parts of the body passes to the kidney and as it does the vessels break up into tiny masses (**glomeruli**) in the **renal corpuscles** (Fig. 14–22). As the blood passes through the glomeruli the urinary products are removed. They pass down a long coiled tubule and finally reach a larger duct, the **urogenital duct,** which in the male (**Wolffian duct**) also carries the reproductive (sperm) cells. The corresponding duct in the female carries urine only. The urine is deposited in the **urinary bladder** which in turn opens into the **cloaca.** Urinary wastes and feces,

ARTERIES   VEINS

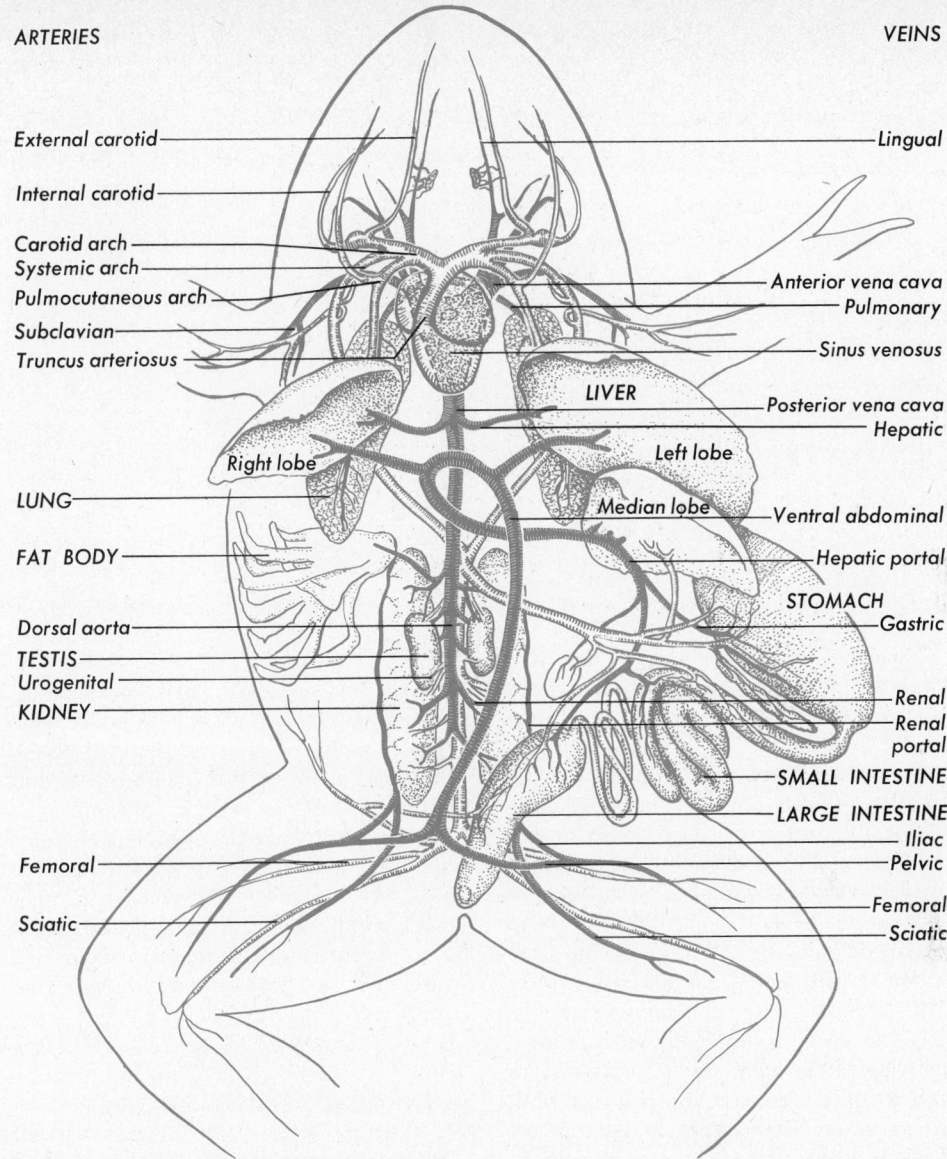

External carotid

Internal carotid

Carotid arch
Systemic arch
Pulmocutaneous arch

Subclavian
Truncus arteriosus

LUNG

FAT BODY

Dorsal aorta
TESTIS
Urogenital
KIDNEY

Femoral

Sciatic

Right lobe

LIVER

Left lobe

Median lobe

STOMACH

Lingual

Anterior vena cava
Pulmonary
Sinus venosus

Posterior vena cava
Hepatic

Ventral abdominal
Hepatic portal

Gastric

Renal
Renal portal
SMALL INTESTINE
LARGE INTESTINE
Iliac
Pelvic
Femoral
Sciatic

Fig. 14–21. Ventral view of the frog circulatory system.

as well as the genital products, all pass to the outside through a single opening, the **anus.**

Reproductive System. The sex organs of the male are the yellowish **testes** located ventral and anterior to the kidneys (Fig. 14–22). They hang in a sheetlike bit of tissue, the mesorchium, through which tiny tubules, the **vasa efferentia,** pass on their way from the testes to the kidneys. Upon entering the kidney, the vasa efferentia con-

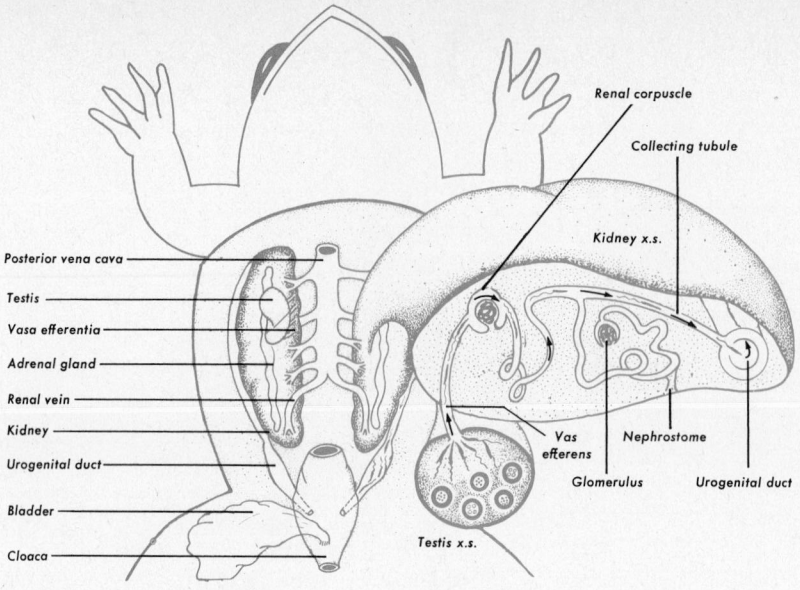

Fig. 14-22. Male frog urogenital system showing the kidney and testis enlarged in a cross section.

nect with the uriniferous tubules which are connected to the renal corpuscles. Therefore, the tubules carry both sperm from the testes and urine from the renal corpuscles. The two products flow to the lateral edge of the kidney where they are poured into a larger tube, the **urogenital duct,** which eventually deposits its sperm load into the **sperm sac.**

In the female, the ovaries lie in the same position on the kidneys as the testes do in the male. Sometime in the summer months when the food is abundant, the residual eggs lying in the walls of the ovaries begin to grow and continue at a rapid pace until the ovaries are tremendous in size, almost filling the body cavity. The eggs develop in tiny pockets in the wall of the hollow ovary (Fig. 14-23) and when the breeding season approaches, the mature eggs burst out in the body cavity. Here they are swept along by the united effort of cilia which line nearly all the walls. Their goal is the **ostium,** the tiny anterior opening of the long coiled **oviduct.** Once inside the opening, the eggs make their way single file through the long oviduct which is also lined with cilia.

During their passage they accumulate a jellylike substance on their exteriors which swells rapidly the moment the eggs become immersed in water. Just before the oviducts join the cloaca they enlarge into thin-walled sacs, the **uteri.** Here the eggs are stored until amplexus occurs, at which time they are laid.

This somewhat detailed study of the frog lays a foundation for an understanding of vertebrates in general, and in particular, man.

LAND CONQUERORS—THE REPTILES

The reptiles (class **Reptilia**) moved one step farther in the long trek to complete terrestrial existence. They spend no part of their life in the water unless they choose to do so. The land **egg** developed whereby the early embryos could exist in a fluid environment. A medium was supplied with sufficient nourishment to carry the embryo through the stages equivalent to the tadpole stage among amphibians. For this reason, reptilian eggs are large, with great quantities

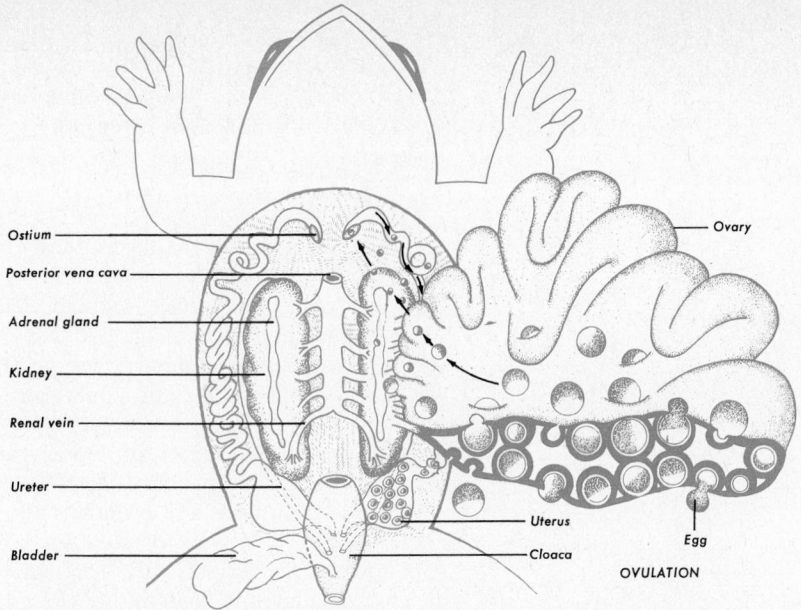

Ostium

Posterior vena cava

Adrenal gland

Kidney

Renal vein

Ureter

Bladder

Ovary

Uterus

Egg

Cloaca

OVULATION

Fig. 14–23. Female frog urogenital system showing the ovary enlarged and in cross section.

of stored food in the form of yolk and albumin (Fig. 14–24). A large, fluid-filled sac, called the **amnion,** develops around the embryo, which not only provides a fluid environment but also protects the developing embryo from injury and desiccation. Shortly after the formation of the amnion, the **allantois** develops from the posterior end of the embryo. This enveloping membrane receives discarded material, including carbon dioxide, from the embryo. It lies very close to the porous, rigid **outer shell** so that a gaseous exchange can readily take place. Therefore, in addition to being an organ of excretion, the allantois acts as a temporary breathing organ during embryonic life. The young reptile need not be immersed in water at any time during its life, and thus the first true land animal has been evolved. This was probably the greatest step forward in conquering the land.

Other changes also occurred that made the reptile better suited for terrestrial existence. Internal fertilization was accomplished by the development of efficient copulatory organs, insuring a direct transfer of the sperm into the genital tract of the female. In addi-

tion, reptilian legs became longer and were usually more ventrally located, making it possible for them to support the body completely off the ground. The heart also began

Fig. 14–24. In order for vertebrates to divorce themselves from water a method of caring for their young during early development was accomplished with the evolution of the land egg.

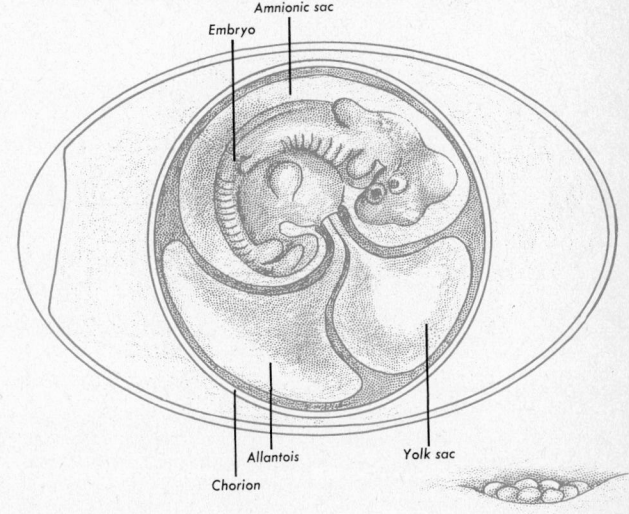

Amnionic sac

Embryo

Allantois

Chorion

Yolk sac

CENOZOIC — Quaternary, Tertiary

MESOZOIC — Cretaceous, Jurassic, Triassic

PALEOZOIC — Permian

Opossum

Primitive bird

mammal-like reptile

Stem Reptile

Fig. 14–25. Primitive reptiles gave rise to all higher groups of animals. The reptiles as a group have been the most successful of all vertebrates, as indicated by their great numbers and variety of form in Mesozoic time and by the fact that they gave rise to the birds and mammals.

to form a complete partition in the ventricle, producing the beginnings of two complete hearts, thereby separating the pulmonary and systemic blood to make a much more effective circulatory system.

Early Reptiles

Sometime in the distant past there were amphibians that gradually took on reptilian characteristics. Such animals have been found in fossil remains and are called the Stem Reptiles. Among these is *Seymouria*, which resembled both amphibians and reptiles. It, or forms like it, probably gave rise to the great variety of reptiles known to have lived during the Mesozoic Era and, indeed, to birds and mammals as well (Fig. 14–25). The Age of Reptiles lasted over 100 million years. By comparison, it is estimated that man is no older than two million years at most.

The Dinosaurs

Dinosaurs had their beginnings in the Triassic Period when they were small unimportant animals. Evolving in both size and numbers during the Jurassic and Cretaceous periods, they reached the pinnacle of walking land vertebrates near the close of the Mesozoic Era, finally disappearing from the earth never to appear again. They ranged in size from the barnyard fowl to the largest of all land animals.

Many of the dinosaurs developed the bipedal method of locomotion, that is, they rose on their hind legs when in haste and propelled themselves entirely by these two appendages. This idea proved successful in some of the largest flesh-eaters, which had powerful hind legs but only short anterior appendages (Fig. 14–26). Bipedal locomotion in four-footed animals is, then, very ancient. Footprints left in various parts of the world by these ancient animals have given paleontologists some interesting evidence as to the nature of the reptiles that made them. For example, in the old mud flats of the Connecticut Valley there are some that resemble bird footprints of today,

Fig. 14–26. A scene from the Mesozoic Period when the great dinosaurs were dominant. The flesh-eating *Tyrannosaurus* is shown attacking the well-armored *Triceratops* (left center) and two tremendous herbivorous *Brontosaurus* are feeding in the swamps (right center). Some reptiles took to the air and are shown flying over their land-dwelling relatives. A primitive bird (*Archaeopteryx*) was present during these early times (lower right).

and hence were first thought to be footprints of giant birds (Fig. 14–27).

The flesh-eaters grew to enormous size and the largest one unearthed, *Tyrannosaurus* (tyrant reptile), reached a height of 19 feet. It must have been an awesome creature in those prehistoric times, perhaps feared by all living animals. Its skull was over four feet in length and the jaws were armed with formidable teeth, which must have functioned well in rending and tearing other animals to bits. It possessed massive hind legs and small front ones. It is thought that its chief source of food was the giant amphibian forms which existed at the same time.

The great amphibious dinosaurs were vegetarians and grew to great lengths and heights. *Brontosaurus* and *Diplodocus* reached a length of 85 feet and a weight of 40 tons or more. They walked on all four feet and their powerful legs were placed in

such a position as to carry the body evenly balanced. The head was much too small for the size of the animal, and it is difficult to see how it could house a brain sufficiently large to govern such a massive hulk. The dorsally placed nostrils have led scientists to conclude that the animal was amphibious and probably remained submerged most of the time with only the nostrils protruding above the surface of the water for breathing air.

In the hip region, the spinal column supported an enlargement several times the size of the brain. Apparently impulses received by the brain from the sense organs were sent down to the large posterior ganglion which operated the posterior legs and perhaps the rear portion of the body.

Horned dinosaurs such as *Triceratops* were the last of the large dinosaurs (Fig. 14–26). The body was relatively bare, but the head was heavily armed with bony

Fig. 14–27. Dinosaurs often left footprints in soft mud that later became buried with fine sand, leaving almost perfect impressions of the feet of these ancient animals. The sedimentary rock in which this one was found separated, so that both the mold (right) and the cast can be seen.

organs of defense. Two great horns protruded anteriorly over the eyes and another over the ridge of the nose. An enormous flare of bone, extending out from the back of the neck, probably functioned admirably in preventing an injurious blow to this vulnerable region.

Why did these great animals become extinct by the end of the Cretaceous Period? There are many answers, but possibly a combination of many factors was responsible for their extinction. Since the flesh-eaters depended on the plant feeders for food, a gradual extinction of the latter meant annihilation of the former as well. Geological changes going on at that time resulted in the gradual rising of the land, culminating in the formation of the Rocky Mountains in this country. This meant not only less water and consequently fewer swamps, the habitat of these animals, but also cooler climates and perhaps much less vegetation. With the declining food supply the great herbivores starved to death, taking the carnivores with them.

Modern Reptiles

At least sixteen orders of reptiles have lived on earth; today only four remain. Of these the most common are the lizards, snakes, turtles, and the large alligators and crocodiles.

Perhaps the most despised of all animals

Fig. 14–28. (Top) One of the largest lizards alive today, *Iguana iguana*, reaches a length of 6 feet. It inhabits tropical America. (Bottom) The Gila monster (*Heloderma suspectum*) lives in southwestern United States and is our only poisonous lizard.

are the lizards and snakes, not because they are particularly harmful to man or because he inherits a fear of them, but because he is taught to be afraid of them, particularly snakes. The group as a whole does little harm to man or his domestic animals, and what harm it does do is offset by its creditable deeds.

Lizards and snakes belong to the order **Squamata**. They are covered with scales which they shed periodically. Lizards are four-legged animals and exemplify the typical modern reptile, that is, they show the least amount of modification in body form of any of the reptiles. A good example of the group is the *Iguana* (Fig. 14–28). The only poisonous lizard is the Gila monster (Fig. 14–28), a highly colored, sluggish, plump

creature found in drier regions of North and Middle America. It rarely uses its venom in killing prey because its diet consists of bird and reptilian eggs as well as nestlings. Its bite is rarely, if ever, fatal to man. Many tropical lizards are very colorful (Pl. 15A).

With the exception of pythons and boas, snakes are without limbs. Many move in an undulating fashion much the same as fish swim. Their eyes are large and without lids (Pl. 15B). Because their sound-recording organs are degenerate, they are deaf. The snake possesses a long forked tongue which has sensory functions useful in tracking down prey. The snake's mouth is equipped with sharp teeth that curve inward and are well adapted for holding its victim (Figs. 14–29, 14–30). Another convenient adaptation associated with food taking is the enormous potential size of the mouth, which can be stretched to accommodate an animal several times its own diameter (Fig. 14–30). Some snakes such as the boas kill their prey by squeezing it to death (Fig. 14–31).

The snakes have received their bad reputation from the poisonous members of the group. Just how these creatures evolved this deadly offensive and defensive mechanism is hard to say. Many different kinds of tooth formations have been produced for inoculating the poison into the wound made by the sharp teeth. In some forms, such as the rattlesnake, the fangs possess a hollow tube through which the venom is injected, as with a hypodermic needle (Fig. 14–30). Others, such as the cobra, have deep-grooved fangs which allow the poison to enter at the base of the tooth and exit near the tip so that, upon striking, the poison will be deep within the wound. The poison glands are located above the angle of the jaw and empty their venom into a flap of skin at the base of the fangs. Once the venom has entered the blood stream, its effect is very rapid, causing a small rodent to become paralyzed in a matter of a few seconds. The venom acts by destroying the red blood cells. Antivenoms have been prepared for all of the common poisonous snakes and have proved valuable in alleviating the effects of

Fig. 14–29. (Top) A rattlesnake (*Crotalus seutulatus*) in striking position. Note that the rattles at the tip of the tail seem blurred because of their rapid vibration. (Bottom) The position and size of the fangs are shown in this picture. Note the pit located anterior to the eye.

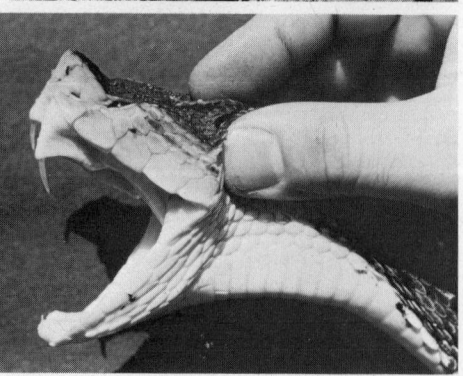

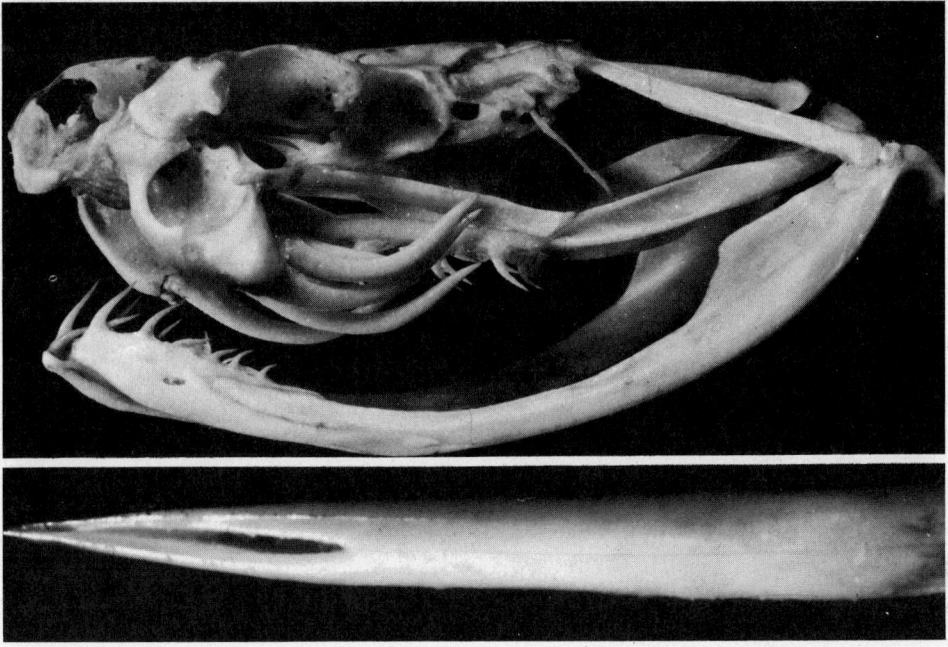

Fig. 14–30. The skull of a rattlesnake. Note how loosely the jaw bones are attached to the skull. This, together with the lack of fusion of the jaw bones in front, makes it possible for the snake to swallow an animal several times its own diameter. One of the fangs has been enlarged in the lower picture to show its hollow construction. It resembles an inoculating needle.

their bites. A person may die from the bite of a poisonous snake if he is not treated. Of the types of poisonous snakes in the United States, the most common are the rattlesnakes. The water moccasin and copperhead have fangs like the rattlesnakes. The coral snake, beautiful though poisonous, has erect fangs much smaller than those of the others.

The turtles, the most odd-looking of all reptiles (Fig. 14–32), belong to the order **Testudinata.** Turtles have existed a long time on the earth, since the earliest ones were contemporaries of the most primitive dinosaurs. The great reptiles came and passed on to extinction, but the turtles have persisted. Even with the advent of mammals, the turtle, concealed in its protective armor, has maintained itself and, who knows, may survive long after the mammals, including man, have passed out of existence.

Turtles live both on land (tortoises) and

in the water. Like many other groups of land animals, they have returned to the water and have so modified their bodies that they are well adapted to an aquatic existence. The great sea turtles, such as the hawksbill and the green turtle (Fig. 14–32), have their appendages modified into flippers. They are never seen on land except during the egg-laying season. Others, such as the common painted turtle, are usually found in water but also frequently are seen on land near bodies of water. Still others, like the high-shelled tortoises, live in certain parts of the world where they have no enemies and grow to enormous sizes, such as those on the Galapagos Islands. These desert forms feed on vegetation alone and rarely if ever take any water.

The surviving members of the great ruling reptiles, the dinosaurs, are the crocodiles, the caimans of the Amazon, the gavials of the Ganges and the alligators of

Fig. 14–31. A California boa (*Lichanura roseofusca*) constricting and swallowing a field mouse.

a four-chambered heart like that of birds and mammals. They also possess the mammalian characteristic of a nearly complete **diaphragm**, which is a muscular separation between the chest and abdominal cavities.

One very interesting reptile is *Sphenodon* (order **Rhynchocephalia**), more commonly known as Tuatara (Fig. 14–34), which carries in its body many anatomical features definitely identifying it with the earliest of reptiles. Tuatara has appropriately been called a "living fossil." Many of its characteristics show definite relationships to the stem reptiles as well as to modern reptiles. It is always interesting to speculate why such isolated members of a once flourishing group were able to survive down to the present day when all its relatives are long since extinct. In the case of Tuatara, its location is probably responsible; the reptile lives on islands off the coast of New Zealand where it has had few, if any, natural enemies. Tuatara living in other parts of the world

Fig. 14–32. A native Australian riding a large marine turtle (*Chelonia mydas*). Note how the anterior appendages are modified into flippers, which are efficient organs of locomotion in water.

Florida (Fig. 14–33). They belong to the order **Crocodilia.** They inhabit the large rivers of the world and are often hunted for their valuable hides. These animals show some anatomical features which place them among the highest reptiles. For example, they have a nearly completely divided ventricle in the heart, resulting in

Plate 1. Portguese man-of-war *(Physalia)* stinging a fish. Note the large gas bag which functions as a float and sail combined. The numerous tentacles perform all of the functions of the animal. Some are heavily armed with nematocysts which are effective weapons against other animals, as this photograph shows. Note the dead fish.

Plate 2A. A sea anemone *(Condylactis gigantea).* These feed on small fish which they capture with aid of their powerful nematocysts.

Plate 2B. A giant jellyfish *(Cyanea capillata)* common on our eastern seaboard. The dome is flattened during relaxation; during its power stroke it becomes more oval in shape.

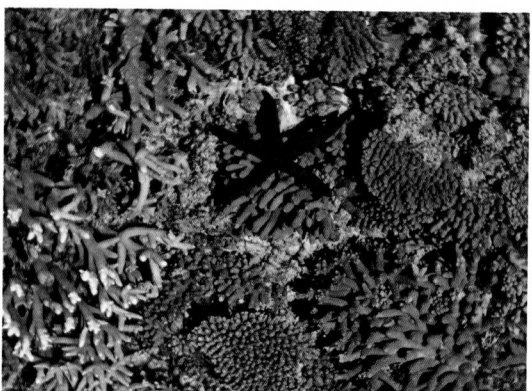

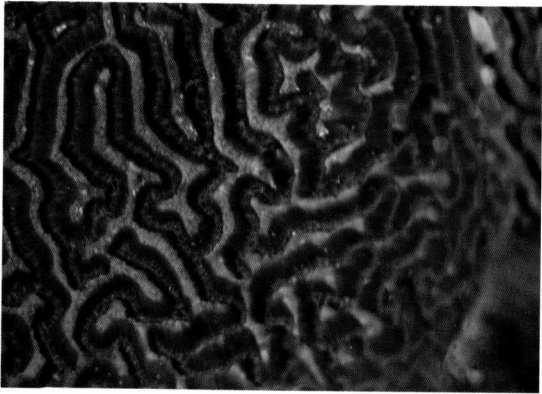

Plate 3A. Living corals from the Great Barrier Reef (Heron Island). The blue starfish *(Linckia laevigata)* is commonly found among the corals. Large specimens measure a foot across.

Plate 3B. Most coral polyps feed at night as shown by the brain coral where the tentacles are contracted.

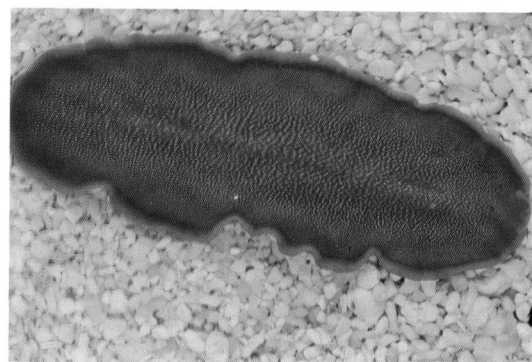

Plate 4. These polyclad turbellarians are from the Great Barrier Reef (Heron Island). They avoid daylight and crawl about on the undersurface of corals. They are extremely fragile, fragmenting immediately when handled.

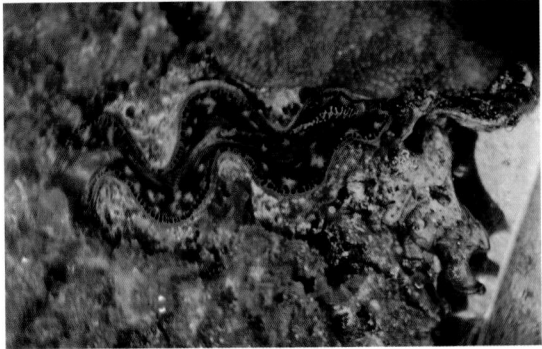

Plate 5A. This conspicuous bivalve *(Tridacna fossor)* lies buried in the coral rock. The highly colored mantle (there seems to be no two alike) is a result of symbiotic dinoflagellates living in it. Some members of this genus reach gigantic sizes (3 feet and over).

Plate 5B. This bright green abalone *(Haliotis)* is popular as a seafood on our western coast. It inhabits the Pacific Ocean, this one came from the Great Barrier Reef.

Plate 6A. This living snail is called a cowry *(Nirigena melwardi)*. Note how the mantle covers the pearly shell.

Plate 6B. This rare snail, called a volute *(Cymiolacca pulchra)*, burrows in the soft sand. Note the matching color design of mantle and shell.

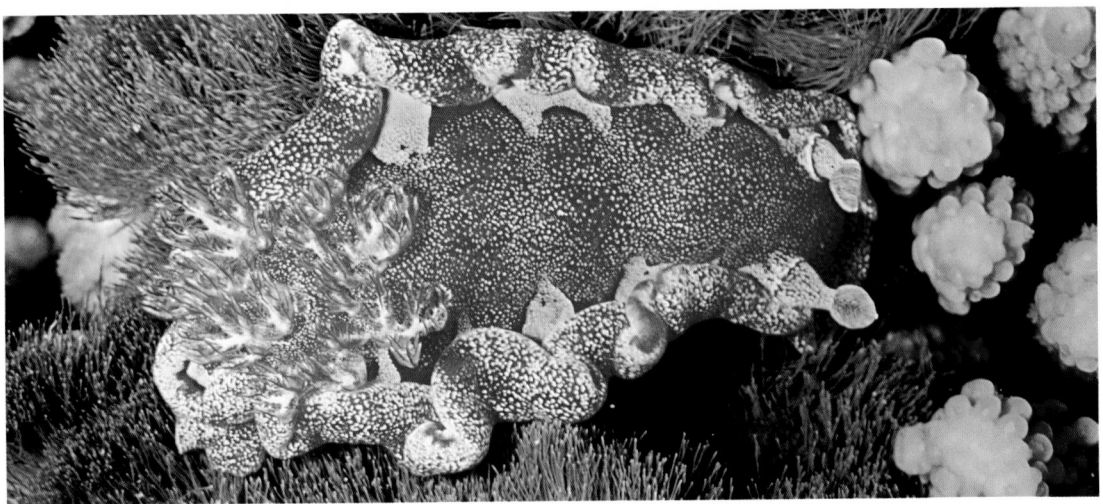

Plate 7. A nudibranch *(Hexabranchus imperialis)* from the Great Barrier Reef. This one is called the "Spanish dancer" because of its agility when swimming. Many nudibranchs only crawl.

Plate 8A. The "sandworm" *(Nereis)* is a common inhabitant of our Atlantic coastal waters. It lives in a burrow in mud or sand from which its head and tentacles protrude. Small animals passing near enough are snatched and devoured. During the breeding season worms leave their burrows and congregate in great numbers near the surface of the sea.

Plate 8B. Aphrodite *(Aphrodita hastata),* the sea mouse, from ventral view which definitely establishes it as an annelid. From the dorsal side it resembles a furry animal, hence its name. It is about 12 cm. long.

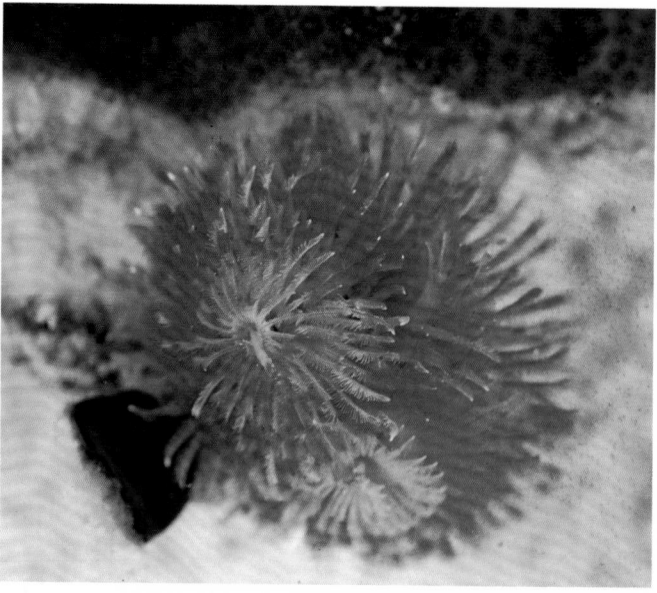

Plate 9A. This annelid *(Chaetopterus)* lives in a tube secreted by its own body. The appendages are used as paddles to keep the water circulating through its tube, thus bringing in oxygen and small animals upon which it feeds. It is luminescent, which seems strange since there is no opportunity for any other animal to appreciate its beauty. This female specimen, which is removed from its burrow, is about 15 cm. long.

Plate 9B. This colorful annelid *(Spirobranchus giganteus)* lives in a tube among the corals on the Great Barrier Reef. The feathery heads *(branchiae)* extend when feeding and are quickly retracted into the tube when disturbed.

Plate 10. (Top) The hermit crab *(Pagurus)* lives in an abandoned snail shell. This habit has resulted in the loss of the heavy exoskeleton characteristic of other crabs. (Bottom) The hermit crab removed from its shell. Its body is unprotected by an exoskeleton.

Plate 11. This ubiquitous mantis shrimp *(Gonodactylus chiragra),* lives among the corals on the Great Barrier Reef. It is a fast, elusive swimmer.

Plate 12A. The praying mantis *(Tenodera sinensis)* has its anterior legs modified into grasping appendages. It gets its name from its posture while waiting for some unwary insect to approach within striking distance. It is one of the larger insects reaching 4 inches in length.

Plate 13A. A basket star *(Georgonacephalus articus)* taken in 420 feet of water. The five principal arms are subdivided into a great many smaller branches. This is a ventral view.

Plate 12B. Cicadas *(Cicada septendecim)* copulating. They were photographed during the summer of 1957.

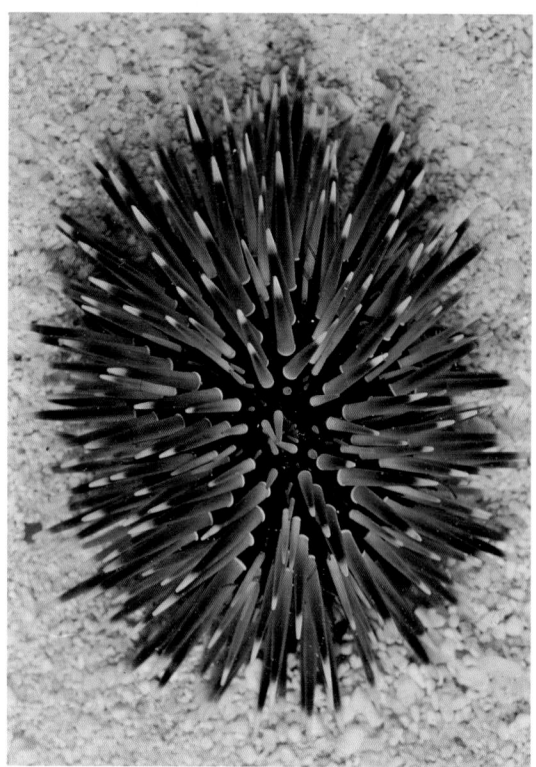

Plate 13B. A sea urchin *(Echinometra mathaei)* from the Great Barrier Reef.

Plate 12C. A Panamanian katydid *(Philophyllia Ingens)*. Note how its coloration matches the leaves, giving it protection from its enemies.

Plate 14. Tree frogs from Panama. Note the adhesive pads on the toes used in climbing to smooth surfaces.

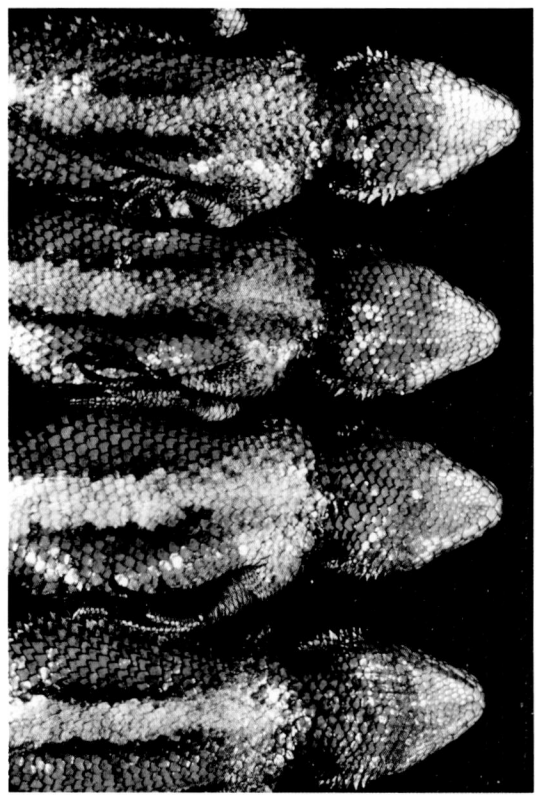

Plate 15B. The design and coloration of snakes are occasionally very beautiful. Note the reflection of the photographer in the eye.

Plate 15A. Lizards can be colorful. These are from Panama.

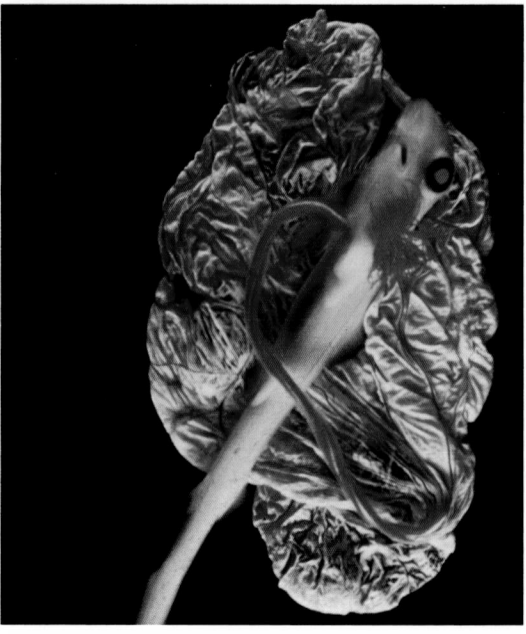

Plate 16A. Birds frequently possess remarkable adaptations for their mode of life. This gallinule has extremely large feet for running on floating vegetation.

Plate 16B. This is a small shark *(Mustelus)* embryo removed from the uterus of the mother. The long umbilical cord is attached to the half empty yolk sac which supplies nourishment for the embryo during its early life. In its later embryonic life there is a ''placenta'' formed in which nutrients and gases pass between the fetal and maternal circulation.

Plate 17. Two examples of symbiosis among animals. (Left) A tiny shrimp *(Conchodytes)* lives in the mantle folds of the Blacklip Pearl Oyster where it is nourished by the organisms brought to it as the oyster feeds. It is difficult to see what benefits the oyster gets from this association so this is probably a case of commensalism. (Right) This small fish *(Amphiprion)* spends most of its time nestled among the tentacles of the sea anemone *(Physobrachia)* coming out when food is available. If pursued by a larger fish, it darts among the tentacles for protection, sometimes completely disappearing into the gastrovascular cavity of the sea anemone. The pursuing fish is paralyzed by the powerful nematocysts of the sea anemone. A feast is thus provided for both the small fish and the sea anemone. This is a case of mutualism.

Fig. 14–33. American alligator (*Alligator mississippiensis*). These grow to 16 feet in length.

Fig. 14–34. Tuatara (*Sphenodon*) from New Zealand. A "living fossil."

were apparently destroyed by the aggressive, more agile mammals. Thus it can be said that isolation has saved one species of animal. Since its environment has remained unchanged, Tuatara, itself, has changed but little through the past 200 million years.

AIR CONQUERORS—THE BIRDS

It is a far cry from the lowly turtle and slithering snake to the most decorated of all animals, the birds (class **Aves**). Yet they have many characteristics in common with reptiles (Fig. 14–35, Pl. 16A). In conquering the air, several body modifications have been necessary. The drastic need for a lighter body framework was met by lighter, hollow bones and a lessening of the weight of the outer covering.

The scales of the reptiles have probably given rise to the feathers of birds, and actually there is little difference between them except in weight and texture. The bird still carries scales on its legs and sometimes around the base of the beak. Because of their loose arrangement, feathers act as excellent insulating material, an extremely important need for birds who live in both tropic and arctic regions. The wings are the modified anterior legs of the reptiles in which the three fingers have fused at the tip and the space between has been spanned by a sheet of skin covered with feathers. The

Fig. 14–35. Birds resemble one another anatomically although they show wide modification of body parts which are used for special modes of life. For example, the beaks of these four birds are modified for food-getting. The pelican's (top left) elaborate beak with its saclike attachment serves as a net in scooping up fish, its sole source of food. The sac is also used during the breeding season for storing fish for the young. The tropical toucan (lower left) possesses a colorful beak which seems overdeveloped for breaking open tree fruits upon which it feeds. Its light, though strong, structure and great length extends the bird's reach in securing fruits on the ends of branches. In addition, the bright color probably is secondarily used in social behavior. The kookaburra uses its powerful beak for killing and eating mice and snakes as this one is doing (upper right). It also has a characteristic throaty cry which has given it the name, "laughing jackass." The spoonbill uses its long broad beak to shovel through the mud in search of food (lower right). This is a rare picture of a mother and young. Both the kookaburra and the spoonbill are found in Australia.

wing spread is due principally to the large quill feathers at the outer edges of the wings. The breastbone is greatly over-developed. It is known as a **keel** and functions as an anchor for the powerful breast muscles which are used to give the power stroke in flight.

Other structural modifications have been essential for the flying animal. The hearts of birds are large and very well developed. The body temperature of birds is constant and slightly higher than in man, enabling the birds to keep active during all times of the year in both temperate and arctic regions.

The swift movement of a bird in flight requires keen eyesight. Birds have eyesight that is several times better than man's. The **cerebellum,** the part of the brain which controls muscular coordination, is proportionately larger in the brain of birds than it is in the reptilian brain (Fig. 16–16). This again is required for flight. The cerebral hemispheres, that portion which controls innate patterns of behavior, or instincts, are also large.

The First Birds

Birds apparently came from reptilian ancestors (Fig. 14–25), a fact that should reveal fruitful intermediate types, animals that are neither reptile nor bird but a combination of both. *Archaeopteryx,* an ancient bird about the size of a crow, was found in limestone about 90 years ago in Germany. These remains were sufficiently intact to enable us to describe the bird rather well; even the feathers were preserved. Discounting the feathers, however, the animal appears very similar to a small dinosaur. The wings were weak and the fingertips bore claws. The beak was lined with teeth and the skeleton was made up of heavy bones, in contrast to the light, hollow bones of modern birds. The fact that the bones were heavy and that the keel was poorly developed indicates that the bird was a poor flyer. In fact, if it could have been observed at that time it undoubtedly would have been seen gliding from branch to branch of trees much the same as the flying squirrel does today.

Some Giant Wingless Birds

There are several species of wingless birds on earth today which have persisted over a long period of geologic time. These birds are all much alike; they are usually very large, some reaching as much as 12 feet in height, with powerful legs, small heads and rudimentary wings which are useless for flight. They depend on their fleetness for security. An interesting fact is that all modern flightless birds live in regions where there are no carnivores. Examples are the ostrich of Africa, the cassowary and emu (Fig. 14–36) of Australia, the rhea of South America, and the kiwi of New Zea-

Fig. 14–36. The emu from Australia is one of the large flightless birds. Its powerful legs carry it swiftly over the ground, and they can be used as weapons of defense.

land (Fig. 14–37). It has been suggested that these birds once were able to fly, but because there were no terrestrial predators they simply ceased flying and remained on the ground all of the time. One reason that birds probably took to the air in the first place was to be better able to flee from their enemies. If the enemies are removed, then the need for flight is no longer present; hence, why fly?

THE CLIMAX ANIMALS— THE MAMMALS

The dominant and most complex animals on earth today are the mammals (class **Mammalia**). Certain characteristics have made it possible for this group of animals to be so successful. To begin with, they possess a unique body covering, **hair,** which functions in conserving heat and as a protection against mechanical injury. Like the birds, mammals are warm-blooded. A new structure, the **diaphragm,** has been added to aid in breathing. This sheet of muscle

Fig. 14–37. A kiwi (*Apteryx mantelli*) probing for worms in leaf mould. These flightless birds are found only in New Zealand, where they are rigidly protected. The female lays very large eggs which are incubated by the male, who also cares for the young. The kiwi is related to the extinct giant moa, which was the largest of all flightless birds.

and connective tissue separates the chest and abdominal cavities and operates like bellows during breathing. Their red blood cells have no nuclei when mature, which enhances their efficiency as carriers of oxygen. The cerebral hemispheres of the brain are more highly developed than in any other group of animals, which probably had much to do with their success. They produce tiny yolkless eggs which must be fertilized inside the body of the female. The young develop within the mother, receiving nourishment from the uterine wall through a special organ, the **placenta.** The young are born alive and receive a special food, **milk,** secreted by special glands, the **mammary glands,** for which the class is named. These are also the characteristics that set the mammals off from the rest of the vertebrates.

When measuring success of a group of animals, we usually think in terms of numbers of species, diversity of habitats occupied, and numbers of individuals. Aside from the arthropods, the mammals are more successful than most other groups of animals and certainly they outstrip all other vertebrates. They show marked variation in size, shape, and habitat (Fig. 14–38). They seem to have invaded every possible portion of the earth, water, and air and are reasonably successful in all environments.

The mammal arose from some primitive reptilian type and the first ones were more lizardlike than mammal-like (Fig. 14–25). Among the earliest groups to show definite mammalian characteristics were the **therapsids** whose skulls possess a tooth pattern similar to that of modern mammals. They had incisors, large canine teeth, and molars with cusps. This pattern is not found among reptiles. The limbs had also shifted to the far anterior and posterior positions, typical of mammals, and were much better developed than in reptiles. Indeed, if one saw one of these creatures, he would think it the result of a cross between a lizard and a cat or dog. Since the fossil remains of these animals include only the bones, we have no way of knowing whether other mammalian characteristics accompanied these skeletal changes. It is highly likely that they did.

These primitive mammals were present on the earth long before the coming of the great dinosaurs but they were small and inconspicuous, and probably kept in hiding throughout the reign of these great beasts. They remained small and insignificant until the dinosaurs met their end. The world was then left free for any animal that was capable of taking over. Even then, while the mammalian characteristics in general persisted, many animals evolved that were not able to survive very long—the mammoth, the saber-tooth tiger, and the giant sloth are examples. Some survived, however, and gave rise to the present mammalian population of the world.

The presence of hair in place of scales is an accessory structure necessary in a warm-blooded animal. Sweat glands aid in regulating body temperature, which is nearly constant. In addition, the heart is a double one, similar to that of birds.

One of the more important characteristics that has contributed to the success of mammals is their method of caring for their genital products and subsequent offspring. Most reptiles lay large, yolk-packed eggs, whereas mammals typically have very small eggs which are fertilized internally in the female and go through a great part of their early life inside the body of the mother. Although internal hatching of eggs is not uncommon in lower forms (reptiles and fish), receiving nourishment from the mother during the process of development is only rarely found (for example, placental shark) in the animal kingdom. Mammals have fewer offspring and retain them longer within the body in order to give them greater protection while allowing them, at the same time, to develop to a more advanced state. Furthermore, since mammals are so highly organized, more time is needed to develop the young to a stage where it can care for itself. The transition took place very slowly and over a long period of time.

The embryos receive nourishment from the fluids of the mother through a **placenta** which is a temporary attachment of the allantois and other membranes to the uterine wall of the mother. At first the placenta must have been a very primitive affair and only partially satisfactory in performing the important function of nourishing the young. Consequently, the young were born in a very immature state and needed extra-uterine care. This was furnished by the use of a belly pouch, called a **marsupium,** in which the young could not only be protected from the cold but also could receive nourishment from the mammary glands through nipples located inside the marsupium (Fig. 14–39). Later in their evolution, more and more time was spent in the uterus and less in the marsupium until the latter was finally discarded.

In certain parts of the world such as Australia, which were isolated during the time the mammals were evolving (Eocene), there survive to this day two different types of primitive mammals which might help to bear out the narrative of the preceding paragraph. They belong to the subclass **Prototheria** and are known as monotremes, which means "one opening," so named because both the urogenital and intestinal tracts open into a common cavity, the **cloaca** (a reptilian character), with only one external opening. The spiny anteater (*Echidna*) and the well-known platypus or

Fig. 14–38. Relative sizes of a few mammals.

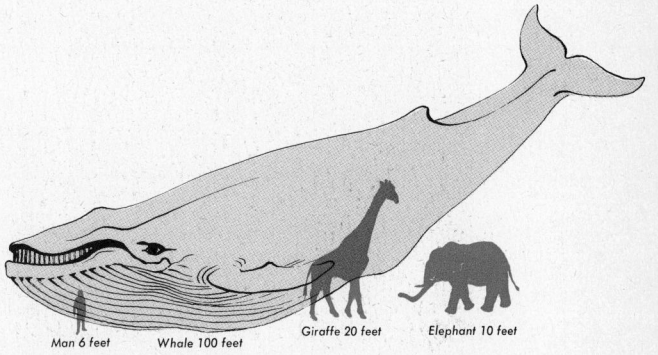

Man 6 feet   Whale 100 feet   Giraffe 20 feet   Elephant 10 feet

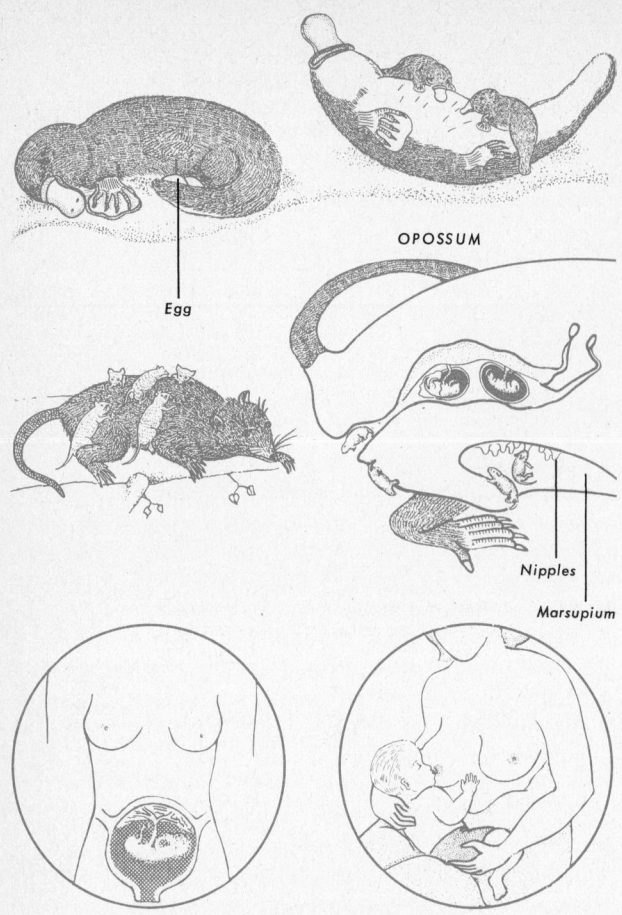

OPOSSUM

Egg

Nipples

Marsupium

Fig. 14–39. Different ways mammals care for their young, from the duckbill to man.

duckbill (*Ornithorhynchus*, Fig. 14–40) are the two examples that remind us of what must have happened many millions of years ago. These animals lay large yolked eggs like the reptiles and birds and they possess broad, flat bills. They are partially warm-blooded and the duckbill incubates its two eggs (Fig. 14–39). When the eggs hatch, the young are nourished for a time from milk secreted by two rows of glands along the belly side of the mother. There are no teats and the young merely lap up the secreted milk. It is noteworthy to mention again that these archaic animals have survived up to the present time due to their isolation. Had they been forced to compete with modern mammals they would long since have vanished from the earth.

Another interesting group of mammals are the **marsupials** (infraclass **Metatheria**), including the kangaroo and koala of Australia (Fig. 14–41) and the opossum of North America (Fig. 25–13). These do not lay eggs like the monotremes but retain them in the uterus where a rudimentary placenta forms. Since this organ is inadequate to maintain the embryos for any great period of time, the young are born in a very immature state (Fig. 14–42). A 200-pound kangaroo, for example, may give birth to offspring no longer than 2 inches. Following birth, the embryos make their way to, or are

Fig. 14–40. Platypus (*Ornithorhynchus anatinus*), an egg-laying mammal. It possesses both reptilian and mammalian characteristics.

placed in, the marsupium where they grasp the nipples of the mammary glands to which they cling during what is equivalent to later embryonic life of a more advanced mammal (Fig. 14–39). After a time they come out of the pouch to feed for themselves, retreating to it only in case of danger. Although the marsupials give us a clue to the past history of mammals, their direct line probably diverged from the reptilian mammal stock very early in geologic time.

The Placental Mammals

These animals have a well-developed placenta and some give birth to their offspring

Fig. 14–42. The young of the opossum (top) are born in a very immature stage and must continue their development in the marsupium. Here they are shown clinging to the teats in the pouch. The young of some true placental mammals are well developed at birth such as the seal (middle). The young seal is shown at the time of birth; the embryonic membranes can be seen clinging to its posterior end.

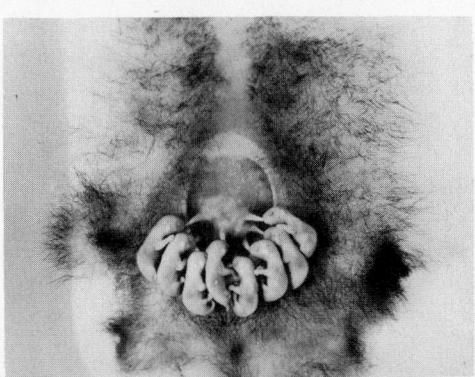

Fig. 14–41. Marsupials from Australia. (Top) Frightened kangaroos (*Macropus*) on the run. (Bottom) The tree-dwelling koala (*Phascolarctos einereus*) or "teddy bear" sleeping in the eucalyptus tree which provides its only diet. It feeds on the leaves.

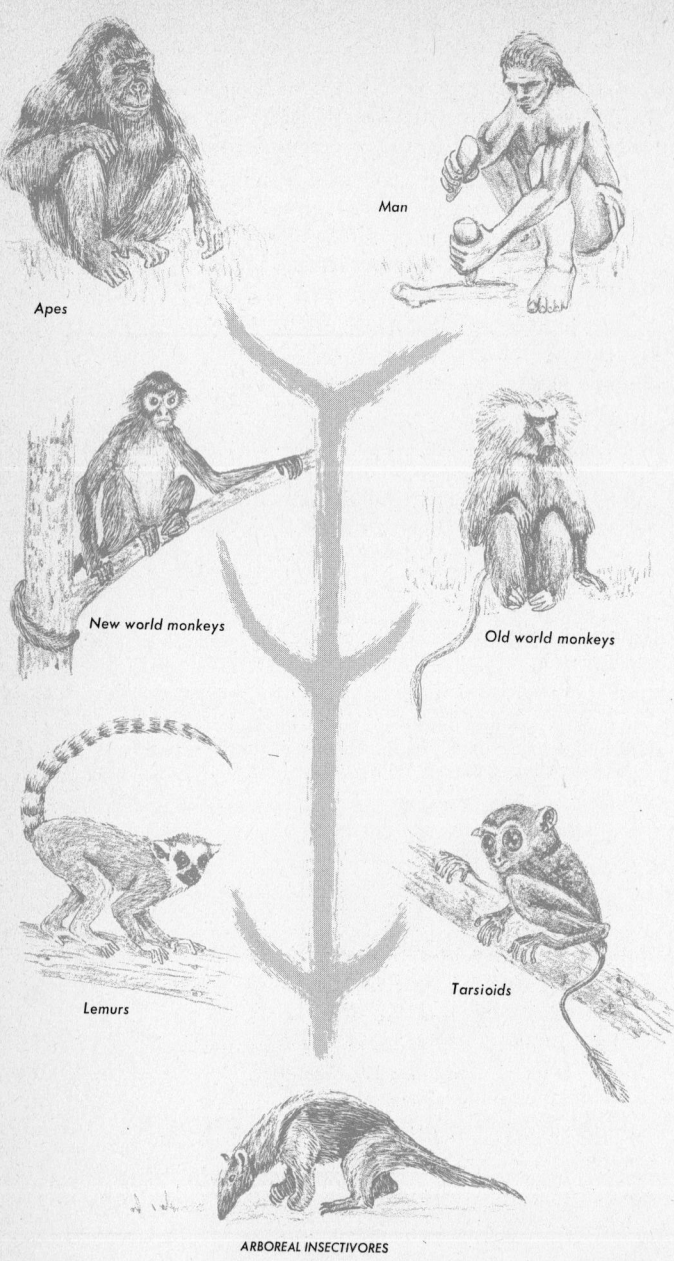

Apes

Man

New world monkeys

Old world monkeys

Lemurs

Tarsioids

ARBOREAL INSECTIVORES

Fig. 14–43. The primates apparently rose from the arboreal insectivores. Some groups retained tree habitats, whereas others, among them man, descended to the ground where they became at home.

in a well-advanced form, e.g., the young seal (Fig. 14–42). These young get along well a few minutes after they are born. Other placental mammals, such as mice and humans, are relatively helpless at birth and require some time before they are able to move about on their own.

Placental mammals show other refinements of ancestral systems. For example, the brain is an even more conspicuous part of the central nervous system and its functions are much more precise. The coordination and precision of operation of all the parts of the body have reached a new developmental high in these animals. This combination of improvements on systems inherited from their reptilian ancestors has been responsible for the success of mammals. The following are representatives of several important orders which belong to the infraclass **Eutheria**.

The most primitive modern placental mammals are the insect feeders (order **Insectivora**). While there are some fairly large, highly specialized forms such as the moles in this group, the most typical are the tiny shrews. These mouselike creatures are probably similar to the stock that gave rise to higher mammals including man himself. Their mouths are armed with needlelike teeth which aid in securing their specialized diet of insects. They are extremely active animals, in fact, so active that it requires a volume of food equal to their body weight each day to satisfy their needs. Most of them live in burrows and are such secretive animals that they are seldom seen by the casual observer. Some, however, live in trees much the same as the ancient forms that gave rise to the tree-living primates (Fig. 14–43).

The bats (order **Chiroptera**) have conquered the air. To be sure, other mammals such as the flying squirrel are able to soar from tree to tree, but they have not approached the bat in any sense as a flying creature. The bat competes very well with the birds; in fact, it can perform some feats that birds cannot. The wing is a modified hand with the fingers greatly extended and covered with skin. Many species of bats live in deep caves, such as the Carlsbad Caverns in this country, where they are forced to fly in complete darkness. Just how these animals can avoid objects has puzzled biologists for a long time. The bat's use of sound as a navigating device is discussed in Chapter 27.

The order **Primates** may seem the most important of all because we belong to it. Other members of the group are the great apes, the circus monkey, the fierce-looking baboon, the lemurs, and the wide-eyed tarsiers, all coming from an arboreal insectivore ancestor (Fig. 14–43). It is difficult to name any especially striking characteristics that set the primates off from all others, though there are several minor variations which distinguish them as a group. First of all, most of them possess a brain and central nervous system that is developed far above that of all others. It must be said, however, that some of the lowest primates are probably not as intelligent as some of the brightest carnivores. The lower mammals depend to a large extent on their sense of smell to orient themselves, for their vision at best is not very good. But the primates, as inhabitors of trees, need good vision to detect their enemies on the ground, this being of greater importance to them than a keen sense of smell. Arboreal life thus appears to correlate with the development of the excellent visual organ of primates. It also has had a profound effect on the skeleton.

As the appendages became well adapted for life in the trees, both the anterior and posterior digits became modified for grasping limbs. The great toes and thumbs opposed the other four digits and the digital tip flattened out into nails. The appendages became well developed for supporting the body weight and for carrying it with considerable speed from limb to limb through the tree tops. When at rest, the hands were free to handle objects and to bring them close to the keen eyes for closer observation. Freeing the front appendages from the burden of supporting the body weight and development of the prehensile hand have *probably* both been responsible to a large extent for the advance in the primate brain.

Other minor characteristics of the primates are the tooth pattern and skull structure. The early primates were apparently omnivorous just as most primates are today, and a generalized mammalian tooth pattern is characteristic of all of them. Most of them possess two incisors, two premolars and four molars in each half jaw, making 32 in all, considerably less than that possessed by the primitive mammals. The jaw is short, so that a relatively short face results. Furthermore, the large nasal chambers essential to lower animals which depend upon smell are much reduced in primates with a corresponding reduction in this sense. Since the brain has grown a great deal in size, it has risen over the face, bringing the latter into a more vertical position. Along with this change has come a gradual shifting of the eyes from the lateral position to a frontal position. This has resulted in overlapping images, producing **stereoscopic vision,** essential to tree-dwellers for judging distances accurately, and to people for driving cars.

Various Primates

The Lemurs. The lemurs are not greatly removed from the insectivores (Fig. 14–43). Their primitive nature is indicated by the fact that their eyes still lie in such a lateral position that their vision does not overlap (Fig. 14–44). They live in the tree tops where they move about so conservatively and cautiously that they are rarely seen by enemies and consequently have been able to survive in the Malagasy Republic up to the present time.

Tarsier. The East Indies are the home of the only hopping primate, which is not much bigger than a rat (Fig. 14–45). Its mark of distinction lies in the fact that zoologists consider it intermediate between the lemurs and the monkeys. Its ratlike tail and long legs are well adapted for leaping, which it performs with great agility. The swollen tips of its digits are useful in catching limbs of trees as it forages for insects and lizards during its nocturnal sorties. The extremely large eyes are directed forward and probably permit stereoscopic vision.

Anthropoids. The highest group of all primates are the anthropoids, the manlike primates. These include the **monkeys,** the **great apes,** and **man.** It is important to point out that man did not descend from monkeys, nor is the monkey a degenerate man. Both had separate beginnings a long

Fig. 14–45. Tarsier (*Tarsius tarsier*) is a rat-sized hopping primate. Note its tremendous eyes and the padded finger tips which aid in grasping twigs.

Fig. 14–44. Lemurs are primitive primates that are found only in isolated parts of the world. This is the ring-tailed lemur, *Lemur catta.*

time ago and have been and are traveling along separate paths in their evolution (Fig. 14–43).

The Monkeys. These creatures are manlike both in physical characteristics and in attitudes. They are primarily arboreal, although they are able to get along on the ground. They walk on all fours, but when at rest usually sit down on their haunches, thus freeing their hands for the job of manipulating food or any other object. Their large and forward-placed eyes are probably as keen as those of humans.

The monkeys are divided into two groups: New World forms of Central and South America and the Old World forms of Asia, Africa, and Europe. Two members of the New World monkeys are of interest because of their specialization: the spider monkey, with its prehensile tail which functions as a fifth hand (Fig. 14–46), and the howler monkey, which has a remarkable voice made possible by modifications of the

Fig. 14–46. This spider monkey (*Ateles paniscus*), a representative of the New World monkeys, possesses a handy adaptation in its prehensile tail which functions as a fifth appendage.

throat into large, bony resonating chambers. Each has specialized in its own peculiar way and these variations set them off from all other members of the group.

The Old World monkeys exhibit a wide variety of form and habits, from the sacred langur of India to the highly colorful ground-dwelling mandrill (Fig. 14–47). Most of them are tree-dwelling, although the baboon lives on the ground most of the time.

The Manlike Great Apes. These predominantly large primates, the gibbon, orangutan, chimpanzee, and gorilla, separated very early from the common stem that also produced the monkeys, probably about the same time a branch separated off on its long course toward man. This appears to have taken place at least 20 million years ago. Among the great apes, the gibbon (Fig. 14–48) is best adapted to arboreal life. Its

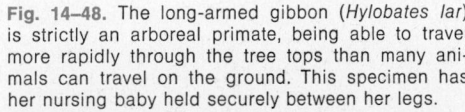

Fig. 14–48. The long-armed gibbon (*Hylobates lar*) is strictly an arboreal primate, being able to travel more rapidly through the tree tops than many animals can travel on the ground. This specimen has her nursing baby held securely between her legs.

Fig. 14–47. The mandrill (*Mandrillus sphinx*) is a colorful member of the Old World monkeys. It is noted for its doglike face and its highly colored cheeks.

long arms permit it to brachiate (swing hand over hand) through the trees with great speed. This type of locomotion in arboreal primates probably had some bearing on the achievement of the upright position in man. Ancestral forms may have begun to assume this position at an early date but before specialization had gone as far as it has in the gibbon, they descended to the ground and started their terrestrial existence.

The closest relatives of man are probably the great African apes, the gorilla and chimpanzee. Although they are at home among the trees, particularly the chimpanzee which brachiates very well, they still spend a good deal of their time on the ground. The gorilla seems to prefer the ground, retreating to the trees only to sleep. The chimpanzee possesses a disposition more favorable to captivity and for that reason a great deal is known about its behavior. In size it matches man very well, being about 5 feet

Fig. 14–49. The gorilla (*Gorilla gorilla*) is the largest and probably the most intelligent of all the great apes. This is a young male that has just been enticed into new quarters which it is thoroughly investigating.

in height and weighing about 150 pounds when full grown. When young it is easily handled, and is a very affectionate, playful, and extremely curious animal. It seems to approach the human type of intelligence as illustrated by its ability to solve problems. Experiments, notably those by Yerkes, have shown that the "chimp" will solve simple problems much like a small child.

The gorilla is perhaps more intelligent than the chimpanzee, but because of its sullen and individualistic disposition, it will not tolerate training or testing of any sort (Fig. 14–49). A two-year-old gorilla responds in a manner similar to a child of about the same age or a little older. Soon, however, it develops a morose disposition and cannot be trusted for close association with man. In size the gorilla exceeds all other primates, and an old male may reach a height of 6 feet and a weight of 500–600 pounds. Because of its herbivorous diet the gorilla is not a predator, and for that reason does not get into much trouble. The gorilla walks rather well on the ground but frequently reverts to all fours, especially when in a hurry. Its short sturdy legs and manlike feet support its weight well, although it is not a swift runner.

At this point we should discuss man because he is a primate, but a discussion of his evolution has been reserved for a later section of this chapter.

Some Other Mammals

One of the more successful groups of mammals are the flesh-eaters, the carnivores (order **Carnivora**). They include cats, dogs, weasels, bears, civets, and the marine seals and walruses. The carnivores have teeth well adapted to the rending and tearing of flesh (Fig. 14–50). The large canines readily tear through tough skin and the shearing molars cut the flesh into pieces sufficiently small to be swallowed. Furthermore, since meat is easily digested, the alimentary canals are short when compared to those of the plant feeders. In general, the carnivores are active and sometimes predatory animals,

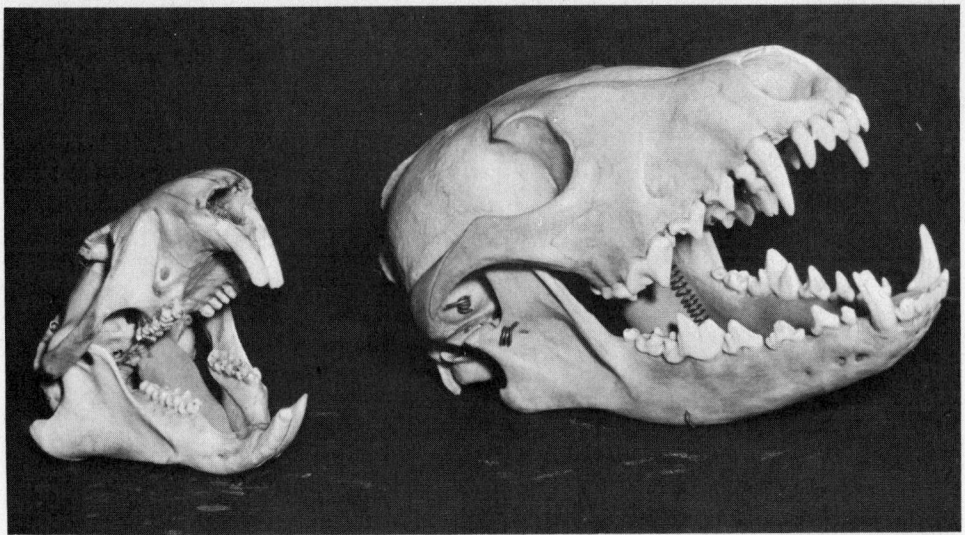

Fig. 14–50. The teeth of mammals are adapted to different diets. The beaver (left) has huge incisor teeth for gnawing and flat molars for grinding whereas a carnivore as the wolf (right) has small incisors but large spikelike canines which are efficient in rending flesh. Its molars have a shearing action for cutting its food.

some hunting in packs like the wolves and others leading more or less solitary lives like the big cats and bears.

Some very interesting and bizarre carnivores are those that inhabit the water, and have practically lost their ability to locomote on land. These are represented by the seals, sea lions, walruses, and others (Fig. 14–51). They are relatively helpless on land, but in the water they compare favorably with the best of the true aquatic forms, including fish. Having been on land long enough to acquire the intelligence and cunning of the mammals before returning to their original environment, they offer considerable competition for the less intelligent fishes. Their appendages have become flippers which are effectively used in locomotion. They swim in a fishlike manner by undulating motions of the body. The tail has dwindled to a useless structure and the two posterior legs function like a tail in locomotion.

In contrast to the carnivores are the herbivores (ungulates), of which the domestic horse and cow are examples. The horse possesses an odd number of toes (order **Perissodactyla**) and the cow an even

Fig. 14–51. A young Atlantic walrus (*Odobenus rosmarus*). Note how well the appendages are adapted for aquatic life but very poorly for land locomotion.

number of toes (order **Artiodactyla**). In their evolution, the former came first and reached large sizes, as illustrated by the giant rhinoceroses that attained a shoulder height of 18 feet. Members of this group have their weight borne on three toes in the rhinoceroses or on one toe in the case of the horse. The even-toed herbivores usually have four toes, of which two bear the main weight of the animal; this is illustrated by cattle, sheep, giraffe (Fig. 14–52), and many others. Some, such as the hippopotamus, bear their weight on four toes. Some of the present-day forms are **ruminants** or "cud-chewers." Their stomachs have several compartments, a condition that permits large amounts of hastily acquired food to be temporarily stored and then brought back into the mouth at a later period to be properly chewed at the animal's leisure. This is a most comfortable and effective adaptation since the animals may feed voraciously during certain periods of the day when grazing is less dangerous and then retire to secluded spots to finish the job of chewing.

Perhaps the most curious of all mammals is the elephant (order **Proboscidea**) with its nose, in the form of a proboscis or trunk, touching the ground. This handy organ makes possible an unusually short neck, because it performs the function of securing grass from the ground. It is also useful in obtaining water. Actually, it is a prolongation of the upper lip, including the nostrils. The teeth have undergone several changes, the most striking of which is the formation of the great tusks, which are overdeveloped incisor teeth. During the Miocene and Pliocene these great beasts spread over much of the northern land masses and survived in great numbers on our continent until ten or twenty thousand years ago when they became extinct.

Some distant relatives of the elephant went into the sea and became adapted for an aquatic existence. These are the sirenians (order **Sirenia**), which include such mammals as the dugong of the Indian Ocean, the sea cows formerly from Bering Straits but

Fig. 14–52. A large mammal from Africa.

now practically extinct, and the manatee from the Atlantic Ocean in the tropics. These beasts feed on the abundant aquatic vegetation along the shores but are unable to come ashore themselves because they lack posterior appendages and the front ones are adapted for swimming only. The sirenians are not related to the whales; fossil records indicate that their closest relatives seem to be the Proboscidea, in spite of the vast difference in their external appearance and way of life.

The whales and porpoises (Fig. 14–53) (order **Cetacea**) are probably the most specialized of all mammals. They are descendants of land carnivores that have returned to the sea and become so successful in this environment that they are found in all of the oceans of the world. They were probably flesh-eating mammals much like the present-day otters, fishers, and minks which, returning to the water in search of prey, gradually became better and better adapted to marine life.

Whales have few anatomical features remaining that are reminiscent of their land life. They have lost their posterior limbs, although they still have useless remnants left buried deep in their bodies, and their anterior appendages have modified into efficient flippers. The adult whale is generally completely lacking in hair; a few species have some around the mouth region, however. The whales' nostrils have moved back to the tops of their heads, which facilitates breathing at the water surface without exposing the head. A tremendous tail is the principal organ of locomotion; it undulates up and down rather than sideways as in a fish.

Even though the first whales were probably all carnivorous, many subsequent species became herbivorous. All whales today are divided into two groups: the **whalebone whales** and the **toothed whales.** The latter have retained their teeth, which may be used in crushing fish and squids, their chief source of food. The more interesting whalebone whales have evolved a sievelike structure called whalebone which hangs from the roof of the mouth and is used to strain small marine organisms. Whalebone, which is derived from ectoderm and is homol-

Fig. 14–53. A group of porpoises, mother and month-old offspring in the foreground. Note the nostril on the top of the head, an adaptation that makes it possible for the animal to breathe at the surface without exposing its head.

ogous to nails and claws, makes up the strainer. The blue whale, also a whalebone whale, has prospered on this diet of minute marine life, and some specimens reach a weight of 150 tons, which exceeds that of any other animal that has ever lived, dinosaurs not excepted. One problem that is difficult to understand is how this animal can dive to great depths and remain under water for 30–45 minutes without being crushed and without getting decompression sickness. It apparently has oxygen reserves and other devices for satisfying its needs under these rigorous conditions. The behavior of porpoises (small whales) has been intensively studied in recent years and

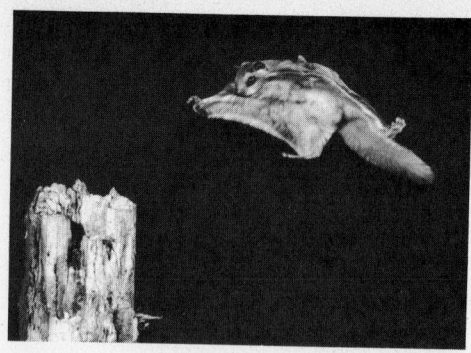

Fig. 14–55. The flying squirrel (*Glaucomys volans*) has adapted itself as a glider. The webbed skin attached to fore and hind legs supports the squirrel as it glides from tree to tree.

Fig. 14–54. South American toothless mammals: (Top) The Giant anteater (*Myrmecophaga tridactyla*) possesses powerful claws for opening ant hills and a long sticky tongue for lapping up the ants which are its sole source of food. (Bottom) the sloth (*Bradypustridactylus*) is remarkably adapted for tree life. Its hooked claws make it possible to hang comfortably from branches.

found to be quite remarkable. We have more to say about this topic in Chapter 27.

A small group of mammals show remarkable reduction in teeth, jaw skeleton, and jaw muscles. These are the so-called toothless mammals (order **Edentata**), represented by the anteaters and sloths (Fig. 14–54).

The most abundant of all the mammals are the rodents (order **Rodentia**), the "gnawers." These include many small mammals commonly known to everyone: squirrels (Fig. 14–55), chipmunks, beavers, porcupines, guinea pigs, and even the pestiferous rats and mice. They are characterized by their large chisel-like incisor teeth which are self-sharpening and which are kept almost incessantly active (Fig. 14–50). The incisor teeth grow continuously and if not worn down could soon become so long that the animal would not be able to close its mouth. Most rodents are small today, although at one time there were giant beavers which grew to the size of a small bear. Rodents have a high degree of intelligence and are usually secretive animals, many living in burrows. One, the beaver, is almost human in its ability to build dams. In some regions where conservation laws have made it possible for them to come back, they have become almost pests because of their habit of damming every small stream in large areas, thus flooding fields and roads in the surrounding country.

The rabbits and hares are often thought to be rodents because of their chisel-like teeth. In other respects, however, they have few rodent characteristics, hence have been placed in a separate order (order **Lagomorpha**). Hares are larger than rabbits and do not construct burrows. In other respects they are very similar. Both rabbits and hares are ubiquitous in distribution and have been very successful, particularly when they have been introduced in new areas. The scourge of rabbits in Australia is a familiar example.

Having now completed a synopsis of animal life, from protozoa through mammals, let us consider the story of man's origin.

The Origin of Man

Fossil remains of early man have been few and only fragmentary to date, a fact which makes the reconstruction of man's evolution a very difficult task. However, certain important discoveries have been made which give some clue as to his origin. The most primitive monkeylike fossil has been found in Egypt in early Oligocene strata. It is named **Parapithecus** and may well represent the basic form which gave rise to the New and Old World monkeys. The most ancient man-ape or ape-man fossils were unearthed by L. S. B. Leakey and co-workers in the Olduvai Gorge near Lake Victoria, Africa. In addition to numerous skeletons, primitive stone tools and other artifacts were found which indicated that these creatures had a primitive culture. These were named **Zinjanthropus**. The skeletons are similar to those found in South Africa and Australia which are called **Australopithecus** (Fig. 14–56). According to the type of tools found with the skeletons, **Zinjanthropus** lived at least one-half million years ago and **Australopithecus** may even be older.

The skulls of these ape-men show characteristics of both the great apes of today and of modern man. The brain case has a capacity of over 700 ml. which is about the size of the largest gorillas today, but smaller

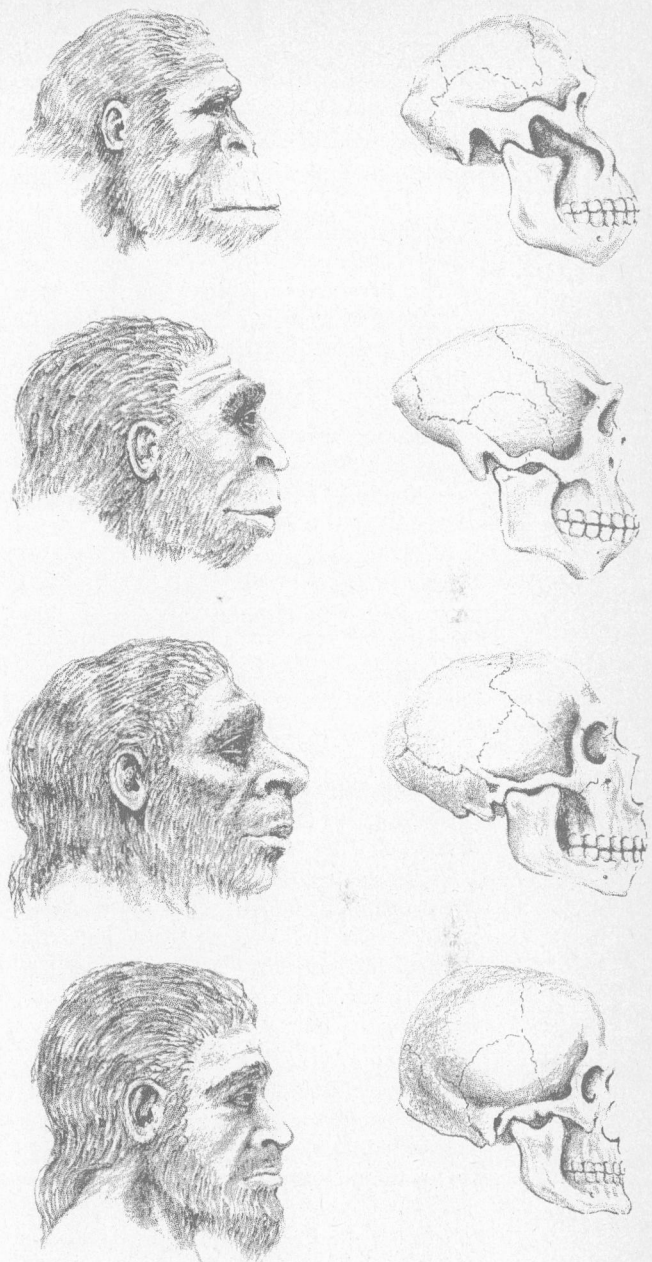

Fig. 14–56. A comparative study of the reconstructed faces of early men together with their skulls from which the reconstructions were made. They are arranged top to bottom in the chronological order, although there may well have been some overlapping. *Australopithecus* appeared first, followed by *Pithecanthropus* and *Neanderthal*. Modern man is portrayed by the *Cro-Magnon*.

than that of the next higher fossil man, namely, **Pithecanthropus** (Fig. 14–56), with a capacity of about 900 ml.

Pithecanthropus has stirred up more controversy since its discovery than any other fossil man. This is understandable, since its discoverer, Eugene Dubois, a Dutch Army officer, set out to find the missing link between man and the apes, probably because of the intense controversy that had been stirred up by Darwin's *Descent of Man.* It is most remarkable that a man should set out to accomplish such a difficult task and actually carry out his promise. Others searched furiously for the next few years, but it was not until 1936 that similar skeletons were found. The original material, which consisted of a skull cap, three teeth, and a femur bone, was discovered in the banks of the Solo River in Java in 1891. Dubois called these finds *Pithecanthropus erectus,* the "erect man." His claim that this was an intermediate form, half ape, half man, is probably not correct. Recent studies seem to indicate that it was a man and not an ape. The man had a very low brow and the size of his brain was about 900 ml., a much greater capacity than that of the great apes though smaller than any modern man (average about 1,500 ml.). The tooth pattern is definitely human, and it is interesting to note that the wisdom teeth show no signs of disappearing, as is common in modern man. He probably lived about 500,000 years ago.

A close relative of *Pithecanthropus* is **Sinanthropus,** commonly known as the Peking man because the remains were found in a cave some 30 miles from Peking, China. These middle Pleistocene deposits stimulated Dr. Davidson Black in the 1920's to investigate them for the possibility of human remains. He found first only a tooth here, but during the next decade over 30 individuals were unearthed, so that the evidence is very complete concerning this primitive man. *Sinanthropus* is so similar to *Pithecanthropus* that they are now thought of as belonging to the same genus (*Homo* [*Pithecanthropus*] *erectus*). However, the

brain case of *Sinanthropus* is somewhat larger on the average (915–1,200 ml.). The leg bones indicate an upright posture. He apparently had a culture which was far above anything the apes would be likely to have. He used fire and made stone tools. There is evidence that he was cannibalistic, based on the fact that so many skulls are present in the deposits, each one crushed at the base. This might mean that bodies were eaten and the brain removed for the same purpose, a probably not uncommon custom for early man.

The **Neanderthal** man (Fig. 14–56) was the first fossil man discovered (1856) and many of his remains since have been uncovered, indicating that he roamed over parts of Asia, Africa, and Europe about 100,000 years ago. Physically he was a rather short man, not exceeding 5 feet 4 inches (the females were several inches shorter), but very powerfully built. His upper arms and femurs were short and the latter were curved forward so that in posture he probably walked with his knees slightly bent forward. Furthermore, his massive head was set somewhat forward over his thick neck and barrel chest, so that he was probably quite gorillalike in appearance. His brain case capacity was 1,550 ml., larger than the average for modern man, although it had different proportions. The forepart of the brain case, the seat of higher intelligence, was rather small, the larger portion being in the back of the head. His skull was heavy, with large protruding brows, and the massive jaws bore teeth much more substantially built than those of modern man. The mandible was chinless just as among the apes; a chin is a more recent development.

He had some type of culture which included religious rites, because he buried his dead—probably the important reason why so many remains have been found. He had learned to shape and use stone implements in the daily business of securing food, and these, together with bones of the animals he hunted, are found in Neanderthal deposits.

There is always great interest not only

among anthropologists but also among laymen concerning the story of the origin of modern man. Which of these early men gave rise to modern man, *Homo sapiens?* Was the Neanderthal man the immediate forerunner of *Homo sapiens?* There has been a great deal of discussion over this point, and there are several possible answers. It must be admitted certainly that Neanderthal showed considerable advance over *Homo (Pithecanthropus) erectus.* Whether or not he gave rise to *Homo sapiens* is questionable. There is evidence to indicate that he was wiped out rather suddenly by another race, probably *Homo sapiens.* If this was the case, then the latter must have been derived from another stock as yet not unearthed.

Whenever *Homo sapiens* made his appearance, he split off into many races which spread over the entire world. One of the outstanding of these is the *Cro-Magnon,* who lived in Europe 15,000 to 40,000 years ago (Fig. 14–56). Most of the remains, about one hundred skeletons, have been found in France, and the records afford a rather complete story of the culture as well as the anatomy of this man. He was tall, many males reaching 6 feet in height, although the females were considerably shorter. His skull was long and massively built with the tremendous capacity of 1,700 ml., which is greater than that of any race today. Furthermore, the shape was definitely modern in that the forehead was high and the back part rounded as in modern man. None of the apelike characteristics observed in *Pithecanthropus* are evident in this man. The mandible terminated anteriorly in a pronounced chin, and the face was short and broad with the eyes set far apart. Undoubtedly, he was a handsome fellow.

In culture these men far surpassed their predecessors. They made their hunting weapons skillfully, producing the spear and axe, and they may have even devised the bow and arrow. They apparently were superior hunters, as judged by the contents of their caves, which are strewn with the bones of large animals they had killed and feasted upon. They probably clothed themselves with skins of animals to protect their hairless bodies from the elements. They evidently possessed great skill in making and mixing paints, and in depicting the life as it existed in their time, for they have left behind on the walls of many caves in Spain and southern France magnificent paintings showing many of the animals which they hunted. These are portrayed in a most realistic manner, even to the natural colors which have resisted the ravages of time. The remains of these men would indicate that they were superior both mentally and physically to any races alive today. What, then, has happened to them? Some think their descendants still live in Europe today. If so, are they degenerate types? Is *Homo sapiens* actually on the decline?

Knowledge of early *Homo sapiens* in other parts of the world is not as complete as in Europe. One might expect to find some early Negroid types in Africa, the home of the Negro. An early skull from the Sahara does show Negroid characteristics. As this part of the world becomes more modernized, the diggings of various sorts which always accompany the process will perhaps throw more light on this. Undoubtedly, rich deposits exist in parts of Asia which when unearthed will reveal an interesting story concerning the origin of the Mongoloids, about whom little or nothing is known today.

Apparently, *Homo sapiens* invaded America comparatively recently, because no remains of very early man have been found. The primates that were here in the early Tertiary times became extinct and no others appeared until modern man made his way from Asia across Bering Straits. It is highly probable that this is the path he took in populating the Americas, because geologists have shown that a drop in the sea level of 100 feet would leave a land connection between Asia and North America. Such rising and falling of the sea took place with each great glacial period. For many thousands of years it was possible for these people, filtering gradually over this narrow neck of land, to migrate southward to the semitropical and tropical regions of the Americas.

Races of *Homo Sapiens*

The criteria that have been used to distinguish the various races are such points as stature, hair, face proportions, skull shape, complexion, eye and skin color, and blood groups. Although there are many difficulties in attempting to separate *Homo sapiens* into races, using these characters, it is possible to define four major groups: Australoids, Negroids, Caucasoids, and Mongoloids.

The Australian aborigines are usually considered to be the most primitive and probably have a somewhat different line of descent from the others. They are referred to as the Australoids. The Negroids include the Negroes, pigmies, Bushmen, and Hottentots, all with heavy pigmentation in their skin and hair. This race is widespread today, occupying much of Africa as well as most of the Pacific islands, from New Guinea to Fiji. The Caucasoids include a wide range of people, indeed all of those that are not Australoid, Negroid or Mongoloid. They possess hair and skin color from light to dark brown, eye color from blue to dark brown. The hair, however, may be straight or wavy, but never kinky or wooly as is typical of Negroids. They include the most aggressive people of the world, representatives of which inhabit Europe, the Americas, and parts of Asia. The Mongoloids are characterized by coarse, black, straight hair, and skin that has varying shades of yellow and brown. Their heads are heavily covered with hair, but the face and the rest of the body is relatively naked. They are also widespread, occupying much of eastern Asia as well as the American continent. They include the Chinese, Japanese, Eskimos, and American Indians.

Man's Present Status. One might think that with all laws of evolution operative through these past 500,000 or more years, present-day man might be an animal physically perfect in all respects and geared beautifully to his environment. This, in fact, is far from the truth. The number of people with defective vision is a glaring example. The occasional inadequacies of the various organ systems is confirming evidence that they do not always function as they were intended, at least under present treatment which itself may be at fault. In spite of these apparent inadequacies, man has done rather well, biologically speaking. He has spread himself over a very large portion of the globe and has reached nearly three billion in numbers, not a large figure, to be sure, when one considers that there are more bacteria in a quart of sour milk! He has managed himself rather well in most respects; he plans for his own food and shelter as well as other comforts of life. He has, by concerted effort, been able to allow himself some leisure time from the endless task of providing the bare necessities of life. He has sometimes used this time creatively, thus improving not only his immediate environment but also his relation to it.

There has been little, if any, improvement in our brain since the Cro-Magnon man; and during the intervening 50,000 years, progress toward civilization as we know it has been extremely slow. Only during the last 5,000 years, and particularly the past 300 years has outstanding progress taken place. Why was man so slow in rising as a social animal, and why, when he started, did he rise so rapidly? It was undoubtedly due to the fact that he acquired the ability to put down in writing what he had learned so that those who followed could profit by his experience. Once this took root it flourished and with it the progress of mankind.

It is well worthwhile, then, to study rather carefully this animal that has made such stellar progress in the past few hundred years, and whose primary mark of distinction is a huge brain.

Books

BLAIR, W. F., BLAIR, A. P., BRODKORB, P., CAGLE, F. R., and MOORE, G. A., *Vertebrates of the United States*, 2nd ed. New York: McGraw-Hill, 1968.

COLBERT, E. H., *Evolution of the Vertebrates*, 2nd ed. New York: Wiley, 1969.

COONS, C. S., *The Origin of Races*. New York: Alfred A. Knopf, 1962.

EDITORS OF TIME-LIFE BOOKS, *The Emergence of Man*. New York: Time-Life Books, 1972–1973.

GANS, C., BELLAIRS, A., and PARSON, T. S., eds., *Biology of Reptilia*. New York: Academic Press, 1969–1970.

GATES, R. R., *Human Ancestry*. Cambridge, Mass.: Harvard University Press, 1948.

GILLIARD, E. T., *Living Birds of the World*. New York: Doubleday and Co., 1958.

GREGORY, W. K., *Evolution Emerging, A Survey of Changing Patterns from Primeval Life to Man*. New York: Macmillan, 1951.

HOOTON, E. A., *Up From The Apes*. New York: Macmillan, 1945.

LORENZ, K., *King Solomon's Ring*. New York: Thomas Crowell, 1952.

NOBLE, G. K., *The Biology of Amphibia*. New York: Dover Publications, Inc., 1956.

OLSON, E. E., *Vertebrate Paleozoology*. New York: Wiley-Interscience, 1971.

ORR, R. T., *Vertebrate Biology*, 3rd ed. Philadelphia: Saunders, 1971.

PETERSON, R. T., *A Field Guide to the Birds*, 2nd ed. Boston: Houghton Mifflin, 1947.

PORTER, K. R., *Herpetology*. Philadelphia: Saunders, 1972.

ROMER, A. S., *Notes and Comments on Vertebrate Paleontology*. Chicago: University of Chicago Press, 1968.

ROMER, A. S., *The Vertebrate Body*. Philadelphia: Saunders, 1971.

TOBIAS, P. V., *The Brain in Huminid Evolution*. New York: Columbia University Press, 1971.

VAN TYNE, J., and BERGER, A. J., *Fundamentals of Ornithology*. New York: Wiley, 1959.

WRIGHT, A. H., and WRIGHT, A. A., *Handbook of Snakes of the United States and Canada*. Ithaca, N.Y.: Comstock Publishing Co., 1957.

YOUNG, J. Z., *The Life of Vertebrates*, 2nd ed. Oxford, England: Clarendon Press, 1963.

Articles

CAREY, F. G., "Fishes With Warm Bodies." *Scientific American*, **228**, 36, February, 1973.

DOBZHANSKY, T., "The Present Evolution of Man." *Scientific American*, **203**, 206, September, 1960.

GAMOW, R. I., and HARRIS, J. F., "The Infrared Receptors of Snakes." *Scientific American*, Vol. **228**, May, 1973.

HOWELLS, W. W., "Homo erectus." *Scientific American*, **215**, 46, November, 1966.

KEETON, W. T., "The Mystery of Pigeon Homing." *Scientific American*, **231**, 96, December, 1974.

KORTLANDT, A., "Chimpanzees in the Wild." *Scientific American*, **206**, 128, May, 1962.

MILLOT, J., "The Coelocanth." *Scientific American*, **193**, 34, December, 1955.

SCHMIDT-NELSON, K., "How Birds Breathe." *Scientific American*, **225**, 72, December, 1971.

SHAW, E., "The Schooling of Fishes." *Scientific American*, **206**, 128, June, 1962.

SIMONS, E. L., "The Earliest Apes." *Scientific American*, **217**, 28, December, 1967.

SIMONS, E. L., and ETTEL, P. C., "Gigantopithecus." *Scientific American*, Vol. **222**, January, 1970.

TINBERGEN, N., "The Evolution of Behavior in Gulls." *Scientific American*, **203**, 118, December, 1960.

WARREN, J. V., "The Physiology of the Giraffe." *Scientific American*, **231**, 96, November, 1974.

WASHBURN, S. L., "Tools and Human Evolution." *Scientific American*, **203**, 62, September, 1960.

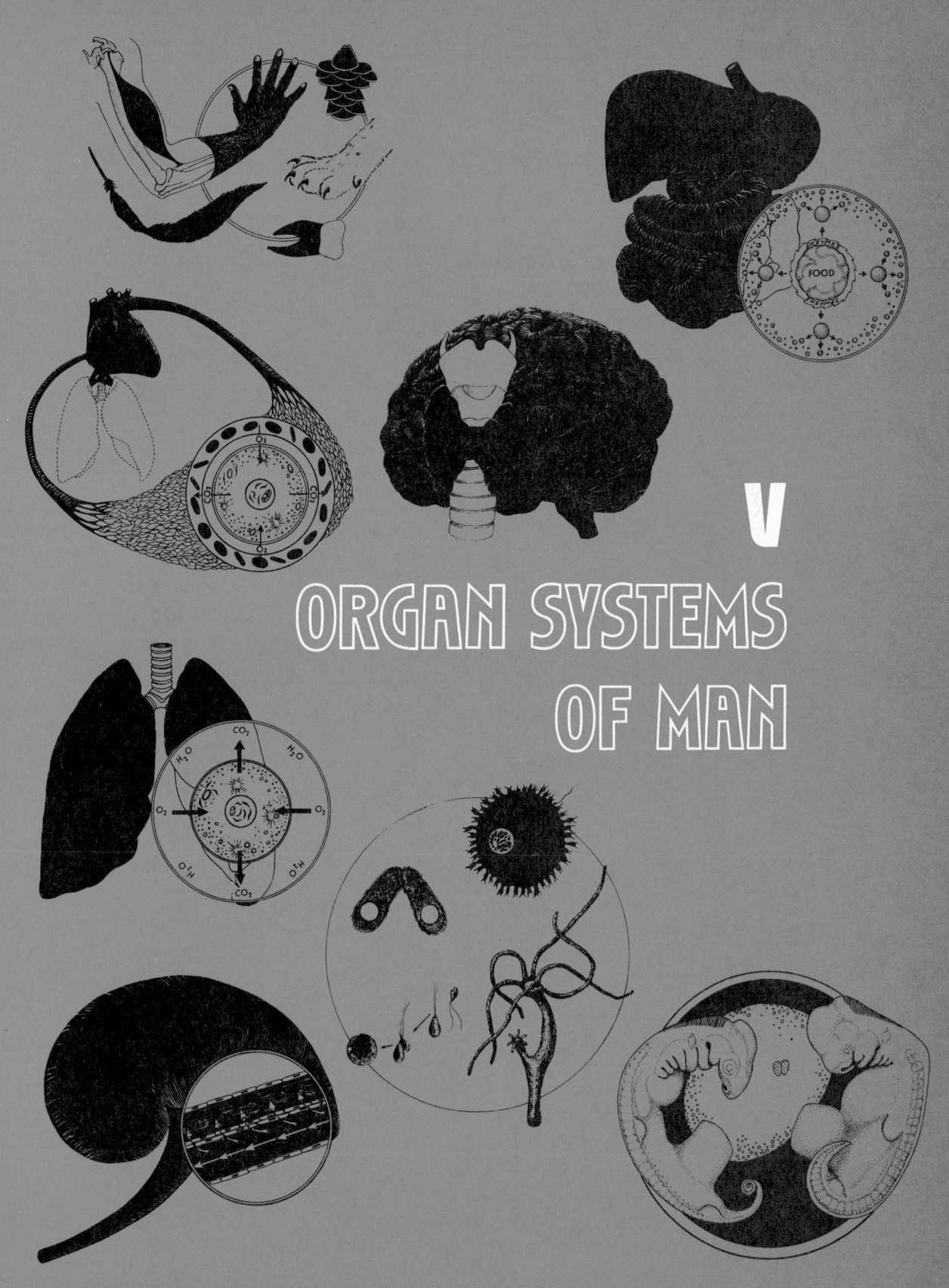

V

ORGAN SYSTEMS OF MAN

15

HOUSING, SUPPORT, AND MOVEMENT

In studying the last group of animals, we have selected man as a typical mammal. A study of the rat, cat, guinea pig, or dog would afford no more information than that of the human body and for the readers of this book certainly no more interest.

MAN—A TYPICAL MAMMAL

Man is a typical mammal, unusual only in the size of his brain. He is not the largest or the smallest mammal, nor is he highly specialized when compared to the whale, for example. In fact, he is a rather mediocre mammal, being poorly endowed with organs of offense and defense. His puny, flat finger nails and short canine teeth are no match for the claw and tooth of the tiger or lion. His hide is not thick, like that of the elephant or whale, and it is completely unprotected by hair, the normal coat for most

mammals. He has no horny outgrowth for defense, like the ungulates, and even his locomotor appendages are only fairly effective in getting him out of danger.

Man is no longer at home in the trees but lives on the ground and employs the bipedal method of locomotion, a method far from new, for the great carnivorous dinosaurs also employed it. Other present-day bipeds such as the ostrich can easily outstrip him in cursorial travel. He is poorly fitted for life in the water where his appendages are not well adapted for locomotion. He can submerge for only very short periods without coming up for air, and in cold water his survival time is very short.

He has, however, one crucial organ that accounts for most of his success, his well-developed brain. This organ, by its intricate disposition of nerve impulses, has made it possible for man to compensate for most of his physical deficiencies. With it he has been able, through the power of speech, to communicate with his fellows and later to put words down in writing. Over a long period of time this type of specialization has finally "paid off," because man today is the dominant species on the earth.

In order to understand man it has been necessary to study other forms of animal life. Man does not lend himself well to experimentation for obvious reasons and, furthermore, he grows too slowly to permit studying succeeding generations. He cannot be kept under the controlled conditions that are possible with rats in a cage. However, most of our information about his functioning has come through the careful study of lower animals. If it were not for these experiments our knowledge would be very meager. This brings up an interesting and important point concerning attitudes.

There are isolated groups of people who oppose any animal experimentation (primarily experiments on dogs, cats, and other pets), sincerely, perhaps, or stirred by some ulterior motive. In any case, these groups are small but usually active and are constantly stumping for legislation that would curtail present-day experimentation. It requires only a slight knowledge of the subject to realize that most of the information about the functioning of the human body can come only through such experiments. Such groups have been responsible for bills that have come before state legislatures and even the Congress. Fortunately, those that would prevent all animal experimentation have been voted down. However a more sane approach to the problem of regulating animal care before the animals reach the laboratory has been formulated into a bill which was passed in Congress and is now law. This law is received enthusiastically by the scientific community and does not, as yet, impair research where animal experimentation is essential.

Now let us discuss the organ systems of man, as illustrative of vertebrate organ systems in general.

Fig. 15–1. Human skin in cross section. Gland and hair development are shown in several stages.

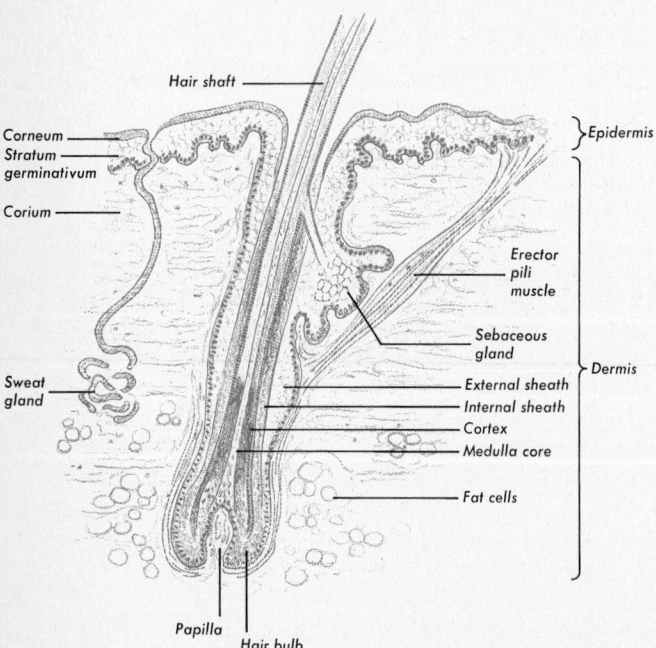

Hair shaft

Corneum
Stratum germinativum

Corium

} Epidermis

Erector pili muscle

Sebaceous gland

External sheath
Internal sheath
Cortex
Medulla core

} Dermis

Sweat gland

Fat cells

Papilla
Hair bulb

OUTER COVERING—THE SKIN

As we have seen in the preceding chapters all animals are provided with an exterior covering that functions as a barrier against the outside world. This is extremely simple in the lower invertebrates but becomes much more complicated in higher forms as, for example, in the vertebrate skin with all of its derivatives.

It is necessary to examine skin microscopically if one is to understand how it performs its many and sundry jobs, and we will take human skin as our example. It is usually divided into two parts, the outer, thinner **epidermis** and the inner, thicker **dermis** (Fig. 15–1). The epidermis is composed of an outermost nonliving covering, the **corneum,** which is the part in immediate contact with the outside world; lying beneath it is a layer of epithelial tissue called the **stratum germinativum** which is composed of actively growing cells. As the cells grow they move toward the outside, die, and eventually become the corneum. These dead cells are constantly being sloughed off in mammals and many of the lower vertebrates. Everyone is familiar with the loss of these cells in unexposed parts of the body such as back of the ears and between the toes. They are particularly noticeable in the hair, where they resemble flaky "scales" and do not have an opportunity to escape readily. These dead cells are spoken of as dandruff and often erroneously considered a pathological condition. The corneum is perforated by many tiny holes through which **sweat** passes from **glands** that lie deep below. The corneum becomes very thick (callused) on the soles of the feet and the palms of the hands, especially in people who perform heavy labor requiring the use of these appendages. Another interesting characteristic of the corneum of these areas is the formation of **friction ridges**. It is the presence of these in the hand and foot which causes fingerprints and footprints. These friction ridges apparently have come through a long evolutionary history, being originally digital pads of four-footed animals. When mammals took to the trees, the pads developed into transverse ridges which functioned in increasing the friction between the hand and the branch, thus preventing slipping. In man the ridges are generally arranged in whorls, although they are transverse to the long axis of the fingers for the most part. The designs appear to be infinite in number and never seem to be repeated on the tip of any digit, either on the hand of one individual or on any other individual. This has provided a convenient means of identification because it positively distinguishes one person from another.

The stratum germinativum produces many structures which, on one side, are sunk deep into the dermis and, on the other side, are an important part of the external covering. The scales of fish and reptiles, the feathers of birds, and the hairs of mammals are such structures. Although these all have similar origins, in the final adult stage they are quite different both in structure and function.

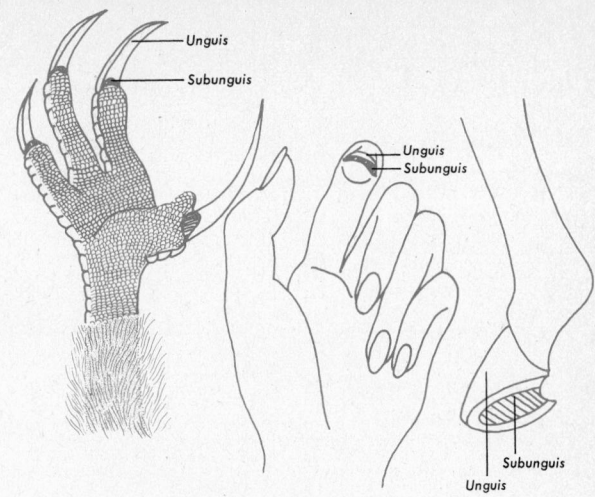

Fig. 15–2. Homologous digital tips: claws, nails, and hoof. Note that the unguis and subunquis are emphasized differently in all.

Digital tips in various vertebrates, such as the carnivore claw, the ungulate hoof, and the primate nail, are likewise produced from this region of the epidermis (Fig. 15–2). They are all homologous, since each has the same origin although each performs a different function.

Derivatives of the Epidermis

Scales and Feathers. Reptilian scales and feathers of birds have a common origin (Fig. 15–3) and they are very much alike both anatomically and functionally. Scales of bony fish, however, are dermal in origin and hence are not homologous with reptilian scales or feathers. Epidermal scales similar to those of reptiles are found on the legs of birds and various parts of the anatomy of some mammals (for example, tail of rat and beaver). Structurally, they resemble overlapping or abutting plates that offer considerable resistance to outside mechanical injury. When overlapping like shingles, they are so arranged as to offer the least resistance to forward motion.

Feathers resemble scales in their overlapping arrangement, although otherwise the

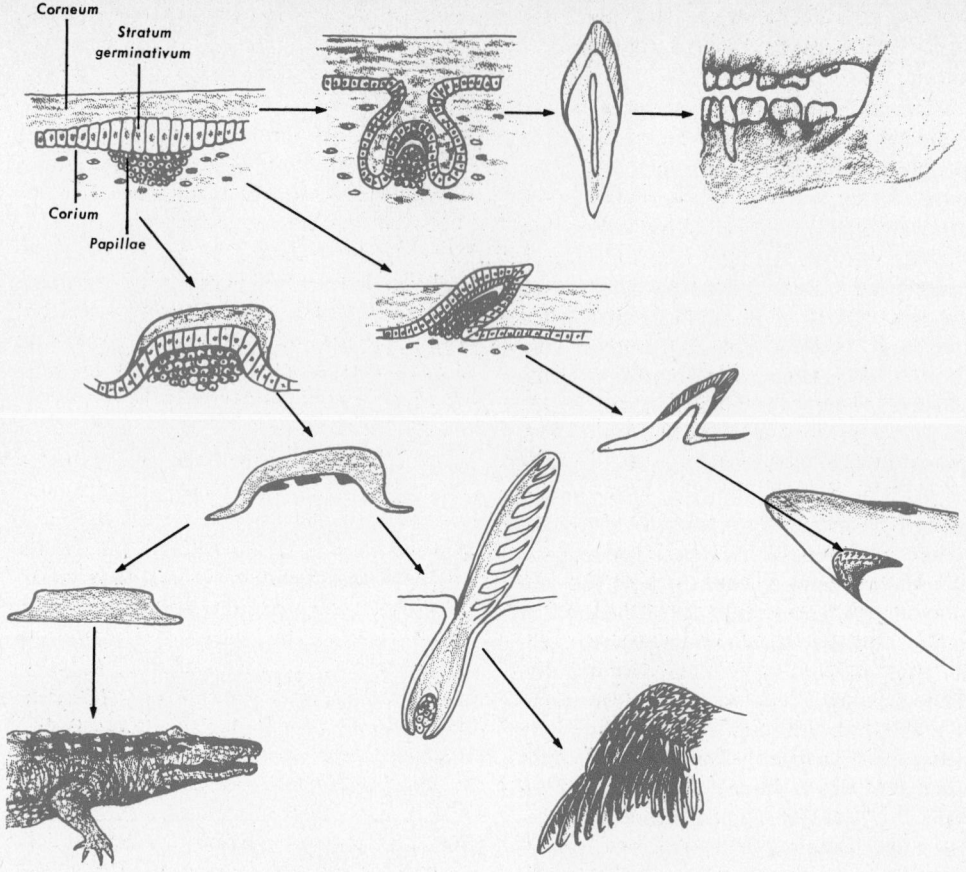

Fig. 15–3. Stages in the development of reptilian scales, feathers and teeth, showing their common origin.

likeness is not so obvious. They are much lighter in construction, possessing numerous tiny filaments that offer resistance to the passage of air through them. The wing feathers of birds will allow air to pass one way but not the other—a beautiful example of adaptation to flight.

Teeth. Teeth are compound structures, the dentine developing from mesoderm and the enamel from ectoderm. Their development is similar to that of the placoid scales of cartilaginous fish (Fig. 15–3). Since the mouth is lined with ectoderm (the germ layer that gives rise to epidermis), we might expect that the mouth could be equipped with any structure that could arise from

ectoderm. The scales in sharks enlarge and grow over the edge of the jaw, producing teeth. Human teeth develop in similar fashion except that the final structures are quite different in that they fit into deep cups or sockets in the jawbone.

Hair. These tiny projections from the skin of mammals perform a protective function against both physical injury and heat loss. The numerous hairs tend to provide a dead air space just above the skin which prevents heat loss much as insulation materials in a house. Feathers also act as heat insulators, a fact readily observed on cold days when birds ruffle up their feathers to improve the insulating properties of their integument. On

302

a hot day, a bird keeps its feathers close to its body to allow as much heat to escape as possible.

Mammals, with few exceptions, are covered with a thick coat of hair. Man has lost most of his hair, probably because he evolved in a warm climate. Today it is present only in the pubic regions, under the arms, and on the face and head. The facial adornment is a male secondary sexual characteristic because it is not found in the female. The rest of the body is usually covered with very tiny hairs which are vestigial, for they perform no function in modern man. This loss of body covering carries with it the obvious disadvantage of rapid heat loss in a cold climate, and man has thus been forced to clothe himself with an artificial covering.

Hairs are arranged in definite patterns in the various regions of the body of man as well as on the body of other mammals. They are not perpendicular to the skin surface but slant, usually in a specific direction. For example, in a dog the hair slants away from the mid-dorsal line, usually in the direction of the pull of gravity, and this probably helps in shedding water. In man the direction of the hair slant in the back region is the same as in dogs. On the arms and legs it also follows a common pattern but on the top of the head it sometimes forms whorls or "cowlicks" which are inherited from generation to generation.

Attached to the base of each hair are tiny muscle fibers which, when contracted, cause the hair to stand on end, producing "goose flesh" (Fig. 15–1). These muscles are under the influence of the autonomic nervous system and are therefore beyond voluntary control. When one is frightened or sometimes under other emotional stress, the hair can be seen to stand on end, particularly along the spine. The common statement, "Chills run up my back," has a physiological foundation.

Pigment granules lying in the lower epidermal layers of the human skin give it color ranging from no color at all, as in the abnormal albinos, to the dense pigmentation in black people. Sunlight has a decided effect, not only on increasing the amount of pigment (tanning), but also on increasing the thickness of the epidermis itself. The purpose of such a response is protection, because ultraviolet light damages unprotected skin. Continual exposure to bright sunlight produces a dark, tough skin, even in white people. It has become the fashion to expose the body to the damaging rays of the sun in order to obtain a "tan," which for some strange reason has been identified with good health. There is no relationship; indeed skin cancer occurs most often in those people who have been exposed excessively to the sun.

Glands. Near the base of each hair is a tiny sebaceous gland which secretes an oily substance designed to keep both the hair and the skin in a soft, pliable condition. These glands, like other skin glands, come from the epidermis, although they are buried deep in the dermis (Fig. 15–1). The tiny, much coiled sweat glands are likewise found deep in the dermis where they function in extracting water from the blood and tissues and spreading it over the surface of the skin for the purpose of cooling the body (Fig. 15–1). The resulting evaporation reduces the temperature of the skin and thereby aids in the regulation of the temperature of the entire body. In man sweat glands are concentrated in the palms of the hands, soles of the feet, and under the arms.

Another very interesting skin gland found only among mammals and fully developed only in females is the *milk* or *mammary* gland, a name that identifies the group. These are modified sweat glands and are confined to areas most convenient for suckling the young mammal, which they supply with a complete food during its early post-embryonic life in a sort of continuation of umbilical feeding.

The Dermis

This layer of the skin is much thicker than the epidermis and is composed of tough connective tissue (Fig. 15–1). Fat tends to store in the deeper parts of the dermis and becomes localized in characteristic regions of the body familiar to everyone. The distribution of this layer of fat is characteristic of the sex. For example, in the female the layer is thicker, giving the skin a more velvety touch. The body contours are smoother and the underlying muscles are less pronounced than is the case in males.

Many tiny bundles of blood vessels project up from below into the dermis where they function in bringing the blood close to the surface for cooling the body when it is too warm. They contract when the body is cold (**vasoconstriction**), preventing the blood from coming to the surface. They are under the influence of the autonomic nervous system and may be influenced in some people by emotions such as in blushing (**vasodilation**). Also, the dermis contains many nerve endings which are receptors for heat, cold, pressure, touch, and pain. With the exception of the pain nerve endings, they are specially designed end organs for special stimuli. Since the skin is in contact with the outer world all of the time, a great deal of information comes to the brain from it. Pain, for example, is a very uncomfortable sensation, ideally designed to make the animal do something about the situation, if possible. This sensation is responsible for preservation of the individual, for without it great areas of the body might be destroyed without the organism being aware of it.

Thus we see that most animals possess some sort of protective covering. Animals also need some internal support, particularly the larger forms and those animals that live on land. This is provided by an **endoskeleton.**

THE SKELETAL SYSTEM

Animals have acquired many forms of support for their bodies. The various structures not only hold the body together but also, in many cases, have an important protective function. Vertebrates have an internal skeleton which functions primarily for support. It affords very little protection to the soft external parts of the animal, although it provides excellent protection for such vital organs as the brain, heart, and lungs. The human skeleton is an example of a vertebrate skeleton.

The skeleton of man is similar to that of other mammals, almost bone for bone, but certain parts are emphasized more or less when compared with similar parts in other mammals. This is because of the upright position man's body has taken. All animals that have adopted bipedal locomotion—dinosaurs, birds, and the kangaroo, for example—have shifted their body weight so that some parts of the skeleton must bear a greater portion of the burden than other parts. In quadrupeds the spinal column and legs resemble a suspension bridge where the column functions as the bridge itself and the legs as the supporting piers at either end. In man and other bipeds the body is elevated at one end until it is in a vertical position which requires more secure footing at the base. In the human skeleton, this shift in weight has also necessitated more rigid connection between the supporting vertebral column and the pelvic girdle to which posterior supporting appendages are attached. This appears to function in a fairly satisfactory manner, although if one may judge by the number of middle-aged people suffering from a "sacroiliac" (the development of a faulty union between the column and pelvic girdle), it is clear that the arrangement could be improved.

The skeleton is usually divided into two general parts: the **axial** region comprising the **skull,** the **column,** and the **ribs,** and the **appendicular** region which includes the **appendages** and their **girdles** (Fig. 15–7). All of the bones are so securely tied together with ligaments that they are torn apart only under great strain.

The Axial Skeleton

The Skull. The skull shows considerable modification when compared to that of lower vertebrates. The greater emphasis on the cranial case compared to the mandible is obvious when the heads of the dog, mon-

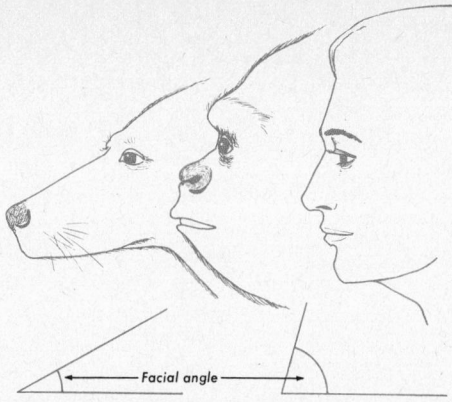

Fig. 15-4. Comparative profiles of the dog, monkey, and man to show the relative shift in the facial angle (see text). Because of the increasing size of man's brain, which has grown over his shortened jaws, his facial angle has increased over that of lower forms.

key, and man, for example, are placed side by side (Fig. 15–4). As the brain has grown forward over the shortened jaws, the facial angle (angle between a line drawn along the forehead and one from the base of the nose to the foramen magnum) has increased and the human face has been formed.

The human skull is made up of 28 bones, 22 of which are joined by jagged-edged **sutures**. The other 6 are the tiny ossicles of the ears. The bone which supports the tongue and larynx, the **hyoid**, is loosely connected with the skull. Although most of the skull is heavy, solid bone, certain portions contain cavities. These are remnants of chambers that formerly had specific functions but which perform no known function today. For example, the three **sinuses** in the anterior and middle portion of the skull once served the sense of smell but do not do so now. There is a pair of maxillary sinuses in the cheek region, a pair of frontal sinuses over the eyes, and a pair of sphenoid sinuses in the posterior part of the nasal chamber. All have small ducts which drain into the nasal chambers, but the arrangement is such that drainage is not good, especially when the membranes are swollen with a cold. Under such conditions the large surface area of the sinus membranes becomes infected, causing

the so-called **sinus** trouble which is often difficult to treat satisfactorily. Another spongy bone, the **mastoid**, lying behind the external ear may also become infected via the eustachian tube and the middle ear. Such an infection can reach the brain because the mastoid is separated from it only by a very thin bone. Surgery, the treatment of choice for mastoiditis some years ago, has largely been replaced by the use of antibiotics.

The brain is exposed to the outside world in only one place, and that is in the nasal chamber. The floor of the brain case where the olfactory nerves leave the brain and pass down into the nasal chamber, is called the **cribriform plate**. It is a piece of bone perforated with many small openings through which the nerves pass and through which, unfortunately, nasal infections can reach the brain.

The cord enters the human brain case ventrally instead of posteriorly as is the case with most mammals. The large opening through which it passes is called the **foramen magnum**. Since the skull is precariously perched on the tip of the spinal column, it might be expected that the cord could be injured at this point rather easily, and such is indeed the case. A severe blow at the base of the neck will stretch the cord at the point where it enters the skull thus causing complete paralysis of all body functions. Other openings into the brain case are the foramina for numerous small blood vessels and for the cranial nerves, including the optic nerves at the base of the orbits.

At birth, several bones from the brain case have not come together (sutured), so that five spaces are left without bony covering. These are called **fontanelles** (little fountains—so named because they rise and fall with each heart beat). This lack of suturing before birth plays a very important function in the birth process, for the head of the child undergoes severe squeezing while passing through the birth canal and needs to change its shape to fit the narrow passage. As the child grows, the fontanelles gradually close, leaving fine jagged lines at the junctures. The age of a skull can be told by the clearness of these lines. They are faint or absent in old skulls.

Injuries to the skull have been common throughout man's history. Early skulls often show evidences not only of natural injuries but also of apparent deliberate removal of small portions. Such drillings (called trephining) seemingly had some religious significance, but the remarkable thing about them is that the patients often recovered, as revealed by the smooth edges of the opening, indicating that the bone healed. Similar operations are performed today for entirely different reasons and with much more satisfactory results.

The Spinal Column. When a comparison is made between the spinal column of man and that of almost any other mammal, certain striking differences are noted (Fig. 15–5). These result from the upright posture man has assumed. In the dog, for example, the column forms a smooth arch between the two pairs of legs; in man, on the other hand, it forms a sigmoid or S-shaped curve. The curved spine of man is admirably designed to give the head a smooth ride. If the pliable spine of a growing child is subject to undue stress, it may ultimately affect the development of the adult skeleton. Much of our posture is dependent on the spine and there is much emphasis today on the desirability of good posture.

The spinal column is composed of 33 articulating **vertebrae** of rather irregular sizes fitted snugly together from the neck to the pelvis. They are securely laced together by many ligaments. This is essential because the column houses the very delicate spinal cord which, if injured even only slightly, may cause dire effects in the operation of the appendages as well as of other parts of the body. The column is more flexible in some regions than in others. For example, the vertebrae of the thoracic region are relatively immovable whereas those in the lower back and neck region have considerable amplitude of movement. This arrangement allows for a large variety of movements of the trunk.

Pairs of small openings between the verte-

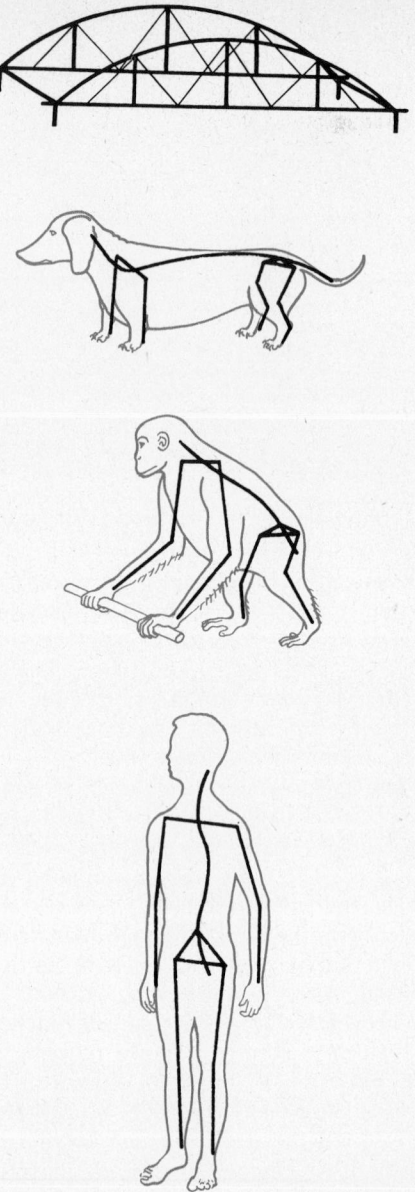

Fig. 15–5. The spinal column is the axial support of vertebrates and is subject to considerable variation among the different groups, depending on the stress and strain put upon it. In the quadruped, such as the dog, the column functions like a bridge with the two supports at either end. When the support is shifted to the two posterior appendages, such as in the ape, a more secure attachment must be effected between the column and the pelvic girdle. This is carried further in man, where we see a huge pelvic girdle, since the posterior appendages must provide the only means of support and locomotion.

brae provide exits for the spinal nerves (Fig. 15–6). Each vertebra has a large cylindrical passageway, and these taken together form the **neural canal** which houses the nerve cord. Five of the lower **sacral** vertebrae are fused into a solid bone, the **sacrum**, which joins the **ilia** (singular, **ilium**) on the dorsal side, thus securely attaching the pelvic girdle to the spine. It needs to be a broad, secure attachment because the whole upper body pivots at this point and the stress is considerable.

The spinal column terminates in several tiny, useless, fused vertebrate collectively known as the **coccyx**. In many vertebrates they give support to a functional **tail**, but in man they are mere vestiges of the past. Undoubtedly, far back in man's early history, long before he was man, he had a tail. It must be remembered that the presence or absence of a tail means nothing from an evolutionary point of view. The bear and guinea pig are without tails, yet they are no more related to each other than either is to man.

The Ribs. The 12 ribs are attached to the **transverse process** and the **centra** on the column side and 10 of them to the **sternum** directly or indirectly on the ventral side. These together form the **thoracic basket,** a convenient enclosure for the vital organs located in the chest region (Fig. 15–7). It is interesting to note that the number of ribs is not always 12. The millions of chest X-rays taken of soldiers brought to light the fact that there is considerable variation, ranging from 11 to 13, the latter number being the most common variation. Incidentally, the gorilla also possesses 13 ribs.

The Appendicular Skeleton

The remainder of the skeleton, consisting of the appendages and their supports or girdles, is called the appendicular skeleton (Fig. 15–7). The **pectoral girdle** to which the arms are attached is located in the anterior region. It consists of two **clavicles** (collar bones) and two **scapulas** (shoulder blades); taken together they form a triangular brace with the arm hanging at the apex. Clavicles are rudimentary or absent in most mammals, but in the primates they are large, functional bones. Life in the trees, where brachiation was responsible for the development of not only the clavicle but also the nerves and muscles of the arms which make them such useful appendages today, is responsible for this difference. While the

Fig. 15–6. The spinal column is composed of interlocking vertebrate that, taken together, form a sturdy, flexible support for the entire body. The large openings in the vertebrae form a bony canal in which the delicate spinal cord is housed. Between the vertebrae are paired openings through which the spinal nerves pass.

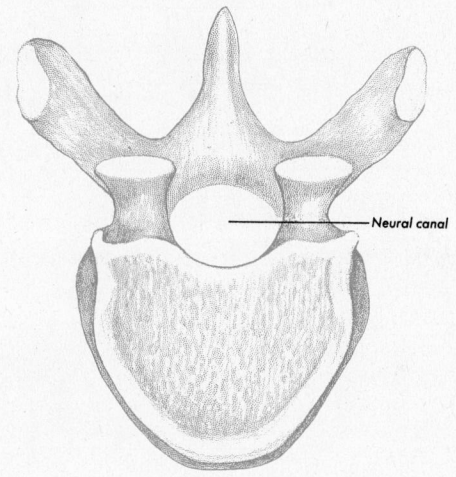

Neural canal

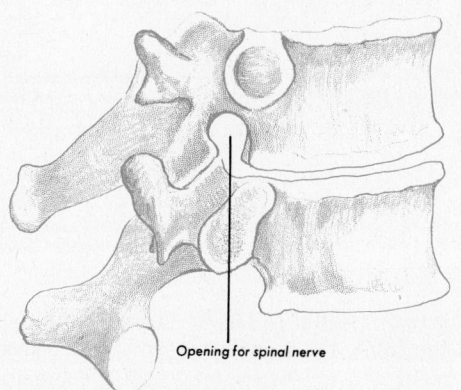

Opening for spinal nerve

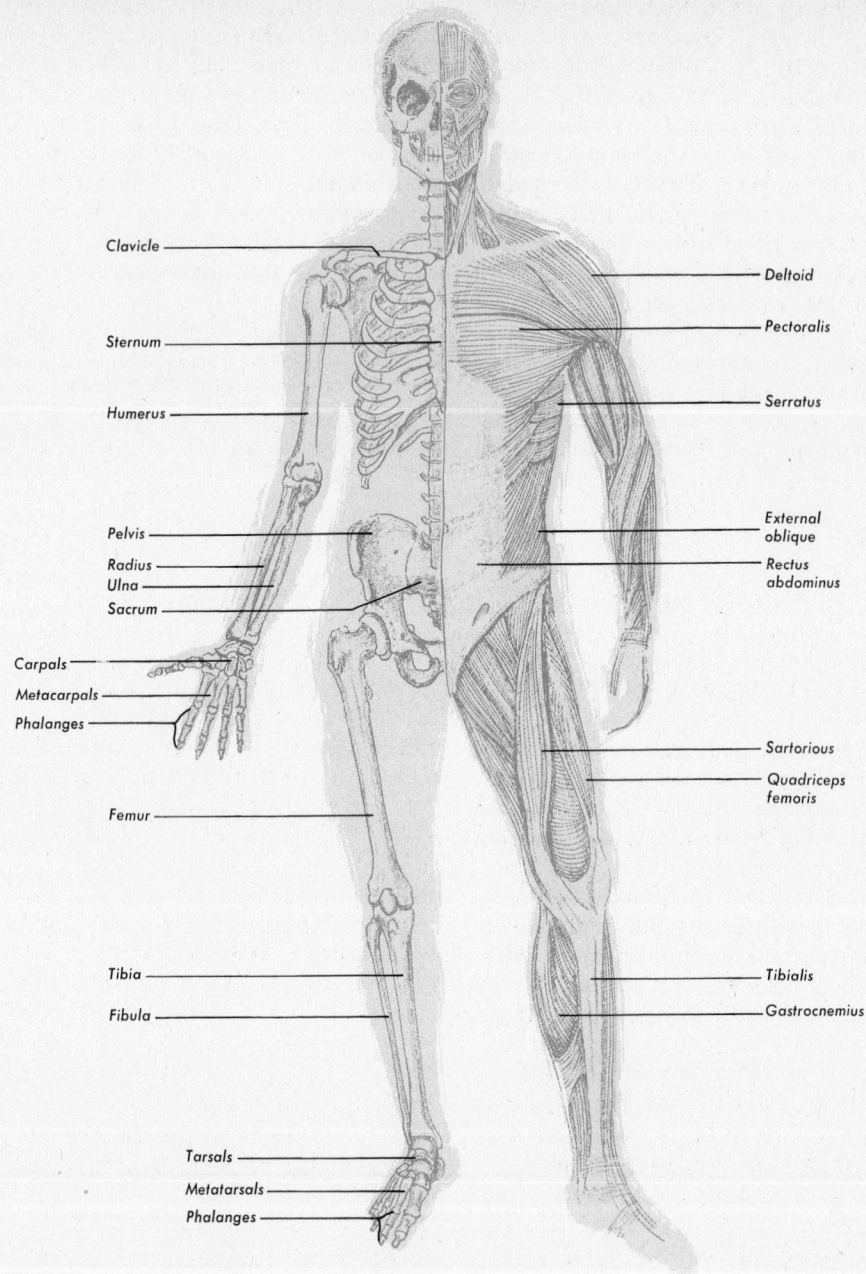

Clavicle

Sternum

Humerus

Pelvis
Radius
Ulna
Sacrum

Carpals
Metacarpals
Phalanges

Femur

Tibia

Fibula

Tarsals
Metatarsals
Phalanges

Deltoid

Pectoralis

Serratus

External
oblique
Rectus
abdominus

Sartorious
Quadriceps
femoris

Tibialis
Gastrocnemius

Fig. 15–7. The human skeletal and muscular systems, front view.

clavicle is firmly attached to the sternum on the front, the scapula has no secure attachment and is loosely slung over the thoracic basket by means of muscles and ligaments. This arrangement permits a great deal of movement in which the shoulders can be freely rolled over the ribs. The anterior appendages have much more freedom of movement than the posterior appendages, whose primary function is locomotion.

The upper arm, the **humerus,** fits into a crude socket made by the union of the scapula and clavicle called the **glenoid cavity.** The humerus is held in place by ligaments at its upper end, but since the attachment is none too secure, under certain stresses it may be forced out of the socket, resulting in a dislocation. Such stretched ligaments allow dislocation more readily under similar subsequent stresses. The advantage of this junction lies in its loose arrangement that allows more freedom of movement for the arm. For example, the arm may be turned in a complete circle as well as rotated in the socket. A dog, on the other hand, could not possibly perform such a feat, for the arrangement of the bones in its pectoral girdle is much more rigid.

The two forearm bones, the **radius** and **ulna,** form a combination whereby hinge action as well as partial rotation can take place. This means that the forearm can be flexed (bent on the upper arm) in a straight pull or it can twist through 180 degrees. At the wrist another hinge is produced by the end of the radius and the **carpals,** the small wrist bones. Actually, this is as much a universal joint as it is a hinge, with the result that the hand can move in all directions with equal facility. The hand with its large, opposable thumb is a primitive but most useful instrument, and it is hard to imagine life as it is lived today without it.

The **pelvic** girdle is the most specialized part of the entire skeleton. A quadruped, running on all fours, does not require as secure an attachment to the column as does a biped, whose pelvis has become correspondingly modified. However, in man the pelvis not only has become an excellent support for the entire body but it has also broadened and flared out so that it functions as a support for the organs of the abdominal cavity.

The pelvis is composed of three pairs of fused bones: the large, flat and cupped **ilia** (singular, **ilium**), the **ischia** (singular, **ischium**), the bones used in sitting, and the **pubic bones** which complete the girdle in front. The fused vertebrae of the sacrum form a complete circle at the back, leaving a large opening through which all mammal offspring must pass in the process of birth. The urinary and digestive tracts pass through here also. The dimensions of this opening are one of the clues used in determining the sex of a skeleton. Not only is the opening larger in females but, in addition, the attachment of the pubic bones is not so broad. Both features are essential to allow such a large object as a fetus to pass through. The ilia also flare outward more abruptly in the female than in the male; this changes the position of the legs somewhat so that the method of walking and running differs in the two sexes.

The **femur** is the longest bone in the skeleton. Its proximal end (end nearer the body) is a pronounced ball which lies at an angle to the rest of the bone and which fits into a deep socket in the pelvis called the **acetabulum.** This is a much more secure arrangement than the one in the shoulder region, although it does not have equivalent freedom of movement. For this reason the hip joint is not nearly so apt to dislocate as the shoulder joint. The leg can circumscribe a narrow cone but not a wide one like the arm. The leg is primarily concerned with the business of carrying the body forward in progression and consequently is constructed to function essentially in a forward and backward motion.

At the distal end (end farther from the body) the femur flattens out, forming a hinge with one of the two lower leg bones, the **tibia** or shin bone. The other lower leg bone is the **fibula,** which is smaller and lies on the outside of the leg. Together with the tibia it affords a point of contact, in turn, with one of the two large ankle bones. The other forms the heel. These two, together with the five other tarsal bones, the **metatarsals** and the **phalanges,** form the foot.

There are two arches in the foot, **longitudinal** and **transverse,** which are primarily supported by stretched tendons that come from muscles in the lower leg. Being always under tension, they possess a resilience that puts a "spring in one's step" and they also take away the shock from sudden contact with the substratum.

The Composition of Bone

If a long bone like a femur is cut in cross-section, it will be found to be hollow with a soft spongy material, the **marrow,** occupying the cavity (Fig. 4–4). The outer portion is very hard and resists breaking. The tubular nature of the bone makes it even stronger than a solid piece of equal weight. The hard part of bone is composed of calcium carbonate, or lime, and potassium phosphate, as well as an organic matrix which resembles cartilage. This can easily be demonstrated by placing the bone in an acid solution which dissolves out the minerals, leaving the matrix. Although the bone retains its original shape, it is very soft and pliable and as such could certainly be of no use to an animal. On the other hand, the organic matrix can be removed by heating the bone for some time so that only the minerals are left. Such a bone also retains its original shape but if disturbed crumbles into ashes. Again a bone of this composition would be of no use to an animal. Minerals and matrix taken together, then, are necessary to produce satisfactory material from which to construct skeletal units.

Bone Growth

It is obvious that the bones of a child, while fully formed and quite solid, must increase in both length and diameter as growth occurs. This is accomplished by a rather elaborate bone-destroying and bone-building process going on within the bone itself. The bone is covered on the outside by a thin cellular membrane, the **periosteum,** which has to do with the increase in the diameter of the bone. At the ends, called the **epiphyses** (singular, **epiphysis**), there is also active cellular growth which causes the increase in length. As bone is produced by both periosteum and epiphyseal cells, a simultaneous bone destruction is going on within the marrow cavity. In other words, as the bone cells produce bone on the outside and at the ends of the bone, similar cells are destroying bone on the inside. Thus the bone gradually becomes longer and increases in diameter.

Although bone may seem dead, it is far from it, as was pointed out earlier (p. 103). The **Haversian system** (Fig. 4–4) consists of a canal in the center containing blood vessels and a nerve, surrounded by concentric rings of bony matrix, and between them scattered tiny spaces, **lacunae,** filled with the bone cells. Very tiny tubes (**canaliculi**) connect the bone cells with one another and the central canal, and it is through these canals that the cells are nourished and kept alive. These bone cells secrete the bony matrix in which they are entombed. It is as if a mason were to surround himself with a concrete wall of his own building and thus be enclosed in a chamber which he could never leave, but in which he would be kept alive by small portals through which nourishment could be supplied.

THE MUSCULAR SYSTEM

One of the most striking characteristics of animals is movement. Since they are voracious feeders they must be on the move most of the time in search of food. Among all but the protozoa and perhaps a few others, contracting muscles are responsible for movement, not only of the body as a whole and its external appendages but of the internal organs as well, such as the organs of digestion and circulation. It is not surprising, therefore, that a man's body has more than 600 separate muscles.

Muscle Structure

Muscle is the principal part of the meat that is bought at the market and it usually makes up about 40–50 percent of the body weight of large animals. Viewed with the naked eye, muscle is seen encased in a sheath of connective tissue which often glistens.

The color of the muscle may vary with the nature of the fibers and with age; young mammals such as calves have lighter muscle tissue than older beef. Finally, muscles of the viscera possess a different texture than those of the skeleton.

The microanatomy of muscle tissue was described in Chapter 4. **Smooth** or **involuntary** muscle is located in the organs of digestion and in the skin, as well as in other places. They are concerned with those movements which are not directly under voluntary control, such as peristaltic movements of the digestive tract. These muscles are slow to respond to stimuli and the response that eventually occurs is of long duration. If one pricks the intestine of a frog with a sharp needle it requires from 1 to 10 seconds before any reaction is noted, but once the contraction starts it lasts for a minute or two, clearly demonstrating the character of smooth muscle action.

The **skeletal muscle** is composed of numerous fibrils (tiny fibers) suspended in the more fluid protoplasm, the **sarcoplasm.** Differences in the relative amounts of sarcoplasm and fibrils make a difference in the appearance of voluntary muscle tissue. Muscle fibers that contain a great many fibrils and relatively little sarcoplasm are light in color, and when the proportion is reversed the muscles are dark. In birds such as ducks, where sustained flight for long periods of time is essential, the breast muscle fibers contain more sarcoplasm and are therefore red, whereas the breast muscles of the domestic chicken which flies only short distances, if at all, are white.

Cardiac muscle functions as a unit because of the nature of its cells (p. 104). As a result of its sustained action, it is dark in color, as one might expect.

Muscle Action

The muscle responds like a rubber band; it can do only one positive thing and that is **contract.** When it is not contracted it is said to be **relaxed.** The function of a muscle, then, is to pull two objects closer together. This means that there must be muscles which pull bones in one direction and those which pull the same bones back again (Fig. 15–8). Muscles working against one another are said to be **antagonists.** For example, by contraction of the large muscle in the front of the upper leg, the bent leg straightens, as in kicking a ball (Fig. 15–7). Once the leg is straight it must be bent again before another step or kick can be executed, and several large muscles on the back side of the leg carry out this movement. To be sure, there is no complete relaxation of one set of muscles during the contraction of their antagonists. Both contract, the resultant action depending on how much each contracts. When bones are bent on one another the action is spoken of as **flexion;** when they are straightened out the action described as **extension.** The example of kicking is a case of flexion and extension of the leg bones. Although there are many other types of muscle action, antagonistic action is the most common in the animal body. Antagonistic muscles are not equally matched in strength. For example, the muscle which raises the jaw is stronger than the one that lowers it. Hence, when both contract violently as they do in convulsions, the jaw is closed tightly (lockjaw).

Muscles vary considerably in size and shape, some being long and fusiform, whereas others are thin and flat (Fig. 15–7). Most of them have a fleshy middle or **belly** part and two tapering ends which terminate in round or flat cords called **tendons.** Tendons consist of tough, fibrous tissue which attaches the muscle to the bone. The two ends of the muscle are identified by the amount of movement that takes place in the bones to which the tendons are attached. The end which moves the bone the greater distance is called the **insertion;** the end which moves the bone the shorter distance is the **origin.** Thus the biceps brachii muscle (Fig. 15–8) has its origin on a point of the scapula and its insertion on the radius, because the latter bone moves the greater distance when contraction occurs.

Tendons act like cables, attaching a muscle to a bone sometimes at a considerable distance from the muscle. This is a very

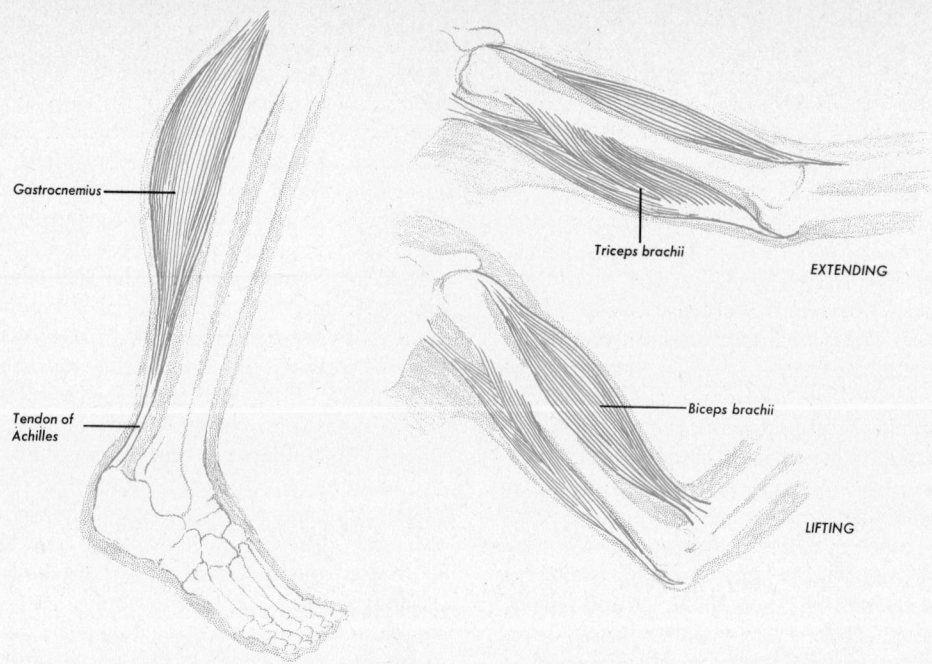

Gastrocnemius

Triceps brachii

EXTENDING

Tendon of
Achilles

Biceps brachii

LIFTING

Fig. 15–8. Muscle action in the human arm and leg. In the upper right figure, the triceps brachii muscle contracts in extending or straightening the arm while the biceps brachii relaxes. In the lower right figure, the opposite action occurs, that is, the triceps brachii relaxes and the biceps brachii contracts. This flexes or bends the arm as in lifting. To rise on the toes as in walking the large gastrocnemius contracts (left).

convenient arrangement because it makes possible the location of muscles some distance from the point where action must occur. For example, the muscles that support and operate the foot are located in the lower leg. The large tendon at the heel, the tendon of Achilles (Fig. 15–8), is like a steel cable when one is standing, particularly if on one's toes. This trend to longer tendons and concentration of the muscle's action farther from the muscle is most handsomely illustrated in the leg of the deer. Its lower leg is little more than skin, bones, and tendons, yet all the power of the strong leg muscles is transmitted efficiently to the tiny digits that contact the ground.

Muscles normally contract as a result of impulses coming to them through nerves. However, an isolated muscle can be made to contract if stimulated directly by an electrical current, even though all the nerves have been destroyed. The nature of the con-

traction can be studied by attaching the muscle to a recording device (Fig. 15–9) and noting its action following stimulation. When the muscle first receives a very brief stimulus there is no visible evidence of anything happening. This period is known as the **latent period** (Fig. 15–10), and lasts about 0.01 second in the frog muscle. Contraction then begins and continues for 0.04 second. This is immediately followed by a relaxation period that lasts 0.05 second, during which time there is a chemical readjustment taking place in the muscle (discussed below). If successive stimuli are increased in their frequency, there will come a time when the contractions will be superimposed upon one another until there is a sustained contraction which is greater than any derived from single stimuli (Fig. 15–11). Such a condition is called **tetanus,** and is what usually occurs in most muscular contractions, however short.

If a stimulus is given to an isolated frog

heart muscle, contraction occurs, provided the stimulus is sufficient to initiate a response. No matter how much the stimulus is increased, the resulting contraction remains the same. This fact has led to the establishment of the **all or none principle,** which means simply that if the heart muscle contracts at all, it will do so to its greatest extent. The question arises as to whether or not this applies to skeletal muscle. Obviously such muscles contract in graded amounts, because one can contract any of his muscles as much or as little as he likes. Here the principle does not apply to whole muscle but to individual fibers or to **motor units** (about 100 fibers). Although there still seems to be some question about it, the available evidence points to the fact that motor units do obey the "all or none" principle. Hence, the force with which a muscle contracts depends on how many motor units are stimulated. A mild contraction would result when only a very few are stimulated; a maximal contraction, when all of the units receive a stimulus.

Chemistry of Muscle Action

The mystery of muscle contraction is gradually being solved although the entire story is still not clearly understood. As a result of the efforts of physiologists, biochemists, and cytologists, particularly those working with the electron microscope, we are obtaining a clearer picture of what happens when a skeletal muscle contracts and relaxes.

It was once thought that the chemistry involved in muscle contraction was rather simple, because when a frog leg muscle (usually the sartorius) was stimulated continuously lactic acid accumulated, which subsequently was oxidized to carbon dioxide and water. Since glycogen disappeared simultaneously from the muscle, it was thought to be the source of energy which brought about the contraction. Further experimentation, however, demonstrated that if glycogen was prevented from breaking down into lactic acid (this can be done by using the specific poison, iodoacetic acid), the muscle continued to contract. In addition, it was discovered that the muscle would contract with the same force if denied oxygen as it would with oxygen present; in other words, it functioned **anaerobically.** What then, was the source of the energy?

Muscle must possess some other substances that released energy in a manner resembling oxidation. A diligent search revealed the presence of a phosphate-containing compound, **creatine phosphate** (CP), which turned out to be energy-rich very much like the phosphate in adenosine triphosphate (ATP), a compound with which

Fig. 15–9. Muscle action can be studied by attaching an isolated muscle, such as the gastrocnemius muscle of the frog, to a lever which can scratch a line of its path on a smoked moving drum (kymograph). When the muscle is electrically stimulated, the nature of the contraction can be recorded on the smoked drum.

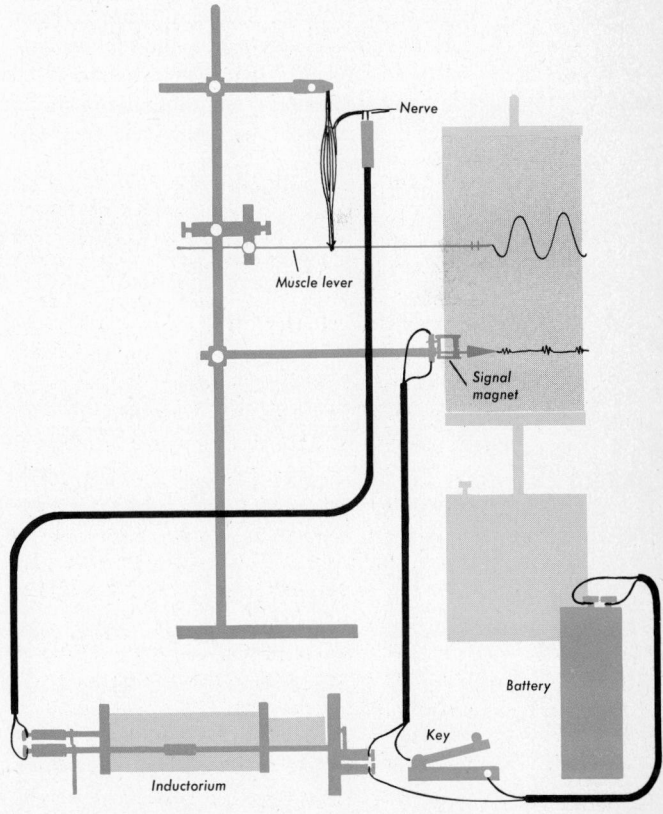

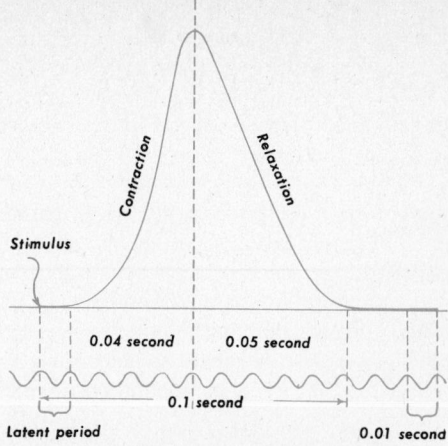

Fig. 15–10. This record was made when the frog gastrocnemius muscle contracted and relaxed, using a recording device as shown in Fig. 15–9. Note the time required for each event to occur.

we are already familiar (p. 87). Both compounds result from glycolysis and the Krebs citric acid cycle and both liberate a great deal of quick energy when their phosphate bonds are broken. This then turns out to be the source of energy for muscular contraction. It is now known, however, that CP acts as a reservoir of high energy phosphate and that the immediate source of energy for muscle contraction is ATP. During glycolysis ATP is generated and when it is in excess

Fig. 15–11. This record shows that by applying stimuli to a skeletal muscle with gradually increasing frequency, contractions merge until there is a sustained contraction called *tetanus*. The contraction is stronger in tetanus than in the single contractions.

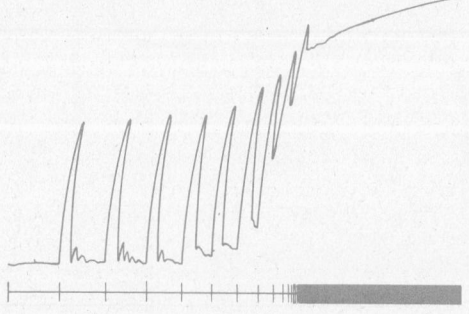

the high energy phosphate is transferred to creatine forming CP, which is stored for use when ATP is exhausted following continued muscle contraction. The CP then comes into action in the following manner:

$$CP + ADP \rightleftharpoons ATP + C$$

thus replenishing the supply of ATP. We see here, then, a very efficient system of chemical reactions which makes possible the sudden expenditure of large amounts of energy without going through the relatively slow process of oxidation. This remarkable achievement in chemical evolution accounts for the swift movement of animals.

In spite of our rather extensive knowledge of the chemical events that occur during muscle contraction we know very little about the mechanism involved in the conversion of the chemical energy in ATP to mechanical work. Obviously it is somehow tied up with the muscle proteins themselves. We do have considerable information about these proteins. Located in the muscle fibers are two proteins, **myosin** and **actin,** which are involved in the actual shortening of the fiber. Both are essential; neither one nor the other can function alone. These have been isolated in test tubes and when they are mixed they form a complex, **actinomyosin,** which can be made into "synthetic" fibers. When ATP is added to these fibers they contract the same as muscle, simultaneously decomposing the ATP to ADP. This experiment, accomplished by A. Szent-Györgyi, is one of the milestones in further understanding the chemistry of muscle contraction.

Adenosine triphosphate supplies the driving force for contraction by its capacity to dissociate actin and myosin. In the presence of ATP, therefore, the two proteins are separated; with the arrival of the nerve impulse the enzymes which convert ATP to ADP (ATPase system) go into action and bring about the conversion. With the destruction of ATP, actin and myosin combine or re-associate which results in simultaneous contraction of the individual fibers.

As was pointed out earlier, we still do not know how the chemical energy in ATP is converted to mechanical work.

We may sum up the chemical steps that take place during muscular contraction and relaxation as follows:

Contraction Phase
ATP → ADP + ~P (high-energy phosphate)

Relaxation or Recovery Phase
Glycogen → lactic acid + ~P (glycolysis)
Lactic acid + O_2 → CO_2 + H_2O + ~P
(Krebs cycle)
ADP + ~P → ATP
~P → Body heat
Excess ~P stored as CP

From this review it becomes clear that during contraction the only reaction is the conversion of ATP to ADP, releasing energy in a sudden but controlled manner. This is an anaerobic reaction, which accounts for the fact that a man can run a hundred yards without taking a breath. When the ATP is exhausted CP comes into action converting ADP to ATP. With continued activity, naturally this reserve supply also is depleted, and when that happens no further contraction can occur until time has elapsed for recovery, or partial recovery, to occur. A person exhausted in this manner must remain quiet while breathing rapidly and deeply to supply sufficient oxygen to permit the next reactions to occur. Actually the ATP and CP are never completely used up in the body; indeed, the moment contraction is initiated, the recovery process starts and these compounds are being restored all during contraction. In recovery glycolysis occurs, during which glycogen is converted to lactic acid, releasing a small amount of ~P. Lactic acid is then oxidized via the Krebs citric acid cycle to CO_2 and H_2O, yielding large quantities of ~P. This step requires large quantities of oxygen, hence the deep breathing following severe exercise (or during, if prolonged). The energy released from this reaction is utilized in several ways; part of it is used to convert ADP to ATP, part to

convert lactic acid back to glycogen, part is converted to heat to keep the body warm, and that which is left over is stored as creatine phosphate. The over-all process results in the most economical method of obtaining the greatest possible amount of energy from glycogen. It means that the animal body is unusually efficient, about 40 percent of the available energy being released in the form of mechanical work, 60 percent as heat. This is a very satisfactory figure when compared with internal combustion engines which rarely exceed 25 percent.

Referring once again to the runner, we know now that during the entire hundred-yard dash his muscles required very little oxygen because only ATP was required, the expenditure of energy being of short duration. Once the race ended he remained quiet and breathed deeply for some time to replenish his ATP. During the run he was building up an **oxygen debt,** which he "paid back" during the heavy breathing at the termination of the race. The obvious advantage of such an arrangement is that muscles are ready to contract with all of their force almost instantly. They contract until their store of ATP has been nearly exhausted; then they must stop until the blood brings sufficient oxygen to oxidize the accumulated lactic acid which releases the energy required to restore glycogen and ATP to their original levels.

Prolonged muscular activity results in **fatigue** which can be painful. From the foregoing discussion it would seem that fatigue results from the accumulation of lactic acid and the exhaustion of ATP and glycogen in the muscles. This is not quite so because in the intact animal fatigue sets in before there is an appreciable amount of lactic acid present in the muscles. Some of the lactic acid slowly diffuses into the blood, passes to the liver where it becomes liver glycogen, and then returns to the muscle and other tissues as blood glucose. The discomfort of fatigue probably is due to a combination of factors. Experimental evidence seems to demonstrate that the site most susceptible to fatigue is the junction between the muscle and nerve, and not the muscle or nerve itself.

The limitations for work are set by the ability of the body to restore exhausted organic compounds in the muscles; this depends indirectly on the functioning of the respiratory, circulatory, and excretory systems which, in turn, the muscles must depend on to receive their quota of burning material (sugar and oxygen) and to carry away their accumulated wastes (urea and CO_2). There are great individual differences among human beings for this capacity. Some get along well with very little sleep—Edison was such an example—whereas others require eight hours or more per day.

Muscles can be developed to considerable

size and strength if they are constantly put to difficult tasks. By lifting heavy weights each day the muscles of the entire body grow disproportionately large and function very well in lifting heavy objects. If this is to be the life work of the individual it is wise to have such a set of muscles, just as it is wise for the man who handles a shovel all day long to have thick calluses on his palm. Modern living requires minimal development of one's muscles, hence recreational activities are helpful in compensating for this deficiency. Man evolved as an active animal and it follows that a certain amount of muscular activity is conducive to good health.

The Mechanism of Muscle Action

Just what morphological changes occur during muscle contraction has come to light only recently as a result of electron microscope studies principally by H. E. Huxley. Whereas it was once thought that the shortening of the muscle fiber was caused by the folding of the actomyosin filaments, it appears now that this is not so. In order to understand the present thinking in regard to the mechanism of contraction let us interpret the morphology of the muscle fiber which has long been known from light microscopy.

Each striated muscle fiber consists of many tiny striated fibrils, the composite giving the typical striated appearance observed under both the light and electron microscopes (Figs. 15–12, 15–13). There is a regular alternation of dense (**A bands**) and light (**I bands**) bands. The A bands also have a less dense region called the **H zone** and the I bands are transected by a dense narrow line, the **Z line**. The entire region between two Z lines is the **sarcomere** (functional unit of myofibril). Under the electron microscope the nature of these structures becomes clear.

Each myofibril consists of two kinds of filaments; thick ones which turn out to be myosin, and thin ones which are actin. It is their parallel and overlapping arrangement that causes the striations. The actin filaments account for the I band; the myosin filaments

Fig. 15–12. These are photographs of rabbit skeletal muscle taken through the electron microscope. The top picture is a longitudinal section of the fibers showing the numerous bands characteristic of voluntary muscle. Note the numerous tiny fibrils that make up each fiber. The fibers are shown in cross section in the bottom photograph. Note the regular spacing of the fibrils. It is from such studies that our knowledge of muscular action is continually increased.

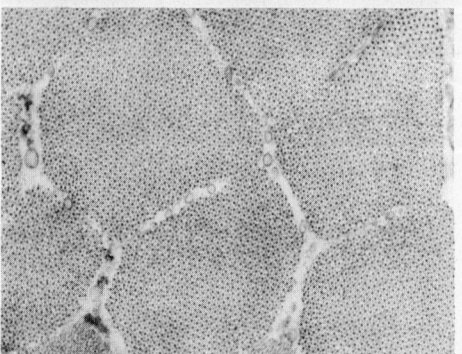

MUSCLE

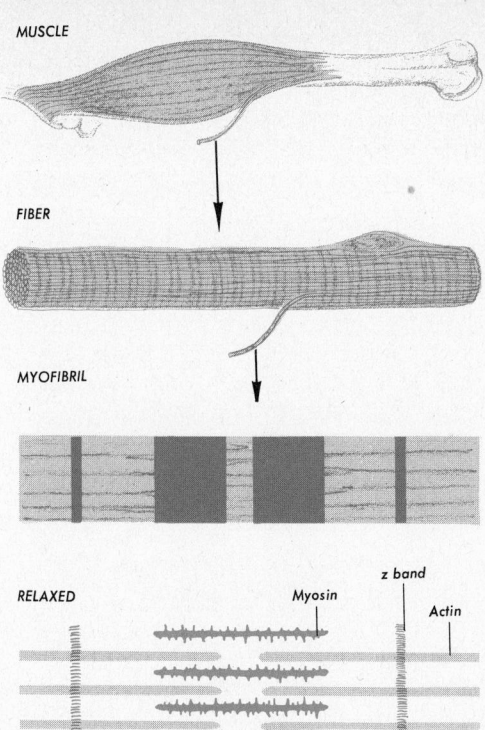

FIBER

MYOFIBRIL

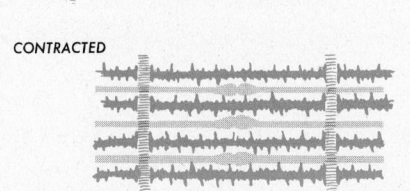

RELAXED

Myosin

z band

Actin

CONTRACTED

Fig. 15–13. This series of sketches portrays skeletal muscle examined from the miscroscopic level to that of the electron microscope. The muscle is composed of many fibers, each of which is made up on numerous myofibrils. The nature of the alternate light and dark bands, easily observed with the light microscope, can be understood from electron micrographs. The protein myosin is arranged in thick frayed fibrils which overlap with the thinner smooth actin fibrils causing the appearance of the dark bands. The actin alone constitutes the light bands. The Z lines mark the boundaries of the contractile unit. When the muscle contracts the actin and myosin fibrils slide past one another until the former touch one.another and the latter touch the Z lines (bottom sketch.) This results in a shortening of the myofibrils and consequently of the whole muscle.

account for the A band; and the region in the A band where the actin filaments fail to meet results in the H zone.

During contraction the length of the A band remains constant and the length of the I band decreases. The actin filaments meet which eliminates the H zone. As a result, the Z lines come closer to one another and the entire fiber shortens. Obviously the actin and myosin filaments slide past one another during contraction, but just how is this accomplished?

Electron micrographs of cross-sections of striated muscle fibrils (Fig. 15–12) show additional structures. The filaments are arranged in a symmetrical pattern, each actin filament being surrounded by three myosin filaments. Each myosin filament contains along its length numerous extensions which act as cross-bridges that unite with specific sites on the actin filaments. By moving back and forth they would cause the actin filaments to move short distances; repetitive oscillations would bring about greater movement of the actin filaments. This type of movement is visualized as ratchetlike. Huxley has been able to separate myosin and actin filaments and then reassemble them into a configuration resembling the original unseparated filaments. He has been able to confirm his original hypothesis of the ratchetlike mechanism of contraction at the molecular level.

During contraction and relaxation of muscle fibers, which usually take place very rapidly, a great deal of metabolic activity must be accounted for on morphological grounds. This has been done by Porter and Franzini-Armstrong in a careful ultrastructural study of striated muscle. They have identified two types of tubular systems which take care of the flow of raw materials into the muscle fiber and metabolic waste products away into the blood stream. One type is called the **sarcoplasmic reticulum** (its counterpart, the endoplasmic reticulum in all cells) which is a system of tiny tubules running parallel to the fibrils with cross connections in the region of the Z lines and H bands. The other tubular system, called the **T system**, consists of invaginations of the sarcolemma (membrane covering the fiber) at the level of each Z line. These two systems provide channels for the flow of materials in and out of the muscle.

Whereas the morphology of skeletal muscle is well known today there is much left to learn about the biochemistry of muscle contraction. Moreover, little is known about the mechanism of smooth and cardiac muscle contraction, nor is the nature of invertebrate muscle contraction understood.

SUGGESTED SUPPLEMENTARY READINGS

Books

BECK, W. S., *Human Design*. New York: Harcourt, Brace Jovanovich, 1971.

BLOOM, W., and FAWCETT, D. W., *A Textbook of Histology*, 9th ed. Philadelphia: Saunders, 1968.

GUYTON, A. C., *Textbook of Medical Physiology*, 4th ed. Philadelphia: Saunders, 1971.

MARSHALL, P. T., and HUGHES, G. M., *The Physiology of Mammals and Other Vertebrates*. Cambridge, England: Cambridge University Press, 1965.

MONTAGNA, W., *The Structure and Function of Skin*, 2nd ed. New York: Academic Press, 1962.

RHODIN, J. A. G., *Histology: A Text and Atlas*. New York: Oxford University Press, 1974.

ROMER, A. S., *The Vertebrate Body*, short version, 4th ed. Philadelphia: Saunders, 1971.

SCHMIDT-NIELSON, K., *How Animals Work*. Cambridge, England: Cambridge University Press, 1972.

THOMPSON, Sir D. W., *On Growth and Form*. New York: Macmillan, 1942.

WOODBURNE, R. T., *Essentials of Human Anatomy*. New York: Oxford University Press, 1961.

Articles

HAYASHI, T., "How Cells Move." *Scientific American*, September, 1961.

HUXLEY, H. E., "The Contraction of Muscle. *Scientific American*, November, 1966.

MURRAY, J. M., and WEBER, A., "The Cooperative Action of Muscle Proteins." *Scientific American*, February, 1974.

PORTER, K. R. and FRANZINI-ARMSTRONG, C., "The Sarcoplasmic Reticulum." *Scientific American*, March, 1965.

SZENT-GYÖRGYI, A., "Muscle Research." *Science*, 1958.

16

COORDINATION

Coordination is obviously a fundamental necessity for all animals, because even protozoa have some method of coordinating their separate parts. The growth and organization of the individual may be likened to the expansion of a telephone system as a small village grows to a great city. As the latter increases in size, the system becomes more and more intricate until a telephone system like that of New York City results, which is about as difficult to understand as the nervous system of a grasshopper or a man.

Among higher invertebrates and all vertebrates the network of tiny fibers connecting all parts of the animal has apparently proven inadequate, because a supplementary system has evolved, namely, an **endocrine system.** In this system an entirely different principle is employed; instead of impulses passing over tiny fibers, specific chemicals produced by special glands are released into the blood

and circulate to other parts of the body where they produce a specific effect.

In studying coordination, we begin with a discussion of stimuli from the external world and the internal environment as received by the **sense organs** and other **receptors;** then we discuss the nervous system which is the intermediary for the transmission and interpretation of the stimuli, the components called the **adjustors;** and then we discuss the **effectors**—muscles and glands —which respond to the stimuli. We conclude with a discussion of the endocrine system.

RECEPTORS

The receptors consist of specialized end-organs which are highly sensitive to certain kinds of stimuli. Conspicuous sense organs such as the eye and ear are familiar to everyone; others, such as the tiny receptors located in various parts of the internal body, are not so well known but are just as important in the proper coordination of the organism.

Skin Receptors

It might be expected that the outer covering of the body would be highly sensitive to the environment around it, and this is true. A pin prick in the skin almost anywhere on the surface of the body results in a pain sensation; this fact indicates that these nerve endings are very numerous and widespread. The same is true of the nerve endings for touch, pressure, heat, and cold. A thin section of the human skin will reveal tiny, oval-shaped **tactile corpuscles** from which nerves lead inward. Any pressure brought to bear on them causes impulses to be discharged from the specialized cells within the corpuscle which travel along nerve fibers to the central nervous system. Other kinds of sensory end-organs which respond to pressure stimuli over larger areas are located in the deeper skin and in many internal organs. Free nerve endings which register pain terminate in the epithelium within the internal organs as well as in the skin. The endings ramify and come into contact with nearly every cell, which explains why pain sensations are felt even if only a small area is stimulated, such as in pricking with a pin.

By marking off specified areas on the skin and using a stiff bristle as a stimulus the appropriate receptors can be located, and they will be found to be quite unevenly distributed over the body. It is difficult, for example, to distinguish two points one-half inch apart in the middle of the back, whereas on the tip of the finger or tongue a distance of one-sixteenth of an inch is perceptible. Likewise, if pointed metal instruments (styluses) are used, the hot and cold end-organs can be detected. There are more cold spots than hot spots and that is why, for instance, one shivers at first if suddenly exposed to a hot shower. When all of the end-organs are stimulated simultaneously, as would be the

case in the above situation, the total response is that of coldness at first because there are more of the cold than hot spots. Later, the proper interpretation of the stimulation is recognized. Pressure end-organs can be found by applying a blunt metal stylus having the same temperature as the skin to various regions. Other sensory nerve endings are located in the tendons and muscles and respond to tension placed on these tendons and muscles. These are important in balance and are discussed under that topic later.

Chemo-Receptors

All of the receptors of the skin have to do with identifying energy changes that occur at or very near the body. In addition, there are chemo-receptors that identify substances dissolved in the saliva of the mouth, giving one the sense of **taste,** and chemicals dissolved in the mucus of the nasal chambers, imparting the sense of **smell.** In the latter case, the organism is made aware of changes in its environment some distance away. The sense of smell is, in this respect, like the senses of hearing and seeing which extend perception to great distances.

Taste. The sense of taste is actually a combination of stimuli coming from the mouth cavity. Stimuli from the end-organs of taste, touch, heat, and cold located in various parts of the mouth cavity give a composite sensation which is called taste. The difference in the "taste" of hot and cold foods is due to stimuli other than those which are caused by dissolved chemicals. The sense of smell is also important to taste, as anyone with a bad cold is well aware. The end-organs of taste are called **taste buds** and are distributed over the surface of the tongue, laryngeal region, and parts of the roof of the mouth (Fig. 16–1). They are oval-shaped bodies made up of several cells which terminate in a slender sensory process on the end toward the mouth cavity.

There are four kinds of taste sensations and consequently there are four different kinds of taste buds, each with a rather specific distribution on the tongue and other mouth parts (Fig. 16–1). The taste buds registering **bitter** are located at the base of the tongue, **salt** and **sweet** on the tip, and **sour** along the edges. These can all be identified both microscopically and experimentally. No matter how these buds are stimulated the resultant sensation is always sweet, sour, salt, or bitter. Some chemicals stimulate two kinds of taste buds, but in each case the taste bud responds as it should according to its predetermined function. There are some classes of substances which have a consistent taste; for example, acidic substances usually taste sour, basic substances bitter. The threshold (a stimulus that is just sufficiently strong to elicit a response) is very low for the sense of both taste and smell. For example, it is possible to taste quinine in concentrations of one part in two million, and much greater dilutions of odorous substances can be smelled.

The taste of some substances varies widely with different individuals and just what taste registers has a genetic basis. For example, the chemical **phenylthiocarbamide** may taste sour to one person, bitter to another, salty to another, sweet to another, and fail to give a taste response at all to another. Taste is inherited in a definite pattern.

Smell. The olfactory end-organs which are responsible for the sense of smell are located in the nasal membranes, and it is through these organs that gaseous chemical stimuli (odors) are received. While in man the receptive area in the two nasal chambers is only about 10 square centimeters, in most mammals it is much more extensive. It will be recalled that the sense of smell is far more important to ground dwellers than to those that live in trees, where keen vision is of more value. It is not surprising, therefore, that when the primates took to the trees the sense of smell diminished and in the present primate is very poorly developed. The dog, on the other hand, receives a great deal of information about the world through its nose.

The olfactory cells give rise to fibers (Fig. 16–1) which coalesce, after passing through

Fig. 16–1. The end-organs for chemo-reception are located in the nasal chambers (smell) and on the tongue (taste). They are shown here in detail together with the nerve pathways that conduct the impulses to the brain.

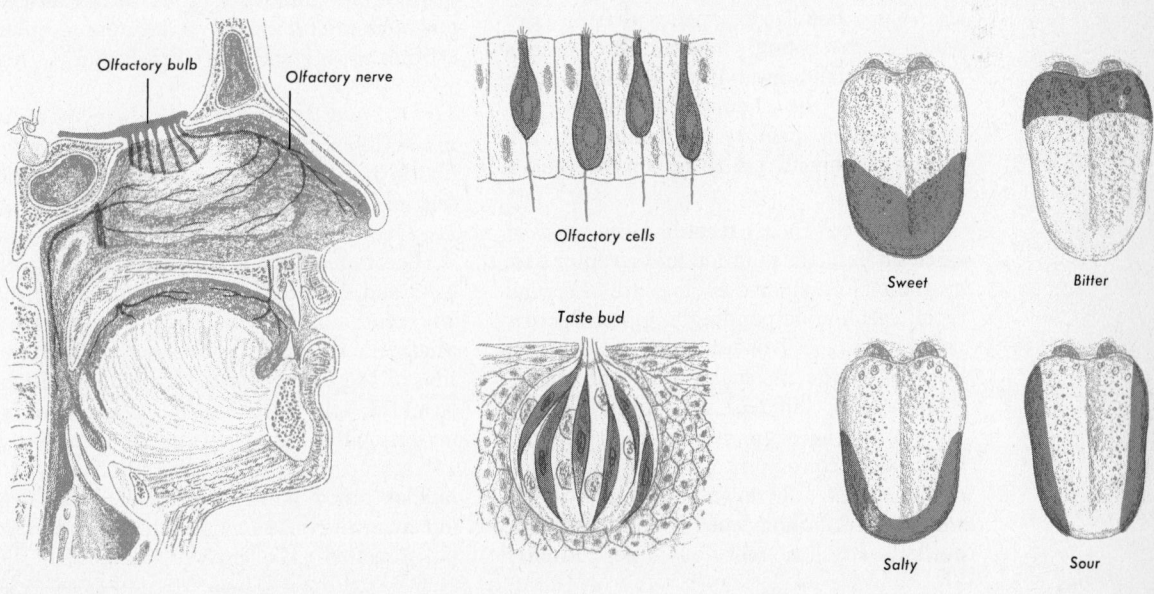

Olfactory bulb

Olfactory nerve

Olfactory cells

Taste bud

Sweet

Bitter

Salty

Sour

the cribriform plate, to form the **olfactory bulb,** which becomes a large nerve leading to the brain. The location of the olfactory end-organs is such as to protect them from the desiccating effects of incoming currents of air during breathing, and the nasal passages are kept continuously moist in order that the incoming odors may dissolve in the fluid bathing them.

It may be necessary for large quantities of air to pass over the receptors before stimulation is possible. As more air passes over them more of the chemical in gaseous form becomes dissolved in the fluid of the nose, thus increasing the concentration to a point where the threshold is exceeded. This accounts for the constant sniffing of the dog, or man, too, on occasion, to bring more air into contact with the nasal epithelium.

In spite of considerable research, our knowledge of the functioning of the sense of smell is still not clearly understood. Odors are referred to by the name of the aromatic substance in question, and there are nearly as many names for odors as there are aromatic compounds. It seems highly improbable that there is an infinite number of kinds of olfactory nerve endings, although when they become fatigued to one odor they seem to respond with normal vigor to another. In general, the nerve endings fatigue readily, and it is a familiar experience that an odor which is very strong when one first enters a room soon fades away until it is unnoticed, not because the chemical in the air has diminished in quantity but because the olfactory end-organs fail to respond beyond a brief period.

One recent theory regarding the sense of smell supports the notation that receptor sites on the olfactory nerve endings are of several types, each one responding to molecules of a particular shape. It is not the chemical composition of the molecules, only the shape; therefore very different substances have the same or similar odors. Five of the seven primary odors correspond with five basic molecular shapes. These are (with typical example): **camphoraceous** (moth balls); **musky** (angelica root oil); **pepperminty**

(mint); **floral** (rose); and **ethereal** (dry-cleaning fluid). The other two odors, **pungent** (vinegar) and **putrid** (rotten eggs), could not be correlated with specific shaped molecules. The fact that putrid molecules have an excess of electrons and pungent molecules a deficiency may mean that their odor-inducing stimuli are dependent on electrical charge rather than shape. Synthetic molecules of specific shapes have been found to have the odor that was predicted for them by their configuration. However, no receptor sites on the olfactory end-organs have been found with specific shapes to fit the different molecules.

Vision

This is the most perfect of the distance receptors. Nearly all animals from the simplest to the most complex are sensitive to light. Not that they all are sensitive to the same wave lengths that are recorded by the human eye, but nearly all have evolved some sort of receptor which is sensitive to light. It has been proved, for example, that the bee sees shorter light waves than we can see. Dogs, on the other hand, appear to be color blind. It is thus abundantly clear that different animals see, hear, smell, and so forth, quite differently from man. Indeed one must carefully avoid an anthropomorphic attitude with respect to all behavior.

The Human Eye. Although there are some minor differences in the eyes of various vertebrates, the human eye serves as representative of the group in our discussion (Fig. 16–2).

The human eye is extremely sensitive to light and registers color better than eyes of any other animal. It is necessary to consider briefly the way light behaves before the function of the various parts of the eye, particularly the lens (see also p. 621), can be understood.

Light consists of small particles or corpuscles called photons, each of which has one quantum of energy. Light, therefore, can be measured in quanta. For example, the

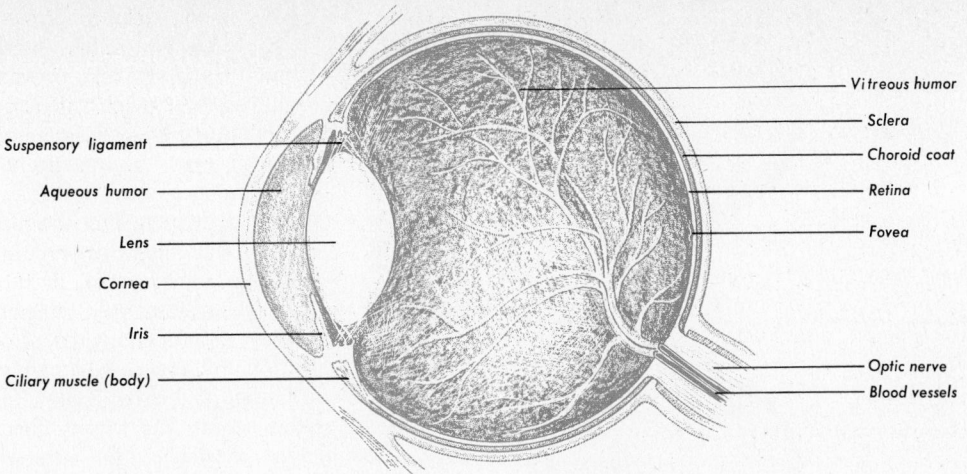

Suspensory ligament
Aqueous humor
Lens
Cornea
Iris
Ciliary muscle (body)

Vitreous humor
Sclera
Choroid coat
Retina
Fovea
Optic nerve
Blood vessels

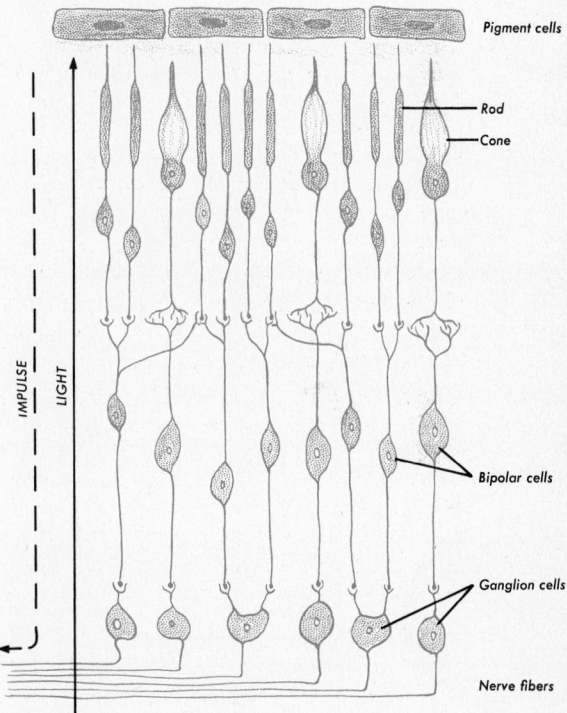

IMPULSE
LIGHT

Pigment cells
Rod
Cone
Bipolar cells
Ganglion cells
Nerve fibers

human eye can detect as little as 6 quanta of light which is equivalent to the light from a candle fourteen miles away. Light travels in straight lines at a speed of 186,000 miles per second in air, but it travels at different rates in other media such as water, glass, and transparent tissues such as the lens. Therefore, when light passes from one medium to another it bends (refracts). It is a familiar fact that when a stick lies at an angle partially in water, it appears bent. Actually, of course, the light is coming to the eye through two different media, water and air, and at different speeds, hence the bending at the juncture of the two media. Light coming through glass is bent in a similar fashion, and when the glass is shaped so that it is uniformly curved on both sides (convex lens) the rays of light are bent toward one another so that they come to a point, or **focus**. The amount of bending increases with the increase in angle between the light ray and the surface of the glass. Therefore, a highly curved surface will bring the rays to focus at a very short distance (**focal length**), whereas a more flattened surface will bring them together at a greater distance. Depending on conditions, therefore, a lens is said to have a long or short focal length. Light comes from an infinite number of points on the object and passes through the lens with the

Fig. 16–2. A longitudinal section of the human eye to show its internal structure (above) and the microscopic details of the retina (below).

result that the image is reversed. With this knowledge of light and convex lenses, let us study the human eye.

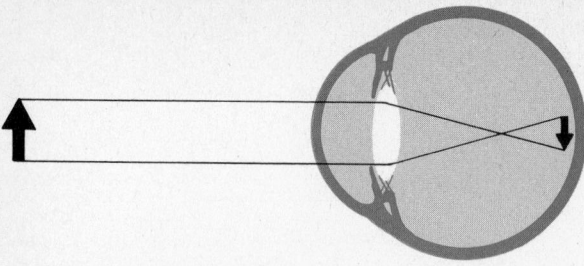

DISTANT VISION: lens thin

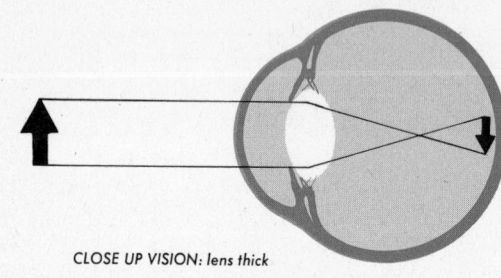

CLOSE UP VISION: lens thick

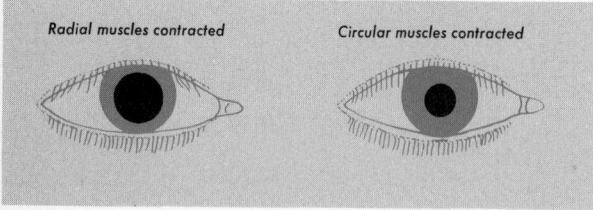

Radial muscles contracted   Circular muscles contracted

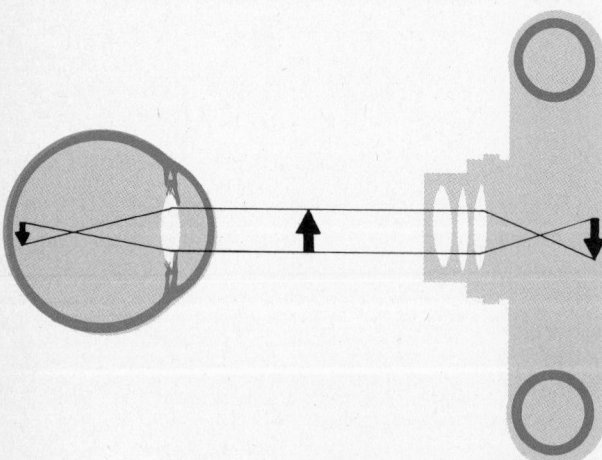

Fig. 16–3. Accommodation is accomplished by changing the thickness and curvature of the lens. The amount of light entering the eye is controlled by the iris diaphragm. A comparison of the camera and the human eye. Note the likenesses and differences. (For names of parts of the eye see Fig. 16–2.)

The remarkable similarity of a simple camera to the eye can help us to understand how we see (Fig. 16–3). Both have convex lenses which bring the rays of light to focus upon a sensitive plate, the film in the camera and the **retina** in the eye. The amount of light entering the chamber is controlled by the **iris diaphragm** in both cases. The housing is a lightproof case which allows the rays of light to pass through unobstructed; in the camera it is made of an adjustable tube or bellows and in the eye it is composed of a tough outer covering, the **sclera,** and a highly pigmented inner lining, the **choroid coat.** In the eye the space behind the lens is filled with a semiliquid substance, the **vitreous humor,** and the cavity in front of the lens is occupied by the **aqueous** (watery) **humor.** These maintain pressure within the eye and keep it from collapsing.

There are some very important differences between the camera and the eye. First of all, the film and the retina function differently. Once a picture is taken, the film is used up, that is, it must be replaced by another, whereas in the retina continuous series of pictures can be recorded without exhaustion of the sensitive cells. Secondly, the method of bringing objects at different distances into focus on the sensitive plate operates differently, although the end result is the same. In the camera, the lens maintains one shape and must be moved forward and backward until the image is in focus on the film. In the eye, however, the lens itself changes its shape, becoming more curved for close objects and less curved for distant objects (Fig. 16–3). This change in shape of the lens is called **accommodation.** Most of the light is brought to focus upon the retina by the **cornea;** only the slight differences that are needed to produce a sharp picture are accomplished by the lens itself. This slight variation, however, makes the difference between clear, sharp vision and poor, fuzzy sight. Accommodation is effected by the action of the **ciliary muscle,** located near the point of attachment of the **suspensory ligament** which supports the lens (Fig. 16–2). When this muscle contracts, pressure changes occur within the eyeball which cause the lens to become more curved, that is, thicker, which brings near objects into focus. When the muscle relaxes, the ligament tightens and the pressures are shifted so that the lens flattens out, causing distant objects to come to focus on the retina. Therefore, the ciliary muscle is active only when one is looking at close objects (under 30 feet), which is the reason why the eyes can be rested by looking out the window at a distant object.

The amount of light entering the eye is controlled by a sheet of circular muscular tissue, the **iris,** which contains the pigment granules responsible for eye color (Fig. 16–3). Both radial and circular muscles are present in the iris and it is the antagonistic action of these muscles that do the job of enlarging or constricting the opening, the **pupil.** They are under the control of the autonomic nervous system and therefore beyond voluntary control. When the eye is exposed to bright light, the circular muscles contract, constricting the pupil, whereas in dim light the radial muscles contract, causing dilation of the pupil. Thus a delicate arrangement is provided to project just the right amount of light on the retina to obtain the best possible picture reception.

Exactly how light rays are transformed into the nerve impulses that pass over the optic nerve to the brain is not understood. The conversion from light energy to nerve energy takes place in the retina, which is composed of numerous cells, some of which are sensitive to light. These are the **rods** (sensitive to white light) and the **cones** (sensitive to colors) (Fig. 16–2). There are about 125,000,000 rods and 4,000,000 cones in the human retina. The cones are crowded around a central region, the **fovea centralis,** where visual acuity is most pronounced. Elsewhere in the retina the cones are mixed with the rods and vision is not so clear.

The retina is so arranged that the light must traverse the layers of neurons (ganglion and bipolar cells) before it reaches the sensitive rods and cones. The latter are functionally interconnected by synapses and all finally leave the eye through one place to form the **optic nerve.** At this point of exit the retina is interrupted so that no images can be formed here and is therefore called the **blind spot.** In order to see clearly, it is necessary to look directly at the object so that the image falls on the fovea—all other vision is peripheral and is less clear. The rods are more sensitive to light than the cones, and detection of weak light sources is best made when looking to one side of the source. Fliers search for beams at night with their peripheral rather than foveal vision.

It is well known that when light falls on the rods in the retina chemical change takes place which is the basis for light reception. The rod cells contain a photosensitive purplish-red pigment, **visual purple** (rhodopsin), which, when struck by light, is bleached by a photochemical reaction forming the compound **retinene,** a derivative of Vitamin A and the protein **opsin.** The important outcome of this reaction is that energy is released which stimulates a neuron setting up a nerve impulse to the brain (Fig. 16–4). Visual purple is restored by the recombining of retinene and opsin with the aid of ATP. Vitamin A is readily converted into retinene by reduction of part of the molecule under the influence of an enzyme. Continued exposure to bright light sometimes causes visual purple to be bleached at a more rapid rate than it can be regenerated; if this occurs **snow blindness** results. A deficiency of Vitamin A may result in **night blindness,** that is, difficulty in seeing in dim light.

The exact nature of color vision is not well understood, although it is known that light brings about a chemical change in the pigment **iodopsin** of cone cells. It is known that cones in some reptiles and birds contain one of four different pigments: red, orange, yellow, and white. The pigment is in the form of a globule resting on the tip of the cone. All light passing through the globule is filtered out except that of the particular globule. The colored cones are distributed over the retina so that color images are possible. Color blindness may be a result of a reduction in certain of the cones.

Eye Defects. The most common defects of the eye are caused by the inability of the focusing mechanism (cornea and lens) to

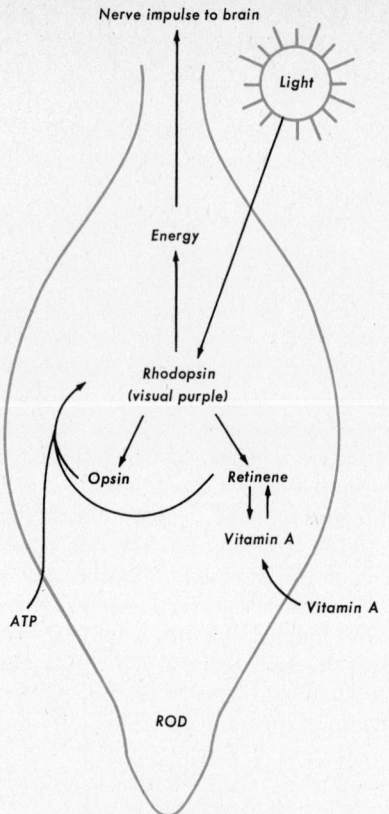

Fig. 16–4. The reactions that occur in a rod cell of the retina. See text for explanation.

form clear images on the retina. If the rays come to focus in front of the retina, the person suffers from **myopia**, or **nearsightedness**; if they come to focus behind the retina, the defect is known as **hyperopia** or **farsightedness** (Fig. 16–5). In order to correct these conditions it is only necessary to place in front of the eye a lens ground so that it bends the rays of light just enough to compensate for the defective focusing mechanism. In the case of nearsightedness a biconcave lens is needed, while in farsightedness a biconvex lens brings about the proper correction. When either the cornea or lens is irregular in its curvature, the defect is called **astigmatism**. This can be corrected by using lenses that are ground in such a manner as to compensate for these variations in curvature.

Sometimes, as a result of disease or for other reasons, either or both the lens and the cornea may become fogged over so that vision is dimmed or completely obliterated. If the lens becomes fogged, it can be removed and a substitute lens placed in front of the eye in the form either of conventional glasses or of contact lenses which fit tightly against the eyeball. If the cornea clouds over it can be replaced by one from a normal eye, restoring the vision to normal. This is an extremely delicate operation but one that is

Fig. 16–5. Common human eye defects and how they are corrected with lenses (spectacles).

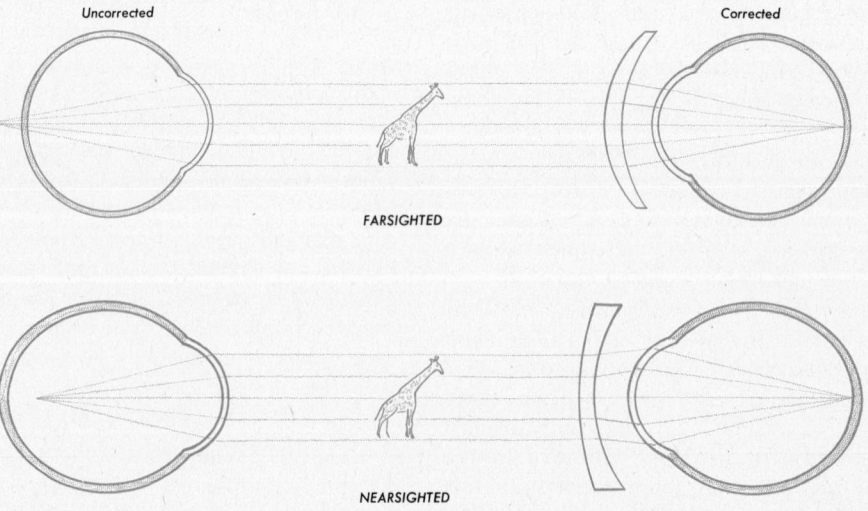

becoming more and more common. People often "will" their corneas to others at the time of death; this custom is prevalent enough now that cornea "banks" have been established.

Hearing

Hearing is another sense that records stimuli coming from a distance. This is a convenient supplement to vision in that sound travels around corners and in the dark, two very important adjuncts in the problem of orientation. Nearly all groups of animals have some means of receiving water or air vibrations (Fig. 16–6). Underwater forms such as fish receive vibrations in the water and can, therefore, hear, although they possess none of the apparatus found in higher vertebrates that is used to receive and amplify sound. They do possess the inner ear structures that are essential for hearing. Apparently these can receive only vibrations set up in their own bodies. Fish not only receive sounds in water, but some even make sounds. Air breathers, both vertebrate and invertebrate, have not only solved the problem of receiving air vibrations but also of setting up such vibrations.

When the vertebrates migrated onto land they, too, started to solve the problem of sound making and sound receiving in air. The frog has a crude ear and emits a very simple sound. It is interesting to note that in providing for this change old structures, the gill arches, which now serve no special function, were employed in making the new organs. For example, the tips of the jaws and the hyoid arch which evolved earlier from gill arches became the **ear bones,** and some of the remaining gill arches became the **larynx** (voice box) as well as other elements of the upper breathing system (Fig. 25–9). Strangely enough, those mammals that returned to the sea, such as the porpoises, have retained excellent underwater hearing. They also produce a variety of sounds which are used to communicate with one another. In man, the production of sound has probably been carried to its greatest perfection, birds

excepted, although the sound-receiving apparatus of some other vertebrates, such as the dog, has a greater range than the human ear.

Intimately associated with and usually considered as a part of the ear in the vertebrates is another organ that is physiologically quite separate from the sense of hearing.

Fig. 16–6. Some animals and the manner in which they produce and receive vibrations in the air and water.

This is the **organ of equilibrium,** which is a receptor for changes in conditions of balance and rotation. In the invertebrates these two organs are quite unrelated. Organs of equilibrium are found in many invertebrates, from the jellyfish to the crayfish.

The Human Ear. The human ear is composed of three portions, the **outer,** or **pinna,** the **middle,** which connects with the throat by means of the **eustachian tube,** and the **inner,** where the **cochlea** (sound receptor) and **semicircular canals** are located (Fig. 16–7). The outer ear is merely a skin-covered, cartilaginous projection from the head, designed to catch and concentrate sound waves. It surrounds the opening which leads through the **auditory canal** and terminates at the **tympanum,** or **eardrum.** The walls of the canal are supplied with glands that produce wax which discourages small creatures such as insects from entering.

The middle ear consists of a chamber connected to the pharynx by the eustachian tube. Bridging across this air chamber is a series of three bones (**ear ossicles**) which conduct vibrations of the tympanum to the cochlea. Although the eustachian tube is advantageous in equalizing the pressure on both sides of the eardrum, it does have a disadvantage in that microorganisms in the mouth can make their way through this tube and infect the middle ear regions. Such infections can be dangerous, sometimes leading to deafness. The tiny bones are named the **hammer, anvil,** and **stirrup** (**malleus, incus,** and **stapes,** respectively) because of their shapes. Together they produce a lever arm which diminishes the amplitude of the tympanic vibrations but at the same time intensifies them. The hammer is attached to the eardrum and the anvil, while the stirrup, attached to the anvil at its opposite end, fits into the **oval window** of the cochlea. Vibrations conducted through the chain of bones are conveyed to the liquid-filled cochlea. Another membrane-covered opening, the **round window,** allows the fluid to vibrate freely without being lost from the closed chambers. Since liquids do not compress, the vibrations retain all of their vigor until they are delivered to the sensory cells which generate impulses that are transmitted to the brain over the auditory nerve.

The highly complex cochlea can best be studied by uncoiling and cutting across it in order to examine its internal structures (Fig. 16–8). It is a long tapering tube which is divided into three chambers, all filled with fluid. The middle chamber contains the **organ of Corti** (named after its discoverer), which consists of a **basilar membrane** composed of tightly stretched connective tissue fibers which are longer at the top of the spiral and shorter at the larger or lower end. Above this membrane are the so-called **hair cells** which send out nerve fibers that make up a part of the auditory nerve. Overlying the hair cells is the **tectorial membrane,** a thin sheet of tissue, which lies very close to the tiny hairs projecting from the hair cells. There are other structures, too, which need not be considered in this discussion.

Fig. 16–7. The human ear dissected in order to show the various parts that have to do with receiving sound and the part that is concerned with balance.

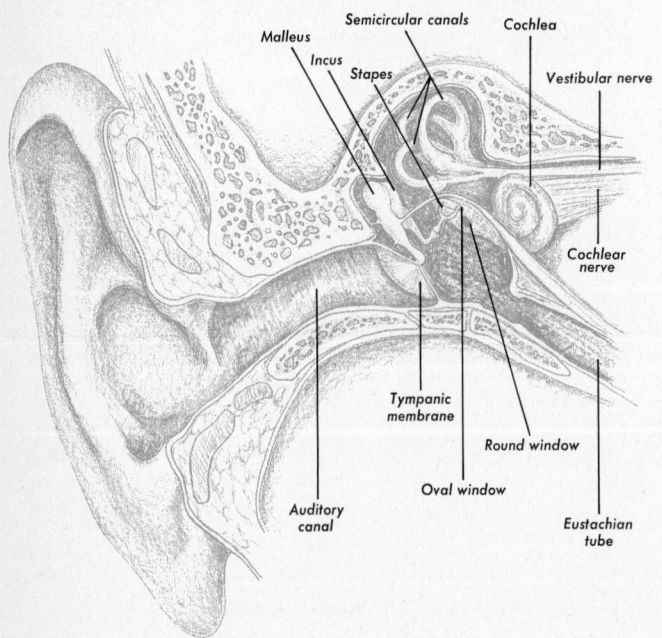

Semicircular canals

Malleus

Incus

Stapes

Cochlea

Vestibular nerve

Cochlear nerve

Tympanic membrane

Round window

Oval window

Auditory canal

Eustachian tube

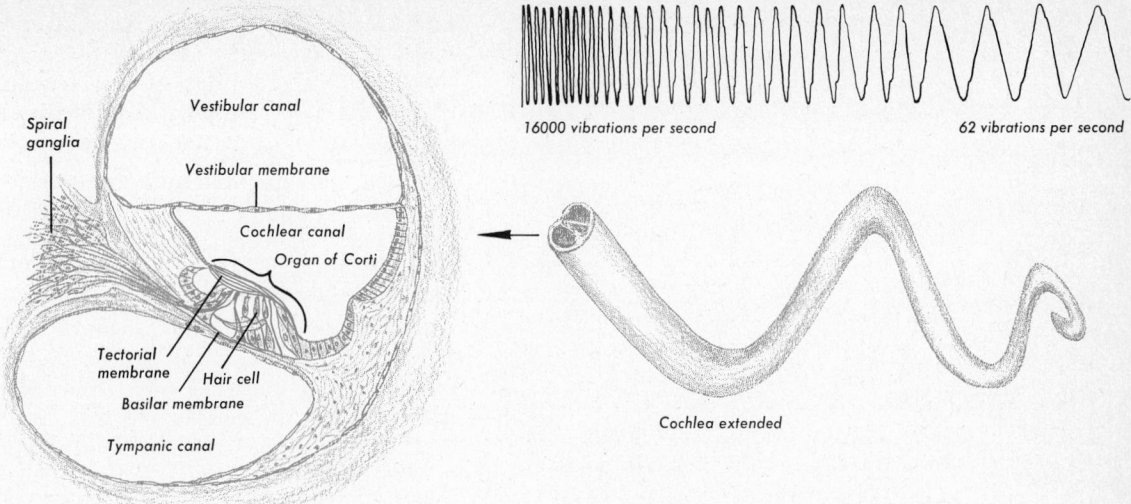

Fig. 16–8. The cochlea uncoiled and cut in cross section to show the organ of Corti in detail. The parallel lines to the left indicate the wave lengths of the various notes and the approximate position in the cochlea where the organ of Corti picks them up.

Sound is perceived when the fluid in the canals of the cochlea is set into vibration by the stirrup entering the oval window. It seems probable, although no one has seen it, that these vibrations set the basilar membrane into sympathetic undulations which cause the hair cells to touch the tectorial membrane, and that such touch stimuli set up impulses in the nerve fibers of the hair cells. The short fibers of the basilar membrane that lie at the large end of the cochlea respond to higher notes, whereas those at the top of the spiral are stimulated by the low notes. This can be borne out by experimentation. If an experimental animal is exposed to sound of high pitch and considerable amplitude (loudness) for a long period of time, the hair cells in the lower end of the cochlea are destroyed and the animal is deaf over this range. The same phenomenon has been observed in people suffering from the so-called boiler-makers' disease where the constant din of a trip-hammer over many years finally destroys that portion of the organ which records the same pitch as that of the hammer. It seems, then, that vibrations coming in stimulate various parts of the organ of Corti according to the frequency of the vibration (pitch). When vibrations of various pitches come in simultaneously the corresponding segments must be stimulated. This is quite remarkable when it is recalled how many different tones can be distinguished when listening to a symphony, for example.

Hearing Defects. The human ear can detect low and high tones ranging from about 20 to 16,000 vibrations per second. A young child can hear even greater ranges; the range diminishes in the upper limits with advancing years.

Deafness is usually caused either by faulty transmission of sound through the middle ear or by some difficulty in the cochlea, rarely by any deficiency in the central nervous system. One of the common causes of deafness is middle ear infections. Infectious bacteria can make their way up the eustachian tube to the middle ear where they can cause damage to the ear bones or drum. With the refinement of electronic equipment, hearing aids have gradually been perfected to a point where they are very

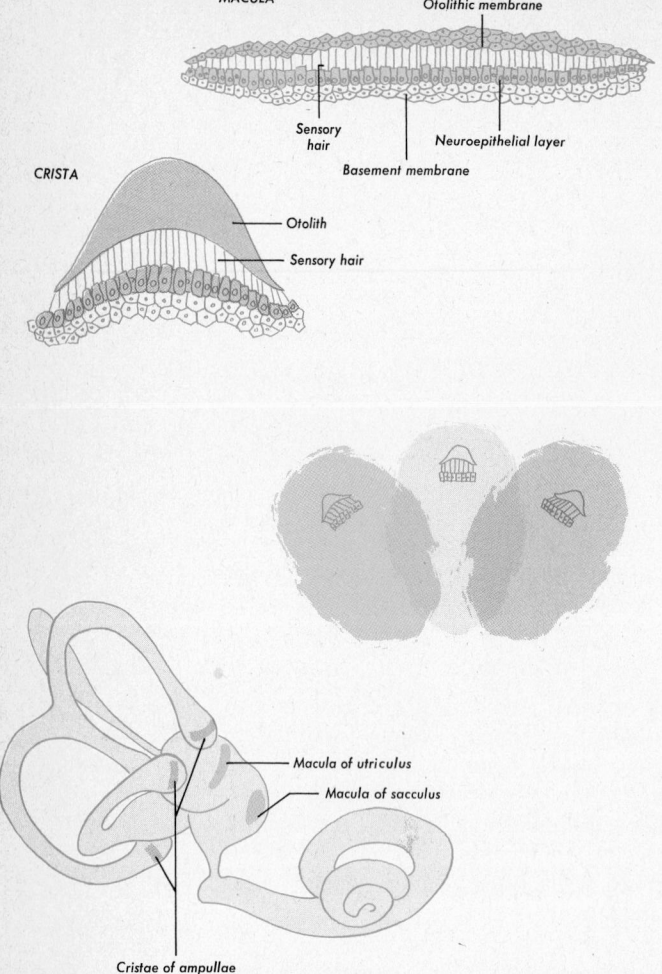

MACULA

Otolithic membrane

Sensory hair

Neuroepithelial layer

Basement membrane

CRISTA

Otolith

Sensory hair

Macula of utriculus

Macula of sacculus

Cristae of ampullae

Fig. 16–9. The nonacoustic labyrinth with the utriculus, one of the semicircular canals and an ampulla cut in cross section to show their internal structures. The lower figures indicate schematically how the otoliths aid in setting up stimuli coordinated with head movements.

able surgery of the ear bones, replacing one or more, in recent years has restored hearing to many people.

Sense of Balance

There are three different organs which have to do with the sense of balance and movement. The eyes, when functioning, operate continually in this capacity. A second very important organ of equilibrium is a portion of the inner ear, the **nonacoustic labyrinth,** while the third organ is the system of **proprioceptors,** which are tiny sensory endings located in the muscles and tendons.

The nonacoustic labyrinth is composed of three **semicircular canals** and two small chambers, the **utriculus** and the **sacculus** (Fig. 16–9). In all vertebrates the inner ear includes the organ of equilibrium. The parts of the nonacoustic labyrinth operate differently. The semicircular canals function when change in rate or direction of movement of the head occurs, and the utriculus and sacculus have to do with the position of the head.

The three canals are semicircular tubes arranged at right angles to one another so that no matter how the head is moved, two of the canals (one in each ear) function. They are fluid-filled and each has a small swelling called the **ampulla** which houses a tuft of hair cells that are sensitive to the movement of the fluid. Any acceleration or deceleration of head movement causes the fluid to flow in definite directions in the canals, thus stimulating the hair cells to send impulses to the brain which results in the sensation of movement. When this occurs in a horizontal plane—the one people are accustomed to—no particularly unpleasant sensation results. However, if the head is moved in a vertical plane, such as in the abnormal movements of flying or riding in an elevator, some very unpleasant sensations are experienced; in fact, they may become so distasteful that nausea occurs, as in seasickness. Just why such movements should affect the stomach and bring about disagreeable

useful to those who have imperfections in the transmitting portion of their ears. Obviously defects in the cochlea cannot be compensated for by hearing aids. These instruments merely amplify the tones so that sluggish or imperfect ear bones will, by sheer force of the vibrations, pick up and transmit them to the inner ear. Some very remark-

feelings is not very clear. Fortunately, one can usually become accustomed to such movements so that eventually there is no more response to them than to horizontal movement.

Head position, that is, static position, is determined by the utriculus and sacculus. These two chambers are also filled with a fluid and, in addition, each contains a tiny lime pebble (the **otolith**) attached to the sensory hair cells. Since the otolith is free to move in the chamber, when the head changes its position the pull of gravity shifts the tiny weight so that the hairs are bent in synchrony with its movement (Fig. 16–9). This is very similar to the operation of the statocyst in the crayfish (p. 317). It is the perception of this movement that makes a person conscious of the position of the body with respect to gravity.

Not only is it important to register the position and movement of the body as a whole but, if coordination of all the complex movements is to be had, there must be some way of bringing about this interrelationship. This is done by the hundreds of **proprioceptors,** tiny sensory endings located in each muscle and tendon. These are sensitive to changes in tension. As the various muscles contract, proprioceptors send out impulses which eventually bring about an interplay of muscles and tendons to perform a coordinated act. This goes on without conscious knowledge and functions without vision, since, for example, a person is able to play a piano in the dark. It is necessary not only to know that a muscle is contracting but also to know where the appendage is at all times. In a way, the muscles can be considered sense organs because sensory impulses are coming from them almost as rapidly as motor impulses are going to them, though the cells involved in the two cases are different. The tendons are also abundantly supplied with proprioceptor end-organs but they receive no motor nerves and cannot contract.

We consider, next, the central nervous system.

The Nervous System

In its evolution it might be expected that the nervous system originated from the external part of the body, because it is this part of the organism that contacts the outside world directly. Embryology and the history of the animal groups both bear this out. As animals became more complex, certain cells in the outside layer became specialized to receive and transmit stimuli. These cells became adapted to receiving stimuli and thus developed into the receptors or sense organs such as the eye and the ear. Transmission of the impulses was left to other cells which combined to form the **sensory nerves.** These, in turn, carried the impulses to a central station, the brain and spinal cord, where adjustment and interpretation took place. Another set, the **motor nerves,** carried the adjusted impulse to the glands and muscles where the final response was executed. With such a system it was possible for an animal to reach great size and complexity and still be intricately coordinated. The analogy of the centralized telephone system may be recalled to good advantage at this point.

It may seem surprising to learn that this very complex system is composed of only one general kind of cell, the **neuron.** Here the telephone analogy breaks down, because, while the organization may be similar, the nature of the wire itself is vastly different from the neurons which go to make up the entire nervous system.

The Neuron. Although all cells may conduct an impulse within themselves, the neuron has developed this characteristic to an advanced stage. Anatomically the neuron is divided into three parts: the **cell body,** containing the nucleus, the **dendrites,** which consist of numerous protoplasmic outgrowths from the cell body, and the **axon,** a single, much longer extension terminating in a brushlike filament (Fig. 16–10). The cell body resembles many other cells in appearance and the dendrites merely extend its surface of contact. The axons are usually covered with a fat-containing **myelin sheath** which is enveloped by a single layer of fat cells, called the **Schwann sheath.** Together they function like an electrical insulator. Only the Schwann sheath is found in nerve fibers of the autonomic nervous system. It has been postulated that the myelin sheath promotes conduction of the nerve impulse owing to its insulating properties. It has been shown that the nerve impulse travels about four times faster in myelinated nerves than in those without the myelin sheath.

Recent studies with the electron microscope indicate that the Schwann cell gives rise to the myelin sheath by the elaboration of its outer membrane as indicated in Figure 16–10. It forms a tightly packed spiral around the axon. Unmyelinated axons, while embedded in Schwann cells, do not possess the elongated spiraled membranes.

The Schwann sheath functions in nourishing the nerve fiber and brings about its regeneration in case of injury. If the axon is severed the distal part from the cell body degenerates. If, however, the Schwann sheath remains intact and heals, a new axon

Fig. 16–10. A motor neuron with dendrites and numerous attached axons from other neurons and its single axon. The top insert is a schematic representation of the synapse based on electron micrographs. The lower insert is also a schematic interpretation of the cross section of an axon based on electron microscopy.

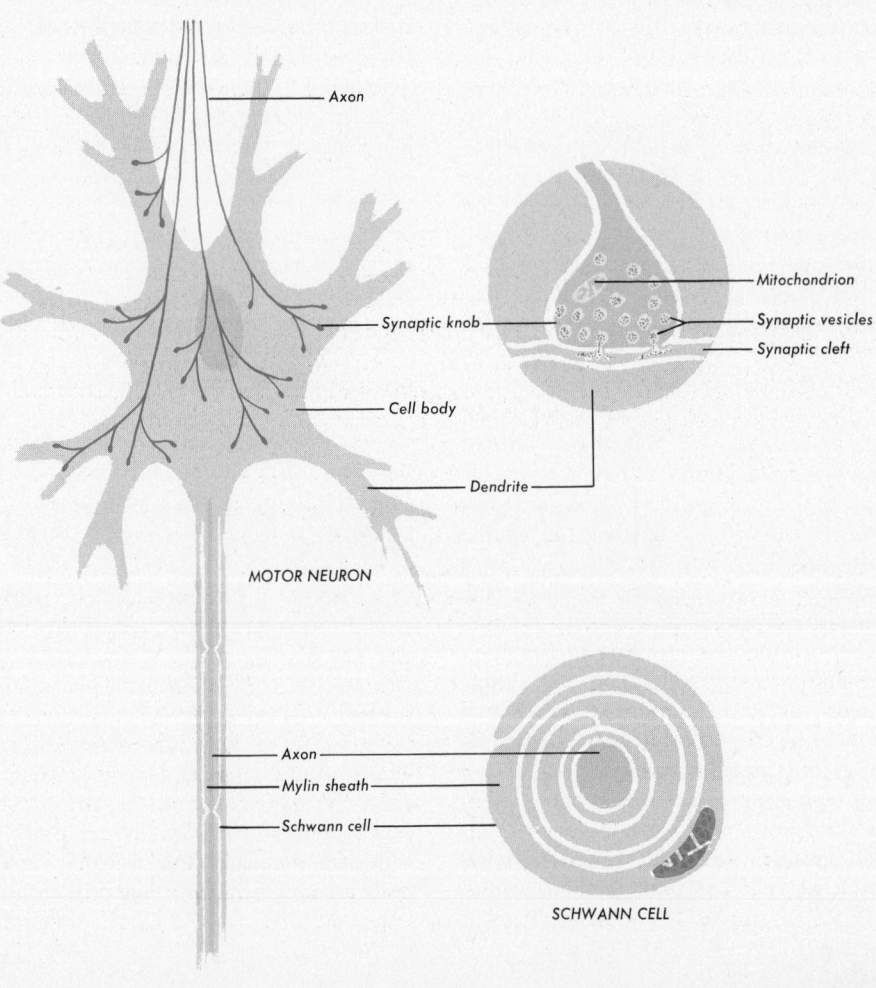

Axon

Synaptic knob

Cell body

Dendrite

Mitochondrion

Synaptic vesicles

Synaptic cleft

MOTOR NEURON

Axon

Mylin sheath

Schwann cell

SCHWANN CELL

replaces the degenerated one. If the laceration is so severe as to make the healing of the Schwann sheath impossible subsequent regeneration is disoriented and the sheath probably never innervates the cells as it originally did. Occasionally this can be corrected by rerouting the Schwann sheaths by surgery.

The axons are usually long. An extreme example is illustrated by the giraffe, in which the axons extend from the tip of the toe to the back, a distance of 6 feet or more. Many axons are bound together to form the nerve trunk, spoken of as the "nerve" by anatomists, which is covered by a tough sheet of tissue. This is remarkably similar to a telephone cable where each wire is insulated from all the others.

There are several major kinds of neurons located in specific parts of the nervous system. In the brain and cord they are highly specialized and occur only in certain locations (Fig. 16–13). These neurons are called **association neurons,** and they function in connecting various parts of the brain and cord. Those that conduct impulses from the distal parts of the body to the brain or cord are called **sensory** or **afferent** neurons and those that conduct the impulses away from the brain and cord are called **motor** or **efferent** neurons. Some nerve trunks are composed entirely of one or the other, in which case they are called **sensory** or **motor nerves.** Most, however, carry both kinds of fibers and are called **mixed nerves.**

Nature of the Nerve Impulse. Up to the present time, attempts to solve the nature of the nerve impulse have been made only with peripheral nerves, that is, those outside the cord and brain. Very little progress has been made toward an understanding of how they work within the brain itself, though there is no reason to doubt that in a general way the functioning is similar.

Once an impulse passes over a nerve fiber there is a short period when the nerve is incapable of transmitting a second impulse, that is, it refuses to accept any further stimulus. This is known as the **refractory period**

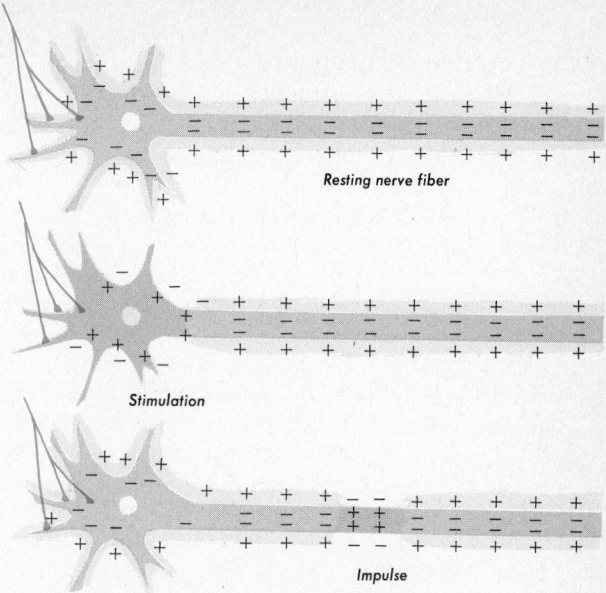

Fig. 16–11. The nerve is normally polarized, being positively charged on the outside and negatively charged on the inside. This polarity is lost coincident with the nerve impulse (indicated in gray).

and is very short in most nerve fibers, lasting 0.001 to 0.005 second. It means that some reorganization is essential before the nerve fiber can once again be stimulated. This, in turn, is due to the physical make-up of the nerve fiber itself. The outside membrane of the nerve fiber is positively charged while the inside is negatively charged (Fig. 16–11), and is therefore said to be **polarized.** This condition is maintained until an impulse passes along which brings about a chemical change; this results in the mixing of the charged ions across the membrane, and in the membrane becoming **depolarized** or neutral in the region of the impulse. It is known that the change is chemical as well as electrical because the nerve fiber consumes oxygen and gives off carbon dioxide, respiring just as all cells do. The depolarized condition is coincident with the refractory period and therefore lasts a very short time. Impulses pass along a nerve in rapid succession and it is highly unlikely that a single impulse brings about a specific action. Im-

333

pulses come in "bursts" like bullets from a machine gun.

By attaching thin wires to a nerve and using electrical measuring equipment, the passage of the impulse can be detected. The study of these currents, called **action potentials,** demonstrates that they differ in frequency, speed, and strength in different nerves. For example, impulses are fired continuously and rapidly in the nerves going to the heart from the cardiac center. The heartbeat is changed with the frequency of the impulses. By contrast, motor fibers to glands are normally at rest; impulses occur only when the gland is actively secreting. Each kind of fiber has its own pattern of impulses.

The fact that the nervous impulse continues throughout its course with equal vigor indicates that something is added to it as it travels. It might be compared to a path of inflammable material where each portion ignites the succeeding part so that the entire trail burns with equal intensity. A minimum amount of heat must be supplied for ignition, but once ignited the flame burns with equal vigor from the point of ignition forward. Furthermore, any excess heat beyond the minimum necessary to ignite the material will not change the situation. It is likewise with a neuron; once stimulated, the impulse travels with equal intensity throughout the course of the cell. The neuron, like the muscle cell, obeys the "all or none" principle. In other words, if a given stimulus elicits an impulse, the impulse starts and continues throughout its course with full vigor. No matter how the minimum stimulus is altered, the impulse travels with its full force or it does not travel at all.

The simplest way to study the nerve impulse is to observe the action of a muscle to which it is attached. The classic setup for such study is the sciatic nerve of the frog attached to the large gastrocnemius (calf) muscle and a mechanical device for recording the contraction of the muscle (Fig. 15–9). Various stimuli can be used to stimulate the nerve, but an electrical one is the best and most convenient. Whenever the nerve is stimulated the muscle twitches, indicating that some change set up in the

nerve has traveled along the nerve to the muscle, causing it to contract. This change is called the **nerve impulse.**

If a special instrument designed to detect minute electrical currents is placed on a nerve over which an impulse passes, there is a definite response, indicating that the impulse has an electrical aspect. If it were electricity, it would travel with the speed of electricity, namely, 186,000 miles per second. An early experiment, however, demonstrated that the impulse travels at a much slower speed, 30 meters, or about 100 feet, per second in the frog and only four times that rate in man. The nature of the stimulation bears no relation to the speed of transmission; whether the nerve is stimulated with heat, pressure, or electricity, the impulse travels at exactly the same speed. Furthermore, once the impulse is started it continues with equal vigor throughout its course, unlike electricity which as it travels along a wire gradually diminishes in intensity the farther it goes.

So far, consideration has been given only to the transmission of the impulses within the neuron itself. How does the impulse travel between nerve cells?

The Synapse. In order for a nerve impulse to complete its circuit, it must pass over more than one neuron; in fact, a great many are probably involved even in the most simple action. Neurons are not directly connected with each other but come in close association only. The region or area where the dendrite of one neuron is in close proximity with the axon of another is known as the **synapse.** This is a very important part of the nervous system because it is here that a selection is made as to whether or not an impulse is permitted to pass on to the next neuron. The impulse can travel both ways within a neuron but where the neurons are in a series, as they always are, the impulse travels **toward** the cell body on the dendrites and **away** from it on the axon. The synapse, therefore, acts like a traffic signal on a one-way street.

In recent years it has been possible to observe the detailed structure of the synapse by means of electron microscopy. This information is important in our understanding of the function of the synapse. The synapse is actually a junction consisting of several parts. The tip of the axon is swollen into the **synaptic knob** which lies close to the dendrite or cell body (Fig. 16–10). The synaptic knob contains many mitochondria, indicating a high rate of oxidative metabolism, and numerous small vesicles called **synaptic vesicles**. These contain the **transmitter substance** which is the compound that is instrumental in the passage of the impulse from the axon to the dendrite or cell body. The transmitter substance may be acetylcholine, epinephrine or norepinephrine. The space between the axon and dendrite or cell body is called the **synaptic cleft** and it is uniform in width (2000 Å). Whereas there are inhibitory and excitatory axons, the synapses of both are identical morphologically. Just how does this rather complex morphology fit into the observed function of the synapse?

When the impulse reaches the synaptic knob it stimulates the synaptic vesicles to approach the synaptic cleft and discharge the transmitter substance (Fig. 16–10). They then return into the knob and become recharged. The transmitter chemical depolarizes the membrane below the cleft (dendrite or cell body) initiating a flow of current in the next neuron. Thus the actual transmission of the impulse across the synapse is by means of a specific chemical. There are two kinds of synapses; the **excitatory** which permits the impulse to travel to the next neuron and the **inhibitory** which prevents the passage of the impulse. These are identical as can be observed in the electron microscope. They differ in the kind of transmitter substance each contains in the synaptic knobs. The two transmitter substances that have been identified we discuss in a later section (p. 345), but there are undoubtedly others. The cell body (or its dendrites) has numerous excitatory and inhibitory synapses lying on it (Fig. 16–10). Whether or not the cell body sends the impulse on through its axon depends on the relative number of both types of synapses that are stimulated at the same time. For example, if more inhibitory synapses are stimulated than excitatory synapses, the impulse does not pass to the axon. In neurophysiological jargon, we say it won't "fire." When the number of excitatory synapses that are stimulated exceeds the number of inhibitory synapses, the neuron fires and the impulse passes along the axon to the next neuron.

The one-way valve action of the synapse can now be understood. The synaptic vesicles are found only in the synaptic knobs of the axons, never in the cell body or dendrites. This means that the stimulation can come only from axons, resulting in a one-way flow of the impulse. The lag in nerve transmission can also be accounted for by the time required for the secretion of the transmitter substance into the synaptic cleft. This new knowledge has been helpful in understanding the regulation of the flow of impulses, but there is much more that needs to be discovered about the transmitter substances themselves.

The Peripheral and Central Nervous System. In an attempt to understand some of the more simple reactions effected through the nervous system, it is useful to divide it into parts, all of which, however, are very interdependent. The **central nervous system** is composed of the brain and cord while the **peripheral system** is made up of the nerves which connect the brain and cord to all parts of the body. The **autonomic system** is discussed later in this chapter. Impulses enter the central nervous system through the afferent fibers of the peripheral nerves. Interpreting the incoming messages and subsequently dispatching them is the function of the brain, the cord acting primarily as a relay. The impulses leave the brain and cord for the muscles and glands on the efferent fibers. The simplest type of this action involving the peripheral and central nervous system is called a **reflex**.

The Cord and Reflex Action. Every person has experienced the operation of a simple reflex in his own body as, for example, the quick withdrawal of his bare foot when it encounters a sharp object. It is difficult to determine whether the foot was removed after or just prior to the sensation of pain. A simple experiment with a frog can show that the reaction could have taken place without the sensation of pain. If the brain is removed and the skin irritated with acetic acid, the frog withdraws its foot by appropriate leg movements (Fig. 16–12). The stimulus can be repeated again and again but the response always is the same. Remember, there is no brain to interpret the message, yet the response is just as effective as if the brain had been intact. Similarly, people with certain regions of their cord injured feel no pain though they respond to the stimulus. It is apparent in these cases that the impulse from the stimulus must have traveled only to the cord, returning from it

Fig. 16–12. The frog has had the cerebrum removed by a cut just back of the eyes. The cord is intact. Acetic acid has just been brushed on the thigh of the right leg (left). The right leg flexes and the toes scratch the region which is irritated by the acid (right).

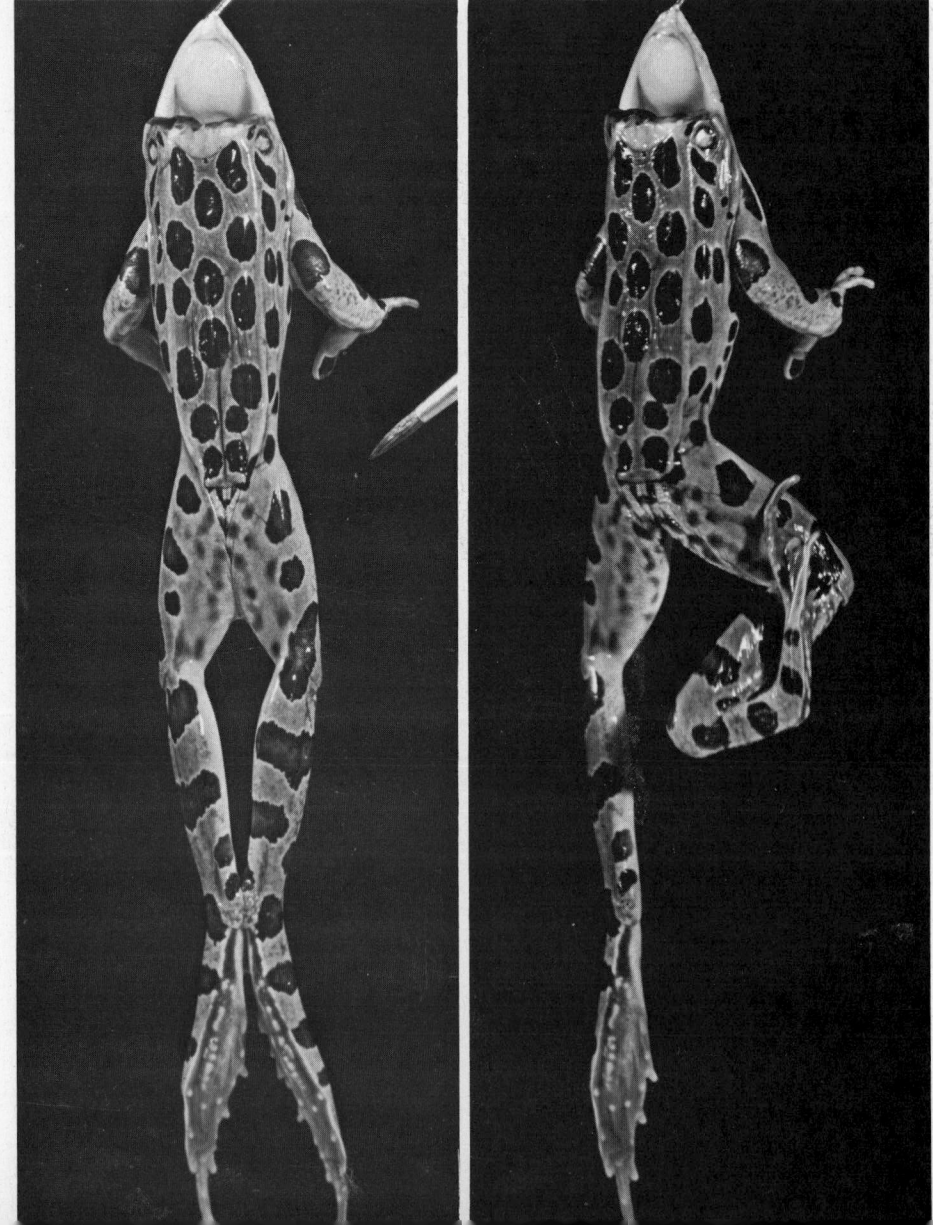

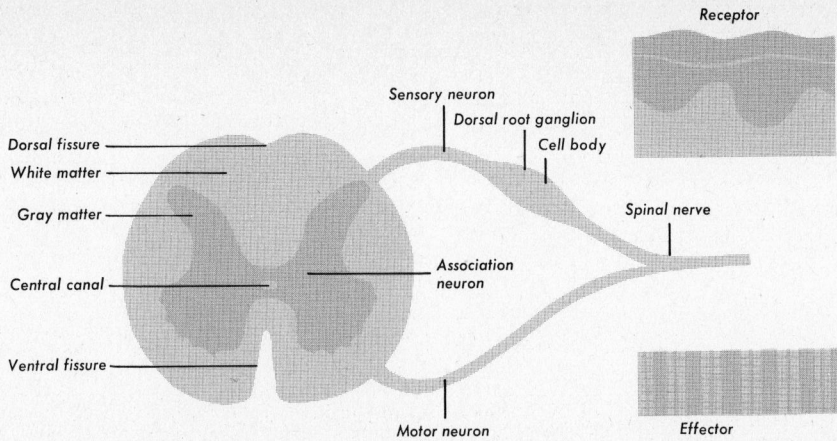

Fig. 16–13. A schematic representation of the vertebrate spinal cord in cross section together with the pathways of the sensory and motor neurons. The impulse originates in the receptor, passes over the sensory neuron, then through the cord (via the association neuron) and over the motor neuron to the effector. These neurons constitute a reflex arc.

to the appropriate muscles for bringing about that response. Various portions of a frog's cord can be destroyed, but the reflex persists as long as the region where the sensory and motor nerves join the cord is intact. The complete action, then, must take place in a very localized region.

In order to understand how a reflex can take place, a little information about the anatomy of the cord is in order. A spinal nerve bifurcates as it approaches the cord to enter on the dorsal side as the **dorsal root** (sensory) and on the ventral side as the **ventral root** (motor) (Fig. 16–13). The dorsal root bears the **dorsal root ganglion,** in which the cell bodies of the sensory neurons are located. The cell bodies of the motor nerves are located in the ventral gray matter of the cord (called the **ventral horn**). The function of these roots can be definitely identified by cutting them and artificially stimulating the stumps. When the ventral root is cut, no impulses reach the muscle from the cord, but when the cut end on the muscle side is stimulated a response is evoked. Likewise, by severing the dorsal root no impulses reach the cord, but when the stump on the cord side is stimulated a response takes place. The dorsal root thus

contains only sensory fibers and the ventral root only motor fibers. The spinal nerve, which is a union of the two, contains both, of course.

A simple reflex begins with an impulse coming from a sense organ and traveling over an afferent nerve fiber to the cord via the dorsal root (Fig. 16–13). Here it comes in close contact (synapse) with one or more association neurons. The impulse is then dispatched to the proper muscle or gland over an efferent nerve fiber via the ventral root. With few exceptions reflexes involve more than the three neurons just described. The incoming sensory neuron usually connects with more than one motor neuron through association neurons in the cord. Thus the impulse may not only elicit the simple response but may also pass along the cord to stimulate other efferent neurons and in turn cause a large group of muscles to contract. Indeed, this is the manner in which it usually happens. When applied to man, the simple reflex accounts for many of our daily movements. These actions do not have to be learned, they are innate—one is born with them.

Not only are different levels of the cord involved but also neurons on the opposite

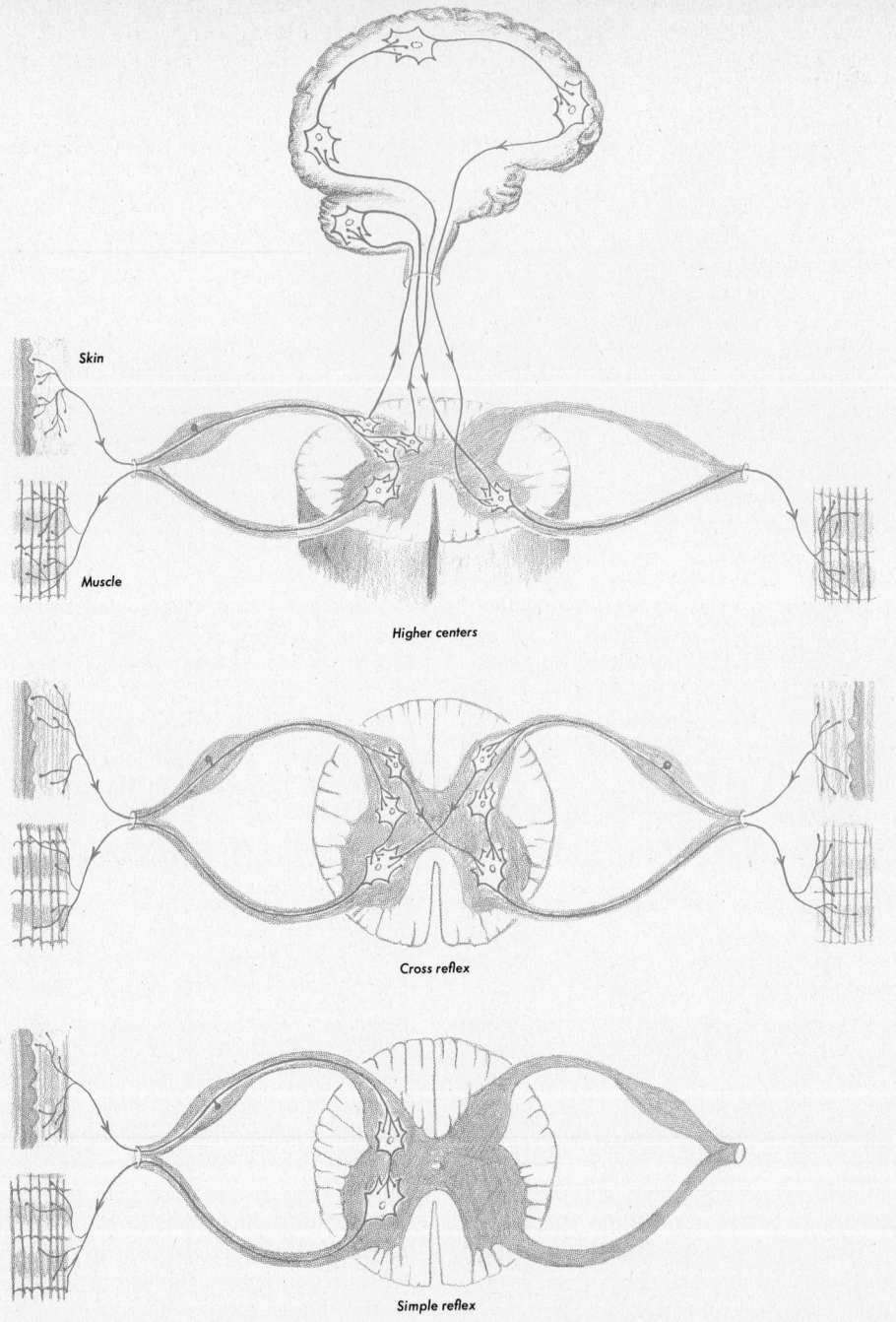

Skin

Muscle

Higher centers

Cross reflex

Simple reflex

Fig. 16–14. A series of sketches showing the nerve pathways in the cord and brain. The lower and middle figures show those where only the cord is involved while the upper figure indicates the pathways to and from the brain.

side of the cord. There is a crossing over of the association neurons so that an impulse may travel to both sides of the body (Fig. 16–14). If, for example, acetic acid is applied to one side of the ventral surface of a frog whose brain is destroyed, the leg on that particular side will attempt to remove the irritating acid while the leg on the opposite side will extend (Fig. 16–12). Presently, the impulse will spread to the front legs so that eventually all of the legs are moving in such a manner as to rid the animal of the offending substance, and the body also begins to make mass movements as it would in attempting to crawl away from the irritation. To be sure, the first pathway is simple, but as the stimulus spreads many other pathways are involved until the whole animal is thrown into movement. Under these conditions the higher centers also are involved (Fig. 16–14).

In man, when the various parts of the brain are included, the impulse becomes a part of consciousness and one is aware of the reflex. When one steps on something sharp, for example, his foot may be withdrawn through simple reflex, but very shortly certain parts of the brain are involved, even to the cortex where reasoning and memory are possible. He may even consider the advisability of wearing shoes, or the injury may be sufficiently severe that he will remember the event for a long time.

It seems that the impulses usually travel over certain paths but these can be altered on a moment's notice and new paths are then followed. In other words, many conditions can determine the pathway over which an impulse may travel. There are many choices; which it chooses depends on a large variety of circumstances.

An illustration of how the reflex may be interrupted may be seen in the sneeze reflex. Sneezing is caused by a stimulus originating in the nasal chambers; its ultimate fruition is an explosive expulsion of air through these chambers to dislodge the irritation. If, when the sneeze sensation begins, a second stimulation is set up by producing pressure on the upper lip, the first pathway will be blocked. In other words, a new set of conditions blocked the original reflex pathway. Neurologists believed at one time that certain pathways were set up not only in the cord but in the brain as well, and that each pathway was traversed by the impulse in exactly the same manner, thus making it "deeper" until it was well established. There seems to be no evidence for this idea, and it is now well known that the impulse may travel over different pathways and may never take the same course twice. Furthermore, pathways may be employed by different reflexes at different times. Which specific pathways are followed, and why, is one of the most important problems in neurology today.

The Brain. This part of the nervous system is the most complex and, of course, the most difficult to understand. Recalling our present scanty knowledge about how the brain functions, one is inclined to doubt the probability of ever understanding it, but to accept such an attitude would be highly unscientific. A great German scientist once said that the speed of the nerve impulse would never be measured, and yet, just six years later, another scientist measured it very accurately. Such statements of finality are unwise, especially for a scientist.

Nerve impulses come to the brain from all over the body on **afferent** nerve fibers like messages coming to the admiral of a fleet. Just as the admiral must make decisions that will be sent by messages to various parts of the fleet in order to accomplish a certain goal, likewise decisions are made in the brain and impulses are sent out over **efferent** nerve fibers to various parts of the body in order to execute certain actions. Decisions are made for the most part in a routine manner, based on inherited principles or on experience gained through having made similar decisions before. Most of the decisions are made by the brain without breaking through to consciousness, so that one is not aware of most of its activity. This great center is indispensable to the harmonious working of the entire body. Aberrations in the brain, so slight that they cannot be detected by any physical change, produce such stark changes in personality and emotional stability that our society is forced to build special institutions such as prisons and asylums to house people so afflicted.

For a better understanding of the human brain it is best to start with a lower vertebrate-type brain, which is relatively simple in structure. In the discussion of invertebrates it was shown that the obvious location for the brain is in the animal's anterior end, which is the part that arrives in any new environment first. Among vertebrates the brain likewise has its beginnings and its subsequent development in the anterior end of the organism, and as we progress from fishes to mammals it becomes increasingly prominent.

The three major regions of the brain, **forebrain, midbrain,** and **hindbrain,** are clearly marked off very early in the embryonic de-

velopment of every vertebrate (Fig. 16–15). These three regions soon subdivide into five regions. The forebrain becomes the **telencephalon** and **diencephalon,** the midbrain remains undivided and is known as the **mesencephalon,** and the hindbrain divides to form the **metencephalon** and **myelencephalon.** Each of these becomes modified and develops other parts as the brain becomes more complex in higher animals. In the human brain the **cerebrum** comes from the telencephalon; the posterior lobe of the **pituitary** (an endocrine gland) and the **optic chiasma** from the diencephalon; the **corpora quadrigemina** from the mesencephalon; the **cerebellum** and **pons** from the metencephalon; and the **medulla oblongata** from the myelencephalon. There is a clear relationship between the size of a particular part of the brain of an animal and its importance in the life of the animal. Detecting odors is one of the most important abilities possessed by fish and frogs, who also have proportionately large olfactory lobes; whereas in birds, where the sense of smell is poorly developed, this part of the brain is proportionately small (Fig. 16–16). In the lowest vertebrates the cerebrum is nonexistent, or almost so, whereas in the birds it begins to dominate the anterior end of the brain and in mammals it overgrows all other parts to become the most prominent part. In man, this trend is carried to the most extreme point of development.

Elaboration of various parts of the brain seems to be associated with the kind of life its owner leads. The part of the brain that controls muscular coordination is the cerebellum, which is large in animals that move in three dimensions, such as birds and fish (Fig. 16–16). Amphibians, reptiles, and land-dwelling mammals (mouse and man), on the other hand, confine their movements for the most part to two dimensions and by comparison their cerebellums are proportionately smaller than those of animals moving in three dimensions.

The cord takes care of the simplest activities in the organism, and that part of the brain which lies at its anterior end, the

Fig. 16–15. The vertebrate brain begins as a hollow sac which subsequently divides first into three distinct regions and later into the many parts shown (left). The embryology of the human brain parallels that of other vertebrates as shown by the series of figures to the right.

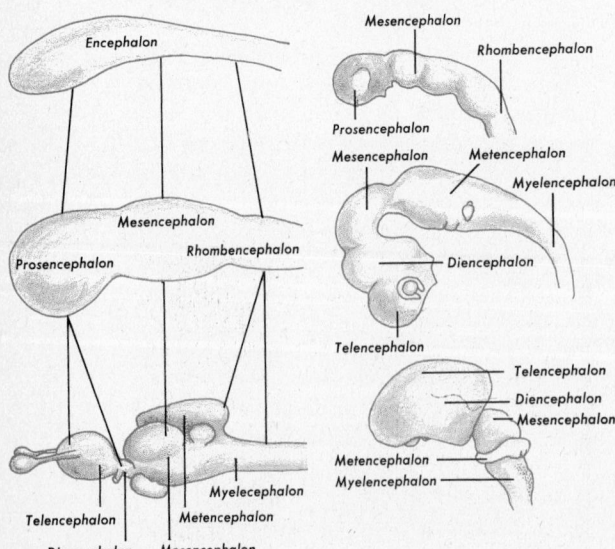

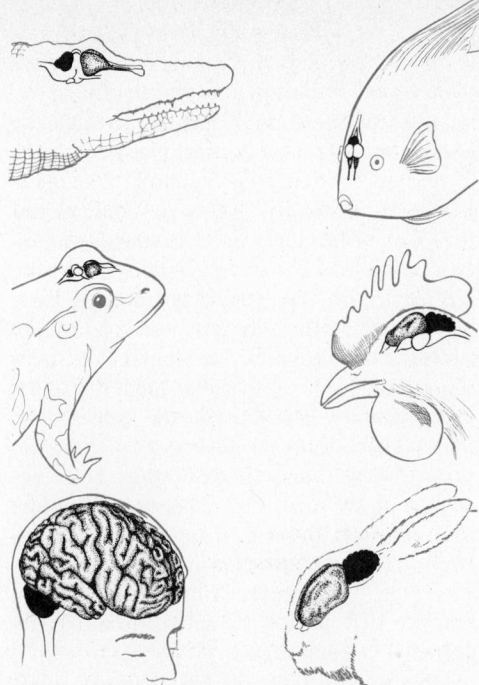

Fig. 16–16. The relative sizes of the various parts of the brains of different vertebrates vary according to activity. Animals that fly or swim have better muscular coordination than do those on land, hence their cerebellums (solid black) are proportionately large. Likewise a well-developed cerebrum (stippled) is essential to mammals, hence this part of their brain is best developed.

medulla oblongata, is the center for the activity of such vital organs as those of breathing and transportation. Even in man, both the cord and medulla function in much the same manner as they do in lower vertebrates. The cerebellum functions in muscular coordination, particularly during highly complex movements, and the cortex of the cerebrum, which evolved last, is the center of such highly complex activities as thought and reasoning.

As the vertebrate brain evolved, there was a gradual shift of function from the lower part, the brain stem, to the higher part, the cortex. This shift can be demonstrated experimentally. When the cerebrum of a frog is removed, the frog's normal activities are influenced only slightly. It jumps normally when stimulated and it can swim in a per-

fectly normal fashion (Fig. 16–17). Even a "decerebrated" reptile shows very little concern about its loss. Such an operation on a bird or mammal, however, brings about striking changes. The ability to locomote is destroyed and all actions which require considerable muscular coordination are lost. This simply means that the higher vertebrates have shifted their nerve centers from the lower brain stem to the cerebral cortex.

This shift has given these higher forms much greater plasticity in the control of their muscle coordination. For example, when certain muscles in man have lost their nervous connection with the brain in the disease known as facial paralysis, functional cranial nerves passing to relatively weak and less useful muscles can be transplanted to the larger more important muscles and eventually become functional through long retraining. In other words, impulses can be sent to a muscle via a wholly new nerve with the result that the muscle can eventually respond in its usual manner. In man, the cortex so completely dominates the body that it is physiologically possible by intense effort to learn to move muscles in almost any manner. This is well illustrated by the many human feats performed, activities that could never be executed by other animals because their nervous systems are constructed in such a way as to make it impossible.

Fig. 16–17. The anterior portion of the head, which includes the cerebrum, has been cut away in this frog. When placed in water it swims in a normal fashion, indicating that this portion of the brain has little, if any, function in performing this act.

As the cerebrum increased in size in reptiles, birds, and mammals, it became necessary to increase the surface area while retaining a reasonable volume. This has been done by the formation of wrinkles or **convolutions.** Starting in the lower mammals with only very few convolutions, the number and extent of these increase as the brain increases in size, reaching a maximum in man. Naturally, as the cortex has increased in size, the entire brain has increased with respect to the cord. The relative weights of these two structures for several common animals are shown in the following figures:

| | RATIO OF | |
	WEIGHT OF CORD	WEIGHT OF BRAIN
Frog	1.0	1.0
Cat	1.0	4.0
Monkey	1.0	15.0
Man	1.0	55.0

In general, the size of the brain is an indication of intelligence, although there are some notable exceptions; for instance, the brain of both the whale and elephant weighs more than that of man. Dogs, gorillas, and men of approximately the same body weights have brain weights of about 140, 450, and 1,350 grams, respectively. Within limits of normal variation in a single species, brain size does not indicate intelligence. For example, woman generally possesses a brain which averages about 100 grams less than man, but what man would be so brave as to imply that she is less intelligent! Likewise, there is a wide variation in the size of brains of different individuals of the same sex, just as there is in stature and body weights. Famous brains have been preserved and weighed, out of morbid curiosity or for scientific purposes, and it has been shown that a variation of 800 grams exists between brains of apparent equal intelligence. So, within limits, brain size alone is not a criterion of intelligence. Intelligence is probably due to such things as the number of cell bodies or the number of association neurons in the cortex of the cerebrum which, in turn, is an indication of the number of pathways over which impulses may travel.

The two cerebral hemispheres are each divided into five lobes, and the nerve cells associated with various functions are localized in these lobes (Fig. 16–18). The center for sight is located in the **occipital** or posterior lobe. Areas for smell, hearing, and taste are located in the lateral or **temporal** lobe, while centers for muscle movements are centered in the anterior or **frontal** lobe. Skin sensations lie in the **parietal** lobe. The fifth lobe (**insula**) lies beneath the frontal and parietal lobes and cannot be seen from the surface view. Large areas of the cortex are spoken of as "silent areas" because it appears that injuries to these regions result in no particular loss of sensory or motor function. These are the great unknown areas of the brain which have stimulated so much research in recent years.

The complex life human beings live in modern society has produced intricate mental "patterns" in man's cerebral cortex, and in a so-called normal person these patterns are satisfactory for the solution of day-to-day problems. There are no physical patterns that can be detected by examining the brain, merely repeatable cortical activity. In certain instances various aberrations in behavior indicate severe mental disturbance. It has been found that by subjecting certain people in specific instances to severe shock by means of insulin, electricity, and other agents, these abnormal responses can sometimes be corrected. More recently certain drugs have been useful in the treatment of certain types of mental diseases.

The medical profession is learning that more and more of the common ailments of mankind are due to an overactive cerebral cortex rather than actual organic disease; that is, people imagine they are sick, and this can be carried so far that actual symptoms from heart disease to stomach ulcers occur. This has resulted in a field of medical research known as **psychosomatic medicine.** With proper treatment, certain neurotic ten-

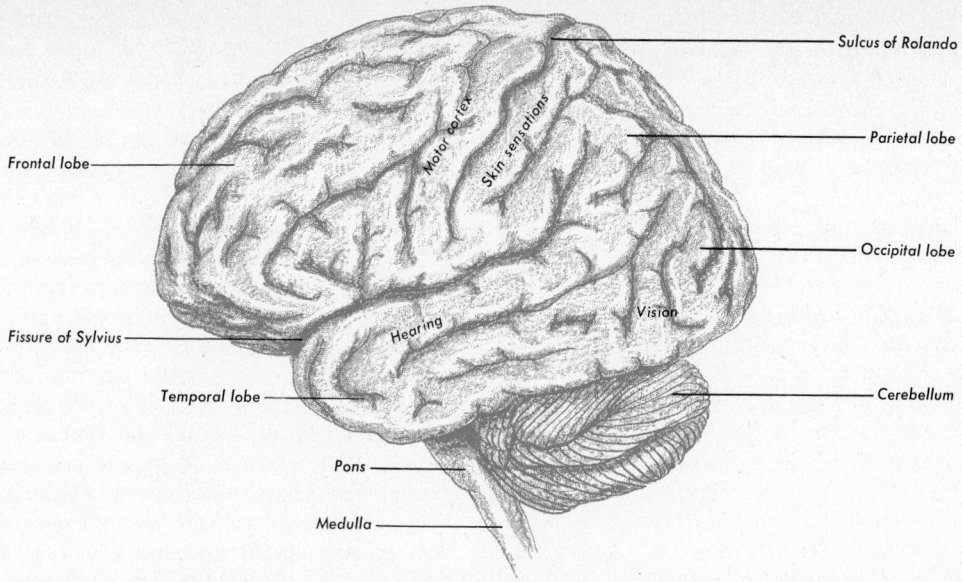

Fig. 16–18. Localized brain areas are concerned with special functions.

dencies can be overcome and the person miraculously "cured." This type of medical research is commanding increasing attention and holds hope for a different approach to the study of certain diseases.

The Autonomic Nervous System

A great many activities of the body such as peristaltic movement in the intestines, breathing, and heart action go on unnoticed and without voluntary control. These are all under the influence of the **autonomic nervous system.** The name implies that the system is a completely automatic one, which is not the case, but it has become fixed by usage.

Anatomically, part of the autonomic nervous system can be seen as two rows of ganglia (lateral sympathetic ganglia) lying on each side of the spinal column in the thoracic and abdominal cavities (Fig. 16–19). The ganglia are secured to the spinal nerves by means of two short connectives, a **white** and a **gray ramus** (Fig. 16–20). Distally they extend as small nerve fibers to the various organs of the chest and abdom-

Fig. 16–19. The spinal column with nerve cord and associated nerves. Note how well these delicate structures are protected from possible injury.

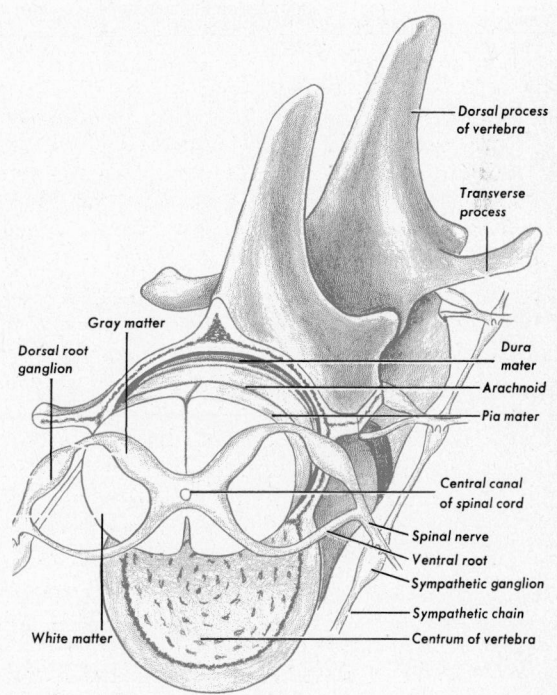

inal regions. The incoming (afferent) fibers pass through the dorsal root just the same as the voluntary nerve fibers, and the cell bodies are located in the dorsal root ganglion. The notable difference between the two systems is that the efferent fibers of the autonomic system have a chain of at least **two neurons** between the central nervous system and the organ innervated, whereas in the voluntary system there is but one. The latter also has a myelin sheath, whereas the former has only the Schwann sheath (p. 332).

The first autonomic cell bodies of the efferent neurons lie in the lateral portion of the gray matter of the cord, while the second neuron cell bodies lie outside of the cord in the ganglia already mentioned; sometimes the second ganglia lie directly on the organ itself some distance from the sympathetic chain. The second neuron may return to the ventral root and pass along with the spinal nerve fibers to such parts of the body as the blood vessels of the skin and sweat glands

which are under the control of the autonomic nervous system. Others may pass through to a large ganglion (**collateral sympathetic ganglion**) outside the sympathetic chain where synapsis occurs with the second neuron. Fibers leading from the central nervous system to the ganglia outside the cord are called **preganglionic fibers,** those leaving the ganglion, the **postganglionic fibers.** Such a system of fibers makes it possible for stimuli to come from the internal organs and to effect a response elsewhere in the body. For example, an impulse can come from the stomach (Fig. 16–20), causing some action to occur in the duodenum, initiating a peristaltic wave, perhaps. Moreover, pain impulses might come from the stomach which, by the intermingling of the two systems (central and autonomic), project through to the cerebral cortex where they are registered as an uncomfortable feeling about which something might be done. Likewise, impulses might arise in the temperature end-organs of

Fig. 16–20. A schematic drawing of the nerve pathways involving both the autonomic and the central nervous systems. Impulses arising from a temperature sense organ in the skin can stimulate sweating. Also impulses arising from the stomach can cause some action in other parts of the viscera.

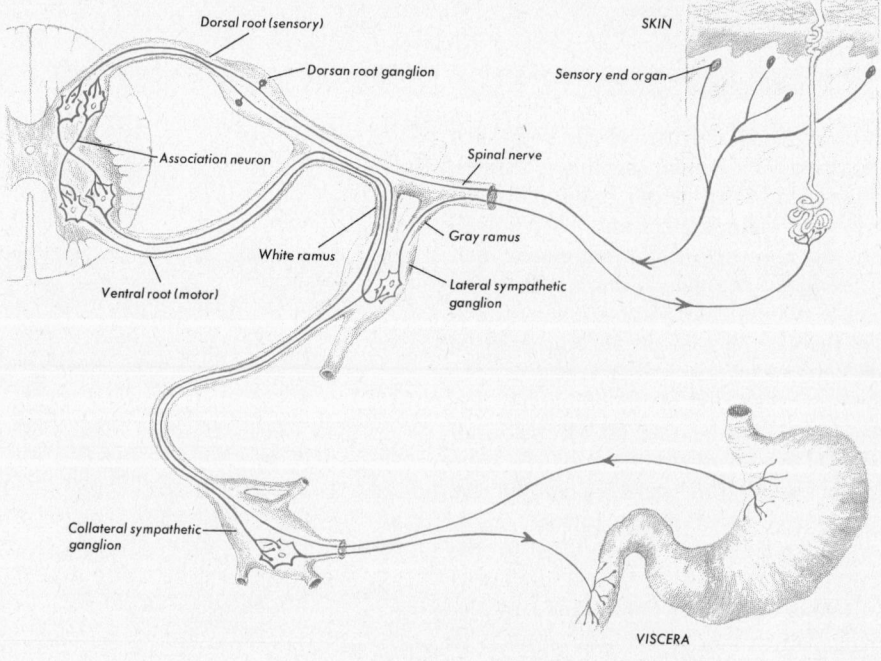

Dorsal root (sensory)

Dorsan root ganglion

SKIN

Sensory end organ

Association neuron

Spinal nerve

White ramus

Gray ramus

Ventral root (motor)

Lateral sympathetic
ganglion

Collateral sympathetic
ganglion

VISCERA

a too warm skin and pass through the circuits (Fig. 16–20), giving rise to efferent impulses that cause the sweat glands to pour their secretion over the skin, and thus cool the body. Thousands of these pathways are continually carrying messages to and from parts of the body.

The autonomic nervous system is divided into two rather distinct parts: the **thoracolumbar (sympathetic)** which is composed of the double chain of ganglia already noted; and the **craniosacral (parasympathetic)** which originates in the posterior portion of the brain (midbrain and medulla) and the **sacral** region (Fig. 16–21). The thoracolumbar system has short preganglionic fibers and generally long postganglionic ones. The opposite is true of the craniosacral system, and in fact, the ganglia lie in some cases, such as the heart, embedded within the organ itself. These two systems send nerve fibers to all of the organs operating involuntarily so that each has a double innervation. However, the two systems produce opposite effects. For example, if the nerve from the thoracolumbar system to the heart is stimulated, the beat is accelerated; but if the nerve from the craniosacral system is stimulated, the beat is slowed down. Stimulation of one nerve may cause excitation in one organ and inhibition in another, thus stimulation of the vagus (craniosacral) inhibits the heart but excites the stomach. The value of such a mechanism is obvious. It is the interaction of these two systems that regulates the flow of blood to various parts of the body, differing with each condition in which the animal finds itself. It also causes the pupil of the eye to dilate or constrict, depending on the amount of light that is needed for vision. These and hundreds of other routine jobs the body does quietly and efficiently and entirely without the knowledge of the owner. This system has certainly taken the drudgery out of operating the body, and has left for the higher centers the job of getting the whole organism in a position to obtain food or to do the many other things that are essential for life.

The puzzling question of why impulses arriving in an organ via either part of the autonomic system cause acceleration or inhibition even though the nerves appear to be identical, was cleared up by some ingenious experiments performed by Loewi in 1921. He removed the hearts of two frogs, leaving the nerve supply intact on one of them, and then placing them in a salt solution in separate containers (Fig. 16–22). He then stimulated the vagus (parasympathetic fibers) nerve of one and permitted the salt solution to perfuse through the other. The rate of beat slowed down in both hearts. He then stimulated the sympathetic fibers and the rate of both increased. Obviously some substance was being released by nerves of the first heart and was having the same effect on the second heart. Further investigations demonstrated that two substances, called **neurohumors** or transmitter substances, were produced. The parasympathetic system secretes **acetylcholine** and the sympathetic system secretes **sympathin**. Thus, in order to complete the mission of delivering a message to a muscle or gland, a physical and a chemical action must take place. As pointed out earlier (p. 335) acetylcholine is a transmitter substance that plays a role in transmitting the impulse across the synapses. An enzyme known as **acetylcholinesterase** is always present, which neutralizes acetylcholine thus preventing cumulative effects which would be disastrous. The action of sympathin is much like that of epinephrine, the endocrine secretion from the medullary portion of the adrenal gland. Indeed, these cells are themselves modified postganglionic sympathetic fibers. There seems to be no inhibitor for sympathin and it must be destroyed by oxidation some time after it is formed. Its slow destruction may account for the slow and sustained response of this part of the autonomic system.

The discovery of specific transmitter substances in the autonomic nervous system led scientists to postulate that similar substances may operate in the central nervous system. This is turning out to be the case although how many there are and their exact nature is yet to be discovered.

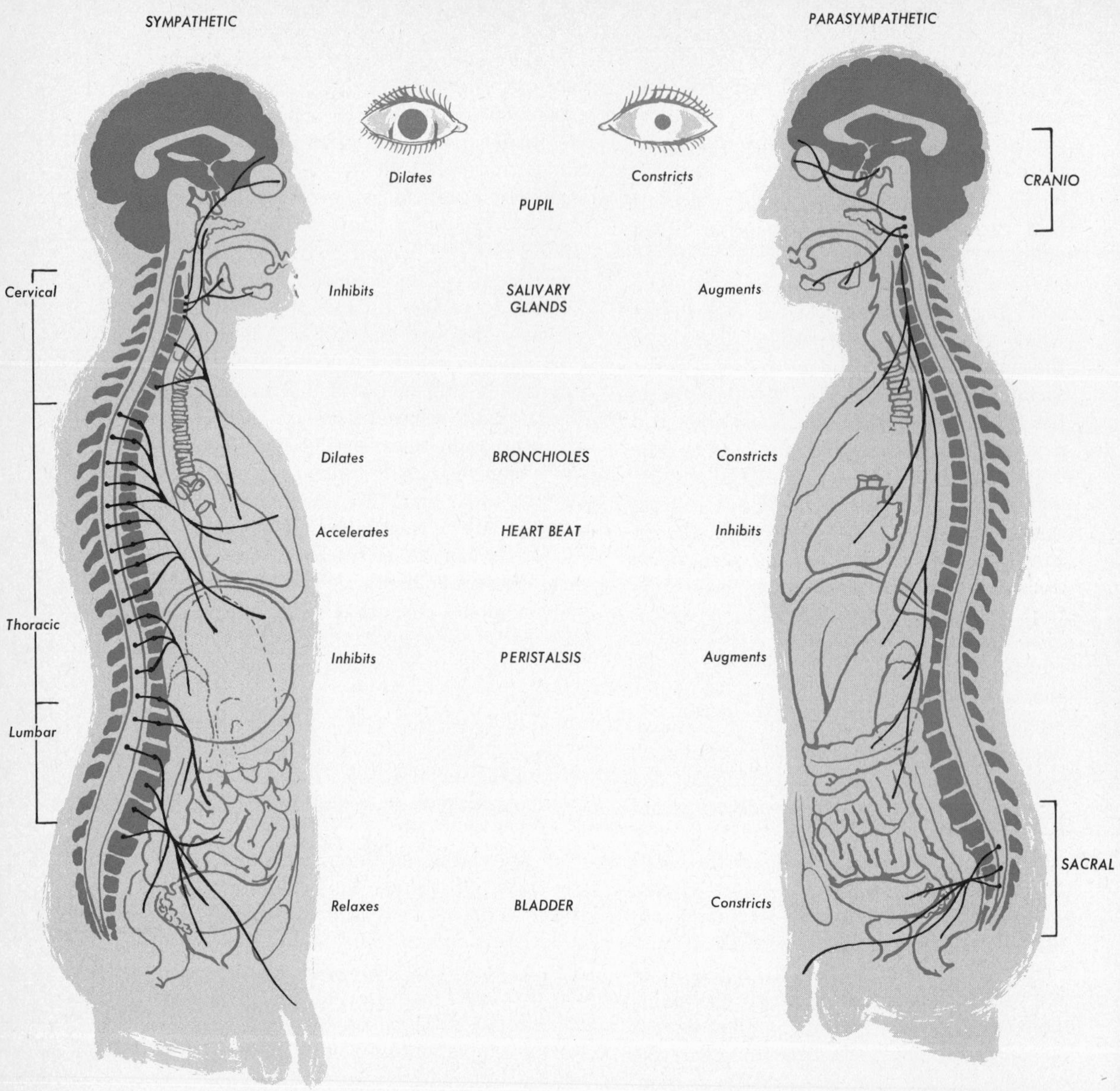

Fig. 16–21. A schematic drawing of the sympathetic and parasympathetic nervous systems, separated for clarity, to show how they function. Note that each organ is supplied with nerves that cause it to be activated or inhibited.

EFFECTORS—THE ORGANS OF RESPONSE

The principal effector organs of higher animals are the **muscles** and **glands,** although there are other types of effectors in use throughout the living world. For example, there are **chromatophores,** found in a large group of vertebrates and invertebrates, that are responsible for rapid color change in the

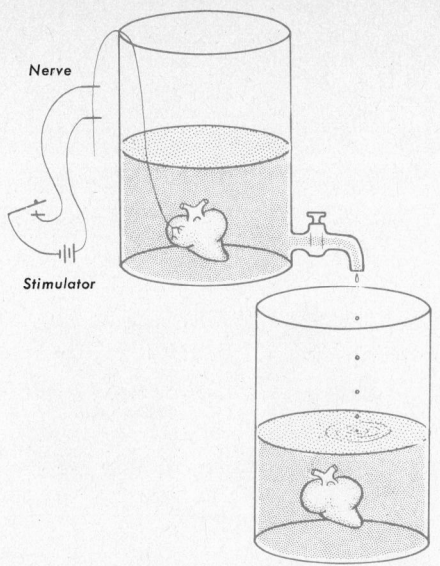

Nerve

Stimulator

Fig. 16–22. A schematic representation of Loewi's classical experiment in which he demonstrated that when the vagus nerve was stimulated in one frog heart, the other was inhibited even though the only connection between the two was by a saline solution.

skin. Some animals possess **luminous organs** which radiate light when stimulated and a few others possess **electric organs** that generate electrical discharges under suitable conditions. Each of these effectors is discussed briefly with the exception of the muscles which have been treated in Chapter 15.

Glands

Skin glands have already been discussed (Chapter 15), but there are other glands in the body which are stimulated directly by the nervous system. Just how a gland cell secretes is not well understood. Its job is to extract the product of secretion from the blood or lymph on one side and discharge it on the other into a lumen or tube that conducts the product to its proper place, either to the outside of the body, in the case of sweat glands, or into the digestive tract, in the case of digestive glands (Fig. 16–23). In such glands as sweat and tear glands no special substances are synthesized; their secretion is merely extracted directly from

the blood. However, in the case of many other glands, the digestive and endocrine glands for example, while the raw materials are extracted from the blood stream, the final product is synthesized within the cell itself.

Glands respond to stimuli from the nervous system and most of them secrete only when stimulated. Some endocrine glands, however, apparently secrete continuously, although the rate may be affected by nervous excitation or by hormones from other endocrine glands. There is considerable energy utilized during glandular activity. Thus the sweat glands produce a fluid with a salt content considerably higher than that of the blood, and energy is required to increase this

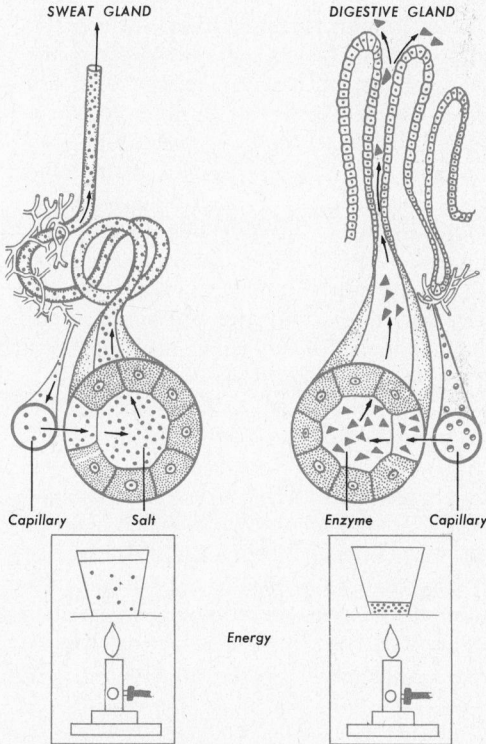

SWEAT GLAND   DIGESTIVE GLAND

Capillary   Salt   Enzyme   Capillary

Energy

Fig. 16–23. Two different types of glands and how they function. Sweat glands merely extract the substance from the blood, concentrate it, and secrete it. Digestive glands, on the other hand, must manufacture the enzyme from raw materials supplied from the blood, and then secrete the finished product. In both instances energy is required.

concentration just as it would be if heat were used to evaporate an equal amount of dilute salt water to a similar concentration (Fig. 16–23). Careful measurements of gland cell demonstrate that some respire at a higher rate than any other cells of the body, even those of the heart.

Chromatophores

Although color change is not found in human beings other than the gradual tanning of the skin as a result of exposure to sunlight, many lower animals are equipped with a very efficient and often spectacular color-changing apparatus. In some animals, such as the squid, the variety and rapidity of change in color are almost fantastic. It can change from pearly white to intense black almost instantaneously. Others such as fish and reptiles change more slowly, but the final product is equally dramatic. Such fish when moved experimentally from an aquarium with a black bottom, over which they become very dark, to one of white sand shortly become a very light color. Furthermore, some, like the flounder, actually produce the mottled effect of colored stones on the ocean floor. Lizards change from the color of the bark on the tree trunk to the intense green of the leaf. The protective value of such a mechanism is obvious.

This color change is accomplished by the movement of pigment either in the effector end-organ in the skin or in parts closely associated with it. In vertebrates, the pigment is confined to single cells which are scattered throughout the skin of the animal. When light falls on the eyes of the fish, stimuli are sent to the chromatophores which adjust the amount of pigment that spreads out on the surface to obtain just the right shade of color to match the background. Blind fish remain one color no matter what the background. On a light background, the normal response of each chromatophore is to concentrate the pigment granules of the cell into compact "pin points." On a dark background the same pigment granules

Fig. 16–24. The leopard frog changes from a light color where the spots are very prominent to a darker color where the spots are obscured. The activity of the chromatophores which are responsible for these varying colors is shown magnified in the enlargement of the portions of skin shown on the right. This activity is controlled by hormones.

spread out at the surface, darkening the entire area.

Although most of the pigment cells in animals are primarily under the influence of the nervous system, some, such as those of Amphibia (Fig. 16–24), are controlled by hormones, although the initial stimulus is via the eyes. Pieces of frog skin, for example, can be stimulated to change color simply by dipping them in solutions containing specific hormones. When in contact with the hormone **intermedin** (from the intermediate lobe of the pituitary), the skin darkens, and the opposite reaction can be induced with epinephrine. However, under normal conditions, stimuli from the eyes excite the endocrine glands whose secretion causes the chromatophores to respond. It is a chain reaction.

Electric Organs

Powerful electric discharges can be produced by a few species of fish. They are "triggered" by stimuli coming from the nervous system. The electric eel and ray possess special "electric organs" that are composed of a great many modified muscle cells so arranged as to accumulate their individual action currents and build up a considerable voltage. In some forms as much as 400 volts have been recorded, which is sufficient to stun or even kill a small fish. When a small light bulb is placed in the circuit, flashes have been recorded.

CHEMICAL COORDINATION— THE ENDOCRINE GLANDS

An important adjunct to the nervous system in bringing about coordination of the vastly complex animal body is the **endocrine system.** This is made up of glands located in various regions of the body which secrete powerful organic compounds, called **hormones,** directly into the blood stream. Their activity is manifest in other parts of the body. While the nervous system is responsible for quick action, the endocrine system functions in bringing about the much slower reactions which may extend over some period of time.

The glands which are known to be endocrine in nature today were described by early anatomists. Thus, Galen, in the second century A.D., described the tiny pituitary gland of mammals, although he could assign no function to it. Indeed, it was not until nearly the end of the last century that actual experimentation began to bring to light the function of these mysterious glands.

The British physiologist E. H. Starling, studying the functions of the duodenum in 1905, named the endocrine secretions **hormones,** and defined their function. As our knowledge of the numerous hormones found in both plants and animals increases it becomes clear that these are specific chemical compounds which are manufactured by special regions of the organism. They either diffuse or are transported in the blood stream to another region of the organism where, in very low concentration, they regulate and coordinate cellular activity. Chemically they are steroids, amino acids, or proteins and are, therefore, very diverse. This wide variation in chemical composition makes it impossible to define them as belonging to any group of chemical compounds.

The Endocrine Glands and Their Secretions

The endocrines evolved after the nervous system, so it might be expected that they would be found in the more highly specialized animals where the nervous system could not take care of the multitudinous jobs assigned to it. However, hormonelike substances from plants and protozoa have been described. Among the invertebrates, endocrine glands have been found in mollusks, annelids, crustacea, and insects where they are important for coordination or other functions. They probably exist in all groups of animals. They are consistently found among the vertebrates, even in such low forms as the cyclostomes. The glands themselves are derived embryologically from various sources and in their evolution have performed different functions. For example, the hormones that control chromatophore activity in the amphibia can exercise no comparable function in birds and mammals because they have no chromatophores. However, the hormone does stimulate pigment synthesis in mammals. The amphibian hormones are present in higher vertebrates as can readily be demonstrated by the injection into a frog of the proper extract. Undoubtedly, such recently acquired hormones as those that stimulate lactation in mammals are derived from similar hormones present in lower vertebrates where they perform a different function. There has been a long, slow, biochemical evolution of these complex compounds which have an intricate interrelationship in the higher animals today.

From the first, experimental research in **endocrinology,** as this branch of biology is called, has required the use of certain techniques in understanding the function or functions of a suspected gland. The simple removal of the gland and assigning whatever changes that appear in the animal thereafter as a result of the hormone produced by that gland may or may not be correct. It so happens, as we soon learn, that the normal functioning of any organ or organ system is due to a number of hormones working together (synergism) or against one another (antagonism); hence symptoms resulting from the removal of a single gland may be caused by more than the mere lack of that hormone. Other hormones, now uninhibited, may be in part responsible for the observed abnormalities. Therefore, the experimental design may involve the removal of a number of glands simultaneously, and the substitution of purified extracts for each one. In this way the function of each may be defined.

A common practice in defining the function of an endocrine gland is to substitute another gland from another animal for the one removed or to feed dried glands, or better yet, to inject purified extracts of the gland. The degree to which recovery is observed gives some evidence of the function of the gland or the extracts. Here again the design of the experiment must be carefully planned if the results are to be meaningful.

A great deal of effort in the past few years has been made to purify hormones so that the chemical formula of each could be written. Since the amount of a hormone in the gland, blood, or urine is extremely small, enormous quantities of these materials are required to find even a trace of the hormone. For example, it requires over two tons of pig ovaries to yield a few milligrams of estradiol. Even so, many hormones are being chemically defined today.

In human beings, the occasional disfunctioning of endocrine glands because of tumors or other abnormalities has given physicians and biologists abundant material for study and, in some cases, for experimentation. Results of experiments on lower animals have also revealed a great deal of information that has been applied directly to the alleviation of many endocrine aberrations in man.

The ultimate goal in experimentation with endocrines is to find the exact chemicals that are involved and to be able to prepare them in the laboratory. Once the chemical structure of these substances is known, they can be synthesized and used in animals with deficiencies to restore normal conditions. This has been accomplished for some hormones, but there is still much to be learned.

The vertebrates have seven clearly recognized endocrine glands: the **gonads, pancreas, duodenum, thyroid, parathyroids, adrenals,** and **pituitary.** Although they are referred to as ductless glands, the first two do have ducts—the endocrine secretion does not leave the gland through the ducts but instead enters the blood stream directly. These two, together with the duodenum, have two separate glandular functions. The gonads function in the production of eggs and sperm in addition to their endocrine function of producing hormones that are responsible for the secondary sexual characteristics. The pancreas produces digestive enzymes in addition to insulin, and the duodenum has several functions besides producing secretin. Moreover, a single gland such as the pituitary produces several hormones, each with a strikingly different function. Other endocrine glands may be found, although this appears much less likely than the possibility of discovering new functions for the glands already known.

The location of these seven endocrine glands is indicated in Figure 16–25; and the functions of their hormones and of other hormones are indicated in Table 16–1.

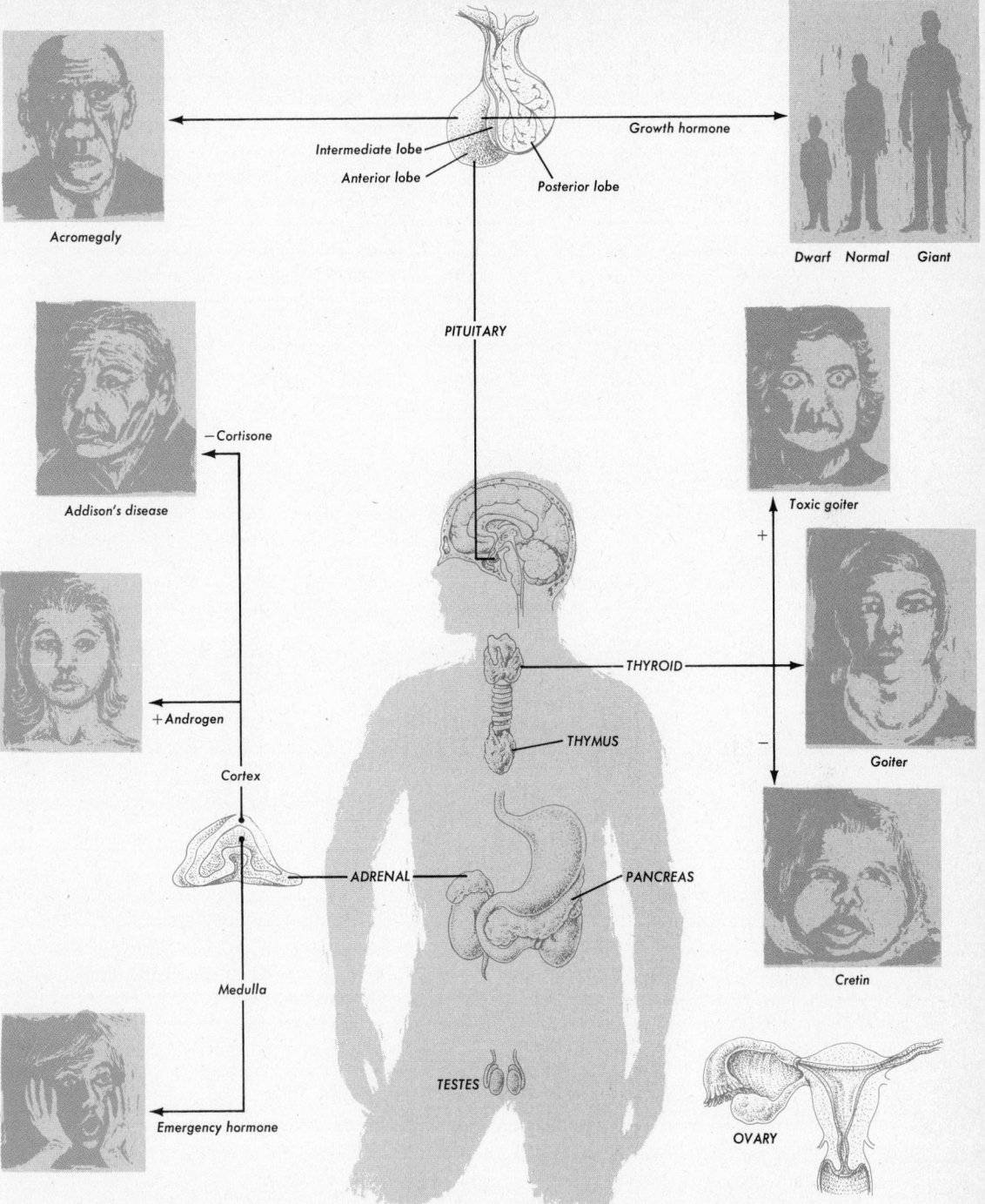

Fig. 16–25. The location of some of the endocrine glands and some of the results of their normal activity and malfunctioning.

TABLE 16–1.   Vertebrate Hormones and Their Functions

SOURCE	HORMONE	FUNCTIONS
Stomach	Gastrin	Gastric juice secretion
Duodenum and Jejunum	Secretin Pancreozymin Cholecystokinin	Pancreatic secretions Frees digestive enzymes into pancreatic juice Controls gall bladder
Pancreas (islets of Langer-hans)	Insulin Glucagon	Lowers blood glucose; stimulates glucose metabolism and protein and fat synthesis Raises blood glucose; enhances glucose release from liver
Testes	Androgens (Testos-terone)	Development of male sexual characteristics

Ovary
  Follicle
 	
  Corpus
  luteum
	Estrogens (Estradiol, estrone, estriol)
	Progesterone
	Relaxin
	Progesterone and Estrogens
		Development of female sexual characteristics
		
		Changes during gestation
		Relaxes cervix and pelvic ligaments during birth

Placenta	Chorionic gonadotropin Relaxin Progesterone Estrogens	Same as luteinizing hormone (see below)
Thyroid	Thyroxin	Controls growth, development (nervous tissue), and basal metabolism (birds and mammals)
Parathyroids	Parathormone Calcitonin	Lowers serum phosphate; raises serum calcium Lowers serum calcium
Adrenal Cortex	Corticosterone	Synthesis of glucose and glycogen, protein degradation, stress, prevents allergic and inflammation symptoms
	Aldosterone Androgens	Controls levels of sodium and potassium; retention of sodium in kidney (see above)
Adrenal Medulla	Epinephrine Norepinephrine	Controls size of arterioles and renules, influences heartbeat; raises blood glucose and stimulates oxidative metabolism Increases blood pressure by constricting arterioles

TABLE 16–1. Cont.

SOURCE	HORMONE	FUNCTIONS
Pituitary Anterior lobe	Somatotropic hormone (STH) Somatotropin	Growth (extremities and skull), protein synthesis
	Thyroid stimulating hormone (TSH), Thyrotropin	Growth of thyroid gland and synthesis of thyroid hormones
	Adrenocorticotropic hormone (ACTH) Corticotropin	Growth of adrenal cortex and synthesis of cortical hormones
	Follicle-stimulating hormone (FSH)	Growth of ovarian follicles in female and promotes spermatogenesis in male
	Luteinizing hormone (LH)	Ovulation, production of corpus luteum, and secretion of progesterone and estrogens in female. Promotes secretion of androgen in male
	Luteotropic hormone (LTH) Prolactin	Synthesis and secretion of milk in mammals
Intermediate lobe	Melanophore stimulating hormone (MSH) Intermedin	Controls dispersion of pigment in melanophores; melanin synthesis
Posterior lobe	Stores and secretes oxytocin and vasopressin	(see below)
Hypothalamus	Unknown number of releaser hormones Origin of posterior pituitary hormones	Controls the release of hormones from the anterior pituitary
	(a) Oxytocin	Controls uterine contraction and release of milk from mammary gland
	(b) Vasopressin	Antidiuretic action
Kidney	Renin	Converts the blood protein angiotensinogen to angiotensin
Blood	Angiotensin	Raises blood pressure and stimulates aldosterone secretion

The Duodenum

Two British physiologists, Bayliss and Starling, found that the highly acidic food passing from the stomach into the duodenum stimulated the walls of the latter organ to produce a substance which circulated in the blood stream to the pancreas, causing it to secrete its products into the digestive tract. This substance they called **secretin.** We speak more about secretin in the chapter on digestion. With their pioneer work, the field of endocrinology was initiated.

Other hormones produced by the intestinal wall are **enterogastrone** and **cholecystokinin.**

The Pancreas

This compound gland has ducts and its most obvious function is producing digestive enzymes which are drained off to the duodenum through the pancreatic duct. The digestive function of the pancreas is discussed later; here we are concerned only with those portions of the pancreas which produce hormones.

Lying scattered among the pancreatic juice-secreting cells are clusters of cells called **islets of Langerhans** which function in secreting hormones. There are two kinds of these cells; the alpha cells which produce **glucagon** and the beta cells that secrete **insulin.** Both of these hormones function in controlling sugar metabolism. Before discussing them let us take a look at the historical background that leads up to the discovery of insulin.

The story of the discovery of insulin is one of the more fascinating sagas in the annals of biological science. As far back as 1890 it was known that if the pancreas were removed from a dog, death followed in a few weeks. The salient point that came out in the early experiments was the appearance of large quantities of sugar (glucose) in the urine of such dogs. The two German workers who first performed these operations noticed that ants were attracted to the cages of these dogs and found that it was the sugar in the urine which was attracting them. This reminded them of human diabetes, a disease that had been known for centuries. A long series of experiments followed by workers all over the world, and it was eventually proven conclusively that the islets of Langerhans produced a hormone called insulin that was responsible for retaining and storing sugar in the body. If these islets were destroyed, as in the case of human diabetes, sugar was no longer utilized in the liver and

other tissues, but poured out through the kidneys into the urine and was lost from the body. This was a great discovery, but what could be done about it now that the cause of this dread disease was known?

The first step was to try to find a substitute for the nonfunctioning islets. Feeding the whole pancreas to depancreatized dogs failed to produce the slightest effect. The hormone was digested in the alimentary tract; consequently, it never got into the blood stream where it could be carried to the liver and tissues in which it would work. The next step was to inject an extract into the body, but because of the difficulty encountered in getting a pure product, no satisfactory results were obtained for 20 years. During this long period experimenters were attempting to obtain a pure hormone from the whole glands of various animals, mostly domestic animals such as cattle, sheep, and hogs.

It occurred to a young Canadian physician, Dr. Frederick Banting, that perhaps the digestive enzymes produced by the pancreas destroyed the insulin before it could be extracted. This was later shown to be true. Banting reasoned that since the embryonic pancreas was known to produce the islet tissue before the enzymes appeared in the pancreas, if such glands were used, perhaps the hormone could be isolated in an active state. In 1922, he and three other men—Best, Macleod, and Collip—working together, set out to isolate the hormone. After a great deal of labor, they eventually prepared a product which caused no ill effects on the dogs when injected under their skin and which alleviated their diabetes. It was a short step to the treatment of humans, where success was immediate. Thousands of diabetics then had, for the first time, some means of staving off an early death from a disease that had always been fatal.

With the ensuing years insulin was produced in more concentrated and purified form, and today its chemical structure is known. Through the brilliant work of the English biochemist, Sanger, insulin is known to be a protein (molecular weight—6,000) consisting of two peptide chains, one containing 21 amino acids, and the other 30. It was discovered that most commercial insulin preparations contain another hormone, called glucagon, which has the opposite effect of insulin; namely, it increases blood sugar levels instead of reducing them as insulin does. Glucagon has been found to be a protein also and is secreted by the alpha cells as already mentioned.

Both glucagon and insulin function primarily in the control of carbohydrate metabolism, although hormones from both the pituitary and adrenal glands also play a part. Insulin has several effects all of which are related. It brings about the transport of glucose across the cell membrane which is reflected in a decreased blood glucose level. It also promotes the storage of glycogen in the liver and muscles, and increases the burning of glucose to carbon dioxide and water. If insulin is produced in insufficient quantities sugar utilization is correspondingly decreased and with this faulty carbohydrate metabolism the metabolism of fats and proteins is seriously interfered with. Glucagon increases blood sugar by its action on the enzyme **phosphorylase** which controls the conversion of liver glycogen to glucose.

Insulin therapy is responsible for the near-normal lives of hundreds of thousands of diabetic men, women, and children. The hormone is either taken by mouth or injected under the skin at rather frequent intervals, depending on the severity of the disease.

The Nature of Diabetes. Without treatment a diabetic suffers from insatiable thirst, excessive urination, a gradual loss in weight, general body weakness, and finally a coma which terminates in death. During this course the sugar in the blood and urine is found, by measurement, to be abnormally high (as much as 8 percent in the urine), the liver loses its glycogen and finally, in the precoma state, acetone and partially degraded fats also appear in the blood and urine. Before death the acetone may reach such concentrations that it can be detected on the breath. All of these symptoms are immediately relieved with the administration of insulin.

The first, most obvious function of insulin is to maintain normal carbohydrate metabolism in the body. For some reason, in the absence of insulin the liver fails to store glycogen, and glucose is oxidized very poorly. Strangely enough, the abstinence from carbohydrates in the diet does very little good in preventing any of the symptoms. In fact, it seems that without insulin the body mobilizes all of the sugar at its disposal and discharges it from the body via the urine. Even the amino acids are deaminized at an abnormally high rate, so that the sugar residue is added to the already heavily sugar-laden urine. Furthermore, the fats are withdrawn from storage and only partially oxidized, leaving the unoxidized fractions in the blood and urine. It seems that all the forces of the body are put forth to produce sugar which is then wastefully thrown away. Death is the inevitable answer to such a course, unless insulin from an external source can intervene.

In some diseases of the pancreas the islets are stimulated to produce more than the normal amount of insulin. The results are the same as when a diabetic gives himself too much insulin. The blood sugar is dropped to such a low level that the brain becomes irritable and finally the person goes into the severe condition called **insulin shock.** Most diabetics are familiar with the possibility of this condition and accordingly carry sugar or some other sugar-containing substance that can be taken quickly to overcome the lowered blood sugar level. Because of his liability to insulin shock or coma, either of which may render him unconscious, it is advisable that the diabetic carry among his possessions a card or tag identifying his disease, so that in event of collapse his condition will not be mistaken for some other malady or even drunkenness.

The Gonads

The testes and ovaries are also compound glands whose primary function is the production of sperm and eggs; in addition, they have very important endocrine functions which have developed in the evolution of the vertebrates. They are concerned with those organs which function in the caring of both the gonadal products (eggs and sperm) and the early embryo. The complete process of reproduction becomes highly complex in mammals where the young are few in number and are cared for both within the body of the mother long before birth and for some time after they have made their appearance in the outside world.

The gonad's primary function of producing eggs and sperm is discussed in Chapter 21 and only such anatomy as is necessary for an understanding of their endocrine function is given at this time.

The Testes. Located among the sperm-producing tubules of the testes is a special type of tissue (interstitial) which produces several hormones, the most active being **testosterone,** a steroid, which is secreted directly into the blood stream. It affects the general metabolism which results in protein synthesis and body growth. It also stimulates the production of the **secondary sexual characteristics** in all male vertebrates, characteristics which are associated with maleness. They are the very obvious traits which separate the male from the female both morphologically and physiologically. The comb and brilliant plumage of the cock, the antlers and massive body of the bull moose (Fig. 16–26), and the beard of man are all secondary sexual characteristics. The absolute proof that these characteristics are associated with the testes can be demonstrated by **castration,** or removal of these glands, an ancient custom practiced by man not only on his domestic animals but in some cases also on his fellow man. The castrated

cock shows none of the comb and wattles of the normal male; the steer is quite different in both its anatomy and behavior from the bull; the gelding has none of the fire nor cantankerousness of the stallion. In ancient times it was customary to castrate male slaves, producing **eunuchs** who would then be docile, subservient beasts of burden and trusted keepers of the harem. When it was desired to retain the soprano voice of a particularly talented youth, castration did the trick, and thus choirs could be produced with remarkable musical qualities.

The onset of interstitial tissue activity is associated with puberty when pubic hair, change of voice, and increased size of the genitalia occur.

If testosterone is injected into a castrated animal or a testis is transplanted into some part of the body where it can grow and secrete testosterone into the blood stream, the male secondary sexual characteristic will be restored. Such injections given to a castrated female will cause her to develop masculine characteristics. A perfectly normal egg-laying hen can be induced to become a functional father rooster with comb, wattles, and crow and all by castration followed by a series of injections of testosterone. By removing the single ovary from a seven-week-old chick a normal rooster will result (Fig. 16–27). This can be accomplished in female birds because they possess a rudimentary testis and no external genitalia. Sex reversal in mammals is limited to the secondary characteristics only, unless treatment with hormones is carried out in early embryonic stages when complete reversal can be produced.

Occasionally the testes in mammals fail to descend normally into the scrotal sac as they should do during the last few weeks of gestation. This condition is known as **cryptorchidism** and males in which it occurs are of low fertility or sterile. If, however, the testes are brought down into the scrotum by surgery, they very soon become functional and produce viable sperm. A cryptorchid is perfectly virile in every way although he is infertile; that is, he possesses all of the normal charac-

Fig. 16–26. Sexual dimorphism in birds and mammals.

Fig. 16–27. The removal of the single ovary in birds results in the subsequent development of a testis and male characteristics. In these photos the chickens on the extreme ends are female and male, respectively. The animal in the middle had its ovary removed at seven weeks of age and shows a well-developed male comb.

teristics of the male including sex drive. This is because his interstitial tissue is unimpaired, so that testosterone is produced in proper amounts to allow for normal development of his masculine characteristics. Upon microscopic examination, these testes will show perfectly normal interstitial tissue but degenerate sperm-producing tubules. Experiments show that if the normal testes of mammals are placed back into the body cavity or heated to the internal temperature of the animal the sperm tubules degenerate. Therefore, sterility of the cryptorchid is due to the higher temperature existing in the body as compared to the scrotal sac. This is difficult to correlate with the fact that the internal testes of birds are fertile and the temperature is even higher than that of mammals. In the long evolution of mammals one fails to discover the advantage of placing these organs, upon which the race depends for its perpetuation, in such a hazardous position when they would be much safer housed within the body cavity as are their counterparts, the ovaries.

Chemically, testosterone is well known and is found to be similar to one of the female hormones, both being steroids. Other related testicular hormones are collectively called **androgenic compounds** and seem to be generally distributed throughout the body. They are probably substances which are utilized in the production of testosterone or they are products of its breakdown, because they are found in the urine. Strangely enough

males produce female sex hormones (estrogens) and females produce male hormones (androgens). It is the relative amounts of each that determine the development of the secondary sexual characteristics.

The pituitary is intimately associated with gonadal hormines. If the pituitary is removed all cells of the testes regress resulting in the loss of androgen secretion and the secondary sex characteristic. Apparently, normal gonadal development requires pituitary hormones (LH, FSH) as well as testosterone.

After the testes had been associated with male vigor, the intriguing idea of transplanting them or injecting their extracts into the body of a senile male caught the imagination of early biologists. Long ago, Brown-Sequard, a famous physiologist, injected himself with testicular extracts which he professed renewed his vigor. This initiated a long series of experiments both on animals and on man himself. The results have been quite successful when hormones are administered to preadolescent boys but not in adult men. It can be concluded that testosterone does initiate and maintain the secondary sexual characteristics and probably contributes to sexual behavior and urge. In man, however, the latter function is so complexly interwoven with psychological reactions that it is difficult to determine just how much effect the hormone really has.

The Ovary. A far more complex battery of hormones is produced by the mammalian female generative apparatus than by the male. This is due to the recently acquired though intricate mechanism of caring for the developing embryo, both before birth and immediately after. These hormones are produced in the ovary, although others occur in various parts of the genital tract.

At birth and throughout the early life of a female, the potential eggs lie dormant in the outer region of the ovary. From puberty on they begin to grow. As an egg grows, there develops about it a fluid-filled space called the **Graafian follicle** (Fig. 16–28). These follicles produce a hormone, **estradiol** (estrogen), which is the counterpart of testosterone in the male. The influence of estradiol in the blood stimulates the onset of changes both in body contours and in the female organs; these changes result in the mature human female. A similar process occurs in all mammals. In the mature female with normal sexual cycles another hormone, **progesterone**, is produced by a special part of the ovary called the **corpus luteum.** After the rupture of the Graafian follicle and the liberation of an egg which moves down the oviduct toward the uterus, the cavity left fills with the corpus luteum which, in turn, produces progesterone. This hormone pre-

Fig. 16–28. A Graafian follicle from a rat, sectioned to show the egg.

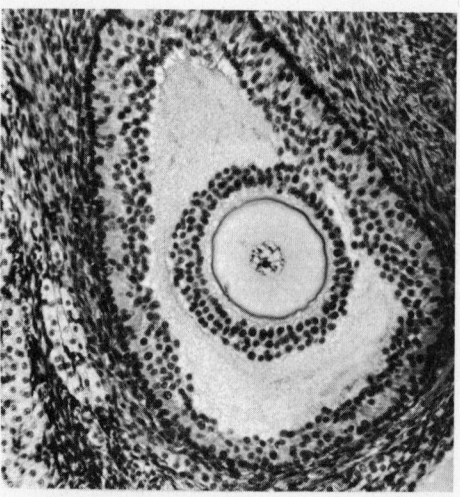

pares the uterus to receive the egg. If no fertilized egg reaches the uterus the menstrual cycle is initiated.

The **menstrual cycle** in humans and other primates has its counterpart in the **estrous** or "heat" cycle in other mammals. It involves a distinct rhythmic cycle of sexual activity during some part of which the female is receptive to the male. Such a cycle is completed every 14 days in guinea pigs, twice a year in dogs, and once every 5 days in rats. In humans and anthropoid apes the menstrual cycle is complex, involving a periodic sloughing off of the highly vascular lining of the uterus every 28 days by menstruation. In other mammals the uterine lining returns to the resting state with no sloughing off or bleeding. The word *menstruation* comes from the Latin *mensis,* which means month. The menstrual flow consists mostly of the epithelial lining of the uterus together with the incorporated gorged blood vessels. The course of events that precede this are rather well known, but just *why* it must occur, since it is not found in the cycles of lower mammals, lacks an immediate explanation.

Since we are dealing with a cycle, description can start at any point in it (Fig. 16–29). The beginning of the menstrual flow may arbitrarily be taken as the starting point for this discussion. About 4 days later a Graafian follicle begins to grow in the ovary and as it increases in size it produces more and more estradiol. Maturity is reached in about 12 to 18 days at which time a rupture occurs in its wall and the egg is released. This marks the end of estradiol production from this structure but not of the hormone itself. The empty follicle is quickly converted into the corpus luteum, which continues to produce estradiol, and in increasing amounts, the closely related hormone, progesterone, from the 15th to the 26th day. These hormones, in addition to bringing on the changes, already referred to, at puberty, are also responsible for the rhythmic menstrual cycle. The production of both estradiol and progesterone has a stimulating effect on the walls of the uterus, causing it to proliferate and to become highly vascular in preparation for the fertilized egg, if and when it makes its way into the uterine cavity. What happens from this point forward depends on whether or not the egg is fertilized.

If the egg is **not fertilized** the corpus luteum retrogresses and progesterone is reduced to zero during the 25th to 27th days of the cycle. This results in the sloughing off of the uterine wall known as the menstrual flow, which continues over a period of 4–5 days. Another Graafian follicle then begins to grow and the cycle is started over again. The pituitary hormones, which are discussed a little later, play an important part in this process. It might seem that a rather elaborate preparation is made each month for the event of pregnancy and that an unnecessary waste results when fertilization fails to occur. One speculates whether a little less preparation might be satisfactory until it is certain that fertilization has taken place.

If the egg **is fertilized** as it passes down the oviduct, the corpus luteum is retained and goes right on producing progesterone until just a few days before the end of gestation. The zygote is implanted in the uterine wall wherever it happens to touch. In fact, the highly vascular wall is so receptive to tiny particles that almost any small object is readily picked up by it at this time. The walls also produce mucus rich in glycogen, which probably acts as a source of energy for the early stages of the embryo until it gains a secure foothold and can withdraw nourishment through its placenta. With the developing stages of pregnancy, progesterone continues to cause further accommodations of the uterine wall for the enlarging embryo. It also causes the mammary glands to increase in size, prevents any further Graafian follicles from forming, and inhibits uterine contractions.

The placenta which develops from the uterus wall and embryonic membranes supports the embryo but also has endocrine functions which are involved with pregnancy. It secretes **chorionic gonadotropin**, a protein hormone which functions in stimulating the corpus luteum to remain active. Chorionic gonadotropin appears early in preg-

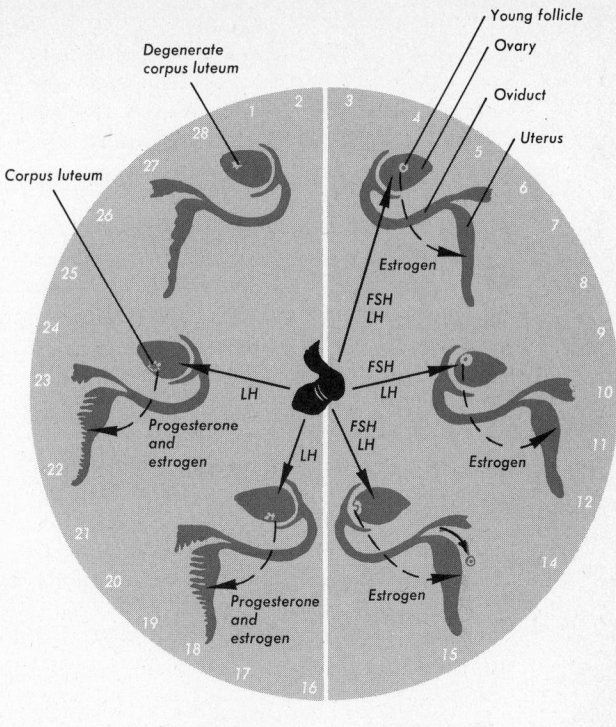

Fig. 16–29. Graphical outline of the events that occur during the menstrual cycle in a human being.*

nancy (about 2 to 3 weeks) and can be detected by injecting urine into a male frog, which will promptly shed sperm, a positive test for pregnancy. The placenta also secretes estrogens and progesterone which supplement that produced by the ovaries. The ovary as well as the placenta produce the protein hormone **relaxin** which relaxes the pelvic ligaments to permit the passage of the fetus through the birth canal.

As the end of pregnancy approaches, the corpus luteum ceases to produce progesterone and the uterine wall, which has during this time become an accessory in producing the hormone, reduces its output. This precipitates changes which are similar to the be-

* According to older accounts prolactin (LTH) is also supposed to be required for corpus luteum function but this is known to be true in only a few species.

ginning of menstruation, that is, in the absence of progesterone the uterine wall begins to degenerate, therefore becoming incapable of nourishing the fetus any longer. Furthermore, without the inhibiting effect of progesterone, the muscles of the uterine wall begin powerful contractions which eventually result in the expulsion of the fetus. The production of milk by the mammary glands occurs after birth due to another hormone, **lactogen,** which is produced by the pituitary gland and also is discussed later.

It takes some time after the birth of the offspring for the hormones to readjust themselves and the menstrual cycle once again to reestablish itself. This usually does not occur until the amount of lactogen from the pituitary subsides, which means, of course, that the offspring has ceased to rely on the mammary secretion as its principal source of food.

The Thyroid

The thyroid, together with the remaining glands, is purely endocrine in function. The thyroid has various shapes in different vertebrates but in man is bilobed and lies on either side and under the larynx (Fig. 16–25). The two lobes are connected by a narrow strip of tissue, called the **isthmus,** passing across the trachea. The presence of the gland can be determined by merely feeling it with the fingers.

This gland was seen by early anatomists, and its importance suspected because they noted that in certain individuals it became enlarged, seeming even to cause their death. At the beginning of the Christian era the Greek physicians prescribed the drinking of sea water as a cure for **goiter** (the term used for the swollen gland). Later, others gave their patients products of the sea, such as dried seaweed leaves, which undoubtedly gave some relief because all sea products are rich in iodine, the important ingredient in the production of the thyroid hormone, **thyroxin.** The exact function of the thyroid was not known until replacement experiments in 1885 demonstrated that the gland did produce a hormone. This was isolated in pure

form in 1916 and synthesized in 1927 by Harrington and Barger, two English investigators. When this substance is administered to an animal deprived of its thyroid, the animal remains perfectly normal in every respect. If it is denied such treatment, stark metabolic changes occur which, if prolonged, may terminate the life of the animal. What is the specific function of this gland?

It is generally agreed that the thyroid gland secretes thyroxin, which controls the level of basal metabolism. The secretion of thyroxin is controlled by the pituitary hormone, **thyrotropin,** and vice versa. That is, if thyroxin drops in the blood, the pituitary sends out more thyrotropin which stimulates the thyroid to produce more thyroxin, a nicely tuned "feedback" system (Fig. 16–30).

The cells of the thyroid are arranged in follicles. These cells accumulate iodide from the blood and incorporate it into the protein **thyroglobulin,** which is subsequently hydrolized by enzymes to thyroxin, a derivative of the amino acid, tyrosine. Thyroxin then diffuses into the blood stream and thence to the tissues where it does its work. Many of the effects of thyroxin are due to stimulation of oxidative processes, but some, such as those on growth and development and on brain function cannot be explained by this effect.

Thyroxin must be produced at a uniform rate in order that these important processes proceed at what is spoken of as a normal level. If more or less is produced, these processes accordingly increase or decrease in speed with accompanying symptoms that are very definite and easily recognized. A diseased thyroid merely produces too little or too much of its secretion. If it continues in either direction too long, illness inevitably results.

Underactivity. When the gland fails to produce the proper amount of thyroxin, the effects are much more pronounced in a young animal than in an adult. For example, if the thyroids are removed from tadpoles or pups the animals do not mature properly. The tad-

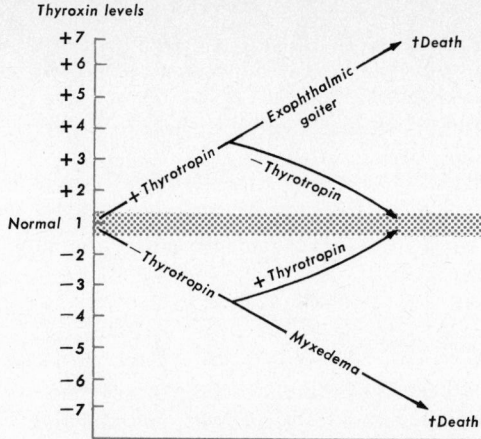

Thyroxin levels

Fig. 16–30. Under normal conditions thyroxin levels in the blood remain relatively constant owing to a carefully controlled feedback mechanism. As a result of thyrotropin stimulation the thyroxin output is increased. If this continues, exophthalmic goiter results, and ultimately death. When the flow of thyrotropin is cut back, less thyroxin is produced and the normal level is maintained. Likewise when too little thyrotropin is produced by the pituitary, the thyroxin level drops eventually, resulting in myxedema. This is normally prevented by an increased output of thyrotropin which causes the level of thyroxin to return to normal.

poles do not metamorphose into frogs, and the pup does not mature into an adult dog. Likewise in human beings, if a child has a deficient thyroid he becomes a **cretin.** Such a child is small and badly formed, with pudgy, puffy skin and swollen tongue, and his mental development is at an almost complete standstill. If given thyroxin in the early stages of the disease, the child responds remarkably well and can grow into a normal adult. Obviously, if a human cretin is allowed to live for twenty years without treatment, thyroxin will do him little good because his body tissues will have completed their development and can be changed but little.

If the thyroid becomes atrophied for some reason and fails to produce an adequate supply of thyroxin in the adult, a familiar disease known as **myxedema** results. The obvious symptoms of the disease are general loss in vigor, reduction in mental activity, increase in weight, and a thickening of the skin to give it a puffy appearance. Less

obvious symptoms are a drop in basal metabolism, the improper burning of food, sometimes as much as 40 percent below normal, a slowing of the heart rate, and a lessening in the sex drive. It seems that the entire machinery of the body slows down. The administration of proper amounts of thyroxin or thyroid extracts restores the rate of metabolism to its normal level, and subsequently all symptoms of the disease disappear. Sometimes in surgery too much of an overactive gland is removed and the patient may then find that he is suffering from myxedema and must take thyroxin all the rest of his life. Fortunately, the digestive enzymes have no effect on thyroid extract, thyroxin, or even the dried gland (in contrast to insulin), a fact which permits administration by mouth, an important detail in the treatment of any disease.

Fifty years ago the presence of an unsightly enlarged thyroid was very commonplace in certain parts of the world. Surprisingly, these regions were rather well defined and in them even the domestic animals had goiters. An examination of the soil and water showed that there was a marked deficiency of iodine. Along with this discovery, the thyroid secretion was found to be remarkably rich in iodine; it was not difficult to fit the two together and conclude that goiter appeared in regions where there was very little iodine available in the food products and water. These areas were spotted over the world. In the United States they are concentrated along the St. Lawrence River and Great Lakes regions. For example, in 1924, 36 percent of the schoolchildren in Detroit showed incipient endemic goiter, but within seven years after the addition of potassium iodide to table salt the incidence had dropped to 3 percent.

The presence of a goiter does not necessarily mean that the gland is under- or overactive. It does mean, however, that there is some sort of disturbance in the thyroid output. It may be compensating for the lack of iodine and does this by producing more thyroid tissue in an effort to supply sufficient thyroxin to keep the body at a normal basal metabolic level. Such compensating action is not uncommon in other parts of the body (for example, an enlarged heart muscle). If the thyroid cannot maintain a normal level of thyroxin, myxedematous symptoms may be evident. If it can supply the proper amount there are no symptoms of the disease, although the individual harbors the greatly enlarged gland on the front of his neck.

Overactivity. For some unknown reason, the thyroid sometimes begins spontaneously to produce more thyroxin than the body needs, and this may be accompanied by a slight enlargement of the gland. It differs from the simple goiter described above because while it may not be enlarged, or only slightly so, its output is far greater. Obviously, abnormally high amounts of hormone in the blood stream increase the rate of burning foods (30 percent or more) and speed up all the bodily activities. More food is consumed, yet there is a wasting away of the body. Profuse sweating occurs, the heart is overworked, and external heat cannot be tolerated. All this produces a highly irritable and nervous individual who is continually on the move but accomplishing very little. The action is like running an automobile at top speed with the brakes set. It is clear that such activity will soon result in the destruction of the organism itself.

This condition, which is called **exophthalmic** (from the fact that it sometimes causes the eyes to bulge out of their sockets) or **toxic goiter,** is controlled by destroying a part of the cells that produce the hormone (Fig. 16–31). This is easily accomplished by surgery and such operations are very successful. In this age of the atom a new method has been discovered which is sometimes employed when for some reason or other it is inadvisable to operate. It is the use of radioactive iodine. Iodine, when subjected to atomic radiations, becomes radioactive itself. Such a form is called an **isotope.** Since the thyroid picks up about 80 percent of all the iodine taken into the body, the exact amount

that will be delivered to the gland shortly after swallowing isotopic iodine can be determined beforehand. Furthermore, radiations are known to destroy thyroid tissue; therefore, by feeding radioactive iodine "cocktails" to the patient, a certain amount of success in destroying a part of the gland has been achieved. Its more satisfactory use, however, is in the treatment of cancer of the thyroid.

The Parathyroids

In the early studies of the thyroid much confusion resulted because in removing the gland the four tiny parathyroids embedded in the thyroid were inadvertently removed also (Fig. 16–25). The symptoms that followed this were the result not only of thyroid deficiency but also of parathyroid deficiency. The small parathyroid glands were discovered in 1891 and many years later their true function was determined.

Fig. 16–31. A case of hyperthyroidism.

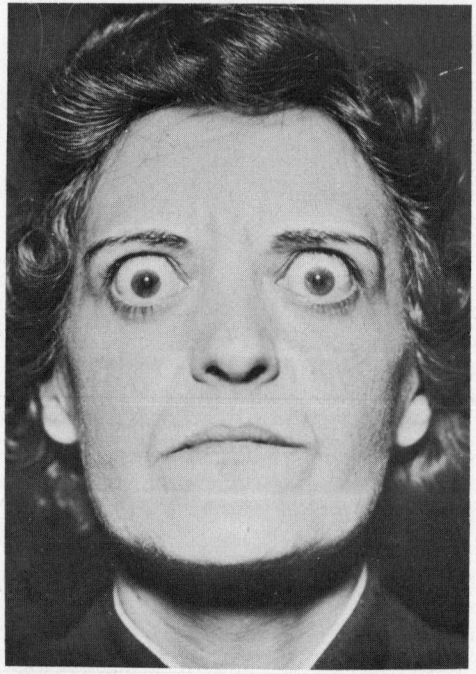

If the parathyroids are removed from an animal, injections of the hormone **parathormone,** produced by the glands, will keep that animal in good health. If no hormones are given, the animal suffers from severe muscular tremors, cramps and finally convulsions. The composite symptoms are called **tetany,** and without treatment the animal passes into a coma and death soon follows. It is now known that the parathyroids actually produce two hormones. The first of these, parathormone, maintains proper levels of calcium and phosphate by regulating kidney excretion of these ions and controlling the amount of calcium and phosphate deposited in bone. The second hormone, calcitonin, suppresses blood calcium by unknown means. When the parathyroids are removed the blood calcium level falls rapidly, which correlates with the symptoms of the disease. Administration of calcium will prevent symptoms of parathyroid deficiency. If the glands produce an overabundance of the parathormone, the calcium level in the blood then rises too high and even the calcium of the bones is sacrificed, so that a weak, twisted skeleton is the result.

The Adrenals

The adrenals are located on the upper inner edge of each kidney, as one might guess from their name (Fig. 16–25). Their combined weight is no more than an ounce, and each is composed of two parts, an outer covering called the **cortex** and an inner dark-colored mass called the **medulla.** The gland is therefore a composite one and each part has a separate origin, the cortex coming from the mesodermal lining of the coelom whereas the medulla is derived from a part of the neural tube. One might expect structures of such different origins to have different functions and they do.

The Medulla. The medullary portion of the adrenal produces two closely related hormones, **epinephrine** (sometimes called **adrenin** or **adrenaline**) and **norepinephrine** (or

noradrenalin). Chemically they are amines derived from tyrosine and their functions are similar. They have also been synthesized from sources other than adrenal glands. Related synthetic compounds such as **ephedrine** produce similar effects when administered to animals.

If the medullary portion of the adrenals is removed from an animal, death does not follow nor is the animal markedly affected by its loss. If injections of medullary extract are given to such an animal or one with intact adrenals, characteristic changes occur rather rapidly. The heart action becomes stronger and the blood vessels to the skin and viscera constrict, sending most of the blood to the muscles, brain, and lungs. The hair "stands on end," the pupils dilate (wide-eyed), and the skin blanches. The spleen constricts, forcing its reserve of blood out into the general circulation, and simultaneously the blood's ability to clot is stepped up. More glycogen in the liver is converted to glucose, so that the total amount in the blood is definitely increased. This chain of events prepares the body for undue stress such as occurs in a fight or a sudden retreat. The body is made ready to function to the maximum of its ability in case a sudden burst of energy is needed. Provision against possible injury is afforded by the increased speed of blood coagulation. This whole series of effects is similar to excitation of the sympathetic nervous system. Thus, both the nervous system and the adrenal medulla play an important rôle in fear and anger. Knowledge of this fact has led to the so-called emergency theory of adrenal function.

Epinephrine has clinical use in cases of asthma where it dilates the breathing passages. It is also helpful in starting a heart that has suddenly stopped beating.

The Cortex. The **cortex** of the adrenal is essential for life, although when even such a small portion as one-fifth of the total gland tissue is present, life is undisturbed. Its products are numerous, in fact, over 30 compounds have been isolated in recent years.

The first crude extract, isolated in 1930, was called **cortin** and was effective in treatment of people suffering from Addison's disease, which is the name identified with a deficiency of this portion of the adrenals. Since that time many compounds have been extracted, the best known and most effective being **corticosterone** which functions in the conversion of proteins to carbohydrates. Another is **aldosterone** which controls sodium and potassium metabolism. A third, **adrenosterone,** influences growth and protein synthesis. It is a weak androgen and in excessive amounts can cause masculinization of women and young children, as sometimes happens with adrenal tumor. Most of the other hormones function as precursors for the synthesis of those just mentioned. The cortex hormones are required in some not-well-understood way for "stress" tolerance (p. 339).

If the cortex fails to function, marked changes occur which are fatal if uninterrupted by treatment. Addison's disease develops and as a result the carbohydrate metabolism is greatly affected, as indicated by a drastic drop in blood sugar, because of the inability of the enzymes to convert the proper amounts of proteins to carbohydrates and then to convert the latter to sugar in the liver. Salt (NaCl) is lost from the blood and tissues at a rapid rate which reduces the entire blood volume and with it the blood pressure. As Addison's disease progresses, the skin bronzes owing to the deposition of melanin and the sexual functions fail due to an actual atrophy of the Graafian follicles and the seminiferous tubules.

If, on the other hand, the cortex becomes overactive as a result of irritation caused by a tumor, changes of a different kind occur. Carbohydrate, salt, and water metabolism is disturbed and the person becomes weak and tends to waste away. In males, the maleness is greatly enhanced, accompanied by excessive hair growth. If it happens to a very young male child, the sex organs may become fully mature (except the testis) within the first or second year of life, and the hair, muscula-ture, and voice resemble those of an adult man. These are very rare cases, fortunately. In females, the situation is even worse. If the overactivity occurs in an adult woman, the changes are all toward maleness; the beard grows (the bearded lady in the circus), the body becomes more muscular, and the voice deepens. Even the female sex organs begin to atrophy and become nonfunctional.

These hormones, particularly corticosterone, are used clinically for suppressing allergies, tissue inflammation, and tissue proliferation in the joints which causes rheumatoid arthritis.

The Pituitary

The last and perhaps the most complex of all the endocrine glands is the **pituitary,** or **hypophysis.** Located approximately in the middle of the head, it lies in a bony capsule and is attached to the base of the brain by a slender stalk, the **infundibulum** (Fig. 16–25). Like the adrenals, the hypophysis is a compound gland, composed of three lobes: the **anterior** and the **intermediate** lobes arise embryologically from an outpocketing of the roof of the mouth; and the **posterior** lobe originates as a solid outgrowth from the floor of the brain (hypothalamus). The point of contact with the brain through the infundibulum is retained while all connections with the pharynx are lost very early in embryological development. The anterior lobe is the larger of the three and has no nerves but is amply supplied with blood vessels. This abundant vascularization provides means for receiving hormones that stimulate it to produce its own secretions and a means for their escape. The posterior lobe is composed of nonmyelinated nerve fibers coming to it from the hypothalamus. It is also supplied with blood vessels but they are distinct from those going to the anterior lobe.

Although it is very difficult to operate on the pituitary in man, experimental animals such as the frog and rat lend themselves to such surgery. By perforating the roof of the mouth the pituitary can be neatly removed and subsequent effects observed. When this operation is performed on a young mammal, growth is inhibited at once, sexual maturity never occurs, and both the thyroid and the adrenal cortex atrophy. With the injection of pituitary extracts, the animal develops normally and the associated symptoms never appear. From these observations it is clear that the pituitary certainly gives rise to more than one hormone and that its influence is far-reaching in the animal body.

The Anterior Lobe. Six well-known hormones are produced by this lobe, and there may be others. The six are named according to the part of the body that they affect, with the addition of the suffix **-tropic**: **somatotropic** (growth), **thyrotropic** (thyroid), **adrenocorticotropic** (adrenal cortex), (two) **gonadotropic** (gonads), and **mammotropic** (mammary). The hormone is sometimes named by adding the suffix **-in,** that is, thyrotropic hormone may be called **thyrotropin.** Each of these is considered briefly in this order.

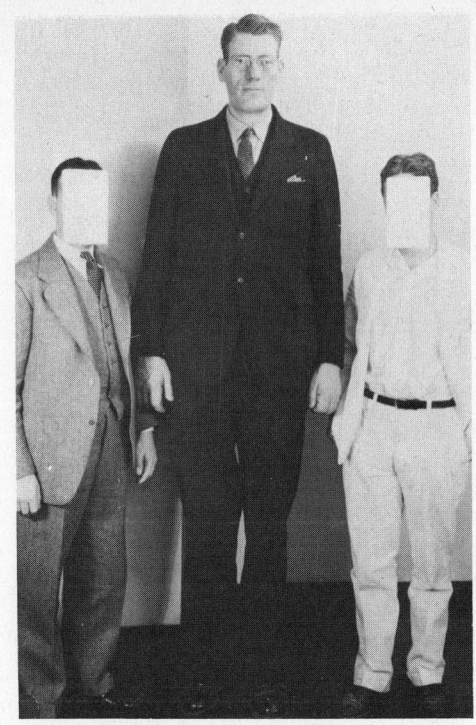

Fig. 16–32. Case of giantism. The men on either side are of normal height.

SOMATOTROPIC EFFECTS. The function of the pituitary was brought to the attention of early anatomists by the fact that an enlarged gland was always associated with giantism. These huge men (Goliath, whom David slew, was undoubtedly one) reach a height of nearly 9 feet and are rather well proportioned, with the exception of the extremities which are longer than normal (Fig. 16–32). An examination of the anterior lobe of the pituitary always reveals a greatly enlarged organ, sometimes reaching the size of a hen's egg. On the other hand, when the gland fails to produce the proper amount of hormone, a midget results. It is to be noted that giants or midgets are produced only when the gland either overfunctions or underfunctions in the young child.

Both dwarfism and giantism can be produced in animals simply by removing the pituitary in the first case and supplementing with grafted pituitaries or extracts (somatotropin) in the second. Dogs can be forced into giants by placing glandular material under the skin or injecting purified extracts of the anterior lobe of the pituitary. If the gland is removed in a pup, growth takes place very slowly, if at all, and the dog becomes a midget. In human beings with overactive anterior lobes, removal is the only remedy known so far. In pituitary midgets, injections of refined extracts of primate growth hormone is sometimes beneficial but it is not sufficiently abundant for clinical use.

If increased activity occurs after maturity has been reached, as it occasionally does when a tumor forms on the pituitary, the person does not then become a giant, although deep-seated changes do occur. Since the body has already ceased growing, the effort to produce a giant by further increase in size is restricted to the regions of the joints and the face. Consequently, a person so afflicted becomes barrel-chested, beetle-

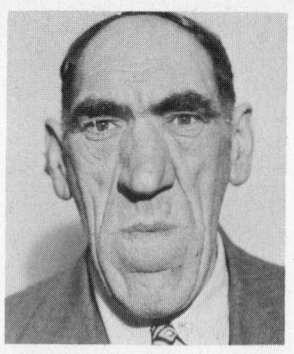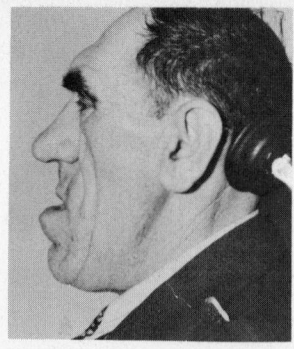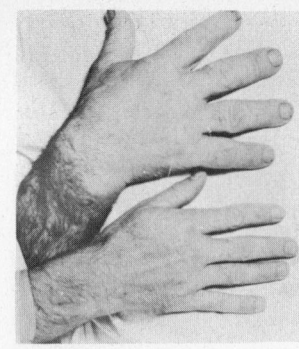

Fig. 16–33. A case of acromegaly. Note particularly the protruding lower jaw, massive brows, and the enlarged hands.

browed, and long-jawed, while the feet and hands grow very large (Fig. 16–33). Such a condition is spoken of as **acromegaly.**

THYROTROPIC EFFECTS. The effects of **thyrotropin** (TSH) are quite well known; it stimulates the enlargement of the thyroid gland, the uptake of iodine, and the synthesis and secretion of thyroid hormones.

CORTICOTROPIC EFFECTS. The adrenal cortex and its production of hormones are directly under the control of a specific hormone from the pituitary called **adrenocorticotropic hormone** (ACTH). This substance has been isolated and its chemical composition determined. It is a peptide containing 39 amino acids. Its action parallels cortisone in every respect, which proves that its action is through the adrenal cortical hormone. Therefore ACTH must control the cortisone output. By increasing ACTH, cortisone is also increased.

An interesting sidelight on the discovery of ACTH came from a common observation, namely, that during pregnancy arthritic women recover from their rheumatism. Shortly after the child is delivered the disease returns. Both cortisone and ACTH have the same remitting effect as pregnancy. Apparently, the added burden of childbearing stimulates a greater output of ACTH which, in turn, stimulates a greater production of cor-

tisone, the combined action of which improves the arthritis. Just what the specific action is on the disease itself is not known.

Both of these hormones have proven beneficial in cases of extensive body injury such as that caused by severe lacerations or burns. They serve to muster all the body's potentials toward regenerating new tissue as well as preventing shock and the other symptoms associated with severe trauma.

GONADOTROPIC EFFECTS. A young hypophysectomized animal never becomes sexually mature, and if the operation occurs after maturity there is prompt atrophy of the sex organs. There is thus a close and important relationship between the pituitary and the sex glands. There are at least two gonad-stimulating hormones produced by the anterior lobe of the pituitary gland: one is called FSH (**follicle-stimulating hormone**) and the other LH (**luteinizing hormone**). Both are complex proteins and have only recently been isolated in pure form. If no FSH is produced, Graafian follicles fail to form in the female and seminiferous tubules cease to function in the male. Without LH, none of the Graafian follicles release their eggs, nor will the interstitial tissue of the testis produce testosterone. Restoration of these two hormones to a hypophysectomized animal allows normal development of the sexual organs and in the female insures normal functioning all the way to pregnancy and full-term development of the fetus. The development and subsequent function of the testis is under the influence of FSH and LH. Both are essential for normal spermatogenesis. LH also influences the interstitial cells of the testis to produce testosterone.

MAMMOTROPIC EFFECTS. As was pointed out in an earlier section on the gonadal hormones, the mammary glands develop under the impetus given them by estradiol, progesterone, and possibly another hormone. Even after they are fully formed, lactation (secretion of milk) does not occur unless still another hormone, **prolactin,** or lactogenic hormone, is produced by the pituitary. This is sometimes referred to as the "maternal instinct" hormone, because it produces certain mothering behavior in animals which normally do not possess such instincts, an old male dog, for example. If such an animal is given this hormone in sufficient quantities over a period of time, it not only will mother pups but will also produce milk to feed them.

The Posterior Lobe. The posterior lobe of the pituitary is no longer considered an endocrine gland in spite of the fact that hormones are released from it. The hormones are actually manufactured by neurosecretory cells in the hypothalamus from which they are released via the posterior lobe. If this structure is removed or damaged the body water balance is greatly disturbed. Water is not reabsorbed by the kidney tubules, which results in great urine flow (from 3 to 10 gallons a day). This disease is known as **diabetes insipidus** and, without treatment, is fatal.

The hormone **oxytocin** has been isolated from the posterior lobe and has recently been synthesized. It is a peptide consisting of nine amino acids. It has a pronounced effect on uterine muscles, causing them to contract with increased vigor; hence this compound is often used to induce labor and to contract the uterus following childbirth. Another interesting function of oxytocin is that in

lactating females it functions in milk letdown. If there is little or no hormone present in the blood of the animal, milk is produced in a normal fashion but cannot be released. **Vasopressin,** another posterior lobe hormone, is chemically very much like oxytocin; it varies only in two different amino acids, yet its action is vastly different. The main effect of vasopressin is in regulation of water balance and when given regularly to a person suffering from diabetes insipidus, normal water balance can be maintained indefinitely. For this reason it is called **antidiuretic hormone** (ADH). When given in very large doses it also brings about constriction of arterioles of the body causing a marked elevation of blood pressure; hence, the name vasopressin. This, however, is probably not a normal function of the hormone.

Control of the Pituitary. Because of the many hormones produced by the pituitary, it was once thought of as the "master gland." It is now known that the release of each anterior lobe hormone is controlled by the quantity of the specific hormone in the blood. For example, as estrogens increase in the blood, FSH and LH are inhibited; thyroxin inhibits thyrotropin, and cortisone represses ACTH (Fig. 16–34). Thus we see a negative "feedback" mechanism always in operation so that the pituitary hormones and their "target" organs are in equilibrium to maintain the so-called normal conditions in the body.

The feedback mechanism does not explain all of the facts about pituitary control of endocrine functions. We know that environmental factors influence the secretion of hormones via the nervous system. It is now becoming clear that this interaction between the nervous and endocrine systems occurs in the pituitary. Just how is this done?

We have already learned that there is a direct morphological connection of the pituitary with the hypothalamus. A great deal of research in recent years on this region of the brain has revealed the fact that the hypothalamus is where neural information is transmitted to the pituitary. Thus, sensory information that is received through the nervous system is conveyed to the hypothalamus which in turn controls the release of pituitary hormones. How does the hypothalamus control the pituitary?

The anatomy of the pituitary and hypothalamus throws some light on this problem. The anterior pituitary has no nerve fibers so neural control is not the answer. However, there is a system of small blood vessels which begins in a network of capillaries in the floor of the hypothalamus, then form small vessels which end in another bed of capillaries in the anterior pituitary gland. This is a portal system reminiscent of that seen in the kidney and liver. This system suggests that the hypothalamus is producing hormones which are delivered via the circulatory system to the pituitary where they control the production and release of pituitary hormones. This hypothesis has been shown to be true by a number of investigators who have made extracts from the hypothalamus and have shown that they cause the release of the various anterior pituitary hormones. Although all of the hypothalamic hormones have not been completely identified, it is thought that there are six of them (Fig. 16–34). They are: corticotropin-releasing factor (CRF), luteinizing-releasing factor, follicle-stimulating hormone-releasing factor, somatotropin-releasing factor, prolactin-inhibiting factor. Note that all of the hormones stimulate the pituitary to produce and secrete the corresponding tropin hormone except the last, prolactin-inhibiting factor, which suppresses the synthesis and release of prolactin.

The origin of the hypothalamic hormones has been cleared up in recent years although the cytological basis for their origin has been

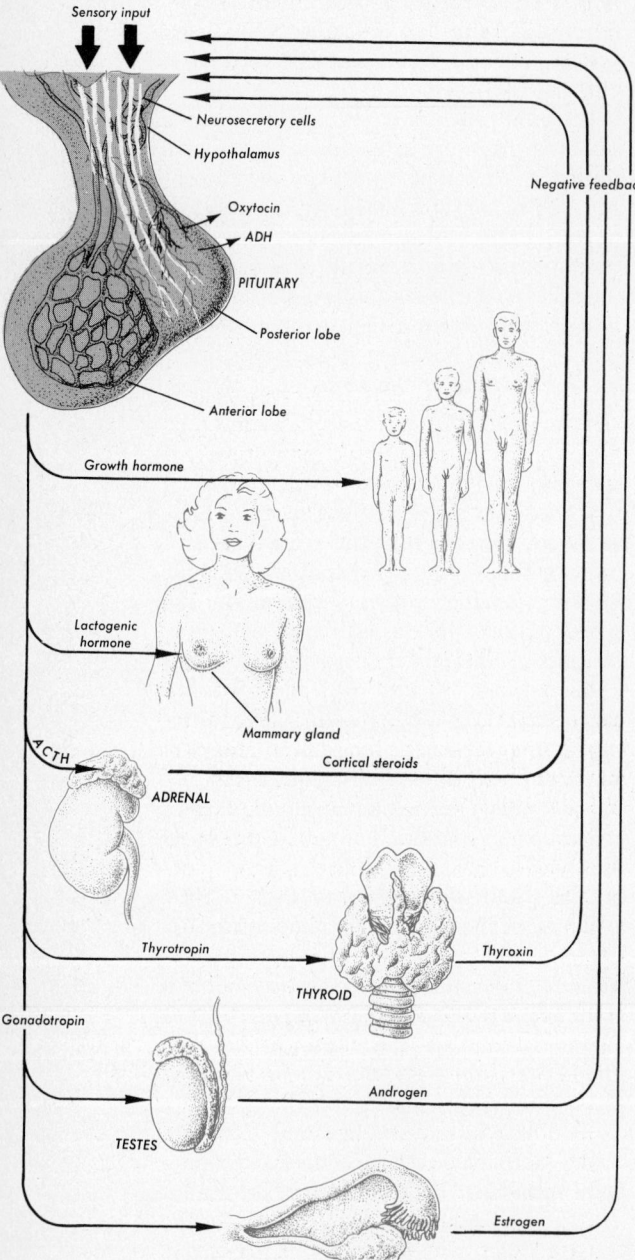

Fig. 16–34. This illustrates the neuroendocrine reflex. Neural sensory signals from the environment arrive at the hypothalamus causing it to synthesize and release the various pituitary hormones. These enter the circulation and pass to the adrenals, thyroid, and gonads, stimulating these glands to synthesize and release their corresponding hormones which have their specific effect on the target organs. Negative feedback from these target gland hormones modify the system. Growth and lactogenic hormones go directly to the target organs from the pituitary. Negative feedback hormones from these organs have not been clearly identified.

known for a long time. The so-called **neuro-secretory cells,** found in the nerves of most animals, were observed by cytologists many years ago, but their function was obscure. They resemble nerve cells except that they contain numerous secretory granules. In the hypothalamus the dendrites of the neuro-secretory cells connect with neural pathways in the central nervous system; in this way impulses coming from the sense organs are relayed to the hypothalamus; thus it responds to changes of the internal and external environments. The axons of the neurosecretory cells terminate in the capillary bed as indicated in Figure 16–34. The hormones are synthesized in the cell body and then transported along the axon to its tip where the hormone is stored in droplets. When the proper sensory stimulus comes along, the hormones are released into the blood stream via the capillary bed. When they reach the pituitary they induce it to synthesize and release its own hormones into the general circulation. These pass to the "target" organs where their specific effects are produced.

We now understand how signals coming from sense organs are translated into hormonal signals which in turn bring about physiological responses of the animal to its internal and external environment. This circuit has been appropriately called the **neuro-endocrine reflex.** Negative feedback from the target gland (thyroid, adrenals, gonads) hormones can modify the system, but sensory stimuli ordinarily take precedence in determining the rates of secretion of hypothalamic and pituitary hormones.

The posterior lobe hormones arise in two groups of hypothalamic neurosecretory cells. The hormones, oxytocin and vasopressin, are synthesized in these cells and then conveyed along their axons to the posterior pituitary lobe where they are released into the general circulation.

Other endocrine glands in the body, such as the **thymus** in the neck region and the **pineal** in the brain, have been suspected of having endocrine functions. There is some recent evidence that the thymus is involved in the production of antibodies. Also a hormone called the **melanocyte-stimulating hormone** (MSH) or **intermedin,** has been isolated from the intermediate lobe of the pituitary which darkens the skin of frogs and man. It is responsible for the increased pigmentation noticed in pregnancy. This hormone is closely associated with hydrocortisone and ACTH.

Stress and Hormones

Considerable attention in recent years has been given to the rôle played by hormones during environmental stress. Han Selye and his collaborators at the University of Montreal have been particularly interested in this problem. During the life of an animal environmental hazards such as cold, starvation, hemorrhage, burns, and lacerations occur more or less frequently and it is under these conditions that certain endocrine glands are stimulated to increase their output and compensate for any damage that might accrue from these stresses. For example, following trauma, particularly with severe bleeding, the adrenal medulla increases its output of epinephrine. The increased epinephrine triggers the release of increased ACTH from the anterior pituitary which in turn releases increased amounts of adrenal cortical hormones and these bring about changes in the carbohydrate and mineral metabolism. These changes are such as to counteract the effects of the injury and permit the tissues to repair rapidly. If the stress is of long duration, such as severe cold or starvation, the tissues become adapted to the stress as a result of hormonal action. However, there is a limit beyond which the body cannot tolerate further strain; if this limit is taxed, the body goes into shock from which it is not likely to recover.

In this chapter we have seen how the many parts of a complex animal are coordinated into an integrated whole. In the next few chapters we examine the organ systems responsible for the energy that keeps the machine going.

ORGAN SYSTEMS
OF MAN

Books

BEACH, F., *Hormones and Behavior*. New York: Paul B. Hoeber, 1947.

BULLOCK, T. H., and HORRIDGE, I., *Structure and Function in the Nervous Systems of Invertebrates*, 2 vols. San Francisco: Freeman, 1965.

CASE, J., *Sensory Mechanisms*. New York: Macmillan, 1966.

COBB, S., *Foundations of Neuropsychiatry*. Baltimore: Williams & Wilkins, 1952.

FRY, B. E., *Hormonal Control in Vertebrates*. New York: Macmillan, 1967.

GARDNER, E., *Fundamentals of Neurology*, 5th ed. Philadelphia: Saunders, 1968.

KATZ, B., *Muscle and Synapse*. New York: McGraw-Hill, 1966.

NOBACK, C. R., *The Human Nervous System*. New York: McGraw-Hill, 1967.

PAVLOV, I. P., *Conditioned Reflexes*. New York: International Publishers, 1941.

PROSSER, C. L., *Comparative Animal Physiology*, 3rd ed. Philadelphia: Saunders, 1973.

RANSON, W., and CLARK, S. H., *The Anatomy of the Nervous System*. Philadelphia: Saunders, 1959.

SELYE, H., *Textbook of Endocrinology*. Montreal: University of Montreal Press, 1949.

TURNER, C. D., and BAGNARA, J., *General Endocrinology*, 5th ed. Philadelphia: Saunders, 1971.

WIENER, N., *Cybernetics*. New York: Wiley, 1948.

WOODBURNE, R. T., *Essentials of Human Anatomy*. New York: Oxford University Press, 1961.

YOUNG, W. C., ed., *Sex and Internal Secretions*, 3rd ed., 2 vols. Baltimore: Williams & Wilkins, 1961.

Articles

BEKESY, V., Von, "The Ear." *Scientific American*, **197**, 66, August, 1957.

BRAZIER, M. A. B., "The Analysis of Brain Waves." *Scientific American*, **206**, 142, June, 1962.

ECCLES, Sir J., "The Synapse." *Scientific American*, **213**, 56, 1965.

GRAY, G. W., "The Great Ravelled Knot." *Scientific American*, **179**, 28, October, 1948.

GRAY, G. W., "Cortisone and ACTH." *Scientific American*, **182**, 30, March, 1950.

HEIMER, L., "Pathways in the Brain." *Scientific American*, **225**, 48, July, 1971.

HENDRICKS, S. B., "How Light Interacts with Living Matter." *Scientific American*, **219**, 174, September, 1968.

KATZ, B., "The Nerve Impulse." *Scientific American*, **187**, 55, November, 1952.

LOWENSTEIN, W. R., "Biological Transducers." *Scientific American*, **203**, 98, August, 1960.

MILLER, W. H., RATLIFF, F., and HARTLINE, H. K., "How Cells Receive Stimuli." *Scientific American*, **205**, 222, September, 1961.

NEISSER, U., "The Processes of Vision." *Scientific American*, **219**, 204, September, 1968.

THOMPSON, E. O. P., "The Insulin Molecule." *Scientific American*, **192**, 36, May, 1955.

WALD, G., "Eye and Camera." *Scientific American*, **183**, 32, August, 1950.

17

THE DIGESTIVE SYSTEM

Procuring food and extracting the energy from it is one of the most important activities confronting animals. This is so because all animal activities require energy which can be obtained only by releasing it from food. All animals, from amoeba to man, have special means for bringing about this conversion. The first steps in this complex energy-releasing mechanism are ingestion and digestion of food.

Let us take a "tour" through the digestive tract of man and examine what happens on the way. In order to elucidate the story of digestion Figures 17–1 and 17–2 have been provided. The former shows the morphology of the human digestive tract and the latter indicates the principal chemical reactions of digestion and absorption and where both occur. It will be helpful to refer constantly to these two figures during the following discussion.

THE MOUTH

Food is taken into the **oral cavity** through the mouth, where it is crushed into smaller particles by the teeth. During this procedure it is thoroughly mixed with **saliva**, a secretion from the three pairs of **salivary glands** (submaxillary, sublingual, and parotid) (see Fig. 17–4). The tongue arises from the floor of the oral cavity where it functions as a handy organ in moving the food about and pushing it to the back of the mouth when it is to be swallowed. The tongue is not only highly sensitive to chemicals, as has already been pointed out, but also to touch. The use of the tongue in forming words needs no further comment. The oral cavity is lined with mucosa which secretes mucus, thus keeping the lining moist at all times; this, together with saliva, aids in lubricating the food so that it can slide

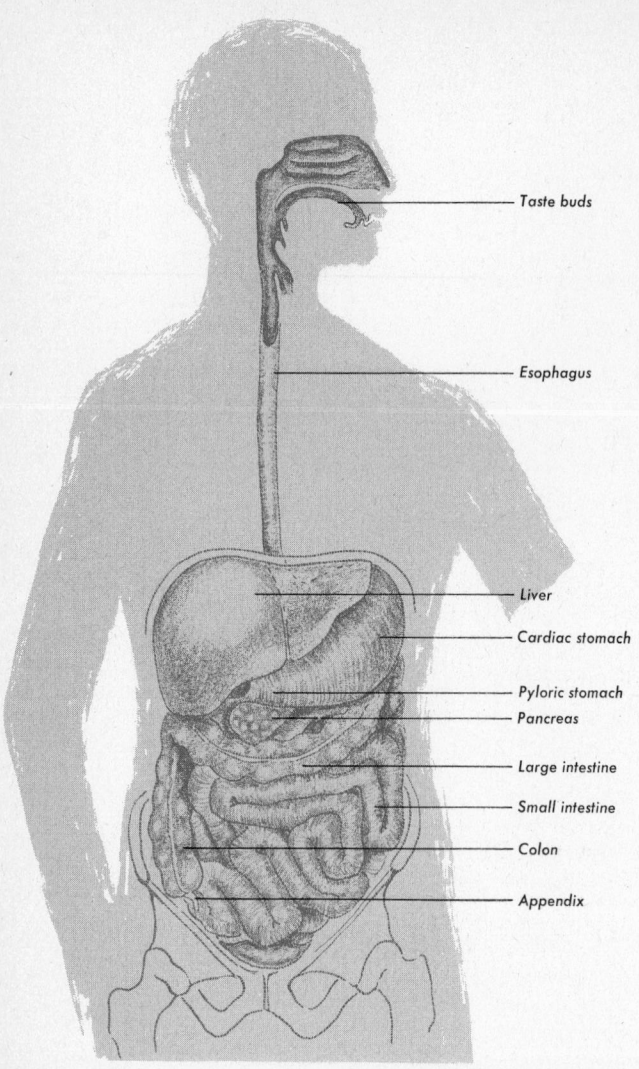

Taste buds

Esophagus

Liver

Cardiac stomach

Pyloric stomach

Pancreas

Large intestine

Small intestine

Colon

Appendix

Fig. 17–1. The human digestive tract. A dorsal view has been drawn in order to show some parts more clearly.

down the esophagus with little friction. The teeth are such important structures from both the utilitarian and esthetic points of view that further consideration is given them here.

The Teeth

The tooth consists of the **crown** which protrudes into the oral cavity, the **neck,** a narrow region where the gum comes into contact with it, and the **root** which is firmly cemented in a socket in the jaw bone. (Fig. 17–3). The crown is covered with hard **enamel** which affords a good grinding surface. Under this is the **dentine,** which is softer, and lying at the center is the **pulp** chamber which contains the nerves and blood vessels. The living tooth is porous and is nourished by the blood stream.

Humans have two sets of teeth during their lifetime. The first set, the **milk** teeth, begin to appear in the first year of life and are fully formed by the eighth year. Before they are all well established, however, the front ones start falling out because of the pressure of the second set coming from underneath. During the first twelve to sixteen years of life the individual is experiencing a continual loss and replacement of his teeth. The second set, when fully formed, is composed of 32 teeth, eight on each side of both the upper and lower jaws. There are two front cutting **incisors,** one **canine** next (proceeding posteriorly), then two **premolars,** used in grinding food, and finally three **molars** which are the heavy grinders. The last molars, the **wisdom** teeth, may appear late in life or not at all.

Teeth that are so perfectly formed in lower vertebrates and primitive man seem to have difficulty withstanding the effects of civilization. It has long been known that the civilized man has notoriously bad teeth while his primitive brother may have perfect teeth throughout his lifetime. A high carbohydrate diet has been thought to be responsible for tooth decay, but natives of many of the South Sea Islands live almost exclusively on a starchy diet and yet their teeth are unusually well preserved. However, when sugar (sucrose) replaces starch in the diet, dental caries (decay) appears. Along with this change to sugar there is a marked increase in the numbers of an acid-forming bacterium (*Lactobacillus casei*) in the mouth. If sugar is withheld from the diet for some time the bacteria disappear and there is no further decay of the teeth. It is rather well confirmed today that it is the sugar in the diet that causes dental caries, and since sugar occurs only in small quantities in nature, it can be considered an unnatural food for all animals including man. Teeth cannot withstand the action of acids produced by the bacteria, hence decay is prevalent among sugar-eating people, which includes much of the civilized world. Fluorine in the water is known to be of value in preserving teeth, probably owing to the fact that in addition to the hardness of the enamel it produces, it retards the growth of bacteria in the mouth.

Digestion in the Mouth

The only part of the saliva that has to do with food breakdown is the enzyme **amylase** which converts starches to **maltose,** a disaccharide. Salivary amylase is a very active enzyme, for if only a few drops of saliva are added to a suspension of starch kept at body temperature, the starch will be converted to maltose in about 20 minutes. This demonstrates one of the properties of an enzyme, for to accomplish the same conversion in the laboratory would require drastic treatment with powerful acids for many hours at high temperatures. Salivary amylase has the power to break the large starch molecules into the much smaller maltose molecules, leaving only one more step to the glucose stage when, as a simple sugar, it can be taken into the cells and metabolized. The way most people eat, amylase has only a momentary chance to function before the food departs for the stomach, although its action does continue during the the swallowing and for a short time in the stomach.

The flow of saliva into the mouth is controlled by nerves which respond to stimuli coming not only from the taste buds on the tongue but also from end-organs in the walls of the oral cavity that are sensitive to the presence of food itself and to the mechanics of chewing. Impulses travel from these end-organs to the brain stem to be routed back along the efferent nerves to the salivary glands (Fig. 17–4) which are stimulated to pour saliva into the mouth through ducts. This is not the only way

salivation (secreting saliva) can be induced, as any hungry person knows who smells, sees, or even thinks about food. Such responses are called **conditioned reflexes** and are acquired by learning.

Swallowing

The problem of getting food from the mouth cavity to the stomach in vertebrates

Fig. 17–2. A partial history of food as it passes through the body.

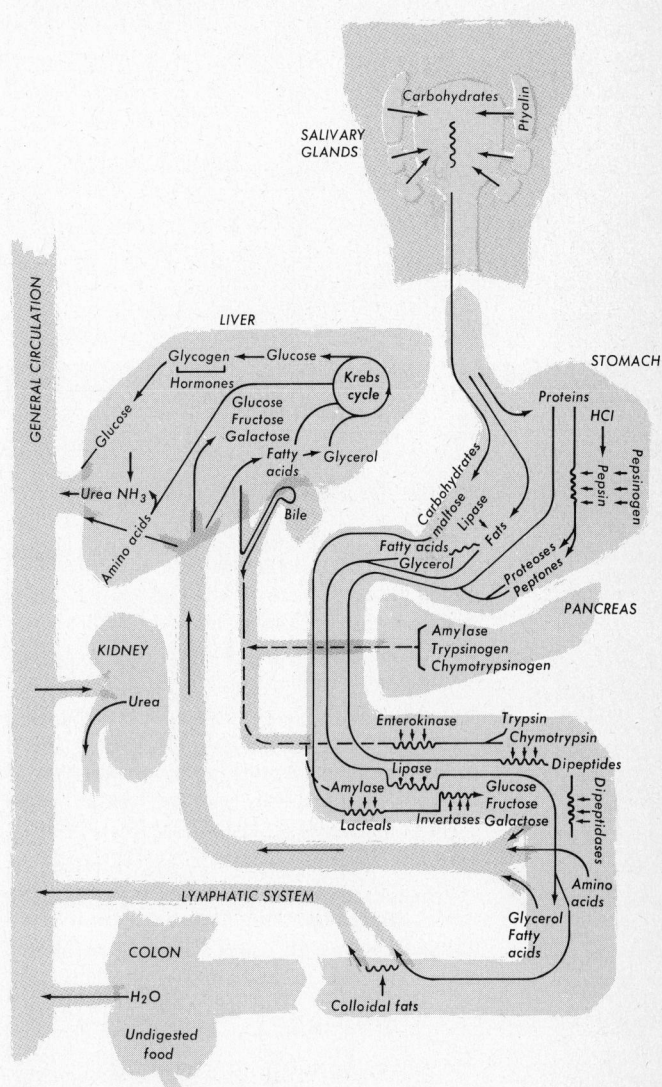

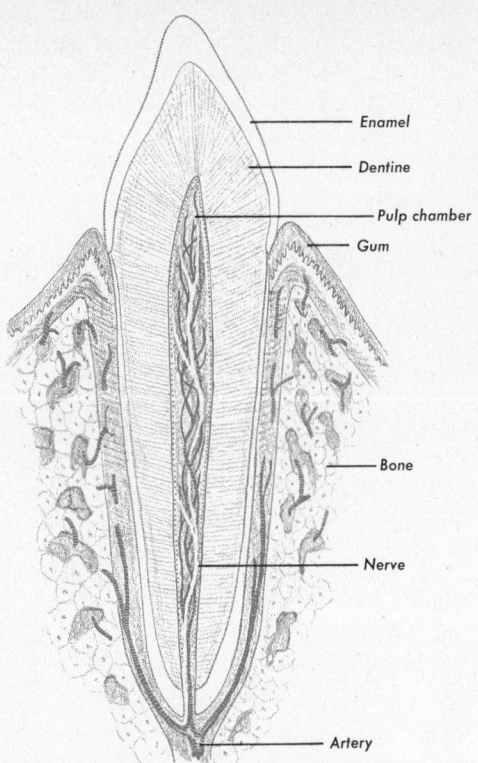

Fig. 17-3. A mammalian tooth in sectional view to show its internal structure. A part of the jaw is also shown to indicate how the tooth is fastened to the bone.

apparatus developed for this purpose is very simple among amphibians but becomes more complex in the higher forms, particularly in mammals. Here elaborate nerves and muscles have evolved to provide a smooth crossing of food and air. The inadequacies of this system are familiar to anyone who has experienced the difficulties resulting when particles of food accidentally enter his air passages.

Let us see how the process of swallowing takes place in man. When the food reaches a pasty consistency in the mouth, it is forced into the back of the oral cavity (**pharynx**) and rapidly propelled to the stomach through the small esophagus. When food or fluids reach the pharynx, a chain of impulses is initiated which brings about the pulling forward of the **larynx** and a tipping of the epiglottis to prevent food from passing into the trachea, together with a simultaneous opening of the upper end of the esophagus (Fig. 17-5). Once all this is started, the food is beyond recall. When it enters the esophagus a peristaltic wave

Fig. 17-4. The salivary glands secrete saliva when end organs in the tongue and mouth cavity are stimulated. The nerve pathway is shown here.

has an interesting evolutionary history. The difficulty arose when animals migrated onto land and the breathing and food paths were forced to coincide. Water, the oxygen-containing-medium, passes into the mouth and out the gill clefts in fish, while the food enters the mouth also and merely continues straight back into the esophagus (Fig. 17-5). No difficulty was experienced with such a mechanism. However, when vertebrates moved out on the land the oxygen-containing medium, air, took a new pathway because the lungs evolved from the floor of the mouth cavity. This meant that both food and air had to pass through the pharynx and this could only be accomplished by providing some way of closing one when the other was functioning. The

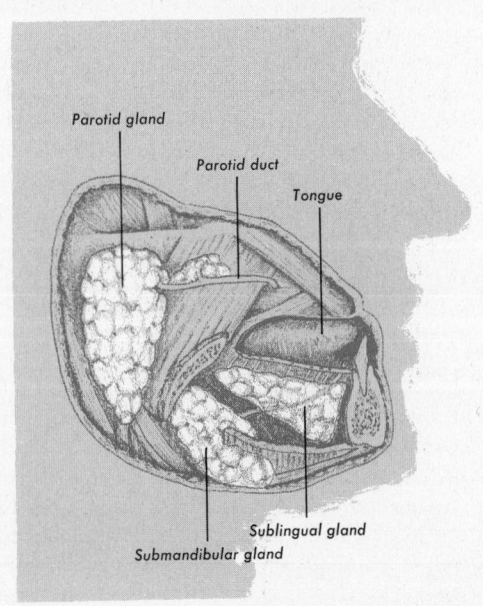

carries it quickly to the stomach. Peristalsis is accomplished by the thick muscles which make up the wall of the esophagus. The presence of food in this collapsed tube causes the longitudinal muscles to relax just in front of the **bolus** (ball) and at the same time causes the circular muscles to constrict just behind it with the result that the food is pushed along the tube (Fig. 17–10). This is a rapid action, requiring only 5 to 6 seconds for food and less for fluids. Furthermore, it has nothing to do with gravity because a horse can swallow uphill, as indeed it must always do, and even a human can drink in an inverted position if he desires. Swallowing is initiated only by the presence of food or fluids in the pharynx; thus it is impossible to swallow twice in rapid succession when no foods or fluids are taken into the mouth. Yet when one drinks fluids one can swallow continuously.

THE STOMACH

After passing through the esophagus, the food drops into the stomach, which is a sac-like expansion of the digestive tube designed not only to store a considerable amount of food (about two and one-half quarts) but to start the digestion of proteins. It is a thick-walled muscular sac lined with **gastric glands.** These are minute, slender pockets in the soft mucosa with tiny openings through which the gastric juice flows into the cavity of the stomach. In addition to the circular and longitudinal muscle layers of the esophagus, the stomach possesses an oblique layer. With this elaborate system the stomach becomes an efficient mixing or churning organ.

The stomach is usually divided into two general regions: the **cardiac region** which immediately follows the esophagus, and the **pyloric region** which is followed by the small intestine. The anatomy and activity of the two portions vary somewhat. The portals of entry and exit to the stomach are guarded by valves consisting of thickened circular muscles which, when strongly contracted,

completely close both of these openings. Such valves in a tube are called **sphincters.** Besides these two in the stomach, the **cardiac** at the entrance and the **pyloric** at the exit, there are two others along the digestive tube: one where the small intestine joins the large intestine (**ileocaecal**) and the other at the end, the **anal** sphincter. These valves are important in retaining the food in its proper place until digestion is complete.

As food is swallowed, the stomach gradually expands until full; the reverse process takes place as digestion is completed and the food is moved along to the small intestine. Fluids such as water pass through the stomach in a few minutes, whereas more solid foods remain from 3 to 5 hours, depending on their nature. Some foods digest more slowly than others, for example, those rich in fats. The food is retained in the stomach and churned until it resembles a thick soup, which is called **chyme.** Peristaltic waves begin in the cardiac region and move

Fig. 17–5. In fish the food and water portal of entry is the mouth. In land-dwellers the air and food passageways cross, as shown here in the amphibian and man. The left lower figure shows the position of the larynx during breathing and the lower right figure shows it during the act of swallowing food.

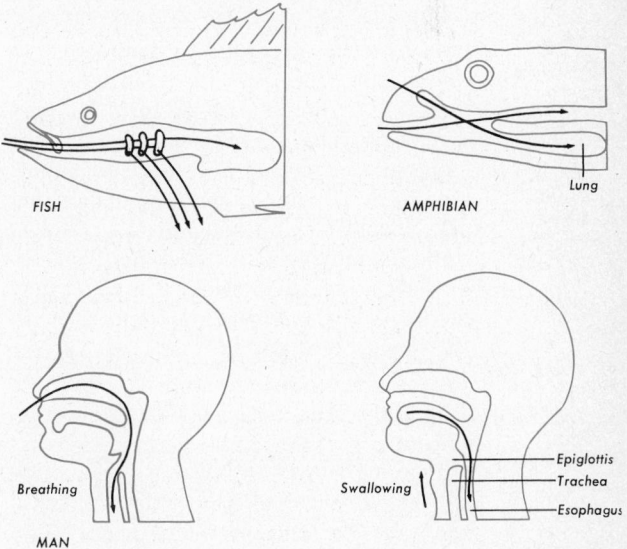

toward the pyloric end, so that the food is constantly being forced into that end. So long as the pyloric valve remains closed, the food cannot pass into the intestine, but must continue to be mixed back and forth until the proper consistency is reached. The pyloric valve then opens intermittently, allowing small amounts of the chyme to pass in spurts into the upper end of the small intestine.

These movements are under the influence of the autonomic nervous system. Parasympathetic fibers reach the stomach through the vagus nerves and their action increases the intensity of the peristaltic waves and augments the flow of gastric juices. The opposite action is brought about by impulses coming through fibers from the sympathetic system. Undue emotional strain during a meal excites the sympathetic fibers excessively, thereby slowing up stomach activity and prolonging digestion unduly.

Most animals have a mechanism for removing anything taken into the stomach that does not "set well." This is particularly true in carnivores, who are apt to eat slightly decayed food that might be toxic. **Regurgitation,** as this chain of events is called, is just the reverse of swallowing, culminating in a complete evacuation of the stomach. Just prior to the event a deep breath is taken in order to hold the diaphragm down; a convulsive contraction of the abdominal muscles then follows which presses the stomach up against the rigid diaphragm. The sphincter leading into the esophagus relaxes and the stomach contents are expelled almost explosively. The action also resembles swallowing in that once it starts it continues until the job is done, regardless of the will of the owner. This safety mechanism is a feature of importance in the survival of a species.

Digestion in the Stomach

The millions of tiny glands in the walls of the stomach secrete **gastric juice,** a highly acidic, watery fluid that contains the enzyme **pepsin** (gastric proteinase). Between 400 and 800 cc of gastric juice are produced during the digestion of an average meal. The flow of gastric juice, like the flow of saliva, is under the influence of the nervous system, at least in part. Everyone has experienced "mouth watering" when hungry and within sight or smell of good food; the stomach "waters" the same way and probably at the same time. Just how this functions was an attractive subject for investigation even for early biologists, and now the entire mechanism is rather well understood.

At the time of the signing of the Declaration of Independence in this country, a clergyman in Italy attempted some experiments on the activities of his own stomach. This amazing Spallanzani swallowed small metal cages containing bits of meat and after these were left in his stomach for varying lengths of time they were retrieved by means of an attached string. He noticed that the tiny bits of meat had disappeared and concluded that they must have gone into solution and therefore were digested. Another notable series of experiments, begun about 40 years later and continued for many years, was performed by an American army surgeon, Dr. William Beaumont, who had the good fortune of treating a man (Alexis St. Martin) with an unusual bit of accidental stomach surgery. This Indian had been shot through the stomach with a shotgun, and the remarkable thing about the accident was that the load had not only torn away the abdominal wall over the stomach but had also taken a part of the stomach as well. Beaumont plugged the wound with a wad of cotton and waited for his patient to die. To his surprise the patient not only recovered but the stomach wall healed to the abdominal wall in such a way as to leave a permanent opening, or **fistula,** as such openings are now called. This gave Beaumont an unusual experimental subject for the study of the functions of the stomach. For years afterwards

Fig. 17–6. Artificial openings into the digestive tracts of two dogs to illustrate how Pavlov performed his basic experiments. The left dog is eating a "sham meal," while the right one is secreting gastric juice from a "Pavlov pouch" into a container.

he kept St. Martin close at hand so that he could observe how digestion progressed under all sorts of conditions.

Since that time there have been many cases where the esophagus had been closed because of the accidental swallowing of some strong chemical, making it necessary to provide the person with a fistula in order that he might receive nourishment. The famous Russian physiologist, Pavlov, produced artificial fistulas in dogs and then studied the rate of flow of the gastric juices under varying physical conditions (Fig. 17–6). In such fistulas a small compartment of the stomach is separated off from the remaining portion, but the circulation and nerves remain intact so that the fistula will respond normally. With such an experimental animal, Pavlov was able to show conclusively what factors influenced gastric secretion. Another operation which supplemented the findings with artificial fistulas involved the severing of the esophagus and the bringing of the cut ends to the surface of the neck (Fig. 17–6). Pavlov showed that if a dog were fed a meal which never reached the stomach, since the cut esophagus led to the outside of the neck, about one-fourth of the normal flow of gastric juice took place. He showed further that this was due to reflexes, because the flow ceased when all of the nerves to the stom-

ach were severed. However, if he placed food directly into the stomach without the dog seeing or smelling it, about one-half of the normal flow of gastric juice took place. While the flow in this case was somewhat reduced by cutting the nerves to the stomach, a considerable secretory activity continued. This could mean only that there must be a hormone produced by the lining of the stomach which circulates in the blood and stimulates the gastric glands, at least in part, to secrete. This hormone, **gastrin,** has since been identified. With this and other types of experiments, Pavlov established the principle of "**conditioned reflex**" about which we have more to say in Chapter 27.

In order that gastric secretion may be turned off as the stomach empties there must be some other hormone to bring this about. It is **enterogastrone,** secreted by the duodenal mucosa. As foods enter the duodenum this hormone is secreted into the blood stream and passes to the stomach where it inhibits gastric secretions. It also reduces the churning action of the stomach.

The contents of the stomach are very acidic, having a pH of approximately 2.0, owing to the high concentration of hydrochloric acid. The ions exist as H^+ and Cl^- in the gastric juice. The acid functions in several ways. It aids in dissolving away the limy material between cells (particularly plant cells), thus making them more readily available for enzyme action. It provides a favorable medium in which the enzymes can do their best work and it has some bacterial effect on detrimental bacteria which may be taken in with the food.

In addition to HCl, gastric juice contains mucus, proteins, water, and the enzymes, **pepsin, rennin,** and **lipase.** Pepsin is manufactured by special gland cells in the stomach lining. It exists in the inactive state when formed in the cells and is called **pepsinogen.** When this substance comes into contact with the acid medium of the stomach, it is transformed into pepsin, the active enzyme. Pepsin, once formed, can itself convert pepsinogen to pepsin. The advantage of producing an inactive enzyme is to spare the lining cells from the powerful proteolytic action of pepsin. Pepsin breaks protein molecules down into **proteoses, peptones,** and **polypeptides,** all of which are soluble in water. The reaction may be expressed thus:

$$\text{Protein} + \text{water} \xrightarrow{\text{pepsin}} \text{proteoses, peptones, and polypeptides}$$

Like all digestive enzymes, pepsin is a hydrolytic enzyme which functions by adding water (hydrolysis); in this case it splits the large protein molecules into smaller molecules. This is the initial stage in protein digestion. Owing to the short time food is in the stomach, not all of the protein molecules are attacked by pepsin; some pass on into the intestine where enzyme action finishes the job.

The enzyme **rennin** is found in the stomachs of young mammals, and it acts on **caseinogen,** the protein of milk. It is secreted in the inactive state as **prorennin** and, when in contact with HCl, becomes rennin which attacks caseinogen splitting it into **casein** and **whey.** The casein then coagulates in the presence of calcium ions forming **curdles** which are then accessible to pepsin for digestion. The young mammal depends on milk as its sole source of food during the early part of its postnatal existence and the efficient digestion of milk protein is advantageous. The formation of curds permits the protein to remain in the stomach long enough for peptic digestion to occur. Without rennin the soluble caseinogen would pass far into the intestine, just as any other fluid does, before protein digestion could start.

The remaining enzyme in gastric juice is **lipase** which converts fats to fatty acids and glycerol. Most tissue fats are not attacked by gastric lipase, only fats in colloidal form such as those found in milk. Moreover, efficient fat digestion requires an alkaline environment and bile salts, a condition found only in the small intestine.

Therefore, fat digestion in the stomach is inconsequential.

Like other parts of the body, the stomach is subject to many ills but it seems to suffer especially from emotional strain. In some people happiness seems to center around the contentment of their stomachs. Everyone is familiar with the difficulty experienced in eating immediately following the reception of good or bad news. This is owing to stimulation coming through the autonomic nervous system. Continued emotional strife can even produce organic damage, such as erosion of small areas of the stomach lining (gastric ulcers). Frequently this malady clears up "miraculously" when the emotional strain is removed. In stubborn cases of ulcers or cancer, parts, or even the entire stomach in extreme cases, may be successfully removed. This type of surgery has progressed remarkably in the last two decades.

THE SMALL INTESTINE

The small intestine is a narrow tube with a length of over 23 feet which extends from the stomach to the large intestine. Although it is possible to get along without a part or all of the stomach, it is absolutely necessary that the small intestine remains essentially intact, because it is here that almost all digestion is completed and **absorption** of the end products of digestion takes place. The small intestine is divided into three parts that differ slightly from one another in their anatomy. The first 10 inches is the **duodenum,** into which the ducts from the **liver** and **pancreas** empty. This is followed by about 10 feet of a region known as the **jejunum,** which is particularly rich in intestinal glands. The remaining portion is the **ileum,** and is characterized by the vast number of tiny fingerlike projections of its lining, the **villi** (Fig. 17–9). These structures increase the surface area of the intestine tremendously, thereby making possible the adequate absorption of digested food substances. The small intestine is thus adapted to retain food for a considerable period of time in order that digestion may be completed and absorption take place. In this respect, it should be recalled that carnivores possess shorter intestines than herbivores because meat is more readily digested than plants (Fig. 17–7).

The entire gut is supported by a dorsal **mesentery** which holds such parts as the stomach and duodenum in place but at the same time allows the rest of the small intestine considerable freedom of movement. In lower vertebrates there is evidence that a ventral mesentery was once present which held the entire digestive tract in a line from mouth to anus. As the gut increased in length this ventral support was lost. Actually, it no longer performed any particular function, since the animal lived in a horizontal position where its gut hung like clothes from a line. However, when man decided to walk on his hind legs, the dorsal mesentery could not continue to support the gut as well as previously, and so the gut had a tendency to slide posteriorly into the pelvis. In youth, the abdominal muscles are adequate to compensate for this deficiency but as age advances the gut tends to sag more and more, resulting in the pot-belliedness of advancing years. It requires a great amount of physical effort to keep it in place.

Specialized gland cells in the duodenum secrete **intestinal juice**, which contains the following enzymes: **enterokinase, dipeptidase, amino-peptidase, invertases, amylase,** and **lipase.** The function of each is discussed later.

The Liver

This, the largest gland in the body, is located on the right side just under the diaphragm (Fig. 17–1). It is tunneled with spaces and with vessels which are filled with large quantities of blood, and because of its construction will not tolerate injury. Fatal internal injuries suffered in car accidents frequently involve the liver. Unfortunately, the steering wheel of the auto-

Fig. 17–7. The length of the digestive tracts of animals varies with the diet, as shown in these figures. The carnivore (dog) has the shortest tract (with respect to trunk length), the herbivore (rabbit) has the longest, while the omnivore (man) has a gut of intermediate length.

mobile is located over this area of the body, and since any sudden stopping of the car throws the driver against the wheel, the liver, being the most vulnerable organ in that area, is most apt to be damaged.

Bile, which is drained from all parts of the liver by tiny tubules, accumulates in one large duct, the **bile duct,** which eventually joins the pancreatic duct just before the two empty into the duodenum. However, there is a storehouse for the continuously secreted bile, the **gallbladder,** which is located in a hollow on the underside of the liver. It is a thin-walled sac connected to the bile duct by a small duct of its own. Between meals the bile accumulates in this sac and when bile is needed the walls contract, forcing the stored bile into the duodenum in large quantities.

In so far as digestion is concerned, the only function of the liver is the production of bile. In a sense, bile is an excretory as well as a secretory product because the residue from broken-down red blood-cells accumulates in it and is thus eliminated from the body, at least in part. Bile also contains considerable quantities of sodium bicarbonate, which functions in greatly reducing the acidity of the chyme in the duodenum. The organic constituents of bile are **bile pigments, bile salts,** and **cholesterol,** each of which has specific functions.

The bile pigments, **bilirubin** (red) and **biliverdin** (green), are responsible for the color of bile. While in man bile is straw-colored, in other vertebrates it ranges from green to red with all intermediate shades. When mixed with the chyme in the gut, the bile pigments undergo further chemical change, turning to dark brown or black, thus contributing the brown color of **stools.** The first sign of faulty bile elimination is the gradual loss of this color, and when the bile fails altogether the stools are gray in color.

Of the numerous constituents making up bile, only the **bile salts** function in digestion. They are responsible for emulsifying the fats in food so that the fat-splitting enzyme from the pancreas can work more effectively. They also seem to **activate** this enzyme, because without them the fats are poorly digested and appear in large quantities in the stools. The bile salts are conserved by reabsorption in the lower end of the small intestine and circulate in the blood back to the liver to be used over again.

The organic compound **cholesterol** is important in the bile because of the trouble it sometimes causes. It does not accumulate if the concentration of bile salts is sufficiently high to keep it in solution. However, if the bile salt level drops, cholesterol sometimes precipitates out in the gallbladder, forming **gallstones.** These are harmless in themselves but if one is forced into the tiny bile duct as the gallbladder contracts in emptying, it may occlude the tube and produce trouble. Peristaltic waves in the tube wall attempt to pass the stone along and this causes extreme pain. If the contractions are successful, the stone eventually passes into the intestine where it will do no more harm. If, on the other hand, it remains lodged in the duct, bile fails to reach the intestine and all the symptoms resulting from an absence of bile in the gut ensue. The stools lose their color, fats fail to digest, and the bile pigments, since they are excretory wastes, accumulate in the blood and eventually in the skin, causing

it to yellow, or **jaundice.** Removal of the stones is a simple surgical operation, and when accomplished the person is usually restored to health very quickly.

The Pancreas

This long, flat, light-colored organ lies between the stomach and duodenum, and by means of its **pancreatic duct** connects with the bile duct. The gland has two functions: endocrine, which has already been discussed, and digestion. It produces several digestive enzymes, all of which are essential to complete the digestion started in the mouth and stomach. In addition, the **pancreatic juice,** like bile, contains large quantities of sodium bicarbonate, which function in neutralizing the acid chyme from the stomach. Even with this large amount of sodium bicarbonate the chyme remains slightly acid during its trip through the small intestine, a fact contrary to earlier beliefs.

The pancreatic juice also contains the starch-splitting enzyme **amylase** (amylopsin); the protein-splitting enzymes **trypsin, chymotrypsin** (secreted as trypsinogen and chymotrypsinogen) and **carboxypeptidase;** and the fat-splitting enzyme **lipase** (steapsin). It should be noted that pancreatic and intestinal lipase has almost the entire job of digesting fats because gastric lipase is ineffective in this operation. The other enzymes are approximately duplicated in the mouth and stomach.

Control of Pancreatic Juice and Bile Flow

The pyloric sphincter relaxes when the content of the stomach reaches a certain consistency, though precisely how this is controlled is not well understood. Once the chyme reaches the duodenum, however, the control of the bile and pancreatic juice flow is rather well known. The wall of the duodenum contains a substance called **prosecretin** which, when brought into contact with hydrochloric acid, is converted to a hormone called **secretin.** Some of this gets into the blood stream where it circulates to the pancreas, causing it to deliver pancreatic juice (Fig. 17–8). The proof of the action of secretin can be had by simply injecting an extract from the wall of the duodenum that has come into contact with hydrochloric acid into an animal which has had all the nerves to the stomach and pancreas cut in order to rule out nervous control. The result is copious secretion of pancreatic juice.

The flow of bile from the gallbladder is caused by a somewhat similar mechanism. The presence of fats and acid in the duodenum causes the formation of a hormone named **cholecystokinin,** which circulates in the blood and causes the gallbladder to contract and deliver its contents.

Digestion in the Small Intestine

As the chyme spurts through the pyloric sphincter into the duodenum, it is eventually mixed with bile and pancreatic juice, together with the intestinal juice from the duodenal wall. Together these function in the final stages of digestion. The chyme is rendered less acid by the highly alkaline nature of both bile and pancreatic juice. Inactive trypsinogen and chymotrypsinogen coming from the pancreas are converted to active trypsin and chymotrypsin by the enzyme **enterokinase,** a constituent of intestinal juice. Actually enterokinase activates the conversion of trypsinogen to trypsin and this enzyme then influences the conversion of chymotrypsinogen to chymotrypsin. Trypsin and chymotrypsin attack whole protein molecules and have survived the effects of pepsin in the stomach, breaking them down to proteoses, peptones, and polypeptides. Further action is necessary before protein digestion is complete.

Further splitting of the proteoses, peptones, and polypeptides is accomplished by amino-peptidase from the intestinal juice and carboxypeptidase from the pancreas.

These two enzymes attack different bonds in these complex molecules, the end result being relatively simple two-amino-acid compounds, the **dipeptides.** The last step in degradation of the protein molecule is brought about by the enzyme **dipeptidase** (intestinal juice) which reduces the dipeptides to amino acids which are then absorbed into the blood stream.

Intestinal carbohydrate digestion follows a similar path. Pancreatic and intestinal amylase take up where salivary amylase

Fig. 17–8. The hormonal control of the secretion of pancreatic juice. The acidic food causes prosecretin (black spheres) in the duodenal lining to be converted to secretin (black spheres with a notch) which circulates in the blood, eventually reaching the pancreas where it stimulates the flow of pancreatic juice.

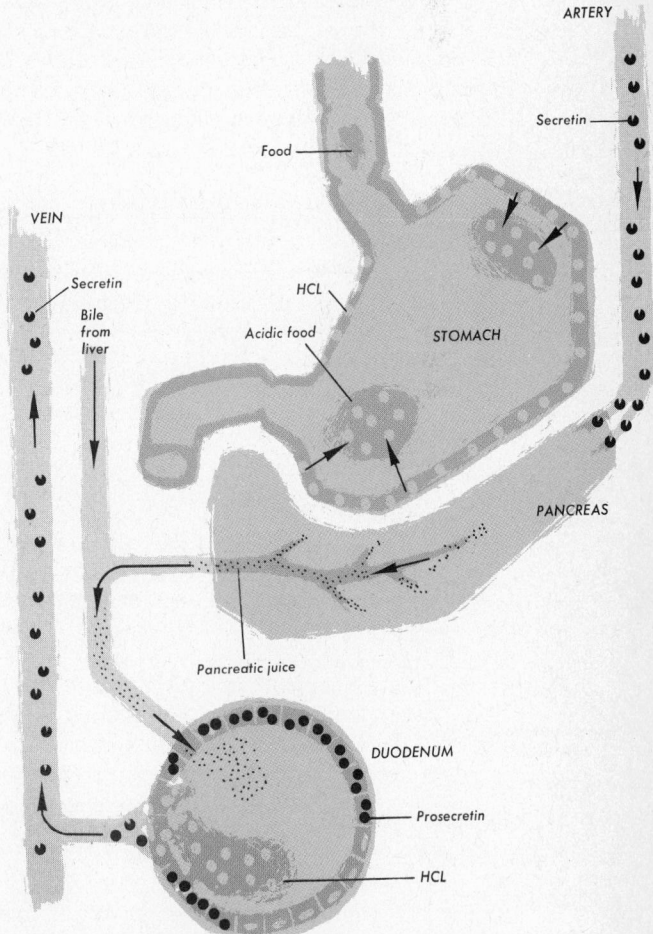

left off since nothing happened to these food molecules in the stomach. Salivary amylase, as was pointed out earlier, does not have a full opportunity to do its work so that considerable amounts of polysaccharides reach the small intestine. These amylases, then, break the remaining polysaccharides to the double sugars, disaccharides, which are then reduced to monosaccharides by the **invertases** in the intestinal juice. There are several of these which are specific for particular sugars; for example, **sucrase** splits sucrose to glucose and fructose, **maltase** converts maltose to two glucose molecules, and **lactase** attacks lactose and converts it to glucose and galactose. These are the end products of carbohydrate digestion and the molecules then are absorbed into the bloodstream.

Fat digestion is continued and completed also in the small intestine. Not all fat is reduced to fatty acids and glycerol; some remains as finely divided fat particles which are absorbed as such. That which is digested is split by intestinal and pancreatic lipase into the end products, fatty acids and glycerol, which are then absorbed. This complete story is shown in Table 17–1.

Absorption

The end products of food digestion are absorbed into the circulatory system and are thus transported to all parts of the body where they are utilized for energy, growth, and repair. With the exception of alcohol, which is absorbed in the stomach, nearly all organic and inorganic compounds are absorbed in the small intestine. Complex modifications of the digestive tract facilitate the process of absorption. In the first place, the small intestine is 23 feet long and, secondly, the **villi**, previously referred to, multiply the surface area with which the chyme comes in close contact, giving every absorbable particle an opportunity to find its way into the blood stream.

The important anatomical unit that

Table 17–1.   Enzymes of Digestion

SUBSTRATE	ENZYME	SOURCE	END PRODUCT
Polysaccharides	Amylase	Salivary glands	Polysaccharides, disaccharide
	Amylase	Pancreas	Disaccharides
	Amylase	Intestinal glands	Disaccharides
Maltose	Maltase	Intestinal glands	Glucose
Sucrose	Sucrase	Intestinal glands	Glucose and fructose
Lactose	Lactase	Intestinal glands	Glucose and galactose
Proteins	Pepsin	Gastric glands	Proteoses, peptones, polypeptides
Casein	Rennin	Gastric glands	Coagulates casein
Proteins, proteoses Peptones, polypeptides	Trypsin Chymotrypsin	Pancreas	Peptides
Peptides	Peptidases	Pancreas and Intestinal glands	Amino acids
Fats	Lipase	Gastric glands	Fatty acids and glycerol
Emulsified fats	Lipase Lipase	Pancreas Intestinal juice	
Colloidal fats			Colloidal fats

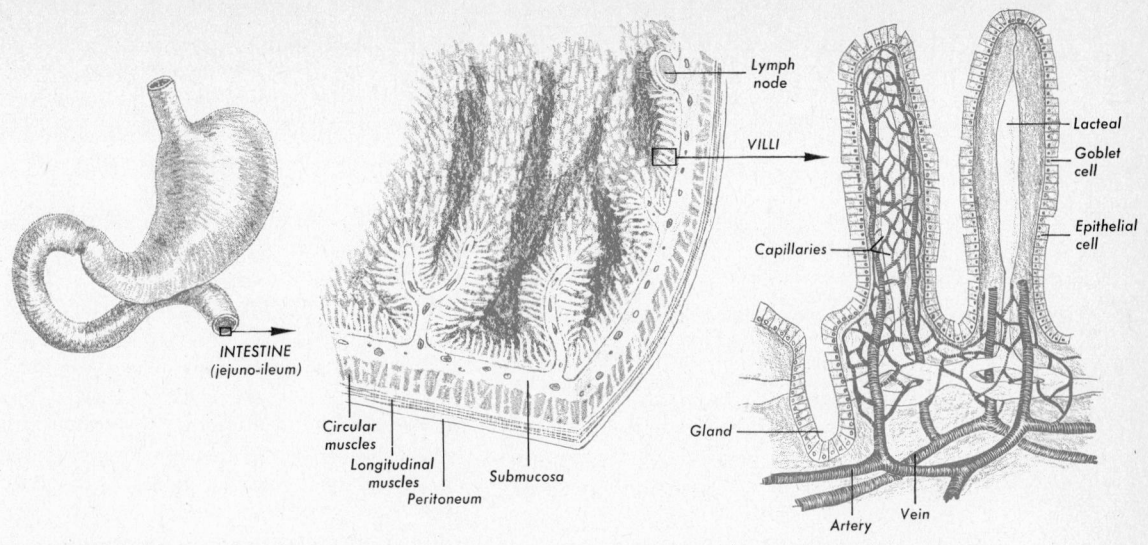

Fig. 17–9. A cross section of the intestine with the villi greatly enlarged to show how absorption takes place. Amino acids, sugars, fatty acids, and glycerol (not shown) enter the capillaries of the villi. Colloidal fats are taken into the lacteals of the villus. One villus is shown contracted while the other is gorged with fat. This "pumping" action aids in moving the lymph.

facilitates absorption is the **villus** (Fig. 17–9). It is a thin-walled projection lined with capillaries and with a central space, the lymph vessel or **lacteal**. As the watery chyme bathes the villi, as osmotic equilibrium is constantly strived for between the blood and the intestinal contents, but since the blood is moving rapidly through the capillaries, the flow of absorbable materials is toward the circulating blood. In addition to this passive absorption, work is required to move some molecules across the mucosal membranes. The fact that such molecules as fructose, glucose, and galactose, all of the same size and chemical composition, are absorbed at different rates indicates that some selection is made. The "active" transporting mechanism involves temporary phosphorylation of the sugars, which "pulls" them into the cell on one side and "releases" them, unchanged, on the other. Other molecules pass through by simple diffusion, as mentioned above.

The end products of protein, carbohydrate, and fat digestion, together with inorganic salts, vitamins, and water, are absorbed into the capillaries of the villi and are transported via the **hepatic portal system** directly to the liver. The colloidal fat droplets, however, enter the lacteals and are carried by the lymphatic system into the general circulation, by-passing the liver. Excess water as well as other molecules that do not enter the capillaries may also find its way into the lymph. The end result is that all food products eventually reach the blood stream, where they are distributed to all parts of the body.

Small Intestine Movements

While the chyme is in the small intestine (5–10 hours), it is in constant motion. The purpose of this movement is to mix it thoroughly and thus ensure proper digestion and absorption. One of the most important sources of our knowledge about these and other movements of the alimentary tract has come through X-ray pictures taken during various stages of digestion. If a person consumes a meal heavily laden with barium (or bismuth), it can be readily followed through the digestive tract because barium is opaque to X-rays. When observed, such a mass appears as black shadows on the screen; and if pictures are taken, the contours of the alimentary canal can be made out very easily.

Fig. 17–10. Food moves along the digestive tract by peristalsis. The motion is rapid in the esophagus, where its purpose is to get food to the stomach as soon as possible. In the stomach and remainder of the gut there are two kinds of movements: one that moves the food along, as shown in the duodenum above; and the other which brings about a mixing of the food (segmental and pendular) as shown in the stomach and lower intestine.

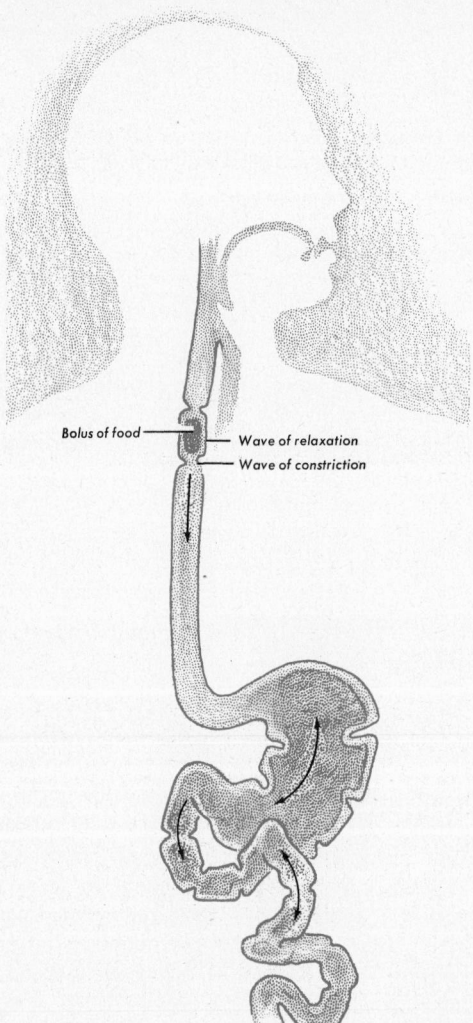

Bolus of food — — Wave of relaxation
— Wave of constriction

There are two types of mixing movements in the small intestine: **segmental** and **pendular**. Both are confined to the limits of a single loop of the gut. The segmental movements started by constricting at regular intervals throughout the loop (Fig. 17–10). Relaxation appears in 10 minutes or so and another series of constrictions occurs which is between the first series. This results in the breaking up and mixing of the masses of chyme. It is supplemented by the pendular movements, which consist of mild peristaltic waves that move the chyme back and forth within the intestinal loop but do not go beyond it. The chyme is moved progressively along the gut by means of peristaltic waves that occur both gently and in sudden rushes. The action is a slow progressive movement with a sudden rush, at times, of the entire contens from stomach to colon. The combined movements result in a continuous progression of the chyme through the small intestine until it reaches the large intestine.

THE LARGE INTESTINE

The small intestine joins the large intestine (colon) in the lower right region of the abdominal cavity (Fig. 17–1). The opening is guarded by the **ileocaecal sphincter.** Chyme that passes into the large intestine is still in a very fluid state, but as it passes through this region of the digestive tract much of the water is reabsorbed. The reabsorption of water is a conservation measure, for if it were retained in the gut the intake of water by mouth would necessarily be much greater. Such demands would impose insurmountable problems for an animal that needed to go any distance for its source of water. Since the chyme moves very slowly through the large intestine (12–14 hours), there is ample time for the water to be reabsorbed. During this time bacteria grow so abundantly that they constitute about half of the weight of feces. Bacteria cause no harm and are responsible for the production of vitamin K in man. In some herbivorous vertebrates, however, bacteria are thought to be beneficial, aiding the digestion of cellulose.

The saclike **caecum** at the beginning of the colon terminates in a finger-like evagination called the **vermiform appendix** because it is wormlike in appearance. There is a remnant of a once functional portion of the caecum. Herbivorous animals possess a very large caecum in which a certain amount of bacterial digestion takes place. From the caecum, the colon extends anteriorly along the right side as the **ascending colon,** crosses to the left side as the **transverse colon,** passes down the left side as the **descending colon,** then bends in an S-shape (**sigmoid flexure**) to become the **rectum.** The digestive tract terminates in the **anus** which is closed by two powerful muscles, the **anal sphincters.**

The chyme, which is now referred to as **feces,** is gradually moved through the colon by mild paristaltic waves as well as by periodic, overall massive contractions. When feces reach the rectum, which is normally empty, a desire to deficate becomes very pronounced. During defication a large peristaltic wave moves over the entire colon, aided by voluntary contractions of the abdominal muscles.

FOOD

Food must provide the body with the proper amounts and kinds of substances from which it can synthesize its own parts. Just as the carpenter or mason must have lumber, bricks, cement, and other materials, so must the body have its building materials. The digestive process breaks down the large, complex molecules to smaller or simpler ones that can be rebuilt into the specific molecules needed for human protoplasm. The process of reconstruction is taken up later. What concerns us now is the kinds of food which, when reduced to the simple state, will provide all of the essential materials in sufficient quantities to build and maintain the body. Re-

turning to the house analogy, if the carpenter does not have enough lumber and nails, or the mason enough bricks and cement, he will be forced to cut corners and produce a house that will not function as it should. Similarly, food must contain all of the essential substances if the body is to be properly constructed. The types and kinds of food have a profound effect on growth (Fig. 17–11). Food must also supply a source of energy to maintain the body at a constant temperature and to permit movement.

Foods for Building

The simplest of these are the inorganic salts. The analysis of any animal reveals a considerable amount of iron, calcium, phosphorus, and other minerals. Moreover, the animal body excretes minerals every day, in man about 30 grams.

The minerals that are most apt to be deficient in human diets are calcium, iron, and iodine. Calcium is essential for building bone and should be maintained at a high concentration in the diets of children and pregnant women. Although calcium is present in small amounts in vegetables and cereals, it is particularly rich in milk—hence, the constant urging for an adequate supply

Fig. 17–11. The value of a complete diet is spectacularly demonstrated by these three litter mates, which were fed on rations that were popular in the years indicated. The diets improved over these 43 years chiefly because of intense research on vitamins and antibiotics.

of this food for children. Iron is necessary for the formation of hemoglobin, the chief constituent of the red corpuscles, and a low iron diet results in anemia. Meat and eggs are rich in iron, as well as many kinds of fruits and vegetables, and ordinarily, under normal circumstances, it is not necessary to supplement the diet for this mineral. Iodine is required for the proper functioning of the thyroid gland, as was pointed out earlier. Near and at the seashore, and in regions covered by the sea in past geological time, this element is in adequate supply in the water and food grown in the region. Sea food in particular is always rich in it.

While sodium chloride (table salt—NaCl) is probably adequately supplied in most meats, it is less abundant in plant products; for that reason, food is usually salted during cooking. Salt is essential for normal body maintenance. Hydrochloric acid, for example, is produced in the cells of the stomach, using the Cl from the NaCl. With severe sweating there is a gradual loss of salts; if prolonged it will result in "miner's cramps," where large muscles go into painful spasms. The victim suffers from extreme thirst, but the more water that is taken the more severe the spasms, because the salt loss is proportional to the water loss.

Since the body is primarily water and since approximately 2,000 cc are normally lost each day, it follows that this important substance must also be replaced continuously. One can survive for many weeks without food but only a few days without water.

In addition to minerals and water, the protoplasm-building proteins must be present in the diet in considerable quantities. Larger amounts of protein are essential for the growing body than the 50 grams per day required by the adult. This amount must include all of the amino acids (about 10) that the human body cannot synthesize from simpler compounds. These are the so-called essential amino acids. There are many others that can be manufactured by the intracellular enzymes, providing the raw materials are available. Some proteins, such as those derived from milk, meat, eggs, and wheat, contain all of the essential amino acids, whereas others, such as those from gelatin and corn, lack one or more.

Foods for Energy

All of the energy utilized by the body must come from the food brought to it. No energy is derived from breaking the large molecules of protein, carbohydrate, and fat down to absorbable size during digestion. The energy comes from further breakdown by glycolysis and oxidative metabolism. When carbohydrates, fats, and proteins are metabolized to the same end products, either in the body or artificially outside the body, they deliver the following numbers of calories per gram: proteins and carbohydrates—about 4 calories; fats—about 9 calories. As far as energy requirements are concerned, one food is as good as another. Carbohydrates, of course, oxidize readily to supply energy. One can also derive all required energy from protein alone by conversion of the amino acids (about 40 percent), following deamination, to pyruvic acid and the subsequent oxidation via the Krebs citric acid cycle. With fats the situation is different. For some reason, when over 50 percent of the caloric intake is fat, oxidation is impaired, and fats burn best when there is an adequate amount of carbohydrate present in the tissues.

Accessory Foods—The Vitamins

The discovery of vitamins has a long and fascinating history. When man first began to isolate himself from natural fresh foods and to live on stored or dried foods, the "deficiency diseases" made their appearance. Among the many diseases that appeared, **scurvy** was the best known, and the first for which a remedy was discovered. That was some 350 years ago, long before anybody knew anything about vitamins, or much else concerning nutrition. The British navy captain Cook discovered that when his men were down with scurvy after many months at sea, their fresh foods exhausted, they recovered almost miraculously if the ship docked where fresh fruits were available. He, therefore, took fruit, particularly citrus fruit, aboard and discovered that small daily rations kept his men from getting the dreaded scurvy. It has been said by some that the use of this knowledge was responsible in part for the success of the British navy.

Vitamins are effective in extremely small quantities, so small that they cannot be considered energy-producing in the sense that proteins, carbohydrates, and fats are. They are specific, relatively simple organic compounds which must be included in the diet for normal health. Most organisms produce some vitamins. Plants manufacture all they need but animals require those in their diet that they cannot synthesize. If vitamins are absent over a period of time, definite symptoms appear which grow progressively worse, terminating in death. Animals vary in their vitamin needs. Rats, for example, can synthesize vitamin C and man can exist without thiamine. In the latter case, however, this vitamin is manufactured by bacteria living in his digestive tract. By trial and error experiments the vitamin requirements have been determined for most laboratory animals, such as the rat, mouse, guinea pig, and rabbit, so that they can be employed in vitamin research.

The role of vitamins is becoming clear as research on the subject continues. Since they are needed in such small quantities in the diet and yet are essential for life of the organism, it was early suspected that they might play some part in metabolism. This has proven true.

Since not all of the needed vitamins are synthesized by the animal, those which it cannot produce must be included in the diet; therefore, animals are dependent ultimately upon plants for the vitamins which they do not make themselves.

Vitamins are usually grouped according to their solubility: those soluble in fats and oils (A,D,E,K) and those soluble in water (B-complex and C). The use of letters to identify the vitamins is disappearing as their molecular structures become known and they are given specific names. There are more than 30 vitamins described. Some of the better-known ones are listed in Table 17–2 together with their sources and the diseases caused by their absence from the diet of man.

Fat-Soluble Vitamins

Vitamin A. This vitamin is derived from the orange-colored pigment of plants called **carotene.** The vitamin exists in two chemical forms. Carotene is present in all green plants and it is most abundant in those that are yellow or red. Carrots, for example, are rich in vitamin A. Some animals have a tendency to store it in great quantities in their livers. This is particularly true of such fish as the shark, cod, and halibut, hence the well-known names—shark-liver oil, cod-liver oil, halibut-liver oil. Note that it is found in the oil of these livers. Its fat or oil solubility is one of its characteristics.

For some unknown reason, the vitamin A concentration in the liver of newborn infants is low. Apparently it is not transmitted through the placenta and therefore must be acquired during the first few weeks of life. Human milk is rich in vitamin A, particularly the **colostrum** which is the milk secreted by the breast immediately following birth.

In severe cases of vitamin A deficiency, rarely observed in this country, the eyes become infected and the cornea becomes very dry and ulcerated (**xerophthalmia**). Another symptom is the inability to see at night, which is called **night blindness.** The retina of the eye cannot synthesize **visual purple** in the absence of this vitamin and consequently does not respond to dim light (see p. 325). It is true of vitamin A, as it is with most of the others, that severe symptoms resulting from a marked deficiency rarely show up among Americans, because

NAME	SOURCE	FUNCTION	EFFECTS OF MARKED DEFICIENCY
Fat-Soluble			
Vitamin A	Yellow and green vegetables, oils, butter	Chemistry of vision; health of epithelial cells	Night blindness, xerophthalmia, infections
Vitamin D	Liver oils, milk, eggs	Ca and P metabolism	Rickets
Vitamin E	Most foods	Gamete formation	Sterility in some vertebrates
Vitamin K	Most foods	Prothrombin formation	Blood-clotting time increased
Water-Soluble			
B complex:			
Thiamine (B_1)		Carbohydrate metabolism	Beriberi
Riboflavin (B_2)		Oxidative metabolism	Skin defects
Niacin (B_5)		Oxidative metabolism	Pellagra
Pantothenic Acid	Whole grains, yeasts, meats, eggs, milk, green vegetables	Coenzyme A precursor	Uncertain in man
Folic Acid Cobalamin (B_{12})		Nucleic acid metabolism	Anemia
Biotin		CO_2 metabolism	Uncertain in man
Pyridoxine (B_6)		Amino acid metabolism	Uncertain in man
Vitamin C	Citrus fruits, fresh vegetables	Amino acid oxidation, maintenance of capillary walls	Scurvy

of our diversified diet. Mild cases, on the other hand, are relatively common but can be corrected simply by supplementing the diet with vitamin A.

Vitamin D. Whereas only a single vitamin is mentioned here, there are about ten related compounds included in the group of D vitamins. Chemically they are all derivatives of **sterols.** The most important is D_2, or **calciferol,** which is formed by the action

of ultraviolet irradiation on **ergosterol,** another sterol. It aids in calcium absorption from the digestive tract and, together with parathormone, controls calcium mobilization in bone formation.

Vitamin D is found naturally in fish liver oils along with Vitamin A. Irradiated fats contain D vitamins in abundance. When vitamin D is deficient in the diet of children there is failure of proper deposition of phosphorus and calcium in the developing bones, causing the disease known as **rickets.** A rachitic child has a malformed skeleton. Fortunately, very few rachitic children have been seen in recent years, compared to several decades ago. In climates where there is a great deal of sunshine the vitamin deficiency is rarely seen. However, in northern climates during the winter months children are very apt to deplete their stores of vitamin D.

Vitamin E. This vitamin is called **tocopherol** and it has three different molecular configurations. It apparently functions in hydrogen transfer during aerobic respiration which is reflected, when deficient, in the degeneration of the germinal epithelium, the production of inactive sperm, and resorption of mammalian embryos. In chickens the liver becomes necrotic, hemorrhages occur, and the testes degenerate when this vitamin is lacking in the diet. Nothing is known about its effect in man, owing primarily to the fact that deficiency is very unlikely because it occurs so abundantly in plant and animal foods.

Vitamin K. This vitamin occurs in two forms, both of which are known chemically. It plays a role in photosynthesis, hence is widely distributed in plants. It is produced by bacterial action in the digestive tract and a sufficient amount of the vitamin is absorbed to prevent symptoms under normal conditions. Newborn infants may sometimes show a deficiency. In mammals it functions in the synthesis of prothrombin (p. 418), an essential factor in blood coagulation. When bile fails to reach the digestive tract, as in gallstone occlusions, the bacteria do not produce vitamin K in sufficient quantities to allow normal clotting of blood, thus making surgery hazardous.

Water-Soluble Vitamins

B Vitamins. This is a group of water-soluble vitamins that occur abundantly in high-protein foods such as meat, liver, nuts, whole grains, and particularly yeast. In the early days of vitamin research, a single B vitamin was described; now a large number, about 12, are known to belong to this group which is sometimes referred to as the **B-complex.**

Thiamine (B_1). This vitamin functions in intermediate metabolism. When it is deficient in the diet, the conversion of pyruvic acid to acetic acid is impaired, and respiration is faulty. Severe thiamine deficiency in man produces **beriberi,** a disease that was common among the rice-eating peoples of the world when modern methods of removing the hulls were introduced. The thiamine was in the hulls, and hence was lost in the processing of the rice. The processing of wheat today is just as impractical from a nutritional standpoint. Nearly all of the vitamins are removed from the wheat grain when it is processed into the white flour. Present-day methods of milling have become a little more scientific, however, for an effort is now made to put the vitamins back in again. It seems rather ridiculous to engage in the labor and expense of removing many nutritious parts of the rich wheat kernel, and then be obliged to restore them, for the sole purpose of keeping the flour white. About all that is accomplished is the production of a white but nutritionally inferior flour at a higher price.

Riboflavin (B_2). This vitamin is necessary for hydrogen transport in respiration. It is converted into a **flavoprotein** which is one of the hydrogen acceptors in the conversion of ADP to ATP (p. 88). A deficiency results in skin disorders in man as well as in laboratory animals. Small amounts taken daily will often restore the skin to normal appearance.

Niacin (B_5). This vitamin also plays an active part in respiration. It functions as a precursor to the synthesis of diphosphopyridinenucleotide (DPN) and triphosphonucleotide (TPN), both hydrogen acceptors. When it is lacking in the diet, the disease **pellagra** results. This is a disease in which the skin and digestive tract are affected; it is fatal in severe cases. Small amounts of yeast, which contains niacin as well as other members of the B-complex, will cure the disease very quickly.

Pantothenic Acid. This vitamin functions as one of the precursors in the synthesis of coenzyme A, which is used in the transfer of acetyl from pyruvic acid to the Krebs citric acid cycle (p. 87). The vitamin is synthesized by bacteria in the colon of man. It is not known what the human requirement is. Its deficiency in rats causes hemorrhages and destruction of the adrenal cortex.

A recently discovered vitamin, **thioctic acid,** also is associated with coenzyme A formation. It, together with thiamine, pantothenic acid, and niacin, is essential in this phase of cellular metabolism.

Folic Acid. This vitamin is required for purine and pyrimidine synthesis. Probably all animals require it in small amounts. In man it is synthesized by intestinal bacteria although a folic-acid-deficient diet leads to anemia.

Cobalamin (B_{12}). This is a cobalt containing vitamin which occurs in natural waters. It functions in protein synthesis and is required by many organisms. A diet deficient in this vitamin causes anemia in mammals because of its importance for the maturation of red blood cells. Man requires about 1 mg per day to prevent anemia.

Biotin. The function of this vitamin involves the metabolism of carbon dioxide, probably through oxaloacetate synthesis. Egg yolk contains it, while egg white is rich in an antibiotin factor which nullifies its action. It is synthesized by intestinal symbiotic bacteria in many animals. If the bacteria are destroyed mammals require biotin

in the diet. The requirement for man is very small, less than 10 micrograms per day.

Pyridoxine (B_6). This vitamin is associated with the metabolism of amino acids. It has been known for a long time that it is required for normal growth of rats and there is some evidence now that it may be essential for health in humans. Rats fed a diet deficient in pyridoxine stop growing and show marked lowering of resistance to disease because of a resulting loss of both white and red cells in the blood. Its role in the human is still not ascertained.

Vitamin C (ascorbic acid). The function of this vitamin is poorly understood. It seems to be involved with oxidation of some amino acids. The fact that it occurs in the mitochondria would indicate that it is probably involved in respiratory metabolism.

Vitamin C is found abundantly in citrus fruits and fresh vegetables. Cooking destroys it rapidly and stored foods soon lose their vitamin C content. The chief symptoms of scurvy are due to the weakening of the walls of the capillaries and small arteries, causing the characteristic bleeding at the joints, gums, and under the skin. There is some evidence that it influences the formation of substances that bind cells together. In scurvy it probably is the failure of this function of vitamin C that brings on the symptoms of the disease. Dental caries is supposed to be due, in part, to a low vitamin C intake.

In a nation where most people are not confronted with the problem of how to get enough food but what kind of food to select, good advice is to eat a highly varied diet. Such a diet should be particularly rich in whole grains, fresh vegetables, and fruits, with an ample supply of meat and lesser amounts of the foods high in carbohydrates, such as potatoes, bread, and so forth, which add weight but have few vitamins.

Books

BEAUMONT, W., *Experiments and Observations on the Gastric Juice and the Physiology of Digestion*. New York: Dover Publications, 1959. (Original was published in 1833.)

BECK, W. S., *Human Design*. New York: Harcourt, Brace Jovanovich, 1971.

CARLSON, A. J., JOHNSON, V., and CAVERT, H. M., *The Machinery of the Body*. Chicago: University of Chicago Press, 1962.

FULTON, J., *Selected Readings in the History of Physiology*. Springfield, Ill.: Charles C Thomas, 1930.

ROMER, A. S., *The Vertebrate Body*, short version. Philadelphia: Saunders, 1971.

WOODBURNE, R. T., *Essentials of Human Anatomy*. New York: Oxford University Press, 1961.

Articles

MAYER, J., "Appetite and Obesity." *Atlantic Monthly*, September, 1955.

WOODWARD, J. P., "Biotin." *Scientific American*, June, 1961.

18

THE BREATHING SYSTEM AND GAS EXCHANGE

In the previous chapter we saw how energy-rich food is broken down and absorbed into the blood stream, ready for distribution to all the cells of the body where it is used for growth, repair, and energy. In order that the energy in food molecules may be released, **oxygen** must be present, because most energy obtained by a living organism is through oxidation. Each cell, then, must not only have a supply of fuel and building materials but also have oxygen to burn the fuel. The function of the breathing system in animals is to provide a means of getting oxygen out of the air or water and into the vicinity of the cell so that it can be utilized. Another function is the elimination of CO_2.

In animals with considerable bulk, where the cells are a great distance from the surface, the circulatory system performs the job of distributing the oxygen, thus working hand in hand with the breathing system to solve this complex problem of oxygen supply.

The definition of **respiration** is not breathing but the **actual oxidation** of the food in the cells, releasing energy and the waste product, carbon dioxide. **Breathing,** on the other hand, is merely part of the mechanism by which cells obtain oxygen. Lungs, gills, and other structures remove oxygen from the air or water and pass it on to some sort of transportation system by which it can reach every cell. It is important that the distinction between breathing and respiration be kept in mind. We deal with the former in the following discussion leaving the latter for another chapter.

The methods employed by different animals to bring oxygen to their cells are highly varied and become extremely complex in large animals. Such a system may be compared to the ventilating system of a large building. The larger and more complex the building, the more intricate the system becomes, and each new expansion of the edifice means the addition of more powerful blowers and an ever increasing number of ducts. A point might even be reached when it would be impossible, or at least impractical, to attempt to provide ventilation for all parts of the building. In animals, likewise, the problems become more complex with increasing size, which has doubtless been one of the factors in limiting the size of animals.

The lungs of man, like those of all vertebrates, arise in the embryo from an outpocketing of the floor of the pharynx. When fully formed, they almost fill the entire **thoracic cavity.** The complete breathing system consists of the **nasal chambers, larynx, trachea, bronchi, bronchioles,** and **alveoli** (Fig. 18–1). The nostrils are the normal portals of entry for air, and they open into the spacious nasal chambers which are especially adapted for warming the incoming air. The surface area is greatly increased by sheets of bony tissue, the **turbinates,** that hang down into the nasal passages. These are covered with a layer of mucus epithelium that is kept constantly wet by the mucus-secreting glands located in this layer of tissue. It is also highly vascularized, giving warmth to the entire nasal chamber. The result is that

Fig. 18–1. A schematic view of the breathing system of man with one alveolus and associated parts greatly enlarged.

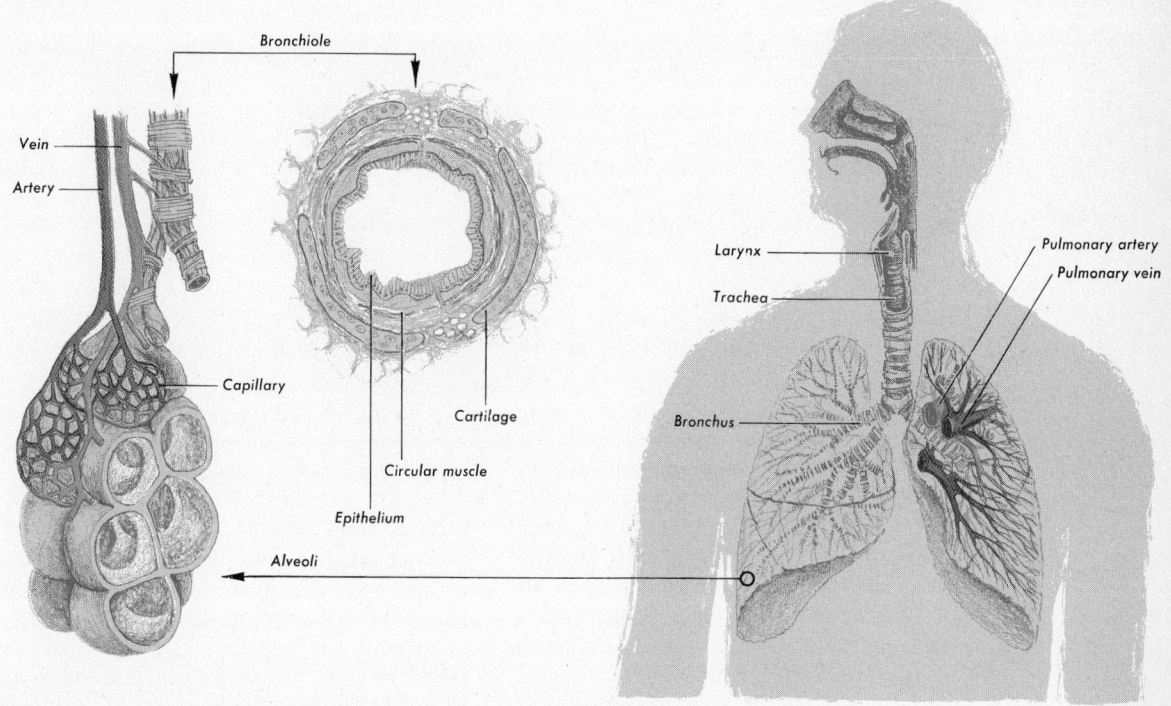

Vein

Artery

Bronchiole

Capillary

Cartilage

Circular muscle

Epithelium

Alveoli

Larynx

Trachea

Bronchus

Pulmonary artery

Pulmonary vein

as air passes through these passages it is not only warmed to approximately body temperature but is also humidified to some extent.

Another important feature is the provision made to filter particulate matter out of the incoming air. The nose is lined with hair which strains out the larger particles, and smaller bits which pass the hair are caught in the mucus secretion and eventually discharged with it. The breathing passages are lined with ciliated epithelium which functions in moving mucus toward the nasal chambers. Proof of the efficiency of this mechanism is noted by observing the color of the nasal secretions after working in dust-laden air. The mucus secretion also has some hygienic value, because bacteria are caught by this mechanism as well, thus preventing them from reaching the lungs where they might cause disease. However, some bacteria and viruses may establish themselves in the nasal areas, as in the case of head colds. Later they may make their way into the deeper portions of the breathing tract where characteristic symptoms appear. The nasal secretions have some bacteria-destroying action, and in head colds the cells of these areas are stimulated to produce large amounts of mucus to prevent the organisms from getting a foothold. This reaction accounts for the "running nose" at the beginning of a cold. Unfortunately, this is frequently carried to such extremes that swelling of the walls closes the air passages, forcing the victim to breathe through his mouth. This makes him liable to infection from the same or other organisms, since the air has not been filtered in the nasal passages as it normally would be.

Once past the nasal passages, the air is carried down into the **pharynx** and into the **larynx** on its way to the lungs. The larynx, or **voice box,** is a cartilaginous structure at the upper end of the **trachea.** It offers a rigid support for the **vocal cords,** which are stretched bands of tough tissue that vibrate to produce a sound when they are tightened and air is forced over them. While the initial sound comes from the vocal cords, the ultimate sound produced is highly modified by the nasal chambers and mouth cavity. These resonating chambers, together with their movable parts (soft palate, tongue, and cheeks), are responsible for the **quality** of the voice, either speaking or singing. Any interference with these passages has a marked effect on the quality of the voice, a familiar phenomenon in anyone suffering from a cold.

Immediately beyond the larynx is the **trachea,** a single tube whose walls are prevented from collapsing by means of cartilaginous rings. The "windpipe," as it is sometimes called, lies at the front of the throat where the rings can be felt through the skin. The trachea passes into the chest cavity where it divides into two **bronchi,** one going to each lung. These subdivide into **bronchioles** which after many divisions terminate in tiny blind sacs, the **alveoli.**

In the alveoli the actual exchange of gases takes place. All of the mechanism up to this point functions in getting the air to and from these alveoli. The combined surface of all the alveoli of the human lungs is over 1,000 square feet, and all of this is in intimate contact with capillaries. The lungs are covered by a thin layer of epithelium called the **pleura** and the spaces between the alveoli and bronchioles are filled with elastic connective tissue. The lungs remain partially inflated at all times. When removed from the thoracic cavity they collapse to only a fraction of their inflated size, because of the constant tension on the walls of the bronchioles. But, even in this contracted condition they resemble a sponge in consistency and readily float in water.

THE BREATHING MECHANISM

The lungs almost completely fill the thoracic cavity, which is lined with pleura identical with that covering the lungs. These two layers normally lie close together with no actual space between them, but when the lungs are deflated, the space that results is known as the **pleural space.** The two surfaces are held together because there is no air in

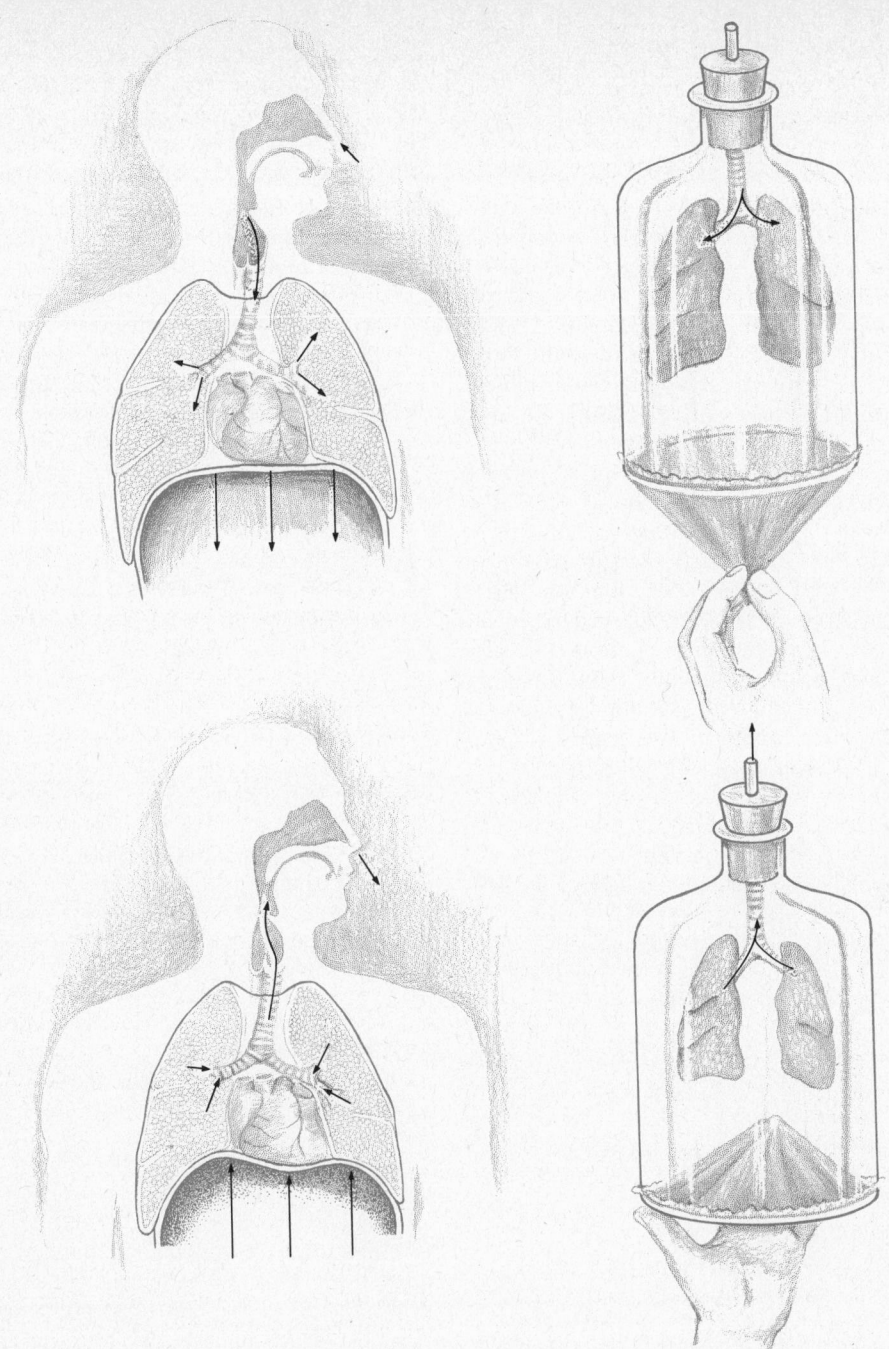

Fig. 18–2. In breathing, the thoracic cavity increases and decreases in size by alternate raising and lowering of the ribs, together with a corresponding elevation and depression of the diaphragm. This can be shown experimentally by placing the lungs and trachea of a dog in a bell jar and covering the open end with a rubber membrane. By pushing up and down on the membrane, simulating the movements of the diaphragm, air passes in and out of the lungs.

the pleural space. In other words, there is a partial vacuum in which the pressure is below that of the outside atmosphere. Therefore, if the walls of this closed cavity move out so as to enlarge the cavity, the walls of the lungs must follow. This creates a decreased pressure in the lungs themselves which, in turn, causes air to rush into the low pressure region deep in the alveoli. The opposite effect results when the cavity is reduced in size. The cavity can be increased and decreased in size by simply raising the ribs (Fig. 18–2) and then allowing them to return. Simultaneous with the raising of the ribs, the dome-shaped **diaphragm** is pulled downward and flattened. The ribs are raised by the contraction of the muscles lying between them (intercostal muscles) and when these muscles are relaxed the ribs fall back into position. At the same time, the diaphragm is returned to its dome-shape by a ligament that keeps it in this position. Therefore, **inspiration** is brought about by the contraction of muscles, **expiration** by relaxation. As a result of these two actions the air in the alveoli is changed.

Under extreme exertion the abdominal muscles also are involved in the breathing movements. On inspiration they relax in order to allow the diaphragm to descend farther, thereby compressing the abdominal organs. In forced expiration the abdominal muscles contract in order to help push the diaphragm upward. These muscles also play an important part in coughing and sneezing. Preceding these actions, the air passages are voluntarily closed and the muscles contract to build up a pressure within the lungs. This is followed by a sudden opening of the passages, resulting in an explosive discharge of air which is designed to dislodge any foreign matter that may have gotten into the air passages.

Accidental puncturing of the thoracic cavity brings about the collapse of the lung because when air within the pleural cavity reaches the same pressure as the outside air, the lung contracts, as pointed out above. Such a lung is nonfunctional and if both lungs collapse at the same time, suffocation follows. Fortunately, there is a partition between the lungs so that one can be deflated while the other functions normally. It even becomes necessary sometimes to collapse one or the other lung artificially in such disease as tuberculosis. This is easily accomplished by simply forcing a hollow needle between the ribs into the pleural space and allowing sterile air to enter (Fig. 18–3). When the needle is withdrawn the lung fills the space again—in about two years—as the air is slowly absorbed into the tissues and eventually carried away in the blood.

Fig. 18–3. Lungs can be deflated by simply forcing air into the pleural space as shown here.

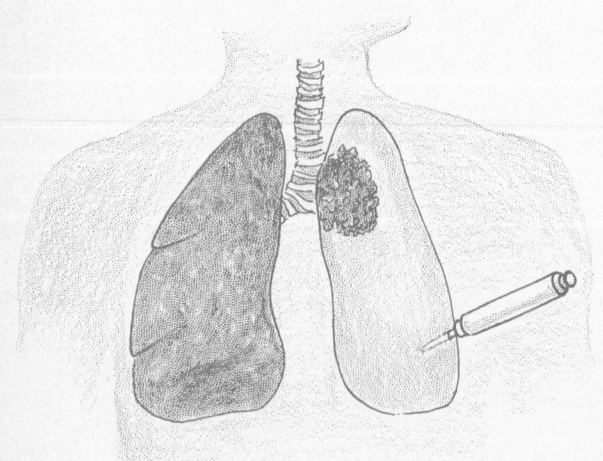

LUNG CAPACITY

It is a familiar fact that breathing is slow and shallow when one is at rest, either sitting or lying down, but rapidly becomes faster and deeper with exertion such as running. It is also well known that when resting one can take in a great deal more air at the end of a quiet inspiration and can force out a great deal more air at the end of a quiet expiration. These facts indicate that the lungs have a great deal more capacity than is used in nonstrenuous work. This constitutes a reserve that is always available when needed.

In quiet breathing, about 500 cc of air is breathed in and out. This is known as **tidal** air, reminiscent of the tides. Upon forced inspiration, 1,500 cc more can be taken into the lungs. This is known as **complemental air.** Forced expiration can deliver about 1,500 cc of air, known as **supplemental air.** Therefore, complemental air together with tidal and supplemental air amounts to about 3,500 cc, the total capacity that can be taken in and forced out under conditions of heavy exertion. This amount is called **vital capacity.** There is some variation in this quantity, depending on the size, age, sex, and training of the individual. Athletes usually have slightly more vital capacity than do non-athletes, women slightly less than men.

Some air, about 1,000 cc, is always left in the lungs even after forced expiration. This **residual air** is present to some extent in lungs that have been removed from the body. If the lungs of an unborn child are placed in water they do not float, whereas those of the newborn who has taken a breath do. This test sometimes has legal significance in suspected cases of infanticide.

CONTROL OF BREATHING

Because the demands for oxygen vary with activity, the rate and depth of breathing vary to meet these demands. It is possible to change the rate of breathing voluntarily, although one does not need to think in order to breathe. One can hold his breath for some time, with considerable distress, but eventually he will be forced to breathe again in a normal fashion. The rate of breathing is altered by a number of conditions arising from within as well as from without the organism. Such conditions include emotional as well as physical strain.

Breathing is a result of the coordinated action of a great many muscles which require a rather precise controlling mechanism called the **breathing center.** This consists of a group of specialized cells located in the **medulla** (Fig. 18–4). Bursts of nervous impulses leave this center over the **phrenic** nerves to the diaphragm and over the **cervical sympathetics** to the rib muscles, causing both to

contract, resulting in inspiration. The rate at which these contractions occur depends on several factors influencing the breathing center. If the carbon dioxide content of the

Fig. 18–4. The rate of breathing is controlled by the breathing center which is sensitive to carbon dioxide in the blood. During activity the increased carbon dioxide content of the blood stimulates the breathing center to bring about increased rate of breathing. The opposite effect prevails when at rest.

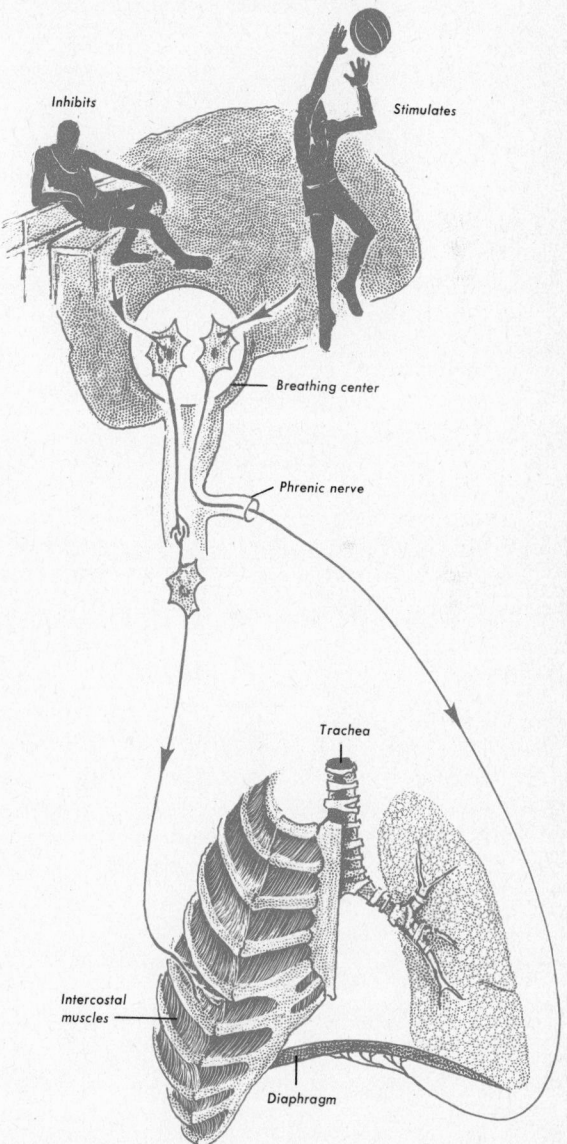

Inhibits

Stimulates

Breathing center

Phrenic nerve

Trachea

Intercostal muscles

Diaphragm

blood rises, as it does in exercise or emotional stress, the impulses are quickened. If, on the other hand, the carbon dioxide is diminished, the impulses will be farther apart. If the phrenic nerve is severed or

Fig. 18–5. The rhythmic movements in breathing are controlled by the breathing center also. Sensory nerve endings lying in the alveoli send impulses to the center which inhibit its action when the alveoli are stretched on inspiration. During the period of inhibition, no motor impulses come from the center to the muscles, hence they relax, resulting in expiration.

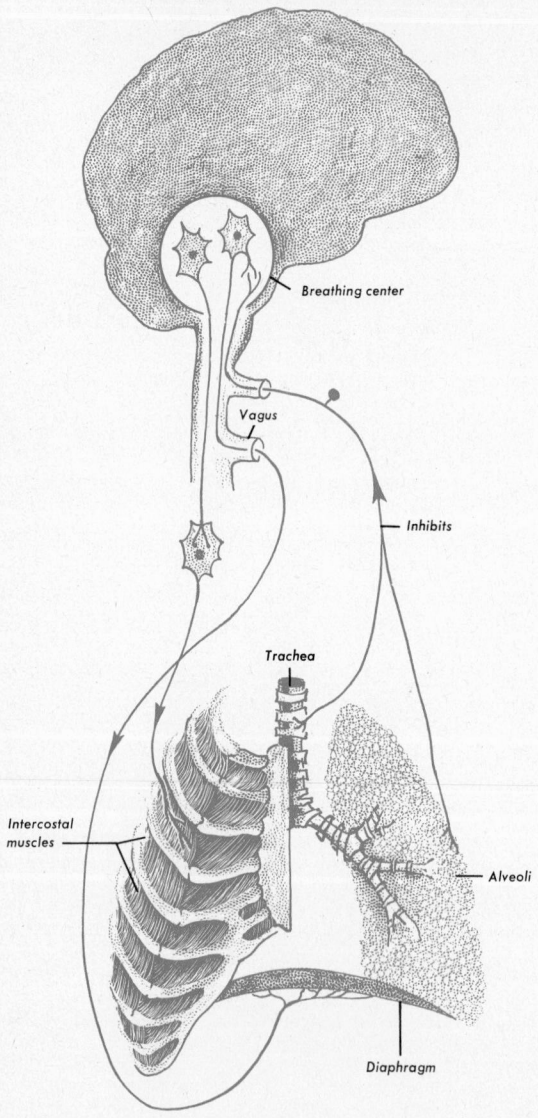

destroyed through diseases such as poliomyelitis, the impulses fail to reach the diaphragm and this muscle ceases its contractions. The same is true of the other motor nerves going to the rib muscles. It is possible by taking a series of deep inhalations and exhalations to reduce the carbon dioxide content of the blood to a point where breathing can be held in abeyance until the gas again reaches a sufficient concentration to stimulate the breathing center. A newborn child, as well, will take its first breath only after the carbon dioxide has reached a certain level. When its placental circulation is cut off the gas begins to build up in the blood. Occasionally, even then breathing will not start. This often can be remedied by forcing air containing 10 percent carbon dioxide into the lungs.

One of the problems of breathing that was not understood for a long time was how the breathing center could send out these bursts of impulses in a rhythmic fashion. It has been discovered by clever surgery that if all of the sensory nerves leading to the breathing center are severed, inspiration will occur but not expiration. In other words, the muscles of the diaphragm and of the ribs will contract but will not relax, because a steady stream of impulses continues to come to them. This means that the breathing center must be inhibited periodically so that the bursts of impulses coming to the breathing muscles are interrupted. This is accomplished by sensory endings in the walls of the alveoli and other parts of the lungs (Fig. 18–5). When air is drawn into the lungs, the stretching of the walls presses on the nerve endings, causing impulses which inhibit the breathing center. Such is the device by which the lungs are alternately filled and emptied.

The breathing center is also inhibited by impulses coming to it from many other parts of the body. Receptors lie in the walls of the larynx and pharynx which are sensitive to foreign bodies or harmful gases. For example, when a particle of food gets into the larynx, breathing ceases at once thus preventing the passage of the particle farther into the delicate breathing tract. This is followed by a closing of the epiglottis until considerable pressure is built up in the chest cavity. It then lets go explosively in the form of a cough as already described. Likewise, if irritating gases such as chlorine are drawn into the pharynx, breathing ceases instantly with the closing of the air passage by the epiglottis. Here is another safeguard to prevent injury to the extremely sensitive and delicate lining of the breathing tract. Severe pain and sudden chilling will also inhibit the breathing center. One also "catches his breath" because of the action of these nerves on the breathing center. Thus we see how carefully this mechanism is tuned to the job of protecting the important breathing organ so that its function may go on uninterrupted throughout life.

GASEOUS EXCHANGE

By measuring the amount of oxygen and carbon dioxide in expired air as compared to that of the atmosphere, it is easy to demonstrate that some exchange has taken place, as shown in the following figures:

	Inspired Air	Alveolar Air	Difference
Oxygen	20.0%	14.5%	5.5%
Carbon dioxide	0.04%	5.5%	5.1%
Nitrogen	79.0%	80.0%	1.0%

Water and other gases have not been taken into account in the above approximate figures. It is noted that atmospheric air has lost approximately 27 percent of its oxygen (difference ÷ total inspired air) and has picked up approximately 5 percent carbon dioxide in passing through the alveoli of the lungs. Blood passing near the alveoli has lost most, but not all, of its carbon dioxide and has taken on a load of oxygen. Just how is this done?

The oxygen molecules in the air dissolve in the watery mucus covering the lining of the alveoli and pass directly through by diffusion to the blood which is flowing very close

to the walls of the alveoli, in fact, only two cells away (one cell layer in the capillary wall plus one cell layer in the alveolar wall) (Fig. 18–6). Oxygen passes from the alveolar air to the blood by simple diffusion, since the concentration of the oxygen molecules is much greater in the alveoli than in the blood. Although oxygen molecules are moving freely across the membranes in both directions, there are more going into the blood than are going into the alveoli, hence the net movement is toward the bloodstream. Conversely, the carbon dioxide moves in the opposite direction for exactly the same reason. The inert nitrogen gas remains approximately the same during breathing. The overall result, then, is that the blood leaving the lungs is

Fig. 18–6. This is a schematic representation of how gaseous exchange occurs in the alveoli and in the tissues. An explanation is found in the text.

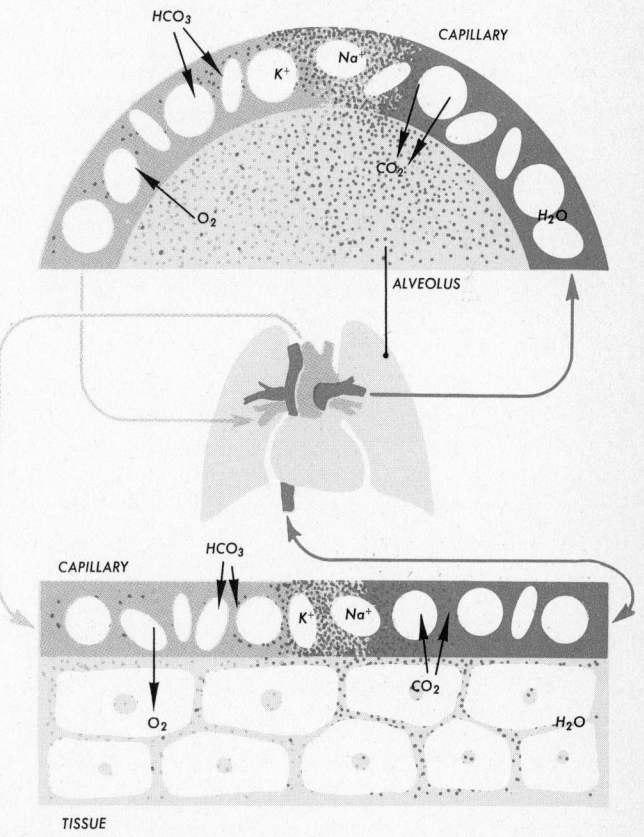

rich in oxygen and low in carbon dioxide, whereas expired air is rich in carbon dioxide and low in oxygen. With the continued breathing movements there is a continual exchange of these gases, and interruptions cannot be tolerated, at least for any prolonged periods of time.

The amount of oxygen in the alveoli depends on the amount in the air, which changes with atmospheric pressure. At sea level it is 760 mm of mercury, of which about 150 mm represents the **partial pressure** of oxygen. There is a gradual drop in pressure as one rises above the earth's surface and with this comes a drop in the pressure of oxygen in the alveoli, even though the breathing rate and depth are increased. At about 14,000 feet, no matter how fast or deep the breathing, symptoms called **moun-tain sickness** appear. These are due directly to a shortage of oxygen in the tissues, resulting from too little oxygen in the alveoli and hence too little in the blood (Fig. 18–7). Airplanes at or above 14,000 feet must, therefore, supplement the oxygen supply. This is done in passenger liners by pressurizing the entire cabin to that of about 4,000 feet, no matter how high the plane goes. People who live at high altitudes (10,000 feet) soon become acclimatized to the rarefied air by producing more red corpuscles, thus increasing the oxygen-carrying capacity of the blood.

In the tissues exactly the opposite condition prevails with respect to the movement of the two gases (Fig. 18–6). The oxygen diffuses from the blood into the tissue cells because the amount of oxygen is higher

Fig. 18–7. There are several physical and chemical factors in the external environment that influence gaseous exchange. Under normal atmospheric pressures (sea level) the gaseous exchange in the alveoli is adequate. At high altitudes (beyond 14,000 feet) the air pressure is so reduced that an insufficient amount of oxygen passes into the blood to maintain consciousness. In shock (a physiological response often caused by serious lacerations or other trauma) the blood pressure drops so low in the alveoli that circulation is inadequate to pick up enough oxygen to maintain normal body activities. In the presence of carbon monoxide (CO) a stable compound (methemoglobin) is formed in the blood, thus rendering hemoglobin incapable of carrying oxygen. If prolonged, breathing of CO can be fatal.

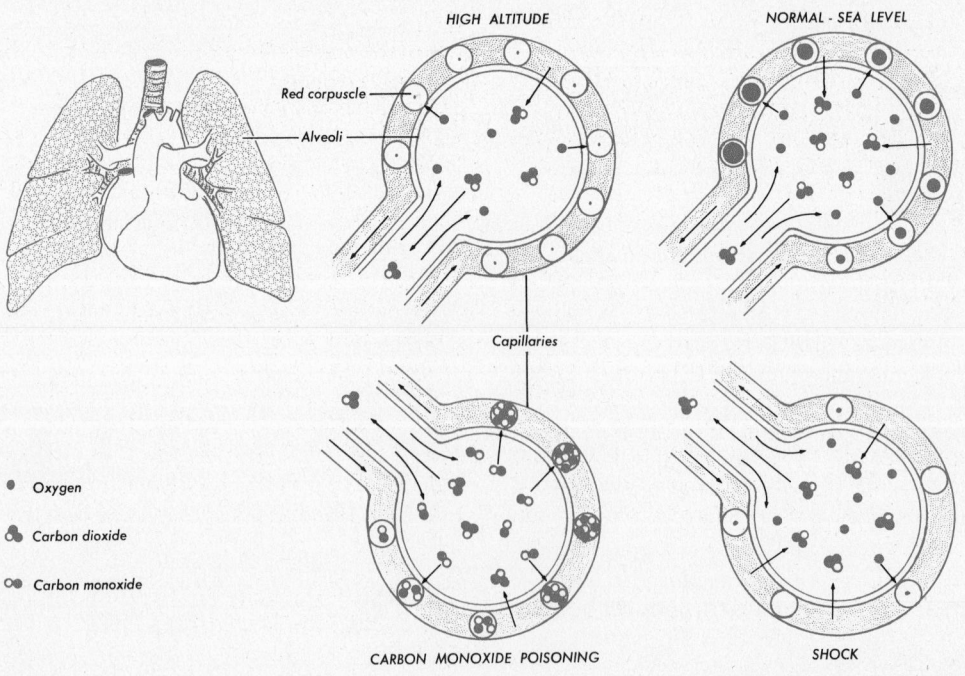

HIGH ALTITUDE

NORMAL - SEA LEVEL

Red corpuscle

Alveoli

Capillaries

- Oxygen
- Carbon dioxide
- Carbon monoxide

CARBON MONOXIDE POISONING

SHOCK

in the blood than in the tissues, and the carbon dioxide diffuses out of the tissues into the blood for a similar reason. Not all of the oxygen leaves the blood at the tissues, however. Only about 7 cc out of a possible 19 cc (per 100 cc of blood) actually diffuse into the tissue cells.

TRANSPORT OF GASES

The value of the blood as a transporting agent for oxygen and carbon dioxide can be determined easily by measuring the amount of these gases that dissolve in equal quantities of blood and water. It will be found that for both O_2 and CO_2 many times more of these substances is picked up by blood than by water. This is primarily because of that amazingly complex organic compound called **hemoglobin** (p. 419). Plasma is little better than water in its capacity to carry these two gases. For example, only 0.2 cc of oxygen and 0.3 cc of carbon dioxide dissolve in 100 cc of plasma, whereas the same amount of whole blood absorbs 20 cc of oxygen and 50 cc or more of carbon dioxide. If it were not for hemoglobin, the blood would need to circulate 35 times faster than it does to accomplish the same job.

About 98 percent of the oxygen is carried in combination with hemoglobin and the rest is dissolved in the plasma. The molecules unite in a one-to-one ratio, that is:

$$\underset{\substack{\text{Hemoglobin} \\ \text{(dull red)}}}{\text{Hb}} + \underset{}{O_2} \underset{\text{(tissues)}}{\overset{\text{(lungs)}}{\rightleftarrows}} \underset{\substack{\text{Oxyhemoglobin} \\ \text{(scarlet)}}}{\text{HbO}_2}$$

Obviously the reaction must be a reversible one, because the oxygen must be as readily released to the tissue cells as it is taken up in the alveoli in the lungs. Hence blood leaving the tissues contains largely hemoglobin, whereas that leaving the alveoli contains primarily oxyhemoglobin. The color difference in these two explains why systemic veins are dull red in color while arteries are scarlet.

Hemoglobin is no less remarkable in its capacity to make it possible for the blood to carry large quantities of carbon dioxide. When this gas leaves the tissues and enters the blood (Fig. 18–6) it immediately enters the red blood cells where, under the influence of the enzyme **carbonic anhydrase**, it reacts with water to form carbonic acid (H_2CO_3), the bubbly ingredient in soda water. Carbonic acid dissociates to form hydrogen (H^+) and bicarbonate ions (HCO_3^-), and since sodium ions (Na^+) are plentiful in the plasma and potassium ions (K^+) in the red cells, the CO_2 is transported to the lungs as sodium and potassium bicarbones. Almost one-third of the CO_2 is carried in the cells and two-thirds in the plasma. If carbonic acid were not immediately transformed to a harmless bicarbonate, it would render the blood very acid because of the release of free hydrogen ions. This we know does not happen, for the blood has a relatively constant number of hydrogen and hydroxyl ions (pH 7.45) at all times. The numerous buffers (inorganic ions, amino acids, and proteins) make this possible. If, however, the alveoli become unable to eliminate CO_2 as in disease (pneumonia), the carbonic acid builds up in the blood which then becomes more acid than usual (it is still alkaline), resulting in a condition called **acidosis**. Tissues cannot tolerate this acid condition and soon die.

In the lungs, the entire series of events is reversed and CO_2 is released to the alveolar air. The rate at which CO_2 is released from carbonic acid is greatly facilitated by carbonic anhydrase. As the carbonic acid is transformed into CO_2 and H_2O, it is immediately replaced by more from the bicarbonates with which it is in equilibrium. The process is simplified in the scheme below.

$$H_2O + CO_2 \rightleftarrows H_2CO_3 \rightleftarrows H^+ + \underset{\text{Alviole}}{\overset{\text{Tissues}}{HCO_3^-}} + \begin{Bmatrix} K^+ \\ Na^+ \end{Bmatrix} \rightleftarrows \begin{Bmatrix} K\ HCO_3 \\ NaHCO_3 \end{Bmatrix} \text{(Blood)}$$

The loading and unloading of the blood with CO_2 is coupled with the oxygenation of the hemoglobin. The chemical equilibria established in the tissues and the lungs are such as to foster the rapid gaseous exchange in each site. The passage of O_2 into tissues enhances the release of CO_2 into blood; likewise, the diffusion of CO_2 into alveolar air in the lungs fosters the uptake of O_2. Very slight changes in the concentration of the gases in both the alveoli and the tissues have pronounced effects on the chemical equilibria which hold or release O_2 and CO_2 at these sites.

For all chemical purposes, nitrogen plays no part in respiratory exchange in man. However, it does produce some effects which become important only under unusual conditions. In probing into the various corners of his environment man has gotten into trouble with this abundant and omnipresent gas. Trouble starts when he leaves the earth's surface very far, either up into rarefied air or down into the depths of the sea with its tremendous water pressures. Gases under pressure remain in solution as long as the pressure is maintained, but the moment the pressure is decreased they come out of solution in the form of bubbles. This familiar fact is observed when the cap is removed from a bottle of "coke"—bubbles of carbon dioxide rise rapidly from within the fluid because of the release of pressure.

At sea level there is an atmospheric pressure of 760 mm of mercury exerted on all parts of the body, inside and outside. Minor differences in pressure, such as those due to altitude changes encountered in driving across the country, have little or no effect, because the change is gradual and the gases in the blood and tissues have a chance to adjust themselves in an undisturbing manner. However, if the change is sudden, that is, if one suddenly ascends to 20,000 feet in a minute or two, these dissolved gases begin to form bubbles in the blood which float about and become lodged in the blood vessels, plugging them and thus cutting off the circulation in parts of the body. This results in violent pain which sometimes causes the person to bend over, hence the name "bends." This discomfort is also experienced by deep sea divers when they begin to surface after being submerged to a considerable depth in the sea. Divers must surface slowly to avoid the "bends," although there are considerable differences in individuals, some suffering more acutely than others. With the advent of high-speed rocket planes in recent years the necessity for preventing "decompression sickness," as it is referred to now, has become more important than ever. Research has revealed that **helium**, another inert gas, leaves the tissues much faster than nitrogen and for that reason has been employed to prevent the sickness, although pressurizing the cabin is more commonly used today.

SUGGESTED SUPPLEMENTARY READINGS

Books

FLOREY, E., *An Introduction to General and Comparative Animal Physiology*. Philadelphia: Saunders, 1966.

FULTON, J. F., *Selected Readings in the History of Physiology*. Springfield, Ill.: Charles C Thomas, 1930.

*GABRIEL, M. L., and FOGEL, S., *Great Experiments in Biology*. Englewood Cliffs, N.J.: Prentice-Hall, 1955.

GIESE, A. C., *Cell Physiology*. Philadelphia: Saunders, 1962.

GUYTON, A. C., *Textbook of Medical Physiology*, 4th ed. Philadelphia: Saunders, 1971.

Hoar, W. S., *General and Comparative Physiology*. Englewood Cliffs, N.J.: Prentice-Hall, 1966.

Romer, A. S., *The Vertebrate Body*, short version. Philadelphia: Saunders, 1971.

Woodburne, R. T., *Essentials of Human Anatomy*. New York: Oxford University Press, 1961.

* Available in paperback.

Articles

Comroe, J. H., "The Lung." *Scientific American*, February, 1966.

Fenn, W. O., "The Mechanism of Breathing." *Scientific American*, January, 1960.

Gray, G. W., "Life at High Altitudes." *Scientific American*, December, 1955.

Green, D. E., "The Synthesis of Fat." *Scientific American*, February, 1960.

19

THE CIRCULATORY SYSTEM

Evolution in the vertebrate circulatory system is one of the many interesting changes that accompanied the transition from aquatic to terrestrial existence as described in an earlier chapter. This can be briefly summed up with the aid of a sketch (Fig. 19–1). In fish, the circulatory flow of blood was between the breathing organs (gills) and the body tissues. The simple fish heart (one auricle and one ventricle) forced the blood through arteries into the gill capillaries where respiratory gaseous exchange occurred. These capillaries coalesced to form arteries which carried blood to tissue capillaries in all parts of the body. The capillaries then united to form veins which conveyed the blood back to the heart—a very simple system.

When vertebrates made the transition to land life, a circuit to care for the newly acquired lungs was established. This meant the construction of a new heart superim-

posed on the old one. The beginning of this development is seen in amphibians where a second auricle, together with a pulmonary circulation, appeared. This was only a partial solution, for the oxygenated and reduced blood became mixed. In a further step forward, a complete pulmonary heart appeared in the higher reptiles. This proved so satisfactory that it was retained in both mammals and birds. Once the two hearts were formed, the vertebrates were fully equipped for land life and evolution was extremely rapid, as evidenced by the tremendous variety of land forms.

The overall design of circulation in man (Fig. 19–2) is similar to that of air-breathing vertebrates in general, differing only in such minor details as the number and location of vessels.

The functional part of circulation is the circulating blood. The elaborate mechanism of the heart, arteries, veins, and capillaries is merely a mechanical device to circulate the blood so that its contents may reach every cell in the body. However, before studying the blood itself, let us examine the machinery of circulation.

THE VESSELS

These consist of a closed system of tubes, large and small, which convey the blood continuously within a circuit. The large vessels carrying blood **away** from the heart are known as **arteries;** the large vessels bringing blood back **to** the heart are known as **veins;** and the tiny intermediate vessels are **capillaries.** The last are the most important, since it is through their walls that the real work of the circulatory system goes on.

The Capillaries

These tubules, often so small that blood cells pass through single file, have walls com- posed of a single layer of cells. These same **endothelial** cells continue as the lining of the larger vessels (Fig. 19–3). The capillaries form a network throughout all the tissues of the body, so vast that a pinprick anywhere usually punctures one, causing blood to ooze out. Because of their thin walls, dissolved substances in the blood can readily pass into the tissues and, conversely, waste substances in the tissues can readily diffuse into the blood and be carried away. In an active ani- mal such as man it is essential that this ex- change be a rapid one. Even in the relatively sluggish frog, whose capillaries can be easily observed in the web of its foot, blood cells race through the capillaries in a fraction of a second and yet this is sufficient time for the important processes of exchange to occur.

The Arteries and Veins

On either end of the capillaries are larger tubules, **venules** (little veins) at the end to- ward the heart and **arterioles** (little arteries)

Fig. 19–1. A schematic portrayal of the evolution of the vertebrate circulatory system. While the simple heart was satisfactory for an aquatic form, it could not handle the problem of transporting oxygen in sufficient quantities for the active air-breathing land dwellers. The second heart (pulmonary heart) had its beginnings in the amphibians and reached its full-fledged condition in the higher reptiles, birds, and mammals.

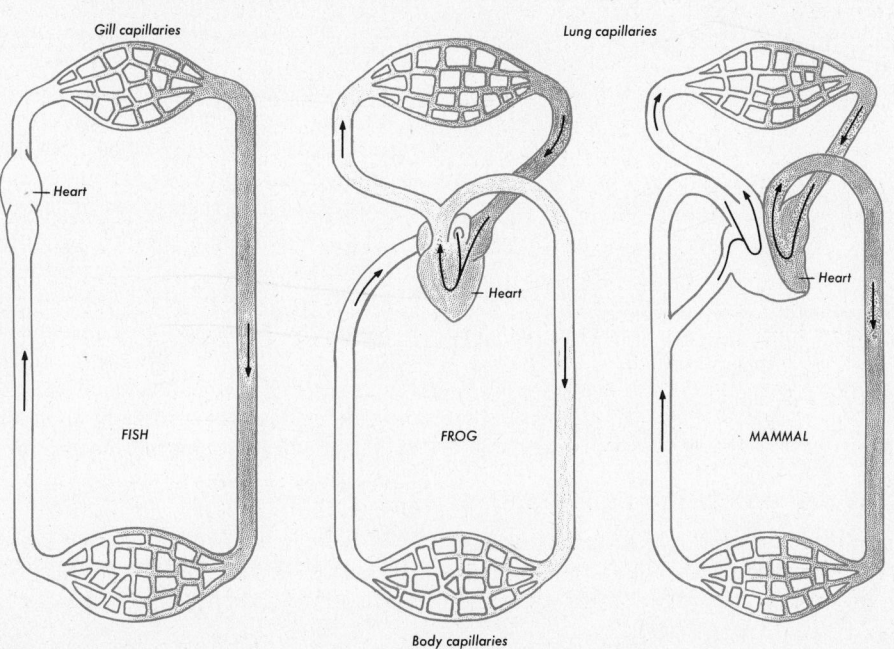

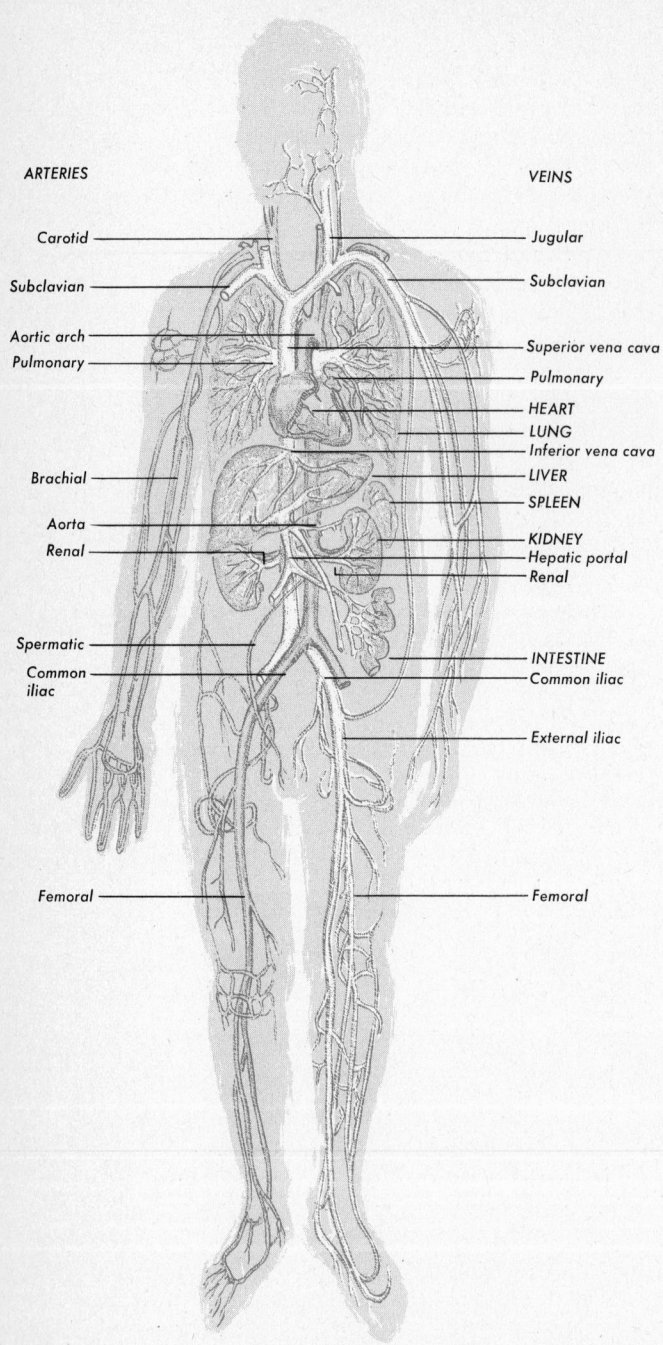

ARTERIES

Carotid

Subclavian

Aortic arch
Pulmonary

Brachial

Aorta
Renal

Spermatic

Common
iliac

Femoral

VEINS

Jugular

Subclavian

Superior vena cava

Pulmonary

HEART
LUNG
Inferior vena cava
LIVER
SPLEEN

KIDNEY
Hepatic portal
Renal

INTESTINE
Common iliac

External iliac

Femoral

Fig. 19–2. A diagrammatic representation of the human circulatory system.

at the end coming from the heart. These venules and arterioles become larger and larger as the heart is approached, where they are designated as veins and arteries. There is no structural difference between the two except that the walls of the arteries are thicker and stronger than those of the veins (Fig. 19–3). Both are composed of three layers of tissue. The inner **endothelium** is the same tissue that makes up the whole of the capillary wall. As a matter of fact, it lines the entire circulatory system. The outside layer of both arteries and veins is made up of tough **connective tissue** so that it readily stretches to permit an increase in diameter, but does not easily bend. Between the two is a **smooth muscle layer** which regulates the diameter of the blood vessel, thereby controlling the amount of blood flowing through it. These muscles (**vasoconstrictors**) are under the influence of the autonomic nervous system and the state of their contraction depends on the need of various tissues of the body for food and oxygen. For instance, following a meal the muscles in the walls of the blood vessels going to and from the viscera relax, allowing more blood to flow to and from these organs. On the other hand, during violent exercise they contract in this region but relax in the muscles and breathing system. By such regulation the various parts of the body are supplied with the proper amount of blood at all times. The veins differ from the arteries in that some have **valves** which prevent the blood from flowing backwards in them. High blood pressure in the arteries makes such valves unnecessary.

The walls of the arteries need to be stronger than those of the veins because the blood is under considerable pressure when it leaves the heart and this pressure alternately rises and falls with each contraction of the ventricles. Since the capillaries need a continuous flow, there must be some means of absorbing these high and low pressure levels, and this is taken care of by the elastic arterial walls. When the pressure rises suddenly, as it normally does, the walls stretch, absorbing the pressure. Between beats, then, the resilience of the walls forces the blood through the capillaries at an even flow. If this were not true the blood would gush through at high pressure and stop altogether between beats when the heart is at rest. Indeed, something approximating this happens in advancing years when the walls of the vessels grow hard. When this occurs, the pressure of the blood rises so high that the brittle vessels are apt to burst and cause damage by cutting off the blood source to vital organs.

THE HEART

The pumping mechanism of animals has had a long history of evolution from the simple pulsating tubes of the annelid to the highly efficient organ of the vertebrate. Much of this history has been portrayed in earlier chapters, and with respect to vertebrates can be summed up briefly by comparing the hearts of several adult forms (Fig. 19–4). The history is correlated with the transition from aquatic to land life. The primitive heart of the shark, as well as the more elaborate one of the salmon, is a two-chambered pump that functions satisfactorily where gills are the breathing organs. With the advent of lungs, a second heart was formed over millions of years of evolution. Today we see how this probably happened in the adult forms of the frog, lizard, and crocodile. The story is also recapitulated in the developing mammalian embryo (Fig. 19–4).

The adult human heart is a muscular organ composed of four principal chambers, two **auricles** and two **ventricles,** and a system of valves to keep the blood moving in one direction (Fig. 19–5). The auricles are thin-walled, saclike chambers whose principal function is to collect sufficient blood from the great veins to fill the ventricles quickly the moment they are empty. The ventricles, on the other hand, are thick-walled chambers

Fig. 19–3. Arteries and veins differ only in thickness of their walls, whereas capillaries are composed only of the layer of cells that lines the larger vessels. Some of the various cellular components of the blood are also shown.

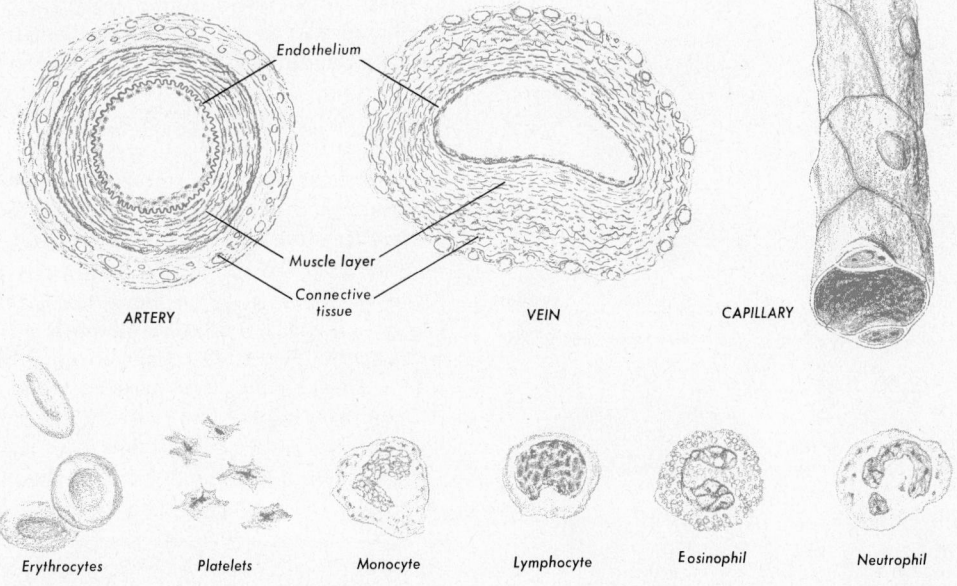

Endothelium

Muscle layer

Connective tissue

ARTERY

VEIN

CAPILLARY

Erythrocytes   Platelets   Monocyte   Lymphocyte   Eosinophil   Neutrophil

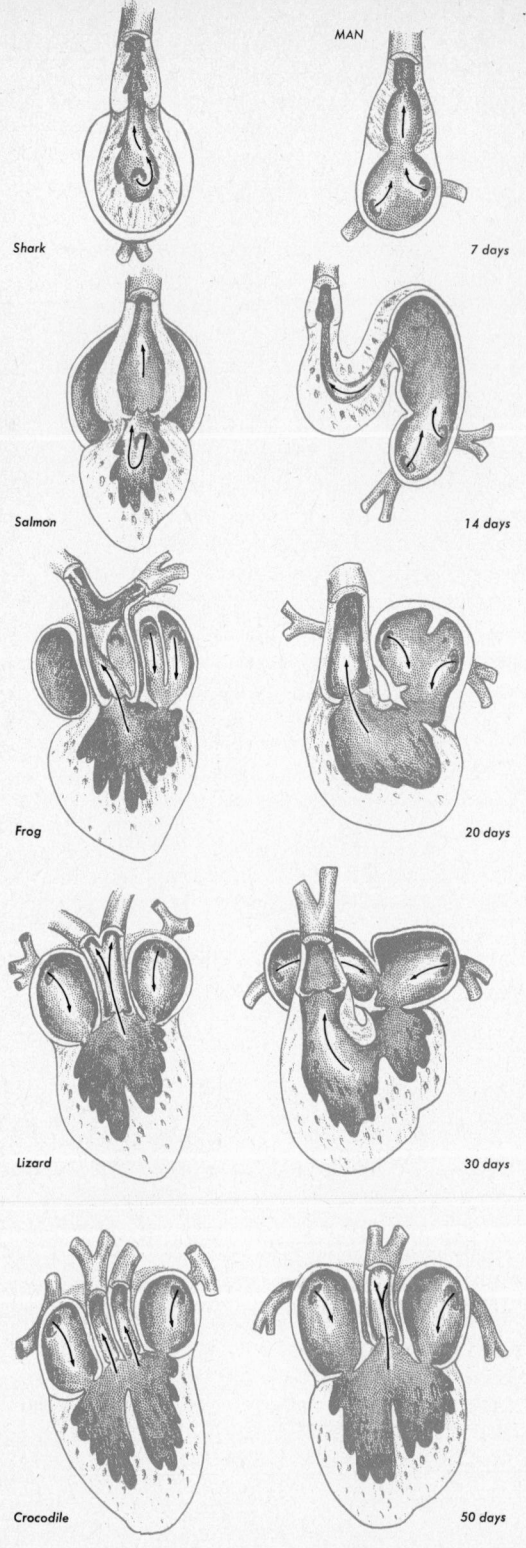

Shark

MAN

7 days

Salmon

14 days

Frog

20 days

Lizard

30 days

Crocodile

50 days

whose sole function is to keep the blood forever on the move throughout the vast network of vessels. Because the left ventricle is thicker and larger than the right, it performs the greater task of forcing the blood throughout the body, whereas the right heart merely keeps the blood moving through the lungs. The two hearts, however, are intimately associated and beat simultaneously even though their circuits are distinct and separate. The left auricle and ventricle are separated by a pair of flaps, called the **bicuspid** (mital) **valve**; similarly, the right auricle and ventricle are separated by three flaps, the **tricuspid valve.** These flaps of tough tissue are held rigidly in place when under pressure by tiny cords that extend to the inner muscular walls of the ventricles.

The ventricles open into large arteries, the left into the **systemic aorta** and the right into the **pulmonary artery.** Very near the openings are half-moon-shaped valves, the **semilunar valves,** which prevent the blood from returning to the ventricles once it is forced out. Each valve is composed of three thin-walled cups that fit very tightly together when under pressure. These, as well as the other valves, are arranged so that the blood can pass only in one direction. This, then, is the pump that keeps the blood circulating continuously throughout life.

The Blood Path through the Heart

Blood passes from the pre- and post-**cavas** into the right auricle, the walls of which then contract (**systole**), forcing the blood through the tricuspid valve into the right ventricle (Fig. 19–5). Actually, when the ventricle empties and begins to relax (**diastole**), the blood requires very little force to flow into the ventricle; hence the muscular walls of the auricles are thin. The ventricle fills completely, then suddenly contracts with sufficient force to exceed the pressure in the pul-

Fig. 19–4. The phylogenetic development of the vertebrate heart from the shark through the reptile is shown on the left, as compared with the embryology of the human heart shown on the right.

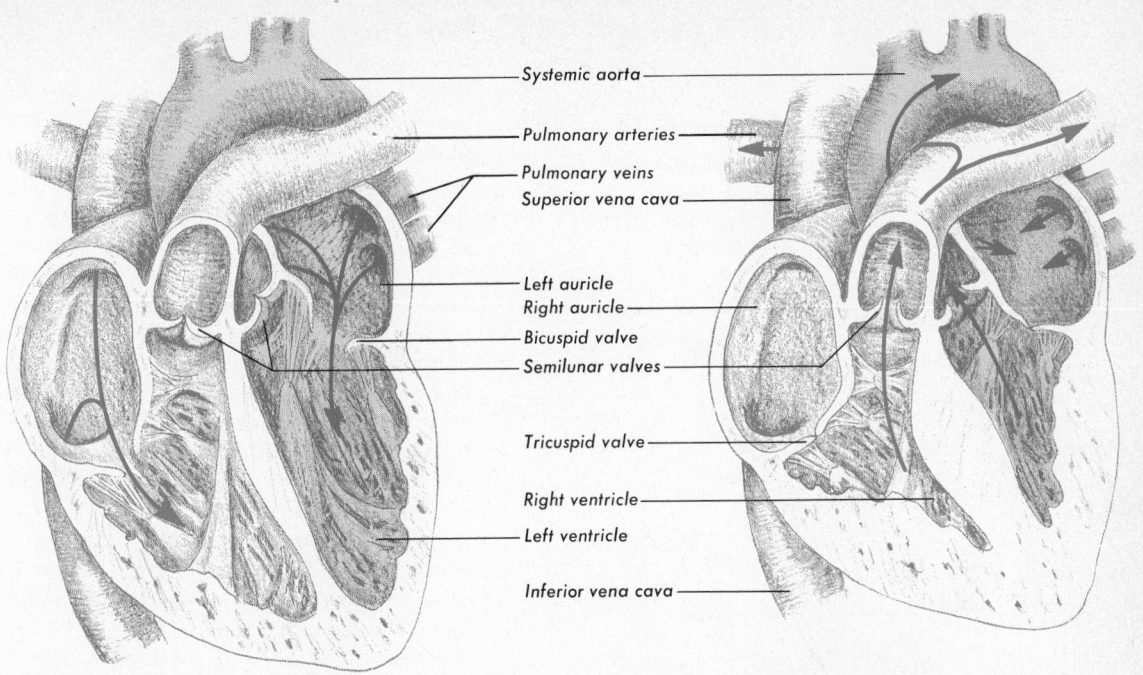

Systemic aorta

Pulmonary arteries

Pulmonary veins
Superior vena cava

Left auricle
Right auricle
Bicuspid valve
Semilunar valves

Tricuspid valve

Right ventricle
Left ventricle

Inferior vena cava

Fig. 19–5. The human heart is shown here cut so that the chambers and valves can be seen. The contracting portions are indicated in solid black, whereas the parts that are relaxed are left white. The two figures demonstrate the way the parts operate during the cardiac cycle.

monary artery, thus opening the semilunar valve which allows the blood to pass to the lung capillaries. The sequence of events during one complete systole and diastole is known as the **cardiac cycle.** The blood then returns to the left auricle by way of the four pulmonary veins. The path taken by blood passing through the left heart is similar to that of the right, although the force with which it leaves is greater because the blood must go throughout the body as a result of the impetus received from the muscle of the left ventricle. It passes over the semilunar valve of the dorsal aorta and out to the capillaries of the body, eventually returning to the right auricle again through the vena cavas.

How the Heart Is Nourished

It is obvious that such an active muscle as the heart must require an enormous sup-

ply of oxygen and food continuously. Even though tons of blood flow through the heart chambers, none of it reaches the heart muscle because there are no direct connections to the muscle from the chambers. The heart has a system of blood vessels of its own, however, called **the coronary circulation.** This rather strange name was given to it because the vessels reminded early anatomists of a crown, since they encircle the top part of the ventricles. The two **coronary arteries** leave the systemic aorta just above the semilunar valves (Fig. 19–6) and pass throughout the heart muscle, ultimately becoming capillaries. Blood returns through a system of veins (**coronary veins**) which coalesce and eventually empty into the right auricle via the **coronary sinus.** Approximately one-fourth of the total blood pumped out by the left ventricle passes through the coronary circulation. This is an extremely important system because the slightest impairment,

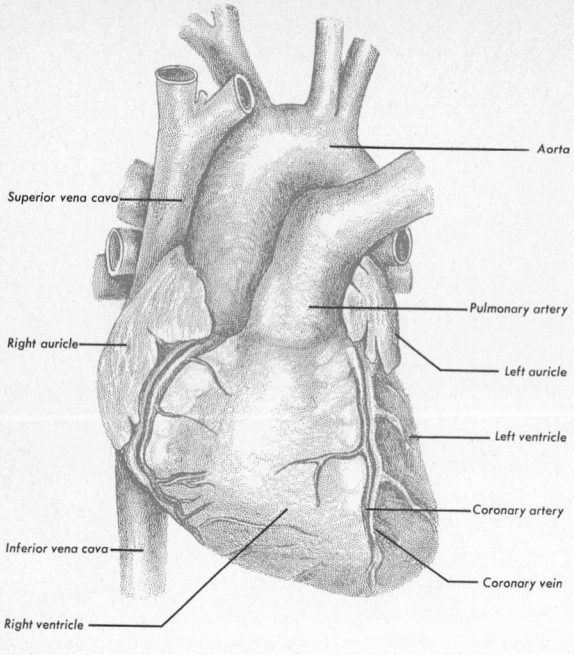

Aorta

Superior vena cava

Pulmonary artery

Right auricle

Left auricle

Left ventricle

Coronary artery

Inferior vena cava

Coronary vein

Right ventricle

VENTRAL VIEW

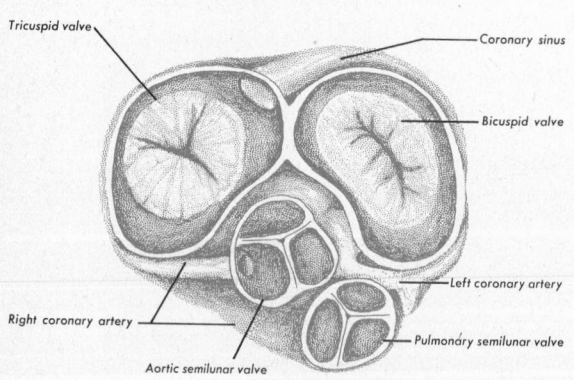

Tricuspid valve

Coronary sinus

Bicuspid valve

Right coronary artery

Left coronary artery

Pulmonary semilunar valve

Aortic semilunar valve

ANTERIOR VIEW
(auricles cut away)

Fig. 19–6. The heart is here shown in ventral and anterior view with the auricles cut away in order to illustrate the nature of the coronary circulation.

such as a tiny clot of blood lodging in it, profoundly affects heart action. This is one form of heart attack. Sudden death results when the stoppage of blood is sufficient to hinder seriously the contraction of the muscle.

Listening to the heart during health and illness led more alert physicians many centuries ago to recognize that certain sounds could be associated with heart malfunctioning. The characteristic "lub-dub," as heard with the stethoscope, is caused by the closing of the various heart valves to prevent the blood from rushing back into the auricles from the ventricles and back into the ventricles from the great arteries. The low "lub" sound is due to the closing of the mitral and tricuspid valves during the initial stages of ventricular systole, whereas the sharp "dub" which follows the first sound very closely is caused by the sudden closing of the semilunar valves in the arteries. There is a brief pause during which time the heart rests and the auricles fill. Any abnormalities in the closing of these valves can be detected by various "murmuring" sounds that are easily recognized by the trained ear of the physician. During some kinds of infectious diseases, and perhaps allergies, the edges of the valves become inflamed and injured; upon healing they frequently do not fit as well as they once did, resulting in the so-called heart murmur. Such an injured heart can, however, increase its size and output so that even with faulty valves it may still be adequate. Such a heart is often spoken of as a "compensated heart" and within limits may function normally for many years.

The work of the heart is almost unbelievable when it is compared to that accomplished by other muscles of the body. Starting long before birth, indeed, when the embryo is no more than 25 days old, it continues its ceaseless contractions until old age; during all this time it does not falter, and its only rest comes between systoles. Beating at 70 times per minute, over 100,000 per day, and nearly 40,000,000 per year, the heart does enough work in one year's time to lift its owner nearly 100 miles above the surface of the earth. It is truly a remarkable organ.

Control of the Heart

The heart muscle functions in an apparently normal fashion when removed from the body that houses it. Histologically, cardiac muscle is unlike either smooth or striated muscle. The nuclei lie deeply embedded in the muscle fiber as in smooth muscle, but cross-striations are present, much as in voluntary muscle. Most important, however, all of the fibers are directly connected to other fibers so that the entire mass of muscle fibers is continuous, forming a syncytium (p. 98).

Skeletal and smooth muscles contract only when they are stimulated by nervous impulses or by hormones. The vertebrate heart, on the other hand, may be removed from the body and if placed in the proper nutrient fluid at the proper temperature will continue beating for hours. Moreover, the rate of pulsation may be quite normal even though it is isolated from any nervous or hormonal

control. This might seem to indicate that the heart is not influenced by either of these types of control, but a check of one's own pulse under varying conditions of excitation or physical exertion quickly shows that such is not the case. This means only that the **intrinsic** nature of the heart muscle is to contract in a rhythmic manner. The rate of this beat, however, may be influenced by outside factors. This can be demonstrated in an excised heart. When placed in a cold fluid it slows down, whereas in a warmer one it speeds up. Therefore, temperature affects its rate. Certain drugs do the same.

The manner in which the heart beats has been determined rather precisely in studies of the excised heart. It has been shown, for example, that if the heart is subjected to reduced temperatures in various parts, its rate of beat is changed only when a specific region is cooled. This region lies near the base of the great veins where they enter the right auricle (Fig. 19–7). There is a discrete bundle of specialized tissue lying here called the **sinus node** or "pacemaker" which is responsible. This is clearly shown by the following experiment. If an excised heart is clamped in the region between the auricles and the ventricles, the former continue their rhythmic beating whereas the latter remain quiet. This demonstrates that impulses pass from the pacemaker through the auricles to the ventricles. We have learned further that in passing from auricles to ventricles the impulses travel over a bridge called the **auricular-ventricular node** (Fig. 19–7). In the contraction of the heart as a whole, the auricles beat first, the ventricles second. Nerve pathways in the heart have been traced from the auricular-ventricular node down through the septum between the ventricles and anteriorly through the muscle, and each beat is synchronized with an impulse that follows this pathway.

Even though the heart beats autonomously in the intact animal, its beat is carefully regulated by two sets of motor nerves which carry impulses directly to the sinus node (Fig. 19–8). Branches from the **vagus**

Fig. 19–7. The sequence of the contraction of the various chambers of the heart is under the influence of the sinus node and the auricular-ventricular node. The beat is initiated in the sinus node, from which it travels over the heart as indicated by the arrows.

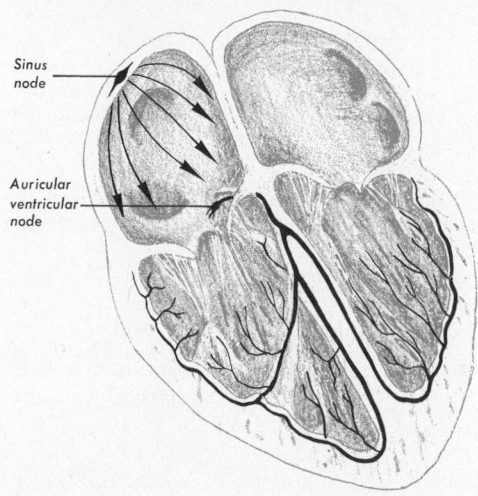

Sinus node

Auricular ventricular node

(cranial-sacral) may inhibit the beat, whereas many branches from the thoracolumbar autonomic system in the cervical region may accelerate it. These are influenced by emotions, a fact familiar to all.

Additional sources of regulation come from three groups of sensory endings close to or in the heart itself. One group originates in the aortic arch and terminates in the **cardiac center** (depressor center) in the **medulla,** and these nerves are called **depressors** because they slow down the heart beat by stimulating the vagus when the pressure is unnecessarily high in the aorta. Another set of nerves comes from the carotid arteries which also record elevated pressures in these vessels. Such action occurs following exercise when the muscles no longer need a large volume of blood. A third group arises in the right auricle and terminates in the **cardiac center** (accelerator center). When the large skeletal muscles are in vigorous action, blood flows into the right auricle in greater amounts than normally. As a result the walls are stretched, stimulating the accelerator nerves to make the heart beat faster with the result that the muscles receive the additional blood they need. Thus, by the combined action of these sets of governors, the heart is able to maintain a rate that supplies all parts of the body with an adequate amount of blood under variable circumstances.

Other Controls of Heart Rate

Many factors other than nerve impulses affect the heart rate. These usually stimulate the heart directly. We have already mentioned that heart rate is influenced by temperature change. The rate is increased with increasing acidity and decreased with increased alkalinity. Carbon dioxide increases the heart rate as well as the rate of breathing, as noted earlier. Hormones, such as epinephrine, have a pronounced effect on heart rate. Actually there is almost no change in the body that does not affect the heart in one way or another. This must be so if it is to do its job properly.

It was brought out in an earlier chapter

that when a muscle contracts, a minute electrical impulse is set up which can be detected with delicate instruments. This fact has been employed in diagnosing heart difficulties. The instrument employed is the **electrocardiograph.** It records slight electrical changes that occur when the various parts of the heart muscle contract. A normal heart

Fig. 19–8. The rate of heartbeat is under the control of the cardiac center in the medulla. The upper figure demonstrates how emotions control the beat. The middle and lower figures show how the rate is regulated according to physical needs.

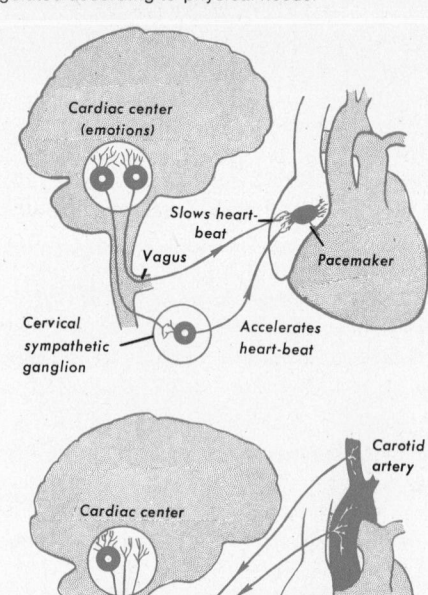

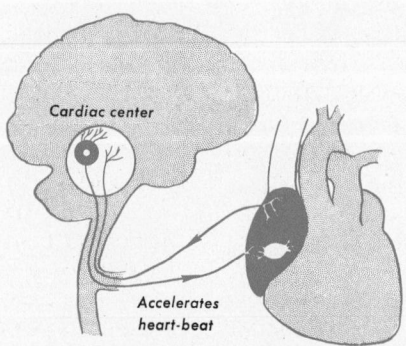

produces a characteristic series of peaks and valleys. Any deviation in this pattern indicates abnormality, the exact nature of which can be discerned rather accurately.

BLOOD PRESSURE

If a vein is severed the blood oozes out in a gentle flow, whereas if an artery is cut it shoots out in spurts which are synchronized with ventricular systoles. It is obvious that the blood is being forced along under pressure, and that the arterial pressure is much higher than the venous pressure. The highest pressure is maintained in the aorta. As the blood passes through the thousands of miles of capillaries, the pressure gradually falls, owing primarily to the friction of the walls of these tiny tubes (Fig. 19–9). By the time it reaches the veins, much of the pressure has been spent and, in fact, when it reaches the large veins entering the right auricle it has a slight negative pressure. In other words, the blood is "sucked" into the heart.

The pressure in the arteries rises and falls with each heart beat. This can be observed by "taking the pulse," which can be felt over the wrist artery or any other artery that is near the surface. The artery swells and collapses in a rhythmic manner. This interested an eighteenth-century English clergyman, Stephen Hales, to the point of experimentation. He placed a crude cannula (a small brass tube), connected to a long glass tube, into the carotid artery of a horse and noted that the blood rose to a distance of over 9 feet. He also observed that the blood rose and fell gently in the tube with each heart beat. He therefore concluded that the pressure in the artery was sufficient to maintain a column of blood 9 feet high, and that the variation in height was due to the contracting ventricle.

Blood pressure is measured even today in a similar manner, although with more refined instruments. For experimental purposes, where blood pressure recordings are desired over some period of time, the blood from an artery is allowed to flow into a tube fitted to

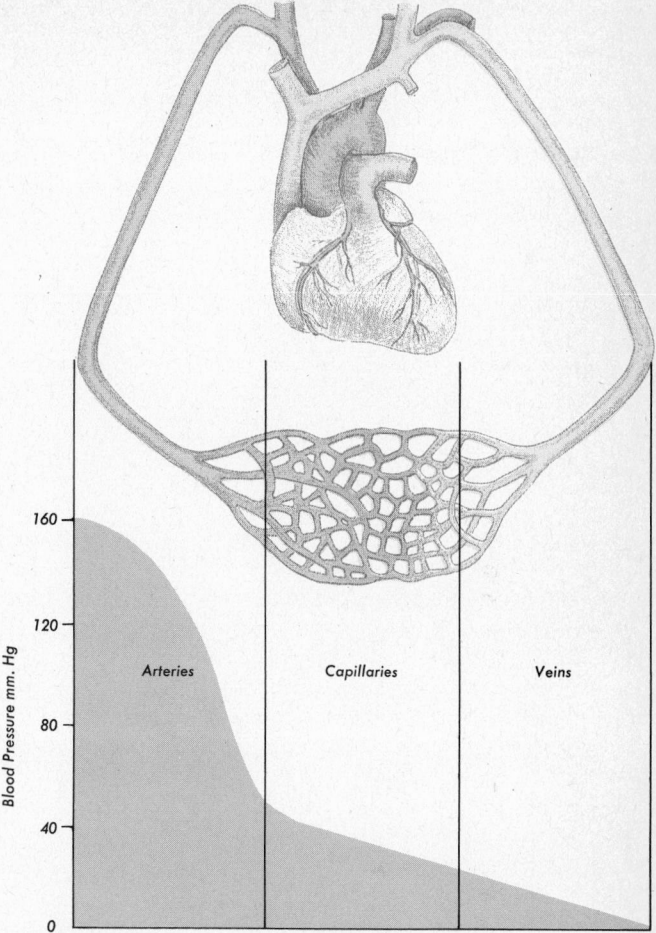

Fig. 19–9. The blood pressure falls rapidly in the arteries but declines at a much more gradual rate in the capillaries and veins. It is highest when leaving the heart and lowest in the large veins that empty into the heart.

a recording device. Continuous recordings can be made in this manner. This would be a rather formidable way to observe the blood pressure of a patient in the physician's office and there would be few who would submit to such treatment. The years following Hales' discovery led to much experimentation, and before the turn of the nineteenth century, other methods had been devised to obtain blood pressure without entering an artery.

413

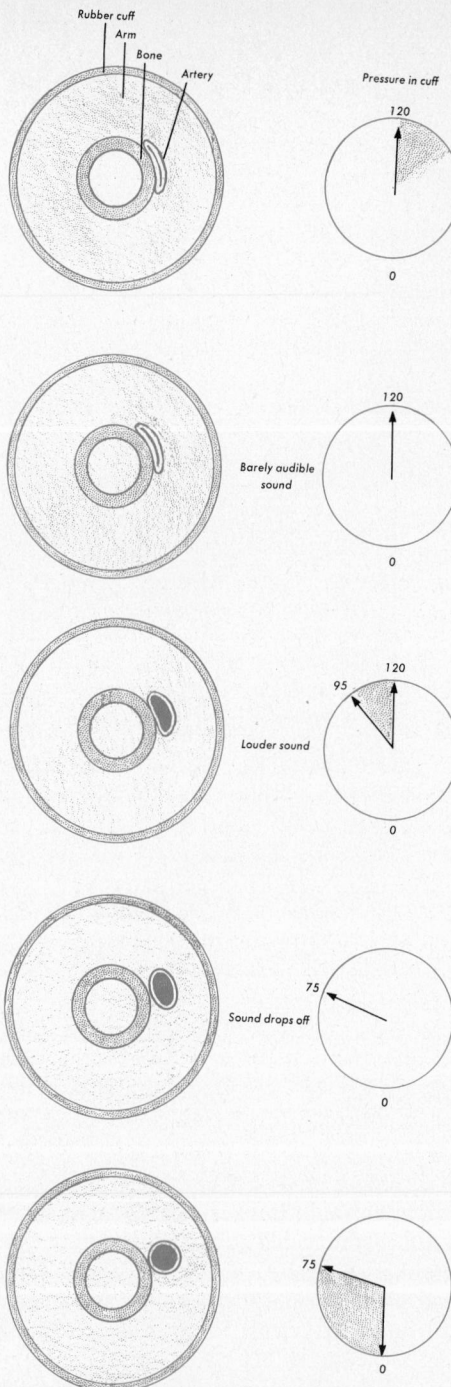

The most successful method is universally employed today. It consists simply of placing a rubber bag that can be inflated around the upper arm where it will squeeze the arm artery until no more blood can be forced through it (Fig. 19–10). The bag is attached to a mercury or air **manometer**, an instrument that records the pressure in millimeters of mercury. Mercury is used because it is a heavy fluid and changes in pressure can be recorded with a small instrument. Water could be used, but the instrument would be very inconvenient because the tube would have to be several feet long. Once the blood is prevented from going through the arm artery, the pressure in the cuff is slowly released, barely allowing blood to pass through at the highest pressure, called **systolic pressure** because it is the point of greatest force due to ventricular systole. This can be heard through a stethoscope placed over the artery at the elbow. As the pressure is released still further, more and more blood flows through the artery and the sound becomes louder and louder, suddenly falling off sharply. The pressure reading just before this point is reached is referred to as the **diastolic pressure,** because the blood is moving during the entire cardiac cycle and it therefore records the pressure which is maintained in the arteries when the semilunar valves are closed. In other words, it is the lowest pressure in the arteries or when the heart is at rest. The normal ranges of blood pressure in man, measured in millimeters of mercury, are: arteries, 120/80 (systolic/diastolic); capillaries, 30/10; and veins, 10/0. A great deal of important information can be obtained by the physician about the condition of the arteries and heart by taking blood pressure readings, and they have become a routine part of medical examinations.

Blood Pressure Control

Blood pressure depends on three major factors in the circulatory system: the blood volume, size of the blood vessels, and the force of the heart beat. Blood volume is controlled by the intake and loss of fluids and

Fig. 19–10. Blood pressure readings can be obtained in this manner. With such a device much information can be gained about the circulatory system.

plays little or no part in rapid changes in blood pressure. It is effective over long periods of time; for example, a gradual loss of blood would decrease blood volume and lower the blood pressure. The remaining two factors are important in rapid blood pressure change, a phenomenon which may be very important to the animal, and which may even save its life.

Blood vessel space is controlled by the changing size of the vascular system. This is accomplished by the contraction and relaxation of the smooth muscles in the walls of the arteries and veins. There are none in the capillaries. The diameters of the vessels are reduced when the muscles contract (called **vasoconstriction**), and increased when the muscles relax (called **vasodilation**). The blood pressure rises with vasoconstriction and drops with vasodilation. The muscles are under the control of nerves which terminate in the **vasomotor center** located in the brain very near the breathing and cardiac centers. Nervous and chemical stimuli influence the vasomotor center, much the same as they affect the cardiac and breathing centers. Under normal conditions there is a "built-in" safety mechanism to prevent the blood pressure from rising too high or dropping too low. The same stretch reflexes operate here also; nerve branches from the aorta and carotid arteries lead to the vasomotor center and when the pressure is high it dictates general vasodilation which causes the blood pressure to drop over the entire body. Vasoconstriction is brought about by dropping the blood pressure in the aorta and carotid arteries.

Blood pressure levels are dependent also on the force of the beat which controls the amount of blood that flows through the heart. As the flow of blood to the heart increases the heart muscle becomes stretched which brings about a reflex action causing the muscle to contract with more vigor. Any muscle works more forcefully when under tension. An upper limit is soon reached, however, when the carotid and aorta are sufficiently stretched to send impulses to the vasomotor and cardiac centers which slow the heart rate and dilate the blood vessels thus maintaining or lowering the blood pressure. When a feeble flow of blood enters the heart, the heart beat is weak and the blood pressure is low. The rate of the heart beat is not necessarily correlated with the flow of blood; a rapid weak beat may not force out as much blood as a slow forceful one. On the other hand, the beat can be fast and forceful, or slow and weak; in other words heart rate is not an index of the efficiency of the heart. Low blood pressure immediately rises when, through exercise, the venous return increases thus stretching the vena cava and right auricle which then sends impulses to the cardiac and vasomotor centers. These in turn bring about increased heart rate (and force) and general constriction of the blood vessels (particularly of the arterioles), both of which cause an immediate rise in blood pressure. We see here then that vasodilation, vasoconstriction, and force of the heart beat are all carefully coordinated to bring about the proper blood flow to supply the demands of the tissues of the body.

Chemicals also have profound effects on blood pressure; indeed this is the basis for the treatment of patients with unusually high or low blood pressure. Normally, carbon dioxide in the blood influences the vasomotor center just as it does the breathing and cardiac centers. As it increases the vasomotor center sends out signals which cause constriction of all the blood vessels, thus raising the blood pressure. The reverse happens when the carbon dioxide content of the blood is low. It is interesting to observe what profound effects can be produced by carbon dioxide. This is reasonable since this gas is a good indicator of the metabolic state of individual cells.

Endocrine secretions also influence blood pressure. Epinephrine, for example, causes vasoconstriction in all of the body vessels except those in the skeletal muscles which dilate. This is understandable when we recall the function of this hormone. During an emergency when maximal strength is needed, the skeletal muscles receive the greatest supply of blood. Many drugs influence the blood pressure through different pathways;

some affect the vasomotor center, some the cardiac center, and some the blood vessels directly. Recently great progress has been made in controlling high blood pressure (hypertension) through drug therapy.

Distribution of Blood

It has already been indicated that blood distribution is not uniform; that is, localized areas may receive more or less blood under varying conditions. For example, following a meal the blood vessels to the viscera dilate, permitting more blood than usual to go to this region of the body. This is advantageous in facilitating digestion and subsequent distribution of food molecules. During this period skeletal muscles receive less blood owing to the constriction of their vessels. This accounts for the lethargy following meals. The opposite action takes place when the digestive tract has completed its job and body action is called for.

When one suddenly moves from a horizontal position to a vertical one, there is an instant shift in the tension on the blood vessels to compensate for the difference in hydrostatic pressure between the two positions. The same is true when one moves against gravity such as in rising in an elevator or airplane. Temperature also induces changes in the size of the blood vessels. Cold brings about constriction of the surface blood vessels whereas heat causes them to dilate, both accounting for the change in skin color under these conditions.

We see here, as we have seen in other systems of the body, "feedback" systems geared to the needs of the body tissues. Wherever blood is needed to permit the organism to function at its best advantage, the interlocking regulatory systems (heart, breathing mechanism, and vasomotor system) come into play to bring about the desired results.

THE BLOOD

It is interesting to recall that not too long ago it was considered sound medical practice to withdraw blood (blood-letting) from the veins during disease, while today such a procedure would be considered "fatal," if not for the patient, certainly for the doctor who attended him! Medical practitioners today conserve the patient's blood, or even add to it by transfusion in certain types of illness and in cases of serious injury. Many thousands of lives have been saved because stored blood is now available for immediate use.

In spite of earlier blood-letting customs, blood has been held in high regard from ancient times, and today still plays a part in rituals of many primitive tribes. Such terms as "blood lines" in breeds of domestic animals or "good or bad blood" or "blue blood" denote the hereditary importance that has been attached to blood. Despite such common beliefs, it is now known definitely that there is no difference of this sort between either the bloods of individuals or various races of man alive today. Any efforts to perpetuate this fallacy are based on emotion rather than fact.

The more we learn about blood, the greater its importance becomes. Let us review its function in nutrition and metabolism by following it on a routine trip through the body (Fig. 19–11). Blood leaving the left ventricle goes to the liver, digestive tract, kidney, and body tissues. The blood coming from the liver contains more urea, that from the kidney less urea, that from the intestines more amino acids and glucose, and that from the general body tissues less food and more waste products. Moreover, all of the blood leaving these organs is low in O_2, which must be replenished in the lungs at the same time that CO_2 is lost. The urea which is formed in the liver must make a complete circuit through the lungs before it reaches the kidneys, where it is extracted from the blood. Fats are absorbed into the lymphatic system which eventually joins the blood system, making possible the distribution of this food. As the blood passes through the endocrine glands, food products are absorbed and converted into hormones which are then secreted into the blood. The performance of all of these functions and many more makes blood truly a most remarkable fluid.

Closer examination shows that blood is a tissue like a muscle, nerve, or bone, even though it exists in a fluid state. The fluid portion of the blood is called the **plasma.** Suspended in the plasma are certain **formed elements;** erythrocytes (red blood cells), **leucocytes** (white blood cells), and **platelets.** In addition, the plasma carries many substances, some of which still are not well understood. Like all tissues, blood is mostly water, about 80 percent in weight; the 20 percent of solids consists of approximately 18 percent protein and 2 percent other chemical substances. When the formed elements are separated from the plasma, they are found to make up nearly one-half of the volume (45 percent). The total amount of blood in a normal person is 5–6 liters, or approximately 8–10 percent of his body weight. One can lose somewhat less than half of this amount and survive. However, a special mechanism for **coagulation** or **clotting** is present to prevent such blood loss.

Blood Clotting

The survival value of any mechanism that prevents the loss of blood is obvious. Blood coagulates even more rapidly in the earthworm, for example, than in man. The blob of jellylike substance on the windshield of a speeding car demonstrates the rapidity with which the insect's blood coagulates. Animals have very short clotting times when compared to the three minutes required for man's blood to clot, for the speed with which the clot forms often spells the difference between life and death.

The exact series of chemical reactions that take place in forming a blood clot is not completely understood. The clot is made up of a mass of threads of a protein called **fibrin,** which enmeshes red blood cells so that a semisolid plug is formed. It starts as a small clot but grows rapidly until it is of sufficient size to fill the opening in the vessel. If the

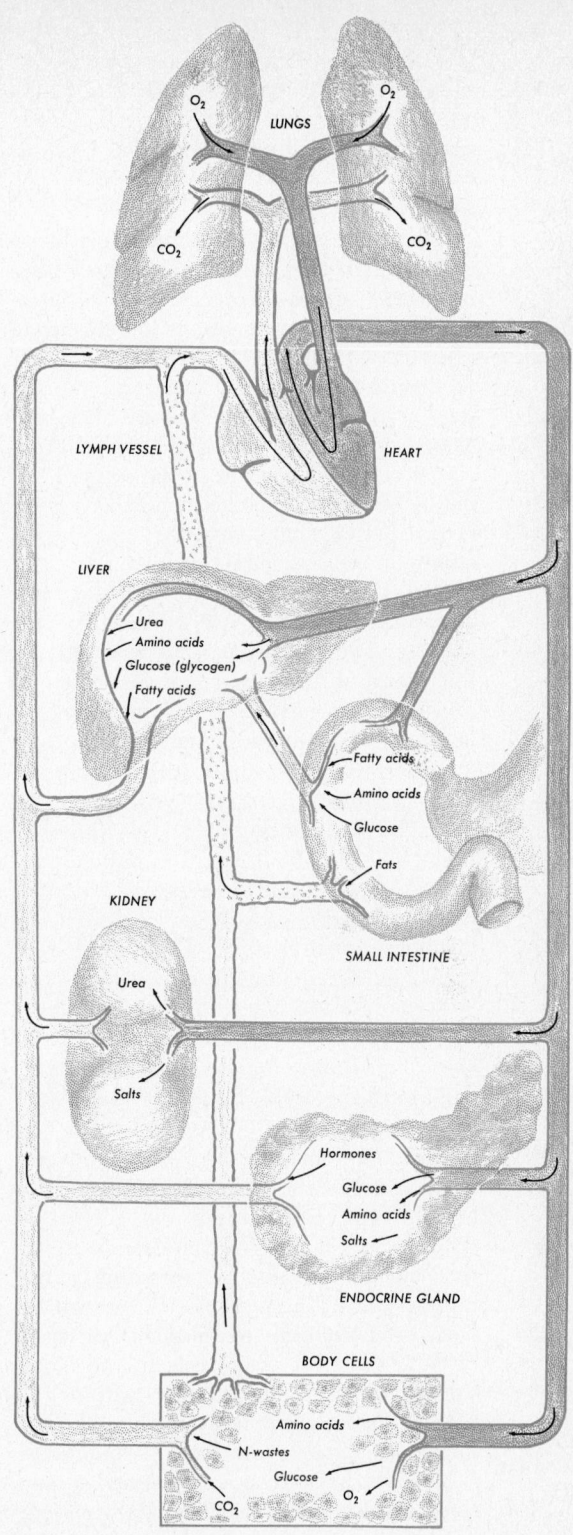

Fig. 19–11. The blood performs many functions as it flows through the body, some of which are portrayed in this sketch. Note what happens to the blood as it enters and leaves each organ.

vessel is too large or the pressure too great from behind, as in the large arteries, the clot fails to stop the blood flow and death of the animal results. Ragged injuries produce better and faster-forming clots than do clean cuts, the explanation of which will follow shortly.

There are several observations on blood that need to be known before the clotting mechanism can be understood. If blood is collected in a vessel containing sodium citrate or oxalate, it fails to clot; and if it is allowed to flow into a paraffin-lined vessel, it clots only very slowly. In the first instance, the chemicals must have blocked the clotting action of the blood, and in the second, the nature of the paraffin surface must have been involved. It is now known that the sodium citrate or oxalate combines with the calcium in the blood, removing it from solution. Therefore, calcium must be essential for clotting. The smooth paraffin is similar to the lining of the blood vessels; therefore, rough surfaces such as often form in ragged injuries must aid in the formation of clots.

If freshly drawn blood is stirred vigorously with stiff bristles, fibrin shortly begins to collect on the bristles, and if the stirring continues all of the fibrin can thus be removed, leaving what appears to be perfectly normal blood except for its failure to clot. This is called **defibrinated** blood. There must have been a precursor protein, much like fibrin, that was in solution in the blood but which became insoluble with the stirring. This is known as **fibrinogen.** The problem is: Why did fibrinogen become fibrin? What initiated the chain reaction?

Something must happen when the blood is taken out of the vessels, because clots do not normally form within the circulatory system itself. Careful chemical analysis has revealed another substance, **thrombin,** in the plasma which is responsible for the conversion of fibrinogen to fibrin. Obviously, thrombin cannot occur freely in the blood stream, or clots would form within the vessels. This happens only rarely and even then only under pathological conditions. Again experimentation has revealed a precursor, **prothrombin,** that is normally present

in the blood. The next problem is: What starts prothrombin to form thrombin?

Another careful search has demonstrated that the platelets and probably other tissues of the body contain an enzymatically active substance called **thromboplastin** (or thrombokinase) which is essential in starting this long chain of reactions. It was stated earlier that calcium was also essential. Just what the relationship is between these substances is not clear at present. Probably when a hemorrhage occurs, the broken cells in the vicinity of the injury, together with the disintegrated platelets, release thromboplastin. Thromboplastin, reacting with calcium and prothrombin in the blood in some unknown manner, produces thrombin and this, in turn, changes fibrinogen to fibrin. Summarized:

Hemorrhage → thromboplastin
Thromboplastin + calcium + prothrombin
$$\rightarrow \text{thrombin}$$
Thrombin + fibrinogen → fibrin
Fibrin + red blood cells → clot

Plasma clots the same as whole blood which means that the cells play no part in the clotting mechanism. After the clot is formed, it shrinks and squeezes out a straw-colored fluid, called **serum,** which does not clot owing to the fact that it has no fibrinogen. In other respects it is similar to plasma.

Earlier in the chapter on nutrition there was mention made that vitamin K had something to do with blood clotting. If too little is formed in the digestive tract, the clotting time increases dangerously, so much so that surgery is inadvisable until large quantities of this vitamin are given to restore the normal clotting time. It is now thought that vitamin K is associated with prothrombin formation, and hence its effect on clotting time.

Clotting can be prevented by precipitating out the calcium, as has already been referred to, and by a number of anticoagulants called **hirudin,** secreted by bloodsucking invertebrates such as leeches, bedbugs, and mosquitoes. You will recall that hirudin permits the blood sucker to obtain a meal without being interrupted by a clot.

A sex-linked blood-clotting defect that

has been known for a long time and has been a particular concern of certain royal families because of their close interbreeding is **hemophilia.** The slightest wound can be fatal to hemophiliacs because their blood clots very slowly. Analysis of their blood indicates that it is normal in all respects except that the platelets do not disintegrate as readily as they do in normal blood. Just why they do not and what can be done about it still remains an unsolved mystery.

Red Blood Corpuscles

Red blood corpuscles are the most numerous and most conspicuous of the formed elements of the blood (Fig. 19–3). A blood smear reveals them as tiny, circular, biconcave disks without nuclei. All vertebrates except mammals possess nucleated erythrocytes; just why nuclei are absent in the latter group is difficult to understand. However, it must be admitted that since the red blood cell lives only about 120 days and never reproduces, it would seem to be a waste of effort to utilize the space in the cell for a nucleus when it could be occupied more profitably by **hemoglobin,** the oxygen-carrying pigment. The erythrocytes possess nuclei when they first form in the **red marrow** of the bones, but lose them just before entering the blood stream. The cells are a light yellow in color under the microscope, although in combined numbers they render the blood a typically red color. Normally there are from 4,500,000 (in women) to 5,000,000 (in men) cells per cubic millimeter of blood. This number varies with disease and nutritional deficiencies. A drop either in the number of red cells or their hemoglobin content results in **anemia,** which can be serious if prolonged because the blood is unable to supply the cells with the proper amount of oxygen.

During embryonic life of the mammal, red cells are formed in the kidney, spleen, and liver; but as the animal matures the red marrow of the bones takes over this task. Red cell production is regulated by the amount of oxygen that the blood carries. If the level drops, red cells are produced more rapidly; and conversely, if the level rises red cell production is inhibited. Simultaneously, the sites of red cell destruction (spleen and liver) are correspondingly stimulated or inhibited. The red cells are formed and destroyed at a tremendous rate, over 10 million per second. Since one of the chief elements of hemoglobin is iron, there must be a continual supply in the diet. The amount is small (0.01 gm per day) and most diets contain an ample amount. However, it is now known that copper is also essential for proper hemoglobin formation. The need for copper lies in the production of an enzyme necessary for hemoglobin formation. Some serious deficiencies in the bone marrow cause a drop in hemoglobin production, resulting in fatal anemias. Recently these have been arrested by administering very small doses of two of the B-vitamins, folic acid and vitamin B_{12} (p. 390). These vitamins are extracted from liver, which was a logical place to find them, since large doses of liver extract were used for years to lessen the symptoms of this particular disease.

The oxygen and carbon-dioxide carrying capacity of erythrocytes has been discussed earlier.

White Blood Cells

White blood cells are called **leucocytes** because they are colorless. They are not so numerous as the red corpuscles, and there are several different kinds, all with nuclei (Fig. 19–3). They can be distinguished by the way they take certain stains, and when counted they serve as an excellent criterion for determining certain kinds of diseases. They usually increase rapidly in number (normal count is about 7,000–9,000 per cubic millimeter) during an infection, because some of them (neutrophils and monocytes) act as scavengers and attack the bacteria by engulfing and digesting them (Fig. 19–12). Such a process is termed **phagocytosis** (p. 83).

White blood cells have the ability to move like an amoeba, and in a localized infection they migrate between the cells of the capillary walls out into the infected region where they do their work. Pus is primarily leucocytes. They probably also aid in healing by transforming into other types of cells in order to repair damaged tissue. Other functions have been assigned to them, but considerable

Fig. 19–12. With a break in the skin, the leucocytes (certain kinds) move out of the blood vessels into the injured area where they engulf and destroy any invading bacteria. Bacteria that are not eliminated by the leucocytes may find their way to the lymph nodes where they are usually destroyed. If, however, they get past these barriers, they are free to invade the circulatory system and the entire organism.

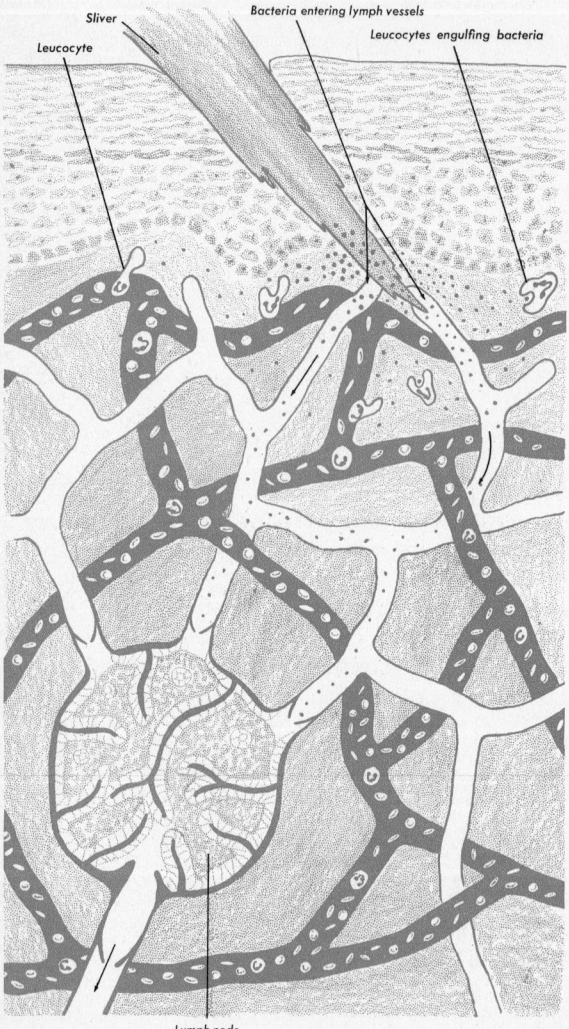

Sliver

Leucocyte

Bacteria entering lymph vessels

Leucocytes engulfing bacteria

Lymph node

information is still needed to understand their complete rôle in the body.

The Plasma

The **plasma** is the fluid portion of the blood and it is chemically very complex. Its composition remains essentially the same although it is constantly receiving and discharging substances. It is composed of about 90 percent water, 7–8 percent soluble proteins, 1 percent inorganic salts, and approximately 2 percent simple organic molecules such as glucose, amino acids, fatty acids, glycerol, urea, and hormones. The principle proteins are **fibrinogen, albumins,** and **globulins,** all of which play important roles in maintaining a stable internal environment. In general the proteins, water, inorganic ions, and glucose maintain rather constant levels in the plasma whereas carbon dioxide, oxygen, hormones, food molecules, and other substances vary from time to time. In addition to its function of delivering food molecules and oxygen to cells and relieving them of their waste products, the plasma also acts as a communicating mechanism between all cells by means of its neurohumors, referred to earlier.

We have already seen the part fibrinogen plays in clotting. The enzymatically active prothrombin also is involved in this process but there are other plasma enzymes whose function is obscure. Albumin functions in maintaining blood viscosity and osmotic pressure, whereas the globulins cause **blood type** differences and are the molecules responsible for **antibodies** which resist infections. The plasma proteins, together with hemoglobin, also function as **buffers** which maintain a constant hydrogen ion concentration.

Antibody formation is an important mechanism in most, if not all, organisms. This fact was first discovered many years ago by Pasteur. Whereas the exact nature of antibodies is still unknown, their importance and mechanism of action are gradually yielding to intensive research. It is best to describe an actual experiment in order to understand the production of antibodies and what they do.

Antibody Production

It was mentioned in the chapter on digestion that proteins must be degraded to amino acids before they can be absorbed into the bloodstream. Any more complex compound of this sort initiates deleterious reactions. For example, if rattlesnake venom is injected into the blood, serious complications follow, hence the value of venom as a defensive and offensive mechanism. If, on the other hand, a very tiny amount of the venom is injected into a large animal, such as a horse, the reaction is not severe, and if the injections are gradually increased at short intervals the horse will eventually tolerate a tremendous dose; enough that, had it been injected in one initial dose, it would easily have killed the animal. Just what has taken place to protect the animal from this poison?

The snake venom is a **foreign protein** and, like any foreign protein, initiates a reaction in the body of the horse. The reaction produces a substance that can combine with the snake venom and neutralize its effect. This substance is called an **antibody** and the snake venom is the **antigen.** Any foreign protein can act as an antigen. In fact, this is a convenient method for separating and classifying specific proteins. The test is more delicate than most chemical tests. There are millions of proteins and each animal possesses its own specific proteins for which there is no chemical means of separation. Using an animal for a test tube, it is possible to determine by experimentation the many different kinds in a very precise manner, even to the point of determining from what animal a bit of blood or tissue originated. This is done by withdrawing the blood from an animal that has previously been injected with a foreign protein, for example, egg albumin, (only proteins are antigenic) and then separating out the cells by whirling the blood in a centrifuge (an instrument for increasing gravity). If some of the original albumin is then added to this clear serum, a white precipitate will form. This reaction will occur for only one protein and no others; it is, therefore, highly specific. This has many practical applications, such as identifying whether or not a specimen of blood is from man or a lower animal.

One interesting use of this reaction was employed by Nuthall of England years ago in the classification of animals. He built up antibodies in experimental animals against the blood of other animals. For example, he might inject the blood of the ape into a rabbit and after a time draw off a sample of the blood from the experimental animal, centrifuge out the cells, and to the clear serum add some of the serum (antigen) of a closely related animal, say a gibbon. He discovered that the closer the animals were related, according to the usual system of classification, the heavier was the precipitate that formed. In this case, the serum of the gibbon would give a rather heavy precipitate, whereas the serum from a pig would give much less, and that from a snake none at all. Interestingly enough, he was able to confirm the classification based on morphological structures.

The graded precipitate of the **precipitin reaction,** as this is called, is a distinct value not only in the relationships pointed out in the preceding paragraph but in determining the closeness of the chemical relationship of different proteins.

One of the great mysteries surrounding immunization has been the source of the cells that produce antibodies. It has long been known that lymphoid tissues and the spleen were involved somehow in the immune response. Removal of the spleen lowers the immune response in experimental animals. It also has been known that phagocytes originate in these tissues and that these large amoeboidlike cells engulf bacteria, viruses, and other foreign material entering the body but they do not produce antibodies. It is now thought that there are three kinds of cells, all interrelated, involved in the production of antibodies.

The first type is the **macrophage** cells, which arise from rapidly dividing promonocytes in the bone marrow, after which they circulate in the blood as monocytes and finally become macrophages in the tissues. The antigen is ingested by the macrophages, and digested, resulting in small non-immunogenic remnants. However, a small amount of the intact antigen remains attached to the outer membrane of the macrophages. This antigen initiates the immune response.

The second type is the **T lymphocyte** cells, which also originate from the bone marrow and exist in the form of precursors, as yet unknown. The precursors move to the cortex of the thymus where rapid cell growth occurs, resulting in T lymphocytes. Some of these move into the thymus medulla and some enter the bloodstream where they make up about 60 percent of the lymphocytes. Others enter the lymph nodes, leave through the lymphatic system, and finally recycle in the bloodstream.

The third type is the **B lymphocyte** cells, which arise from bone marrow much the same as the others, but very little is known about where they migrate. They do circulate in the blood and finally locate in the centers of the lymph nodes. The B lyphocytes are the precursors of **plasma cells,** large cells which manufacture great quantities of antibody.

What is needed then to produce an antibody for a specific antigen is a macrophage with antigen on its outer membrane, plus a T and B lymphocyte, both specific for the antigen. The T lymphocyte stimulates the B lymphocyte to divide rapidly and become an active antibody-producing plasma cell. This cell produces immunoglobulins which are responsible for neutralizing or otherwise destroying the destructive material that has entered the body. This complex story is only now being unravelled and there is a great deal more to be learned about the mechanism. The above account is much abbreviated; a more detailed account would be beyond the scope of this book.

Body Defense

In addition to the phagocytic action of the leucocytes already referred to, the blood and tissues of the body employ the antigen-antibody reaction to destroy invading microorganisms that produce disease (Fig. 19–13). Bacteria as well as animal parasites are protein in nature and therefore induce antibodies when they get into the bloodstream and other tissues. The parasite is toxic to the host, so **antitoxins** (antibodies) are produced that either destroy the parasite or render it helpless so that the leucocytes may engulf and destroy it more easily. Often the parasite gives off a protein product which is also toxic to the host, and again an antitoxin is built up against this product, so that it is neutralized and can no longer harm the body. The end products of this reaction are removed from the body through various channels.

The production of antitoxins is spoken of as **immunization.** Once the antibodies have been produced, they are active for some time against a second invasion of the parasite. How long they are active seems to be specific for the parasite. For example, immunity against typhoid fever may last a year or two, whereas one may expect never to

Fig. 19–13. A schematic representation of the antigen-antibody reaction. When a harmless antigen is inoculated into the blood, an antibody forms on its surface and is then released into the plasma. If, at some later time, a harmful antigen of similar configuration enters the blood, it is inactivated by combining with the antibody, thus protecting the tissues of the body.

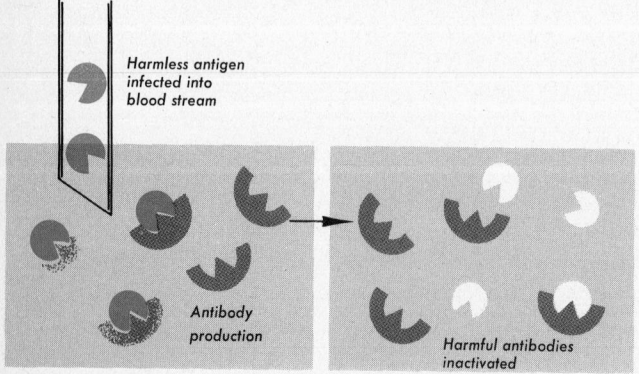

Harmless antigen
infected into
blood stream

Antibody
production

Harmful antibodies
inactivated

have a second case of whooping cough, the antibodies for which last throughout the lifetime of the individual. Immunity built up by the actual participation of the parasite in question is called **active immunity.** A similar immunity can be produced artificially by introducing a weakened or even a dead strain of the parasite into the body so that it will bring about antibody formation but will not cause the disease or, if it does, only in a very mild form. In the case of immunization against smallpox, a weakened strain of the virus (developed in cows) is injected which brings about active antibody formation but causes only minor discomfort itself. This type of active immunization lasts a long time and has been very effective in stamping out certain diseases.

A remarkable human experience was undertaken in the United States in the fight against poliomyelitis. This disease is caused by a virus (actually the several forms of the disease result from infection with different viruses) against which active immunization has been known for several years. Since all people who are exposed to the virus do not get the disease, because of varying amounts of natural or acquired immunity, the effectiveness of using a vaccine (the killed virus) could only be revealed by immunizing human populations on a large scale. The experiment involved the cooperation of many state and national health agencies who helped to get the general public to participate in such an undertaking. This, the largest human experiment ever attempted, was remarkably successful; the final tabulations indicate clearly that the vaccine had been effective in protecting children against poliomyelitis.

This project followed the tenets of a good scientific experiment. The population was divided into two principal portions, each containing representative numbers from all walks of life (habitats). One-half was offered a solution containing the vaccine, the other was offered a *placebo* (the solution with no vaccine present). No one, except those in charge, knew which was which. The effectiveness of the vaccine became evident only when the data were accumulated and arranged in tabular form. Poliomyelitis is now under control, as smallpox is, as a result of the knowledge acquired through this experiment.

Another method of bestowing active immunization in a person is by employing an altered toxin, such as that used in building up active immunity against diphtheria. The toxin taken from the diphtherial organism is treated so that it has lost none of its antigenic properties but is no longer toxic to tissues in the body. This substance is called **toxoid.** When a small quantity of toxoid is injected, antibodies (antitoxins) are produced, so that if at any subsequent time the diphtherial organism enters the body its effect will be neutralized at once and it will be unable to obtain a foothold. This type of treatment has made cases of diphtheria very rare and they could be nonexistent if everyone were thus protected.

Sometimes, as in the case of advanced tetanus, it is necessary to build up the supply of antibodies immediately. There is not sufficient time to allow the body to produce them in the usual slow manner. It is possible, then, to add them directly by injections of antitoxin that has been previously produced in a horse. The type of immunity received from this treatment is called **passive immunity,** because the person himself contributes nothing toward the production of the antibody. Passive immunization is short-lived, which is its chief disadvantage, but in certain diseases it can save a life. It is wiser to prevent the appearance of the disease by active immunization rather than attempt to cure the disease once it has struck.

Blood Types

From the foregoing discussion on the relatedness of animals, it might be expected that all animals of the same species have the same specific proteins. In general this is probably true. However, there are minute variations of which we have become aware primarily through our efforts to transfuse blood from one person to another. They have also been called to our attention in

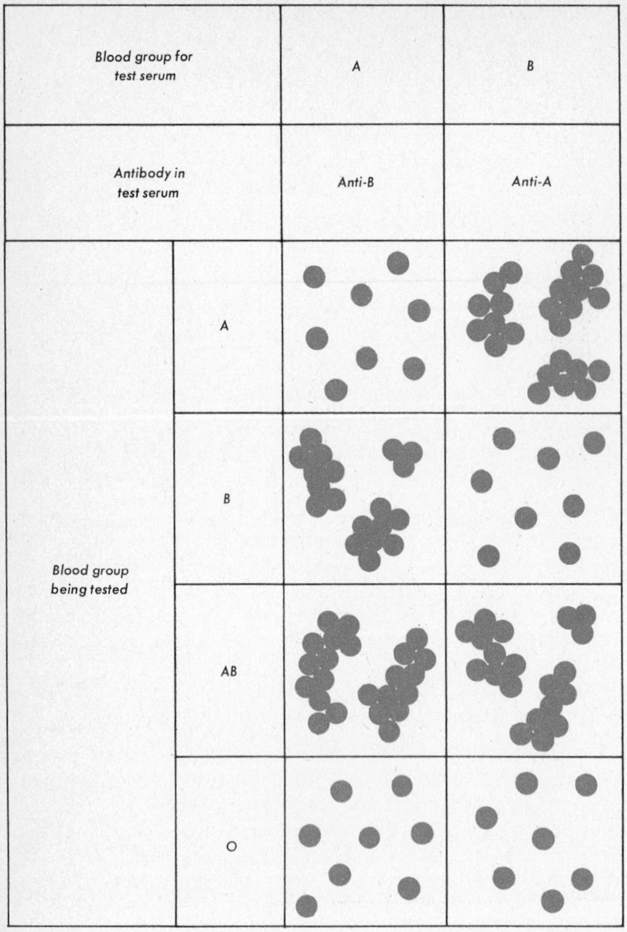

Blood group for test serum	A	B
Antibody in test serum	Anti-B	Anti-A

Blood group being tested: A, B, AB, O

Fig. 19–14. A schematic representation of how blood types (AB, A, B, O) can be determined. Agglutination (clumping of red cells) occurs when sera containing a specific antibody is mixed with blood containing red cells of a certain type. See the text for explanation.

plastic surgery, where attempts are made to graft tissues from one person to replace the destroyed tissues of another, as in cases of severe burns, for example. One wonders why the healthy, intact organs of people dying in accidents could not be saved and transplanted in the bodies of those who are dying because the same organs are no longer functioning properly as a result of some organic or infectious disease. So far we have been limited to blood, veins, kidneys, some bones,

and corneas of the eyes in this regard, primarily because of the specificity of proteins of individuals. In other words, the proteins of one person are slightly different from those of another, so that such transplants are incompatible and will not "take." The spectacular heart transplants have had very limited success, and until the problem of protein incompatibility is solved, the success of such surgery will remain questionable.

A, B, O Blood Types. At the turn of the century, Karl Landsteiner, an American Nobel Prize winner, gave us the first explanation of why people sometimes suffer severely when they are transfused with blood from another person. These occasional catastrophes made the blood transfusion business a rather risky procedure to be used only as a last resort. Landsteiner showed that the red cells of the blood contained two proteins, called **A** and **B**, and that they existed in people singly, in combination, or not at all. Persons could accordingly be classified into groups, depending on the nature of their blood: those with protein A were placed in blood **Group A**, those with B in blood **Group B**, those with both proteins in blood **Group AB**, and those with neither protein in blood **Group O**. The serum of each of these groups contains the antibody for other groups but not for its own, that is, the serum of Group A has anti-B but not anti-A, Group B has anti-A but not anti-B, Group AB has neither antibody, and Group O has both. These antibodies are naturally present in the serum of people, and their kinds must be known in any case before blood can be transfused without possible ill effects.

The groups can readily be determined by mixing sera and whole blood from the various groups in the manner indicated in Figure 19–14, where it immediately becomes apparent that the only safe transfusions are between people of the same blood group. In practice, however, this is not true. Persons with Group O have been called **universal donors** because it is possible to transfuse their blood into people with any of the other blood groups without ill effects. The complete answer for these observed facts is not entirely clear, although two points help in the solution. In the first place, the erythrocytes have no antigenic proteins, so there can be no reaction between them and the antibodies in the recipient's plasma. Secondly, introduced anti-A and anti-B antibodies are diluted so rapidly by the recipient's plasma that they have little opportunity to cause the red cells to agglutinate. If the blood is added too rapidly or in too large amounts, some agglutination might occur. People possessing blood group AB are spoken of as **universal recipients** because they may receive blood from any of the other groups with no harm done. The explanation for this situation is similar to that for universal donors. It seems that, in general, the presence of antibodies in the donor's plasma is unimportant, whereas the presence of the antigens in the donor's red cells and the presence of antibodies in the recipient's plasma determine the safety of transfusions.

Like other blood groups, described below, these groups are inherited in a definite manner, which is considered in a later chapter.

The Rhesus Factor

Another group of red cell proteins has been discovered, designated as the **Rhesus** or simply the **Rh factor** because it was first discovered in the Rhesus monkey. A survey of various populations showed that it occurs in about 85 percent of the people. If a person possesses the protein he is said to be **Rh positive,** whereas if his red cells do not contain the factor he is **Rh negative.** Normally there is no anti-Rh in the serum, and of course no difficulty is encountered unless for some reason or other the blood of an Rh positive and an Rh negative are mixed. Under these conditions the protein containing the Rh factor acts like a foreign protein in the Rh-negative person, giving rise to the anti-Rh antibody. If, then, at some subsequent period such a person receives another transfusion of Rh-positive blood, agglutination of the donor's red cells occur because the

anti-Rh is present in the recipient's serum. Such clumped blood cells produce serious reactions and even death. For this reason, before blood transfusions are given, the Rh condition of the blood is determined and a history of any previous transfusions is important.

The Rh factor also explains the cause of a disease of newly born infants called **erythroblastosis fetalis.** This disease results in the death of a small percentage of babies shortly after birth, and was heretofore a complete mystery. It is now known that the Rh-positive factor is inherited as a **dominant,** and the Rh-negative factor as a **recessive** trait (p. 501). This means that the positive factor appears in either half or all of the offspring of an Rh-positive father and an Rh-negative mother. When the developing fetus of this combination inherits the Rh-positive factor from its father, it is possible that trouble will follow (Fig. 19–15). Normally the blood of the mother does not come into direct contact with the blood of the fetus, but occasionally a very small number of red cells apparently do get into the maternal circulation, perhaps by the accidental breaking of small capillaries in the placenta. Once Rh-positive red cells get into the mother's circulation, Rh antibody is produced. Since the antibody is present in the serum, it can easily diffuse through into the fetal circulation, causing damage to the red cells of the developing fetus. Such a child, when born, is highly deficient in red blood cells, which results in a severe jaundice, and unless immediate treatment is given will die. The only treatment until recently was to replace essentially all of the fetal blood with massive transfusions before birth or shortly thereafter. However, during the past decade a blood extract called Rh-immune globulin has been developed which when injected into an Rh-negative mother carrying an Rh-positive child will prevent the formation of antibodies by the mother, thus saving the child. This is routine practice in countries where there is a high percentage of Rh-negative women.

The situation is not as serious as it might

appear, as attested by the small number of babies born with this disease. This is due, perhaps, to the fact that not all cases of pregnancy result in a mixing of the fetal and maternal blood. Furthermore, not enough antibody is generally produced to cause

trouble on the first pregnancy so the condition does not usually show up until the second and subsequent pregnancies. Since only 15 percent of the population is Rh negative, the chance combination of an Rh-positive man with an Rh-negative woman is not great. Although it is important to know the Rh condition of the mother before and during the pregnancy, it should not cause alarm on the part of an Rh-negative woman who is contemplating marriage to an Rh-positive man.

It is important, however, to know the Rh-factor condition of a young girl about to receive a transfusion. Suppose an Rh-negative girl receives large amounts of Rh-positive blood in a transfusion. She may develop such a high concentration of anti-Rh that it might be impossible for her to ever bear an Rh-positive child. Furthermore, if at some subsequent time another transfusion were necessary and she were given Rh-positive blood again, she would suffer a severe, perhaps even fatal, reaction. Concern for the type of transfusions used is not confined to the female alone. Rh-negative males can also suffer severe reactions if transfused intermittently with Rh-positive blood. As a result of this information, blood is routinely typed for the Rh factor.

It has been shown recently that there are at least 35 different proteins in human blood which are sufficiently distinguishable to be identified and to have their inheritance followed. This is a very active field of research today, and there is no doubt that much fruitful information will come from it in the years ahead.

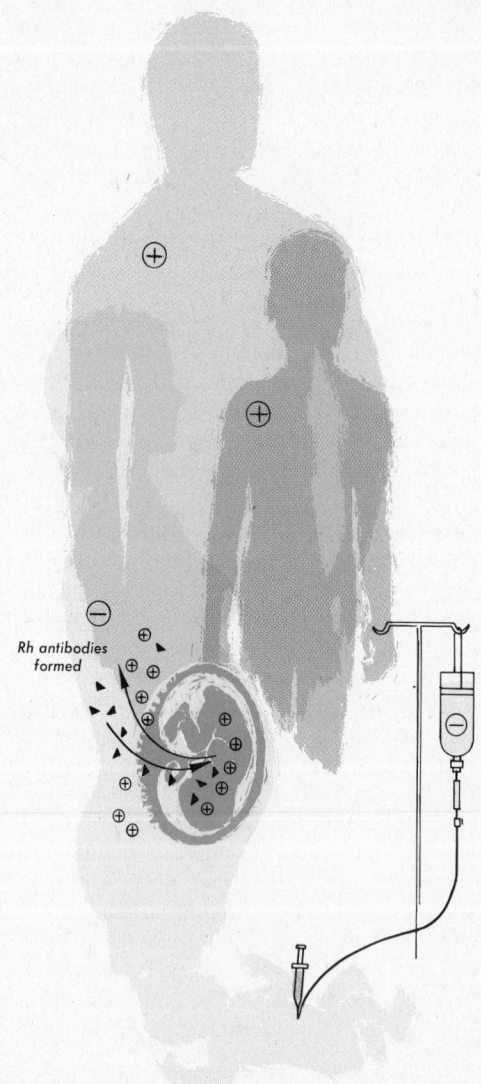

Fig. 19–15. Complications may arise when an Rh-positive man marries an Rh-negative woman, as indicated in these figures. The explanation is given in the text.

Rh antibodies formed

Other Functions of the Blood

Among the many functions of the blood in addition to body defense, are pH and temperature regulation, and water balance.

Acid-Base Balance. All of the cells of the body are very sensitive to the amount of acid or base that is present in their environment, and can withstand only very slight changes in the concentration of hydrogen ions. Since the blood controls the internal environment, it follows that it, too, must be very constant. Such is indeed the case. In whatever part of the body the **hydrogen ion concentration** or _p_H of the blood is measured, it is found to be remarkably constant, being slightly alkaline (_p_H 7.45). This may seem difficult to understand in view of the many compounds being "dumped" into and withdrawn from the blood continuously. It is made possible by substances in the blood, appropriately called **buffers,** which maintain a constant _p_H. They combine with both acids and bases so as to prevent any important change in the relative number of hydrogen and hydroxyl ions. The most important buffers in the plasma are the proteins, amino acids, phosphates, and bicarbonates. Acids that form in the cells as a result of metabolic activity are passed through the cell membrane into the blood where the buffers absorb the extra hydrogen ions, thus preventing an excess and hence an increase in acidity. Carbon dioxide coming into the blood from all the cells would form carbonic acid if it were not for the sodium ions which convert it to bicarbonate, thus maintaining a constant _p_H. The carbon dioxide is lost in the lungs, an important factor in removing acid conditions from the blood. Likewise, acidic substances are removed from the circulation in the kidneys. All of these factors work together in order that the blood can remain constant as far as the acid-base balance is concerned.

Water Balance. Water is constantly being added to and withdrawn from the blood because it is the only way that this important compound can be delivered to and taken away from the body cells. This is extremely important, since all life processes are maintained in a water medium within the cells. Therefore, they must have the proper amount at all times. Water is taken into the blood from the digestive tract and lost through the skin as sweat or through the kidneys as urine. Excess water is normally lost through the urine, so that a delicate water balance is maintained in every cell of the body at all times.

Temperature Regulation. Because the body is exposed to widely varying external temperatures, several different mechanisms have been provided to regulate the body temperature. The added muscular contraction brought about by shivering in mammals causes a greater burning of sugar, thus raising the temperature of the entire body. When the body is exposed to low temperatures, the skin becomes pale and cold, due to vasoconstriction, which prevents blood from coming close to the external surface where it would be unduly cooled. The opposite effect, flushed skin, is noted following violent exercise or during particularly hot weather. This is caused by vasodilation in the skin which allows the warmer-than-normal blood to come near the surface where it can lose its excess heat. This control of the size of the small blood vessels in the skin is a very important temperature-regulating mechanism. In addition, the sweat glands pour out water which, by evaporation, provides an important cooling device.

Certain special provisions are present in some animals for increasing surface area to make the cooling process more effective. For example, elephants and rabbits are thought to employ their large ears for this purpose as well as for collecting sound waves. Bats are thought to rely on the circulation in the skin of their wings for heat regulation. Man, of course, covers his body with various fabrics, the color and texture of which are changed depending on the temperature. Hair and feathers are excellent insulators against temperature change. Thus the bodies of the warm-blooded birds and mammals are reasonably well suited to withstand the varying temperatures they encounter in their particular environments.

THE LYMPHATIC SYSTEM

A swelling following a blow on any portion of the body is gorged with a fluid called **lymph,** which is much like plasma except that it does not contain so much protein material. It does, however, contain some

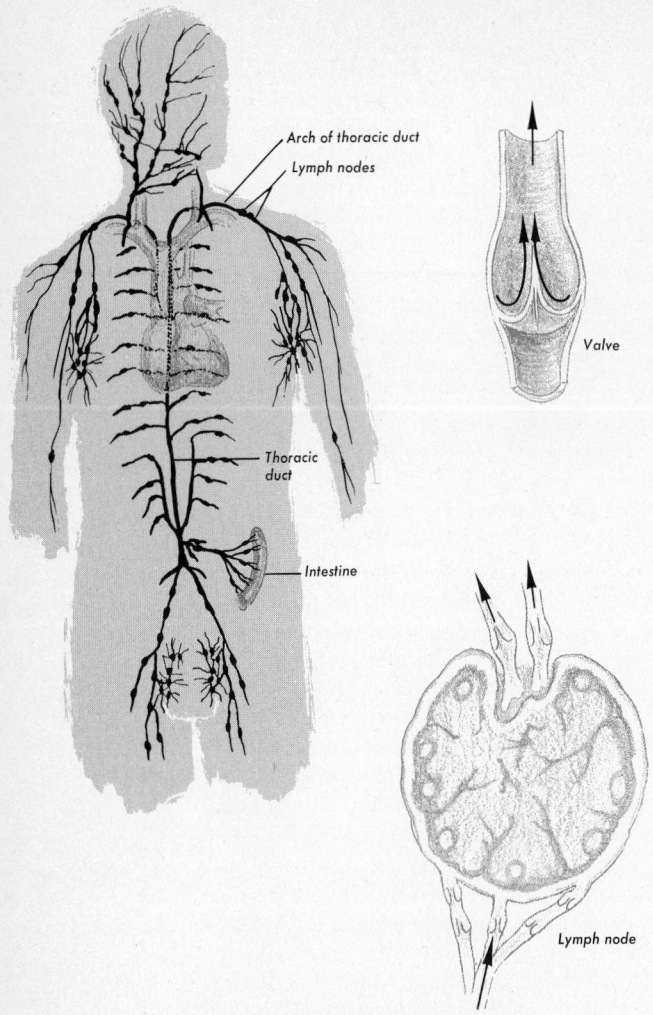

Arch of thoracic duct

Lymph nodes

Valve

Thoracic duct

Intestine

Lymph node

Fig. 19–16. The lymphatic system of man. The vessels are provided with valves which keep the lymph flowing in one direction. A detailed drawing of the lymph node is shown.

white cells, principally **lymphocytes.** Lymph fills all of the spaces between and around cells and thus bathes every cell of the body. It functions as a medium between the capillary and the cell, a continuum through which food, oxygen, and wastes can enter and leave the cells.

Lymph passes out through the capillary walls around the cells, and from there moves slowly on into a system of vessels which coalesce and eventually reach the circulatory system in the neck region where the large lymph vessels join the large neck veins (Fig. 19–16). This is as intricate as the venous system and resembles the latter in many respects. It merely is an alternate route by which water and wastes from the cells can reach the general circulation. In other words, it resembles a "sludge pump" in its action.

The lymph glands which produce the lymphocytes of the blood lie along the lymph channels, through which they are dumped into the blood. The lymph glands have another function, namely, as filters. Foreign particles such as dust, debris, and bacteria float in the lymph and finally make their way to the lymph glands where considerable phagocytic activity goes on. This can be so active as to produce noticeable swelling of these glands, particularly under the arms if the source of infection is in the hand, or in the groin if the difficulty arises from some portion of the leg or foot. They function in stopping the infection before it gets into the general circulation where it may do a great deal of damage. Carbon particles are inert, so that those entering the body from air laden with coal smoke will lodge in the lymph glands, contributing a dark color to them if the person has lived in cities where smoke is abundant.

There is no pumping station in the lymphatic system of man, although such mechanisms are present in some of the lower animals. The movement of the lymph is due to the continual massaging of the lymph channels by the contracting muscles, both visceral and skeletal. The contraction of the villi in the intestinal walls, as well as the negative pressure created at the point of entry into the large neck veins, aids in bringing about lymph movement. Any stoppage of this movement results in swelling, as is the case when certain roundworms invade the lymph glands, thereby clogging them so that the lymph cannot pass through (p. 182).

Books

FULTON, J., *Selected Readings in the History of Physiology*. Springfield, Ill.: Charles C Thomas, 1930.

HOAR, W. S., *General and Comparative Physiology*. Englewood Cliffs, N.J.: Prentice-Hall, 1966.

ROMER, A. S., *The Vertebrate Body*. Philadelphia: Saunders, 1962.

WOODBURNE, R. T., *Essentials of Human Anatomy*. New York: Oxford University Press, 1961.

Articles

BURNET, M., "The Mechanism of Immunity." *Scientific American*, **204**, 58, January, 1961.

CLAMAN, H. N., "The New Cellular Immunology." *Bioscience*, **23** (10), 576, October, 1973.

HARARY, I., "Heart Cells in Vitro." *Scientific American*, **206**, 141, May, 1962.

LAKI, K., "The Clotting of Fibrinogen." *Scientific American*, **206**, 60, March, 1962.

WIGGERS, C. J., "The Heart." *Scientific American*, **196**, 74, May, 1957.

WOOD, J. E., "The Venous System." *Scientific American*, **218**, 86, January, 1968.

ZWEIFACH, B., "The Microcirculation of the Blood." *Scientific American*, **200**, 54, January, 1959.

20

METABOLISM AND DISPOSAL OF WASTES

We have seen in the last several chapters how complex food molecules have been rendered soluble and reduced to the proper chemical configuration to be transported to the individual cells of a complex animal. In addition, we have seen the method by which oxygen has been made available for oxidation of these energy-rich molecules. The next step is to consider the processes by which energy is released and how this energy is utilized by the cells in doing all of the things that living things must do, such as, grow, move, produce heat, and many other things. This most important series of processes is called **metabolism**. Once metabolism is completed certain toxic end products are left over that must be disposed of. In this chapter we consider both metabolism and the mechanism the body has for ridding itself of wastes. Both of these processes involve some complex chemistry and physics, much of which is fairly well understood.

METABOLISM

In Chapter 2 we learned that metabolism goes on within the individual cells. We also learned the chemical steps involved. Once again we must refer to this information and, in addition, describe how metabolism takes place in a complex animal.

Metabolism involves two processes: releasing energy and the utilization of this energy. In the former, the food molecules are oxydized, thus providing the energy to propel the machinery of life. These two processes complement one another. The first makes the second possible; protoplasm can do all of the things it does only if energy is available. Conversely, without the fabric of protoplasm there would be no means by which energy could be released. In other words both processes are interdependent and must proceed simultaneously.

430

We learned earlier that the only source of energy in organisms is through respiration, and this occurs in each individual cell. The raw materials are fats, carbohydrates, and proteins, each of which goes through a complex series of steps in degradation before the energy is finally stored in the form of ATP. The immediate energy is derived when ATP is converted to ADP. If this were completely so then one could compute the amount of energy in any food molecule in terms of ATP molecules, and when an animal consumes these molecules the energy stored as ATP should equal the amount of energy taken in. In this case the efficiency of the organism would be 100 percent, and we know this is not so. Then how can we determine the energy transformations that go on in an animal; that is, how can we study intake as compared to outgo? Obviously the energy is taken in as potential (stored) energy in the form of food molecules, and it comes out in the form of chemical work, mechanical work, and many other forms of activity. In order to do this we need to convert all of the intake and outgo energy into a single unit. The easiest to measure and one into which all other forms of energy can be converted is heat, expressed in **calories** (Cal.). By definition **one calorie** is the amount of heat required to raise the temperature of one kilogram of water through one degree centigrade. With this unit one has a means of comparing different forms of energy.

It is very simple to measure the number of calories in food molecules. They are merely burned and the heat given off is expressed in calories. For example, 1 gram of glucose or protein equals about 4.0 calories and 1 gram of fat equals 9 calories. This has been done with most foods and lists are available showing the number of calories contained in each kind of food. One can use them in computing the exact number of calories which are consumed in any specified time. This is useful information when it is desired to control weight. Does all of this stored energy become available in the form of useful energy, that is,

in the form of ATP? In other words does all of the potential energy in foods show up as ATP as a result of respiration? This can be answered by comparing some experimental data. For example, when 1 gram of glucose is oxidized in cells, the equivalent of 1.7 calories of energy is captured as ATP. When compared with the 3.8 calories a gram of glucose actually contains, we see that respiration has an efficiency of 44 percent, which is very high when compared to machines built by man. The remaining energy is lost in the form of heat.

In order to determine the amount of energy derived through respiration it is necessary to use some other criterion rather than that of simply measuring the number of calories consumed, owing to the fact that not all of the food is converted to energy. Some may be unused, some used in growth, and some stored. Moreover some organisms are more efficient in food utilization than others. A far more reliable method of determining energy utilization of animals is to measure the **oxygen consumption** because oxygen is used only in respiration. It is a simple matter to measure the amount of oxygen consumed by any organism per unit of time. This is done routinely with man and the results often have medical significance.

Basal Metabolism

Most oxygen consumption data are taken from records on man. The oxygen consumption, and therefore the energy utilization, varies widely among human beings. It depends on age, sex, weight, body configurations, and activity. Therefore, a standard procedure must be followed if any abnormalities are to be detected. The rules are that a person must be completely at rest and must not have eaten for at least twelve hours. The energy utilized then is just sufficient to keep his heart beating, to keep him breathing and to maintain body temperature. The rate at which he consumes oxygen under these conditions measures his **basal metabolic rate** (BMR). For a young mature male this is about 1,600 calories per day; for a young woman about 5 percent less. By taking thousands of these records standard tables have been established which can be referred to when measuring the BMR of anyone. These are useful in diagnosis of certain diseases which may be reflected in changes in the BMR.

A person pursuing a sedentary occupation requires about 2,500 calories per day. Heavy physical labor boosts this total to as much as 6,000 calories per day. One's weight remains constant providing the intake of calories is equal to those burned. With age the tendency is to continue eating at the same rate as when young, but muscular activity is reduced. This inevitably results in an increased waistline, the bane of middle age.

On the other hand, a caloric intake which does not satisfy the energy requirements forces the body to draw on its reserve food stores. The first reserve foods to be used are the carbohydrates (glycogen) of the muscles and liver. The next to go are the stored fats, which in an average person will supply enough energy for about six weeks. The last to go are the proteins, starting with the least used parts, such as certain muscles. Finally the essential organs such as the heart and brain are reduced to a point of failure, which then terminates the life of the organism. It is interesting to note that one can tolerate starvation for long periods without apparent harm.

Metabolism of Carbohydrates

Starches and sugars are the chief sources of energy for herbivorous and omnivorous (man) animals. Carnivores, however, do not rely on these foods for their primary energy source, but on proteins and fats.

It will be helpful during the following discussion of metabolism to refer to Figure 20–1. Once the complex carbohydrates (polysaccharides) are digested they are absorbed into the hepatic portal system as the three simple sugars (monosaccharides), glucose, fructose, and galactose. When they reach the liver they are converted to **glycogen** which may be stored or immediately reconverted to glucose, depending on the energy requirements of the organism, which change from moment to moment. The glucose level in the blood must be maintained at a relatively constant level because it is the chief source of fuel for respiration in the cells. This level must not drop below about 60 mg per 100 ml of blood. If it does, stark changes become immediately evident in the animal, starting with mental confusion, followed by convulsions and coma, and finally ending in death. Similar symptoms occur following an overdose of insulin (p. 355). These dramatic effects are a result of the sensitivity of brain cells to lowered glucose levels. Since they store little or no glucose a minimal level is required at all times to maintain normal functions. They are equally sensitive to the lack of oxygen, and the symptoms are the same.

Fig. 20–1. A schematic representation of the metabolism of the end-products of digestion.

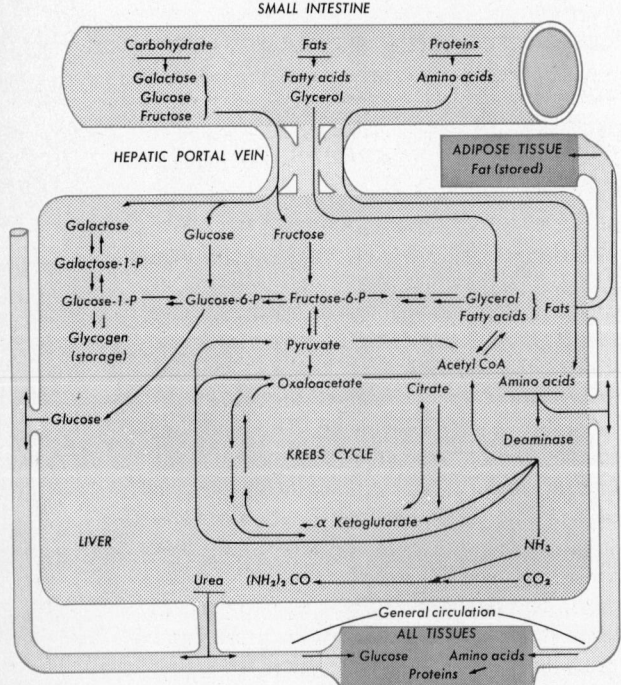

As was pointed out earlier, the liver releases the proper amount of glucose at all times. This is under the influence of hormones, primarily insulin from the pancreas, although those from both the medulla and cortex of the adrenals, as well as somatotropic hormones from the anterior pituitary, are involved. Moreover, the two-way reaction:

$$\text{Glucose} \underset{\text{G-6-phosphatase}}{\overset{\text{hexokinase}}{\rightleftharpoons}} \text{glucose-6-PO}_4 \rightleftharpoons \text{glycogen}$$

takes place only in the liver because of the enzyme **glucose-6-phosphatase.** Muscle cells convert glucose to glycogen, just as liver cells do (both have the enzyme hexokinase), but the reaction does not go the other way because muscle cells have no glucose-6-phosphatase. Glucose going to muscle cells is stored as glycogen and ultimately metabolized there. Muscle glycogen does not supply free glucose to the blood stream as liver glycogen does. In evolution it apparently was of more survival value to have an energy reserve in the muscles than in the brain, although just why this is so is unclear.

Glucose can be stored as fat as well as glycogen. It is common knowledge that an excess of carbohydrates in the diet leads to fat accumulation, in man or beast. The metabolic pathways by which this happens are well known. During glycolysis fructose-1, 6-diphosphate is converted, through several intermediates, to glycerol. The fatty acids are derived from acetyl-CoA. The fatty acids and glycerol are transported by the blood stream to adipose tissue where they are converted to fats and stored. In addition, adipose tissue itself can convert glucose to fats.

Metabolism of Fats

Fats and oils accumulate in both plants and animals and, therefore, appear in the diet of animals. As was pointed out earlier, they are rich in energy, containing over twice as many calories per gram as either carbohydrates or proteins. Their high energy yield is accounted for, in part, by the fact that they contain little water and oxygen. When they oxidize they require an abundance of oxygen. They are excellent materials for the storage of energy, which is one of their chief functions. Fats also have an important function in the construction of membranes within and surrounding cells. They, together with protein, make up the framework of these membranes. Another function of fats, which may be secondary, is to form an insulating layer beneath the skin to prevent heat loss and also to cushion internal organs from shock with body movements. Another incidental function of fats is that they contain the fat-soluble vitamins.

Fats are essential for life but since they are present in most foods, there is little likelihood of one suffering from a fat deficiency. The body is able to synthesize most of the fatty acids, although some are **essential** and must appear in the food. However, the diet would need to be extremely low in fats before a deficiency in the essential fatty acids would be evident.

Fats which are split to fatty acids and glycerol in the digestive tract are absorbed into the hepatic portal system. Colloidal fats by-pass the liver and go, via the lymphatic system, into the general circulation (p. 373). The fatty acids that reach the liver are those present in the original fat that was taken into the digestive tract. That is, they may have come from corn oil, animal fat, or from many other sources. These then are reassembled in the liver to form human fat, which is then absorbed in the blood stream and carried to adipose tissue where it is stored. Or it may be utilized in the manufacture of cell membranes.

It has been known for many years that fats oxidize readily only when in the presence of carbohydrates. The reason for this observation is now known. Oxaloacetic acid, which results from carbohydrate metabolism, is needed to combine with the acetyl coenzyme A which is derived from fatty acids, in order for fat metabolism to proceed. Whenever carbohydrate metabolism is interrupted, fat metabolism is also involved. Diabetes, for example, is a disease in which carbohydrates are not properly metabolized as a result of the loss of the hormone insulin (p. 355). Symptoms of the disease are the accumulation of fat in the liver and the appearance of so-called acetone bodies in the urine. Improper metabolism of fatty acids in the liver cause these abnormalities. Other hormones influence fat metabolism. Those from the adrenals, sex glands, and pituitary are involved but just how is not fully known.

Metabolism of Proteins

Every cell of the body must rebuild or replace itself from time to time. During the growing stage of the whole animal, and in some cases during adulthood, cells also duplicate themselves. This means that protein must be synthesized in each cell more or less continuously, and this is done by the utilization of the amino acids that come to the cell from the blood stream. The amino acids are assembled in the proper sequence to form proteins that are specific for the particular species of animal. This is done under the influence of specific enzymes, and just how this is accomplished in cells is described in Chapter 23. The source of an amino acid, such as glycine, does not matter since it is the same whether it be from a cow, oyster, or plant. As it combines with others, however, the resultant protein is highly specific. This relationship may be compared to bricks in a house. The red bricks are all alike no matter from which brick yard they came, and they only become distinctive when they form a part of a particular house. They are then a constituent of a pattern which in this case is Smith's house, not Jones' or Brown's. Likewise with the construction of proteins from amino acids, the numerous amino acids (about twenty) are precisely selected by cellular enzymes, and built into the final protein structure.

It, therefore, becomes necessary that all of the amino acids be present in the circulating blood. During the course of evolution animals have lost the capacity to synthesize all of the amino acids that are required for growth and repair. This has come about by the loss of specific enzymes, probably because the amino acids were present in the diet making it unnecessary to manufacture them. The amino acids that the body can manufacture are the **nonessential** amino acids and are not needed in the diet, although they can be utilized if available. The amino acids that the body is unable to synthesize are the **essential** amino acids and must be included in the diet. Man requires ten of these and this requirement varies with other animals. For example, if a person ate nothing but gelatin as a source of protein he would starve to death, because it lacks three essential amino acids. Corn lacks two, lysine and tryptophan. Meat, eggs, and milk contain all of them and for that reason are considered complete proteins. Usually one eats a variety of proteins so that amino acids lacking in one are present in another. However, there is always the matter of quantity; monotonous diets consisting of a single protein which may be low in one essential amino acid would require that the individual consume large quantities in order to satisfy his minimal needs. If, on the other hand, the amino acids were perfectly balanced one could take in as little as 25 grams of protein a day to maintain normal health. Actually from 80 to 100 grams are customary in most diets.

One must have a continual supply of proteins because they are not stored in the body, hence during starvation the tissue proteins are sacrificed to replace the other nitrogenous compounds lost from the body. It is possible to measure quantitatively the nitrogen ingested in the form of protein and that excreted as urea and determine whether or not the body is maintaining a protein "balance." This is useful information, particularly during such disturbed conditions as disease or following severe injury. At these times the body often loses more nitrogen than it takes in and is said to be in "negative nitrogen balance." This results in a wasting of the body tissues and, if continued, terminates life.

Just what part hormones play in protein metabolism remains obscure, although there is no doubt that they exert a profound influence. Certainly the growth hormone from the anterior pituitary and the adrenal cortical hormones influence the synthesis of protein. We have already seen that insulin is important in carbohydrate metabolism and probably also influences protein metabolism. Other hormones such as those from the gonads also influence protein metabolism but exactly how is unknown.

Fate of Excess Proteins. Excess proteins taken into the body are converted to carbohydrate or fat and can then be used for energy or stored. But proteins contain nitrogen, which is not true of fats and carbohydrates. The nitrogen must be removed by a process called **deamination** before amino acids can be utilized as fat or carbohydrate. This occurs in the liver and requires energy which is evidenced by an increase of as much as 30 percent in the BMR shortly following protein intake. This can be proven by a simple experiment. Such a rise does not occur following protein intake in an animal whose liver has been removed.

The amino groups of the amino acids are removed and combined with carbon dioxide to form **urea** which is then removed from the blood in the kidney (p. 439). The process by which this is accomplished involves cyclical chemical reactions (Fig. 20–2). The first step in this cycle is stripping the amino acids of their amino groups ($-NH_2$). This is accomplished by numerous specific enzymes called **deaminases** (transaminases). This results in the formation of **glutamic acid** (from α-ketoglutaric acid, a Krebs cycle compound). The glutamic acid is then decomposed by the enzyme, **glutamic dehydrogenase**, to form α-ketoglutaric acid and ammonia (NH_3). The latter toxic substance is removed by means of the **ornithine cycle.**

Three amino acids, **ornithine, citrulline,** and **arginine,** are involved in the ornithine cycle, and it is the sequential conversion of one to the other that provides a means by which ammonia is continuously eliminated from the body. One molecule of ammonia and carbon dioxide first combines with ornithine producing citrulline. A second molecule of ammonia transferred from the amino acid, aspartic acid, combines with this amino acid to produce arginine which then splits into urea and ornithine under the influence of the enzyme **arginase.** This completes the cycle; ornithine starts the cycle over again and the urea, now quite harmless, finds its way to the kidney where it is excreted. Note that the ornithine cycle is dependent on the Krebs cycle for its operation. Starting from the amino acids the net result is:

$$2\,R - CH\,NH_2COOH + CO_2 + O_2 \longrightarrow$$
$$CO(NH_2)_2 + 2\,R - \overset{\overset{\displaystyle O}{||}}{C} - COOH + H_2O$$
Urea

Many aquatic animals do not produce urea, but are able to tolerate ammonia in the blood until it is excreted in the urine. Insects, birds, and reptiles excrete their

Fig. 20–2. A schematic representation of the ornithine cycle. See text for explanation.

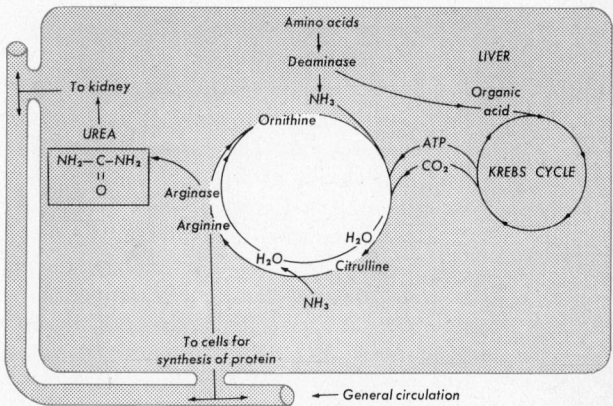

nitrogenous wastes as **uric acid,** which has the advantage of requiring little or no water to flush it out of the body since it is in crystalline form. For this reason these animals can survive better in desert regions. We have more to say about this a little later.

Interconversion of Food

From the foregoing discussion it is clear that fats, carbohydrates, and proteins can be used as energy sources, but only protein is essential for building protoplasm. One can survive on protein as the sole source of energy but not on carbohydrates or fats. By consulting Figure 20–1 again you will be able to follow the interconvertibility of the three types of food molecules.

The monosaccharides, glucose, fructose, and galactose, which result from poly-saccharide breakdown, enter the liver and are converted into glycogen and stored, or they may be transformed into glucose and released into the general circulation, or they may be converted to fructose-6-phosphate and undergo glycolysis, ultimately entering the Krebs cycle where their stored energy is converted to ATP. Some may be converted to glycerol and combine with fatty acids to produce fats, which then enter the general circulation and are stored in adipose tissue. Considerable fat is also synthesized in this tissue. Hence we see how carbohydrates are converted to fats.

Fats are digested to fatty acids and glycerol in the small intestine and pass to the liver where the fatty acids may enter the Krebs cycle and supply energy just as carbohydrates do. They may also recombine in a specific pattern to form fats characteristic of the organism.

The amino acids from protein digestion also reach the liver where they follow several pathways. Some may immediately return to the general circulation and pass to the cells where they are synthesized into proteins under the influence of cellular enzymes. Others may lose their amino groups (deamination), and the remaining organic acids may take part in supplying energy via several pathways. Some fragments enter the Krebs cycle directly whereas others enter via pyruvate. Thus proteins are a source of energy.

This elaborate biochemical mechanism makes possible the most efficient use of all foods, for the maintenance and growth of the body. Whereas plants can manufacture all of their needs from very simple compounds (carbon dioxide, nitrates, salts, and water), animals cannot and therefore have come to depend on plants for many of their food molecules. As was pointed out much earlier, there is a satisfactory balance between the two so that neither overruns the earth.

Molecular Turnover

As one observes a plant or animal over a period of time it seems to remain about the same as far as its general appearance is concerned. It was thought until recently that the molecular composition of the framework of organisms was stable, that is, they remained relatively constant throughout life. It was thought that either food molecules were burned to supply energy or they were utilized in the production of new cells or to replace those that wore out. In other words there was very little turnover of the molecules that composed the organism.

With the advent of tracer techniques with which food molecules could be labelled with radioactive or heavy isotopes and followed during their circuit through the body, a whole new concept emerged. It was shown that labelled amino acids are rapidly incorporated into the proteins of the body even though there is no obvious growth. Likewise, labelled fatty acids become a part of the body fat without increase in the bulk of fat. Even bones become labelled very quickly indicating that these stable structures are changing constantly. Obviously molecules making up the body framework are constantly being replaced by others, the framework being built and destroyed continuously. This is a dynamic equilibrium, in that the body remains essentially the same while this turnover is going on.

The rate of incorporation of labelled atoms into man indicates that one half of his tissues are replaced about every 80 days. Not all tissues are replaced at the same rate; some such as those of the liver turn over much faster than those of the skeletal muscles. Eighty days is an average figure for all tissues of the body. We now know that our bodies change from day to day, as far as their chemistry is concerned. The incorporation of radioactive isotopes into our bodies, as well as into those of all organisms has considerable bearing on the problem of radioactive fallout and should be of great concern to everyone.

DISPOSAL OF WASTES

Once the food molecules are metabolized, the resulting waste products must be removed promptly from the body. Retained, they act as toxic substances, causing death within a short time. In this discussion we are primarily concerned with the removal of the end products of protein metabolism which contain nitrogen, although carbon dioxide is also an excretory product which was discussed earlier. The nitrogenous waste products are urea, uric acid, and ammonia.

At this point we should be certain that there is no confusion about the term, waste products. Undigested food which accumulates in the colon and is periodically removed (defecation) is not metabolic waste. Such materials (feces) have actually never been inside the body, since they follow a one-way passage from mouth to anus without having ever entered the cells where metabolism takes place. **Metabolic wastes**, on the other hand, result from chemical change that occurs within the cells of the body. Carbon dioxide and nitrogenous end products are such metabolic wastes.

In the evolution of kidneys, the first simple kidney consists of a system of tubules ramifying the body tissues, as illustrated in planaria. Each tubule drains the tiny flame cells which selectively pick up the nitrogenous wastes from the neighboring cells, and the entire system conveys those products to the outside through many pores.

The next great step is taken by the annelids, where nephridia replace the flame cells. Not only do the cells at the funnel (nephrostome) pick up coelomic fluid which contains many substances beside nitrogenous wastes, but as this fluid passes down the tubule selective reabsorption occurs along the way. This principle established in the annelids is retained throughout all higher groups.

In the vertebrates the origin of the excretory system is intimately associated with the reproductive system. Primitive cyclostomes have kidneys that are not greatly different from those found among invertebrates. They are long, thin, paired structures lying in the dorsal wall of the body cavity, one on each side of the vertebral column (Fig. 20–3). Coelomic fluid is drawn into the ciliated nephrostomes which lead into a long tubule much like the earthworm's. The tiny tubules coalesce forming larger tubes, the **urinary ducts,** which terminate in the cloaca. In this animal, the eggs and sperm are shed into the body cavity and find their way out through two openings from the posterior end of this cavity into the cloaca. This is an extremely simple method reminiscent of some of the lower invertebrates. The excretory system and the genital system are thus distinctly separated. Perhaps because this system of getting rid of the germ cells was so fortuitous, it became necessary to provide tubes for this purpose. By chance, or perhaps by proximity, the urinary ducts took over that task, and an intimate relationship became established between the excretory and reproductive systems. This was done very gradually through several groups of animals over a period of many millions of years.

Among the fishes and amphibia, the testes became connected with the upper end of the kidney by means of tiny tubules, the

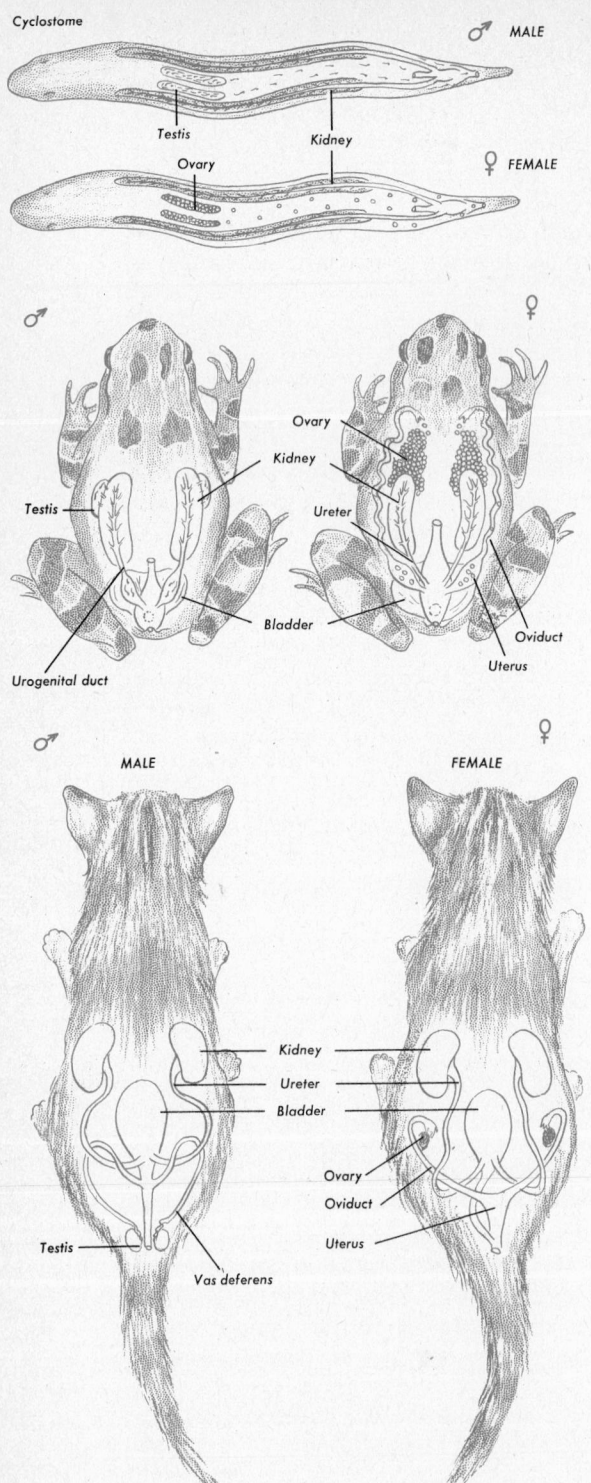

Cyclostome

♂ MALE

Testis   Kidney

Ovary

♀ FEMALE

♂   ♀

Ovary
Kidney
Ureter

Testis

Oviduct
Uterus

Bladder

Urogenital duct

♂   ♀
MALE   FEMALE

Kidney
Ureter
Bladder

Ovary
Oviduct
Uterus

Testis

Vas deferens

vasa efferentia. Sperm cells then made their way out of the body through the same tubes as the urine, namely, the **urogenital ducts** (Fig. 20–3). New ducts, the **oviducts,** were formed for conveying eggs out of the body. The lower ends of these ducts became large and saclike to form the **uteri** (singular, **uterus**) in order to accommodate the great numbers of eggs that accumulated before deposition. The uteri, too, opened into the cloaca. This seemed to be a very satisfactory arrangement, and not until mammalian evolution was well underway did further radical changes take place.

In a mammal, such as a cat, we find that the kidneys have become "kidney-shaped" and much more compact than the long thin organs of the fishes or even the thicker structures of the amphibians. As in the cyclostomes, the tubes which convey urine away from the kidneys have no affiliation with the genital system. Indeed, these ducts, the **ureters,** are new tubes which formed very late in evolution. They connect with a **bladder** (urinary), thence through a tube, the **urethra,** to the outside. The old urogenital ducts of the frog have lost their urinary function and have been taken over completely by the genital system. Their sole function in the mammal is to carry sperm cells. These tubes, the **vasa deferentia,** connect with the urethra, which is urogenital throughout the rest of its course to the outside of the body. The terminal portion is modified into a copulatory organ, the **penis.**

The path of the eggs in mammals is not greatly modified from that of the frog. They pass into the oviducts from the ovaries and then into the uteri, which may be paired as in the cat or fused as in man. The eggs of mammals are much smaller, of course, but they follow essentially the same

Fig. 20–3. The urinary and reproductive systems were separated in lower vertebrates (cyclostomes) but became intimately associated, in higher vertebrates (fishes, amphibia, birds, and mammals). Three forms are shown here to show how this came about. See the text for details.

path as in the amphibians and fishes. A new structure, the **vagina,** has been added which receives the penis of the male in sperm transfer, an essential for land animals.

The embryological development of the urogenital system of mammals follows basically the same course as its evolution. That is to say, at one time the kidney resembles that of a cyclostome, and a little later, that of a frog. Finally, some time before birth, the true mammalian kidney and associated organs are formed. It must be remembered that, although the excretory and reproductive systems are anatomically intimately related, they bear no relationship to one another functionally. The job of reproduction and excretion are two separate and distinct functions.

The Human Kidney

The two kidneys in man are about 4 inches long and are located near the mid-dorsal line just below the stomach. The ureter and blood vessels emerge from a depression on the medial side. The kidneys lie in a capsule of peritoneum which excludes them from the coelom. If a kidney is sliced lengthwise, it will be seen to consist of an outside layer, the **cortex,** and an inner capsule, the **medulla** (Fig. 20–4), both of which are visible to the naked eye. At the point where the ureter leaves, there is a large cavity, the **pelvis** (not to be confused with the pelvis of the skeleton), which is a depository for the urine as it comes from the millions of tiny tubules of the kidney. The entire internal kidney is tied together with connective tissue and interlaced with blood vessels, and it is a very complex and extremely delicate organ.

The cortex is made up of **renal corpuscles,** which are the tiny units where the excretory process begins (Fig. 20–4). Each consists of a minute ball of capillaries, the **glomerulus,** surrounded by the double-walled, cuplike sac, **Bowman's capsule.** The inner wall of the capsule closely adheres to the glomerulus in order that substances may filter readily from the blood stream into the

cavity and thence through the long tubule to the pelvis of the kidney. The medullary portion of the kidney consists almost exclusively of these tiny tubules which play an important function in excretion.

Urine Formation. The function of the kidney has long been known but just how urine forms has become clear only in recent

Fig. 20–4. A schematic view of the human kidney cut to show its internal structure and with a single excretory unit (renal corpuscle and tubule) enlarged to show its detailed anatomy.

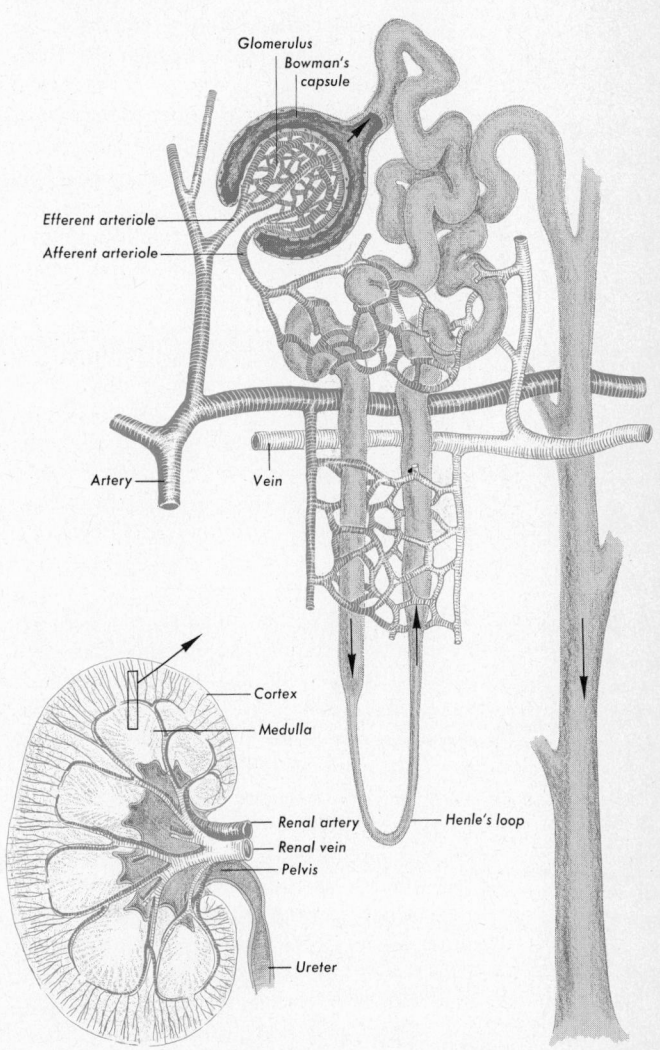

years. Following through the process as it is now believed to take place, the blood passes into the kidney through the renal artery which immediately breaks up into smaller and smaller vessels until these eventually form the glomeruli of the renal corpuscles. As the blood passes through, most of the substances in the blood except the cells and proteins pass through Bowman's capsule by **filtration.** This is purely a physical process which depends entirely on the pressure in the blood vessels, and the amount of fluid passing into Bowman's capsule rises and falls with that pressure. The vessels coming from the glomerulus are slightly smaller than those going to it, so that the pressure remains high in these vessels. About one percent of the blood volume is lost to the **capsular filtrate** as it passes through the kidney.

By some ingenious experiments with tiny micro-needles, Prof. A. N. Richards was able to examine the capsular filtrate and found that it contained urea, sugar, amino acids, salts, and so forth, in about the same concentrations as those in the blood plasma. Obviously, if all of these valuable products remained in the capsular filtrate, the animal would shortly drain its body of the essentials of life. Therefore, these products must be selectively **reabsorbed** into the blood while the fluid passes through the extremely long tubule on its way to the pelvis. This is possible because, when the blood leaves the glomerulus, instead of forming a vein, as is customary in other organs, it becomes another set of capillaries, this time surrounding the tubule that leaves Bowman's capsule (Fig. 20–4). As the capsular filtrate passes down the tubule, the valuable portions, such as the amino acids, sugars, salts, and so forth, are reabsorbed into the blood again. This requires considerable work on the part of the cells lining the walls of the tubules, as indicated by the fact that the kidney requires more oxygen than the heart when calculated on equivalent weights. The secretion is against a diffusion gradient and therefore is called **active transport** (p. 54). This accounts for the large amount of work that is necessary. If the

kidney is denied oxygen, reabsorption stops, although filtration proceeds normally.

With this arrangement it is apparent that the kidney functions as an organ which selects what substances shall remain in the blood and what shall be removed. If there is too much sugar in the blood, for example, the tubules reabsorb some of it but leave the surplus in the capsular filtrate. Under these conditions the urine shows sugar, which is what happens in diabetics. The same is true of other substances.

The Kidney as a Regulatory Organ. Because of its ability to secrete substances selectively, the kidney is a very important organ in maintaining the proper composition of the blood and other body fluids. As we have seen, the various end products of metabolism are injurious if allowed to accumulate. The kidneys remove just the right amount of each to prevent harmful effects, and yet leave enough to maintain a proper balance of ions and molecules. The kidney figures prominently in the retention or release of hydrogen and hydroxyl ions to maintain their proper balance, and thus hold the pH of the blood constant. If the salt concentration of the blood should rise too high or fall too low, there would be a harmful movement of water which might destroy cells. This too is prevented by the elimination of exactly the right amount of salts through the tubules of the kidneys.

Many substances that are used in medicine are eventually eliminated through the urine. Such compounds as antibiotics, aspirin, and many others are removed from the body via the kidneys. Hence testing the urine is an important criterion of the effectiveness of a drug in treating a specific disease. If the drug appears in the urine shortly after it is given, it can be of little value. Antibiotics, for example, must remain in the body long enough to have an inhibitory effect on pathogenic bacteria. Frequently when searching for a new drug, it is found that the drug does the intended job very well but is lost too rapidly through the urine to be effective.

The kidney is able to return a certain

amount of various substances to the blood from the capsular filtrate, but if the amount appearing in the blood is above a certain critical level, which is spoken of as the "threshold" level, the kidney no longer is able to prevent the substance from appearing in the urine. In diabetics for example, the blood sugar glucose reaches such high levels, owing to the lack of insulin, that a large amount of it appears in the urine. If, in man, the amount of glucose per 100 cc of blood exceeds 170 mg, glucose appears in the urine. In other words, that is the threshold for glucose. Other substances have different thresholds.

Blood volume is also regulated by the kidney. Following a severe hemorrhage the blood pressure drops, thus slowing up urine production and conserving body fluids. Similarly, if the blood contains too much water, the blood pressure rises and more fluids are eliminated as a consequence. The amount of urine that is excreted depends not only on the amount of fluids taken into the body but also on the amount of salt that is to be eliminated from the blood. On a salty diet the urine output is greater because more salts must be removed from the blood if a normal balance of these ions is to be maintained. Also it requires more water to pass the high proportion of solids that appears in the urine after a heavy intake of salts. In the case of diabetes there is too much glucose in the capsular filtrate and this produces a high osmotic pressure. For this reason, not much water can be reabsorbed into the blood and so a large urine output results.

Urine volume is also controlled by a hormone (ADH) secreted by the posterior lobe of the pituitary. This hormone controls the rate of water reabsorption in the tubule. If it is insufficient or completely lacking, a disease known as **diabetes insipidus** results. People suffering from this diease may have a urine output of 30–40 liters per day instead of 1.3–1.5 liters, which is average. As might be expected, they also suffer from an insatiable thirst.

A normal kidney restores all of the utilizable substances in the capsular filtrate to the blood except in cases where the retention of such substances might be harmful. However, urea and other substances are highly concentrated as the capsular filtrate flows down the tubule toward the pelvis of the kidney. Once in the bladder it is known as urine. Its concentration varies, of course, with the amount of water included with it. This can be measured by determining its specific gravity, a routine procedure in diagnosis.

Like most organs of the body, the kidney has a tremendous latitude within which to operate, and can withstand considerable abuse and still do its job satisfactorily. Actually, only one-half of one kidney, or one-fourth of the total kidney tissue, is necessary to handle the normal business of living.

Kidney Function in Adverse Environments

Animals live in a wide diversity of environments, many of which call upon the kidney to assist in keeping the balance of molecules commensurate with that needed for life. Not only the kidney, but often special organs have evolved to keep this balance, particularly among animals that have returned to the sea after having evolved in fresh water or on land.

Most marine invertebrates have little trouble with ion exchange due to the high salt concentration because their body fluids have approximately the same osmotic pressure as sea water, although they may not be in the same proportion. Some, however, such as the shore crab, *Carcinus*, can resist osmotic change as when it moves from the sea into brackish water (mouths of rivers). It is said to have the power of **osmoregulation**. When it moves into dilute sea water its own salt concentration exceeds that of its environment causing an inflow of water which, if continuous, can be fatal. The kidney (green gland), however, excretes water; nevertheless, this alone is not adequate to maintain the high salt concentration of its tissues, hence the gills actually secrete salt into the blood thus keeping it from being diluted. This requires work since the movement of the ions is against a gradient; that is, the salts tend to be lost to the more dilute environment, but by forcing them in via the gills the osmotic pressure of the blood can be maintained. Thus we see again in this marine organism how active transport is required to maintain stability of the ions.

The marine fishes also have osmotic problems because they evolved in fresh water and, therefore, their body fluids have an osmotic concentration of about one-third that of the sea. Since their thin-walled gills are constantly in close proximity with sea water they lose water and must counter dehydration. They drink sea water but cannot eliminate the salt through the kidneys (Fig. 20–5). Instead, the gills concentrate the salt and excrete it back into the sea. Again this requires energy and work on the part of the gill cells (active transport).

The marine mammals, such as whales, porpoises, and seals, also have had to adapt to the high salt concentration of the sea, which far exceeds that of their own tissues. Most whales eat fish which have a salt concentration about one-third that of the sea. However, being mammals they excrete nitrogenous wastes as urea, which requires water to flush it out of the body. They also lose some water through breathing, so there is a constant water loss which must be made up. This is accomplished by excreting a highly concentrated urine. The water gained through oxidation of their food replaces that which is lost; consequently they need not drink at all. During certain seasons of the year female marine mammals lose water through another channel, namely, when they are nursing their young. Interestingly, little water is lost this way because their milk is ten times as concentrated (fat) as is cow's milk. This is another remarkable adaptation to prevent water loss.

Marine birds, such as petrels, gulls, penguins, and albatrosses, never drink fresh water, so must also have an osmoregulatory problem. In spite of their ability to excrete uric acid, they still must counteract the high salt concentration of the sea water which they drink. This is accomplished by means of a special gland in the head which excretes salt at a concentration about twice that of sea water. Ducts from the glands lead into the nasal cavity, thus conveying the salt to the tip of the beak. It is a common sight to see gulls at rest dripping salt water from their beaks (Fig. 20–5).

Marine reptiles, such as the giant turtles, solve their salt problem much the same as birds. Their salt glands, however, are located near the eye and the salt is excreted in the form of tears. This becomes obvious when the females come out on land to lay their eggs. They are said to "weep" during this important activity.

Desert animals have a water problem of quite a different sort. Their problem is to get along on little or no water, other than that taken in with their food. They have evolved no new organs to solve their problem but have improved on their kidneys' capacity to concentrate salts and urea in the urine.

The kangaroo rat, a desert rodent, has been studied rather extensively in regard to its capacity to live on dry seeds without drinking at all. When fed on rolled oats in the laboratory it maintains a constant water balance for months without consuming

Fig. 20–5. Various methods by which animals maintain water balance when drinking sea water.

water. It has no sweat glands, being nocturnal, hence loses no water for heat regulation. It conserves water by producing very dry feces and, most important, by excreting highly concentrated urine, far more concentrated than any other mammal. Indeed, its urine is nearly five times as concentrated as man's, and has a salt concentration twice that of sea water.

The camel's reputation for going many days on the desert without water is well known. Just how does it do this when man and most other mammals would perish in a fraction of the time? It has a problem that the kangaroo rat did not have: namely, it works in the heat of the sun and therefore must control its body temperature by heat regulation through water evaporation from the skin. Whereas it has a large compartmentalized stomach, as other ruminants do, is has no extra "water sacks" for a reserve water supply as is sometimes thought. More-

over, the hump contains fat and no water. The camel then seems to have no special devices by which it can store water, and must then conserve what it takes in.

One interesting adaptation is the camel's wide body temperature variation. At night its body cools to about 34°C and during the day it will rise to 41°C before it begins to sweat to prevent a further rise. Man begins to sweat at slightly over 37°C and loses water continuously through this channel. It requires a good many hours in the hot sun before the camel loses water through sweating. Another means of conserving water is the camel's tolerance to dehydration. It can lose as much as 40 percent of its body water without ill effects whereas other mammals are in serious trouble if they lose 20 percent. When water is again available the camel can replace this loss in ten minutes. This animal has merely become extremely efficient in water conservation.

SUGGESTED SUPPLEMENTARY READINGS

Books

BALDWIN, E. B., *An Introduction to Comparative Biochemistry*, 4th ed. Cambridge, England: Cambridge University Press, 1964.

CARLSON, A. J., JOHNSON, V., and CAVERT, H. N., *The Machinery of the Body*. Chicago: University of Chicago Press, 1962.

FLOREY, E., *An Introduction to General and Comparative Animal Physiology*. Philadelphia: Saunders, 1966.

FULTON, J. F., *Selected Readings in the History of Physiology*. Springfield, Ill.; Charles C Thomas, 1930.

GABRIEL, M. L., and FOGEL, S., *Great Experiments in Biology*. Englewood Cliffs, N.J.: Prentice-Hall, 1955.

GIESE, A. C., *Cell Physiology*. Philadelphia: Saunders, 1962.

HOAR, W. S., *General and Comparative Physiology*. Englewood Cliffs, N.J.: Prentice-Hall, 1966.

PROSSER, E. L., *Comparative Animal Physiology*, 3rd ed. Philadelphia: Saunders, 1973.

ROMER, A. S., *The Vertebrate Body*. Philadelphia: Saunders, 1962.

WOODBURNE, R. T., *Essentials of Human Anatomy*. New York: Oxford University Press, 1961.

Articles

COMROE, J. H. Jr., "The Lung." *Scientific American*, 214, 56, February, 1966.

SMITH, H. W., "The Kidney." *Scientific American*, 188, January, 1953.

21

REPRODUCTION

In the preceding chapters we have considered in some detail the problems of structure and maintenance in the individual animal body. We now turn to the problem of reproduction of the individual. All animals reproduce, from the simplest protozoan to the most complex mammal and, furthermore, elaborate provisions are usually made for this important event. The methods employed by animals today must have had a long and interesting history.

EVOLUTION
OF THE REPRODUCTIVE SYSTEM

Undoubtedly, the first forms of life duplicated themselves by some sort of fission, perhaps much as many bacteria and protozoa do today. This is one form of **asexual** reproduction, so called because there is no sex involved; the cell simply divides into two parts, usually equal in size (Fig. 21–1). When animals became many-celled, some form of asexual reproduction was still retained by many of the lower forms. Hydra, for example, forms buds (Fig. 21–1) which develop into miniature hydras. Planaria, a more complicated animal, still employs fission. With increasing complexity of body structure in higher forms, this method of reproduction was lost entirely.

Sexual reproduction must have been introduced very early in the evolution of living things, because we find it well established even among the single-celled plants and animals. There is a series of single-celled green forms living today that shows a graded sequence from the union of cells that are similar in size and activity to those that are quite unlike in these respects (Fig. 21–1). At certain times, individual cells of *Chlamydomonas* fuse, forming a **zygote** which overwinters and continues its asexual method of reproduction the next spring. Since these uniting cells are equal in size, they are called **isogametes** and their union, **isogamy**. A closely related form, *Ulva*, undergoes a similar process, although in this case the cells are unequal in size. The smaller of the two is more active than the larger and in this respect resembles a sperm. Such - unlike gametes are called **anisogametes** and their union, **anisogamy**. Among the parasitic protozoa, such as plasmodium (see p. 000), two totally different cells are produced, a large immotile **egg** and a tiny active **sperm cell**. Here there is not only a size difference but also a physiological difference—one moves, the other does not. Once the sex cells reached this relationship in their evolution, they then maintained it throughout all higher forms. The smaller sperm cell is usually motile and is able to maintain sustained movement, while the egg is large and immotile. Such gametes are referred to as **heterogametes** and their fusion, **heterogamy**. After sexual reproduction became established it was retained, and while we see a rather wide range in sizes and shapes of both eggs and sperm, the fundamental plan remains unchanged in all animal groups.

The methods of bringing eggs and sperm together is relatively simple among both the lower invertebrates and the lower vertebrates. The union is purely fortuitous, although some arrangement, such as seasonal aggregation, usually occurs so that the animals are in the immediate vicinity of one another. As long as the animals remain in a fluid environment, all that is necessary is to discharge the sex cells into the water where by sheer chance they are brought in proximity to each other. However, when animals invaded the land the whole process became much more complex. It must have taken a long time to accomplish

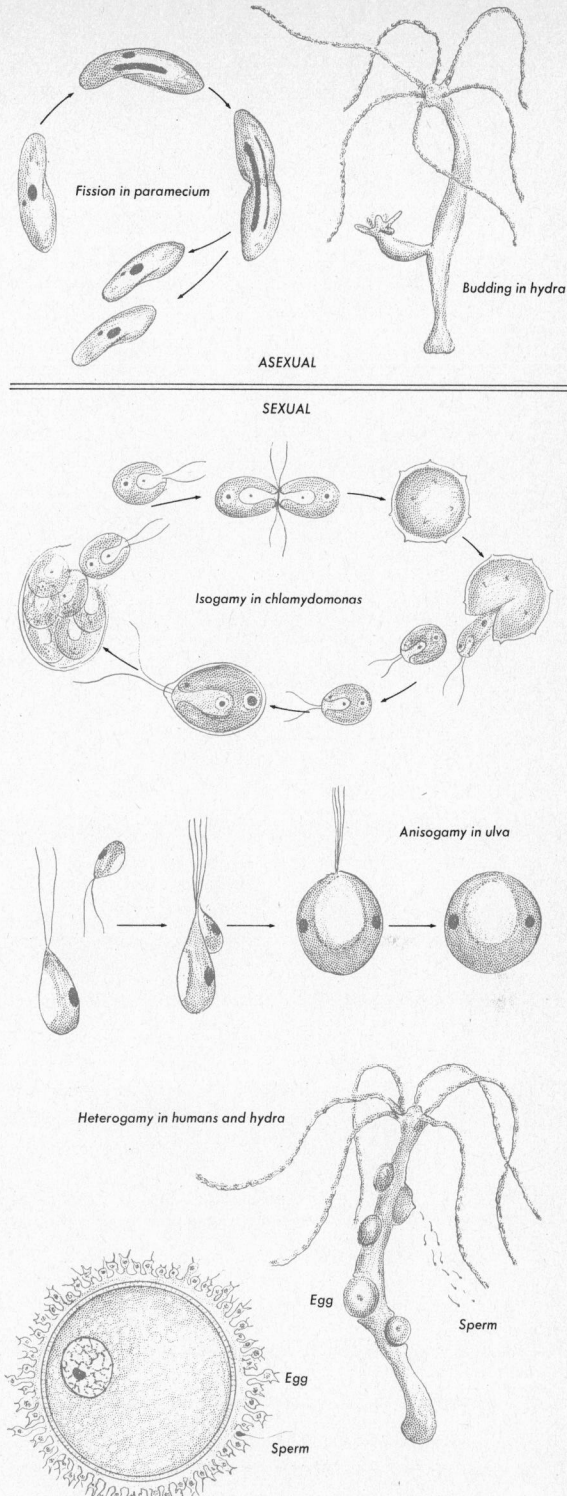

Fission in paramecium

Budding in hydra

ASEXUAL

SEXUAL

Isogamy in chlamydomonas

Anisogamy in ulva

Heterogamy in humans and hydra

Egg

Sperm

Egg

Sperm

Fig. 21–1. Asexual reproduction occurs among the protozoa and lower invertebrates. Sexual reproduction was initiated among some of the simpler protozoa where it progressed from simple fusion of two similar cells to the union of highly diverse reproductive cells, the sperm and egg. The steps which brought this about may have been similar to those outlined here, using animals that are living today.

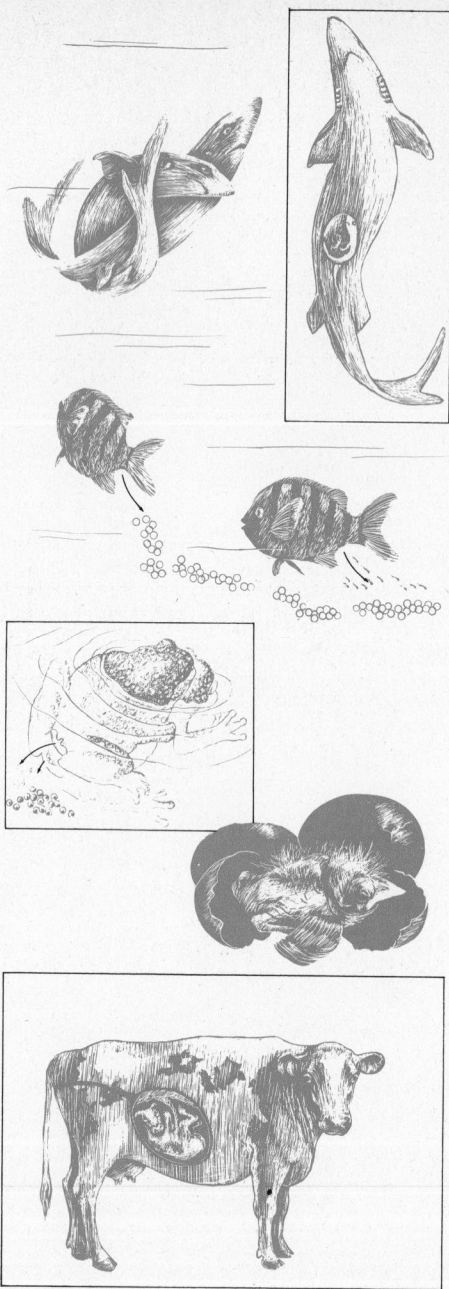

Fig. 21-2. Some vertebrates, such as the fish, the frog, and the bird, lay eggs (oviparous) which hatch outside the body. Others, such as the shark, retain the eggs within the uterus (ovoviviparous) where they hatch, and the young are born in an active condition. Still others, such as the mammals, produce young from small eggs (viviparous) and the young receive nutrients from the uterine wall.

this transition along with the many others resulting from the pronounced change of habitat. Perhaps the most interesting modifications came about among the vertebrates, some of which we have already discussed.

Care of the Young

Among the vertebrates there are three ways in which young are cared for in their early development (Fig. 21-2). Some are hatched from eggs that are laid, as in the case of most fishes, amphibia, many reptiles (Fig. 21-3, Fig. 21-4), and all birds. These are called **oviparous** forms. Others retain the eggs within the uterus until they hatch, and the resulting young are therefore born in a relatively advanced and active stage of development. These are said to be **ovoviviparous.** Some fish and some reptiles (snakes) are of this type. Still other vertebrates (mammals) produce small eggs without yolk that develop in the uterus, and the young receive most, if not all, of their nourishment from the uterine wall of the mother. These are said to be **viviparous.** Young born thus are, of course, more or less advanced in development.

In general, fishes and amphibians give their young little or no care whatever, and consequently no provisions in the way of accessory structures are found in these animals. However, when the reptiles moved onto land, certain anatomical modifications were essential if the young were to survive in a dry environment. For one thing, the egg became very large and abundantly supplied with reserve food for the developing embryo, thus providing a means for the embryo to reach a rather advanced stage before it had to shift for itself. Besides this food reserve, the egg had to supply a fluid environment in which the embryo could develop. In other words, a tiny bit of its ancestral aquatic world had to be incorporated into the egg. This was supplied by the introduction of the **amnion.** Moreover, as the embryo advanced in its development it required more oxygen than could be supplied by diffusion to the embryo directly. The development of the

Fig. 21–3. Hatching pythons (*Python molurus*). Note the wrinkled shell which is soft in reptiles.

Fig. 21–4. Special membranes have evolved in the eggs of vertebrates that have come out of the water and live on land (reptiles, birds, and mammals). The amnion and allantois make possible embryological development on land. Reptile and bird eggs contain sufficient food (yolk) to permit the embryo to develop to an advanced stage before hatching. The mammalian contains little or no yolk, hence the embryo receives its nourishment from the circulating blood of the mother. In addition to the allantois and amnion present in reptile and bird eggs, mammals have evolved a special organ, the placenta, which makes this possible.

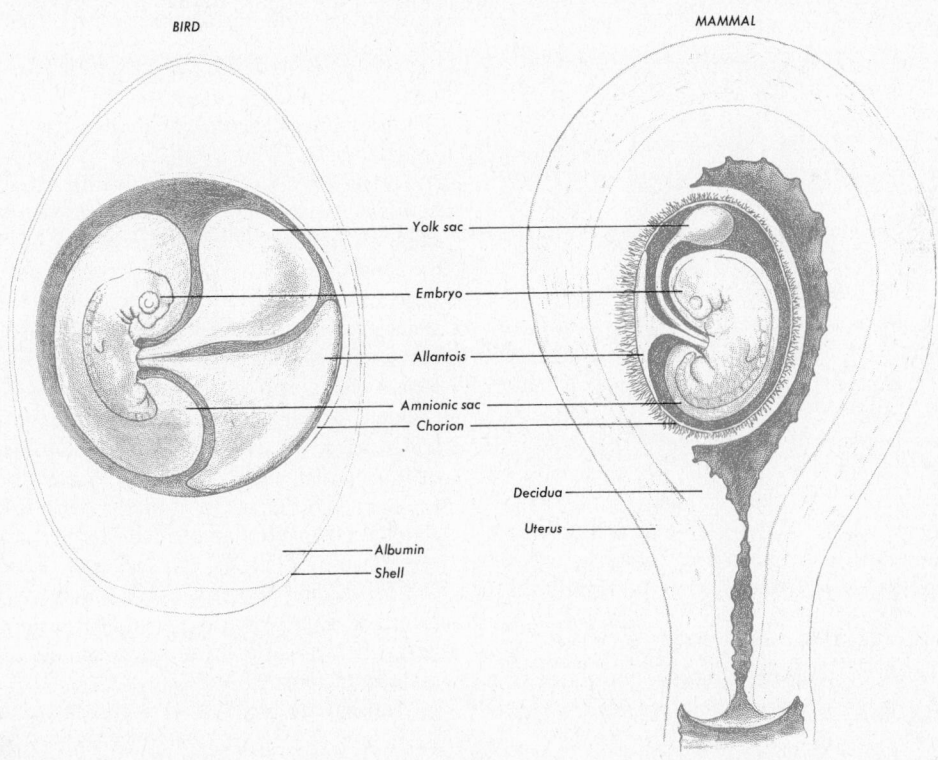

allantois met this need (Fig. 21–4). Large eggs, together with their extraembryonic membranes, served very well for the reptiles and birds. The former reached great heights as dominant worldwide animals during the Mesozoic Era and the latter are a dominant form today.

However, the keen swift-moving mammals that were to follow developed a new approach to this problem. When their embryo began to receive nourishment from the uterine wall via a special organ, the **placenta**, the large food reserve of the reptilian egg was no longer necessary, although the membranes of the latter were retained. Gradually the yolk disappeared but the yolk sac still remained (Fig. 21–4).

The **amnion** is an outfolding of the body wall of the embryo and is lined with peritoneum, the lining of the **coelomic cavity**. This membrane continues to grow around the embryo until the latter is completely enveloped and lies in the resulting **amniotic cavity**. The membranes fuse at their point of juncture, so that the cavity is a closed, fluid-filled sac reminiscent of the aquatic environment that was the home of all earlier embryos. Because the amnion is double-walled, an outer layer, called the **chorion**, is formed. In birds and reptiles the chorion comes in contact with the inner layer of the egg shell, whereas in mammals it comes in contact with the uterine wall and ultimately becomes a part of the placenta.

The **allantois** arises as an outpushing from the posterior gut and forces its way into the extraembryonic cavity to become a large sac-like structure, the outer layer of which fuses with the chorion. The resulting intimately fused membrane then lies in close proximity with the inner surface of the shell in birds and reptiles but in mammals eventually becomes a part of the placenta. The allantois functions as a respiratory organ in picking up oxygen and giving off carbon dioxide for the developing embryo; in addition, it absorbs food materials from the large egg (yolk) of birds and reptiles and acts as a repository for nitrogenous wastes.

In higher mammals the allantois, together with the chorion, comes into temporary contact with the uterine wall where the chorion sends out fingerlike projections, the **choirionic villi**, deep into the wall's soft tissues. This region of contact is richly supplied with capillaries from the **umbilical artery.** Embryonic blood is sent to this region under the impetus of the fetal heartbeat and returned to the embryo via the **umbilical vein.** Simultaneous with the development of the chorionic villi, the uterine wall in the same region becomes highly vascularized, forming many blood spaces into which the fingerlike villi dip. The tissues contributed by both the fetus and the uterine wall constitute the **placenta,** which is shed at birth. It must be remembered that there is no blood connection between the embryo and the mother; each has its own circulation which is kept distinct at all times. The placenta acts like the attachment organ of a fungal parasite— a means of extracting nourishing fluids from the "host." Through this organ oxygen is obtained, carbon dioxide is eliminated, food is absorbed, and nitrogenous wastes (urea) are discharged.

Evolution of External Structures

We usually think of the need for external genitalia to facilitate the union of the sex cells as directly related to the change to life on land. To be sure, such accessory structures were essential for land life, but we must not forget that copulatory organs did evolve among fishes that never left the water. The sharks again are a striking example. The **claspers,** which are modified pelvic fins in the male shark, are utilized as an intromittent organ for carrying the sperm to the cloaca of the female (Fig. 21–2). Many bony fishes also have converted their fins into a copulatory organ. These are unusual cases among the fishes, and most of them have so little external sexual dimorphism that even the best ichthyologists have difficulty in telling male from female.

Among the strict land vertebrates—reptiles, birds, and mammals—a special copulatory organ, the **penis,** has evolved. Among the reptiles it is a pair of elongated masses of erectile tissue with a groove between, the entire structure originating from the floor of the cloaca. In mammals the groove is closed to form a tube which is a direct continuation of the urethra. The terminal portion of the female tract has followed a parallel evolution by becoming transformed into a tubelike receptacle, the **vagina,** for the penis and seminal fluid.

Along with these changes has come the introduction of a large complex of chemical regulators (hormones) and certain nervous modifications which has made copulation not only a necessity for the continuance of the race but a highly gratifying experience. During the breeding period of large mammals such as cattle, both sexes will go to great lengths, even exposure to death itself, in order to tend to the business of bringing about the union of their sex cells. Owners of female dogs are well aware of the semiannual heat cycle of their charges. It is easy to understand how those animals with the greatest sex drive were more apt to become the parents of the next generation, whereas those indifferent in this regard might never have become parents at all, so that the intensity of the sex drive in animals is actually a contributing factor to their survival.

Evolutionary Significance

Viewing the vertebrates in a general way, it is true that giving birth to active young is more advanced than egg laying. There are, however, a number of very interesting primitive vertebrates, the sharks for example, that have progressed a long way toward producing and caring for young in much the same way mammals do. Some sharks inhabit the deep sea and never come into shallow waters to deposit their eggs in safe places. In such species, the eggs are retained in the uteri where they can be better cared for. In one small shark (*Mustelus*) tiny projections, like villi in the intestine, protrude from the

yolk sac and penetrate the uterine wall from which the embryo derives nutritious secretions (Pl. 16B). This condition certainly approaches the placenta of mammals; indeed, these sharks are called "placental" sharks. Another interesting device for extra-egg nourishment of the developing embryo is found among the rays, close relatives of sharks. In some of these animals, glandular teats grow out from the inner uterine wall and by contractions force their secretion into the mouth or spiracle of the embryo, thus resembling extrauterine feeding of mammals. In spite of these rare cases, it is generally true that evolution of the reproductive system among vertebrates has been from the egg-laying forms to those that retain and nourish their young within the uterus of the mother.

We can make one further generalization, constantly keeping in mind the occasional exception which prevents one from drawing final sweeping conclusions. Most primitive animals make headway by sheer weight of numbers; millions are produced but only a small fraction of a percent survive to maturity. As the animals become more complex, fewer and fewer offspring are produced but the early mortality rate drops also. With smaller numbers greater parental care is given, so that the percentage of survival is much greater (Fig. 21–5). The fish deposits its eggs in a scooped-out hollow on the stream bottom and immediately leaves them to the ravages of other fish. A large mammal like a cow bears young only once a year but gives the offspring great care, even to the point of sacrificing her own life. In each case the chance for race survival may be the same. Immediately we can think of exceptions to this generalization. The codfish, for example, lays millions of eggs each season, whereas the shark may have no more than a dozen in an equal period of time, yet the shark is more primitive than the cod. Frogs and toads give their young no care whatever, but the bluegill (fish) protects its nest of eggs viciously against all intruders. These are good examples to remind us of the caution one must exercise in stating a generalization.

Fig. 21–5. A mother flying squirrel (*Glaucomys volans*) carrying her young to a new home. Mammals generally give exceedingly good care to their young.

THE HUMAN REPRODUCTIVE SYSTEM

Although there are minor differences in the reproductive apparatus among various mammals, they are all essentially alike, man included. One rather striking variation has undoubtedly had considerable influence on man's habits and indeed his entire social structure; that is the lack of a seasonal or periodic breeding season. The sexes are mutually attractive throughout the year. This fact probably has had some bearing on the establishment of permanent unions that could provide for offspring which required such long periods of time for growth to maturity. This is the basis of the family which, in turn, underlies our whole social order as we know it in civilized nations today. It should not be implied, however, that man is by nature fitted perfectly into the straitjacket of civilization he has fashioned for himself. In fact, he is constantly in trouble with those individuals who tend to break with established convention and to follow basic instinctive urges. This applies to sex urges as well as others. It is important that we recognize these basic urges as normal and regulate our social order accordingly. This attitude has not always been maintained in the past, but as knowledge grows concerning human behavior we ought to reach a better understanding of what curbs can be placed on basic instincts that benefit society and thus make for a more successful civilization.

Male

The sperm-producing glands are the **testes,** which are very similar in their anatomy to the same organ in other mammals. Sperms are produced by rapidly dividing cells located in the walls of the long, tiny, coiled tubules (seminiferous) which, taken together, make up most of the gland itself (Fig. 21–6). The paired testes lie in a pendent sac, the **scrotum,** which is located in the pubic region. Sperm cells make their way out of the testes into the **epididymis,** a long convoluted tubule, where they mature and are stored for a time. They then pass into the **vas deferens,** a thickened sperm duct that passes anteriorily through the **inguinal canal,** finally entering the **urethra** and thence through the **penis** to the outside. Near the region where the two vasa deferentia enter the urethra each receives secretions from the glandular **seminal vesicles.** Also at this junction the **prostate gland** surrounds the urethra and pours its secretions into it. Two small glands, **Cowper's glands,** located at the base of the penis, secrete a mucoid fluid which, together with secretions from the prostate gland and seminal vesicles, constitute the transport medium for the sperm cells.

The penis is composed of a special tissue with large capillaries called **erectile tissue** which, when gorged with blood, causes the organ to become much enlarged and turgid. In such condition it becomes a satisfactory organ for transferring sperms to the vagina of the female. Erection is accomplished when, under sexual excitement, the small arteries leading to the penis relax, allowing blood to fill the large capillaries; at the same time the veins constrict where they leave the organ, thus trapping blood within it. The organ becomes flaccid by a reversing of the process. The penis ends in a caplike structure, the **glans,** which is partially covered by loose skin, the **prepuce.**

Because of the anatomical arrangement of the male generative organs, certain difficulties may be experienced, especially by older men. For example, if the prostate becomes enlarged, which it not infrequently does, elimination of urine becomes difficult because the urethra is squeezed shut. Another difficulty arises from the fact that when the testes descend through the inguinal canal into the scrotum some weeks before birth, a weakened area is left at the point where the spermatic cord perforates the abdominal wall. If, at some later time, sufficient pressure is brought to bear on this area, a small segment of the gut may be forced into this canal resulting in an inguinal **hernia.** This sometimes occurs during birth when the child is apt to be squeezed unduly while passing through the small birth canal. If further constriction occurs the circulation to the gut may become so impaired as to kill the region, thus resulting in gangrene which can be very serious.

Female

The egg-producing **ovaries** lie deep in the lower body cavity where they function much as they do in lower vertebrates (Fig. 21–7). Their function is to produce eggs, and since the chances for fertilization in mammals are very good only a comparatively small number are actually generated. The cells destined to become eggs lie in the periphery of the ovary and as they mature migrate into the deeper portions (Fig. 21–8). Several potential egg cells mature in a cluster, only one of which exceeds all others in size and eventually matures into an egg. The rest becomes folli-

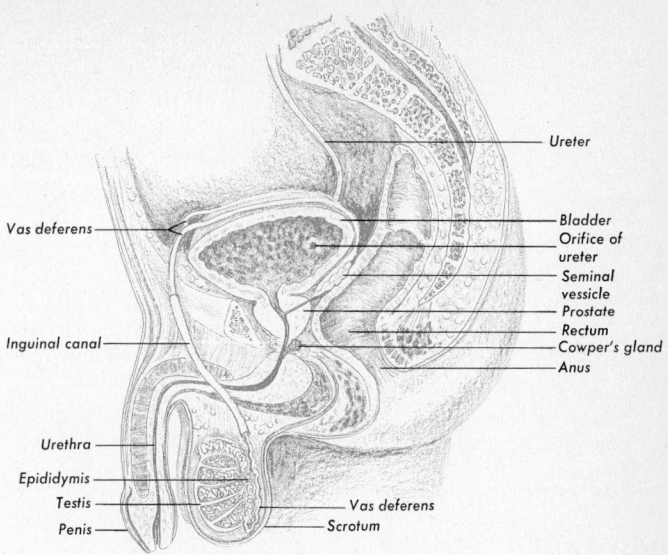

Fig. 21–6. The human male reproductive system.

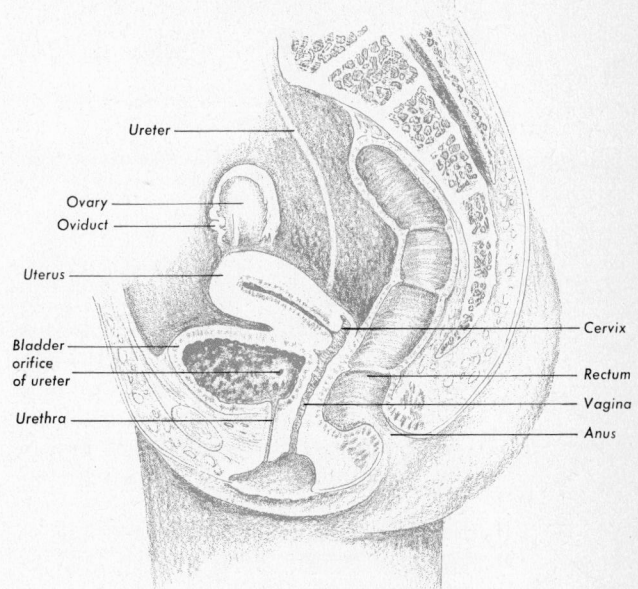

Fig. 21–7. The human female reproductive system.

cular epithelial cells which surround the developing egg. Several layers of follicular cells form around each egg at first; later a liquid-filled cavity appears, in which the egg floats, being attached by a thin stalk to the

451

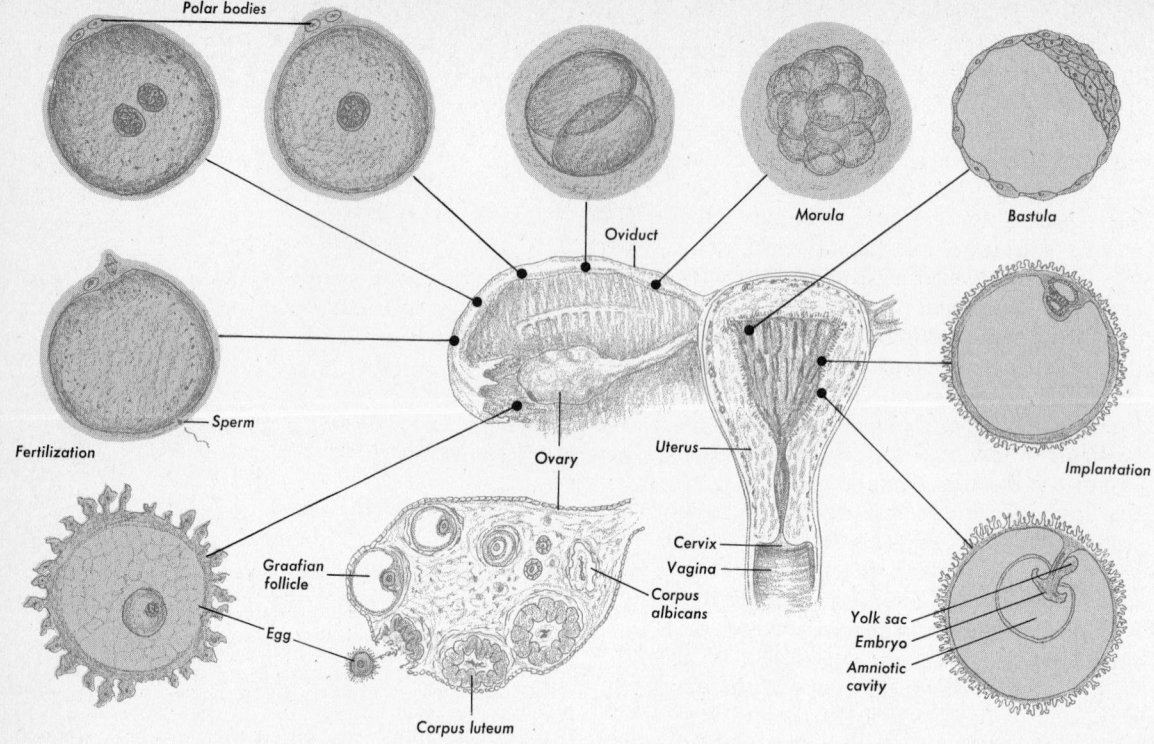

Polar bodies

Oviduct

Morula

Bastula

Sperm

Fertilization

Ovary

Uterus

Implantation

Graafian
follicle

Cervix

Vagina

Corpus
albicans

Yolk sac

Embryo

Amniotic
cavity

Egg

Corpus luteum

Fig. 21–8. The human ovary and uterus cut in such a way that the pathway of the fertilized egg can be followed. Note that fertilization occurs in the upper end of the oviduct. The stages of early development of the embryo from fertilization to implantation are shown enlarged.

inner wall. The entire structure is now known as a **Graafian follicle** (p. 000). When mature it is over 1 centimeter across and is located near the periphery of the ovary once more. The region nearest the outer edge of the ovary thins out and finally splits, allowing the follicular contents to be extruded into the body cavity, a process known as **ovulation.** The tiny egg (about 140 microns across) is carried out with the fluid and is subsequently drawn into the funnel-like terminal end of the oviduct (**Fallopian tube**), which is lined with beating cilia that create a current directed into the oviduct. Occasionally the egg gets "lost" in the coelom and never reaches the oviduct. If, however, it should be fertilized while outside the confines of the oviduct, which sometimes happens, the embryo may become attached

to any convenient organ, including the ovary itself, and develop for some time. It usually is unable to go to full term, however, and aborts, causing severe internal hemorrhages that may be fatal if immediate care is not given.

Fertilization

The union of the egg with a sperm must occur at a rather specific time because neither is long-lived. Fertilization is most apt to occur if sperm are in the vicinity of the egg within a few hours after ovulation, and it usually does not occur if more than three days elapse before the union is possible. Ovulation is definitely timed with respect to the menstrual cycle, usually occurring about the fourteenth day following the onset of menstruation. There may be rather wide variations in unusual cases, but as a rule a day or two before or after the fourteenth day finds the egg in the oviduct where it may be fertilized if sperm are in the vicinity. Sperm deposited in the upper end of the vagina make their way through the **cervix** (Fig. 21–8) and **uterus** into the oviducts in a matter of several hours. It is interesting to note that it seems to be essential that millions of sperm be supplied at once in order that only one reach its destination, even though only one is necessary for fertilization. Apparently the cooperative effort of many is needed to digest their way through the external layers of the egg. The sperm cells move at random in the female genital tract; they are not attracted in any one direction. They reach the egg primarily as a result of the activity of the genital tract itself.

Once the egg is fertilized, it divides several times en route down the oviduct, and by the time it reaches the uterus it is a ball of cells (Fig. 21–8) and is ready for **implantation,** that is, to become attached to the uterine wall. During the time these things have been happening to the egg, the uterine wall has been preparing itself for the reception of the young embryo. When the embryo reaches the uterus, it is actually drawn into the

Fig. 21–9. A human embryo (about six weeks old) floating in the amnionic cavity. The spherical-shaped body in the lower left is the yolk sac. The feathery nature of the chorion is shown. The part nearest the viewer has been cut away.

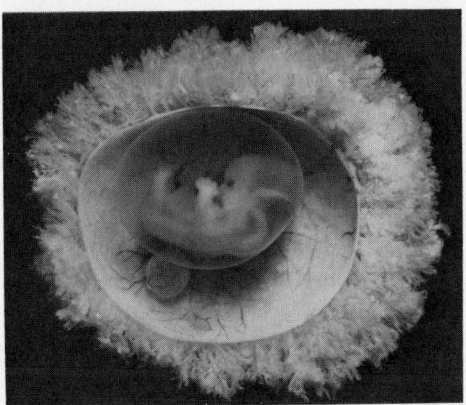

uterine wall because of the receptive nature of the lining itself. Once implanted, the embryo grows rapidly (Fig. 21–9), starting as a tiny reddened spot and finally, after approximately 40 weeks, becoming a full-term **fetus** (Fig. 21–10). This entire process is intricately linked with a complex battery of hormones.

During intrauterine life the embryo must develop a complete set of organs—lungs, digestive system, excretory system—and all except the circulatory system function only slightly but must be ready to carry on full activity within a matter of minutes after birth. All of its nourishment and oxygen have come to the embryo through the umbilical vein from the placenta, and all of its waste products (carbon dioxide and urea) have been disposed of through the same organ. Its lungs have never breathed, its kidneys and digestive tract have functioned but little, yet within a few minutes after it is born all of these organs perform their jobs to full capacity and, almost without exception, they function perfectly from the very start. The most dramatic change occurs in the circulation.

The blood path through the fetus is quite different from that in the adult (Fig. 21–10). Obviously, circulation of blood must start very early in the developing fetus because it is only through the circulation that nourishment and the elimination of wastes can take place. The blood must be pumped to and from the placenta at a rapid rate during **development;** hence the circulatory mechanism is one of the first organs that is well developed in the fetus. The beating fetal heart is easily detected with a stethoscope long before the child is born. Since there is no point in sending blood through the nonfunctional lungs, the heart has modified the main blood flow through itself, and functions essentially as a single organ rather than the double one of the adult. This temporary detour has meant that the fetal heart must be constructed on a slightly different plan than the adult heart. There is an opening between the auricles, the **foramen ovale,** which allows incoming blood from the great

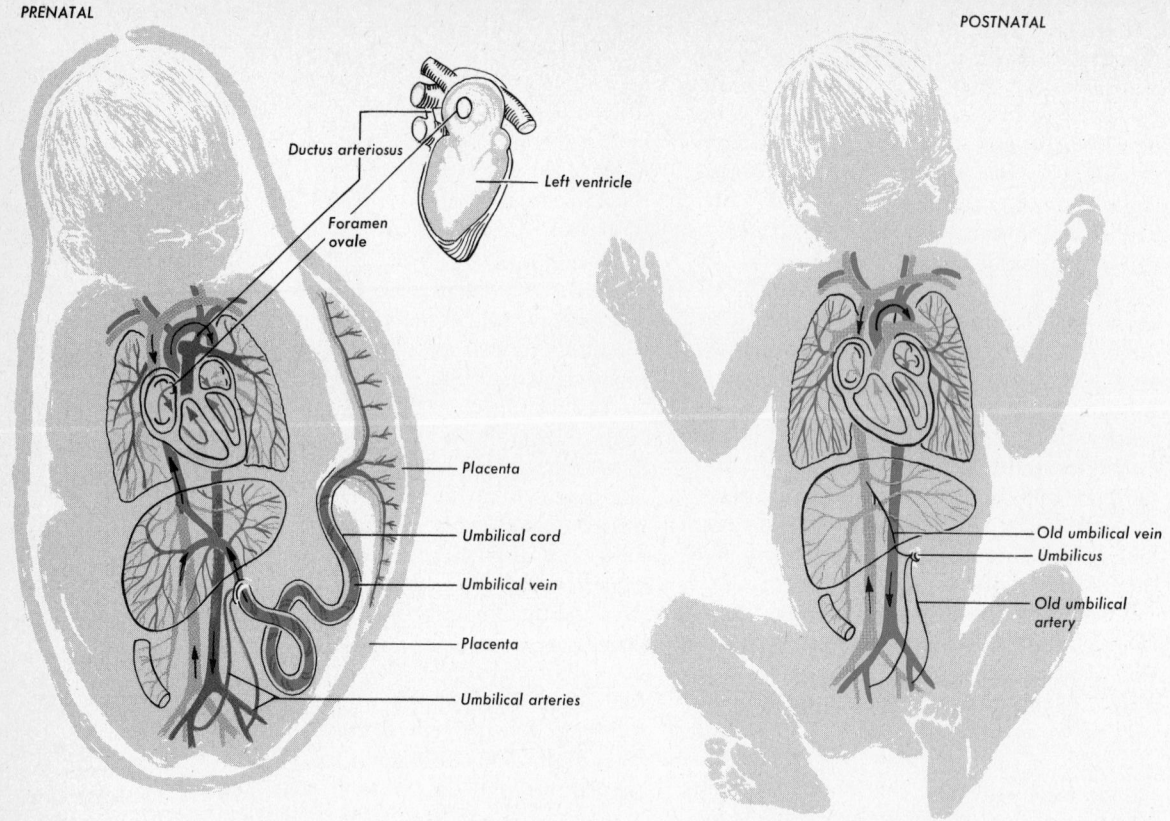

Fig. 21–10. The blood pathway is quite different during fetal life than it is after birth, as indicated in these sketches.

veins to fill both auricles simultaneously. When the blood is forced out from the ventricles, that coming from the right ventricle, which in the adult goes to the lungs, is short-circuited through a large, short connection, the **ductus arteriosus,** to the aorta. Very little blood goes to the lungs, so very little comes back to the left auricle. Almost the entire output from both ventricles therefore passes out into the general circulation. This is necessary because the blood must be forced through the paired **umbilical arteries** to the placenta, which is a highly vascular organ requiring large quantities of blood. The blood flows in this path throughout fetal life, while the organ systems are being developed so that at the moment birth occurs they function properly.

The basic structure of the human embryo is well delineated by the end of the twelfth week of intrauterine life, although there is a great deal of growth that must take place before the child is ready for the rigorous life that awaits him in the outside world. The uterine wall keeps pace with the growing embryo and the added burden gradually changes the entire body contour of the mother (Fig. 21–11). As the fetus grows it slips posteriorly, with its head low in the pelvis, and by the fortieth week of its life it is ready to be born.

Birth

The process of birth is one of the more remarkable events in the biological world

(Fig. 21–12). As the fetus passes through the tight-fitting birth canal, the umbilical cord remains attached to the placenta, and until the moment the child takes his first breath, nourishment and oxygen are received from this source. Upon exposure to the external world, stimulation, coming from being chilled and from the high CO_2 in his blood, causes the child to take his first breath. The partial vacuum created in the chest cavity causes the blood to be diverted from the ductus arteriosus into the pulmonary artery and lungs. This is aided by a powerful sphincter muscle surrounding the ductus arteriosus which contracts at this moment and never relaxes. The large volume of blood then makes its way back to the heart through the pulmonary veins, filling the left auricle. Since this chamber is now filled from a new source, it is no longer necessary for an opening to exist between the auricles, hence a flap of tissue closes the foramen ovale. It requires a few minutes for this adjustment to take place, during which time the child is bluish in color, but gradually he becomes pink as the circulation is separated into its pulmonary and systemic parts. It is indeed amazing that with all its complications this process so seldom fails.

However, if the foramen ovale does fail to close or something else goes wrong with the mechanism for separating the blood, the baby remains bluish in color, a condition given the term **blue baby**. He is bluish in color (cyanotic) because the oxygenated and reduced blood are mixed, just as they are during fetal life. Some remarkable heart and vascular surgery has been accomplished in

Fig. 21–11. Changes occur in both the uterus and the mother to accommodate the developing fetus.

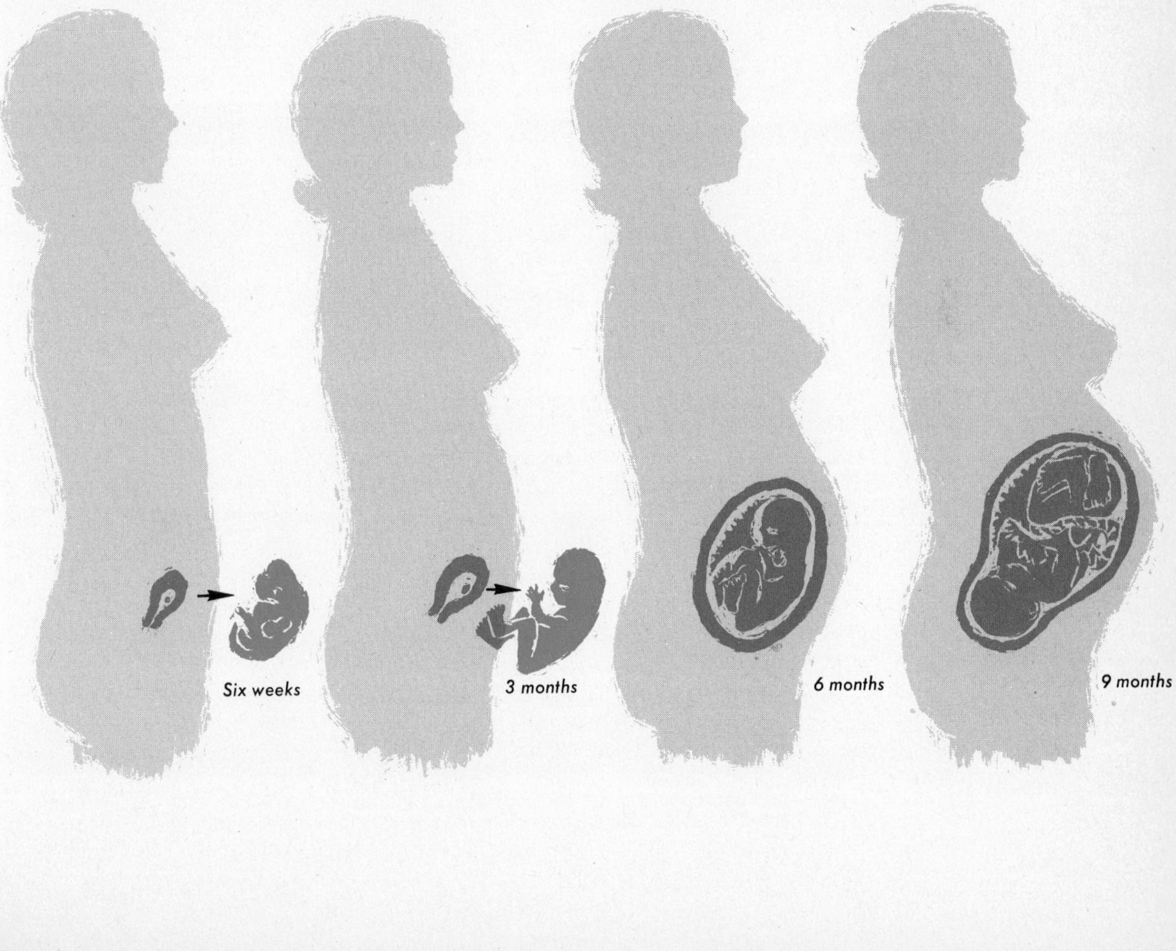

Six weeks   3 months   6 months   9 months

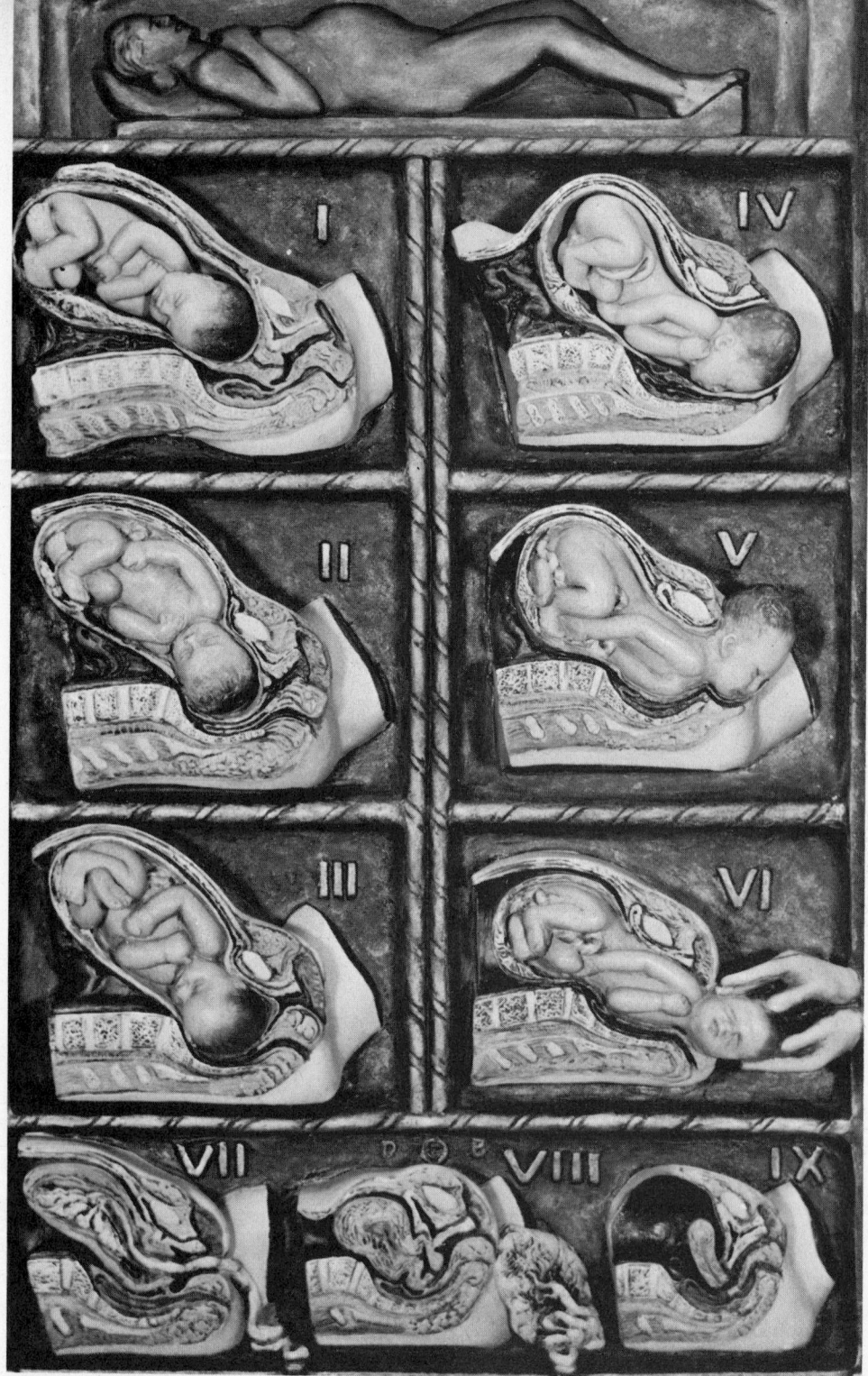

Fig. 21-12. A series of models showing the various stages in the delivery of a baby. The placenta remains attached to the uterine wall for some time after the child is born. It is then shed and the uterus returns to its normal size.

recent years to correct such occasional defects.

We have now completed discussion of the organ systems of man. Now let us turn our attention to the development of an individual from its very beginning and some of the problems involved in this very complex though fascinating story.

SUGGESTED SUPPLEMENTARY READINGS

Books

BULLOUGH, W. S., *Vertebrate Reproductive Cycles*. New York: Wiley, 1961.

*CORNER, G. W., *The Hormones in Human Reproduction*. Princeton University Press, N.J.: Princeton, 1942.

HOAR, W. S., *General and Comparative Physiology*. Englewood Cliffs, N.J.: Prentice-Hall, 1966.

VILLEE, C. A., ed., *The Control of Ovulation*. London: Pergamon Press, 1961.

WOODBURNE, R. T., *Essentials of Human Anatomy*. New York: Oxford University Press, 1961.

* Available in paperback.

Articles

GRAY, G. W., "Human Growth." *Scientific American*, October, 1953.

PINCUS, G., "Fertilization in Mammals." *Scientific American*, March, 1951.

TYLER, A., "Fertilization and Antibodies." *Scientific American*, May, 1955.

ZAHL, P. A., "The Evolution of Sex." *Scientific American*, April, 1949.

22

DEVELOPMENT OF THE EMBRYO

When the sex cells unite to form a zygote which then undergoes millions of mitoses directed specifically toward the formation of a new organism, the development of the embryo has begun. **Embryology** is one of the most fascinating, and yet in its fundamental aspects one of the least understood, of any of the fields of biology. To realize the magnitude of the problem, consider how such a tiny object as a fertilized egg can carry within itself the potentialities of producing a living organism as complex as man.

The question has stimulated speculation among thinking men for many centuries. Aristotle, for example, believed that the male element, or semen, was the seed that gave rise to the new individual. Indeed, the word **semen** means seed. The female was thought of as the earth (Mother Earth) in which the seed was nourished so that it might grow. This was a natural deduction, since it was known that castrated animals

were sterile; without semen there was no offspring. Furthermore, this was during the time when the male sex was considered all important, the female being passive in her social life as well as her biological life. The appearance of characteristics in the offspring was apparently ignored. Moreover, the finding of the placenta attached to the embryo by the umbilical cord furthered the concept that the female functioned only in nourishing the fetus. The idea that both sexes contributed to the offspring came much later.

With the actual discovery of the human sperm by Ludwig Hamm in 1677 the importance of the male retained its place. This was further emphasised by the work of Hartsoeker, who not only claimed to have seen the human sperm first but who also drew a miniature fetus within the head of the spermatozoan and thus advanced the idea of the **homunculus** (Fig. 22–1). This led to the establishment of the **preformation** school which included those who believed that the human embryo was completely formed within the sperm and merely unfolded during its development. Impetus was given to this idea by Malpighi (1672), who concluded that the chick was fully formed, though in miniature, at the time the egg was laid. This was probably due to an error that could easily arise from a study of an egg that had remained several hours (10–15) within the oviduct where development of the chick could proceed, and which might slowly continue development in the warm Italian climate even though unincubated. However, because of the obvious existence of large eggs, such as those of birds and reptiles, there developed another school which believed that the egg was the center of development. This meant there were two schools of thought among preformationists, the **spermists** and the **ovists.**

An interesting aspect of this controversy was the development of the **encasement** theory. One group of followers believed that the child was derived completely from the egg and that the egg contained a minute human being within it. If the egg in ques-

tion contained a diminutive woman, within the eggs of her own ovary were still more minute women and so on, each encased in the preceding. From this point of view the first woman contained all of the future generations wrapped one within the other and the race would come to an end when all of the encased homunculi were exhausted. The absurdity of this theory became apparent with a few mathematical calculations such as Hartsoeker made on the size that an original female must have been to house all future generations. The theory was finally laid to rest with the discovery of the mammalian egg by Von Baer in 1827.

Not all intellectuals of these early times believed in preformation, for some thought that the adult parts were not present at the beginning but developed during embryonic life, that is by **epigenesis.** Actually Aristotle believed this, as did Harvey some two thousand years later. In 1759 the German biologist Wolff published his careful observations on the developing chick, which dealt a severe blow to the theory of preformation. He showed that adult structures do actually make their gradual appearance throughout embryonic life, and his observations have been confirmed many times since; hence epigenesis is the view held today by biologists. As we see a little later, this does not mean that the egg is completely without organization; but the idea of a fully formed miniature adult is no longer held by anyone.

The various theories of preformation seem strange to us now, but when they were advanced they performed a necessary function in that they stimulated investigators to search more intensively for an explanation of development.

Fig. 22–1. One of the early schools of embryology was that of the spermists, who believed that the child was preformed in the sperm and merely unfolded within the confines of the female uterus. Among followers of this concept was Hartsoeker, who published a drawing in 1678 similar to the one shown here, and advanced the idea of the homunculus. The human sperm as it appears under a modern microscope is shown for comparison.

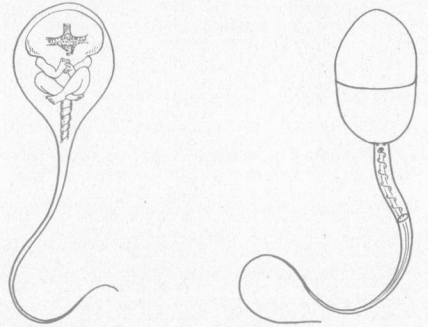

THE PATH OF DEVELOPMENT

It might seem more meaningful to select human development for our study, but because of the obvious difficulties in obtaining suitable material we must resort to lower animals. Because the development is relatively simple and yet typical, we select the eggs of *Amphioxus,* a primitive chordate (p. 248) for our study. Eggs of starfish and sea urchins also provide excellent material and are frequently used for this study.

Cleavage

In some animals the sperm cell enters the egg before the latter is completely mature, that is, before the formation of the second polar body (*Amphioxus* and vertebrates— (Fig. 22–2) or even before the formation of the first polar body (starfish). The fertilizing sperm must penetrate the membranes surrounding the egg before fusing with the egg surface. After being drawn inside the egg, the sperm nucleus enlarges to become the male pronucleus and then lies at rest in the cytoplasm while the egg completes its last maturation division. After the egg pronucleus has formed the two pronuclei unite and the first division of the zygote follows shortly. In frogs the first cleavage occurs in about 3 hours. The split divides the zygote into two equal parts. Cleavage divisions follow in rapid succession, resulting in a hollow ball of cells, the **blastula** (Fig. 22–2), the total being no larger than the original zygote.

Formation of Germ Layers

The ball of cells indents on one side until the two layers of cells almost touch, giving the appearance, somewhat, of a rubber ball which has been pushed in with the finger.

Fig. 22–2. Fertilization and early embryology of *Amphioxus* is shown in this series of figures.

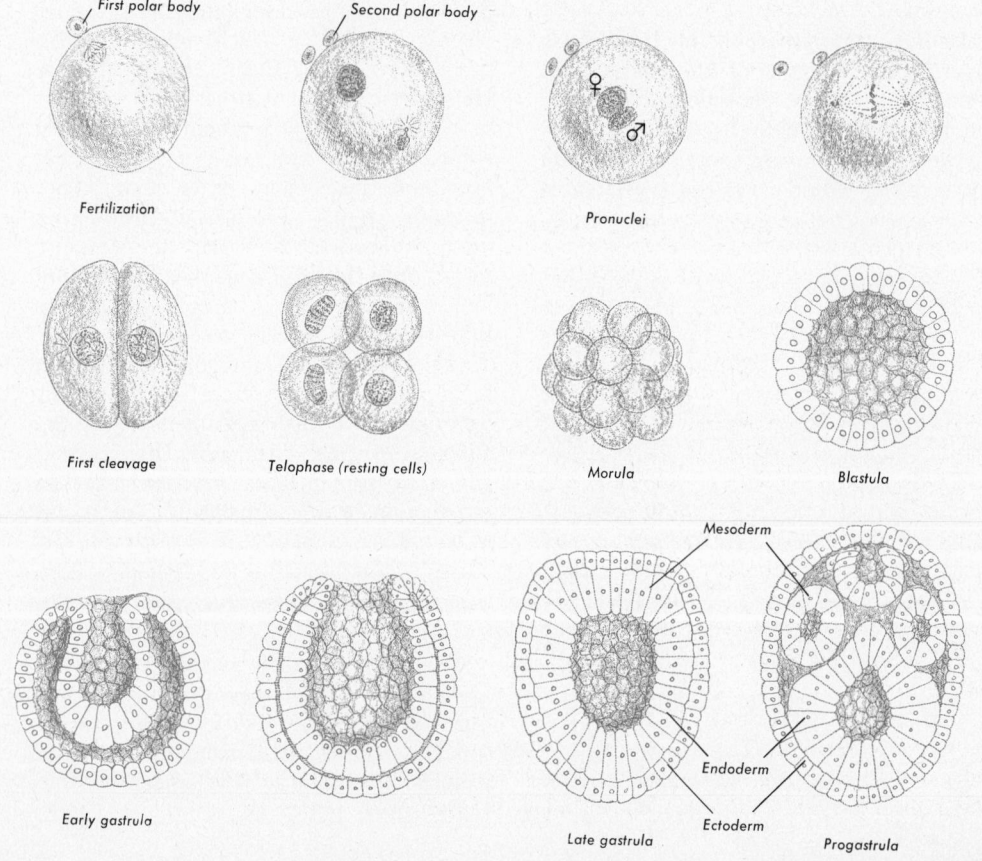

This is the **gastrula,** and the single opening formed at one end is the **blastopore.** The newly formed cavity is called the **archenteron,** which eventually becomes a part of the digestive tract of the adult animal. The two layers of cells that make up the gastrula are called **germ layers;** the one on the outside is the **ectoderm** and the inner one, the **endoderm.** Presently, tiny pouches push out from the archenteron and pinch off, forming a double layer of cells, the **mesoderm,** with a cavity, the **coelom** (enterocoel, p. 115), between (Fig. 22–2). The coelomic pouches fuse, and the coelom thus formed finally becomes the body cavity in the adult. In higher vertebrates such as mammals the coelom gives rise to the pericardial, pleural, and peritoneal cavities of the adult.

The three germ layers are laid down in somewhat different ways in various animal groups, owing primarily to the presence of yolk in the egg. In the case of *Amphioxus,* where there is very little yolk, development proceeds in the manner described. However, when the egg contains a great deal of yolk, as in the case of the bird's egg, the formation of the three germ layers is modified, although the end result is essentially the same. In the chick, which is best known and most frequently studied (Fig. 22–3), the embryo forms on top of the yolk mass. The two-layered embryo or gastrula is produced by a splitting of the single-layered embryo, and the mesoderm subsequently appears between the ectoderm and endoderm, spreading laterally from a central axis called the **primitive streak.** The embryo becomes organized anterior to the primitive streak. Segmental blocks of mesoderm, the **somites,** appear first in the anterior trunk region of the embryo and continue posteriorly as growth occurs. By the 33rd hour of incubation the notochord and three divisions of the brain can be seen, and the heart is laid down. From these beginnings the embryo body is formed.

With the formation of these three germ layers, a complex metazoan body can be built. The layers grow, fold, and differentiate into the adult structures. In general,

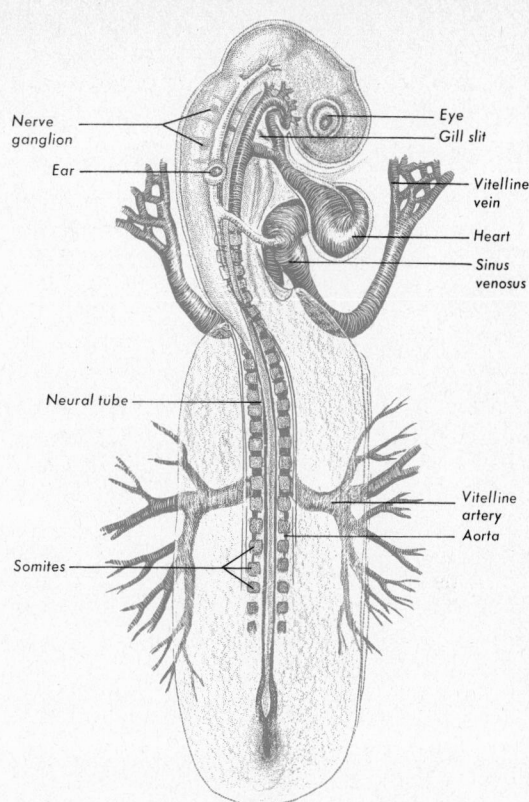

Fig. 22–3. A chick embryo at 48 hours of incubation. Note the developing eye, ear, heart, and other parts. The large blood vessels carry food-laden blood to and from the yolk for growth and differentiation of the embryo.

the endoderm becomes the epithelial lining of most of the digestive tract and its derivatives. The ectoderm gives rise to the outer layer of the skin, the nervous system, and the sensory cells of the adult. The two layers of the mesoderm in the trunk region form different parts: the inner portion becomes associated with the endoderm to form the gut wall, while the outer becomes associated with the ectoderm to give rise to the body wall, thus producing the familiar "tube-within-a-tube" body plan discussed in earlier chapters. Between the two tubes is the cavity of the mesoderm, the coelom. Derivatives of these layers in man are portrayed in Figure 22–4.

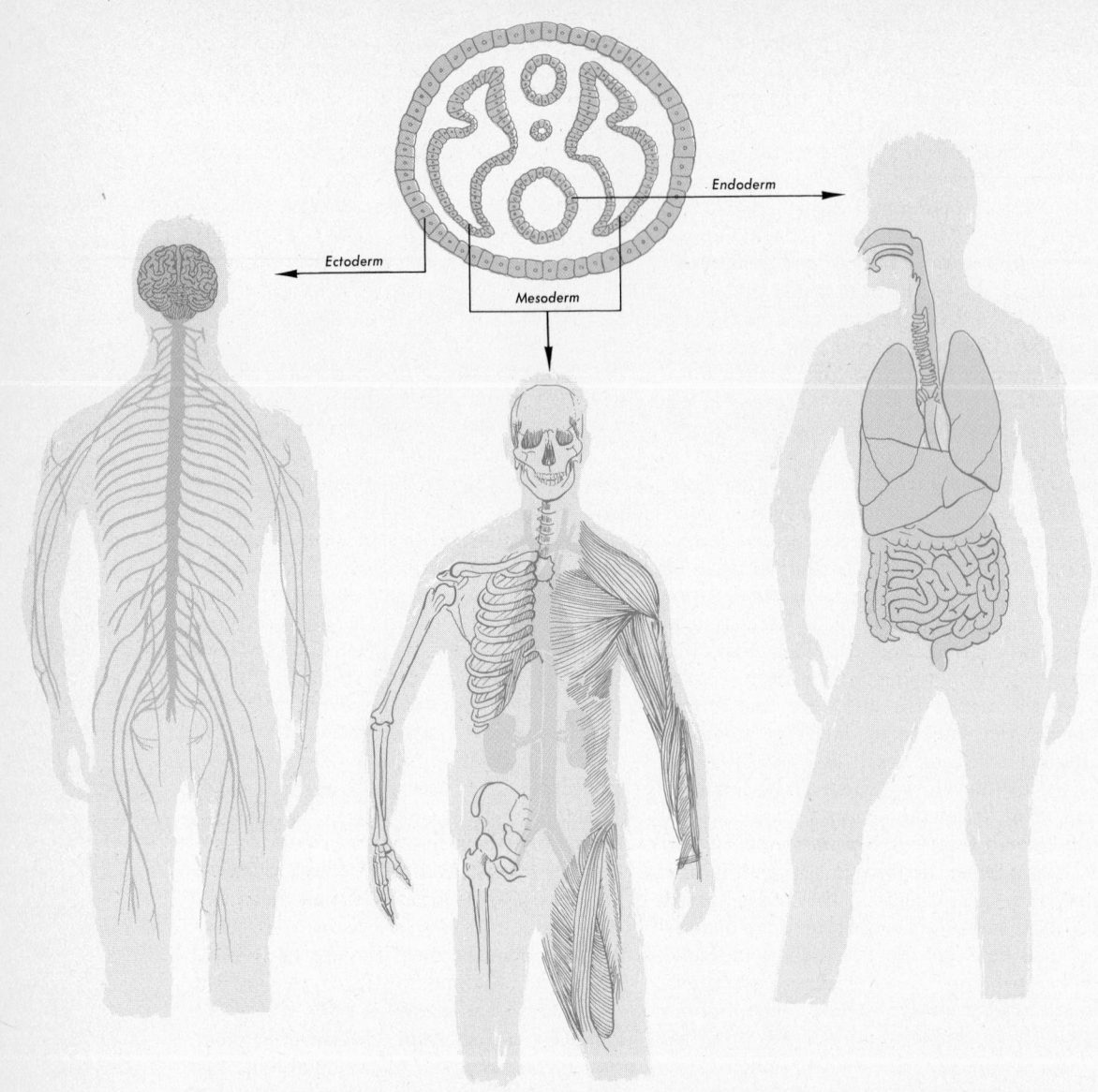

Ectoderm

Endoderm

Mesoderm

Fig. 22–4. Derivatives of the three germ layers in man are shown.

Organ Formation

It is a long, complicated story from the simple germ layers to the full-fledged functional organs which, taken together, constitute the complete animal. The formation of the functional organs is accomplished by several methods. One of the most common is the folding of the layers, which is well illustrated in the formation of the nervous system of *Amphioxus* and of the chordates in general (Fig. 22–5). It starts by a pushing in of a groove along the dorsal side of the embryo about the time the mesoderm is

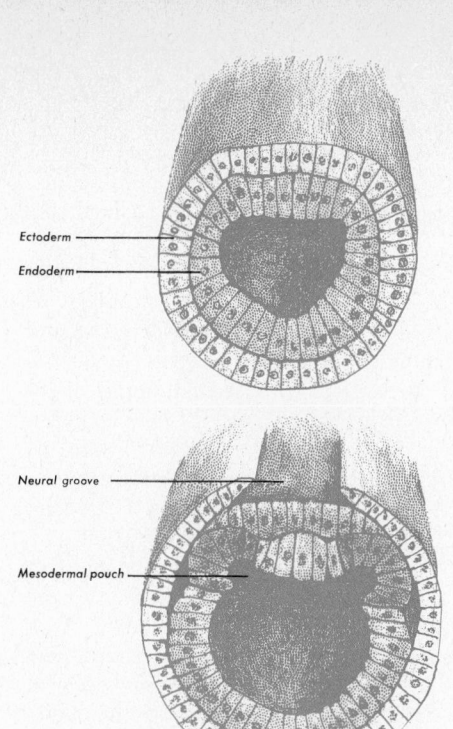

Ectoderm

Endoderm

Neural groove

Mesodermal pouch

Nerve cord

Coelom

Notochord

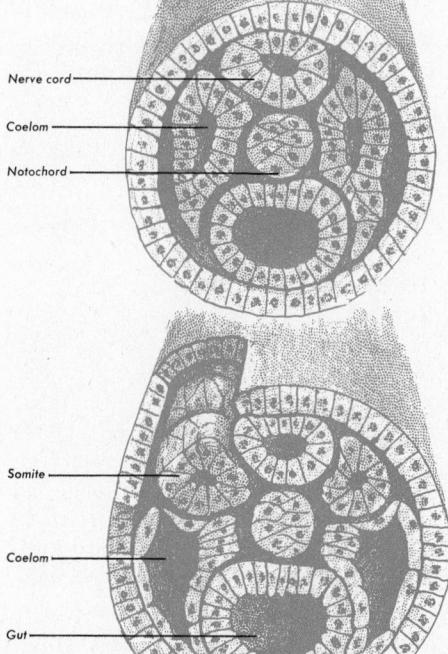

Somite

Coelom

Gut

established. The groove closes over on the dorsal side and sinks below the surface to form a tube which gives rise eventually to the entire nervous system and its associated parts. The retina of the eye forms from an evagination or outpocketing of ectoderm from the brain and the lens from an invagination of the overlying ectoderm of the head. Evaginations from the wall of the digestive tract give rise to the lining of the lungs, most of the liver, and the pancreas, as shown in the human embryo (Fig. 22–6).

Another method of morphogenesis during development is cell migration. Cells once formed do not always remain in their original positions, and during early development individual cells as well as groups of cells migrate a great deal. The mesodermal cells especially, either as single cells or as small groups, break loose and wander through the tissues to set up housekeeping in new locations where they differentiate into new structures, such as muscles, blood vessels, and connective or supporting tissue. What forces are operative in directing migrating cells to their proper locations and in causing them to differentiate there into the right kind of cells to do a specific job? For example, how do the tiny processes of nerve cells that migrate out from the primitive nerve tube find their way peripherally to the muscle cells of a developing limb bud? There is a question of time as well as space. If the nerve cells arrived and differentiated before the muscle cells differentiated, would functional neuromotor endings develop? Migration of cells must occur at precisely the right time and to the right location if the final organ is to form and function properly.

Still another essential phase of development must take place after these cells reach their destination. The cells must differen-

Fig. 22–5. The nervous system of *Amphioxus* is formed from a groove along the top side of the embryo, as illustrated by these figures from top to bottom. The mesoderm gives rise to the somites and lining of the coelom.

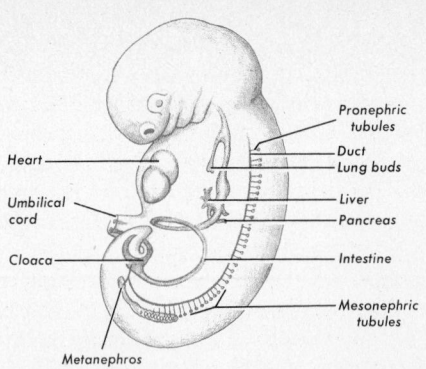

Fig. 22–6. The very young human embryo shows the beginnings of several organ systems. The digestive tract is heavily shaded. Note how the lungs and pancreas are forming as outpocketings from the primitive gut. The nervous system and the kidney are very primitive at this stage of development (under five weeks).

tiate into specific kinds of cells in order to produce the definitive organ. In general, such differentiated cells are specialized for particular functions; and it is probable that once they are formed, they do not revert to other cell types. Some cells, however, even in the adult may retain the ability to differentiate into other kinds of cells, for example, the primitive connective tissue cells that fill in the space resulting from a wound.

Final differentiation of cells occurs at the end of a long sequence of changes collectively called the developmental history. Consider, for example, the mesenchymal cells in the interior of the limb buds which eventually form the appendages of vertebrates. These cells all look alike at first but they gradually differentiate into types that are easily distinguished. At an early stage of development, five bumps, the ultimate digits, push out around the distal margin of the limb bud. Soon thereafter cartilage appears in the center of the bud and extends out into each digit. Skeletal muscle, fibrous connective tissue, bone, blood, and blood vessels all differentiate from mesenchyme, while nerves grow in from the spinal cord. Growth and development of a typical pattern of arrangement of cell types con-

stitute the later stages of development leading to the form we refer to as the adult.

Differentiation of the germ layers follows a similar pattern in all animals, and the final organs derived from them can be predicted fairly closely from the story of evolution. Layers that first formed ectoderm and endoderm in evolutionary history might be expected to give rise to the covering and the lining of present-day animals. Ectoderm, since it contacts the external world, might also be expected to give rise to the organ system that keeps it in touch with its external environment, namely, the nervous system and its associated organs. Endoderm lined the gut of the first metazoa and still performs that same function in the embryos of higher animals, giving rise as well to the epithelial components of the lungs, liver, pancreas, thyroid, thymus and parathyroid glands. The last layer to appear both in evolution and in embryology is the mesoderm, which produces all of the remaining organs such as muscles, blood vessels, skeleton, kidneys, and gonads. Although the three germ layers are destined in normal development to produce specific tissues, deviations can be induced experimentally. Much of the research in the field of **experimental embryology** is concerned with the potentialities of the germ layers under varying environmental conditions.

FACTORS INFLUENCING DEVELOPMENT

Development does not always follow the "normal" pathways, as attested by the occasional appearance of malformed (**teratologous**) offspring among all groups of animals. This means that something has gone wrong along the way in the formation of the various systems. At some critical time chemical or physical factors became sufficiently modified to shift the pattern of differentiation into "abnormal" pathways. Since certain kinds of malformations, such as cleft palate or development of extra limbs, can be induced experimentally in certain organisms, the experimental embryologist can frequently make fairly accurate guesses as to the nature of **teratogenic** factors.

Fertilization

The logical place to start in an experimental study of development is fertilization. Is it necessary that a sperm enter the egg? Aside from contributing a complete set of choromosomes, what other functions does the sperm perform?

Even Aristotle knew that the drone bee was a result of an unfertilized egg. Many animals possess **natural parthenogenesis,** which means development without fertilization, during some stage of their life history. It even occurs occasionally among some animals (insects, fish, amphibians, reptiles, and birds) whose eggs normally require sperm for development. How does this fact fit into the general assumption that all eggs must be fertilized? The situation in bees as well as other animals led biologists very early to investigate the necessity of the sperm for development. Artificial parthenogenesis can be produced in many animals by simply subjecting the eggs to an appropriate chemical or physical change, such as immersing them in certain chemicals or pricking them with a needle in the presence of blood (Fig. 22–7). This would seem to indicate that the sperm serves two functions: one to contribute a haploid set of chromosomes (p. 491), the other to initiate development of the egg. Apparently the potentialities of development are locked up in the egg, waiting to be released at the right time by the entrance of the sperm. If other suitable physical or chemical factors are applied, they too can perform this second function of the sperm very satisfactorily. The action is probably tied up with enzymes in some sort of trigger mechanism which cen be set off by a variety of stimuli.

It is known that mammalian sperm possess an enzyme, **hyaluronidase,** which is capable of digesting the intercellular cement of the follicle cells surrounding the egg, thus making fertilization possible. Perhaps this is why so many sperm cells are essential to effect fertilization. A few cells would not produce enough enzyme to penetrate the follicle cell of the egg. Eggs of echinoderms are surrounded by a jelly coat containing a substance called **fertilizin.** This substance, a glycoprotein, causes sperm of the same species to agglutinate and also activates sperm cells to become motile. The surface of the sperm possesses an acidic protein, called **antifertilizin,** which is thought to interact with fertilizin in a species-specific manner during fertilization.

Theories of Development

The most important problem in embryology today is: How does one type of cell differentiate into another type? The full impact of this question becomes obvious when we realize that all of the potentialities of the adult animal are locked up in the fertilized egg. All of the different types of cells that compose the adult are derived from this single cell. Obviously from the very beginning of development cells must

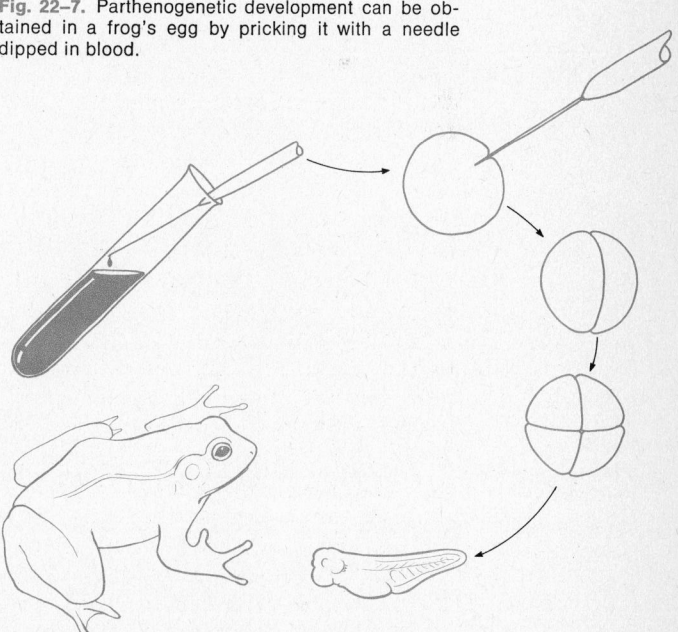

Fig. 22–7. Parthenogenetic development can be obtained in a frog's egg by pricking it with a needle dipped in blood.

constantly differentiate from one type to another. Since their genes are probably all the same and since all of the information that is needed to bring about developmental changes came from them, we must look to the subtle environmental changes in the cytoplasm of cells or in the intercellular matrix surrounding them in order to explain development.

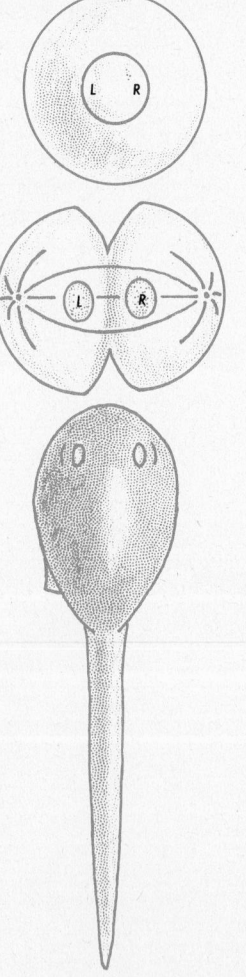

Fig. 22–8. An idea that convinced some early embryologists of the mosaic theory of development is represented schematically. The egg nucleus (top) was supposed to contain the determinants for the two halves of the embryo. When the egg cleaves (middle), the determinants would segregate to the two cells, each of them giving rise to its corresponding half of the embryo (bottom).

Mosaic Development. One of the first attempts to solve the problem of development was made in the latter part of the last century by Wilhelm Roux, who worked with frogs' eggs. He noted that the first cleavage usually coincided with the long axis of the embryo and that the **gray crescent**, a light area near the equator of the egg, usually appeared opposite the point of sperm entrance. The plane of the first cleavage usually cuts through the middle of the gray crescent, thus dividing the egg into two cells, one becoming the right half and the other the left half of the future animal (Fig. 22–8). About the same time August Weissmann postulated that the zygote nucleus contained **determinants** which were segregated to the daughter cells during the first cleavage. At each subsequent division, according to his hypothesis, determinants were passed to the resulting cells until finally cells contained only those determinants responsible for differentiation of the various types of cells found in the tissues of the adult.

In order to test Weissmann's hypothesis Roux killed one of the cells of the two-celled embryo with a hot needle. The uninjured cell cleaved in normal fashion but gave rise to only a half embryo (Fig. 22–9). This experiment seemed to be dramatic proof that the egg is a mosaic of cells, each possessing the capacity to give rise to only a particular portion of the adult animal. This idea was upheld by other investigators experimenting on a variety of embryos, including annelids, mollusks, and tunicates. The experiments were refined to the point where in some annelid embryos a single cell in the 64-cell stage could be observed to give rise to all of the mesoderm in the adult. The mosaic theory of development seemed to answer all the questions of development. However, when the other embryos of other types of animals were studied the theory ran into difficulties.

Regulative Development. Sea urchin embryos provided the first clue that the mosaic theory might not always hold. If the cells of the two-celled embryo were completely separated by shaking, Driesch showed that each gave rise to a perfectly formed embryo though half the size of a normal embryo (Fig. 22–10). This was also true if the cells were separated in the four-celled stage. For this reason the sea urchin embryos were said to be **regulative.** When later investigators finally succeeded in completely separating the first two cells of amphibian embryos, they found that each gave rise to a complete embryo just as in the case of the sea urchin embryo (Fig. 22–11). Separation was effected either by withdrawing one cell with a pipette or by tying a fine thread of hair around the embryo and pulling it tight to separate the two cells. Thus it could be concluded that the presence of the dead cell in Roux's experiment caused the development of a half embryo from the cell left alive. Evidence today seems to indicate that the nuclei of all cells contain a full complement of genes and thus that specific "determinants" are not segregated to determine the many kinds of cells found in the adult. How the nucleus exercises its control of specific cytoplasmic differentiation has become the focus of embryologists today.

A rather interesting outcome of the experiments outlined above was an explanation for the striking anomaly of Siamese twins in humans. Spemann, a famous German embryologist working with constricted amphibian embryos, showed that if a loop of hair was sufficiently loose so that some part of the first two cells was allowed to remain in contact, a partially double embryo resulted (Fig. 22–11). It seems possible that similar anomalies in fish (Fig. 22–12) or humans may be formed in an analogous fashion. Perhaps cleavage between the two cells starts but is not completed, thus producing embryos attached to a greater or lesser degree. Partially double embryos could also result from splitting of the body axis at the gastrula stage. Occasionally in man one embryo grows normally and incorporates the abnormal one inside its body, forming a tumor, known as a **teratoma.** Some teratomas contain skin and hair, indicating that they might have started as partial twins. Teratomas often give trouble in adult life and need to be removed.

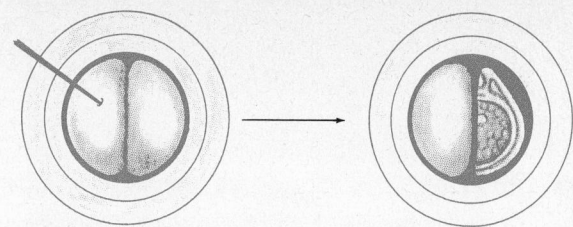

Fig. 22–9. If one cell of a frog's egg is killed when the embryo is in the two-cell stage, only a half embryo develops.

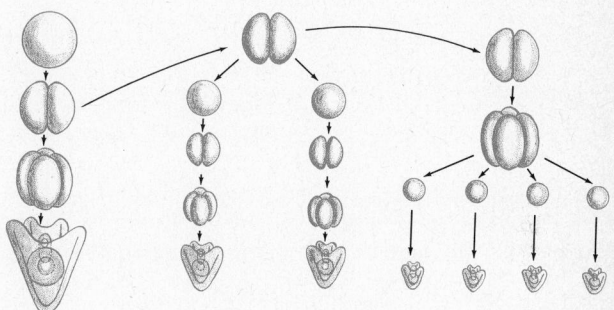

Fig. 22–10. When sea urchin eggs are separated in the two- or four-cell stage, they give rise to fully formed though proportionately smaller embryos.

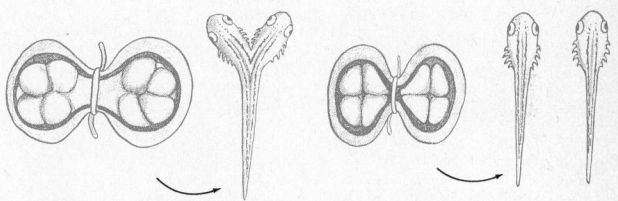

Fig. 22–11. If a hair loop is tied in such a way as to separate the cells partially in a two-cell amphibian embryo, a partially double ("Siamese twins") embryo will result. If the hair loop is pulled tightly enough to separate the two cells completely, two normal, half-size embryos result.

The Organizer. Perhaps the most important embryological contribution made in this century was the organizer theory of Hans Spemann. This theory was the outcome of many ingenious experiments on amphibian embryos performed by Spemann and his collaborators. They demonstrated that by transplanting cells from the dorsal

Fig. 22-12. "Siamese twins" in fish. These are probably formed as a result of an incomplete separation of the early embryo, perhaps in the two-cell stage.

lip of the blastopore to an area in a second embryo that would normally form epidermis, a second embryonic axis was established in the new site, even though the host embryo had one of its own (Fig. 22–13). The transplanted cells sank beneath the surface and induced a nerve cord and

Fig. 22–13. If a few cells from the region above the blastopore of an amphibian embryo are removed and placed on some other region of a second embryo, they produce a nerve cord just as they would have done had they remained in their original location. The operation is performed in the two top figures and the results are shown on the lower left (early) and lower right (later).

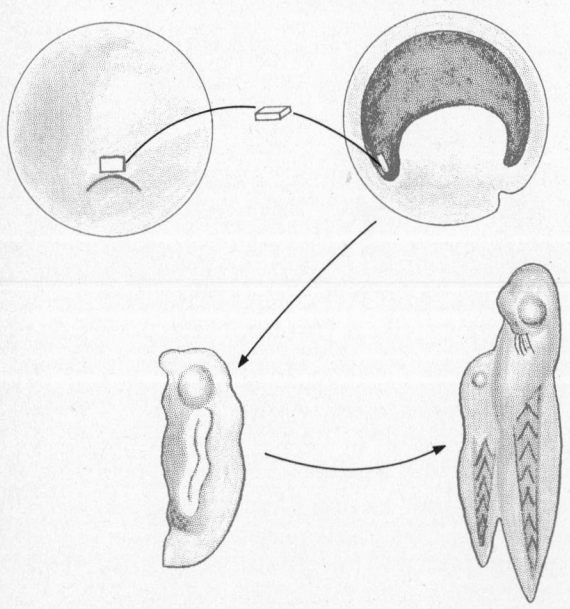

in some cases almost an entire secondary embryo. From these experiments Spemann came to the realization that the cells of the dorsal lip were the organizing center of the axial structures of the embryo, and hence he referred to the lip cells as the **organizer.** Since the organizer was thought to induce a change in the development of other cells the action was called **induction.**

Tissue culture experiments with embryonic cells can be explained in the light of the organizer theory. When cells that normally become nerve cord are removed from an early gastrula and grown in physiological saline away from other cells they form epidermis. However, if the same cells are removed from the late gastrula and grown separately in culture, they differentiate into nervous tissue. Obviously the cells in the early gastrula had not yet been influenced by the organizer, whereas in the later gastrula the organizer had induced the overlying cells to become nervous tissue. The cells of the early gastrula are said to be **undetermined** as regards their capacity to form a nerve cord. They are said to be **determined** after the organizer has acted upon them. Determination is thus a result of induction. If a piece of embryonic tissue is grown in culture and does not differentiate, it is considered to be determined at the time of isolation.

The action of the organizer can be further illustrated in the formation of the vertebrate eye. As the bulb of ectodermal tissue that pushes out from the brain to form the cup comes near the external ectoderm, it induces the latter to form a lens. This happens even if foreign ectoderm is placed over the eye cup. In this case the eye cup is an organizer. Possibly some chemical compound elaborated by the organizer accounts for formation of the lens in this case and the nerve cord in the previous experiments.

Light was thrown on this possibility when it was discovered that even dead organizer tissue could induce neural structures. Dorsal lip cells could be killed with chemicals or heat yet still would cause undetermined ectoderm to form neural tissue. It should be possible to isolate such a stable chemical compound by proper techniques. From a long series of experiments only confusion resulted, because it was discovered that cells of the gastrula that were noninductive when alive became so when killed. Even adult tissues, such as kidney and liver, and various substances completely foreign to organisms can induce. Efforts to purify the organizer have not been satisfactory. Although there is some recent evidence that the initial neutralizing substance (organizer) may be ribonucleoprotein, determination of its exact nature must await further research.

Mosaic and regulative eggs may now be reconsidered. Perhaps the main difference between them lies in the time at which the cells of the embryo are determined, that is, the time when an organizer exerts its influence on specific cells. Mosaic eggs are determined much earlier, evidently when they are formed in the ovary, whereas regulative eggs are determined during embryonic development.

Consider this old idea of preformation in a new light. Modern genetics has shown that the egg is actually preformed in the sense that all the genes have all of the information that is necessary to bring about the development of an adult organism. The genes are derived from the parent during meiosis and are incorporated in the nucleus of the egg. Once cleavage starts and development proceeds, the genes "give the orders" for each stage of development. They undoubtedly are responsible for the formation of the initial organizer, and perhaps for all of the subsequent organizers that must form in sequence as each new structure is formed. Spemann theorized that organizers are produced one after the other as development proceeds. One organ develops as a result of an organizer and this organ, in turn, produces another organizer which induces the development of another organ and so on. To what extent development proceeds by specific induction or alternatively by specific inhibition of differentiating tissues is another of the major questions confronting embryologists.

As was pointed out earlier the question of change in DNA during development does not lend itself well to experimental proof. However, recently some elegant experiments involving transplantations of nuclei in frog eggs suggest that some sort of nuclear change may occur during early embryological development. First, the spindle for the second maturation division is removed from the egg and then replaced with one from a cell taken from a blastula or gastrula (Fig. 22–14). The egg, then, contained a nucleus from a cell whose cytoplasm had already undergone differentiation. When such an egg was permitted to develop, some very interesting results were observed.

A normal embryo resulted when the enucleated egg received a nucleus from a blastula cell. However, nuclei from late gastrula cells usually produced grossly ab-

Fig. 22–14. These sketches show the process of removing the nucleus from the frog egg and the results obtained when older nuclei are introduced into them. See text for further explanation.

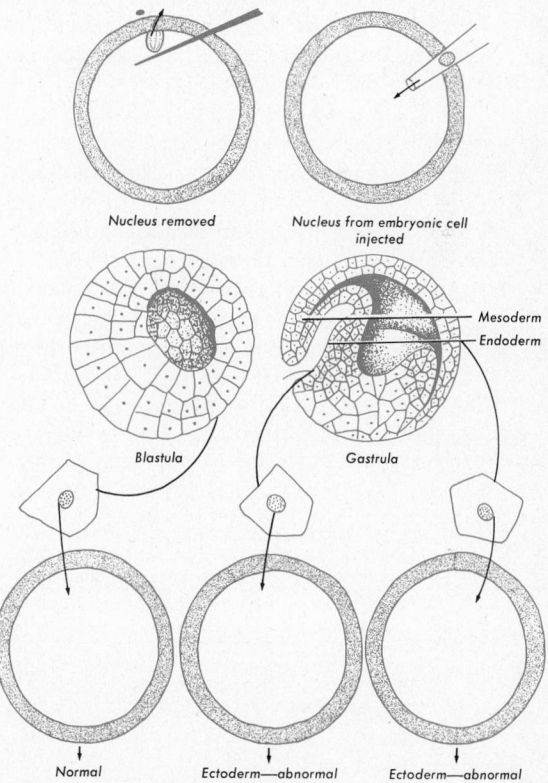

Nucleus removed

Nucleus from embryonic cell injected

Blastula

Gastrula

Mesoderm
Endoderm

Normal

Ectoderm—abnormal

Ectoderm—abnormal

normal embryos. The type of abnormality proved to be equally interesting. A nucleus from an endoderm cell when transplanted to the enucleated egg might give rise to an embryo with normal endoderm but with ectodermal and some mesodermal derivatives abnormal. Ectodermal and mesodermal nuclei of the late gastrula occasionally promoted normal development. These experiments tend to support the idea that the loss of differentiative potentialities by the nucleus is somehow related to the part of the embryo from which it originated. This implies that the activity of nuclear DNA somehow becomes altered differentially in cells with different developmental potentialities.

Regeneration

In earlier chapters we considered the phenomenon of regeneration among the invertebrates, especially in hydra and planaria where it is particularly striking. Can the observations made on these forms, and others as well, be explained in the light of the information available from experimental embryology?

Planaria, when cut into several pieces, gives rise to an equal number of whole animals, each with mouth, digestive tract, and nervous system (p. 169). In planaria the situation is much more complicated than in the coelenterate, because of the greater variety of structures, but unequal distribution of head-producing substance along a concentration gradient diminishing toward the posterior end may account for the observed pattern of regeneration in this organism.

Another aspect of regeneration has come to light through metabolic studies of a number of lower animals including planaria. It has been shown that the metabolic rate, as demonstrated by oxygen consumption studies, is greatest at the anterior and least at the posterior end and, indeed, that there is a definite **gradient** running in an anterior–posterior direction. This is confirmed by applying poisons of various sorts to the animals, in which case the anterior end is always injured first, the effect gradually diminishing in a posterior direction. This same gradient is observed in cut pieces of planaria, for the anterior part of each piece has a higher metabolic rate than the posterior part. The head always develops at the more active end and the tail at the other. This principle has been stated by Child in his **axial gradient theory,** which has been utilized to explain the origin of some abnormalities appearing in various kinds of embryos, including man. Suppose two centers of high metabolic activity somehow became established along the anterior–posterior axis of a developing embryo. The embryo might subsequently develop with some sort of twinning, partial or complete. Separation of high metabolic centers would have the same effect as the mechanical separation performed in Spemann's hair loop experiment, and the end result would be the same.

Study of the regeneration of parts in adult animals is a natural outgrowth of work on embryos. Many interesting questions raise themselves in the field of regeneration. For instance, why do limbs of adult salamanders regenerate completely whereas those of the closely related frogs do not? Further, why have the limbs of higher animals lost most of the power of regeneration? Some interesting experiments have thrown light on these problems.

When a limb of a larval salamander is amputated, the epidermis quickly grows over the cut surface while a concomitant series of changes goes on underneath, resulting eventually in a new limb (Fig. 22–15). The cells of the cut muscle, con-

Fig. 22–15. An amputated limb of a larval salamander will regenerate completely as shown here. Why does a mammalian appendage fail to regenerate?

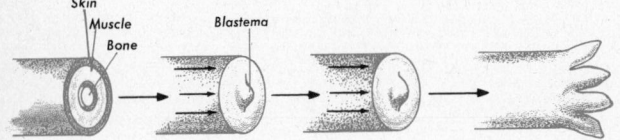

Skin
Muscle
Bone

Blastema

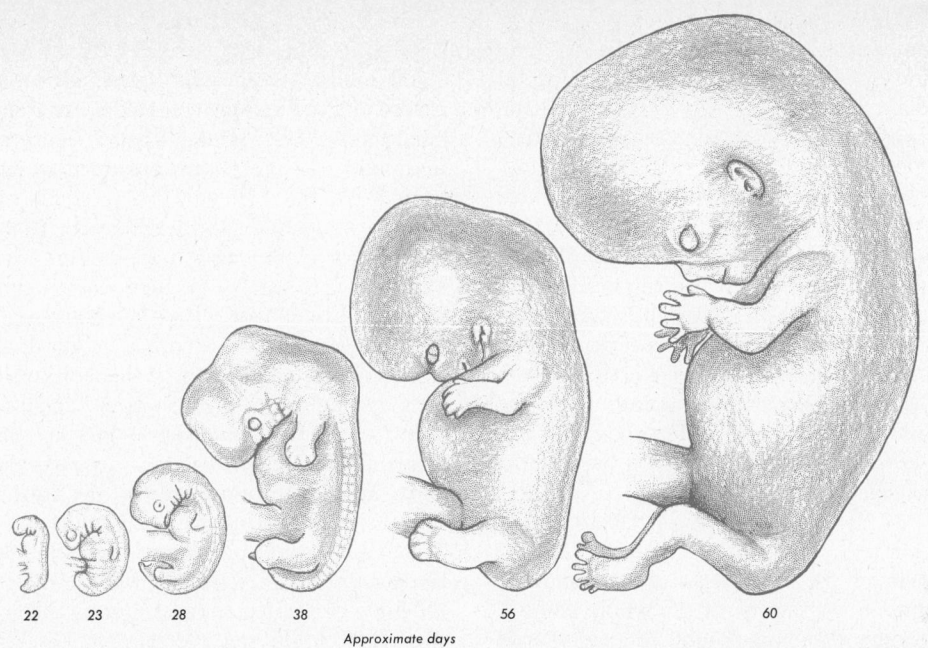

Fig. 22–16. A human fetus undergoes considerable change in body proportions during gestation. These six are compared as to relative size and body shape. Their ages are 22 days, 23 days, 28 days, 38 days, 56 days, and 60 days.

nective tissue, and bone change in appearance. They appear to **dedifferentiate,** that is, lose most of their specialization, and eventually form a small region of special cells called a **blastema.** Once the blastema is formed, further differentiation ceases. The cell mass then begins to **redifferentiate** into the various tissues of the limb and eventually normal form is restored. Why do these events fail to materialize in frogs and higher vertebrates? A simple experiment performed by Rose gives us a partial answer.

If the cut limb of a frog is placed in saline (salt solution) it actually will partially regenerate. Apparently the saline prevents the normal closure over of the wound by the dermis (connective tissue) of the skin and brings about the formation of a blastema. Once the blastema is formed, regeneration proceeds just as it does in salamanders. It appears that normally in the frog the entire skin including the dermis grows over the wound and blocks the formation of the

blastema which would bring about regeneration. On the other hand, in the salamander the wound is closed simply by the epidermis; a blastema is formed and regeneration can proceed. It seems, therefore, that in the frog the ability to regenerate appendages is not lost but merely blocked. The possibility of regeneration of appendages or other parts in reptiles, birds and mammals is still under active investigation.

Since we observe that the embryo grows continually while it is differentiating, let us consider next the problem, growth.

GROWTH

By the end of the eighth week of human life, the embryo has increased to 2 million times its original mass and during the next eight lunar months it increases another 4,000 times (Fig. 22–16). This tremendous accumulation of protoplasm we call growth.

In Chapter 28, we study growth of protozoa in a culture medium and discover the nature of the **growth curve.** By removing a small bit of embryonic chick heart, for example, and placing it in a flask containing adequate nutrients, growth of the cells occur according to the same curve as that demonstrated for the protozoa. At first the cells divide very slowly, then increase to a uniform rate, finally, they fall off until no more divisions occur. If a small bit of tissue is transferred to fresh medium, the cycle is repeated; and this can be continued indefinitely. If cells of isolated tissues grow according to the same patterns as do protozoan cells, how do cells grow when they are a part of the intact embryo?

Just as we note limiting factors (nutrients and accumulation of wastes) in the growth of protozoa, we also note limiting factors in the growth of the whole embryo. We know that animals grow to a certain size and stop; they do not grow forever. Something about the community of cells limits the number of divisions that occur. The individual organs also fail to grow after a certain number of cell divisons or after a certain time has elapsed. For example, if bits of tissue are removed from embryos at various ages it is found that those from older embryos do not grow as well as tissues from younger ones; in fact, some, like nerve tissue, grow little, if at all, when removed from adult animals. Others, like epithelium, retain the capacity for growth throughout the life of the organism. Still others, like some cancer tissue, grow readily no matter what their source or age. From these experiments we can conclude that cells, when confined to an organism, are subject to factors that limit growth. These factors may be the same as the ones that limit growth in a culture medium. What all of these limiting factors are is a fertile field for research.

In the growth of an embryo, the sizes of the various organs increase at varying rates. The human heart, for example, approximates the growth of the body as a whole, whereas the brain grows tremendously during the first five years of life and thereafter grows very little. Because all of the organs grow at different rates, the general overall proportions of the human body gradually change. The change is radical during the gestation period (Fig. 22–16), but after that it progresses slowly throughout life (Fig. 22–17).

Fig. 22–17. Various parts of the body grow at different rates during the life of an animal. Here in the human being many changes occur from prenatal life at the left to the adult at the right. Note particularly the proportionate difference in the head and body size.

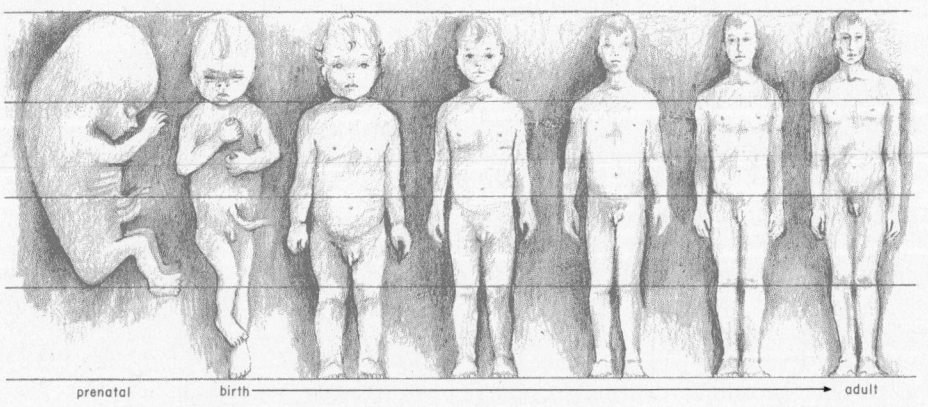

prenatal   birth —————————————————————————————————→ adult

Books

ABERCROMBIE, M., "Cellular Interactions in Development," in *Ideas in Modern Biology*, J. A. Moore, ed. Garden City: The Natural History Press, 1965.

BALINSKY, B. I., *An Introduction to Embryology*, 3rd ed. Philadelphia: W. B. Saunders Co., 1970.

*BARTH, L. J., *Development: Selected Topics*. Reading, Mass.: Addison-Wesley, 1964.

BERRILL, N. J., *Growth, Development, and Pattern*. San Francisco: Freeman, 1961.

CORNER, G. W., *Ourselves Unborn*. New Haven, Conn.: Yale University Press, 1944.

*EBERT, J. D., *Interacting Systems in Development*. New York: Holt, Rinehart & Winston, 1965.

*EDD, M. V., JR., "Animal Morphogenesis," in *This Is Life*, W. H. Johnson and W. C. Steere, eds. New York: Holt, Rinehart & Winston, 1962.

GROBSTEIN, C., "Differentiation of Vertebrate Cells," in *The Cell*, J. Brachet and A. E. Mirsky, eds., Vol. I. New York: Academic Press, 1961.

MARKERT, C. L., "Mechanisms of Cellular Differentiation," in *Ideas in Modern Biology*, J. A. Moore, ed. Garden City, N.Y.: The Natural History Press, 1965.

Molecular and Cellular Aspects of Development, E. Bell, ed. New York: Harper & Rowe, 1965.

PAGE, E. W., VILLEE, C. A., and VILLEE, D. B., *Human Reproduction*. Philadelphia: W. B. Saunders Co., 1972.

PATTEN, B. M., *Human Embryology*. New York: The Blakiston Co., 1953.

RUGH, R., *Experimental Embryology*. Minneapolis, Minn.: Burgess, 1962.

*SUSSMAN, M., *Animal Growth and Development*. Englewood Cliffs, N.J.: Prentice-Hall, 1960.

* Available in paperback.

Articles

MOSCONA, A. A., "How Cells Associate." *Scientific American*, September, 1961.

WADDINGTON, C. H., "How Do Cells Differentiate?" *Scientific American*, September, 1953.

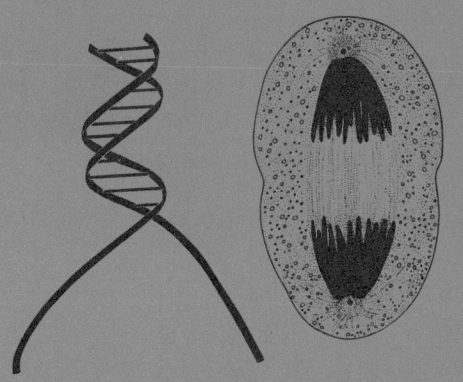

VI
CONTINUITY
OF LIFE

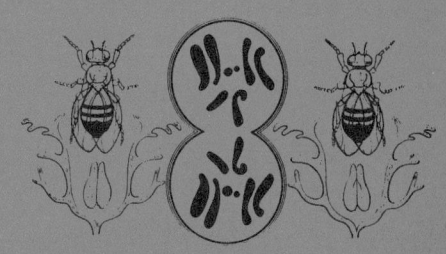

23

CONTINUITY OF CELLS

One of the most obvious facts in biology is that living things grow by increasing their number of cells. The details of the process of growth were worked out many years ago with the aid of the light microscope. We must, however, investigate a lower level of organization, far beyond the resolution of the most powerful microscopes, if we are to learn the initial stages in this duplicating mechanism. In Chapter 2 we learned something about the gene and hinted at its significance in the duplication of cells. At this time we return to the gene to understand its part in cell division, after which we examine that part of the process which can be seen through the microscope.

COMPOSITION OF GENES

The first approach to a study of the composition of genes is to analyze the chromo-

somes chemically. This is because once we can describe a substance in chemical terms we have taken the first step toward understanding how it functions. To that extent we know the chromosome is composed of four major kinds of molecules: two proteins, one called **histone** with low molecular weight (about 2,000) and another which is more complex with a much higher molecular weight; and two nucleic acids, one **deoxyribose nucleic acid** (DNA) which has less oxygen than the other, **ribose nucleic acid** (RNA). These latter two compounds were described earlier (p. 70). The four kinds of molecules are combined in some fashion not clearly understood to form chromatin, the stainable material of which chromosomes are composed. Of these, DNA carries the genetic information and it is about this molecule that we know the most, probably because it has been the most exciting to study.

Workers in both Europe and America

learned in the 1940s that the nuclei of germ and somatic cells of any one species contained the same amount of DNA for each set of chromosomes. This evidence pointed to the fact that DNA was the essential molecule in genes. This idea was confirmed sometime later when it was discovered that it was possible to change one type of *Pneumococcus*, the causative organism of pneumonia, to another (there are many types) by exposing one to DNA extracted from the other. Such transformations have been accomplished with other microorganisms which definitely establishes DNA as the molecule which carries genetic information. It is interesting to note that whereas we have only very recently understood what DNA does in the cell, it was first isolated by Miescher in 1868, long before we knew anything about genes.

It has been shown that DNA is a long unbranched molecule (sometimes called a "macromolecule" because of its large size) of variable molecular weight. Most DNA molecules are composed of four repeating nucleotides. The deoxyribose and phosphoric acid are the same in all DNA molecules; the molecules differ only in the arrangement of their bases (purines and pyrimidines). The purine nucleotides are **deoxyadenylic** and **deoxyguanylic** acids; the pyrimidine nucleotides are **deoxythymidylic** and **deoxycytidylic** acids. The chemical structures of the bases are given in Figure 23–1. The configuration of the DNA molecule was discovered as a result of the following observation. When DNA was extracted from different cells and the relative proportions of the various nucleotides measured, it was discovered that the amount of purine always equalled the amount of pyrimidines in any given cell type. Even more astonishing was the fact that the number of deoxyguanylic acid molecules equalled the number of deoxycytidylic acid molecules; likewise, the number of molecules of deoxyadenylic and deoxythymidylic acids were equal. This specific 1:1 relationship formed the basis for an understanding of the architecture of the molecule itself.

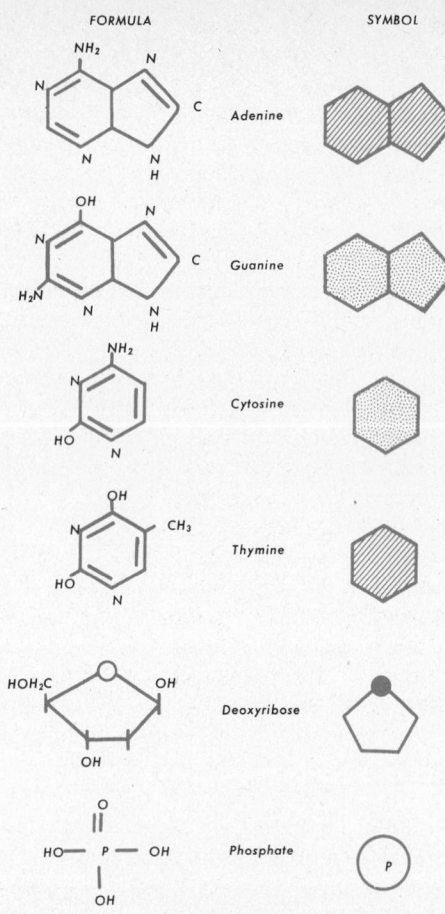

Fig. 23–1. The formulae and the corresponding symbols are shown here for the purine (adenine and guanine) and pyrimidine (cytosine and thymine) bases, as well as those for deoxyribose and phosphate. The symbols are used in Figure 23–2 to show how the DNA molecule is constructed.

James Watson of the United States and Francis Crick of England, using this as well as other information, were able to construct a model of the DNA molecule. They suggested that it is composed of two strands that spiral about one another, sometimes called a double helix (Fig. 23–3). The "backbone" of each strand is made up of deoxyribose and phosphate with the bases (adenine, guanine, thymine, cytosine) pointing toward the middle of the spiral and being held together by hydrogen bonds (Fig. 23–2). The sequence of the bases is such that adenine in one strand always pairs with thymine in the other strand and guanine with cytosine. The salient features of the model are its double strands and their complementary nature (Fig. 23–3). This makes it possible, as suggested by Crick and Watson, to understand how information can be stored in the DNA molecule. The sequence of the base pairs can occur in any number of combinations which means that an infinite number of DNA molecules are possible. The molecule may be compared to a tape upon which an alphabet of four letters could be encoded. By employing each letter as often as desired it would be possible to make an almost infinite number of tapes with no two being identical. With the Watson–Crick model from which to start, it has been possible to formulate theories on the replication of DNA as well as to speculate on the way in which the coded information stored in the DNA molecule is transferred to RNA and subsequently to the synthesis of specific proteins.

Replication of DNA

The complex DNA molecule with its double spiral configuration appears to be most difficult to duplicate, which of course it is. It seems that it somehow unwinds and then forms a second strand in the immediate vicinity of each of the parent strands. But the surface of such strands would be extremely complicated and how could replicas be made of them? Methods used in industry can help to solve this problem. Very irregular and complex surfaces of objects have been faithfully copied for a long time by craftsmen by employing the **template** principle. A template is made from the original which is then used as a mold to produce the copy (Fig. 23–4). It is possible that the surface of the DNA molecule is replicated in the same manner. Indeed, as we soon see this same principle may well explain RNA and protein synthesis.

Evidence for the correctness of this hypothesis, as well as for the reality of the DNA model, comes from several sources. One is the actual synthesis of DNA in a test tube without the presence of cells. A. A. Kornberg placed the four nucleotides and an enzyme **DNA polymerase** in a test tube and to this mixture added a small amount of DNA called "primer" DNA. The result was that long chains of DNA were formed and more importantly, the synthesized DNA had the same physiochemical structure as the primer. The structural identity held when DNA from different sources (virus, bacterial, and animal) was used. Obviously, according to the Watson–Crick model, the double helix must unwind or uncoil and a new complementary chain form along each chain.

By means of autoradiography it is possible to follow the distribution of DNA from one generation to the next in both plant and animal cells. The growing cells are exposed to thymidine which has been previously labelled with tritium (H³), a radioactive iso-

Fig. 23–2. Using the symbols indicated in Figure 23–1, the configuration of the DNA molecule is represented here. The dotted lines connecting the two strands represent hydrogen bonds. Note that the two strands are aligned in opposite directions.

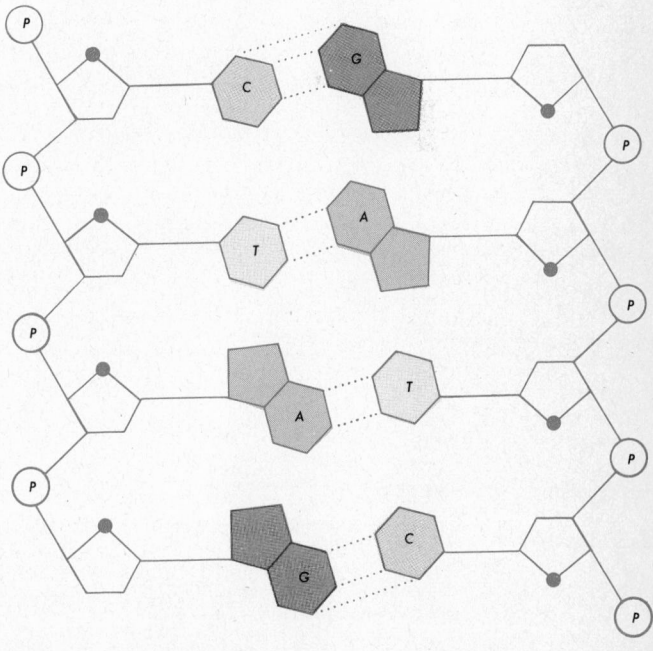

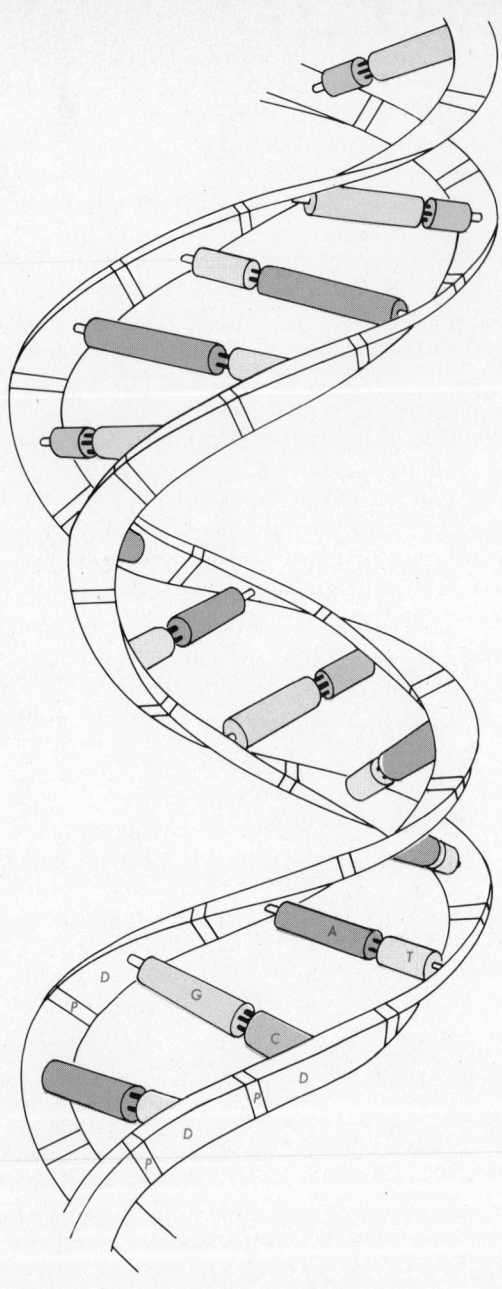

Fig. 23–3. A schematic representation of the double helix of a portion of the DNA molecule. The letters represent the following compounds: A—adenine; G—guanine; C—cytosine; T—thymine; D—deoxyribose; P—phosphate.

tope of hydrogen. Because the radioactive thymidine is incorporated into DNA as it is synthesized, one can actually follow, in the light microscope, how the DNA is distributed in several generations of cells. The DNA in each chromosome doubles during interphase (p. 488) and can be visually observed during metaphase. The experiment is done by placing cells in a solution containing tritiated thymidine for a period of time sufficiently long to permit doubling of the DNA. The cells are then washed and allowed to continue growing in a solution of colchicine (a substance that stops cell division but not chromosome duplication). Autoradiograms of such cells show that both members of the pair of chromatids contain DNA (show up as black dots over the chromosomes on the photographic plate). When the chromosomes duplicate again in the absence of radioactive thymidine each one gives rise to one labelled and one nonlabelled chromatid. In the next duplication of the chromosomes about one-quarter of the chromatids are labelled. Obviously the DNA, as represented by the labelled and nonlabelled whole chromosomes and chromatids, has duplicated during interphase in a manner which is consistent with the Watson–Crick proposal.

According to the present interpretation, DNA replication in the nucleus follows a sequence of events best described with a diagram (Fig. 23–5). We may represent the DNA molecule in simplified form, emphasizing only the relationships of the bases. Adenine is bonded to thymine, and guanine to cytosine (Fig. 23–2). The first step in replication requires that the hydrogen bonds break and the helices uncoil and separate. The separated strands can do one of two things; they can form a complementary strand of DNA or they can form a strand of RNA. Replication of all of the DNA occurs during interphase; RNA is also synthesized during interphase, but only on selected portions of the DNA templates, depending on which genes are active. The RNA is responsible for coding for specific synthesis of proteins. The overall reaction as previously indicated is called the **Central Dogma of Molecular Biology** and is diagrammatically expressed by the following:

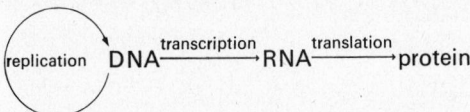

We consider the replication of DNA first. Assume that all of the DNA nucleotides and DNA-polymerase are present in nucleoplasm. The complementary base pairs are attached to one another by DNA polymerase so that a new strand of DNA is synthesized, thus restoring the DNA molecule to its original double-stranded condition. The important point to note is that the new strand is an exact replica of the original strand. Replication of the DNA molecule can thus go on endlessly and always retain its specificity, an absolute requirement if genetic continuity is to persist as we know it does.

The Watson–Crick proposal for DNA replication, as just stated, has general support among molecular biologists. Its importance in understanding the continuity of cells cannot be overestimated. This concept offers a firm foundation on which future investigations can be built in attempting to understand the many problems left unanswered.

Origin of RNA

It is generally accepted that RNA is responsible for protein synthesis and that it originates in the nucleus from DNA. We must also assume that it receives information from the DNA so that specific proteins can be constructed. Just how is this information transferred from DNA to RNA? This problem has been essentially solved in prokaryotic cells and it probably has general application to all cells.

When considering information transfer, we need some specific terms to describe the events. The synthesis of DNA on a DNA template, just described, is called **replication.** The synthesis of RNA on a DNA template is

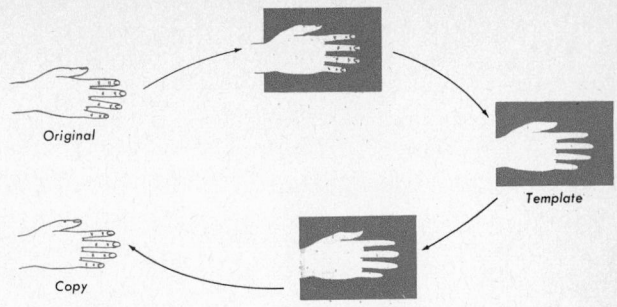

Fig. 23–4. The template principle gives us a possible explanation of how a structure with complex surfaces may be accurately copied. It has been given as a possible explanation for DNA duplication.

known as **transcription.** The last step, the synthesis of protein from RNA, is called **translation.** Let us now consider transcription and later translation. All of these events are schematically portrayed in Figure 23–5.

There are three kinds of RNA: **transfer** (tRNA), **messenger** (mRNA), and **ribosomal** (rRNA). They are all synthesized on DNA templates in the nucleus and make their way into the cytoplasm, just how is not as yet very clear. Their transcription takes place under the direction of the polymerizing enzyme, **RNA polymerase,** in the presence of double-stranded DNA and four nucleotides. Divalent ions, Mg^{++} and Mn^{++}, are also essential. The three types of RNA are synthesized on the DNA templates, frequently at different times, by base-pairing. The RNA nucleotides are similar to the DNA nucleotides with certain exceptions. First of all, the sugar is **ribose** instead of deoxyribose which simply means that it contains more oxygen, and secondly, deoxythymidylic acid is replaced by **uridylic acid.** The assembling of the RNA strands is identical with that for DNA replication except that now uridylic acid pairs with adenylic acid since it has replaced thymidylic acid. The RNA strands now possess the same information as the DNA strand. They then separate from the DNA strand and are free in the nucleoplasm, each of the three probably following different pathways into the cyto-

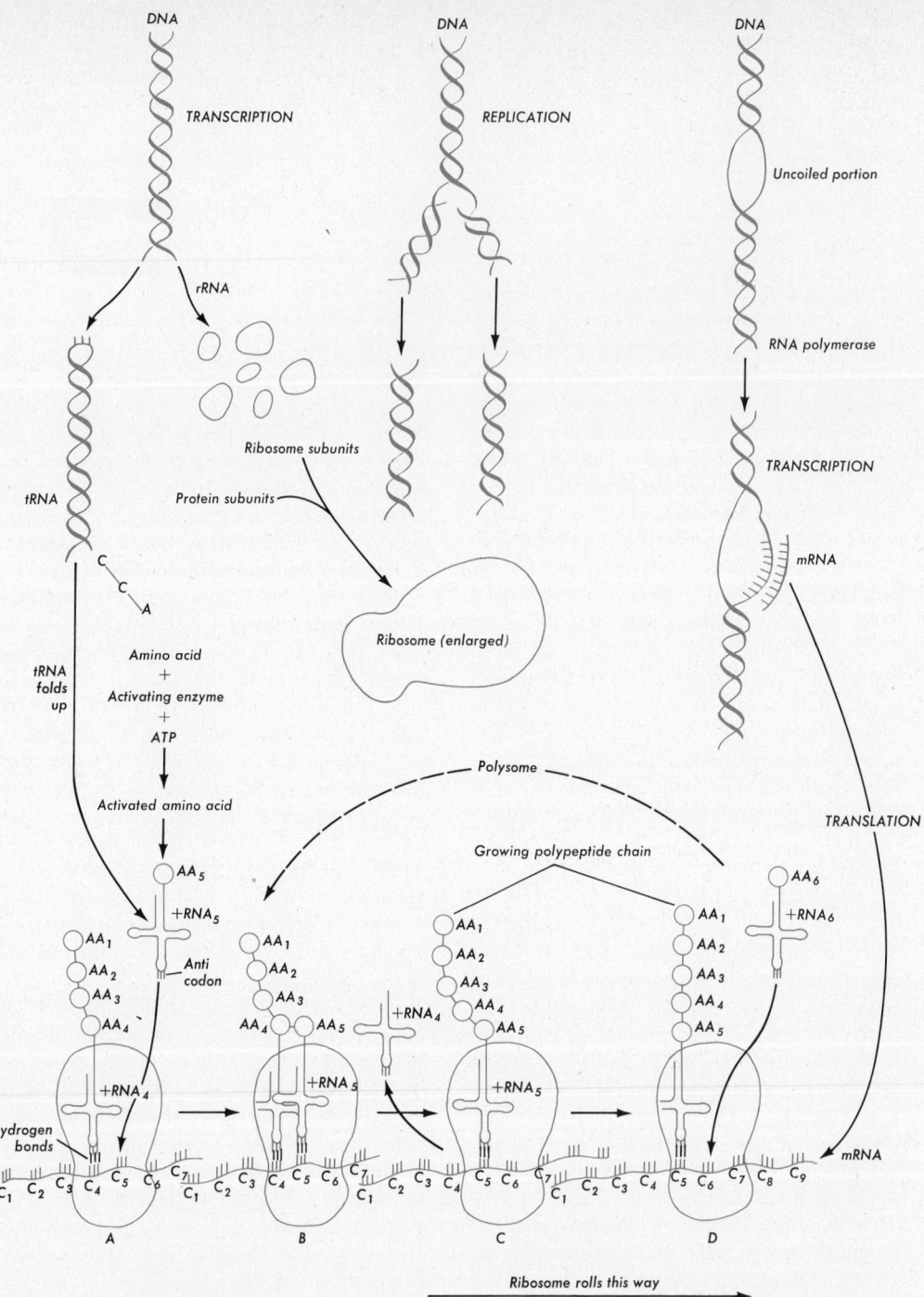

Fig. 23–5. A schematic representation of DNA replication, RNA transcription, and protein synthesis. (Bottom) A series of stages (A–D) in the translation of a mRNA strand consisting of codons C_1, C_2, C_3 ... C_n into a peptide strand with corresponding amino acids AA_1, AA_2, A_3 ... AA_n. For detailed explanation see text.

plasm where they perform their specific jobs in protein synthesis. Ribosomal RNA may be stored for a time in the nucleolus before moving into the cytoplasm; messenger and transfer RNA probably enter the cytoplasm shortly after being synthesized.

It has been demonstrated that transcription occurs repeatedly, that is, the DNA strands are used over and over again in synthesizing the different RNA strands. A given amount of DNA transcribes as much as 60 times its weight in RNA which suggests that the DNA acts as a template in synthesizing a great many molecules of RNA. It is assumed that all three types of RNA are synthesized on specific regions of the DNA molecule. The amounts of each transcribed is different; for example, in some bacteria, about 0.3 percent of the total DNA is complementary to tRNA, about 0.4 percent is complementary to rRNA, and about 99.3 percent is complementary to mRNA. Since the major portion of RNA in a cell is ribosomal, a small section of the DNA molecule must be used repeatedly to produce the large quantities of rRNA.

This proposal for the origin of RNA is based on certain crucial experiments, most of them done with microorganisms, but the evidence is sufficiently abundant now from other sources as well to convince biologists that it has general application.

Many autoradiographic studies have shown that the different kinds of RNA move from the nucleus into the cytoplasm, but just how this occurs is not clear. Once in the cytoplasm, the various RNA molecules follow different pathways. Experimental evidence suggests that the tRNA is freely distributed in the cytoplasm and the mRNA becomes attached to the ribosomes. Several ribosomes are linked together by the long mRNA molecules forming groups which are called **polysomes** (Fig. 23–5). The tRNA becomes attached to the amino acids and carries them to the proper site on the mRNA (lying on the ribosome) where the transcription of the polypeptide (protein) is made. Just how this is done involves a coding system which we now discuss.

Enzymes and perhaps all proteins are under gene control, and since the business part of the gene is DNA, it follows that DNA controls the arrangement of amino acids in proteins. Proteins contain 20 amino acids arranged in a multitude of combinations, hence the difference between them. Then the question arises, how can DNA with only four bases—a four-letter alphabet—be so coded that it can spell out the sequence of the 20 amino acids in a protein? It would seem that a 20-letter alphabet would be required. Just how this is done has been elucidated within the past few years.

The combination of nucleotides that specify one amino acid is a code word or **codon.** The simplest code would be one in which each nucleotide would represent one amino acid. This would code for only four amino acids which, of course, is inadequate. Using two nucleotides, a doublet code, 16 (4 × 4) amino acids could be specified, but this still would not be adequate. A triplet code would specify 64 (4 × 4 × 4) amino acids; this is more than adequate but it has been accepted as the most likely code because more letters would be wasteful and nature is conservative. According to this scheme, each amino acid would be coded by three letters, for example, GGU would specify glycine. Obviously, since there are 64 possible combinations of letters to specify 20 amino acids, some must be identified by more than one triplet, which is the case.

Convincing evidence for the triplet code was provided by Dr. M. W. Nirenberg working at the National Institutes of Health. He passed mononucleotides, dinucleotides, and trinucleotides through special filters that were designed to permit tRNA to pass through, but retained ribosomes. Transfer RNA and ribosomes were mixed with an amino acid which was previously made radioactive. These were then exposed to the three kinds of nucleotides and forced through the filter. It turned out that only the trinucleotides with their amino acid remained on the filter with the ribosomes; both mono- and dinucleotides went through. This means that the tRNA with three nucleotides and its attached amino acid became affixed to the mRNA on the ribosomes. Dr. Nirenberg and others have used this technique to assign codons to all of the amino acids. These are shown in Table 23–1.

The understanding of the genetic code has been one of the major "breakthroughs" in biology of this century. Only time and much experimentation will determine whether or not this code operates in all plants and animals.

What are the steps in protein synthesis once the different kinds of RNA and amino acids are available in the cytoplasm? The first thing that happens is that each tRNA molecule must recognize and become attached to the amino acid for which it is coded. For example, the one carrying the code UUU picks up phenylalanine and all others do likewise. In order for this to happen the amino acids must be activated by specific enzymes and ATP. There is a specific activating enzyme for each amino acid. Once the amino acid is activated it is transferred to the specific tRNA molecule (Fig. 23–5). This complex (tRNA with its amino acid) then finds its way to the mRNA which apparently lies on the surface of the polysomes. The mRNA had received a coded message for the particular protein to be synthesized from DNA (Fig. 23–5). Each molecule of tRNA finds its proper site on the mRNA by base-pairing; this pairing insures the flow of information from DNA to the protein during its synthesis. The most recent finding suggests that there are two sites of attachment of tRNA, lying side by side, on the mRNA. One tRNA molecule is attached to the growing polypeptide chain and the other brings in the next amino acid. Once the amino acid joins the growing polypeptide chain, the ribosome moves on the mRNA, thus bringing the next site for tRNA into position. As the ribosome rolls along the mRNA, the amino acids are added to the polypeptide chain in a specified sequence. As the polypeptide chain of amino acids grows in length it begins to fold in a manner characteristic of the protein being synthesized. Once the protein is fully formed it is released in the cytoplasm and performs whatever function it is designed for.

This remarkable story explains how genetics works at the molecular level and

TABLE 23–1.   Assignments of RNA Codons to Amino Acids

Alanine	GCU, GCC, GCA, GCG
Arginine	CGU, CGC, CGA, CGG
Asparagine	AAU, AAC
Aspartic Acid	GAU, GAC
Cysteine	UGU, UGC
Glutamic Acid	GAA, GAG
Glutamine	CAA, CAG
Glycine	GGU, GGC, GGA, GGG
Histidine	CAU, CAC
Isoleucine	AUU, AUC
Leucine	UUA, UUG, CUU, CUC, CUA, CUG
Lysine	AAA, AAG
Methionine	AUG
Phenylalanine	UUU, UUC
Proline	CCU, CCC, CCA, CCG
Serine	AGU, AGC, UCU, UCC, UCA, UCG
Threonine	ACU, ACC, ACA
Tryptophan	UGG
Tyrosine	UAU, UAC
Valine	GUU, GUC, GUA, GUG

how the most complex of all molecules are formed. The proteins, thus produced, not only make up the architecture of the cell itself but are also the enzymes which are instrumental in the synthesis of many other molecules, such as carbohydrates, fats, sterols, purines, pyrimidines, amino acids, hormones, pigments, and many others. As has already been pointed out, much synthesis goes on in the nucleus; therefore, enzymes formed in the cytoplasm must make their way back into the nucleus to supervise the building of other compounds. Here again we see a feedback system in operation, in other words a very carefully tuned cycling of events which constitutes life itself.

THE CELL CYCLE AND CELL DIVISION

The Cell Cycle

The most visible evidence that DNA is present and active occurs during the condensation of chromosomes and their separation during cell division. It turns out that the full attention of the chromosomes must be given to this division process, and that they do little else during division. Accurate and reliable chromosomal apportioning to daughter cells is of extreme importance. Most of the interesting molecular events, however, occur during **interphase,** and the cell surface reflects some of these activities. The development of the scanning microscope and of the specialized technique of critical point drying was necessary before we could see such interphase features.

The **cell cycle** can be divided into four phases. After cell division, each daughter cell begins to grow: in a typical mammalian cell in tissue culture this first growth period (G_1) lasts about 6–8 hours.

At that time DNA synthesis (S) in the nucleus begins and continues for about 6 hours; meanwhile the rest of the cell continues to grow. After the amount of DNA in the cell has been doubled, the second growth period of about 5 hours occurs. The cell then enters into mitosis which represents only about 5 percent of the whole cell cycle. At the end of G_2 the cell is about twice its original size, and this mass is fairly accurately proportioned between the two daughter cells. Figure 23–6 shows the corresponding cell surface structure and shape which accompanies these cell cycle events.

Mitosis and Cytokinesis

The first visible evidence that DNA, which makes up the genes, and its associate proteins have doubled, occurs in the cell nucleus and ultimately in the entire cell, which cleaves into two identical daughter cells. Nuclear division was first seen by Flemming in 1882 and was named **mitosis** by him. In most cells, nuclear division is immediately followed by a division of the cytoplasm (cytokinesis). In a few tissues, such as skeletal muscles, nuclear division is not followed by division of the cytoplasm; such tissues are said to be **syncytial.**

The most significant aspect of cell division is the remarkably equal distribution of the chromosomes to the daughter cells. This is essential because the genes control the future metabolism and ultimate success or failure of the cell. The importance of the nuclear material is illustrated by removing small portions of the cytoplasm of an egg and following the nucleated part in subsequent development. Such an egg will produce a normal embryo. If, however, a chromosome is removed from the nucleus of an egg, it usually fails to develop normally. Even though it

$G_1 = 8$ hrs	$S = 6$ hrs	$G_2 = 5$ hrs	$M = 1$ hr

Growth period

G_1

G_1

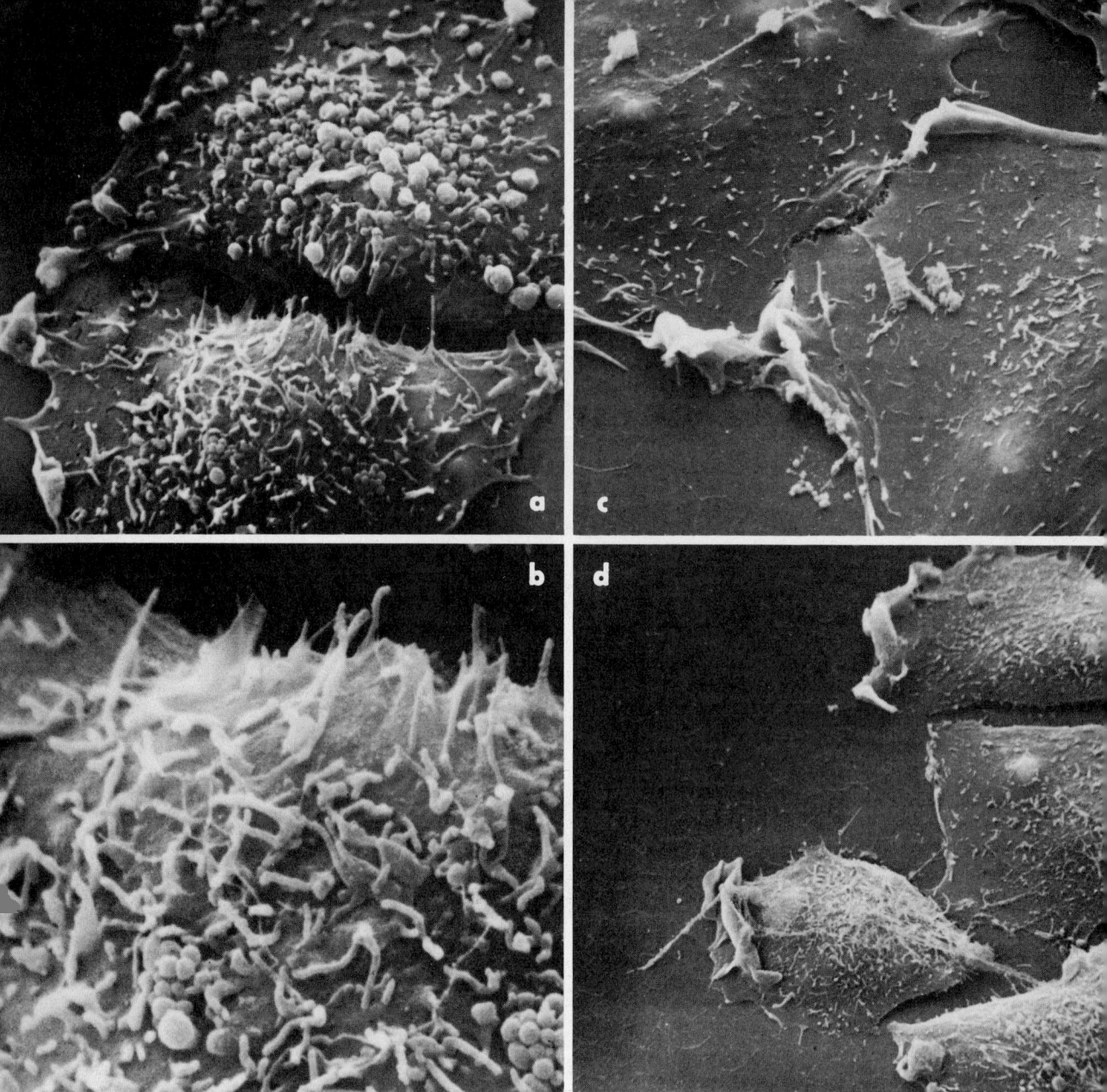

Fig. 23–6. Hamster ovary cells grown in tissue culture. (a) Cells in G_1 are flattened and their surfaces are covered with small hairs (microvilli) and blisterlike spheres (blebs). These two cells are in contact with each other via their microvilli and may be daughter cells. (b) Higher magnification of the lower cell in (a). The blebs appear to be subdivided into little spheres, while the fingerlike microvilli show curls and kinks. (c) In S phase the cells are even flatter, and there are no blebs and fewer microvilli. Verticle ruffles (light areas) are present at some of the edges of the cells. (d). In G_2 the cells are developing more microvilli and blebs and look more like cells in G_1; also, ruffling is more pronounced. The meaning of all these changes is obscure, but the more convoluted surfaces at G_1 and G_2 suggest that the cell membrane is taking in more substances from the medium and exchanging wastes. (For a picture of a cell in division, see Fig. 3–11). Magnification. (a) $\times$ 2,000; (b) $\times$ 8,000; (c) $\times$ 2,600; (d) $\times$ 1,000.

should develop to some extent, the resulting embryo would be quite abnormal. Chromosomes are irreplaceable, because the genes of which they are composed are vital to the life of the entire cell.

The process of cell division is no less precise than replication of DNA or protein synthesis. It is equally complex and many aspects are not as well understood. A major problem is to coil up about 6 feet (175 cm) of DNA (human cells) into 46 metaphase stages, each less than 1 μm in length. Figure 23–7 compares a DNA strand in interphase, which is in the process of making ribosomal RNA (rRNA), with a metaphase chromosome.

We can understand why it is necessary to make small packages for distribution during mitosis, and we know that DNA synthesis has occurred so that these packages are actually double. Cytologists have talked about the "splitting" of the chromosomes and their subsequent segregation into the daughter cells. It is now known that chromosomes do not split; rather, a second chromosome which is an exact replica becomes synthesized in the immediate vicinity of the parent chromosome. The two chromosomes lie so close together that they appear as one. They can be distinguished as two only long after mitosis is under way.

The process of cell division can best be seen in large cells that have been cut into thin slices and stained with different dyes that will differentiate the chromosomes from other regions of the cell. Excellent preparations have been made of a large variety of dividing cells. Among them the early whitefish embryo affords some of the best material (Fig. 23–8). The various stages in cell divi-

Fig. 23–7. (Left) O. L. Miller and his coworkers developed a technique for isolating intact nucleoli from amphibian oocytes, spreading the contents and staining them for electron microscopy. The nucleolar preparations contain long stretches of DNA about 30 Å wide. At several points rRNA genes are simultaneously producing rRNA which forms a fan shape. Each fan represents many strands of rRNA, each one being translated by an RNA polymerase enzyme attached to the DNA. Near the starting point of the gene the RNA molecules are short, increasing in length until the complete rRNA molecule is built, after which they presumably become free in the nucleoplasm. Many such genes are visualized "in action" in these small segments of DNA [× 14,000]. (Right) A scanning electron micrograph of a fully condensed Chinese hamster cell choromosome, about 0.8 μm long. The individual fibers seen coiled up in the chromosome are about 200 Å in diameter [× 100,000].

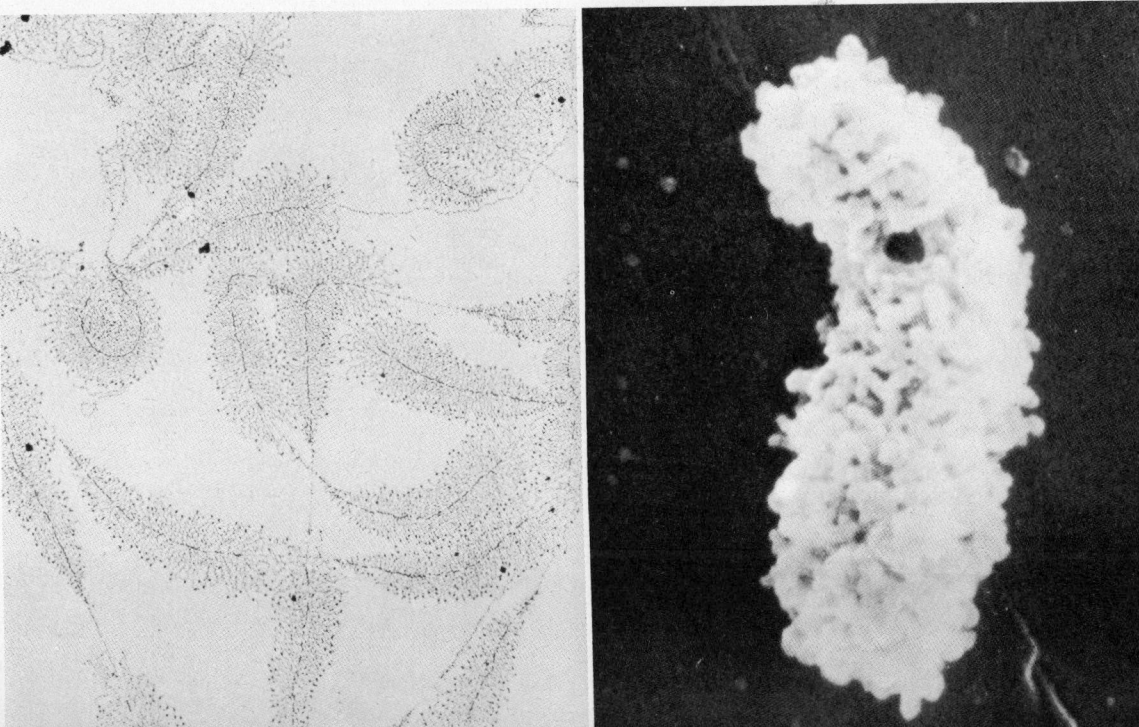

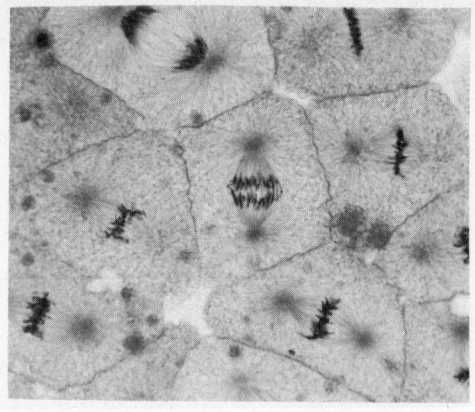

Fig. 23–8. Mitosis in the whitefish early embryo.

sion are shown in a generalized cell (Fig. 23–9).

Biologists have agreed, for convenience, on five steps or phases which cells pass through during the duplication process. (It must be remembered that there is no hard and fast line of demarcation between these phases because the process is a continuous one. They merely denote approximate stages.)

Interphase. The so-called **"resting" phase**, or **interphase**, is the stage when there is little or no apparent activity in the nucleus. Actually, however, it is during this phase that

Fig. 23–9. Schematic representation of a generalized animal cell undergoing division, including approximate time in minutes for each stage.

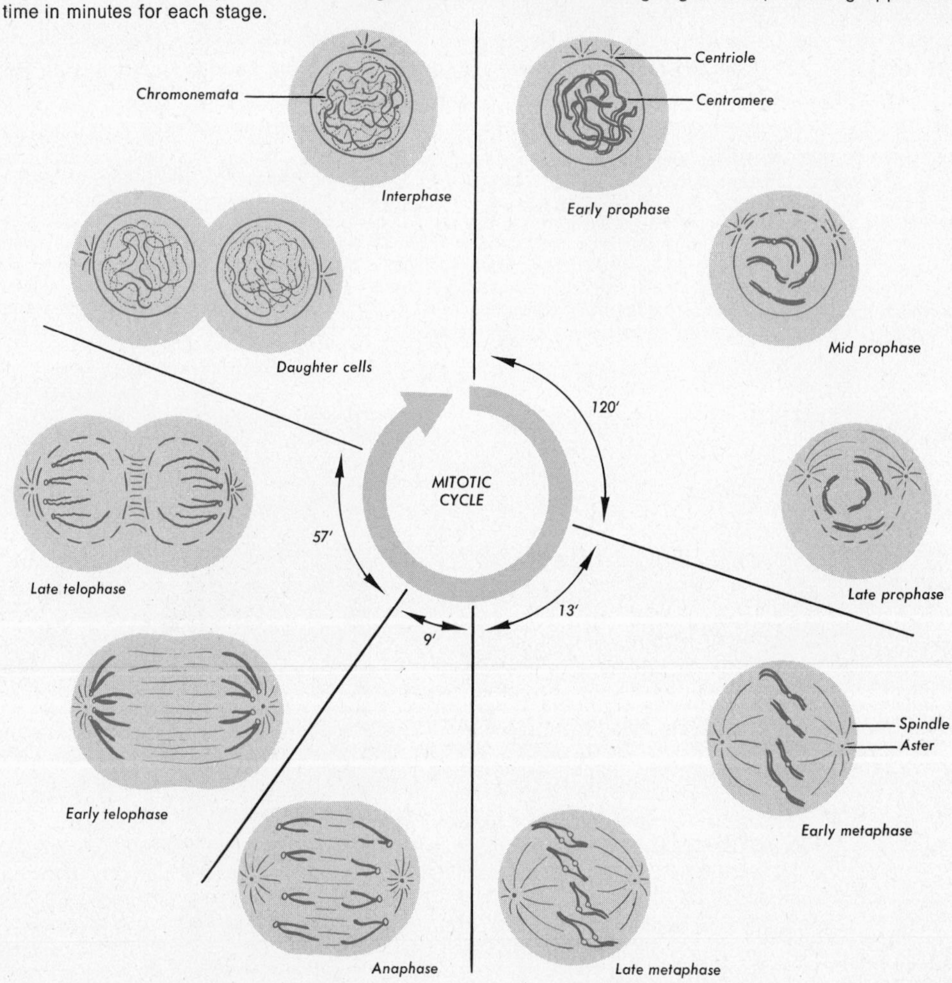

Chromonemata

Interphase

Centriole

Centromere

Early prophase

Mid prophase

Daughter cells

MITOTIC CYCLE

120′

Late prophase

Late telophase

57′

13′

9′

Spindle

Aster

Early metaphase

Early telophase

Anaphase

Late metaphase

growth is going on at a rapid rate and metabolism is at its peak. The chromosomes at this stage appear as long threads and it is essentially impossible to identify them. The chromonemata (p. 71) are uncoiled, which results in the long threadlike appearance of the chromosomes. Actual duplication of the chromosomes occurs during this stage. The long uncoiled paired chromonemata would have difficulty in becoming sufficiently untangled to be easily separated and passed to the daughter cells. Hence in the stages that follow they contract (coil) to only a fraction of their length which facilitates separation and migration.

Prophase. The first sign that cell division is imminent is the separation of the **centrioles** (not seen in the usual whitefish preparations) and the linear contraction of the chromonemata, resulting in thickened chromosomes that now become dimly visible. This phase is often spoken of as **early prophase.** The **mid-** and **late-prophase** stages follow in which the coils in the chromonemata become tightened, resulting in discrete, well-defined chromosomes that stain readily, appearing as black strands. The double nature of the chromosomes is clearly indicated at the termination of prophase, and the centromeres appear as single bodies holding the strands together. The centrioles with their accompanying **asters** have moved toward the opposite poles of the cell, and the nuclear membrane begins to distintegrate.

Metaphase. By the time the nuclear membrane has disappeared, the chromosomes line up at the equatorial plate of the cell. This is called **metaphase.** The centrioles have reached positions opposite one another and a **spindle** is formed, which in fixed preparations consists of fibers that pass from one centriole to the other. Spindles have been isolated from dividing cells and their composition studied (Fig. 23–10). They are complex structures which make up about 15 percent of the cytoplasmic proteins.

Once the spindle is fully formed the centromeres divide and separation of the chromosomes begins. Each centromere is attached to a spindle fiber which seems to play some part in subsequent movement of the chromosomes to the poles. The separation of the chromosomes marks the end of metaphase.

Anaphase. This phase, usually relatively brief, includes the movement of the chromosomes to the poles of the spindle. The chromosomes, now known as **daughter chromosomes,** takes the form of V's and appear to be "pulled" toward the poles by the spindle fibers. Just what forces are involved in this movement is unknown.

Telophase. The daughter chromosomes now at the poles of the spindle come together, maintaining a parallel arrangement. Shortly, the nuclear membrane appears and the chromonemata begin to uncoil, finally attaining the threadlike appearance seen in the resting cell. All of the stages in nuclear reconstruction reenact, in reverse order, the events seen in prophase. At the same time the cytoplasm and cell membranes cleave forming the daughter cells. Thus the interphase is reached and, following a period of growth, the entire process is repeated in cells that are proliferating. Some cells in the adult

Fig. 23–10. These spindles at metaphase have been isolated from sea urchin eggs. Note the complexity of the entire structures.

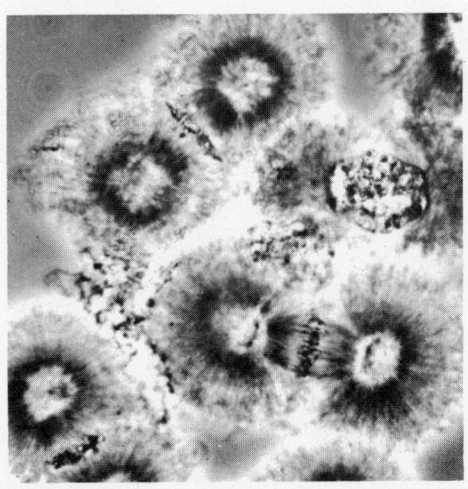

body rarely divide whereas others, such as those generating blood cells and spermatozoa, are constantly undergoing multiplication.

Although there are individual variations among animal cells, the process of mitosis is essentially the same in all cells. Many plant cells do not have centrioles and they separate the cytoplasm in a manner different from animal cells, but division of the nuclear elements is the same.

Factors Regulating Mitosis

Cells divide at varying rates in the different tissues of the body, some undergoing mitosis at extremely short intervals whereas others divide only rarely. For example, sperm cells are produced continuously at a tremen-

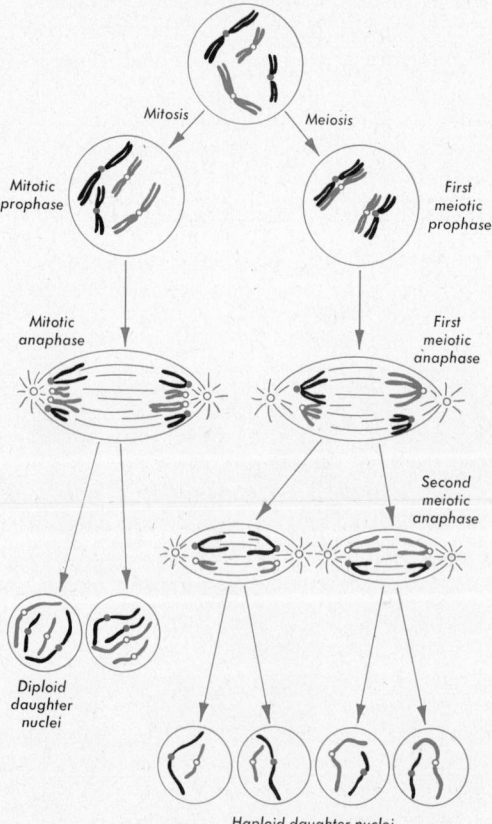

Fig. 23–11. A comparison of mitosis and meiosis, in which dark chromosomes may be considered to be of paternal origin, light ones of maternal.

Mitosis

Meiosis

Mitotic
prophase

First
meiotic
prophase

Mitotic
anaphase

First
meiotic
anaphase

Second
meiotic
anaphase

Diploid
daughter
nuclei

Haploid daughter nuclei

dous rate in the mammalian testis, millions every day. Likewise, red blood cells are manufactured at the rate of 10 million per second, meaning that an equal number of mitoses must occur. On the other hand, some cells such as those in the bones and nervous systems are rarely replaced; hence mitoses occur only rarely in these tissues. All of the cells in an embryo are undergoing rapid cell division, but as the organism matures, the tempo gradually slows down. Biologists have been and still are puzzled over the problem as to what controls this rate of cell division.

It is common knowledge that cells in the region of an injury divide more rapidly than those in the uninjured area. Obviously, the increased rate is essential if the wound is to be closed. Some biologists have been able to demonstrate the presence of a "wound hormone" produced by injured and dead cells which is thought to stimulate mitoses. This has been demonstrated in plants and it is thought that a similar mechanism may operate in animals.

The significance of mitosis should be apparent when it is recalled that all the cells of our body have arisen by mitosis from the original fertilized egg and that every one of the quadrillions of cells contains the same numbers and kinds of chromosomes and genes that were present in that original cell!

FORMATION OF THE GERM CELLS—
MEIOSIS

It has been implied that all cells divide in the manner outlined in the preceding section. There is a very special kind of nuclear division, called **meiosis,** which does not follow this procedure in every detail. Meiosis consists of two nuclear divisions with but one duplication of the chromosomes. Consequently, during meiosis the number of chromosomes is reduced by half. Mitosis and meiosis are compared in Figure 23–11. Note that the end products of the latter consist of four cells with one half the number of chromosomes present in the products of mitosis.

When fertilization occurs the chromo-

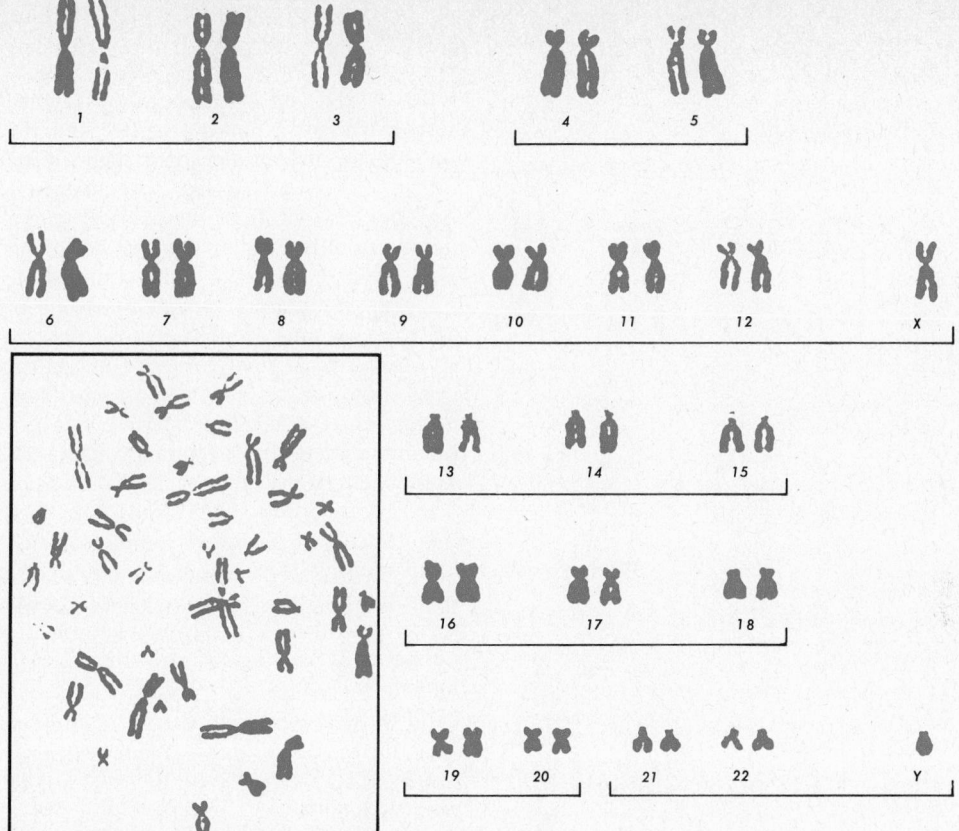

Fig. 23–12. These are stained chromosomes from blood cells taken from a normal human male. Those in the box are randomly distributed as they occur in the dividing cell (flattened for better observation). They are enlarged, arranged in pairs and identified by numbers in the rest of the photograph: There are 22 homologous autosomes and X and Y chromosomes. Each chromosome has duplicated itself but the chromatids remain attached at the centromere.

some number doubles. Obviously at some time during the life cycle the number of chromosomes must be halved. If this were not true, the number of chromosomes would double with each generation, which is not the case. It is during meiosis that the number of chromosomes is reduced from the **diploid** number (2N), which is the number found in body cells, to the **haploid** number (N), just one-half the diploid number, which is the number found in the sex cells. As a matter of fact, all body or soma cells contain duplicate sets of chromosomes, one from the paternal parent and one from the maternal parent. Man, for example, has 23 pairs of chromosomes, 46 in all (Fig. 23–12). The

germ cells contain one full set, or 23 chromosomes; in other words, only one complete complement of genes or chromosomes, not two sets as in the body cells.

The formation of the **germ cells** (gametes or sex cells), namely, eggs and sperms (Fig. 23–13), is called **gametogenesis—spermatogenesis** for sperms and **oogenesis** for eggs.

Spermatogenesis

The formation of sperm cells takes place in the testis, of course, and varies only in detail among various animals. Let us consider the case of mammalian spermatogenesis.

Sperm have their beginning in the peri-

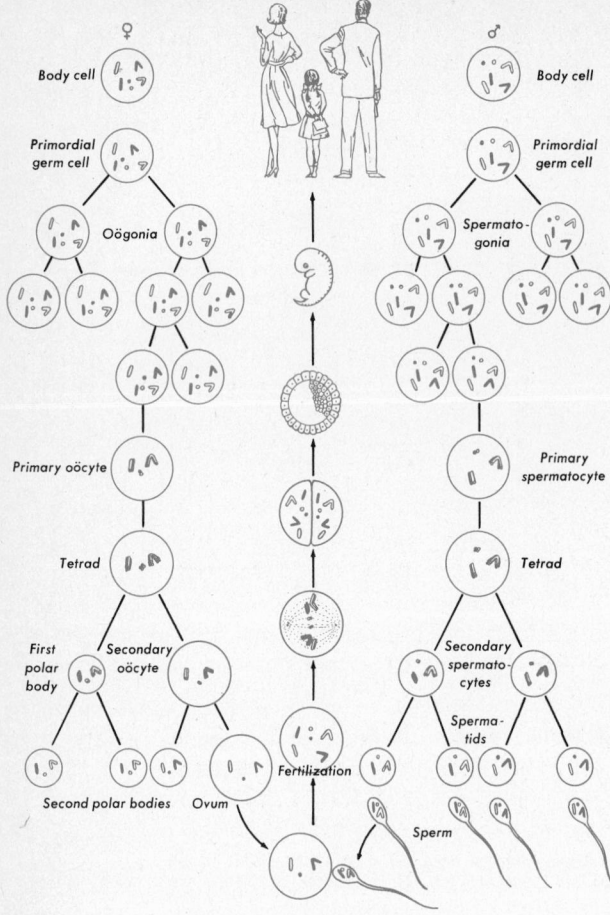

Fig. 23-13. Meiosis.

mitotic divisions before launching into the two divisions (meiosis) which precede maturation and the production of mature sperm. At the end of the mitotic divisions, the spermatogonia increase in size and the **homologous chromosomes** pair. The pairing occurs only between like (homologous) chromosomes, one from the paternal parent and one from the maternal parent. While in this paired condition, each chromosome duplicates itself; thus each chromosome of the pair becomes double. Duplication of DNA occurred at some earlier period, but it is visible only at this time. The association of the four threads is called a **tetrad**; the cells containing them are the **primary spermatocytes.** These nuclei then divide, carrying two of the threads, now called **dyads**, into each daughter cell, or **secondary spermatocyte.** A second division occurs immediately, during which the two threads of the dyads separate so that each daughter cell, now known as a **spermatid,** gets a single thread from each of the original tetrad associations. Each spermatid thus contains only one set of chromosomes instead of two that were present in the

Fig. 23-14. This is a section of the mammalian testis (rat), showing a seminiferous tubule in detail. Sperm cells begin their development at the outer edges of the tubule and as they mature they move into the cavity. The dark rod-shaped bodies nearest the cavity are mature sperm.

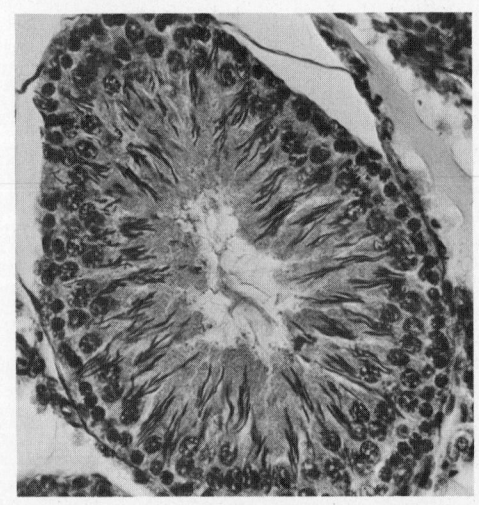

phery of the **seminiferous tubules** (Fig. 23-14), where the cells are very similar to those of any other body cells, namely, diploid with respect to chromosome number. They undergo successive mitoses to build up their population and move toward the **lumen** (cavity) of the tubule to become mature sperm. It is during this period that their number of chromosomes is reduced to the haploid condition. By following these cells it is possible to determine how this reduction occurs. This is done schematically in Figure 23-13.

Because a great number of sperm cells are produced, the **spermatogonia** undergo many

spermatogonium; in other words, the spermatid has the haploid number. The spermatid becomes reduced in size, develops a tail, and matures in other ways to become a full-fledged sperm capable of fertilizing an egg (Fig. 23–15).

During this process it will be noted that the maternal and paternal chromosomes (indicated by black and white in Figure 23–13) may be distributed in several ways so that by the time they reach the spermatid there may be any combination present. There is, however, always a complete set. The mature sperm cell gets only one member of each homologous pair of paternal and maternal chromosomes, never both. This chance distribution of the members of each chromosome pair is very important in the understanding of the mechanics of heredity.

Oogenesis

The process by which eggs are formed varies only slightly from the preceding account for sperm. The principal difference begins when the **primary oocyte** (comparable to the primary spermatocyte) divides. Instead of forming two equal-sized **secondary oocytes,** it produces one large one and one very small one, the latter being known as the **1st polar body.** This discrepancy in size is owing to the fact that virtually all of the cytoplasm goes to one daughter cell. The next division operates on the same principle and results in a single large **ovum** and a **2nd polar body.** Because of their scanty cytoplasm, the polar bodies are nonfunctional and soon disintegrate. The apparent value in this unequal division is to conserve the cytoplasmic contents in order to retain sufficient stored foods for energy during the early stages of embryonic development. This is not necessary for sperms, because once they have contributed their load of chromosomes to the egg in fertilization their job is done. Furthermore, millions of sperm cells are needed to insure fertilization, whereas only comparatively few eggs are necessary to maintain the race.

Eggs vary widely in shape and size (Fig. 23–16). The relative size has little to do with the size of the animal that produces it. For example, the eggs of mouse, man, and elephant are about the same size, and none is as large as the egg of the humming bird. We learned why this is so in an earlier chapter.

The zygote formed at fertilization thus possesses the full diploid number of chromosomes, and all subsequent divisions are mitotic until the mature individual is formed. Then some of its cells are set aside to undergo meiosis in the production of gametes, which again unite with others, and so the process continues from generation to generation. We have seen that in the formation of gametes, the chromosomes may be distributed in a variety of ways. The greater the number of chromosomes (genes), the larger the variety. It is this perennial recombination which accounts for the great variation seen in all living things.

Significance of Sexual Reproduction

The most obvious fact about sexual reproduction which has come from the discussion of gametogenesis is the great variety of germ cells that can be produced by one individual. A germ cell receives one single-thread chromosome from each tetrad. The assortment of

Fig. 23–15. Sperm cells vary widely in shape, as indicated by these sketches.

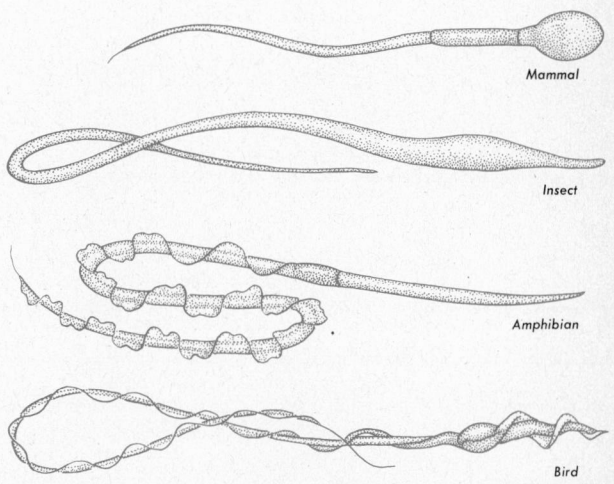

Mammal

Insect

Amphibian

Bird

maternal and paternal chromosome threads to the germ cells is independent for each of the tetrads, so that it is possible that different germ cells may have a different combination of maternal and paternal chromosomes. This is further complicated by the fact that while the chromosomes are paired and in the four-strand stage, they often exchange homologous segments, a phenomenon known as the **crossing-over** (p. 523). This exchange may occur in several places along the paired threads. Thus the threads which are finally assorted to the germ cells are not all like the original maternal or paternal chromosome; some have segments of both. On the average, chromosomes break and exchange genes in

about three places, which means that the original mixing on the basis of chromosome number is increased by 2^3, or 8 times. Let us consider, for a moment, the possibilities in man: there are 23 chromosome pairs in the body cells and, of course, 23 single chromosomes in the germ cells. The possible chromosome combinations then become 2 raised to the 23rd power, or 8,388,608. If we multiply this figure by 8 we arrive at the number of kinds of germ cells it is possible for one person to produce. Providing it were possible, one couple could become the parents of all the people in the world and no two of their offspring need be exactly alike.

Because of this wide variability among germ cells, the offspring resulting from their union are different from each other and from either parent. They are new and different individuals, each with its own respective parts assembled in a slightly different manner from any other members of the species. This makes for great variability among species, which is the fundamental reason why evolution has gone on the way it has. Without variation there would be no way for animals to become better suited to their environment because they would all be exactly alike; hence no one of them would be able to explore a new situation any better than another. This is what happens when **asexual** reproduction is the only means of reproduction. While it is generally confined to the lower animals, it is known to occur among some plants under man's cultivation. The seedless orange trees, for example, are all grown from grafts, so that actually all that is done is to increase the size of the original tree. Since all of the cells carry the same genes, they have no opportunity for variation and must therefore all be exactly alike, barring an occasional mutation. Most of the lower invertebrate animals reproduce asexually as well as sexually. In the former case they likewise have no opportunity for variation. Sexual reproduction evolved very early among animals and is probably largely responsible for subsequent evolution of more complex forms.

Fig. 23–16. Animal eggs vary widely in shape and size.

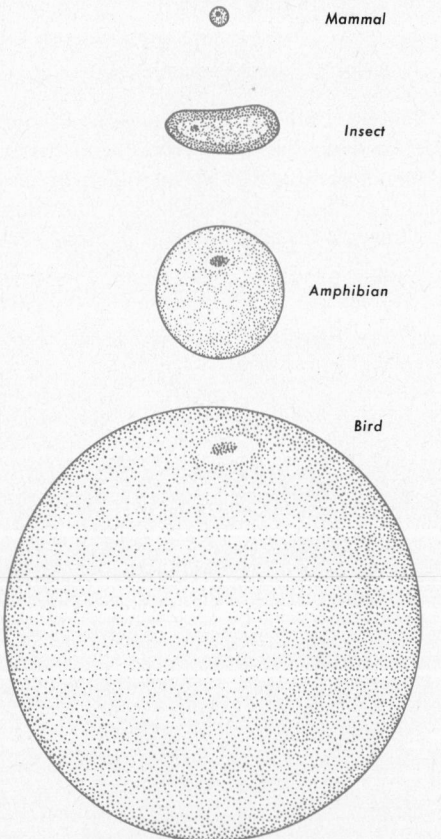

Mammal

Insect

Amphibian

Bird

Books

DeRobertis, E. D. P., Nowinski, W. W., and Saez, F. A., *Cell Biology*, 4th ed. Philadelphia: Saunders, 1970.

Mazia, D., "Mitosis and the Physiology of Cell Division," in *The Cell*, J. Brachet and A. E. Mirsky, eds., Vol. **III.** New York: Academic Press, 1961.

Rhoades, M. M., "Meiosis—In the Cell," in *The Cell*, J. Brachet and A. E. Mirsky, eds., Vol. **III.** New York, Academic Press, 1961.

Articles

Crick, F. H. C., "The Genetic Code: III." *Scientific American*, October, 1966.

Mazia, D., "The Cell Cycle." *Scientific American*, January, 1974.

Taylor, J. H., "The Duplication of Chromosomes." *Scientific American*, June, 1958.

24

CONTINUITY OF THE SPECIES

The statement that "like begets like" is so obviously true that we could easily overlook some far-reaching questions about the physical basis of similarity between parents and offspring. Revealing questions could well be put about the physical, chemical, or organizational basis of the material that perpetuates species characteristics. Our experience, furthermore, shows us that although like begets like, variation may occur. Offspring are not necessarily identical to their parents. Although a basic theme continues through successive sexual generations, variations upon the theme occur. An elephant produces only elephant offspring, but these offspring may differ from the parent in growth rate, size, disease susceptibility, or other traits, Any hypotheses about the nature of the material that maintains continuity must also account for the opportunity for variations between generations and among individuals within a generation.

Genetics is the branch of biology that is concerned with heredity and variation. Geneticists ask such questions as the following: What is the genetic material that is transmitted from parent to offspring? What is its chemical and physical structure? How is the genetic material distributed to the offspring? What regularities underlie the distribution of genetic material within families and within populations? How does the genetic information act? How does it come to expression as the structure and function of the individual? How might the expression of a given genetic element be modified by the presence of other genetic elements and by fluctuations of the environmental system within which the genetic element acts? How do variations arise among individuals of a population?

Descriptions of specific animals and plants show that specialized reproductive cells, **gametes**, bridge the generations. In all cases of sexual reproduction, plant or animal, a

496

necessary part at some point in the life cycle is the fusion of gametes and of their nuclei to form a **zygote**. Reproduction is thus really a cellular phenomenon. We must, therefore, look to the content and to the organization of the gametes and of the zygote for the genetic basis of continuity of a species.

In the description of cell division already given we see that cells arise from preexisting cells by division. We note further that during division an orderly distribution of the chromosomes takes place. During mitosis the distribution of the chromosomes is qualitatively and quantitatively identical in each of the daughter cells. At meiosis, one, but only one, of each kind of chromosome is contained in each meiotic product. In either case there is a controlled and orderly distribution of the chromosomes.

In descriptions already given of early embryology we see that development is **epigenetic**; development proceeds stepwise, acquiring structures and complexity with each succeeding step. The mature structures do not exist as such within the fertilized egg. We cannot look at a fertilized egg and predict the appearance of the mature individual. The fertilized egg of rabbit and of man are quite similar in appearance.

The fertilized egg, therefore, contains as part of its organization the potential for specific development. The genetic information comes to the cell as a code, which is subsequently expressed by cellular activity during development. Earlier chapters have shown that cell control resides in the nucleus, which acts by directing cellular metabolism. The chromosomes, containing the protein-DNA complexes, are the visible, structural organelles of the nucleus that carry the genetic elements, the genes. The DNA of the chromosomes, as described in Chapter 23, encodes the genetic information in a sequence of nucleotides along the DNA molecule. When this information is to be expressed, the code of the DNA is transcribed into a complementary sequence of ribonucleotides to make up a messenger-RNA molecule. The information that has been thus transcribed into the RNA molecule is trans-

lated into a specified sequence of amino acids to make up a specific unique polypeptide chain. The proteins so specified are the biologically functional molecules that determine the pattern of cellular biochemical reactions. Thus the genes act by determining the structure of biologically functional molecules, which in turn influence the course of the development and the behavior of the cell. The genetic information is transmitted from generation to generation by deoxyribonucleoprotein molecules. The appearance of an individual is a consequence of the kinds of nucleoprotein molecules received and the conditions under which the information comes to expression during development.

Genotype and Phenotype. **Genotype** is a term used to indicate the total genetic material. Genotype refers to the sum total of all the genes. The genotype therefore represents the material which has been received by the individual from its parents. An individual does not inherit the characteristics as such from its parents, but rather the potentiality for their development.

Phenotype, on the other hand, is a term used to refer to the total characteristics of the individual. Whereas the genotype of an individual is described in terms of the genes it contains, the phenotype is described in terms of the appearance or the characteristics of the individual. The features or traits by which we may describe an individual make up its phenotype. If we, for example, describe a person as having brown eyes, black curly hair, light skin, good mental ability, AB blood type, and so on, we are describing the phenotype of the person.

Because in genetic analyses we often must infer the genotype from a study of the phenotypes of an individual and its relatives, we may be misled, unless cautioned, into using the terms interchangeably. The terms stand for different ideas. In interpreting all genetic phenomena we must keep clearly in mind the distinction between the genotype, the genetic material itself, and the phenotype, the end result of the expression of the genotype in a given set of circumstances.

The phenotype is a consequence of the interaction of all the genetic elements with each other and with the environment within which development is taking place. The phenotype usually observed is removed by few or many reaction steps from the gene itself. Each of these reaction steps is a point at which the usual sequence of events may be blocked or redirected, depending upon the type of reaction sequence taking place or the type of environmental change affecting the developmental step. Some genetically controlled reaction sequences are less susceptible than others to modification by the environment. Blood types, for example, are little, if at all, changed by the usual environmental fluctuations. On the other hand, the reaction sequences leading to other traits may be easily modified by environmental modifications. Weight of an individual may be markedly changed by the amount of food intake.

It is well to recognize, furthermore, that closely similar phenotypes may not necessarily have the same genotypes. The observation that two individuals are similar in appearance in no way insures that two individuals are identical in genotype. And, conversely, identical genotypes may give rise to different phenotypes. Identical twins have the same genotypes; yet if they are reared under different nutritional conditions (one half-starved and the other well fed) they will differ in stature and perhaps in other ways.

HISTORICAL BACKGROUND OF MODERN GENETICS

A fusion of three lines of investigation is included in the rise of modern genetics. These three fields of investigation were originally distinct and independent. One line, **hybridization studies,** was concerned with explaining and controlling the results of experimental breeding of both animals and plants. A second line, **cytological studies,** was concerned especially with chromosome mechanics as related to cell division, to

Fig. 24–1. Careful selection for thousands of years has produced a wide variety of shapes and sizes in dogs.

gamete formation, and to fertilization. A third line, **biochemical studies** of the cell, was concerned with the structure and the function of the nucleic acids and their relation to protein synthesis.

Hybridization Studies

Hybridization of animals was known by early man; he apparently understood that certain traits could be emphasized and improved by selective breeding. Records indicate that one of the first attempts to utilize this idea was with the wolflike ancestor of the dog. The first association was undoubtedly casual, but through selection the animal became more and more the kind of beast that best suited man's personality and habits. Thousands of years of selection were required, intentional or accidental, to produce an animal so well adapted by size, aggressiveness, and behavior to the whims of primitive man, to say nothing of the idealistic requirements of modern man (Fig. 24–1).

Other animals, such as cattle, horses, and sheep, as well as numerous plants, were domesticated later, and in most of them intentional selection for improvement was practiced. This was accomplished simply by applying the rule that "like begets like." Even though we know now that there are certain discrepancies in this "principle," the fact remains that after thousands of years of selection, plants and animals were gradually molded into the kinds that could best be utilized by man.

It is interesting to note that practically all of the plants and animals capable of domestication were domesticated long before the principles of genetics were known, indeed most of them before writing was established. Very few wild organisms have been domesticated in modern times; all that has been accomplished is improvement of the established types, and in some instances this has been remarkable. For example, modern cattle have become veritable factories for milk and beef production; hens have been known to produce an egg a day for as long as 354 days. An outstanding example among plants is hybrid corn in which the application of the principle of hybrid vigor (p. 515) has resulted in a 30 percent increase in yield. Even greater accomplishments are anticipated in the future.

In spite of the fact that information concerning selective breeding of domestic plants and animals was accumulating throughout the eighteenth century and the first half of the nineteenth, no explanation was given as to how specific characteristics were passed from one generation to the next. Inspired by Darwin's *Origin of Species*, Sir Francis Galton initiated a series of studies on inheritance in about 1859 which resulted in significant conclusions concerning variation within a species. Regarding human inheritance he, like many others, had studied certain complex characters such as height and intelligence and tried to follow them in succeeding generations. The results turned out to be hopelessly complex and he could formulate no generalizations that were meaningful.

This confused situation continued throughout the nineteenth century, in spite of the fact that the basic principles of heredity were discovered and reported by **Gregor Mendel** in 1865 (Fig. 24–2). This modest Austrian monk, as a result of his efforts, is now known as the "father of genetics." In order to better understand how this remarkable man came by his discovery let us review briefly the story of his life.

Mendel grew up on a small fruit farm where he developed a love from his father for growing plants. He learned how to graft and hybridize plants at an early age. His education included preparatory schooling and two years of college, following which he entered Altbrünn Monastery where he was encouraged to follow his interest in plant hybridization. He carried out his famous pea experiments in the monastery gardens during the years 1856 to 1864, during which time

Fig. 24–2. Gregor Mendel (1822–1884), working with plants, laid the foundation for modern genetics, although his great work was not recognized by the scientific world until 1900.

he also attended the University of Vienna where he studied science and mathematics. It was here that he acquired technical skill and an understanding of the mathematical principle of probability which was essential in making his later discoveries. Since Mendel's accomplishments were not recognized during his lifetime, little information is available concerning his personal life. Much about how he came to his original concept concerning hereditary mechanisms remains obscure although his interest in living things, his keen intellect, and his devotion to detailed observation contributed to his success in making the revolutionary contribution that bears his name.

Prior to Mendel a great deal of data had been collected but it had very little meaning since it was undigested and therefore incomprehensible. Data were accumulating in other areas of biology that had important bearing on genetics, a science that had not yet been born. The morphology of the cell was being understood and it is highly probable that cytologists would have uncovered Mendel's discoveries through the study of cells, but at a much later date. During this period sexual reproduction was known at the cellular level, that is, meiosis was deciphered and explained. Being at the heart of recombination, this knowledge was an absolute essential before the mechanism involved in hybridization could be understood. Another discipline that was instrumental in solving the mysteries of inheritance patterns was mathematics. During the early eighteenth century the area of probability became popular, primarily as a result of a need for more information by gamblers and insurance companies concerning predictability. By the beginning of the present century information in these divergent fields was brought to bear upon the problems of heredity and the science of genetics emerged.

Going back to 1865, Mendel presented his work in the form of a short paper before a scientific society of his day. He could hardly be heard above the heated arguments that had been precipitated by Darwin's forceful presentation of the theory of evolution published six years previously, yet in this paper Mendel had portrayed one of the principal keys to evolution. No one understood the significance of this work because he was 35 years ahead of his time. The significance of his monumental work was not appreciated until 1900, 16 years after Mendel had died, at which time three botanists working in different parts of Europe uncovered the same principles that Mendel had so clearly pointed out many years before. They were DeVries in Holland, Correns in Germany, and Tschermak in Austria. Mendel's paper, which had gathered dust through the years, was republished and only then did this humble man receive the acclaim due him. With this rediscovery of **Mendel's Principles** or Laws the science of genetics was born.

Mendel succeeded where others had failed for several reasons. First he chose to study single characters instead of the whole individual as others had done before him. He was fortunate, also in his selection of material for study, although this may not have been mere chance. Mendel knew a great deal about plants and probably was drawn to those which possessed the characters he wished to study. It is now known that he worked on over twenty different characters of the pea but chose to report only those that were similar and which gave well-defined ratios. The probability of selecting at random seven characters, each on a separate chromosome (the pea has seven chromosomes) is highly unlikely; rather Mendel probably selected them because they behaved similarly. This is no discredit to him but is another indication of his appreciation of experimental design. Had his work been accepted in 1865 he would probably have reported his complete work in subsequent publications.

Following his spectacular work with the garden pea, Mendel crossed other plants such as different varieties of hawkweeds, but failed to confirm his earlier results. These findings must have shaken his faith to some extent in his work with peas and may have been one of the reasons why he did not press for recognition. He may have felt that the pea was a special case and that what he had learned had little significance as a general principle that could be applied to all organisms. We can now explain the results he obtained with hawkweeds and they in no way detract from his conclusions drawn from his work with peas.

Mendel's choice of the pea as an experimental tool was a good one owing to the purity of the plants for particular characters such as flower color. They always bred true, that is, the offspring retained the character of the adult from generation to generation. Once Mendel had pure breeding strains he could cross them and follow the trait in subsequent generations. This he did by keeping copious notes with actual counts recorded in tabular form. From these figures his ratios for each experiment stood out strikingly, and they were essentially the same for seven different characters in the pea. The uniformity of these ratios led him to formulate his final conclusions which were consolidated in the famous lecture which he gave in 1865. These were later incorporated by others into what are now known as Mendel's Principles.

During the first ten years of the twentieth century, the science of genetics became firmly established. Some of the great biologists of this period, already deep in genetic problems, were stimulated to apply Mendel's ratios to other plants and animals. Among these was the Englishman William Bateson (1861–1926), who immediately set about to test Mendel's Principles on plants other than the pea, as well as on poultry and other animals. The full impact of Mendel's discoveries became known when Bateson and others found that the ratios applied to the organisms they worked with, and this was, therefore, a general principle for all organisms, a concept which Mendel himself believed.

By 1906, the new science had received the name of **genetics** from Bateson and continued to grow in recognition up to the present. Organisms from viruses to mammals are being used in a large variety of experiments including those concerned with the molecular nature of the gene to those concerned with the inheritance pattern of cancer in man. The information coming from laboratories all over the world has great significance not only because of its practical importance but also because it contributes to our understanding of the processes of nature.

Mendel's chief contribution—that which provided the key for analyzing heredity—was his supposition of a particulate nature for the hereditary material. He regarded hereditary material to be an aggregate of discrete, independent, particles (later called **genes**); he regarded these particles, or factors, to be transmitted to offspring by the gametes according to rules of probability. Mendel did not speculate upon the chemical or physical nature of the factors, nor did he suggest anything about their intracellular localization. He, of course, knew nothing about any relation between these factors and the chromosomes. At the time of Mendel's paper not much was known by anyone about the chromosomes and their distribution. Mendel's conclusions are especially impressive when we realize that the hereditary factors were abstract, hypothetical entities to him. Nevertheless, on the basis of a penetrating and original analysis of carefully collected and selected breeding results, he predicted their existence and formulated two principles of their distribution—**segregation** and **independent assortment**.

MENDEL'S PRINCIPLES

First Principle—Segregation

Mendel worked primarily with plants, but what he learned in plants has since been found to be equally true of animals.

If guinea pigs of the same color are crossed with one another, as for example, a pure-breeding black with a pure-breeding white, the offspring will be black (Fig. 24–3). From this cross one outstanding fact is observed, namely, that black is **dominant**, white **recessive**. This means that when two such characters are brought together in the production

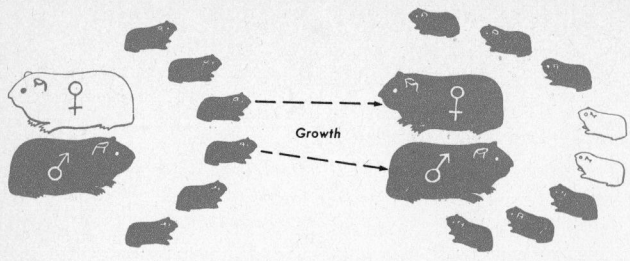

Fig. 24–3. Mendel's first principle—segregation—can be shown by breeding guinea pigs as outlined here. Black and white animals (parents) give rise to only black pigs in the F_1 generation. When these are crossed the black and white hair color segregates out so that the ratio is three black pigs to one white in the F_2 generation.

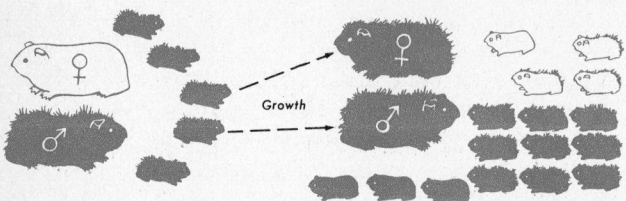

Fig. 24–4. Mendel's second principle—independent assortment—can also be demonstrated with guinea pigs. In a crossing of a black rough pig with a white smooth one, the F_1 generation consists of only black rough pigs. If any two of these pigs are crossed, both characters (coat color and texture) reappear, but in different combinations. There will be 9 black rough, 3 black smooth, 3 white rough, and only one white smooth. From this it can be deduced that the characters are independent units.

of a hybrid, one character completely masks the other. In the above case, the factor for black hair completely dominates the factor for no color, or white. Consequently, all of the offspring are black.

The hybrid is known as the **first filial** or **F_1 generation.** In order to determine whether or not the white is completely lost in the F_1 pigs, they can be bred together, producing another generation, the **F_2 (second filial)**. When many crosses are made, so there are large numbers of offspring, about three-fourths will be black and one-fourth white. This is called the 3:1 ratio which Mendel was able to obtain so many times with his plants. From this experiment it is obvious that the character for white was not really

lost in the F_1 generation. It was merely temporarily hidden or latent, because when the hybrids were crossed the character for white **reappeared unchanged** and in a definite ratio. This means that the character was a **unit** and remained as a unit, even though it was unable to express itself when in the presence of the dominant character. This is Mendel's **Principle of Segregation.**

Second Principle—Independent Assortment

The above crosses involve only one character, namely, color of the skin coat. What would happen if two characters are followed through two generations? Will the characters again be lost, will new combinations be formed, or what will happen?

There are thousands of pairs of contrasting characters in any animal, but by selecting any two, preferably each pair that demonstrate a striking dominant and recessive condition, it should be possible to determine the fate of the characters as they pass from one generation to the next. Employing guinea pigs again we may select another pair of characters in addition to black and white color, such as **rough** and **smooth** coats (Fig. 24–4). Ordinarily the coats are smooth, with all of the hairs pointed in one direction. However, there is a breed in which the hair grows in whorls in various places on the body, giving the animal a roughened appearance. The character is easily seen and is dominant over smooth. When a **rough black** guinea pig is crossed with a **smooth white** one, all of the hybrids (F_1) are **rough black.** When these are inbred new combinations are seen. There are **nine** rough black, **three** rough white, **three** smooth black, and **one** smooth white. This is called the 9:3:3:1 ratio. The essential fact obtained from this experiment is that each of the two characters retains its identity absolutely **independent** of the other. They are all combined in the F_1, though only the dominant ones, namely, rough and black, are visible. In the F_2 each character went its own way independent of the other and showed up in the offspring in new combinations that were not present in either parents or grandparents, that is, rough white and smooth black. This is the **Principle of Independent Assortment.**

The significance of Mendel's work is difficult to overestimate because it was the starting point from which geneticists were able to probe more deeply into the precise nature of the mechanism of inheritance. With the accumulation of experimental evidence since 1900 the basic ideas of Mendel have been extended, made more precise, and have been modified to include what at first appeared to be apparent exceptions. Some modifications (for example, linkage) to the Mendelian principles are described later in the chapter. Some of these modifications become understandable when we consider the implications of the chromosome theory of heredity, which resulted from a merging of the fields of chromosome cytology and of Mendelism.

Cytological Studies

During the latter part of the 1800s cytologists actively pursued microscopical studies of chromosomes and their distribution during cell division. The object of these investigations was a clarification of the mechanics of cell division, gamete formation, and fertilization. During the period from about 1875 until the end of the century, the details of mitosis and of meiosis (as described in Chapter 23) were accumulated. By 1900 there was general acceptance that a zygote nucleus, arising from fusion of egg and sperm nuclei, has two sets of homologous chromosomes; a single set contributed by each of the two parents. Thus, *each* chromosome is present in duplicate in the fusion nucleus and in all nuclei descended by mitosis from the zygote nucleus. By 1900 it was also established that during meiosis homologous chromosomes separate from each other so that each meiotic product receives only one of each pair of homologous chromosomes; each meiotic product, of course, receives one representative of all the chromosome pairs. Soon after the rediscovery of Mendel's paper it became apparent that there were similarities in the behavior of the chromosomes and of the Mendelian particles. In 1902 an American, Walter S. Sutton, and a German, Theodor Boveri, independently published papers listing parallelisms in the observed distributional patterns of the chromosomes and the inferred distribution of Mendelian factors. Their suggestion that the Mendelian factors are borne on the chromosomes began the intimate relationship of Mendelian genetics and of cytology, the combined field known now as **cytogenetics.** Additional early suggestions of a relationship between genetic traits and chromosomes came from McClung's (1902) and E. B. Wilson's (1905) observations detailing correlations between sex and particular chromosome constitutions.

The overwhelming evidence for the theory that genes are carried on the chromosomes grew out of the remarkable series of publications, starting about 1910, based upon the breeding experiments devised by the American biologists, Thomas Hunt Morgan (1866–1945), and his colleagues, with the fruit fly *Drosophila melanogaster.* They could explain their total experimental data sensibly only by supposing that each gene had a fixed and unique position, or locus, along the length of a particular chromosome; that all the genes were arranged along chromosomes in a specific order; that genes located on one chromosome tended to be transmitted together (thus were said to be linked); and that linked genes could recombine by reciprocal interchange of homologous segments of homologous chromosomes. The **chromosome theory of heredity** was strongly supported by numerous examples of parallelisms of chromosome and of gene distribution. If accidental abnormal chromosome distributions occur, they are matched by similar abnormal distribution of the genes associated with the chromosomes.

Biochemical Studies

The nucleic acids were discovered in 1869 by the Swiss biochemist, **Friedrich Miescher** (1844–1895). Miescher's research began as an attempt to determine the chemical composition of the cell nucleus. From a series of extractions of cell nuclei, from pus cells and fish sperm, he obtained a nonproteinaceous substance that was rich in phosphorus. This finding came as a surprise and stimulated intensive research to characterize the substance. Because it was extracted from nuclei, Miescher named it "nuclein"; one of his students, noting the acidic properties of nuclein, renamed it **nucleic acid**. Additional studies by Miescher, by his students, and later by other biochemists, uncovered the chemical composition of the nucleic acids, which has already been described in the preceding chapter. We note here that during the late nineteenth and early twentieth centuries nothing was suspected about the genetic function of nucleic acid. Rather, nucleic acid research was concerned with determining its composition, its intracellular distribution, and its distribution among different species of animals and plants. This research showed that nucleic acids were distributed universally among all organisms, animal and plant. They also drew a distinction between the composition, the intracellular distribution, and the relation to protein synthesis of two kinds of nucleic acid, **RNA** and **DNA**.

During the early 1940s various kinds of indirect evidence accumulated indicated that nucleic acids were somehow associated with protein synthesis. Some tentative suggestions began to be made about a possible involvement of nucleic acids in genetic information transfer. But the direct demonstration of a genetic function for DNA came in 1944, seventy-five years after Miescher's discovery, with the publication by Avery, MacLeod, and McCarty that the bacterial transforming principle was DNA. The phenomenon of bacterial transformation, whereby a recipient cell exposed to extract from a donor cell acquired some genetic properties of the donor cell, had been known for nearly twenty years. Avery, MacLeod, and McCarty demonstrated that the active principle by which the genetic properties were transferred was protein-free DNA. This meant,

then, that pure DNA had the capacity for storing genetic information, for transmitting it from one cell to another, and for directing metabolic activities of a host cell.

These studies are convincing with respect to bacteria, but what about multicellular organisms? Although the evidence comes from varying sources it is all consistent with the conclusion that DNA is the genetic material. Note that DNA is a stable compound; its amount is constant per chromosome set per species and its distribution to daughter cells is precise. All these are characteristics expected of genetic material. Moreover, the relationship between DNA, RNA, and protein synthesis appears to be essentially the same for bacteria and for multicellular organisms.

The physical and chemical structure of DNA as proposed by Watson and Crick in 1953 (p. 000) supports the conclusion that DNA is the genetic material because its structure is such that it can meet the two essential requirements of a genetic material —information for directing its specific self-replication and information for directing cellular activities.

There seems little doubt now that the specificity of a gene, its specific information, resides in a unique sequence of nucleotides along a segment of DNA.

The discovery that DNA transcribes RNA, which in turn translates the sequence of amino acids in a protein, which then controls a specific biochemical reaction, makes understandable the relationships between a Mendelian gene, its localization on a chromosome in the nucleus, its control of protein synthesis in the cytoplasm, the control of a biochemical reaction by the protein, and the consequence for the phenotype of the metabolic activity of the cell. The connection between the Mendelian gene and its expression in the phenotype is no longer as mysterious as it once was.

We have only glimpsed the consequences that have resulted from the fusion of biochemistry and biology during the last several decades of genetics research.

To date one of the most powerful methods for analyzing genetic differences associated with differences in traits is that of experimental breeding. Such studies are often referred to as recombinational genetics, formal genetics, Mendelian genetics, or transmission genetics. Whatever the name, the method involves hybridization of strains selected for the differences to be analyzed; the progeny of several generations after the cross is classified for the kinds and relative numbers of the different characteristics. From these data inferences are made about the number, localization, and other properties of the genetic differences responsible for the trait differences. Several examples are given below of the hybridization method of genetic analysis.

Before the illustrative crosses are discussed, the terminology and symbolism need to be defined. The two members of a single pair of genes are known as **alleles** of each other; alleles reside at **homologous loci** on homologous chromosomes. If two alleles are identical, the pair is said to be **homozygous.** If the two alleles are not identical, the pair is said to be **heterozygous.** A nucleus may, of course, be homozygous for some gene pairs and heterozygous for others; or, it may be homozygous or heterozygous for all gene pairs—depending upon the genetic constitution of the gametes that fused to form the individual. Ordinarily, when we use the words homozygous and heterozygous, we refer only to that allelic pair, or those allelic pairs, under investigation.

Genes may be symbolized by any one of several systems. All systems attempt to distinguish not only between different members of an allelic pair but also between nonallelic genes. In one widely used system letters are used to represent the genes; alleles are represented by the same letter (or group of letters). A capital letter is used to represent one form of a gene and a small letter its allele. For example, a gene leading to black coat in guinea pigs is symbolized as W and

its allele, leading to white coat, as *w.* Unless a gene exists in more than one form it cannot be detected and its effect analyzed. If more than two forms of a gene are known they constitute a **multiple allelic series;** the members of a multiple allelic series may be distinguished by adding superscripts to the basic gene symbol. Nonallelic genes are, of course, represented by different letters. In the guinea pig, the gene leading to rough coat might be symbolized as S and its allele (for smooth coat) as *s.* A homozygous black pig is symbolized as W W; a homozygous white pig as *w w.*

But what about the appearance of the heterozygote, W *w?* Its phenotype depends upon what interaction there might be between the alleles. What type of **allelic interaction** takes place between any particular pair of alleles must be experimentally determined. In the case of the coat color alleles being discussed here, the heterozygote, W *w,* is black. This type of interaction is described by saying that black coat is **dominant** to white coat; or alternately, that white coat is **recessive** to black coat, as we have already discussed. We see that other types of interaction can also occur; that is, dominance and recessiveness are not the only types of allelic interaction.

In observing ratios in actual practice one might think that they are not very precise. For example, in one litter of guinea pigs as a result of a mating between the F_1 hybrids in the above experiment there may be nearly all combinations of the expected 3:1 ratio. If only four pigs were born they may all be black or they may all be white, although the latter possibility is remote. It is more probable that there would be some combination of black and white, two of each color or even the expected 3:1 ratio. If, on the other hand, the litters of many such matings were used, say 1000 offspring, the 3:1 ratio would be approximated closely. This simply means that all genetic ratios follow the laws of probability and in small samples these laws, while operative, are not very reliable. In the above cross there are always three chances out of

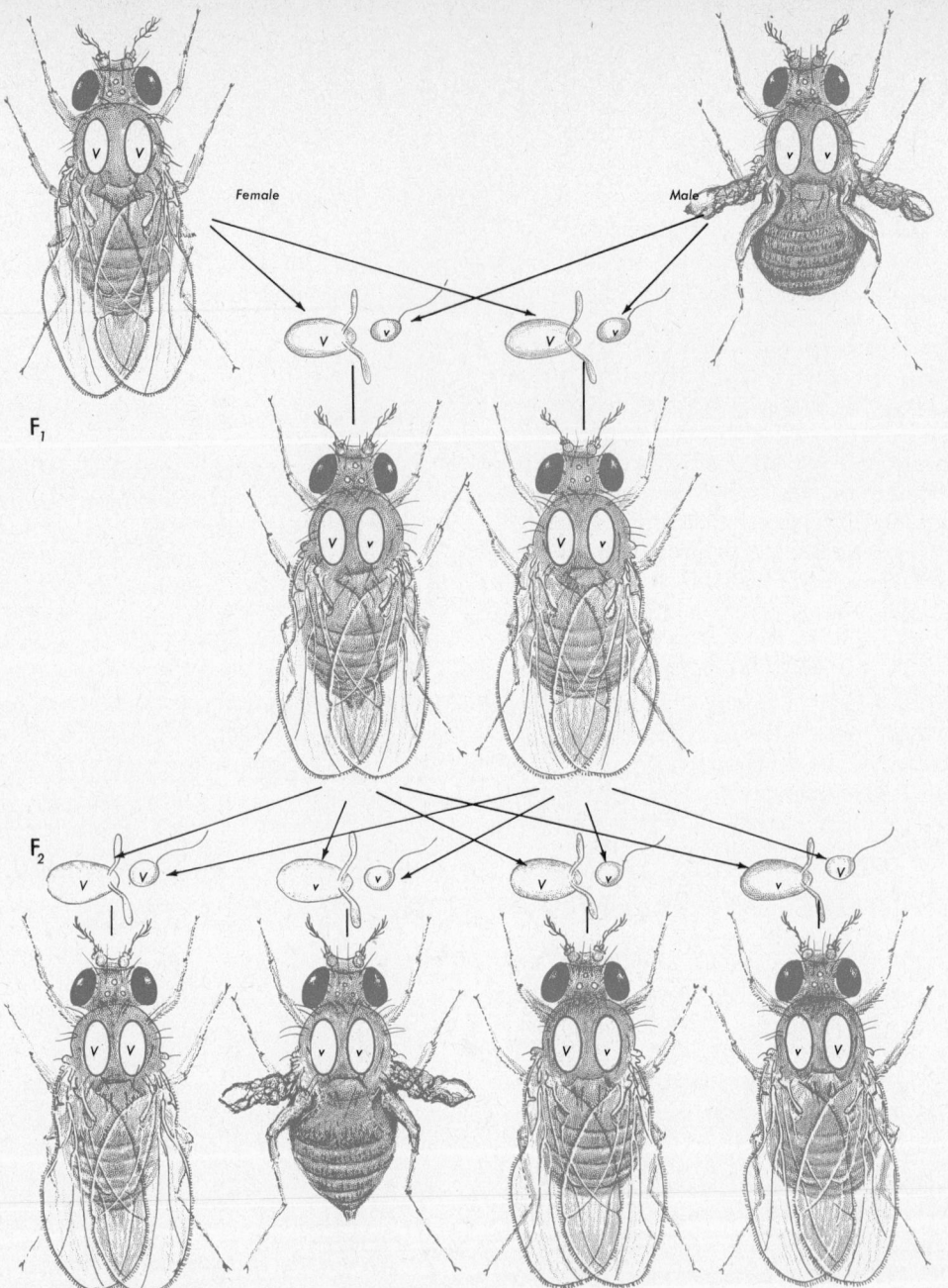

Fig. 24–5. A simple monohybrid cross in *Drosophila* shows the behavior of the genes.

every four that the pigs will be black and only one chance that they will be white. Likewise, in the matings of heterozygous brown-eyed people, each child has three chances of being brown-eyed and only one of being blue-eyed, and this holds true no matter how many children there are in the family. If there are three brown-eyed children in a family and a fourth one is expected, there is no more likelihood that he will be blue-eyed than there was for any of the other children. It must be remembered that at every union of the egg and sperm the chances are three to one for brown eyes no matter what previous unions have produced.

A Simple Monohybrid Cross

In tracing through the simple Mendelian ratios, we use the same organism that Morgan first employed because much more is known about the genetics of *Drosophila melanogaster* than of any other animal. This tiny fly possesses four different chromosomes, all of which have been partially mapped (Fig. 24–21). By the use of X-rays hundreds of gene changes (mutations) have been produced, so that there are a great number of alleles, many of which have been investigated. Through the study of hundreds of thousands of generations of this animal has come much of our knowledge of animal genetics.

In *Drosophila* cultures, occasionally a fly appears with very much reduced wings, called vestigial wings (Fig. 24–5). If this individual is crossed with a normal fly, what might we expect in the offspring? First let us examine the nature of the germ cells of these two flies, assuming that the female is normal-winged and the male has vestigial wings. In this cross it makes no difference which sex has vestigial or normal wings. During meiosis in the male each sperm cell receives a single gene, v, for vestigial wings because it is homozygous; likewise, each egg produced by the female

contains a gene, V, for normal wings. The flies resulting from the union of these eggs and sperm cells are heterozygous (Vv) and normal-winged (Fig. 24–5), because normal wing is dominant. Without making this cross we have no way of knowing which of these two traits is dominant. If these F_1 hybrids are crossed with one another (inbred), we already know the ratio will be 3 normal to 1 vestigial-winged flies. But why?

During meiosis of both male and female hybrids, the two alleles are distributed to the sperm and eggs in a random manner. Half of the germ cells of each contain a gene for vestigial wing and half for normal wing. Upon fertilization these unite according to the laws of chance, and since there are four possibilities, we would expect them to occur in equal numbers for each combination. There is no special attraction between eggs and sperm of different or the same kind of gene. One-fourth of the offspring should be homozygous for normal wing (VV), one-fourth should be homozygous for vestigial wing (vv) and the remaining one-half should possess both genes, that is, they should be heterozygous (Vv), and since normal is dominant, they will all have normal wings just like the parents. It does not matter whether we designate the genes as one-fourth Vv and one-fourth vV, since the end result is one-half heterozygotes Vv, the order of the genes making no difference in the end result. See Table 24–1.

These germ cells unite at random. This can best be shown by employing the "checkerboard" or Punnett, square, named after the British geneticist who first used it.

TABLE 24–1. Symbols Used in a Typical Monohybrid Cross

FEMALE (♀)		MALE (♂)
Parents:		
Normal wings	$VV \times vv$	Vestigial wings
One kind of egg	V v	One kind of sperm
F_1 *Generation:*	Vv	Normal wings (Heterozygotes)
Crossing two hybrids	$Vv \times Vv$	
Two kinds of eggs	V & v V & v	Two kinds of sperm

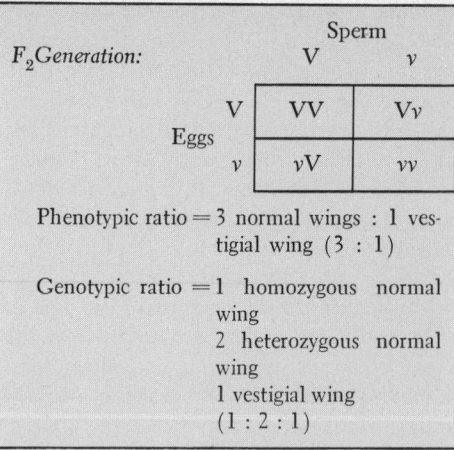

F₂ Generation:	Sperm	
	V	v
Eggs V	VV	Vv
Eggs v	vV	vv

Phenotypic ratio = 3 normal wings : 1 vestigial wing (3 : 1)

Genotypic ratio = 1 homozygous normal wing
2 heterozygous normal wing
1 vestigial wing
(1 : 2 : 1)

Back Crosses

It is sometimes important to know the genetic constitution of animals showing the dominant trait but suspected of being heterozygotes. This is particularly true in the breeding of livestock. In the above F₂ flies, three-fourths of them are normal-winged, but the question arises as to which of these are heterozygous and which are homozygous. By mating each of them with the homozygous recessive grandparent this can be told. This kind of cross is known as a *back cross*. If the normal-winged fly is heterozygous the offspring are half normal and half vestigial-winged, but if it is homozygous the offspring are normal-winged. This is made clear as follows:

If the fly is heterozygous for vestigial wings (Vv) the cross is:

Vv × vv (recessive)		
Two kinds of gametes	V & v v	One kind of gamete
Possible offspring	Vv vv	
Ratio	1 : 1	

In other words, one-half of the offspring would show the recessive trait.

If the fly is homozygous for the dominant trait (VV) the cross is:

VV × vv (recessive)		
One kind of gamete	V v	One kind of gamete
Possible offspring	Vv	

(All offspring would show the dominant trait.)

A study of the inheritance of the abnormality, albinism, in man also demonstrates this type of inheritance. An albino is lacking a gene to produce pigmentation in the skin, hair, and even in the eyes. This is inherited as a recessive, hence it must be homozygous (aa) to appear. If, then, a normal man (AA) marries an albino woman (aa), their offspring will be normal, although heterozygous for albinism (Fig. 24–6). If this offspring later marries an albino, his children will have a one-to-one chance of being albino, and definitely establishes him as heterozygous. If, however, he marries another heterozygote like himself, their children will have one chance in four of being an albino. This is simply an F₂ generation of a monohybrid cross. If he marries a normal person, that is, a homozygote for normal pigmentation (AA), all of their offspring will be normal, but there is a one-half chance that they will be heterozygous (not shown in Fig. 24–6).

Such back crosses have proven of great value in the establishment of "pure" breeds of domestic animals such as cattle for beef or milk production and chickens for egg laying. Rather than pick out the best phenotypes as was the custom for centuries, breeders now make such back crosses until they establish the genotype for a particular trait. Animals possessing this trait are then used as breeding stock. This has been particularly profitable to farmers who wish to build up their herds in respect to milk production. It is possible through recent techniques in artificial insemination to fertilize thousands of cows with sperm from a single bull that is known to produce offspring with high milk-giving qualities. With the use of **progeny selection,** the output of cows is increasing each year until today some record cattle have produced as much as 19,000 quarts of milk a year. This is a far cry from unselected cows which give no more than 350 quarts in a year!

Dihybrid Cross

So far, we have explained what happens when only one pair of contrasting genes is considered. Now let us attempt to explain what Mendel did not know when he crossed two pairs of contrasting characters. The F_1 generations of such crosses are called **dihybrids.** It sometimes is essential for geneticists to follow three and four or more characters simultaneously; this becomes increasingly

complex with each added character. We confine our attention to the two-character cross and leave the others for advanced books in genetics.

Using normal and vestigial wings in *Drosophila* again, we can also follow body color. Ebony or dark-colored individuals occasionally occur in stocks of the normal gray-bodied wild type, and these dark individuals are recessive to the gray-bodied wild type (Fig. 24–7). By crossing a gray vestigial female with an ebony normal-winged male the F_1 offspring are all gray and normal-winged. Note that each of the parents was recessive and dominant for different traits. We could as well have crossed a gray normal-winged individual with an ebony vestigial fly and obtained the same normal-appearing heterozygous offspring. The sexes could also have been switched and it would have made no difference in the outcome. When these two hybrids are crossed, 16 combinations are possible in the F_2 generation. To follow these possible combinations it is convenient to employ the Punnett square again (Fig. 24–7).

Each of the hybrids can produce four different kinds of gametes during meiosis, each bearing the two different genes. These genes must necessarily be located on different chromosomes, for if they were on the same chromosomes the typical ratios would not occur, for reasons that we discuss a little later (linkage, see p. 523). By arranging the four different kinds of sperm along the top of the checkerboard and the eggs along the left side and placing in each square the gene combinations that result from the union of those particular eggs and sperm, the 16 possible combinations become obvious. Remembering which are dominant and which recessive, it is clear that there will be 9 gray normal, 3 gray vestigial, 3 ebony normal, and 1 ebony vestigial. This is the 9:3:3:1 ratio that Mendel verified many times with plants. We see here an explanation for both genotypes and phenotypes. The heterozygotes can be checked for their genes by employing back crosses just as was done for the single-pair genes. If, for example, a gray-bodied, long-winged heterozygote (*VvEe*) is crossed with a double recessive (*vvee*), the offspring

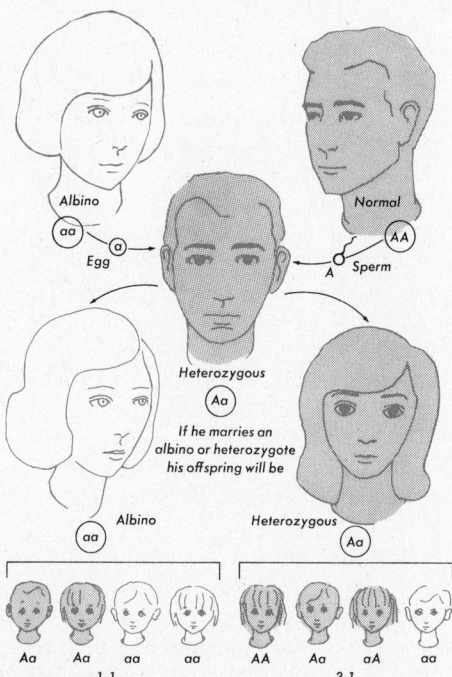

Fig. 24–6. Albinism in humans is recessive and appears only when homozygous. The manner in which it is inherited is indicated here.

TABLE 24–2. Symbols Used in the Backcross of a Hybrid (Heterozygote)

Long Gray VvEe × vvee Vestigial Ebony			
Four kinds of eggs: VE, Ve, vE, ve			ve Only one kind of sperm
Progeny $\frac{1}{4}$ VvEe,	$\frac{1}{4}$ Vvee,	$\frac{1}{4}$ vvEe,	$\frac{1}{4}$ vvee
(long, gray)	(long, ebony)	(vest., gray)	(vest., ebony)
Ratio: 1	: 1	1	: 1

fall into four groups; ¼ long gray, ¼ vestigial gray, ¼ long ebony, and ¼ vestigial ebony. This is portrayed in Table 24–2.

INTERACTION OF GENES

When mention was made above of dominance and recessiveness reference was made to the fact that other kinds of allelic interactions were known. One such interaction is commonly called **incomplete dominance,** or **blending inheritance.** In incomplete dominance, the expression of both the alleles interacts to bring about a phenotype that is intermediate to the two homozygoes. Many examples of incomplete dominance are known, such as roan cattle, described below. In another type of allelic interaction, called co-dominance, both alleles are expressed but their activities do not interact; the consequence is that the phenotype exhibits fully the characteristics of both the alleles. AB blood types of man illustrate co-dominance (p. 423). When an individual has an allele for antigen A and also an allele for antigen B, both these alleles are expressed so that the red blood cells contain both antigens, A and B (and not an intermediate antigen).

Interactions may also occur, of course, between nonallelic genes. The discussion below about the inheritance of comb in poultry is based upon an interaction of two pairs of nonallelic genes. Indeed it is likely that most traits are the end result of coordinated working of many genes. We detect the interactions and they become a part of the analysis only when the strains that enter the cross differ in more than one pair of genes affecting a given trait.

We should also be aware that a single gene may affect more than one trait; this is referred to as multiple effects of single genes, or **pleiotropy.** One gene in early development may start a series of reactions that ultimately alter or control several traits. A gene that we usually think of as controlling eye color in *Drosophila*, for example, may have had its share of influence along the way in the development of a host of other vital processes, though we see its effect most strikingly in eye color. An interesting example of this multiple effect of a single gene in humans is illustrated by the rare anomaly known as **phenylketonuria** and discussed later (p. 536).

Blending or Incomplete Dominance

Occasionally a gene does not demonstrate complete dominance, so that the heterozygote shows a mixing or **blending** of the dominant and recessive traits. This may be due to a cumulative effect where the recessive allele is negative and the dominant allele manifests itself according to whether or not it is single or double. The heterozygote would then show half the effect that the dominant homozygote would. For example, in hair color in cattle, a red bull (WW) may be mated to a white cow (ww) and the hybrid (Ww) will be roan, which is a mixture of red and white (Fig. 24–8). Actually the hairs are still red or white, but they both appear, thus giving the coat its roan color. The appearance of one dominant red gene pro-

Fig. 24–7. A schematic representation of the gene behavior in a typical dihybrid cross using *Drosophila*.

duces only half as many red hairs as the two genes will produce. When the hybrids are mated (Ww × Ww), the F_2 generation shows a ratio of 1 red, 2 roans, and 1 white. In other words, the genotypes and phenotypes are identical. In such matings it is impossible to get a pure breeding roan stock, because reds and whites will continue to show up in future progeny no matter what selective procedures are employed.

MULTIPLE GENE INHERITANCE

So far we have been considering characters that are clearly defined, where the phenotypes are sharply set off by color or some other trait. But there are a great many characters which, instead of demonstrating clear-cut differences, show continuous variation from the two extremes; this is particularly true of such human traits as skin color, height, weight, special abilities, and intelligence. In matings between individuals possessing the two extremes of these traits, children do not resemble one parent or the other but are intermediate between the two. This is like blending inheritance in which more than one pair of genes is involved. This means that several genes, known as **multiple genes** or **factors,** influence one trait, and their ultimate effect depends on additive action. This can be illustrated by a consideration of skin color in humans, which is the best-known example of this type of inheritance.

In certain parts of the world marriages between blacks and whites are common, and this has afforded geneticists sufficient information about the inheritance of skin pigmentation to understand it genetically. It is known that two pairs of genes on separate chromosomes are involved. These may be called *Aa* and *Bb* where *A* and *B* are responsible for pigmentation. In the F_1 generation resulting from a mating between a homozygous black (*AABB*) and a homozygous white (*aabb*) we would expect the hybrid (*AaBb*) to be either all black if the genes for pigmentation were completely dominant, or half black and half white (skin pigmentation condition which we will call mulatto in accordance with the classic scientific studies in this field), if blending occurred. The latter proves to be the case. The genes for pigmentation show incomplete dominance and

the offspring are dark-skinned, approximately intermediate between the white and black parents. This resembles exactly what one would expect in a single gene pair trait. The double gene-pair proof comes when two mulattoes mate (*AaBb* × *AaBb*). If there were but a single gene pair, the typical 1 white : 2 mulatto : 1 black ratio would be expected, but instead an entirely new ratio appears, that is, 1 black : 4 dark brown : 6 mulatto : 4 light brown : 1 white. Since there are sixteen possibilities, there must have been four genetically different sperm and four genetically different eggs, or two pairs of genes, each located in different chromosomes. This is simply the F_2 generation of a dihybrid cross. Why, then, should this peculiar ratio result instead of the typical 9:3:3:1?

If we assume that the genes *A* and *B* influence the production of equal amounts of pigmentation even though located on different chromosomes, and that blending occurs between each with its recessive, *a* and *b*, then the 1:4:6:4:1 ratio is exactly as one would expect. Moreover, the intensity of the skin pigmentation, that is, the phenotype, would give a clue as to the genetic constitution or genotype of an individual. The darker the skin, the more capital letters would appear in the genic constitution. For example, a homozygous black would have the genetic formula of *AABB*; a dark brown would be *AABb* or *AaBB*; a mulatto would, of course, have one of each gene, *AaBb*. However, it would be impossible to tell whether the formula was *AAbb*, *AaBb* or *aaBB*, since each dominant gene has the same influence in producing pigmentation. The light brown would have but one dominant gene, *Aabb* or *aaBb*, whereas the white would be double recessive (*aabb*), naturally. Thus, in this F_2 generation there would be one person resembling each of the grandparents and all others would be graded in respect to pigmentation.

Can such information be of assistance in predicting what one might expect in offspring of matings where the skin pigmentation varied in the two parents? Obviously

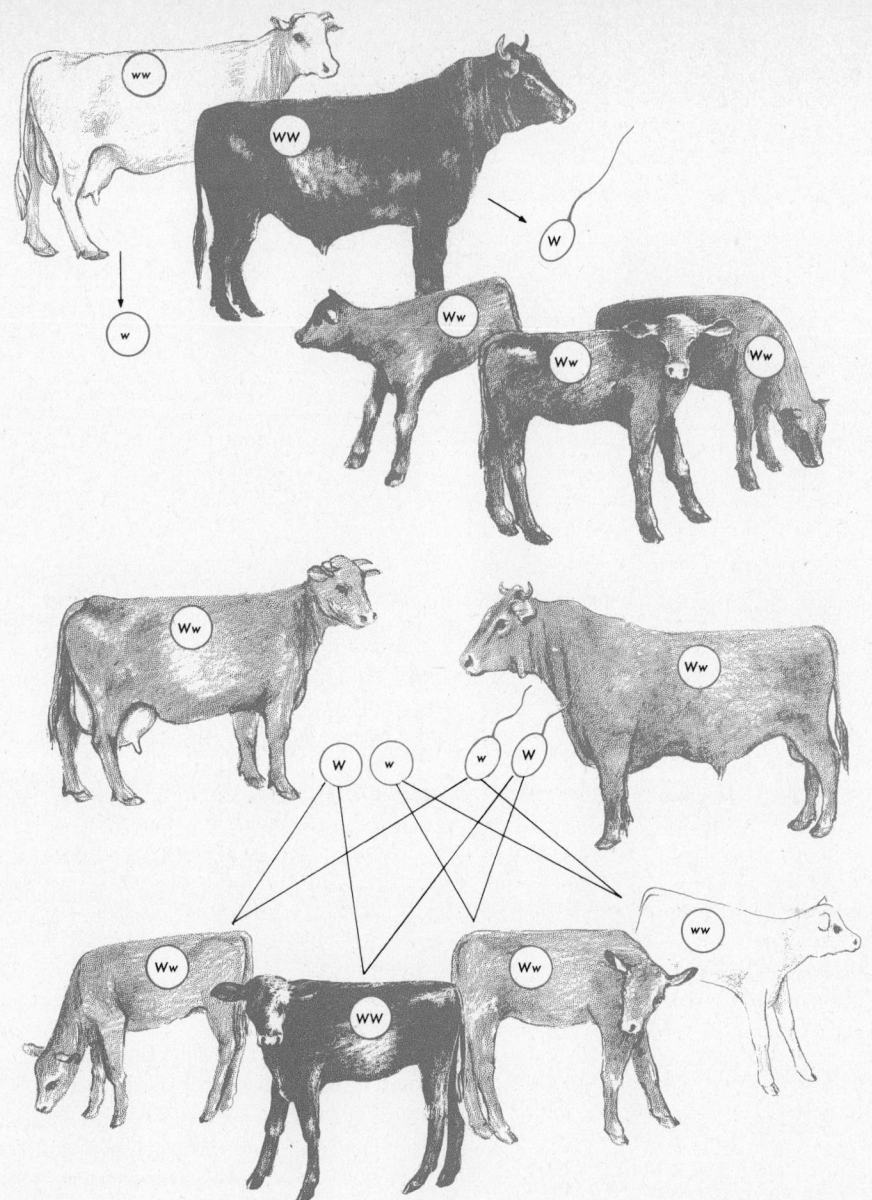

Fig. 24–8. Blending dominance can be illustrated with red and white cattle. Here the genotype can be distinguished by simply observing the coat color.

matings between the double recessive white and a person with any degree of pigmentation will produce offspring lighter than the dark parent. Matings between individuals carrying only one gene for pigmentation and appearing as light as many nonblacks *may* produce children darker than either parent. On the other hand, they have an equal (1:4) chance of producing children with no genetic trace of pigmentation. Such white children are double recessives (*aabb*) and therefore can never transmit pigmentation to their offspring although other characteristics may appear.

Fig. 24–9. Many gene pairs are involved in the control of height in humans. Three generations are represented here where only two gene pairs are considered, leaving space for any number of others to be added. The tall (dominant) and short (recessive) grandparents produced heterozygous children of average height. Crossing of two such heterozygotes results in offspring of a whole range of heights, which turns out to be a normal curve in the grandchildren (F_2). The number of dominant genes received by each individual (dose) and the number of individuals in each category (ratio) appear at the bottom.

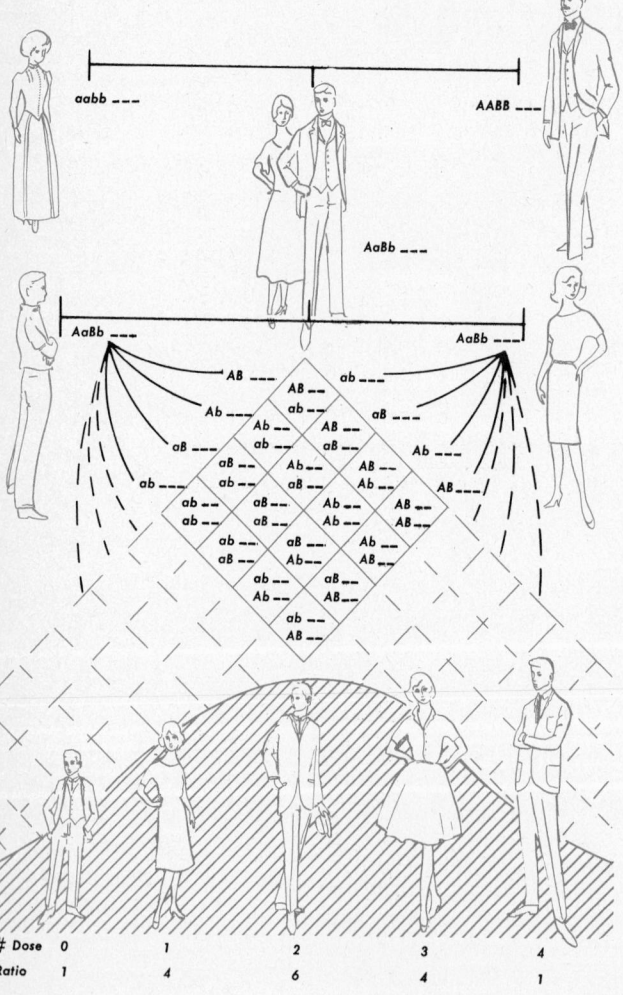

# Dose	0	1	2	3	4
Ratio	1	4	6	4	1

Normal Frequency and Multiple Genes

The character of skin color involving two gene pairs in man is a relatively simple case of multiple genes. As a rule such characters involve many more gene pairs, as, for example, height in man, where ten or more are influential. With three gene pairs influencing one trait, the F_2 comes out in the ratio 1:6:15:20:15:6:1. With each added gene pair the shape of the distribution approximates more and more closely a normal bell-shaped frequency curve (Fig. 24–9). When ten gene pairs influence a single trait, the distribution coincides with the normal distribution curve. In other words, ten gene pairs are sufficient to produce a normal population in respect to one trait whether it be height, weight, length of neck, or degree of intelligence. If one measures the height of 10,000 adult white males in America, he would find a range from about 55 to 85 inches with an average falling around 68. There would be very few as short as 55 inches and very few as tall as 85 inches but a great many around the average of 68 inches. By plotting the number of people at each height against inches, a normal distribution curve results (Fig. 24–10). This is exactly what is obtained when the distribution is computed for height on the basis of ten gene pairs being involved.

Quantitative Studies of Populations

Long before Mendel's Principles were known, some effort was made to understand inheritance by a careful analysis of a single trait in large populations. The most outstanding investigator in this field was Sir Francis Galton of England, referred to earlier, who studied the inheritance of many traits of British people, including such intangible ones as intelligence. He claimed that high intelligence seems to "run in families," as shown by the frequency that names of members of famous families appeared in the British "Who's Who." Such traits as height, which, as we have seen, is inherited through multiple genes, led him to formulate his

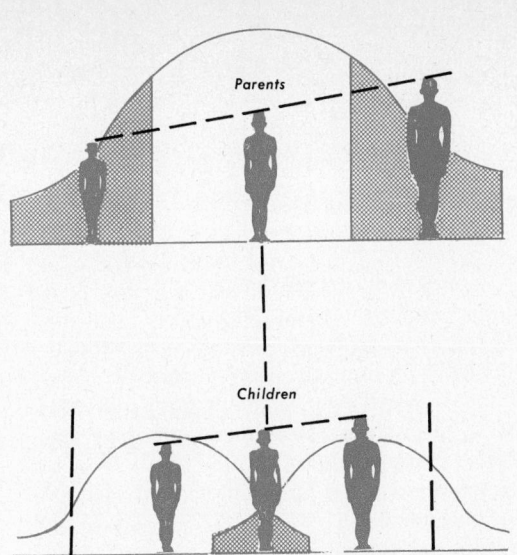

Fig. 24–10. Height in a population of humans shows a normal bell-shaped frequency curve as shown in the upper figure. There are a few about 55 inches tall and a few about 85 inches tall; most people are between these extremes, the average being about 68 inches. When these heights are plotted against numbers, a bell-shaped curve results, which is what is expected when ten gene pairs are involved.

The lower portion of the figure shows how Galton's law of filial regression operates. Children of short and tall parents are never as short or as tall as their parents. They tend toward the average height of the population.

filial regression law. This law merely states that offspring of tall or short parents, while taller or shorter, respectively, than the average British subject, are still not as tall or as short as their parents. This means that at the extremes of height offspring tend to regress toward the average (Fig. 24–10).

Although Galton's studies contributed virtually nothing to the mechanics of inheritance, they did give later workers, not only in biology but also in sociology, economics, and other fields, a tool by which a quantitative description of a naturally or artificially occurring series of variables could be stated. It has been used extensively in biology in determining a normal population in respect to one or more traits. This is important to animal breeders who are trying to improve their stock with respect to a particular

character such as egg-laying in chickens. By carefully selecting for breeding stock those birds that produce the greatest number of eggs, the whole population can gradually be improved along that line. This means a selection toward all dominant or all recessive genes. When the stock becomes homozygous for a particular trait there is nothing more to be gained by further selection. One needs only to inbreed and keep the desired qualities which have thus been attained.

Hybrid Vigor or Heterosis

One of the strange and unexplained outcomes of multiple gene inheritance is that the hybrid resulting from two parents with contrasting traits is frequently more vigorous, bigger, and apparently better in every way than either parent. This condition is known as **heterosis** or **hybrid vigor** and is particularly well known in corn. This observation has also been made on a great many animals. The mule, which is a result of a mating between a jack and a mare, is notorious for its strength and endurance as well as its stubborness (Fig. 24–11). Every farmer is familiar with the value of hybrid corn, which demonstrates heterosis very markedly. Some striking cases of heterosis among

Fig. 24–11. Cross-breeding of a jack and a mare results in a mule which is usually sterile. The reciprocal cross (female burro with a stallion) results in a more horselike hybrid, the hinny.

Jack

Mare

Mule

human beings have occurred where two distinctly different races have crossed. The most celebrated instance is that of Captain Christian and his fellow mutineers of the *Bounty* who took native Polynesian women from Tahiti and settled on tiny Pitcairn Island. The first generation of children were more vigorous, taller, and more fertile than either the white or the Polynesian stock from which the parents came. Incidentally, this is also one of the few cases in human history where a nearly perfect inbreeding experiment was carried on. Studies of half Indian–half white individuals as well as hybrids from Hottentot and Boer crosses, have likewise revealed heterosis to a marked degree, although the condition does not seem to be evident in certain black-white crosses.

The second generation following the original crosses usually results in a gradual loss of the early hybrid vigor, and the offspring in succeeding generations slowly return to an intermediate type. This might be explained by considering that each member of the first cross possesses certain dominant genes for desirable traits which were lacking in the mate. When crossed, a now full complement of dominants would produce their desirable effects, thus producing the vigorous

offspring. In subsequent hybrid matings the dominant genes would segregate out, as one might expect, and eventually the recessives would once more reduce most of the race to the condition of the original parental stock.

Supplementary Genes

Sometimes two pairs of genes interact and influence the same trait, so that the expected 9:3:3:1 ratio in the F_2 is significantly altered; in fact, an entirely new phenotype may appear. When two dominant genes interact so as to produce a new phenotype, they are called **supplementary genes.** One of the simplest cases of this interaction occurs among combs in chickens. Three kinds of combs are well known: **single, pea,** and **rose** (Fig. 24–12). Both pea and rose are dominant to single. What happens when pea and rose, two dominants, are crossed? The result is a new kind of comb, called **walnut!** This demonstrates the combined action of two dominant genes, namely, that each supplements the other and the result is a comb different from that each would produce alone. The proof of this conclusion can be shown merely by crossing two walnut-combed animals. The offspring appear in the ratio of 9 walnut : 3 rose : 3 pea : 1 single, which means that wherever the genes for rose and pea come together their supplementary action produces a walnut comb. Note that the genes for rose and pea are not alleles of each other; this is a case of interaction of genes located at different loci on the chromosomes. It is possible to produce a pure breed of walnut-combed chickens simply by making back crosses with the double recessive single comb to find out which birds are homozygous dominant for both pairs of genes. Once this is done, the mating of two such homozygous birds results in all walnut-combed offspring and the pure breed is established. Remember that this procedure is impossible with those showing incomplete dominance.

Another example of supplementary gene action can be demonstrated in guinea pigs. The gene for black coat (*B*) is dominant to

Fig. 24–12. The inheritance of combs in fowls is due to two dominant interacting or supplementary genes. In a cross between pea- and rose-combed animals a new type of comb, walnut, is produced. Crossing walnut-combed fowls gives the typical 9:3:3:1 two gene ratio, but the interaction of the genes shows a distribution of comb types as shown here. The double recessive is a fourth kind, single comb.

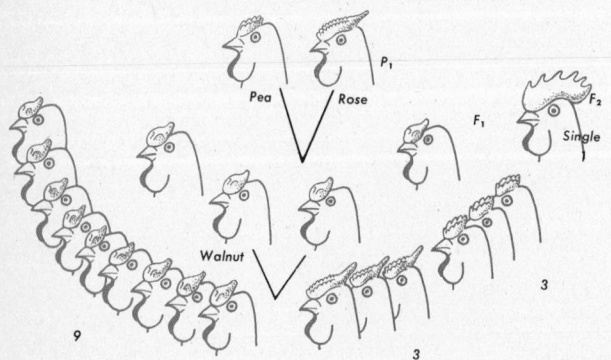

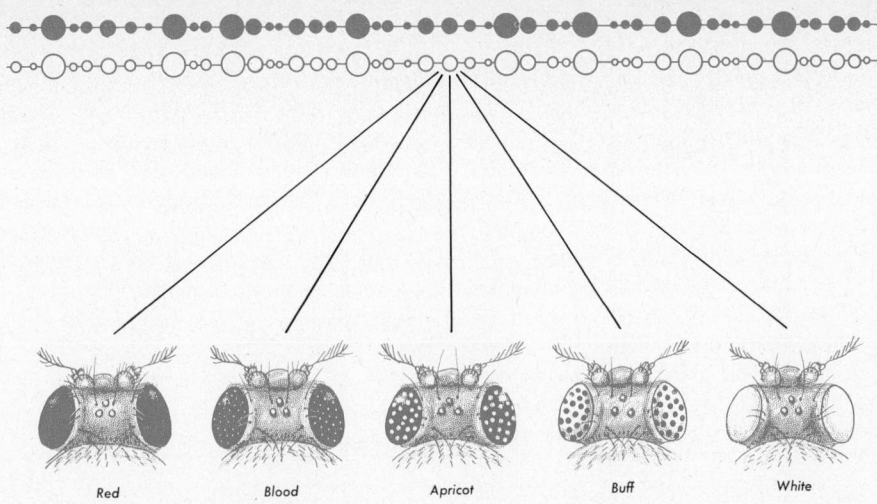

Fig. 24–13. A schematic representation of five of the fourteen alleles for eye color in *Drosophila*. The gene locus on the X-chromosome is indicated at the top of the drawing and the five phenotypes below.

its allele (*b*) which produces brown coat. Another locus controls whether there will be any color at all: thus the gene (*C*) must be present for any color and is dominant to its allele (*c*) which when present in homozygous condition (*cc*) leads to an albino. If a brown guinea pig (*CCbb*) is mated to an albino with the genotype of (*ccBB*), the offspring is black (*CcBb*) which might seem surprising until with further breeding the interaction of these genes is revealed. Whenever the gene for color (*C*) is present the coat color may be black or brown according to whether (*C*) is in combination with (*B*) or (*b*). But if the recessive (*cc*) is present neither (*B*) nor (*b*) can be expressed. Here we see two separate gene pairs interacting in such a way that one dominant produces its effect with or without the other, but the second dominant can produce its effect only in the presence of the first.

Multiple Alleles

It was pointed out earlier that a gene may exist in one or more forms. These are called **multiple alleles.** By the use of X-rays, hundreds of multiple alleles have been produced in **Drosophila,** so that for one locus, eye color for example, there are fourteen known alleles, all different (Fig. 24–13). Their influence on eye-color production ranges from dark red to white with all shades in between. Coat color in rabbits likewise is controlled by multiple alleles. There is a gene (*C*) which controls coat color, whereas the homozygous genes (*cc*) cause no color or albinism. Two alleles, (*c^{ch}*) and (*c^{h}*), when homozygous, produce "Chinchilla" and "Himalayan," respectively. The former is steel gray, while the latter is white all over except the tips of the ears, toes, tail, and nose (Fig. 24–31). When arranged in the series, C, *c^{ch}*, *c^{h}*, *c*, each gene is dominant to those following it and recessive to all of those preceding it. For example, a rabbit with the genes *c^{ch}c* is a "Chinchilla," one with *c^{h}c^{ch}* is also "Chinchilla," whereas one with genes *c^{h}c* is "Himalayan."

One of the best-known characteristics in man that is inherited by multiple alleles is the A, B, AB, and O blood groups which we discussed in Chapter 19 (see p. 424). Three alleles are responsible for the various groups: genes A, A^B, and *a*, of which the first two are dominant to the last. Gene A controls formation of antigen A (agglutinogen); gene A^B controls the formation of antigen B; and gene *a* is without effect in

517

that it causes no antigen formation. Neither A nor A^B is dominant to the other, so when both are present in homologous chromosomes the blood group AB results. Since these genes are inherited according to Mendelian Principles, the knowledge of blood types has some value in addition to that needed in transfusions. In cases of questionable paternity the knowledge of blood types can be used and may rule out certain males as possible fathers of a child. It cannot be employed to determine whether a certain man is the father but only that he could be. The manner in which this operates can be seen from a consideration of all of the genotypes:

Blood Group	Genotype
A	Aa or AA
B	A^Ba or A^BA^B
AB	AA^B
O	aa

A child with blood group A (Aa or AA), for example, and a mother with group B (A^Ba or A^BA^B) must have a father with group A or AB. A male with group O or B could not possibly be the father of such a child. Do you see why?

The Rh blood antigen discussed earlier (p. 425) was thought for a long time to be simple. It was postulated that a single pair of alleles, R and r, accounted for the difference between Rh-positive and Rh-negative blood and these were inherited in Mendelian fashion. However, as more antisera were studied additional genes were needed to account for the complex inheritance patterns. Eight alleles and even more have been postulated by some investigators.

How a single gene can mutate in so many ways to give a series of multiple alleles is easy to understand when we visualize the physical gene as being a segment of DNA containing some 1000 nucleotides in linear order. A change of a nucleotide sequence through rearrangements, substitutions, gains, or losses of nucleotides could occur at any point along the sequence and modify the message of the gene accordingly. Therefore a gene contains many mutational sites. Changes at different sites could, of course, result in different amino-acid sequences in the specified polypeptide. The altered proteins might well have different degrees of activity according to the kind and degree of change in the amino acid sequence; if these changes are detected by changes in the phenotype, the molecular bases for multiple allelic series are easily visualized.

INHERITANCE OF SEX

In many species the two sexes appear generation after generation in a 1:1 ratio. Geneticists were struck by the similarity of this ratio to that obtained from a cross of a hybrid (Ww) with a double recessive (ww), which also gives a 1:1 ratio. They wondered whether sex was genetically determined by some similar type of mechanism. Cytologists had noted that chromosome complements of male and female (of some species) were alike, except for certain special chromosomes. The chromosomes for which the sexes were consistently different were referred to as **sex chromosomes**. For example, male and female *Drosophila* chromosome complements are identical for three of their four pairs of chromosomes. One pair however, differs in the two sexes; the female has a homologous pair (represented by the symbols XX) but the male has just one X chromosome, whose pair member (the Y) is visibly different. The X and Y chromosomes, for which the sexes differ, are the **sex chromosomes**, and the chromosomes for which the sexes are alike are the **autosomes**. This does not mean that sex chromosomes alone influence sex or that they influence nothing but sexual characteristics. It simply indicates those chromosomes for which the sexes differ. Thus, nuclei of female *Drosophila* have three pairs of autosomes plus an XX pair of sex chromosomes, while the nuclei of male Drosophila have the same three pairs of autosomes plus an XY pair of sex chromosomes. The human female has 22 pairs of autosomes and one pair of X chromosomes in each body cell, and the male has 22 pairs of autosomes plus one X and one Y chromosome. This sex chromosome situation exists in some plants and in most animals, although there are some notable exceptions among animals (Fig. 24–14). In some insects and birds the arrangement is reversed, that is, the female has the heteromorphic pair. Sex differences in the hymenopterans (bees, ants, wasps) is determined not by special sex chromosomes but by whether the individual comes from a fertilized egg and is diploid (female) or from an unfertilized egg and is haploid (male).

At meiosis two kinds of sperm are produced, each containing one set of autosomes plus either an X or a Y chromosome (Fig. 24–15). Upon fertilization the egg, which contains only a single X, may receive either a sperm containing an X or one with a Y. In the former case the offspring will be female (XX) and in the latter, male (XY). Since fertilization occurs at random, there is a 50–50 chance of either kind of sperm fertilizing the egg, hence the equal numbers of both sexes. In other words, the male is heterozygous for sex, the female homozygous. It would seem that sex determination is a very simple matter, but experimental work over a long period of time has brought to light certain conditions which indicate that the process is not as simple as was once thought.

Genic Balance and Sex Determination

One of the first experiments that extends the theory of sex determination by the X and Y chromosomes was the appearance in about 1922 of a fly in Dr. Calvin **Bridges'** *Drosophila* cultures at Columbia University which possessed three sets of autosomes (9 chromosomes instead of only 6) and 2 X chromosomes. Considering the number of X chromosomes alone, this fly should have been a female, but actually it was an intermediate between male and female, or an **intersex**. Apparently the presence of an extra set of autosomes upset the "genic balance" in such

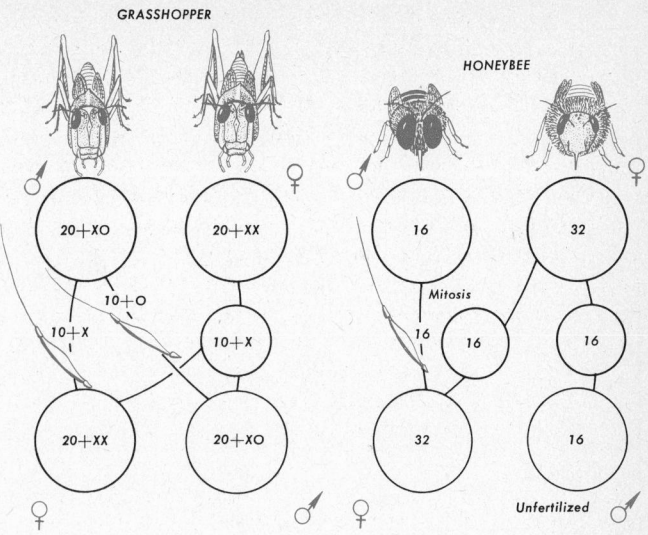

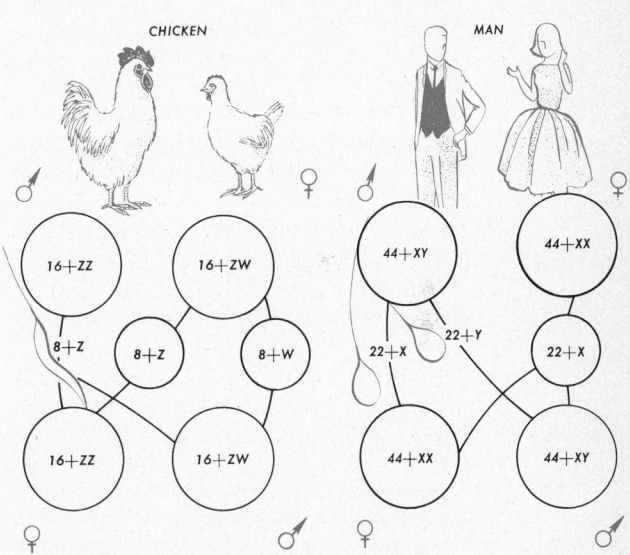

Fig. 24–14. Sex determination varies in different animals. Many insects, such as the grasshopper, have no Y chromosome; males have one X and females two X chromosomes. The drone bee is produced from an unfertilized egg, therefore requires no meiosis to produce haploid gametes. Males are haploid and females are diploid. In chickens, as in all birds, the female produces two kinds of eggs (Z and W) which determines the sex. The opposite situation exists in most animals, man included, where two kinds of gametes (X and Y) occur in the male. The fertilized egg containing a Y chromosome is male.

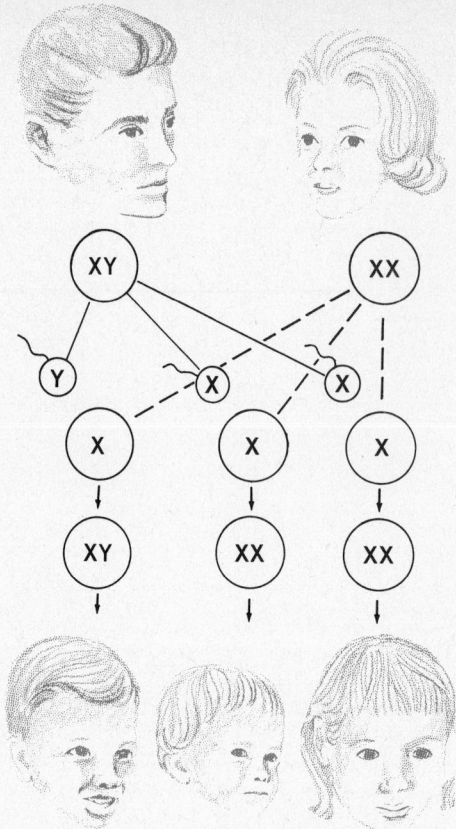

of the genes in the autosomes and X chromosomes seemed to be the deciding factor, as indicated in Table 24–3.

From these experiments it is evident that the X chromosomes carry a preponderance of genes for femaleness while the autosomes contain a preponderance of genes for maleness; how much "maleness" or how much "femaleness" the offspring exhibits is determined by how great a dose of either one or the other it receives. If the dose is intermediate, the offspring displays the external characteristics about evenly divided.

Bridges' explanation of sex determination emphasized the concept of balance of sex determinants. Although in *Drosophila* the Y chromosome appears to play no part in sex determination, in man it does. It seems that sex determination in man depends upon a balance between the number of X's and the number of Y's, with the Y chromosome carrying the male determinants. In recent years chromosome complements of some human individuals showing aberrant sex development have been studied. Some individuals have two complete sets of autosomes and one unpaired X chromosome (XO condition). These individuals manifest what is known medically as **Turner's syndrome**, a certain type of underdeveloped female. Others have two complete sets of autosomes, two X's, plus a Y (XXY condition). These individuals show **Klinefelter's syndrome**, a condition of underdeveloped, defective, maleness. Other combinations of X and Y chromosomes manifesting imbalance have been observed; they are all consistent with the conclusion that in human inheritance the X chromosomes carry female determining factors, the Y carries male determining factors, and the balance is such that XX leads

Fig. 24–15. Sex is determined in animals at fertilization, depending on the genetic composition of the sperm which enters the egg. In humans if it contains an X chromosome the offspring will be female; if it contains a Y chromosome the offspring will be male.

a way that the two X chromosomes could not completely control the situation and produce a female. This and subsequent experiments made it clear that other genes scattered throughout the autosomes were also influential in determining the sex. The distribution

TABLE 24–3.   Genic Balance Between Autosomes and X-Chromosomes

2 A (autosomes) plus	1 X =	Normal male characters
2 A "	2 X =	Normal female characters
3 A "	1 X =	Over-developed male characters
2 A "	3 X =	Over-developed female characters
3 A "	2 X =	Intersex

to normal female development and XY to normal male development.

Genes may be regarded as determining elements that set into motion a sequence of processes culminating in the trait. Many developmental steps intervene between the primary action of the gene and the final character. Ordinarily, development proceeds regularly along the pathway determined by the genes. However, the course of development could be altered by some unusual, non-genic agency, and its alteration bring about an unexpected end result. The genes we have been discussing in this section are sex determiners; the realization of their potentialities as male individual or female individual is a process of sex differentiation. The process of differentiation might be altered without changing the genes in any way. Earlier (p. 357) we saw that the sex of a hen chicken could be reversed when the ovary was destroyed because it received male hormone from an activated residual testis. Certainly the genetic balance was not changed.

Sex Linkage

In addition to carrying genes for sex the X chromosomes contain genes which have nothing whatever to do with sexual characteristics. For example, in *Drosophila*, the X chromosomes carry a gene locus affecting eye color. In *Drosophila*, the Y chromosome is virtually without genes, that is, it seems to be mostly a genetic blank, lacking the alleles of the genes of the X chromosome. Some insects (*Anasa*—squash bug) have no Y chromosome at all and get along satisfactorily. Since the X chromosome carries certain genes that control traits other than sex, it follows that these traits are tied in some fashion to sex distributional patterns. Genes located on the sex chromosomes are said to be **sex-linked** and, of course, follow the distributional pattern of the sex chromosomes. Since a female contains two representatives of a sex-linked gene, one on each of the X chromosomes, it may be homozygous or heterozygous for the alleles. A male with only one X chromosome contains only one representative (**hemizygous**) of the gene. The pattern of inheritance for sex-linked genes thus differs from that in which both parents have two alleles, as in autosomal inheritance. In man the genes responsible for **red-green color blindness** and the blood abnormality, **hemophilia,** are carried on the X chromosome, so that one might expect these anomalies to be associated with sex. This is true and has been recognized for centuries, particularly with respect to hemophilia. Because this condition has occurred in royal families whose histories are well known, it was learned very early that the dread disease passed from mothers to sons but never from fathers to sons. For example, Queen Victoria of England apparently carried the defective gene because she herself did not have the disease but transmitted it to one of her sons and through two of her daughters to succeeding generations.

This great enigma in man's inheritance remained a mystery until a mutation producing white eye color occurred in some of T. H. **Morgan's** fly cultures. When the inheritance of this trait was worked out it paralleled the transmission of hemophilia in man. From this white-eyed stock of *Drosophila*, Morgan was able to show that the gene for eye color was located on the X chromosome and that its inheritance was definitely linked with this chromosome and consequently with sex. As a result of many experiments, it has been found that the inheritance of this sex-linked trait follows a specific pattern (Fig. 24–16). The gene for white eye color is recessive and is therefore expressed only when homozygous. Since the male has but one X chromosome, white eyes will appear whenever a single X chromosome carrying a mutant gene is a part of the genetic make-up of the male fly.

In a cross between a white-eyed female, whose X chromosomes must each carry the defect, and a normal red-eyed male, all the female offspring will be red-eyed, although heterozygous, and all the males will be white-eyed because their one X chromosome carries the gene which induces this defect (Fig. 24–16). When these F_1 flies are allowed to

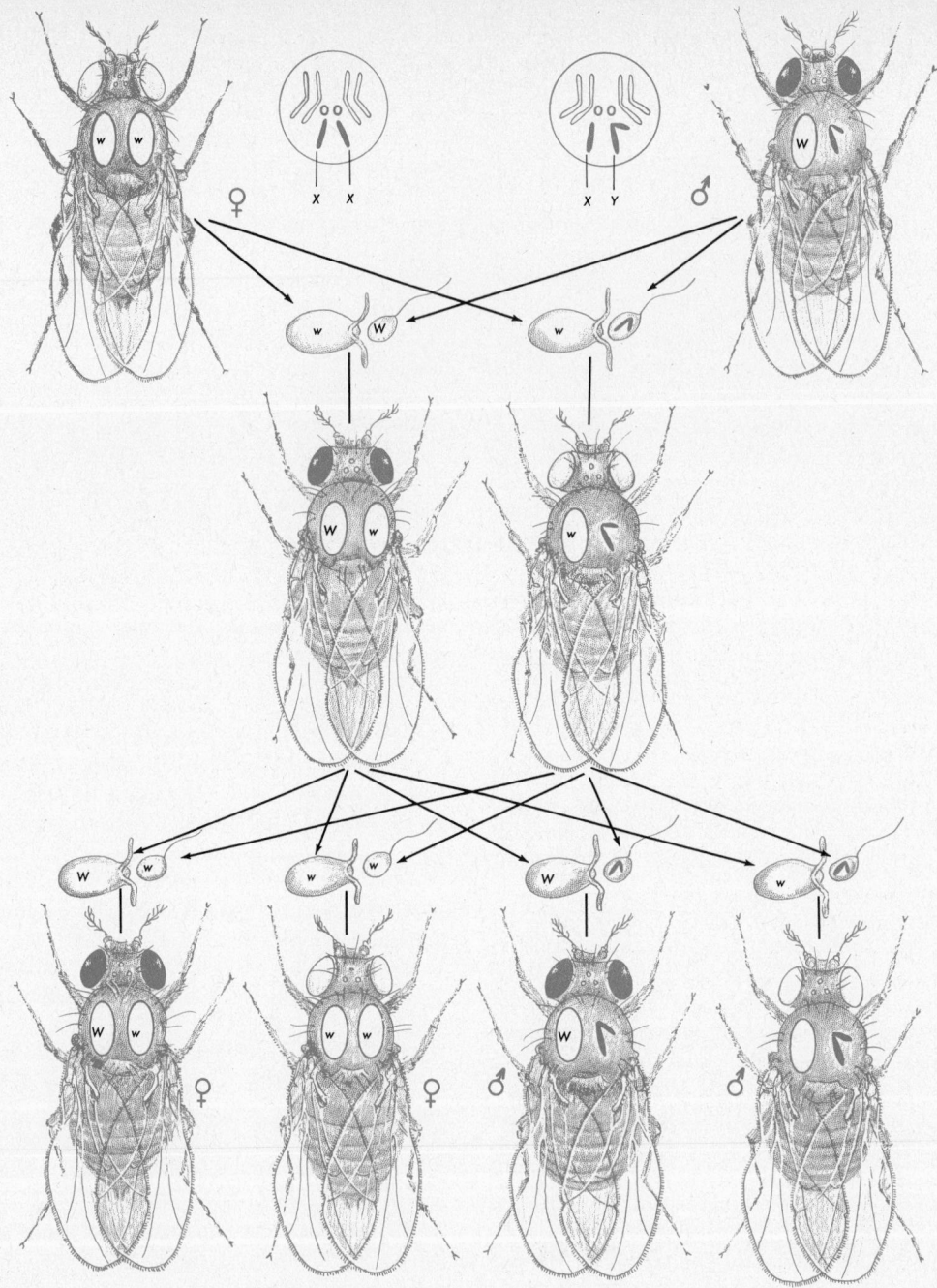

Fig. 24–16. Genes that are located in the sex chromosome (X) are associated with the sex of the animal, as indicated here in *Drosophila*. The gene for eye color (white) is located on the X chromosome and behaves in the manner shown here.

interbreed, the effect of the linked condition is further demonstrated. The female can produce two kinds of eggs; one-half containing an X chromosome with a normal gene and one-half bearing a mutant gene. The male produces two kinds of sperm as usual, that is, half containing an X and half a Y chromosome. The sperm containing the X chromosome, however, carries the mutant gene (w). Random fertilization results in half of the female offspring being homozygous and therefore white-eyed; the other half is red-eyed but heterozygous. Half of the males are white-eyed and half are red-eyed because the eggs from which they came contained one-half mutant and one-half normal genes for eye color. The Y chromosome, of course, plays no part. Therefore, the ratio is 1:1 for each of the sexes and also for the entire F_2 generation.

The explanation of sex-linked genes in *Drosophila* suffices for hemophilia and red-green color blindness in man. For example, in the latter instance, if a woman heterozygous for color blindness marries a normal-visioned man, all the daughters of this combination will have normal vision but half of the sons will be color-blind (Fig. 24–17). One-half of the daughters, however, will be heterozygous for the defect, whereas the normal sons will show no trace of the anomaly and therefore never transmit it to their children. The heterozygous daughters can have color-blind sons while the homozygous daughters will never pass the trait on to either sons or daughters. Obviously, this type of **crisscross** inheritance explains the early observation concerning hemophilia in royal families. Many other organisms, both plants and animals, exhibit this type of inheritance.

Characters other than the sex organs are controlled by genes on the autosomes but, even so, they may be associated with the sex. Their effect is probably on the production of hormones which in turn control the appearance or absence of the trait. Such traits are said to be **sex-limited;** they do not follow the pattern of sex-linked genes. The presence of horns in one sex of certain animals and not in the other is controlled by such autosomal genes.

LINKAGE AND CROSSING-OVER

During the early studies of Mendelian inheritance it was soon discovered that there were more genes than chromosomes and this fact led workers to conclude that more than one gene existed on each chromosome. If this were true, then what happens during meiosis? As long as the genes are on separate chromosomes the ideal ratios of Mendel come out beautifully, but if they are located on the same chromosome, the ratios are different, though in each case quite repeatable. Morgan and his students showed that some genes entering a cross tended to stay together in subsequent generations and proved that they were linked together on the

Fig. 24–17. Color blindness is sex-linked in man and is inherited, as white eye color is in *Drosophi'a*. In these figures the solid black retina indicates normal vision, the broken retina represents the heterozygous condition in the female, and the white retina indicates red-green color blindness.

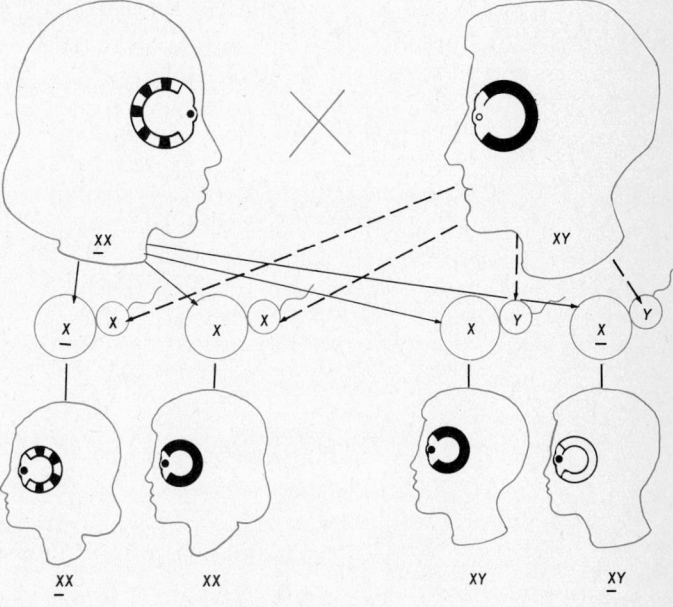

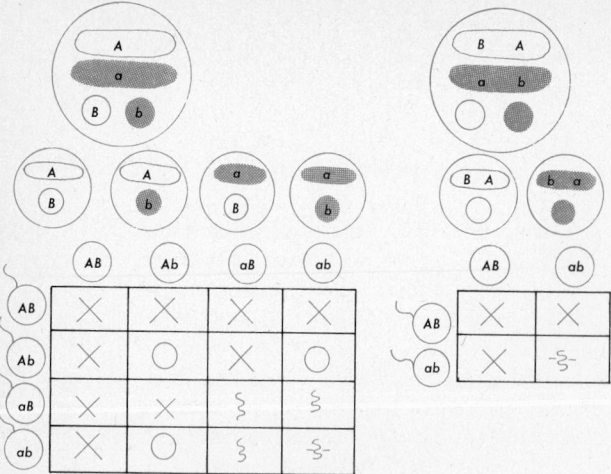

Fig. 24–18. Genes that are located on separate chromosomes give the typical Mendelian ratios, as in the case of four genes located on four chromosomes (two homologous pairs) in the left above. The typical 9:3:3:1: ratio results in the F₂. If the dominant genes are located on one chromosome and the recessives on its homologue they behave as single genes, as shown in the figures to the right.

same chromosome. Therefore, genes located on the same chromosome are said to be **linked.**

There should be as many linkage groups as there are chromosomes, which is the case in those plants and animals where sufficient information is available. Corn, for example, has ten pairs of chromosomes and ten linkage groups. *Drosophila* has four pairs of chromosomes and four linkage groups. Moreover, the linkage groups even approximate the chromosomes in size. There is a small pair of chromosomes and a small linkage

group, a middle-sized pair and a middle-sized linkage group, and there are two pairs of large chromosomes and likewise two large linkage groups. These facts, together with information that is to follow, certainly associate genes with chromosomes.

Completely linked genes respond in a manner similar to single gene inheritance. For example, if genes A and B are linked in one chromosome (Fig. 24–18) and their homologous recessives, a and b, are linked on the homologous chromosome, they segregate in the formation of germ cells as if they were single genes and the subsequent typical one-gene ratio will result, whereas if the genes were located on separate chromosomes, the typical 9:3:3:1 ratio would follow. Such linkages obviously decrease the possible number of combinations and thus definitely limit variation.

Strangely enough, genes once linked do not always remain so. During meiosis, you will recall, the homologous chromosomes come together and wind about one another just before the tetrads are formed. Also, the homologous genes lie opposite each other, that is, those for the same trait are drawn together by some mutual attraction. During early prophase, sequences of genes in the chromosome break and exchange places with their homologous counterparts so that when the chromosomes separate during subsequent meiotic divisions, groups of genes from one chromosome become a part of the other chromosome in some such manner as indicated in Figure 24–19. The resulting chromosomes, containing a mixture of genes, seem to heal perfectly because for the most part they contain their full complement of genes. This process of **crossing-over**, as it is called, operates purely fortuitously, so that in it we see another opportunity for a further juggling of the genes, compensating in part for the limitations placed upon variation through linkage.

The ratios that resulted from crossing-over which initially puzzled geneticists were a blessing in disguise because they made possible the ultimate construction of **chromosome maps.** Sturtevant, in 1913, reasoned

Fig. 24–19. Homologous genes have a tendency to exchange places during synapsis in some such manner as that shown here.

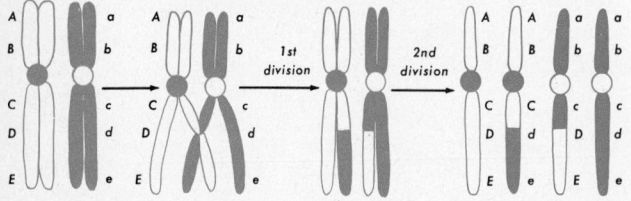

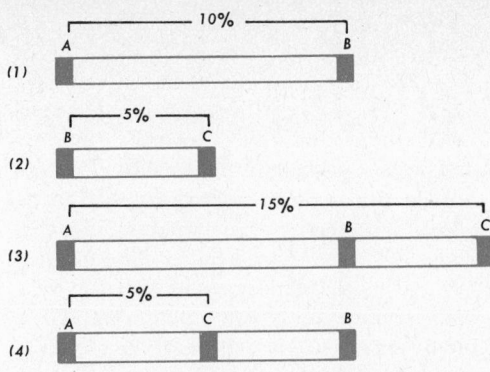

Fig. 24–20. Chromosome maps are based on crossover percentages. See text for explanation.

that crossing-over should occur more frequently between genes that lie farther apart in chromosomes than between those lying close to one another. For example, in Figure 24–19 there would be a greater chance for crossing-over to occur between genes A and E than between A and B, owing to the distance between them. The frequency of crossing-over can be taken then as a criterion of the distance between genes, providing information in chromosome mapping.

If genes are arranged in a linear fashion on the chromosome then let us suppose that the frequency of crossing-over between A and B is 10 percent and between B and C is 5 percent. This means that B and C are one half as far apart as is A and B. This does not tell us the sequence of these genes—that is, are they arranged A, B, C, or A, C, B? The answer to this question can be determined by obtaining the percentage of crossing-over between A and C. If this turns out to be 15 percent, the linear arrangement is A, B, C; if, on the other hand, it is 5 percent, the sequence is A, C, B. These two possibilities are given in Figure 24–20.

Cross-over maps reflect recombinational frequencies of linked genes. They are used by geneticists wishing to calculate probabilities of obtaining recombination of some specified loci. The maps indicate that the gene loci lie in linear array along the chromosome; the maps reveal the relative order of the genetic loci. The crossover maps do not necessarily

reflect, however, the actual physical distance between loci. Remember that cross-over percent was plotted as a unit of length, but no specification was laid down as to what should be the actual physical length of each map unit. Moreover, it is not strictly true that cross-over exchange points are equally probable along the entire length of the chromosome pair. If, for example, crossovers were slightly more frequent at one point as compared to another point, then apparent distance would be greater between loci separated by higher frequency cross-overs. In certain special cases, however, it is possible to localize a particular gene to a physical site in a chromosome. Figure 24–21 compares a cross-over, or genetic, map with a cytological, or physical, map for a portion of a *Drosophila* chromosome.

MUTATIONS

In its most general and widest possible sense, the term **mutation** refers to a heritable change in the structure of genetic material. A mutation might be a change either in the structure of a gene or in the structure of a chromosome. Each kind of change could lead to a phenotypic change. One frequently has no operational way of distinguishing a gene

Fig. 24–21. The linear arrangement of the genes on the chromosomes has been worked out for a few animals, among them *Drosophila*. This has been accomplished in two ways, one by employing crossover techniques, and the other by actual observation, the latter method being possible only because of the tremendous (1,000 to 2,000 times normal) size of the chromosomes in the salivary gland cells. A portion of the left end of the X-chromosome is shown here with a few genes indicated as to observable position on the chromosome and as computed by cross-over studies. Of course the genes cannot be seen.

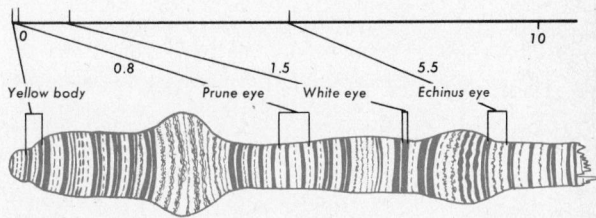

mutation and a chromosome mutation; in this case the general term of mutation is applicable. Sometimes, however, the two general types of mutations can be distinguished; they are then labelled as **gene mutation** (if the change is intragenic) or **chromosome mutation** (if there is detectable change in structure or in number of chromosomes). In any case, the process of change is referred to as **mutation;** an individual that manifests a phenotypic change as a result of a mutation is called a **mutant.**

Gene Mutation

Molecular changes within a gene may come about as a result of an accident during DNA replication (p. 479). Although DNA replication is usually precise, mistakes occasionally occur, changing the sequence in the linear nucleotide array of the gene. Such mistakes could involve the substitution of a wrong nucleotide, the loss of one or more nucleotides, or the addition of one or more nucleotides. Such accidental changes modify the information coded by a nonmutated nucleotide sequence. The extent to which the phenotype manifests such mutations depends upon how much the mutation significantly changes the DNA message and how much the changed DNA message affects cell function.

Gene mutations can be detected only when they produce visible changes in the organism, and since most mutations are recessive, they only become effective when homozygous. A great many mutations may occur but because they are recessive they remain hidden in heterozygotes, perhaps for many hundreds of generations. Eventually, by pure chance, they show up, which probably accounts for the apparent spontaneous occurrence of anomalies in many plants and animals, including man. Strange as it may seem, most mutations are harmful to the species and soon after they appear are eliminated, because the animals possessing them are not as well fitted to cope with their environment as the unchanged type. Through centuries of selection the un-

changed type probably already possesses the best possible combinations of genes to fit it for its particular environment, so that any change that might be made is more likely to result in an organism less suited for survival.

Undoubtedly gene mutations occur in somatic cells, but when they do, there is no effect on subsequent generations because the mutation is not passed on to the off-spring unless the change occurs in the sex cells. Aside from the results in the individual possessing them, somatic mutations are of no importance in evolution.

Occasionally a mutation may reverse itself, that is, mutate back to its original condition. Such **back mutations,** as they are called, rarely occur, but when they do the resulting organism is apparently in no way different from the original stock. Such reverse actions might well be expected in the light of similar actions in protein molecules. Similarly, certain mutations seem to occur again and again in a stock of animals. This might mean that under certain environmental conditions genes are receptive to change, and such changes are more likely to occur when these conditions are met. Moreover, similar mutations have been known to occur in closely related species of organisms under similar conditions. Such mutations are referred to as **parallel mutations.**

Mutation Rates. Obviously, genes must be very stable, because if they were not, it would be difficult to understand how a species could maintain itself for any length of time. If gene changes responded readily to minor alterations in the environment, the species would be markedly unstable. No single mutation rate can be accurately given for a species, because mutation rates of different genes are not the same. Some genes mutate as frequently as 1/10,000 (regarded as a high mutation rate); some mutate as infrequently as 1/1,000,000, or even less. How then can such an infrequent process furnish the raw materials for evolutionary change? The answer seems to be: lots of time and lots of individuals. Moreover, when we point out the low mutation rate, we are referring to change of a particular gene. The total mutation rate for a whole nucleus is, of course, higher. It is equal to an average mutation rate multiplied by the number of genes in the nucleus. For example, suppose that the nucleus of man has about 10,000 genes; suppose further that the average mutation rate in man is 1/100,000; then, one in every ten gametes would have at least one mutation! These figures are only estimates, to be sure, but they appear to fit closely to what is observed.

Since mutations are the tools with which the geneticist works, it is understandable that in the early days of modern genetics a continued effort was made to find some means of bringing about these gene changes artificially. For a long time the work was fruitless until Muller, a Nobel prize winner in 1946, found that by exposing fruit flies to a blast of X-rays just short of the lethal dose he was able to increase the normal mutation rate 150 times. Apparently the X-rays were able to penetrate to the genes in the sex cells and bring about a change in the structure of their DNA. Many of the mutations that appeared in Muller's cultures were no different from those appearing spontaneously. They were also recessive and usually harmful or lethal. Here, then, was a technique of bringing about gene changes artificially, which was a boon to genetic research. Other radiations such as ultraviolet and radium as well as certain chemicals have since been successfully employed. Even high temperatures may be effective if used at particular times in the life cycle of the organism.

Chromosome Mutation

The organization of the chromosome complement plays a part in the functioning of the genetic material. Any alteration of this organization can lead to a marked change in expression of the separate genes. Cases have been documented in which activity or expression of a particular gene changes with a rearrangement of its position relative to other genes (**position effect**). Cases are known in which a change in the

proportion of some part of the complement is accompanied by a marked phenotypic change (see Down's syndrome, p. 542). Thus any change in chromosome structure or in chromosome number threatens an ordered functioning of the elements making up the genetic material.

Structural Chromosome Changes. Changes in which segments of chromosomes are lost are called **deletions;** nuclei containing such deleted chromosomes of course lack all the genes of the deleted segment. The phenotypic effect of a deletion depends upon its size and the kinds of genes associated with the lost fragment. **Translocation** refers to the abberration in which a segment of a chromosome breaks away from its usual place and becomes attached to another chromosome where it does not belong. No genes are lost by this change. The phenotypic effects depend upon whether the genes near the translocation point show position effect. Translocations change linkage relations of the genes in the translocated segment. As might be expected, unusual synaptic relationships arise at meiosis in a translocation heterozygote; these are often associated with a high degree of sterility. Sometimes a chromosome segment may be present in the complement in excess; the segment is said to show a **duplication.** The expression of the genes in a duplication often shows phenotypically as an exaggeration of the characteristics influenced by the genes. If two breaks occur in a chromosome, and if the segment between the breaks rotates through 180 degrees with rejoining of the broken ends, the resulting aberration is known as an **inverson.** Inversions, as could be expected, frequently lead to difficulty in synapsis of homologous chromosomes at meiosis; they are often associated with sterility.

Chromosome Number Changes. One might expect that occasionally during meiosis the number of chromosomes going to the gametes might vary in number. This happens when a pair of homologous chromosomes fails to separate at reduction division. One gamete then has one extra chromosome (*n* plus 1) while another is short one (*n* minus 1). If fertilization occurs with normal gametes, one offspring possesses an extra chromosome in each cell while another is short one. The latter probably will not develop whereas the former may show variations in the resulting trait because the genes are doubled. There have been many cases of **heteroploidy,** as this condition is called.

This phenomenon of extra chromosomes includes the addition of complete sets, so that gametes possess a diploid instead of the usual haploid number, and in some cases (particularly in plants where the phenomenon is better known) even 3, 4, and 5 extra sets. Organisms possessing more than the diploid number are called **polyploids.** They may be larger and more vigorous than the usual diploid. Polyploid flowers may be larger than the normal ones. Leaves of polyploid tobacco plants may be larger. Efforts to induce polyploidy artificially have been successful. The drug **colchicine** prevents division of the cells but does not interfere with the dividing of chromosomes; hence the gametes possess the diploid number of chromosomes instead of the normal haploid number. Upon fertilization, the resulting offspring then have a double set of chromosomes, that is, they are **tetraploid. Triploids** result from the union of a diploid gamete with a haploid gamete. Experimental breeding of these various polyploids has resulted in the introduction of new strains of plants that have proven valuable in increasing our food output. Very little is known about polyploidy in animals probably because it occurs so infrequently.

GENES AND THEIR ACTION

At several places in this book reference has been made to the physical and chemical nature of genetic material. At this point, on the basis of these data, we review some current reevaluations of the meaning of the terms gene and gene action. The genetical-biochemical probing of the nature of the gene and intensive recombinational analysis of selected microorganisms have brought into sharper focus some basic concepts of genetics. Microorganisms have been especially profitable to use as tools in these investigations: numerous enzyme mutants have been recovered from them; it is feasible to recover, test, and select large numbers of mutants and recombinants, making possible the detection of rare recombinants. High-resolution analyses of linkage regions are thus possible. These analyses demonstrate a linear arrangement of mutational sites within a gene and show that these sites are separable (though rarely) by intragenic recombination.

Then, too, more sophisticated techniques are being applied to a study of the structure and the functioning of chromosomes. In recent years **giant chromosomes,** such as those originally described in 1881 by Balbiani (Fig. 24–22), have been studied cytochemically, with radioactive tracer elements, and with both light and electron microscopes. The dark bands of these chromosomes contain DNA and are therefore known to be the location of the genes. The bands are remarkably constant in their location and number in cells of a particular species. Because of the great size of these chromosomes

Fig. 24–22. The chromosomes from the salivary glands of *Drosophila* are extremely large and for that reason show their constitution. Note the black disks arranged on a nonstaining strand. The genes are thought to correspond to the position of the disks.

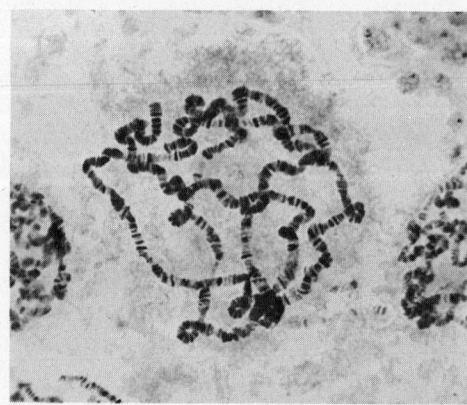

they are useful in checking for differential activity of segments correlated with developmental stages of the organism.

The newer knowledge about genes has forced a revision of our ideas about the definition of a gene. The evolution in our thinking has come about as a result of more and more critical experiments, particularly with *Drosophila*, corn, the bacterium *E. coli*, the fungus *Neurospora*, and the T2 and T4 bacteriophages (p. 532) of *E. coli*. The gene has for many years been described from three points of view: (1) a unit of segregation and recombination; (2) a unit of mutation; and (3) a unit of function. It is now clear that these three units are not physically identical. To underline their differences bacterial and viral geneticists have found convenient the terms, **recon, muton,** and **cistron.** The recon refers to a unit of recombination; that is, the smallest segment that can be defined by a recombinational event. The muton refers to a unit of mutation; that is, the smallest element within a gene, that is able to change, so altering the message of the entire gene. The muton could, theoretically, be as small as one nucleotide within a gene. The cistron refers to a functional gene; that is, it is a segment of DNA, all of which is necessary for the specification of a unique polypeptide chain. The cistron is, physically, the largest unit. Although cistrons are likely to differ in their nucleotide length, they probably are made up of hundreds of nucleotides.

Regulation of Gene Activity

Do genes carry out their predestined function all of the time or are there some controlling mechanisms which cause them to function at one time and not another; that is, can they be turned "on and off"? It seems quite obvious that the latter situation must exist because all cells of a higher organism come from a fertilized egg containing a specific set of genes. As development of the embryo proceeds cells **differentiate;** that is, they take on different functions and are certainly not all alike. They change in response to their changing environment and these alterations are induced by cellular chemical reactions initiated by enzymes. Since enzymes are proteins which are synthesized from coded mRNA it follows that the genes (DNA) must be responsive to their environment. Each cell contains the genes essential in manufacturing all of the proteins but obviously all cells do not produce all of the proteins all of the time. Some must be active at one time in certain cells and inactive at other times. They must be turned "on and off" at very precise periods in the life of individual cells in order that differentiation can occur. Many observations demonstrate that gene activity is precisely controlled and in recent years the way this is done has been elucidated.

To illustrate how genes can be controlled by their environment, we can cite two experiments with bacteria. The colon-inhabiting bacterium, *Salmonella*, when grown on an artificial medium in a test tube, synthesizes the amino acid, histidine. If this amino acid is added to the medium no more histidine is produced. The enzymes (called **repressible enzymes**) responsible for the synthesis of histidine are turned off. This small molecule in the environment is sufficient to inhibit histidine synthesis. The opposite effect can be demonstrated by growing bacteria (*Escherichia coli*) in an artificial medium using glucose as the carbon source. The enzyme **galactosidase** which splits the disaccharide lactose into glucose and galactose, is present in the culture medium at very low levels. This enzyme appears in high concentrations very shortly after lactose is added to the medium. The enzyme appears when the environment includes lactose; in other words, the disaccharide somehow induces the synthesis of galactosidase. Such enzymes are called **inducible enzymes.** These two experiments illustrate the sensitivity of genes to their environment. They also demonstrate the basic aspects of cellular steady-state control. All cells are able to maintain their steady states by regulating the synthesis of their enzymes.

In recent years, as a result of the efforts of many workers around the world, it is now

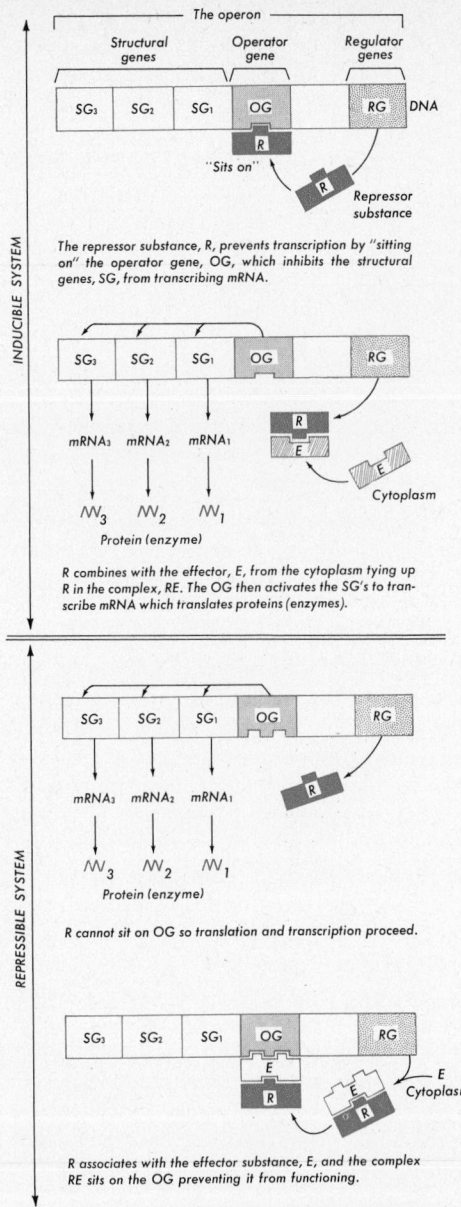

The repressor substance, R, prevents transcription by "sitting on" the operator gene, OG, which inhibits the structural genes, SG, from transcribing mRNA.

R combines with the effector, E, from the cytoplasm tying up R in the complex, RE. The OG then activates the SG's to transcribe mRNA which translates proteins (enzymes).

R cannot sit on OG so translation and transcription proceed.

R associates with the effector substance, E, and the complex RE sits on the OG preventing it from functioning.

Fig. 24–23. A schematic representation of how genes regulate cell functions. See text for additional explanation.

possible to state a hypothesis as to how repressions and inductions might operate. This is the **operon hypothesis,** or **model,** postulated by Jacob, Monod, and Lwoff of the Pasteur Institute in Paris for which they

received the Nobel Prize in 1965. It states that there are two types of genes, **regulator genes** (RG) and **structural genes** (SG); the former control the synthesis of specific enzymes which influence the latter (Fig. 24–23). Structural genes transcribe mRNA which codes cellular proteins, some of which are enzymes that promote metabolic reactions. Those that control a specific reaction lie close together on the chromosome forming a region called the **operon.** This also includes a gene, the **operator gene** (OG), which is active when the structural genes are active. The operon represents the portion of a chromosome that regulates all the steps in the synthesis of an enzyme or other protein.

The inducible and repressible systems are assumed to work like this. During a repression the repressor substance, R, produced by the regulator gene, has no effect on the operator gene, OG. The latter functions in transcribing mRNA which then translates the protein (Fig. 24–23, bottom). As the end product builds up in the cytoplasm, it is thought to combine with R, forming a complex which inhibits the functioning of the operator gene. Once this happens the structural genes cannot transcribe mRNA, and further synthesis of protein ceases. The level of the protein in the cytoplasm determines whether or not the operator gene would again function. Apparently as the protein is used up, the operator gene functions once more. By this feedback system the structural genes are turned on and off.

During an induction the repressor substance (R) normally associates with the operator gene (OG) and no transcription takes place (Fig. 24–23, top). However, if another substance called E (for effector) is introduced into the cell, E would combine with R, forming the complex RE which would remove the inhibition of OG and transcription of mRNA would occur and proteins would be synthesized. When all of E is used up, OG would be inhibited once more and transcription would be turned off. When induction occurs a repression is removed (called **derepression**), whereas in repression this is not the case.

The difference between the inducible and repressible systems lies in the form of the repressor substance that combines with the operator gene. Histidine and lactose, in the two experiments cited earlier, are effectors which combine with the repressor substance, changing its capacity to associate with the operator gene. It is only when the repressor substance is not associated with the operator gene that transcription occurs.

The operon control mechanism vastly improves the cell's efficiency. It permits the cell to quickly respond to its environment from moment to moment. Food molecules entering the cell may function by inducing (induction) enzyme production for its own metabolism. Likewise, accumulated end products may act as specific stimuli for repressing (repression) their own synthesis. Thus, the operon mechanism permits a cell to be in constant readiness to respond to its changing environment. Such a switch mechanism is imperative for the maintenance of the cell's steady state. It may also be important in explaining differentiation, as we see later.

Genes and Hormones

Hormones in very minute concentrations have a profound effect on cells. How this is accomplished has been a perennial puzzle among biologists. However, during the past decade, it is becoming clear how some hormones exert their influence on cells. The German biologist, Beermann, working with the dipteran insect, *Chironomus*, described "chromosome puffs" which had been observed by Balbiani many years before. These are expanded regions along the giant chromosomes which can be readily seen with the light microscope. The chromosomes are polytenic, that is, they consist of a great many strands (single chromosomes) which in concert make observations of individual chromomeres very clear (Fig. 24–24). By observing the chromosomes of the salivary glands of this fly Beermann noted that there was a correlation between the puffing of certain regions of particular chromosomes and the functions of the cells in which they were located. He did this by hybridizing two species, *C. tentans* and *C. pallidivittatus* and studying the offspring. In the former parent all of the chromosomes of the salivary gland cells are alike, whereas in the latter, four cells, which produce a granular secretion, are different from the other salivary gland cells. One chromosome in these cells has a puff not present in all of the other cells. The hybrids have the four cells which produce

Fig. 24–24. This sketch shows a region of a chromosome where RNA is being synthesized as indicated by the "puff." See text for further explanation.

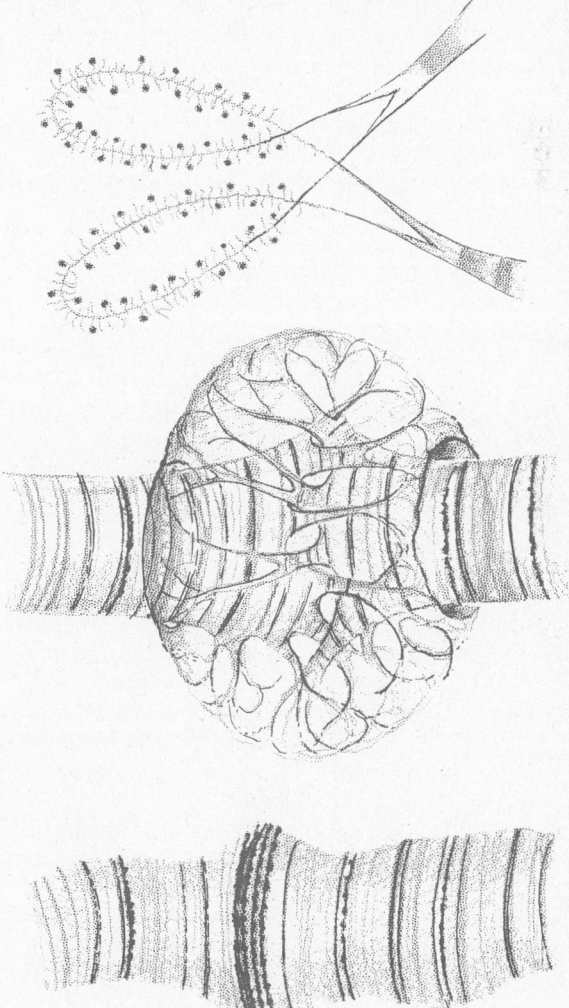

the granular secretion the same as the *C. pallidivittatus* parent, although in smaller amounts. Their chromosome also has the puff in the same place as in the parent. Obviously, the puffed region of these chromosomes was responsible for the production of the granular secretion. This demonstrates correlation of chromosomal morphology with specific function.

A striking demonstration of the effect of hormones on gene activity was accomplished by Clever of Purdue University. He injected the molting hormone, ecdysone into *Chironomus* larvae and was able to observe puffs appearing on two of the chromosomes within an hour. The larvae molted shortly thereafter and the puffs disappeared. Apparently, the hormone induced the genes to produce the materials required for molting, an elegant demonstration of the effects of hormones on genes.

Compound Loci

Since it is impossible to see the gene, there is no present cytological method that separates it into its parts. Hence biochemical and genetic tests must be relied upon to detect the physical dimensions of the gene. Highly refined techniques have been devised to detect crossing-over of small sections of the chromosome. With this technique, some areas of *Drosophila* chromosomes have been shown to be divided into several parts. These subdivisions are now considered to be genetic units which lie very close to one another and affect a single character. Ordinarily they stay together during recombination but sometimes they separate, producing combination effects. A number of loci which were once considered to be single genes have been found to consist of more than one unit or recon.

The first inkling that the gene might be a compound unit came when, over 20 years ago, C. P. Oliver crossed *Drosophila* with lozenge eyes and found in the progeny a few wild-type red-eyed flies which were not supposed to be there. The parents carried the genes lz^g and lz^s which were thought to

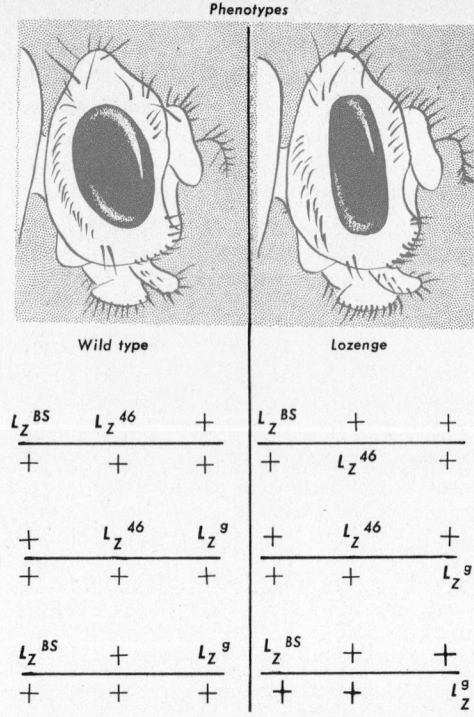

Phenotypes

Wild type   Lozenge

Fig. 24–25. Lozenge eye in *Drosophila* has been shown to be controlled by three recons rather than by one gene, as was once thought. Wild type appears when the recons are arranged as on the left, lozenge when they are arranged as on the right.

be alleles occupying exactly the same locus on the chromosome pairs, that is lz^g/lz^s. All the progeny were expected to be lozenge-eyed; instead a few red-eyed flies appeared. This was explained by speculating that the two genes were not located in exactly the same position on the chromosomes, but slightly to one side of one another thus, lz^g+/+lz^s. It then had to be assumed that crossing-over occurred which would place the two alleles side by side on the same chromosome, lz^g lz^s/+ +, each being heterozygous to the wild type. This, of course, would explain the red-eyed flies.

The lozenge-locus was further investigated by other workers who showed that at least three recons (lz^{BS}, lz^{46}, lz^g) existed which could be separated by crossing-over. These are shown in Figure 24–25, where the phenotypes are indicated. This locus, which was once thought to be occupied by a single gene is now known to consist of several recons, and is identified as a **compound locus.** Several other examples of compound loci have been demonstrated in *Drosophila.* In the white-eye cistron, as many as five sites have been identified.

Compound loci have also been identified in lower organisms, particularly bacteria and molds. Indeed, these organisms lend themselves to this type of research owing to the fact that large numbers of organisms are easily obtained, and selective media make the identification of mutants relatively easy. By subjecting the common baker's mold, *Neurospora*, to high-energy radiations, great numbers of biochemical mutants have become available for study. From these studies further evidence for compound loci has emerged. This is called **complementation.** It has been demonstrated that mutants within a given locus or cistron will complement each other and produce wild types. This is possible without observing recombination. The test involves the mating of different mutant types and observing the phenotypes in the zygote. The resulting data suggest the presence of compound loci. For example, the final step in trytophane synthesis is controlled by the cistron, *Td*, which is composed of several subunits arranged linearly.

Perhaps the most sophisticated analyses of compound loci are those carried out with the viruses infecting bacteria, called *bacteriophages.* The most celebrated virus in this respect is the T4 bacteriophage which infects the colon bacillus, *Escherichia coli.* This virus has a "head" which contains a single long strand of DNA (molecule), and a tail which is used in attaching to the bacterial cell wall prior to infection (Fig. 24–26). Once attached the DNA molecule is "injected" into the bacterium where it controls the reorganization of the cell machinery to produce 100 or more identical copies of the virus. This is a complex process because each copy consists of six proteins in addition to the DNA. The entire operation is under the control of the original virus DNA molecule. During this process errors are occasionally made (mutations) and these are the tools the geneticist uses to investigate the gene.

The existence of a virus gene as a compound locus came from a study of the so-called rII mutant. When it invades bacteria spread in a thin layer over an agar plate, a specific type of plaque (clear space) is formed which can be easily identified. Other mutants produce different types of plaques. The mutant rII can be divided into two different cistrons, called rIIA and rIIB, each producing plaques unlike rII. However, when both rIIA and rIIB infect the host bacterium simultaneously, recombination occurs and the resulting phages produce rII plaques. The two subunits complement each other and the region can be identified as a compound locus.

The rII cistron continues to be investi-

Fig. 24–26. This series of sketches shows how a virus infects the bacterium and replicates its own DNA from materials supplied by the host. When two mutated viruses infect a bacterium, recombination may occur.

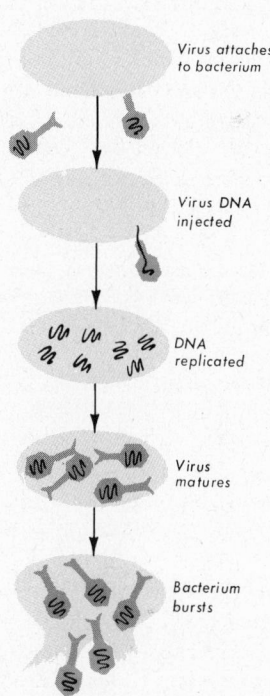

Virus attaches
to bacterium

Virus DNA
injected

DNA
replicated

Virus
matures

Bacterium
bursts

gated by S. Benzer and his collaborators and the outcome is approaching a rather complete analysis of this one region of the DNA molecule. They have found some 80 segments occupying the rII region, all of which are apparently able to cross over among themselves as well as complement each other.

From these analyses we have genetic units of different sizes. The smallest are the mutons and recons; the next in size are the two cistrons, A and B; and the largest unit is the rII region which includes both cistrons. One may well ask, which of these may be defined as the gene? The earlier definition of the gene could apply equally well to all of them. And it makes little difference if one is referring to the higher level of integration where no discrepancy exists between the units referred to. It only becomes important when the work involves the more specific operational aspects of the unit and then the more precise terms must be used.

Genes and Enzymes

In the past genes were designated by the effect they produced as expressed in the altered phenotypes. Today, we are learning what they do chemically and this field has come to be known as **biochemical genetics.** We still rely on phenotypic expression although in a few cases it has been possible to

identify the chemical steps involved. One example of an error in metabolism may be taken from man.

Today very small amounts of various chemicals can be identified by paper chromatography, a simple process employing the principle that various molecules move through paper at different rates and thus can be separated. The mixture of molecules is placed in one corner of a large piece of filter paper which is then dipped in special solvents. The solvents move over the unknown molecules by absorption and carry them at different rates, ultimately depositing them in isolated spots which can then be visualized by chemical reagents. Employing this technique urine samples of many people, both normal and abnormal, have been examined. Most normal individuals excrete only six amino acids (Fig. 24–27). One type of abnormality, called **cystinuria,** is identified by the excretion of large amounts of the amino acid, **cystine,** as well as arginine and lysine, none of which is excreted by the normal person. While some people possessing this abnormality show no symptoms, others form cystine-containing kidney stones which often cause damage to the kidney. This abnormality is due to a recessive gene which is observed in the individual only when it is homozygous.

Most of our information about biochemical genetics, however, has come from the study of viruses, bacteria, and molds, some of which has been transposed and applied to long-known aberrations in higher animals.

Since we know that chemical reactions in organisms are under enzyme control, and that genes are responsible for enzyme production, then there must be some connection between the genes and chemical reactions. A great deal of knowledge has been gained in this regard by a study of *Neurospora*, initiated by two American workers, G. W. Beadle and E. L. Tatum who received a Nobel Prize in 1959 for their efforts. They found that this mold normally thrives on sugar, ammonia, salts, and one vitamin, biotin. It is able to synthesize all other compounds

Fig. 24–27. Chromatographs of the urine from a normal person (A) and one suffering from cystinuria (B). Note that in the latter three amino acids (black) appear which are not found in normal urine. See text for further explanation.

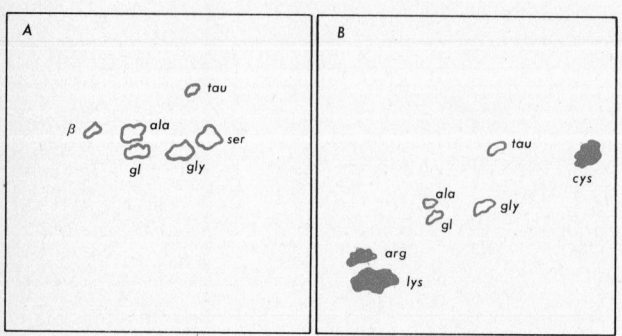

essential in the manufacture of its protoplasm. By showering the spores of this mold with X-rays the workers produced a large number of strains that required more than the minimal diet. The genes mutated so that they could no longer control the production of a certain nutrient, say an amino acid or a vitamin, and therefore required that substance in their diet. These men were able to demonstrate, by employing a large number of strains of the mold, that each step in the production of a nutrient is controlled by a single enzyme and that this enzyme is controlled by a single gene. Therefore, there is a one-to-one relationship between the gene, the enzyme and the ultimate chemical reaction. More recently this idea has been refined to mean that one cistron is responsible for one polypeptide. This generalization has added much to our knowledge of gene action.

The pigment that is responsible for brown, gray, and black coloring is melanin. The degree of coloring depends on the quantity of melanin and this is controlled by genes. Blacks possess genes that provide large amounts of melanin, whereas white people have genes that produce less melanin. The distribution of pigment to the eyes is controlled by separate genes, and the various coloring of the iris of the eyes is dependent on the quantity of brown pigment present. Blue eyes have no pigment and the color is caused by the physical properties of the iris itself. Brown or black eyes contain sufficient pigment to completely obscure the natural blue.

One of the best-known examples of the relation between a gene and a specific character is pigment formation in man. The complete absence of pigmentation results in albinism in which the skin, eyes, and hair are devoid of coloring. The eyes are very light blue or pink owing to the appearance of the blood vessels within the eyeball which can be seen through the iris. This condition appears when the single gene (*a*) is in the homozygous condition (Fig. 24–6). The heterozygous individual (*Aa*) does not show the trait nor, of course, does the homozygous normal (*AA*). The series of chemical reactions that lead to pigmentation are well known.

The chain of reactions that produces melanin begins with the essential amino acid, phenylalanine, which is oxidized to another amino acid, tyrosine, under the control of a specific enzyme (Fig. 24–28). This compound is further oxidized to dihydroxyphenylalanine (dopa) under the influence of a second enzyme. Two more steps are involved, each controlled by enzymes, which convert dopa to melanin, the end product. Obviously, one or more blocks in this series of reactions prevent the formation of melanin which would then appear as albinism in the organism. It is now known that the gene *a* can prevent the oxidation of dopa, providing it exists on both chromosomes, that is, is homozygous. In such individuals dopa accumulates because there is no enzyme for oxidizing it to melanin; hence they are albinos. Albinism can be explained by this genetic aberration in all organisms, from

Fig. 24–28. This series of biochemical reactions demonstrates what happens when certain enzymes are absent or defective. Each enzyme is presumed to be gene-controlled. If the gene does not function properly, the enzyme is not formed, and the reaction stops at that part of the sequence. The diseases resulting from these blocks are shown in the boxes. See text for further explanation.

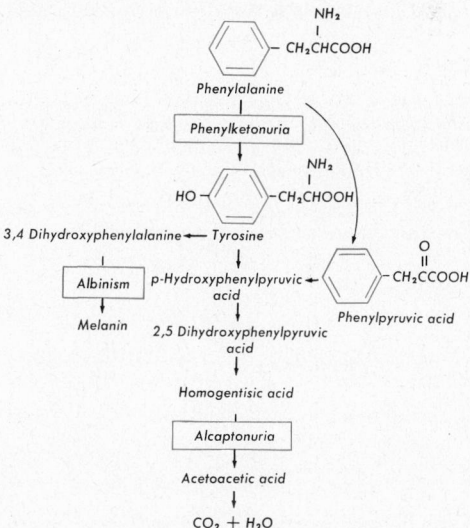

mice to men. An obvious cure for the defect would be to supply the enzyme exogenously. Some progress has been made in this direction with *Neurospora* but so far, there has been no method discovered by which an enzyme can be introduced into a cell in a metazoan animal. It seems that the enzyme must be built by the cell itself in order to be effective. Further research may overcome this obstacle.

Another hereditary human abnormality can also be explained by a block in the metabolism of phenylalanine. This disease is called **alcaptonuria,** which is characterized by the urine turning black upon standing. Another symptom is the darkening of cartilage which can be seen in the ear, elbow, and wrist. The blackening of the urine is caused by the presence of **alcapton** or **homogentisic acid** which in normal people is converted to acetoacetic and fumaric acids and then to carbon dioxide and water (Fig. 24–28). Alcaptonurics apparently lack genes which produce the enzymes responsible for these conversions. The amount of alcapton that appears in the urine can be controlled by the amount of phenylalanine in the diet, which is not the case in normal people.

A far more serious disease in humans, called **phenylketonuria,** is caused by another block in the normal metabolism of phenylalanine. People suffering from this disease excrete large quantities of phenylpyruvic acid

in their urine and, what is worse, are seriously retarded mentally. The amount of the acid in the urine is proportional to the amount of phenylalanine in the diet, which, again, is not the case in normal people. Normally phenylalanine is converted to tyrosine but in phenylketonurics the enzyme responsible for making the conversion fails to function; hence a block occurs and phenylpyruvic acid accumulates and is excreted in the urine (Fig. 24–28). Again a single recessive defective gene is responsible for the block. Normal people possess the genes to convert phenylalanine to tyrosine and tyrosine into p-hydroxyphenylpyruvic acid which can also be produced from phenylpyruvic acid. Phenylketonuria results when the block occurs between phenylalanine and tyrosine. It will be noted that this occurs much earlier in phenylalanine metabolism than the block which is responsible for alcaptonuria and it has far more profound effects. It seems that the earlier in the chemical series a block occurs, the more drastically the person is affected. Another associated trait is the blond complexion of all phenylketonurics. This is caused by the failure of the production of the normal amount of tyrosine which is a precursor to melanin formation.

The gene control of metabolism is well illustrated by this series of steps in phenylalanine metabolism. Clearly the phenotypes are markedly different; yet the chemical processes involved are very closely related. They proceed in a stepwise manner, each controlled by an enzyme, and each enzyme is produced under the influence of a gene. Blocks in other parts of this chain of reactions, would probably cause other defects that would be expressed phenotypically. While this is one of the best-known series of reactions, it is generally believed that most, if not all, proceed in somewhat the same fashion.

Complex Consequences of a Single Gene Defect

Defective genes have one primary effect which is usually very obvious. However, there are often many other minor abnormalities

Fig. 24–29. The frizzled chicken on the left has defective feathers, a gene-controlled abnormality. Owing to a lack of insulation, profound physiological and morphological abnormalities appear in these animals.

which result indirectly from the primary defect. This is well illustrated in chickens where, owing to a single abnormally functioning gene, the trait called *frizzle* appears. Such chickens possess brittle, curly feathers which are few in number. Chickens with this trait are easily distinguished from normals (Fig. 24–29). In addition to this primary trait they possess a whole galaxy of other abnormalities which result from the primary defect. They lay few eggs and the number of chicks hatching is low and those that do hatch have a high mortality rate. The young, as well as the adults, have high metabolic rates although they are less active than normal chickens. They consume large quantities of food, their hearts are larger and beat faster, the spleen is enlarged, and all of the digestive organs are unusually enlarged, presumably owing to their large food consumption. All these characters are associated with the defect in feather formation. Because the insulation from temperature change cannot be regulated in the frizzy chicken its organs have become overdeveloped to take care of the body needs. Under normal temperature fluctuations these chickens fare poorly, but when the temperature is uniformly high, they survive better than normal chickens.

Another example of the profound effects of what appears to be a simple abnormality can be drawn from the study of human blood. When the red blood cells of certain people are placed under reduced oxygen conditions, they become distorted into a characteristic "sickle" shape; the phenomenon is called "sickling" (Fig. 24–30). This condition is caused by a single gene and when it exists in the heterozygous condition (Ss) the person suffers no ill effects. However, when a person carries both recessive genes (ss), a number of serious abnormalities appear which together have been called "sickle-cell disease." These defects include severe anemia, poor body development, and paralysis, all of which apparently results from the sickling condition of the red cells in the blood vessels. The red cells do not sickle in heterozygotes; consequently such people appear normal.

In sickle-cell disease the hemoglobin mol-

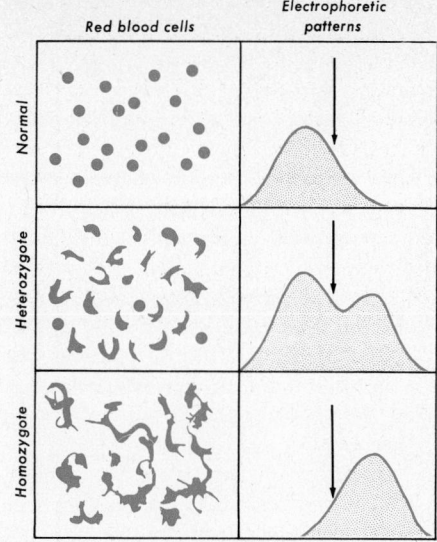

Fig. 24–30. Sickle-cell anemia is an abnormality of the red blood cells, as shown at the left. It is gene-controlled; under normal oxygen tensions the heterozygous condition is harmless, but homozygotes are seriously ill. The electrophoretic pattern on the right indicates that the hemoglobins are slightly different, owing to the substitution of one amino acid for another.

ecules are chemically different from those in normal blood, and this difference was first noted by observing that the electric charge on the two hemoglobins was not the same. Further analysis demonstrated that the two hemoglobins differ by one amino acid out of about 500. One of the glutamic acid molecules in normal hemoglobin is replaced by the amino acid, valine, in exactly the same place in the sickle-cell hemoglobin. Valine has a charge opposite to that of glutamic acid, which accounts for the difference in electrophoretic patterns in the two hemoglobins (Fig. 24–30). At present there seems to be no other difference in the two hemoglobins. One would guess that so slight a change would have little, if any, difference in the large hemoglobin molecule. Quite the contrary is true when one observes a person suffering from sickle-cell disease.

A single gene controls the substitution of valine for glutamic acid in the hemoglobin molecule, and this simple switch of one amino acid for another has far-reaching effects in the victim of sickle-cell disease. Their anemia weakens them physically and retards them mentally; the clumping of the cells in the blood vessels brings on heart failure, paralysis, and kidney obstruction. Thus we see a whole chain of events brought about by one small genetic change at the molecular level. Undoubtedly other abnormalities of genetic origin will be explained in this manner as we gain more information about disease at the molecular level.

An interesting outcome of this abnormality has been observed in parts of the world, such as Africa, where malaria is common in the population. Those persons who are heterozygous for sickle-cell disease are more resistant to malaria than are normals. Just why this is so is not known. For this reason they have a selective advantage and the percentage of the population carrying the gene for sickle-cell anemia is higher in regions where malaria is common than in other parts of the world where the disease does not exist. Therefore, a trait, which may under some conditions be deleterious to a race, actually is beneficial under other conditions.

HEREDITY AND ENVIRONMENT

The age-old question of which is more important, heredity or environment, stimulates the most passionate arguments among people in various walks of life, from educator to the man on the street. What are the best unbiased answers to the questions rising from such discussions that have come to light so far?

One should first be concerned with the problem of what is really inherited. Is a child unmusical because his parents have no musical skills or is he unintelligent because his parents possess a low order of intelligence? Is a starved child short and underdeveloped because his parents are likewise starved and poorly developed? A careful examination of some data should give us an inkling of the answer to these questions. A child placed in an environment in which every opportunity is afforded him to study music as well as other cultural subjects may show unusual artistic talents and intellectual achievements even though he might not appear to have inherited such ability. Likewise, well-fed children of starved, stunted parents often exceed them in height and physical vigor by a considerable margin, as has been demonstrated so many times when children of immigrants are larger and more robust than their parents, who matured on inadequate diets. In these instances was it the environment or was it heredity that played the more important part?

It is clear that the children inherited a set of genes which provided the **capacity** to reach these goals. The environment was merely the factor which determined whether or not the genes would be given a full opportunity to express themselves. Without the substantial set of genes in the first place, no amount of encouragement from a satisfactory environment would have brought them to heights beyond the basic design established by the genes. Of course, the desired goal in any society is to give the genes in each person ample opportunity to express themselves to their greatest capacity. Therefore, while good genes are essential for good stock, they are not enough. Good genes in a good environment is the ideal goal.

There is experimental evidence to demonstrate further the intricate relationship between environment and heredity. In studying the life span in fruit flies, Raymond Pearl found that in a closed milk bottle the vestigial-winged forms lived only half as long as the wild type. If he increased the number of flies of the wild type he cut their lifetime to that of the vestigial-winged flies. If both types were starved they lived about the same length of time. Certainly the environment limited the capacity of the genes. Similarly, in "Himalayan" rabbits (Fig. 24–31) the environment is important in determining the coat color pattern. The intense black fur that occurs at the tips of the ears, paws, and tail is controlled by temperature. These tips being the coolest parts of the animal, the lowered temperature induces the formation of heavy pigmentation. If the hair is re-

Fig. 24–31. The normal Himalayan rabbit has black extremities (top). If it is placed in an incubator at elevated temperatures, the black areas become lighter (left). If its ears and feet are clothed and an ice pack placed over a shaved area on its back for some period of time (right), the new hair will be white on the feet and ears, and a black patch will appear on the back (bottom). From this we can conclude that environment does influence the expression of genes.

moved from a portion of the body that is white and this region kept cool, the hair that grows in is black. If the rabbit is chilled in a refrigerator, it becomes totally black; conversely, if it is warmed up to a relatively high temperature it becomes totally white. It is quite obvious in these cases that the genes lay the pattern for capacity, whereas the environment determines whether or not that full capacity will be realized.

The close interaction of these two forces, environment and heredity, can be still further emphasized by following the development of an organism, say a human being, from fertilized egg to young adult (Fig. 24–32). Beginning with the fertilized ovum, we have a cell consisting of cytoplasm, controlled by the **genes** (G) and surrounded by an environment. Very soon, as cleavage starts, the surrounding **environment** (E) contributes to the cytoplasm, thereby changing it slightly. Moreover, the immediate community of cells has its effect upon the cytoplasm of each of the other cells. The cytoplasm is therefore influenced from within by the genes (G) and from without by the environment (E), even at this early stage. As

development proceeds, the cytoplasm of the many different cells continues to be influenced by surrounding cells, yet the genic constitution remains essentially the same in all of them. All of this brings about the development of organ systems, and ultimately the full-fledged fetus, but in each step the genes control the cytoplasm which, in turn, is influenced by surrounding cells (organizers) so that the cells are channelled into a pattern directed toward the final organized individual. Another environmental factor that is influential is the **nourishment** (N) received by the fetus from the uterine wall. If food, as well as oxygen and hormones, that comes to it from the placenta is deficient in any respect, the fetus is directly affected, perhaps to a point where normal development is impossible. This is a very important environmental aspect of development. The genes strive toward the production of a normal offspring, but they can accomplish this end only if the environment is adequate.

When the child is born, another factor, the **psychological factor** (P), enters the picture. All the other factors still operate, and the addition of this new factor definitely

Fig. 24–32. The individual is a result of two forces, one due to the inherited genes and the other to the surrounding environment which begins with the zygote and continues throughout the life of the individual. The various steps sketched here are explained in the text.

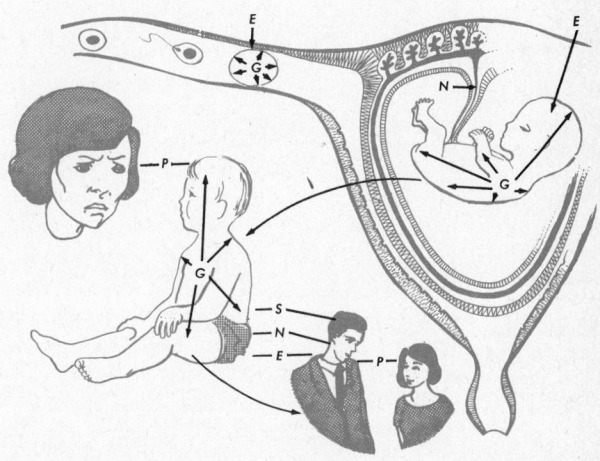

controls further mental development and probably physical development as well. Lastly, as the child matures, the **social factors (S)** become more and more a dominant part of his environment. The genes are still at work and will continue to be throughout life but the environment has a marked molding effect on their action. Thus, we see that it is impossible to have heredity without environment and vice versa. The two forces work hand in hand to produce a normal, well-adjusted individual.

Fig. 24–33. Identical or monozygotic twins are a result of a single fertilized egg; therefore they are genetically alike, as shown in the upper figure. Fraternal twins are a result of two zygotes and may be as genetically different as any other members of the family. Furthermore, each developing fetus has its own placenta, whereas this is usually not the case with monozygotic twins.

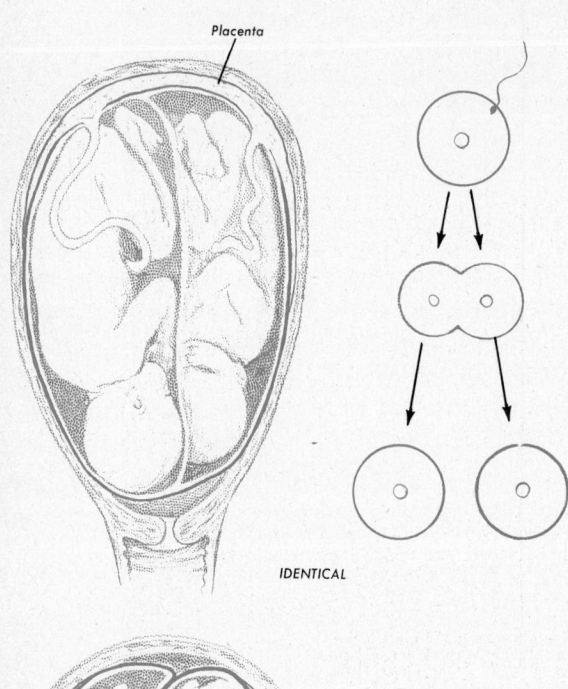

Placenta

IDENTICAL

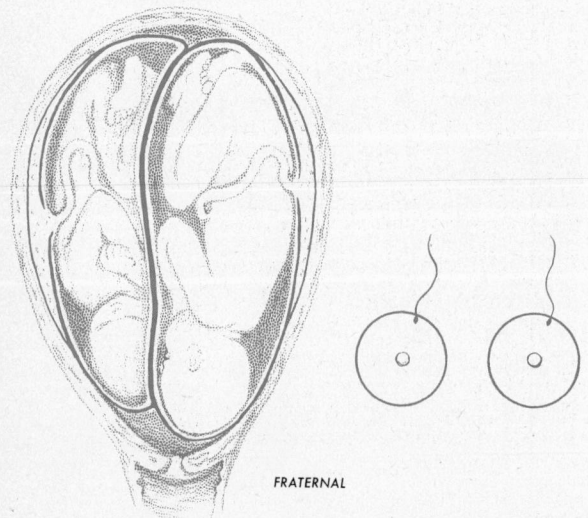

FRATERNAL

The occasional appearance of two offspring as a result of one fertilized egg (**identical** or **monozygotic twins**) has provided almost perfect material for the study of the relative effects of environment and heredity (Fig. 24–33). Monozygotic twins have the same sets of genes, so that if heredity were solely responsible for all of their characters they should be exactly alike. A study of a large number of identical twins by Newman and others has shown that they are indeed remarkably, almost unbelievably, alike (Fig. 24–34). Such traits as height, weight, general appearance, fingerprints, and tooth formation show very close similarity as might be expected, but when such intangible traits as susceptibility to disease (both organic and infectious), food preferences, and even attitudes of mind show marked correlations it seems quite incredible. Intelligence tests usually show that they vary no more than either individual might on two different tests. With such material as this, it should be possible to learn some interesting facts about the effects of environment.

Newman found in his studies that identical twins reared apart were over 90 percent alike in height, nearly 90 percent alike in weight, but only about 70 percent alike in intelligence as measured by standard I.Q. (intelligence quotient) tests. Identical twins reared together had about the same scores for height and weight but were also nearly 90 percent alike in regard to intelligence. Two-egg twins (Fig. 24–33) of the same sex were only about 65 percent alike in all of these traits. When one monozygotic twin is bright, the other is also, but when their educations vary considerably their I.Q.'s also vary correspondingly. Better schooling definitely raises the I.Q. rating. However, under similar cultural environments identical twins, although reared apart, differed only 5 points

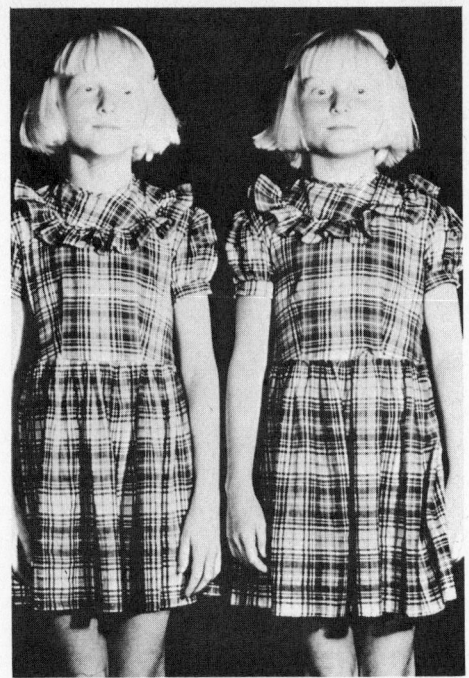

Fig. 24-34. A pair of identical or monozygotic female twins who happen to be albinos as well. They are the result of a single zygote, hence their likeness. They are alike physically, mentally, and even emotionally. Studies of such genetically identical individuals has added much to our knowledge concerning the relative importance of heredity and environment.

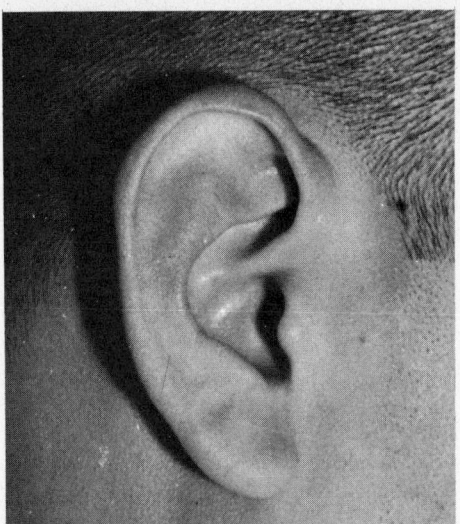

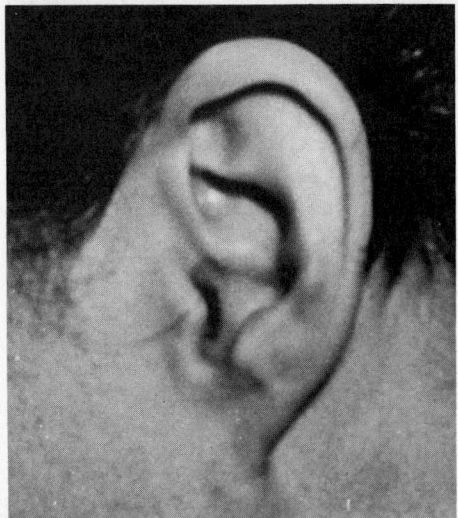

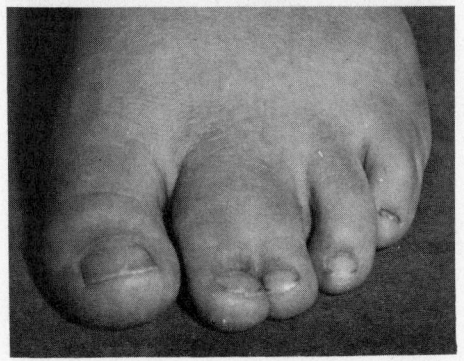

Fig. 24-35. Lobed and nonlobed ears (upper), as well as fused toes (lower), are inherited in a definite fashion.

in their I.Q. ratings. This is within the normal variation that might be expected from day to day with the same person. From these studies, environment seems to have a considerable influence on intelligence as it is measured with I.Q. tests, although this may not be a valid measure of true intelligence. On the other hand, in regard to such physical traits as height and weight, the inherited genes are strikingly more important.

Fortunately, because of the stability of genes, the environment has little or no effect upon them. Excellent genes may be carried along for many generations in starved, poorly developed bodies, but as soon as the environment becomes completely adequate they express their full capacity by producing a vigorous, healthy body. In a world that has always suffered periods of famine intermingled with times of abundance, only stable genes could have made possible a race that has been able to survive up to the present.

HUMAN HEREDITY

Human beings, like all species of animals, show wide variations in height, weight, eye color, hair color, skin color, facial configuration, mental ability, and many other traits (Fig. 24–35). Within limits, these variations are highly desirable because they allow evolution to take place. However, at the extremes of each one there are usually either mildly or highly undesirable conditions which do not permit the individual possessing them an equal chance with others to become a self-sustaining member of his own social group. Some defects such as albinism render the individual only slightly more handicapped than his normal fellows. Most of us are pretty apt to agree, however, that a combination of genes that produces idiots, blindness, or deafness is undesirable. Although we are concerned with the inheritance of defects in order to breed them out of our people, we should be even more concerned with the positive approach, that is, retaining the quality of our present stock and improving it where possible.

Some Inherited Defects

We have previously discussed the slight defects of color blindness (p. 000) and albinism (p. 000), as well as the more serious anomalies, hemophilia (p. 000) and phenylketonuria (p. 000). Some others are listed in Table 24–4 (see also Fig. 24–36). **Down's Syndrome,** formerly known as "mongolism," is characterized by a group of symptoms: retarded physical development (short stature, loose-jointedness, stubby fingers), distinctive facial features ("slanting" eyes, large tongue, broad skull), and very low mentality.

Mongolism is a common defect in children caused by an increase in the number of chromosomes, called **trisomy.** The cells of such children contain an extra autosome which apparently upsets the metabolic balance resulting from extra genes contributed by the additional chromosome. The cause of this anomaly is an error in oogenesis (p. 493). During meiosis in either the first or second division, one pair of autosomes stays together (nondisjunction) instead of separating as they normally do. Thus, the

TABLE 24–4. Some Genetic Diseases of Humans

DISEASE	CONDITION	TYPE OF GENE
Congenital cataract	Opaque growths in eye	Dominant
High grade myopia	Extreme nearsightedness	Recessive
Astigmatism	Irregularity of cornea	Dominant
Polydactylism	Extra digits	Dominant
Lobster-claw	Split hand and foot	Dominant
Achondroplasia	Short appendages	Dominant
Diabetes mellitus	Improper sugar balance	Not clear
Amaurotic idiocy	Degeneration of nervous system	Recessive
Deafness	Atrophy of auditory nerve	Dominant
Harelip and cleft palate	Malformed lip and palate	Recessive
Allergy	Allergy	Recessive
Epilepsy	Convulsive seizures	Dominant
Congenital hemolytic jaundice	Increased red cell fragility	Dominant
Pernicious anemia	Blood deficiency	Dominant
Schizophrenia	Split personality	Recessive
Sickle-cell anemia	Abnormal red blood cells	Recessive

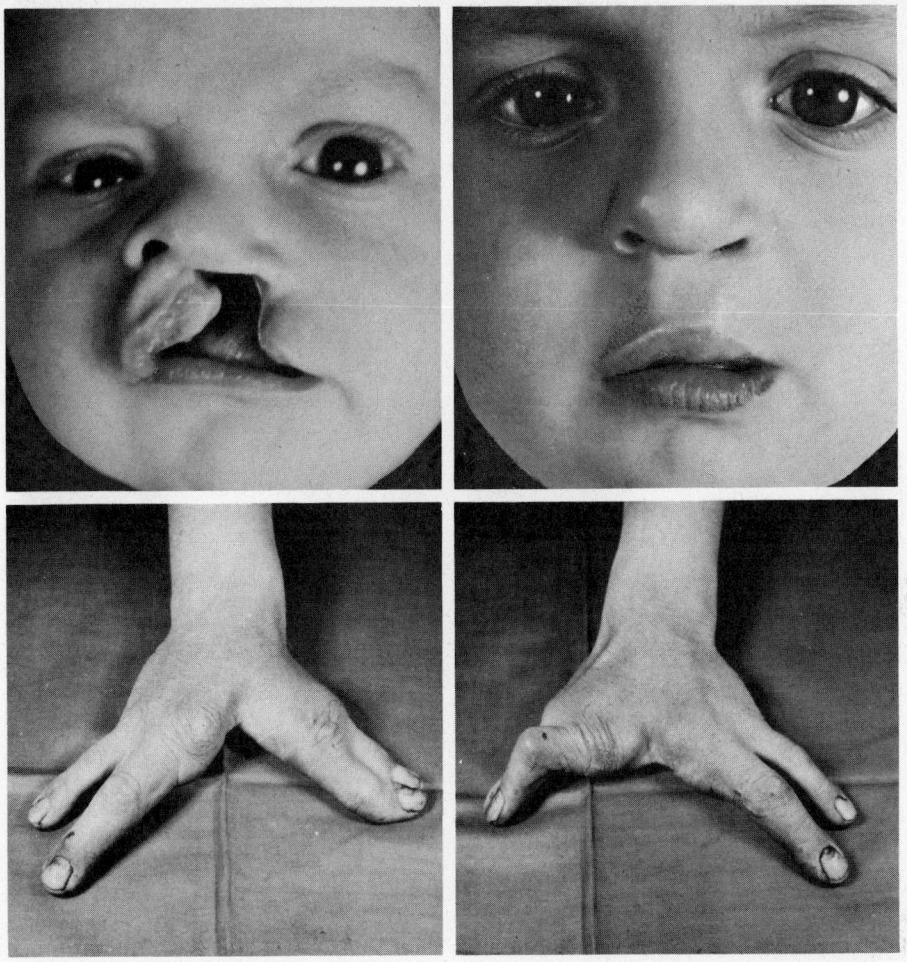

Fig. 24–36. There are a great many human anomalies that have been traced to a genetic origin. Certain types of harelip are inherited. One of the common types is shown here (upper left). Modern surgery has worked wonders in repairing these unfortunate abnormalities (upper right). Many deformities of the hands and feet are inherited. Here are two cases of "split-hand" (lower). This trait is usually dominant and occurs once in about 90,000 births. More often than not the gene responsible for this abnormality is also responsible for others. Both of these cases were abnormal in other respects.

resulting egg has 24 chromosomes instead of the normal 23 (haploid). When fertilized by a normal sperm cell, the zygote containtains 47 chromosomes instead of the normal 46. Approximately one child in 600 births is a mongoloid in the over-all population. The number of mongoloids increases with the age of the mother. After 40 years of age such defective births increase to one in fifty. This is a genetic accident about which

nothing at present can be done. Approximately 95 percent of mongoloids are of the nondisjunction type and the trait is not inherited. The remaining five percent are familial, that is, the trait is inherited. Thus, when a mongoloid child appears in a family, the chances of having a second one are greatly increased (about one in three) if the trait is inherited. It is impossible to determine by examination whether a mongoloid is of the nondisjunction or inheritable type because all mongoloids resemble one another.

Ichthyosis, sometimes called fish-skin disease, is a congenital disease. Those suffering from this malady have dark skin, ¾-inch thick, which is cracked in such a way that it resembles scales, hence the name, fish skin. It occurs over the entire body except the palms of the hands and the soles of the feet. It is fatal, the child dying shortly after birth. The disease was seen for the first time in England in 1731 where it appeared in the son of a laborer. Another type of ichthyosis (nonlethal) has been traced through five generations and has successively appeared in the sons but never in the daughters. Furthermore, the daughters never transmit it to their offspring, which definitely places the responsible gene for the trait in the Y chromosome. This is one of very few cases where a gene has been definitely located in the Y chromosome in man.

A type of blindness caused by the clouding, or fogging, of part of the transparent portions of the eye is called **cataract.** Pedigrees of families in which this disease occurs bear out the contention that the trait is carried by a dominant gene, although it usually appears in the heterozygous condition. That is to say, a parent may have the disease but usually only certain of his children, regardless of sex, will show it, indicating that the parent must have been heterozygous. Had he been homozygous for the trait, all of his children would have the defect. Cataract is also acquired by excessive exposure to heat, radiations, and contusions, but in these cases heredity, of course, is not involved.

Aside from obvious physical defects there are hereditary traits that affect the nervous system and alter the mental behavior of the individual radically. There is no doubt that many mental illnesses, like physical defects, are gene-controlled, so that a careful study of them should yield some valuable information to aid in guiding our future actions.

One kind of nervous disorder, called **microcephalic idiocy,** is inherited as a recessive. It usually springs up in families of normal parents, as one might expect. However, a large proportion of the children resulting from the marriage of an idiot to a normal person would be idiots. Fortunately such marriages are rare and the line usually dies with the defective person.

Another very serious congenital disease is **Huntingdon's chorea,** in which there is a marked degeneration of brain tissue, resulting in poor muscular coordination, twitching, and jerking movement. In addition, there are other associated defects which contribute to produce socially and mentally deranged individuals. The most unfortunate part is that the disability does not became apparent until middle life (25–55 years of age) after most, if not all, of the children have been born. Thus the trait is passed on to the next generation before it is recognized in the parent. Cases of this disease in this country have been traced back to three brothers who arrived here from England in 1630. The malady has shown up hundreds of times in the several thousand descendants of these men, and studies have revealed that about one-half of the children of afflicted persons develop the disease some time after they are 25 years old. A person who knows that he might develop or pass it on to others is under terrific mental strain, which in itself is often sufficient to bring on nervous disorders. Some recent studies indicate that electro-encephalographic patterns can identify these people during adolescence.

MEDICAL GENETICS

From such studies as those indicated above, the medical profession seems to have found some information of real clinical value. Data accumulated by the theoretical geneticists for the past fifty years are just now reaching a point in development where some practical use can be made of them by physicians. This has resulted in the young flourishing science of **medical genetics,** which is not to be confused with **eugenics** (see later), a subject that has enlisted the active participation of people outside the field of medicine.

Many illustrations of the use of genetics have been cited in the diagnosis, prognosis, and even prevention of a wide variety of diseases. For example, in attempting to diagnose what was causing a patient to vomit blood, the doctors decided it must be one of two things, ulcers or liver disease. Elaborate laboratory tests revealed that neither of these seemed to be the cause. One of the physicians recalled that the father of the patient had shown a tendency to bleed due to the fragility of his capillaries in local lesions. Because this condition is controlled by a dominant gene, the likelihood of a similar condition appearing in the son was good. By an exploratory examination of the stomach the suspected diagnosis was verified, the region of capillaries giving the trouble was removed and the case was cured. Here a knowledge of genetics was very helpful in solving a problem which probably would have continued to be a mystery to the physician and a source of discomfort to the patient.

Of equal and perhaps greater importance is the use of genetics in the realm of prognosis. Young married people or those about to be married often wish to know what the probabilities are for them to have normal children. This is particularly important to those who have some so-called "taint" or "skeleton in the closet" in their family. Sometimes advice is sought for on the positive side, that is, they wish to know what the chances are for them to transmit certain talents to their offspring. One case will illustrate the importance of such information. In a family of five children, two boys were severely crippled by serious muscular disease; the others, two girls and a boy, were normal. The problem that concerned the parent was the chance of this disease appearing in the next generation. After careful analysis of the situation it was evident that the gene causing the difficulty was sex-linked. Therefore the father must have had a normal X chromosome (AY) while the mother must have carried the defect in one of her X chromosomes (Aa).

The boys showing the affliction must accordingly have been aY, whereas the normal son was like his father (AY) and the daughters might or might not have had the defective gene (Aa or AA). The normal son would, of course, be unable to transmit the defect to his children. The daughters, on the other hand, had an even chance of picking up the defective gene from their mother. Such information is highly valuable to intelligent people because they can make their decisions in the full light of scientific fact rather than blind faith.

It is now possible to cultivate the cells from the amniotic fluid of the fetus which makes it feasible to detect genetic abnormalities during the early stages of pregnancy. Analysis of the chromosomes of the fetus not only determines its sex but also demonstrates the presence or absence of nondisjunction which results in an abnormal child. Other tests will undoubtedly be devised to determine genetic defects in a child long before it is born. Such information could be of great value, particularly if some treatment is available to correct the defect. If not, it will be up to society to make a decision as to whether or not the pregnancy should be terminated.

The spectacular success that molecular biologists have had in recent years may ultimately be useful in correcting genetic defects. Such possibilities as replacement or removal of genes and DNA transformations are not feasible today but they may be in the future. Medical genetics is progressing rapidly and has become a well-established branch of medical science.

EUGENICS

Eugenics is the study of race-improvement or the "science of being well born." Since planned matings are not possible in humans, scientists who are interested in eugenics are forced to collect and analyze family histories in an effort to account for

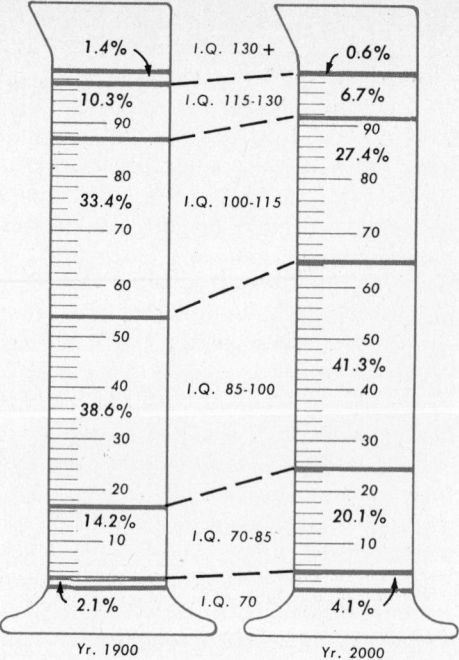

I.Q. 130+ 1.4% → 0.6%

I.Q. 115-130 10.3% 6.7%

I.Q. 100-115 33.4% 27.4%

I.Q. 85-100 38.6% 41.3%

I.Q. 70-85 14.2% 20.1%

I.Q. 70 2.1% 4.1%

Yr. 1900   Yr. 2000

Fig. 24–37. There is substantial evidence that our intelligence, as measured by I.Q. tests, is steadily dropping with each generation. When these figures are projected to the year 2000, the shift in intelligence may reach the levels indicated in the above figures.

our status now and to predict what is likely to happen to us genetically in the future. It is relatively easy to collect vast amounts of data regarding family histories, and certain trends seem to be indicated, some of which are disconcerting. Some authorities estimate (Fig. 24–37) that by the year 2000 the percentage of the population with an I.Q. of 115 or better will have been reduced from the current 12 percent to 7 percent, and the present 2 percent below 70 (moron, etc. group) will have doubled. (One, of course, must consider the relative merits of I.Q. scores as an indication of success or failure in our socio-economic system.) Even though there might be only a small loss in our intellectual heritage, it would seem that every effort should be made to combat this downward trend. What, if anything, can be done about it?

Suppose we could breed the ideal person, what would we breed toward? What kind do we want? What are the desirable traits and who is to be the judge? These are difficult questions for which no scientist and only a highly conceited layman would have an immediate answer. Probably most people, however, would not favor the continual production of defectives such as microcephalic idiots. Fortunately, very few of the hopelessly defective individuals reproduce.

Present Dysgenic Practices

Instead of improving our people genetically it would appear that the operating forces in our modern civilization are doing just the opposite. Our way of life has vastly improved in the past two hundred years, but paralleled with it may be a decline in our genetic heritage. The features that have made our lives more safe and pleasant and have been responsible for our tremendous increase in numbers have simultaneously introduced factors that interfere with the agencies of natural selection, which through the ages have tended toward the building of a sturdy body and mind in a very demanding environment. When that point was reached, it seems that decline has followed. Modern man has interfered with natural selection in several important ways which are only now showing their effects.

Birth rates compared to death rates readily reveal whether a population is increasing or decreasing. Whether the number of people in a population such as we have in America is increasing or decreasing is not as important as the more serious problem of what groups are reproducing the population. Are the qualities that we agree are desirable being perpetuated or are they gradually being sacrificed for less desirable qualities?

One of the potent forces bringing a shift in the quality of people is the difference in the rates at which children are born to groups of various abilities. If different groups reproduce at different rates, as we know they do, the overall reproductive rate of a country is a composite of all of these groups. For example, if one group reproduces at a rate 20 percent higher than another group, the latter will gradually be replaced by the faster multiplying group. This is important only if there is a concomitant dropping in the quality of the people as a whole.

During colonial days in America, the families of the more prosperous citizens were larger than those of the less prosperous. Statistics taken from an early study of New England showed that the upper quarter of the more successful families had on the average one more child than the lower three-quarters (that is, 7.2 to 6.2). The years since this period have seen a gradual decline in the size of the more prosperous group. Thus today persons belonging to the professions and business class are not replacing themselves by 15 to 25 percent. The largest families occur in people in the lowest occupational levels. In other words, there is an inverse relationship between occupational level and size of families. The birth rate among unskilled laborers is almost double that of people in business or professions.

Modern medicine, while certainly alleviating the miseries of mankind during the past two hundred years, has also prolonged the lives of many and saved the lives of a multitude of others who, in the normal course of events, would never have survived. The genes responsible for these weaknesses would have been kept to a low frequency under natural selection, but with the intervention of modern medicine they have been saved to be perpetuated and thus increased in subsequent generations. Let us hasten to point out that saving these lives from a humanitarian point of view can only be praised, but in the eyes of the geneticist it simply means that one of the most potent factors in building a strong race is prevented from functioning. The results could lead to inferior stock.

Another dysgenic factor in our present civilization is modern warfare. In centuries past the strongest, cleverest, and most intelligent men went into battle and the best of these survived to come home and become the fathers of the next generation. This was natural selection at work. Since the advent of gunpowder and subsequent deadly weapons, the strong and able are cut down equally with the less well endowed. Furthermore, in all wars the best equipped both physically and mentally are selected for duty first, thus eliminating many of them from the opportunities of becoming parents. The military can hardly be blamed for wanting the best men they can get to man the complicated instruments of modern warfare, but these same men are also the best stock we have and their chances of becoming parents of the next generation are much reduced when in service as compared to life at home. For the past several hundred years, then, we have been following a policy that is definitely contrary to natural selection and may have had some influence in reducing the quality, not the quantity, of our stock.

The human population, like any other Mendelian population, has various undesirable phenotypes resulting from genetic variability. These are maintained by various selective factors in the environment and in the past have been held at a constant level. Deleterious mutants appear in the population but they are eliminated at about the same rate as they occur. However, if the selective factors in the environment are altered so that defective genes survive and are passed on to succeeding generations, they tend to increase in the population. A number of such factors have recently been introduced into human populations, particularly in the more advanced societies such as ours. Let us examine some of these.

The level of radiations to which humans, as well as all other living things, are being exposed has increased rather dramatically within the past three decades. To be sure life evolved in an environment with a low level of cosmic radiation, but with the advent of atomic and hydrogen bomb explosions, other types of radiations have increased on a worldwide scale. The fallout from these blasts is being carefully monitored and its accumulation in foods, domestic animals, and even human beings is being measured. Studies with experimental animals show that radiations even at low levels over long periods of time are deleterious. They cause an increase in the mutation rate and such mutations are almost always less desirable than the normal. We do not know the exact level that induces mutations in man, but it would seem to be a wise course to maintain the level of radiations as low as possible. It is true that mutations occur spontaneously and probably always have, but when this number is stepped up and these are preserved in the population through the remarkable success of modern medicine, they tend to increase in the population as a whole. This possibility has created concern among such geneticists as the late Professor H. J. Muller who believed that some steps should be taken to preserve our genetic heritage. He suggested that sperm from healthy males should be frozen and stored for use at some future time to artificially inseminate women, thus increasing the possibility of healthy offspring in the generations to come.

It seems that while our present civilization has brought about a higher standard of living, at least in some parts of the world, it has likewise introduced several environmental and other factors that may cause deterioration of our genetic heritage. This could have far reaching effects on the future of man. Is there no way out of this dilemma? It would seem that the application of what we already know about human genetics would tend to curb the apparent decline that seems to be upon us.

Possible Solutions

Many thinking people have faced this problem and there is some concerted effort, feeble though it is at present, to halt this tide of ever-increasing defectives and the dilution of our precious germ plasma that has taken millions of years to produce. The problem is extremely difficult because of our incomplete knowledge of human inheritance. Even more difficult is the problem of convincing people that something should be done about it.

To stem the tide of defectives, sterilization is being practiced in some states in the United States and many countries in Europe. Just what effect does sterilization have in reducing the number of defectives in future generations? It has very little, unfortunately. It is estimated that most defects which are carried as recessives would probably require 2,000 years to reduce their numbers by 50 percent, employing the most rigid sterilization laws. (See Hardy–Weinberg Law, p. 573.)

Isolation and sterilization will have little influence on the ominous drop in general intelligence brought on by differential birth rates. Birth-control measures have been gaining a footing in recent years and seem to be providing a partial answer to the problem. Workers in birth-control clinics are well aware that people in all walks of life are very much interested in spacing their children so that they bring into the world only the number they can adequately care for.

Is there any way, then, to reverse the present dysgenic trends that are threatening many nations of the world? The most comprehensive studies of this problem have been made by the Scandinavian countries where they have settled on a solution which seems to have more merit than any others. They argue that since the nation's people is its prime asset, the burden of providing a continuing stream of high-grade protoplasm is the responsibility of everyone. Therefore, all should share in the care and education of all children. Consequently, the burden of educating and feeding during the maturing years falls upon the state. The present movement in the United States toward federally supported scholarships to young people who can profit by a higher education should also have gratifying results in the years to come.

Books

*BARRY, J. M., *Molecular Biology: Genes and the Chemical Control of Living Cells*. Englewood Cliffs, N.J.: Prentice-Hall, 1964.

BEADLE, G. W., *Genetics and Modern Biology*. Philadelphia: American Philosophical Society, 1963.

BONNER, D. M., and MILLS, S. E., *Heredity*, 2d ed. Englewood Cliffs, N.J.: Prentice-Hall, 1964.

EHRLICH, P., *The Population Bomb*. New York: Ballantine Books, Inc., 1968.

EHRLICH, P., and EHRLICH, A. H., *Population, Resources, Environment, Issues in Human Ecology*, 2d ed. San Francisco: Freeman, 1972.

EHRLICH, P., EHRLICH, A. H., and HOLDREN, J., *Human Ecology, Problems and Solutions*. San Francisco: Freeman, 1972.

HARDIN, C. J., ed., *Population, Evolution and Birth Control: A Collage of Controversial Reading*, 2d ed. San Francisco: Freeman, 1969.

*LEVINE, R. P., *Genetics*. New York: Holt, Rinehart & Winston, 1962.

NEWMAN, H. H., *Multiple Human Births*. New York: Doubleday, Doran and Co., 1940.

*PETERS, JAMES A., *Classic Papers in Genetics*. Englewood Cliffs, N.J.: Prentice-Hall, 1959.

SRB, A. M., OWEN, R. D., and EDGAR, R. S., *General Genetics*. San Francisco: Freeman, 1965.

STERN, V., *Principles of Human Genetics*, 3rd ed. San Francisco: Freeman, 1973.

The Biochemistry of Human Genetics, G. E. W. Wolstenholme and C. M. O'Conner, eds. Ciba Foundation Symposium. Boston: Little, Brown, 1959.

The Metabolic Basis of Inherited Disease, J. B. Stanbury, J. B. Wyngarden, and D. S. Frederickson, eds. New York: Blakiston, 1960.

VOELLER, B. R., *The Chromosome Theory of Inheritance*. New York: Appleton-Century-Crofts, 1968.

*WATSON, J. D., *Molecular Biology of the Gene*, 3rd ed. New York: W. A. Benjamin, Inc., 1975.

* Available in paperback.

Articles

BENZER, S., "The Fine Structure of the Gene." *Scientific American*, January, 1962.

MULLER, H. J., "Radiation and Human Mutation." *Scientific American*, November, 1955.

VII

ORGANIC
EVOLUTION

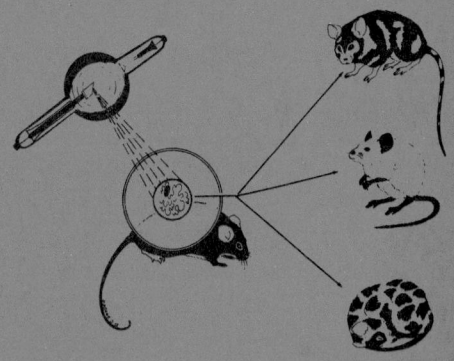

25

EVOLUTION-PAST AND PRESENT

Throughout this book constant reference has been made to the gradually increasing complexity of life from protozoa to mammals; in fact, the underlying theme is this basic concept of change from simple to complex. The word **evolution,** in its broad sense, means unrolling or unfolding and, used in terms of the living world, simply implies that all plants and animals alive today have descended by slight modifications from simpler preexisting forms. Our discussion of the various animal groups has been based on this idea and, even though we have accepted it as a basis for our thinking, let us explore the rise of animal life and examine the theory of organic evolution as a logical explanation of what has taken place.

EVIDENCE FROM ANCIENT ANIMAL LIFE

Geologic Time

In tracing life from the beginning, the course of events has afforded us a convenient way of reckoning time. The earth's thin crust has provided "clocks" that tell time in millions of years. Geologists are becoming more and more adept at reading these clocks, although a reading with an error of a few millions of years one way or the other is still considered satisfactory. Even though the exact times may not be accurate, the sequence of events and the orders of magnitude of each event are correct. By reading these geologic time clocks it is possible to approximate dates of the origin of various forms or animal and plant life.

One of the most accurate methods of reading time which has recently come into prominence employs radioactivity. Radioactive elements disintegrate at a remarkably uniform rate, extending over millions of years in some elements. Uranium, for example, decays or disintegrates at an extremely slow rate into a special kind of lead (atomic weight 206 as compared to 207.2, the average atomic weight of ordinary lead) and helium gas. It requires 2 billion years for one-quarter of a sample of uranium to decompose into these constituents. In determining time by this method, a rock containing a mineral impregnated with uranium is analyzed for its lead content. Since the mineral was incorporated into the rock when it was formed, the relative proportions of lead and uranium would be an indication of the age of the rock. If one-fourth of the uranium had been converted to lead, the rock would be 2 billion years old. It is from such data as these and others that the earth is thought to be at least 4 billion years old.

In the study of animals that have lived in past ages it is more convenient to use terms which denote sequence of animal life rather than numbers of years. The entire history of the earth's crust is divided into **eras, periods,** and **epochs,** each succeeding one being a subdivision of a previous one, as shown in Figure 25–1. Although the dates are only approximate and may vary one to several million years either way, the sequence of events is rather well established.

How Fossils Are Formed
and Preserved

Under very special conditions parts or even whole animals have been preserved in many different ways. From these remains we can learn not only something about their anatomical features but also how they lived. It must be remembered that fossilization occurs only under ideal conditions and that only a very small fraction of the animals existing at any one time died under these conditions. Most of them, of course, disintegrated completely, just as they do today, leaving no clue to anyone in the future that they had ever lived. There are many ways in which fossils are formed, some of which we shall consider.

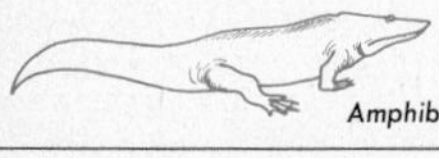

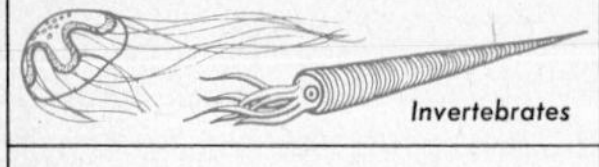

The ideal fossil is the whole animal preserved intact so that its entire anatomy can be studied in detail. This has occurred in the case of insects of the Oligocene, when they became embedded in the sticky pitch of the coniferous trees of that period. These animals show bodily structures in the finest detail. Even scales on the wings of butterflies are as perfectly preserved as if they had lived only yesterday. In certain cold regions such as Siberia the wooly mammoths often fell into crevasses in the ice where they were quickly frozen and thus preserved—flesh, skin, and all—for at least 20,000 years.

Animals that have fallen into petroleum surface deposits have been preserved in their entirety, although more frequently the flesh disintegrates, leaving only the bones in perfect condition. The most famous case is that

Fig. 25–1. Geologic time is represented here both in approximate numbers of years and in terms that denote sequence of animal life. Representative animal types are shown in the approximate periods they lived.

Dominant animal life	Millions of years	Epochs	Periods	Eras
	025	Recent	Quaternary	Cenozoic
	1	Pleistocene	Quaternary	Cenozoic
	11	Pliocene	Tertiary	Cenozoic
	16	Miocene	Tertiary	Cenozoic
	11	Oligocene	Tertiary	Cenozoic
	19	Eocene	Tertiary	Cenozoic
Mammals	17	Paleocene	Tertiary	Cenozoic
			Cretaceous	Mesozoic
	130		Jurassic	Mesozoic
Reptiles			Triassic	Mesozoic
			Permian	Paleozoic
	75		Pennsylvanian	Paleozoic
Amphibians			Mississippian	Paleozoic
	80		Devonian	Paleozoic
Fishes			Silurian	Paleozoic
	145		Ordovician	Paleozoic
Invertebrates			Cambrian	Paleozoic
First life	1500		Precambrian	Proterozoic
Formation of earth's crust	?			Archeozoic

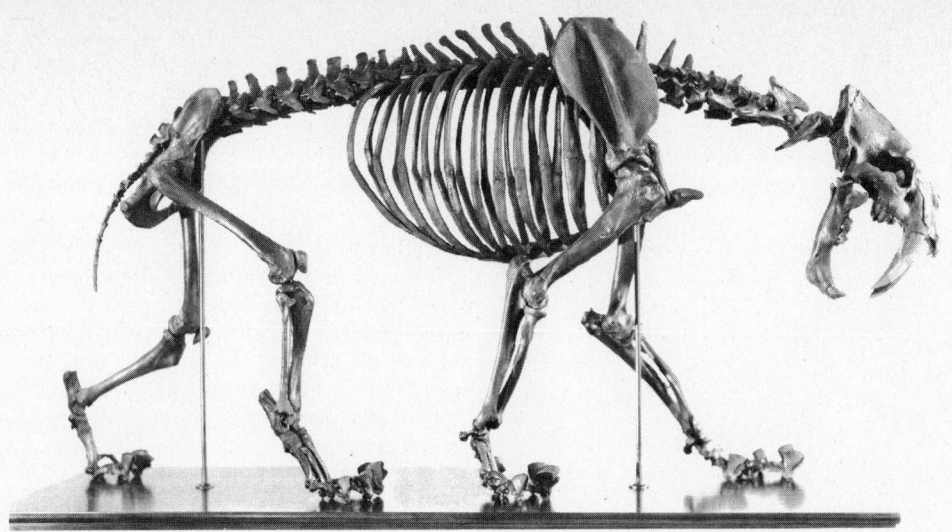

Fig. 25–2. A reconstructed skeleton of a saber-tooth tiger taken from the La Brea tar pits in Los Angeles. These are actual bones which have been preserved by the action of tar. Note the large canine teeth which undoubtedly made this big cat one of the more formidable animals of its day.

of the La Brea tar pits in Los Angeles where elephants, antelopes, bears, lions, horses, and the famous sabertooth tigers (Fig. 25–2) have been found in great abundance.

The hard parts of animals become fossilized very readily, as attested by the large numbers found in various parts of the world. Such parts as shells and other exoskeletons of invertebrates (Fig. 25–3) and the endoskeletons of vertebrates are most commonly preserved. The animal must be buried shortly after death, usually by the sinking of its body into the soft mud bottom of a stream or other body of water, and then be quickly covered by silt which subsequently becomes rock by the cementing action of minerals in the water (Fig. 25–4). Thus the original shape of the animal is maintained even though the organic parts completely disappear at some later time. Ground water containing carbonic acid dissolves away the shell, leaving **a mold** which is later filled with minerals that precipitate out as the ground water seeps through. The **cast** that is formed is an almost perfect replica of the original shell or other hard part. More commonly the replacement is accomplished a little at a time: the most soluble parts are

filled in first, the least soluble portions last. This type of replacement often reveals the minute detailed internal anatomy and is referred to as **petrifaction**.

Animal products such as eggs and excrement have been fossilized, and even tracks have been preserved. Tracks reveal whether the animal walked on all fours or hopped on its two hind legs. This information, together with the fossil remains of the animal itself, gives a rather full picture of animal life in the past.

You may wonder how it is possible to know the age of a fossil animal, what plants and other animals were associated with it, and even what the climate was like at this early time. It is a result of over a century of very careful observations and recently more accurate inferences can be made from new sources of information. One of these methods is called **carbon dating**.

As the earth's atmosphere is bombarded with solar radiations, some of the nitrogen is converted into radioactive carbon (isotope) known as carbon 14 (C^{14}) and becomes oxidized to CO_2. Since this is chemically the same as ordinary CO_2 with carbon 12 (C^{12}), it is taken up by plants at a

Fig. 25-3. Invertebrates with rigid outer coverings were fossilized in great numbers. Here are two examples. The ancient spider (left) was found in coal from southern France. It lived 200 million years ago. The insect (right) is embedded in its entirety in amber. Note how closely it resembles modern hymenopterans.

rate which is proportional to its relative abundance in the atmosphere. All radioactive isotopes disintegrate (decay) into

nonradioactive elements with the emission of detectable particles which can be measured. Radioactive carbon is produced and disintegrates at a constant rate so that there is an equilibrium built up in the atmosphere. Photosynthesizing plants also maintain the equilibrium concentration. We know that this equilibrium has not changed from ancient times because it can be checked. For example, the known age of some Egyptian artifacts is about 5,000 years and this agrees with present calculations which means that ancient trees incorporated about the same proportion of radioactive carbon as do trees of today. By measuring the amount of C^{14} in a piece of charcoal or wood its age can be estimated rather accurately. The half-life (time required for one half of the C^{14} to disintegrate) of radioactive carbon is about 6,500 years. Therefore, if one finds one half the C^{14} in an old piece of wood compared to a modern tree, he could conclude that the ancient tree was about 5,600 years old. Bits of wood have been dated back as far as 50,000 years using this method.

Ancient Animal Life

We have concrete evidence that life took hold in earnest only during the last billion years of earth's history. As was pointed out at the beginning of this book, millions of years was required to produce the first living thing, but once it was formed evolution went forward at a rapid pace. It probably took longer to produce the first simple cell than it has taken to evolve man from that first cell. Evidence of living things appears for the first time in the Cambrian rocks and, strange at it may seem, occurs here in great abundance. This must mean that once evolution started it went forth with a burst of speed from the very beginning, because fossils are found in all periods following the Cambrian. There is very little proof that animals lived in pre-Cambrian times, although had they possessed soft body parts which do not fossilize they could have lived and died without leaving recognizable imprints in the rocks.

Most, if not all, of the major phyla have left fossil remains in the Cambrian, which lasted 60 million to 90 million years, certainly a sufficiently long period of time to allow for evolution of such a large variety of forms. Each of the principal phyla represents a distinct level of anatomical organization which had its beginnings then. Apparently by happy coincidence, specific anatomical and physiological characteristics appeared that possessed great evolutionary potentialities. These became evident in the diversification of species within each phylum that followed. The course that each has taken from the Cambrian up to the present is portrayed in Figure 25–5 in graphic form. Note that each of the principal phyla today started in the Cambrian and through the subsequent periods has had its "ups and downs." Moreover, in the past geologic periods the relative proportions of the various groups differ from those of today (Fig. 25–6). The most striking fact that emerges from this observation is that all of the original phyla have living representatives today and that all of them are more numerous than in Cambrian times. Not only are there more individuals, but there is also a much greater variety of species.

The Story of Vertebrates

The Chordates undoubtedly had their beginnings in the late Cambrian because they are well established as vertebrates in the Ordovician (Fig. 25–7). There are eight classes represented, of which four are swimmers and four possess legs adapted to movement over solid surfaces. Of the classes of swimmers, all but the placoderms have living representatives today. The jawless fishes (cyclostomes) had rather modest beginnings and all but died out in the late Devonian. In recent times they have shown signs of increasing in numbers and varieties, as indicated by considerable numbers in the oceans. Some have also invaded fresh water. The placoderms had the first true jaws which were hinged to the skull, a feature that was retained in all subsequent forms. They also possessed heavy bony skeletons.

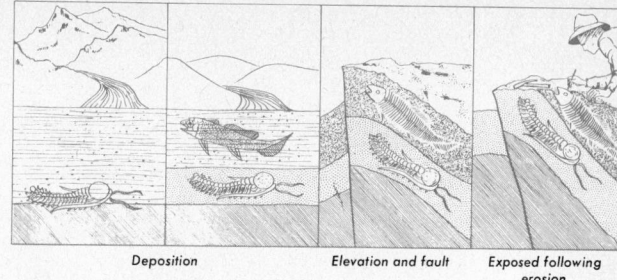

Fig. 25–4. The hard parts of ancient animals become fossils as indicated.

Deposition   Elevation and fault   Exposed following erosion

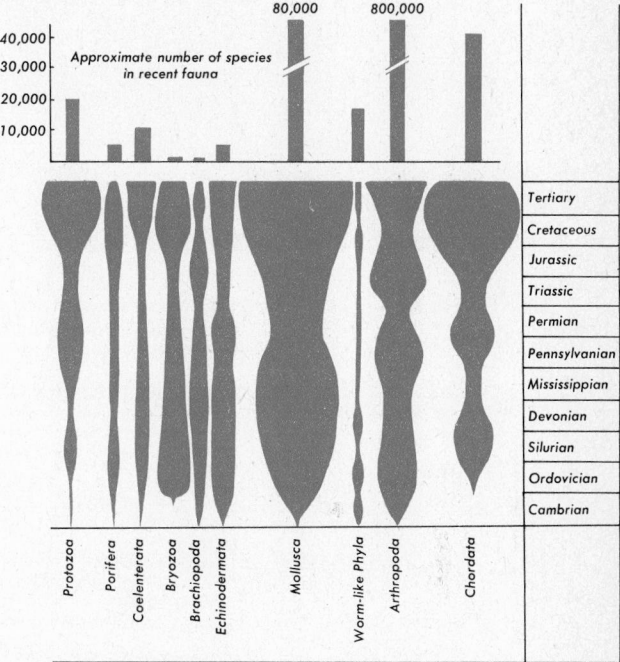

Fig. 25–5. The historical record of life is here shown in graphic form. The various phyla extend from the Cambrian or Ordovician up to the present, and the width of the black pathway indicates the relative numbers of genera in each phylum over this period of time. The upper figure represents the variety of species in each phylum in fauna of the present time.

Both the cartilaginous and bony fishes were offshoots from the placoderms, the former sacrificing bone for cartilage and the latter retaining the bone. Today we have both represented: the cartilaginous fishes in the sharks and rays, and the more successful bony fishes which abound in all waters of

Fig. 25–6. This is a reconstruction of the ocean floor during the Cretaceous Period in the region of Tennessee (U.S.), showing the abundance of invertebrate life. Although none of these species is alive today, it is not difficult to recognize them as mollusks. The two squidlike forms in the upper right are very similar to our present-day forms. The large coiled ammonite in the lower right is very much like the chambered nautilus of today. The numerous cephalopods with shells resembling cones are all extinct in any form. Our present-day clams and snails seem to be well represented on the floor of this ancient ocean.

Fig. 25–7. The historical record of the vertebrates; the width of the pathways indicates the relative variety of known forms in the various vertebrate classes.

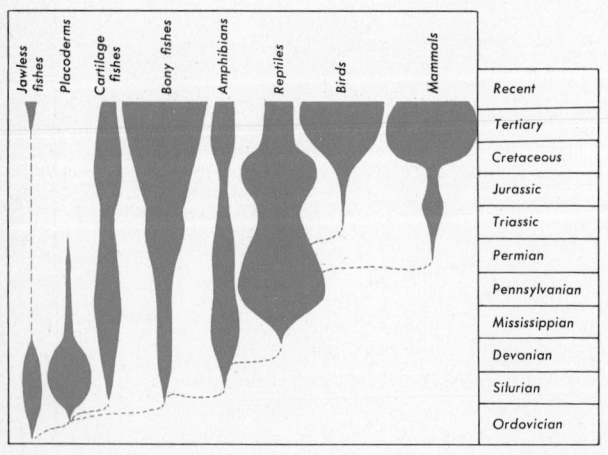

the globe. The cartilaginous fishes have held their own rather well through this long geologic time while the bony fishes have increased at a tremendous rate, particularly since the Permian.

The most primitive land-dwellers, the amphibians, arose from the bony fishes some time during the Devonian, reaching a peak which they maintained through the Pennsylvanian and Permian, but then suddenly lost ground probably because their chief competitors, the great reptiles, became the dominant land life. With the passing of the great dinosaurs, the amphibians made a slight comeback which they have maintained up to the present. The reptiles came from the amphibians in the Pennsylvanian and very soon reached a peak which they continued to hold, with a setback during the Triassic, until the end of the Mesozoic (beginning of the Tertiary), when all of the large dinosaurs suddenly disappeared leaving only remnants which have continued on to the present. The reptiles have been the most successful of all animals living in the past and it is problematical whether any group in the future will reach such peaks as they did.

Both the mammals and birds were derived from the reptiles some time during the Triassic but good records are available only of the early Tertiary. Both started out very slowly, while the great reptiles ruled the earth. With the decline of these beasts, both the birds and mammals came into their own, increasing in numbers and variety at a tremendous rate. One can imagine that through the Triassic and Jurassic small birds and mammals occupied the secluded niches in the environment, keeping well out of the way of the numerous reptiles. All the while their bodies were evolving to a stage where, when the opportunity for survival improved, they were ready to set forth on the road that led them to the dominant position among present-day animals. Certainly the mammals have become the most diversified of all animals, whereas the birds have clung pretty much to a common pattern that seems to serve them well for life involving flight. Mammals reached their peaks in the Tertiary and have since shown a steady decline in numbers of species even though man, a member of this group, is considered to be the dominant form of life on earth today.

All of the evidence from studies of ancient life confirms and elaborates the theory that living things today descended from similar but different forms living in past ages; in other words, they evolved.

EVIDENCE FROM RECORDS OF LIVING ANIMALS

From Embryology

As early as 1821 Meckel had noted that embryos of higher animals passed through successive stages which resembled more or less completely the lower animals. Haeckel

in 1866 summed up the evidence that had accumulated at this time in his statement, "Ontogeny (the development of an individual) is a brief and rapid recapitulation of the phylogeny," and assigned to this formula the imposing name of the Biogenetic Law. He claimed that animals recapitulate in their early embryological stages the phylogenetic history of the race, which we have already seen in various animals. Haeckel's all-inclusive statement has since been attacked from many sources and today stands only in its skeleton form. Although there is definite resemblance between the embryos of related species (Fig. 25–8), they are not clear-cut and absolute as was once thought, and this is as one might expect. Since evolution has come about by a series of small changes (micromutations) in the genic constitution of a species, it might be expected that gene changes would occur affecting all stages of the individual, that is, gene changes might affect the embryo

Fig. 25–8. The early embryos of all vertebrates are very similar during some stages of their development.

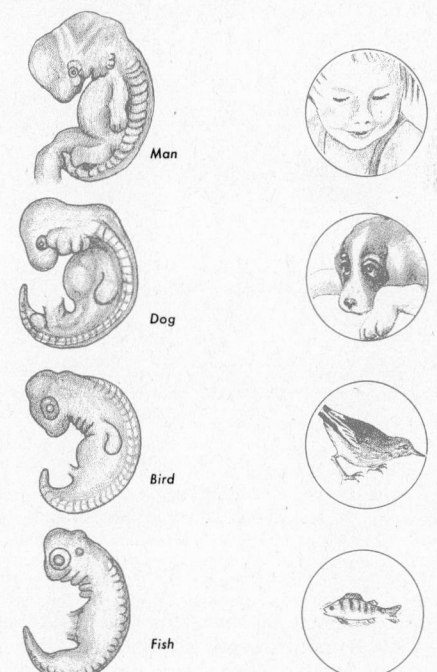

Man

Dog

Bird

Fish

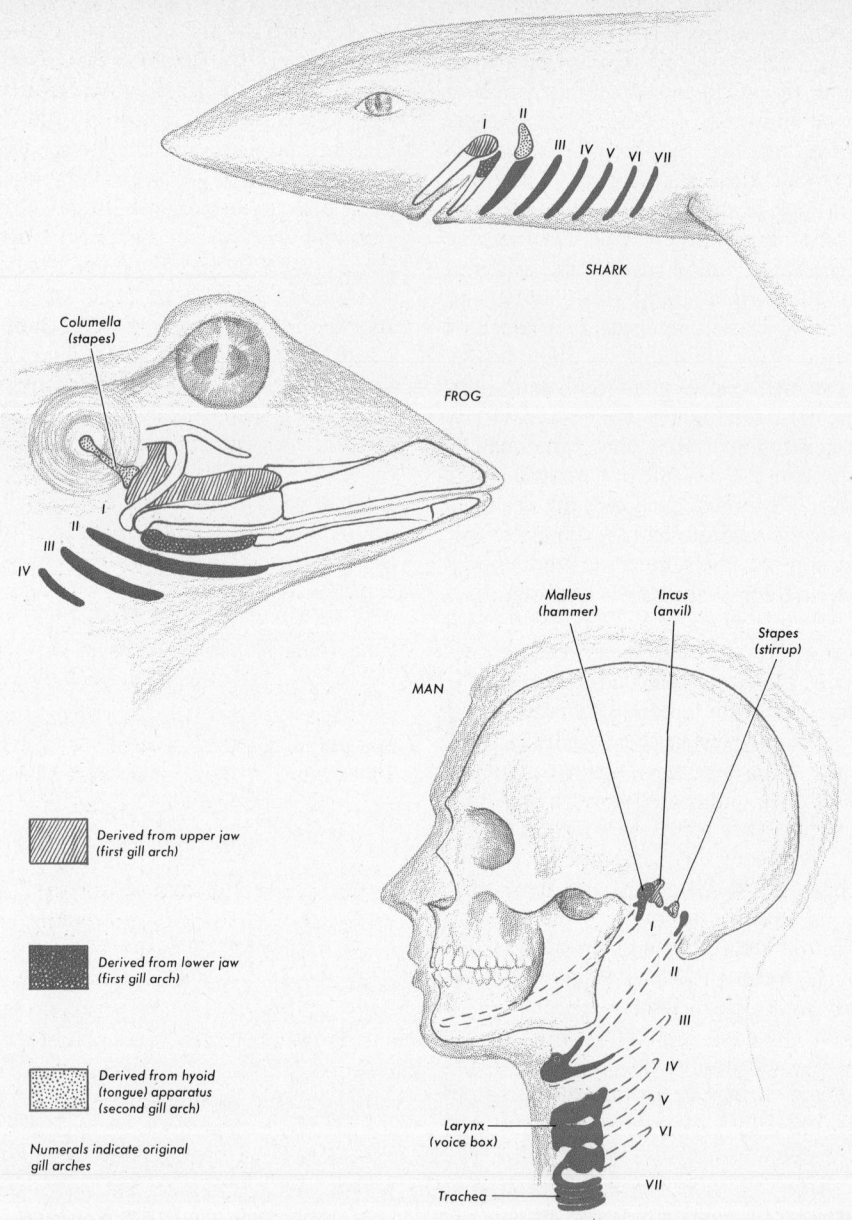

Fig. 25–9. The history of the visceral arches indicates that many changes took place between the shark and man, yet evidence of the transition are clearly present. In the shark the arches gave support to the gills used in breathing water. In the frog the beginnings of the sound-making apparatus have developed and the sound-receiving apparatus has been further perfected. A portion of the hyoid (II) has become the columella, a forerunner of the stapes in the mammalian ear. The frog has also made use of the remaining gill arches in the formation of the laryngeal apparatus for the purpose of making sound. In the mammal the sound-receiving and sound-making apparatuses have developed to a high degree of perfection, but the origins of the ear bones and the larynx are still evident.

itself so that "short cuts" could be taken in producing a certain organ system, for example. Similarly, structures which no longer serve a purpose in the adult animal might be retained simply because they were not "in the way" during the course of development and because no specific gene change occurred that ruled them out. Therefore, one might expect numerous remnants lingering on in the bodies of some animals but not in others, even closely related species. This is exactly what we observe.

Obviously, the higher embryos possess features that are functional in the adults of the lower forms. For example, the human embryo has a set of folds in the neck region in the fourth and fifth week of life. The fish embryo in a comparable stage of development possesses a similar set of folds; and, in fact, the embryos resemble one another so closely at this early stage that it is difficult to distinguish one from the other (Fig. 25–8). In the fish, the folds give support to functional gills in the adult, whereas in the human they undergo a whole series of modifications, ultimately becoming a part of the hearing and soundmaking apparatus as well as a part of the upper breathing tract (Fig. 25–9). We are reminded once again of the fact that, in the long evolutionary history of animals, structures which no longer find a use in the body often take over new tasks and by slight or great modifications become adapted to the new function.

There are many other features that appear in the developing human embryo that are tell-tale evidence of our lowly origin. For instance a well-developed tail (Fig. 25–12) makes its appearance in the first few weeks of life but soon vanishes without performing any function other than to indicate to us that man came from stock which once possessed a tail. We have already noted the presence of a yolk sac which could hardly have a function since it is devoid of yolk (food) and is not needed because nourishment is provided by the placenta. Here again it tells us that some time far back in

evolutionary history our ancestors were egg-laying animals in which the embryos did receive their nourishment from a yolk sac. The human embryo, in the fifth month, develops a coat of hair called the **lanugo,** which is shed before birth, usually in the eighth month of gestation. Such hair could not possibly perform any conceivable function but does lead us to believe that at one point in our evolution we possessed a coat of hair as adults, extending back, probably, to some common ancestor of both man and apes. The entire study of both invertebrate and vertbrate embryos supports the theory of evolution; no other explanation serves so well.

From Comparative Anatomy

One of the most striking observations that impresses anyone who studies a large variety of vertebrates is the fundamental likeness of the body architecture. The similarities are evident no matter what system is examined—appendages, muscles, digestive systems, circulatory systems, or any other. By comparing the appendicular skeletons of the horse, man, and the frog (Fig. 25–10) it is obvious that the bones are similar, varying only in emphasis of specific parts of certain bones. In the horse, the forelegs are extended by elongating the equivalents of the wrist and hand bones of the frog and man. This provides the horse with an appendage well adapted for traveling at high speeds over soft turf. The same bones modified in different ways have given man a maneuverable appendage which is handsomely adapted to the kind of life he leads. An appendage not greatly different from that of man serves the frog satisfactorily in its way of life. Similarly, the hind appendages have been modified to perform specific jobs. The important point to note here is that the fundamental plan is the same; that is, the appendages are **homologous,** which means that they must have had a common origin.

A study of the muscles in any two vertebrates, reveal the same sort of homologies noted in the skeletons. The large, major muscles in both animals are similar, most of them performing the same function in both. They vary only in size due to use, which is reflected in the activity of the animal. Facial muscles of both man and monkey are likewise similar, although we have more of them and consequently can register more emotions in our faces (Fig. 25–11). The digital tip of a bird, a man, and a horse, referred to earlier (Fig. 15–2), can be compared in the same way. Other examples were mentioned in the discussion of the origin of scales, teeth, and feathers (p. 301). All of these comparable structures, with modifications in various species in response to function and many more, support the thesis of common origin. Common origin implies evolution.

From Vestigial Structures

Many animals possess structures which are vestiges of some past functional organ. The appendix in man is such a structure. It

Fig. 25–10. The appendages of the vertebrates are homologous, as shown by comparing the skeletons of these three. It would be difficult to explain these similarities on any other basis than organic evolution.

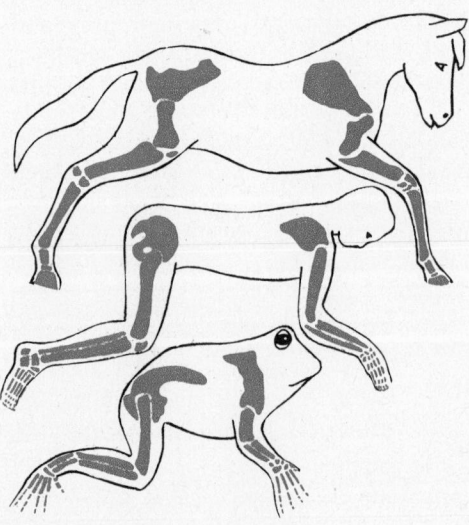

apparently performed some function when man's diet was not quite what it is today. Perhaps at one time he was more of an herbivore.

Another human vestigial structure is the ear muscle. Man possesses a complete set of muscles, similar to those in other primates, for moving the ears in all directions (Fig. 25–11). However, it is only with extreme difficulty that he can move his auditory appendages and even then the movement is ineffective in improving his hearing. Man also possesses a rudimentary tail together with a complete set of muscles for wagging it in all directions. These are even more useless than the ear muscles; yet they have persisted for thousands of generations. Since there is wide variation in the number of segments in the tail (coccyx) one might believe that it is disappearing, and this is probably true. A similar situation exists with the last set of molars (wisdom teeth). Perhaps a few thousand generations hence people may be born without wisdom teeth, an appendix, or a tail.

Vestigial structures in other animals often tell us something about their life history. The porpoise, for example, has a tiny pelvic girdle buried deep in its hip region where it could not possibly perform any function whatever, yet it is faithfully produced in the body of every porpoise. This vestige supports other evidence that this animal was a four-footed land animal at one time in the far distant past.

Anomalies in development occasionally occur which give some hint as to the history of certain structures. They are caused by the retention of embryonic structures in the adult and are called **atavisms** or **reversions**. For example, it is not uncommon for a child to be born with a well developed external tail, reminiscent of a similar structure in the embryo (Fig. 25–12). On very rare occasions rudimentary gill slits occur in the neck region. Sometimes, additional nipples occur along lateral lines that coincide with the milk ridge in the embryo. In animals with multiple births this arrange-

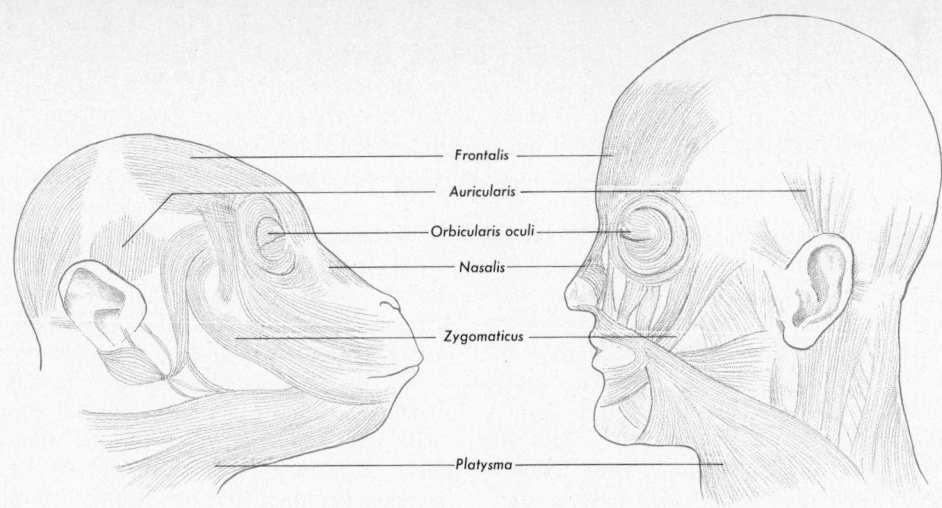

Fig. 25–11. Vestigial structures such as ear muscles in man (auricularis) are similar to those in the monkey. In the latter they have some function in moving the ears which perhaps aids in directing the sound into the ear; in man they are vestigial and serve no purpose, yet they persist.

ment of the mammary glands is customary. Such "mistakes" in development throw light on the history of the body.

From Comparative Physiology and Biochemistry

The very fact that all animals possess essentially the same elements and chemical compounds would indicate a common origin. Even protoplasm is very similar among all animals. The various enzymes that are found in the digestive tract of man can be identified in other animals, even down to the protozoa. Other intracellular and extracellular enzymes can likewise be identified in all animals which have so far been studied. It appears reasonable to assume, then, that the chemical constitution of two closely related species is more nearly alike than that of two distantly related species. Blood is most commonly studied because it is most easily obtained and most easily handled in the laboratory. By utilizing the antigen-antibody technique described earlier (p. 421), it has been possible to verify the relationship between many animals. It is gratifying to note that the

Fig. 25–12. Occasionally an embryonic structure persists in the adult, such as a well-developed tail or numerous nipples in humans. These structures are found in relatives of man and may have existed in his ancestors. The only feasible explanation is that of common origins.

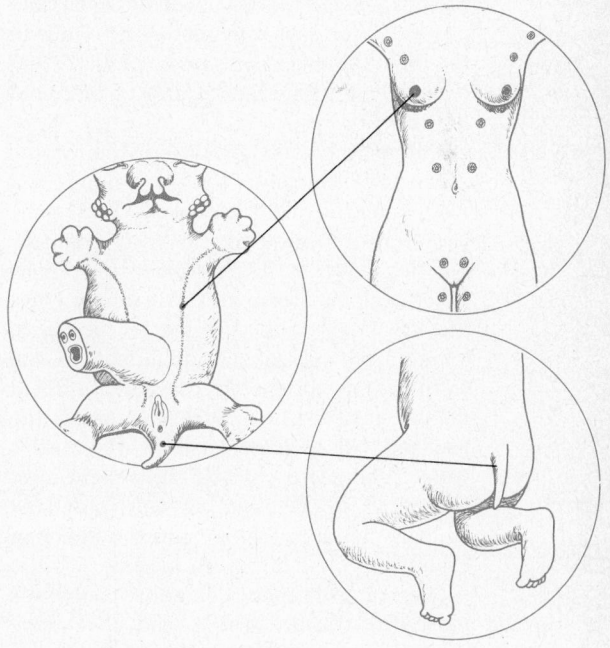

relationships established by these blood tests corroborate those established by our system of classification based on homologous structures. For example, it is possible to show that among vertebrates, seals and sea lions are more closely related to one another than to other mammals. Man's nearest kin among the primates has been shown to be the great apes and his most distant the lemur. All physiological and biochemical data that have been carefully examined supplement those gathered from other disciplines supporting the theory of evolution. Many other substances present in the bodies of higher animals can be traced to their origins in lower forms, which certainly lends support to the thesis that they would have a common origin. It would be difficult to explain them any other way.

From Animal Distribution

There are many places on the earth's surface that are idyllic spots for a large variety of animals, yet none occupies these regions. If animals were created simultaneously, why are not all of the favorable places in the world occupied by those that are fitted for such environments? Only in the light of evolution, both physical and organic, can a satisfactory answer be found for this situation.

Interplay of two evolutions, one of the earth itself, the other of the animals and plants upon it, has taken place. The young earth changed violently at first, gradually quieting down as it grew older and finally becoming inhabited with living things. Once life originated and new species began to appear, any subsequent alterations of the earth's surface or climate drastically affected the life then existing. Such changes as the rise and fall of large areas of the earth's surface and sharp climatic shifts were most important in determining not only the animals that survived but also the direction evolution took in any given locality.

There is little doubt that entire continents move (continental drift), and that there have been many "ups and downs" within the various continents themselves. The presence of fossil remains of sea animals on the top of mountains is unequivocal evidence that these regions were at the bottom of some ancient ocean. The rising and falling of these areas has occurred again and again in certain regions. Coincident with these shifts in the surface of the earth, the climate changed dramatically from one extreme to another. The great coal deposits in the northern sections of the globe are a result of luxuriant plant growth in some ancient period, which is substantial proof that such regions were tropical during these times. Areas that once received a tremendous rainfall later became dry plains. These changes have come about because of the gradual trend toward a more arid climate. The sudden appearance of a mountain range must have had a profound effect on the amount of precipitation that fell in surrounding regions, which in turn altered both plant and animal life. Undoubtedly the success of a species depends a great deal upon the coincidence of gene mutations with changes in the climate and topography of the region. The proper timing of genetic changes in the animal with climatic and environmental changes might mean the success or failure of a species. These constant shifts in climate and topography have influenced the evolution of animals to a marked degree.

Closely related species of animals are usually found in adjacent **ranges** (region occupied by a species), just as one might expect. If they had split off from one another only recently, they would not be separated by very great distances, and more than likely some barrier would exist between them such as a mountain, canyon, or body of water, which may have been responsible for the new species in the first place. The more recent the separation of the two species, the more closely related they are, and the more ancient the separation the greater the differences between them. The greater differences constitute genera, families, orders, classes and phyla. When great land masses

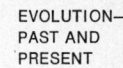

Fig. 25–13. The opossum is the oldest living mammal, having originated in the Cretaceous and maintained itself unchanged up to the present. Note the handlike front appendage in the young specimen.

are separated for millions of years one would expect the entire fauna to be quite different on each of them. Such is found to be the case when, for example, the animals in Australia and in the continent of Asia are compared. The explanation of the occurrence of primitive animals in Australia has direct bearing on the history of this large land mass.

The primitive mammals, which probably resembled our present-day monotremes and marsupials, spread over these land bridges to Africa, South America, and Australia. Later the placental mammals evolved in the northern continents and, because of their superior ability to cope with the environment, were able to drive out or destroy all of the primitive mammals in these northern areas. However, before the placental mammals made their way too far south, the sinking of the land connection between North and South America below the surface of the oceans prevented them from getting into the southern continent. Likewise, the land connections between Asia and Australia and to a lesser extent between Europe and Africa gave way, thus cutting off the possibilities of further migrations southward. The primitive mammals were thus isolated from their more aggressive relatives and were able to survive up to the present.

The fact that primitive mammals could not withstand the onslaught of their more recent placental relatives has been demonstrated within historical times. Rabbits introduced into Australia multiplied so rapidly that they soon outstripped the local marsupials at a tremendous pace. They are well under control today, however, owing to man's ingenuity. A disease, fatal to rabbits, was introduced which has kept them in reasonable numbers for the past several years. One interesting exception to this apparent superiority of northern fauna is the case of the opossum (Fig. 25–13). This marsupial made its way back north over the recent land bridge between the two Americas and it has competed rather successfully there with its aggressive descendants.

The theory of evolution is the most logical explanation for the distribution of animals over the earth.

SUGGESTED SUPPLEMENTARY READINGS

Books

BERRILL, N. J., *The Origin of Vertebrates*. London: Oxford University Press, 1955.

BLUM, H. F., *Time's Arrow and Evolution*, revised. Gloucester, Mass. Peter Smith, 1969.

*COLBERT, E. H., *Evolution of the Vertebrates*. New York: Wiley, 1955.

*DOBZHANSKY, T., *Mankind Evolving*. New Haven: Yale University Press, 1962.

DODSON, E. O., *Evolution: Process and Product*. New York: Reinhold, 1960.

* Available in paperback.

MERRILL, D. J., *Evolution and Genetics.* New York: Holt, Rinehart & Winston, 1962.

MOODY, P. A., *Introduction to Evolution.* New York: Harper & Row, 1962.

Ross, H. H., *A Synthesis of Evolutionary Theory.* Englewood Cliffs, N.J.: Prentice-Hall, 1964.

WALD. G., "Biochemical Evolution," in *Trends in Physiology and Biochemistry,* E. S. G. Barron, ed., New York: Academic Press, 1952.

Article

GLAESSNER, M. F., "Pre-Cambrian Animals." *Scientific American,* March, 1961.

26

THEORIES AND MECHANISM OF EVOLUTION

The question as to *how* the present colossal array of living things came about has puzzled man from the dawn of human intelligence. Civilizations of all kinds have pondered the problem and have come up with some sort of answer which usually has been incorporated into their philosophy and religion. The most gratifying answer was a simple one, and to the western mind special creation as portrayed in the Book of Genesis was also the most satisfactory. The idea that all living things were created for a purpose which they now fulfill left little room for argument and was satisfactory for large segments of the populations of the civilized world until about 100 years ago. The belief that the earth was much older than the allotted 6,000 years ascribed by Christians did not obtain a secure footing until geologists began to unearth very convincing evidence confirming the antiquity of the earth. Once the earth was shown to be very old,

the history of life on it became an immediate and very important problem. The constant discoveries of fossil remains of plants and animals led to the concept that life on earth was likewise very ancient, much older than anyone had imagined. With this start, the whole matter of the origin of living things, both past and present, became a problem demanding careful scientific analysis. Several theories have been advanced, the more important of which we consider after presentation of a brief historical background.

HISTORICAL BACKGROUND

Rudiments of the theory of organic evolution can be found in writings of the early Greeks, especially Anaximander (588–524 B.C.), Empedocles (495–435 B.C.), and Aristotle (384–322 B.C.). Of these, **Aristotle** overshadowed all others because of his command of the natural history of the world of his day. He, like others of his time, believed that living things progressed from simple to more complex forms, reaching the pinnacle in man. In this sense his writings had the germ of the evolution idea, but beyond that they showed little advance over primitive mysticism. Unfortunately, Aristotle's ideas were proclaimed as dogma by the church during the Middle Ages hence the possibility of scientific reasoning was impeded for many centuries. His tenet was idealistic in that he claimed each animal was specially created in its present form. Since that fitted perfectly into the doctrines of the churches there was no quarrel at this early period. Historical geology had not yet been born, so there could be no conflict from that quarter. It was only when fossils were unearthed and needed explaining that

Fig. 26–1. Charles Darwin in his later years. He was born February 12, 1809 and died in 1882.

serious clashes occurred between those adhering to special-creation dogma and those willing to accept historical facts of organic descent.

Certainly during the long period between Aristotle and the Renaissance there must have been some men who questioned the historical hypothesis.

Among these was **Lamarck** (1744–1829), a French biologist, who not only accepted the theory of evolution as an explanation of the existence of all living things but also proposed a theory which seemed to explain evolution. He believed that all living things are endowed with a vital force, distinguished from a physical force, that controls the function of all their parts and ultimately makes it possible for them to inhabit the environment where they now reside. Furthermore, he believed that any traits acquired in the lifetime of an organism were transmitted to succeeding generations. Any such acquired traits could be enhanced or depressed by "use and disuse," and the more an organ or part was used the better suited it became to do the job. Such acquired advantages would then be passed on to the offspring. This is a captivating idea and was readily acclaimed by many as the ideal answer to the question of how evolution occurred.

Lamarck's theory was the outcome of much thought about the problems that had to be solved, namely, random and oriented evolution. He knew that a general theory, if it were tenable, had to explain both. In addition, the great problem of adaptation had to occupy the center of the stage in any explanation, because it was at that time and for many years later the most difficult to interpret. His theory encompassed these features in a manner that would be completely satisfactory if his major premise could be borne out by fact. It has been thoroughly established that acquired characters cannot be transmitted because they do not influence genes which are the only means of transmission from one generation to the next. Any traits acquired in the lifetime of an individual influence only his soma cells and have no effect on the germ cells. Thus the very heart of Lamarck's theory is faulty. Since genetics has firmly established this fact, his theory has no following among modern biologists.

Although interested in natural history as a child, **Charles Darwin's** (Fig. 26–1) college training was in medicine and theology. Neither of these professions suited his taste, so when the famous naturalist, Professor Henslow, suggested that he might go on the ship *Beagle* for a five-year cruise around the world, Darwin grasped the opportunity to continue his earlier studies of geology and biology. The trip took him to the east and west coasts of South America, to Ecuador and west to the Galapagos Islands. It was while he was making a comparison of the flora, fauna and geology of these islands with those of South America that he became convinced that living things were not static nor specially created as he had thought up to this time, but rather that they were undergoing constant change and that this change was intricately linked with the movements of the earth's surface.

During the trip he gathered voluminous notes which, when assembled back in England in the course of the next twenty years, so thoroughly convinced the world of the validity of evolution that it has never been seriously questioned since. This one effort alone placed Darwin head and shoulders above his contemporaries or predecessors. Not satisfied with this accomplishment, he sought an explanation for evolution which eventually was resolved in his theory of natural selection.

Alfred Russell Wallace, a friend of Darwin's, working on the flora and fauna of the Malay Peninsula and the East Indies, hit upon an explanation of evolution which was in essence identical with Darwin's theory of natural selection. He conveyed his ideas to Darwin in the form of a letter. This occurred in the year 1858 after Darwin had completed his data and was about to consider publishing his findings. Rather than rush into print to obtain priority, Darwin presented a short draft of his work simul-

taneously with that of Wallace before the Linnean Society in London. The following year, 1859, he published his classical work, *On the Origin of Species by Means of Natural Selection*, in which he formulated his theory backed up by abundant evidence. Even though both men had arrived at similar explanations of evolution, the credit has gone to Darwin, as it justly should.

Darwin's book convinced the scientific world of the soundness of evolution, and his theory of natural selection provided food for thought for those interested in how it came about. Because the book was written in a manner comprehensible to laymen and because it was staunchly defended by such eminent biologists as T. H. Huxley, the theory of evolution came to be generally accepted by the intelligent world. Natural selection as an explanation as to how evolution came about was immediately attacked by scientists and has continued to be under fire up to the present, although today it is generally accepted, with modifications, as the most tenable of the explanations advanced so far.

THE THEORY OF NATURAL SELECTION

In essence, Darwin's theory of evolution or origin of species through natural selection involves the following steps in reasoning which are borne out by fact.

First, he considered variation an innate property of protoplasm because he encountered it in all groups of plants and animals. Any species of living thing varies widely, a fact that he could not account for but for which we have since found an explanation, to be considered presently. This variation within a species provides raw material with which evolution can proceed. Without it there could be no evolution.

Second, since each organism produces many more offspring than can possibly survive and yet populations remain fairly constant, there must be a continual "struggle for existence" within and between groups as well as with the environment; those best suited to survive under the existing conditions prevail. This could be a passive struggle, such as that engaged in by plants and animals resisting the desiccation of desert regions; or it could be extremely active, as in predation. In any case, only the most fit would survive, and for this Darwin used the phrase, "survival of the fittest," suggested to him by Herbert Spencer. Thus the struggle for existence has a selective effect in removing the unfit and preserving the fit. Moreover, since only the most fit survive, they alone can perpetuate the traits that made them best suited and hence pass them on to their offspring.

Modern Interpretation

The theory of natural selection as an explanation of evolution has stood the test of time although it must be interpreted in the light of discoveries made in the past century. Information has come from cytology, genetics, and many other sources, all of which have contributed to our understanding of how evolution takes place. The modern, so-called **Neo-Darwinian** view includes five principal factors which are responsible for evolutionary change. First, variation can now be explained by **mutation,** and mutated genes recombine during sexual reproduction; second, **natural selection** promotes favored traits and eliminates undesirable ones; third, chance, or **genetic drift,** is effective in survival; fourth, **geographical** and **genetic isolation** are responsible for the origin of new varieties; and fifth, sexually mature larvae (**neotony**) contribute significantly to evolutionary change.

Mutation. The science of genetics has made it possible to understand variation, something Darwin recognized but could not explain. Variations provide the raw material with which natural selection operates, and hence were the heart of Darwin's theory. We know now that variation results from gene or chromosomal changes collectively called **mutations.** Any variations due to environmental changes during the life of the individual, in the embryo or later, have no effect on the genes, hence are of no significance in evolution. Mutations may involve only a single gene (micromutations), which is the most usual, or they may include a group of genes (macromutations), which accounts for sudden large changes such as short legs in sheep or extra toes in cats. Most biologists believe that species have originated from the accumulation of a great many micromutations.

Mutations have been noted in a large number of plants and animals under experimental conditions and they are undoubtedly occurring continuously among organisms in their natural habitats (Fig. 26–2). Over a thousand mutations have been recorded for *Drosophila* alone. A great many gene changes probably go unnoticed which, in nature, might have considerable survival value. For example, a mutation might occur in an animal living in a region where some element in the diet was becoming deficient, which would so affect the enzyme systems of the animal that it could get along on a minimum amount of the element. Such a mutation might show no visible difference in the animal itself and yet be most important in survival. Animals with the favorable gene would be more apt to live to produce progeny than those without it; hence selection would be toward the favorable gene. While we have a great deal of information concerning anatomical mutations we know almost nothing about biochemical mutations (p. 534). Nevertheless, such physiological changes must have been and must continue to be very important in their evolution.

Mutations occur at random both in nature and as a result of mutagenic agents (chemicals or high-energy radiations). They bear no relationship to the needs of the organism and they are unpredictable, at least in our present state of knowledge. Almost 100 per cent of the mutations are harmful, a fact that is understandable since

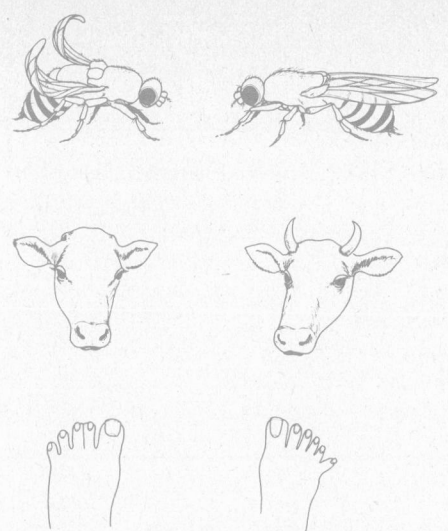

Fig. 26–2. Three mutations occur commonly in animals. Curled wings in *Drosophila* is recessive to normal wings. Hornless (polled) condition in cattle is dominant to horned. Fused or webbed toes (syndactyly) and extra toes (polydactyly) in man are dominant to normal toes.

all organisms alive today have gone through millions of years of selection. Only the best traits are preserved; hence any further change is likely to be less desirable than those already in existence.

Under different circumstances a mutation could be unfavorable and be preserved —for example, a mutation that rendered an amphibian partially or totally blind. In its normal environment death would be certain. However, if it wandered into a cave where vision was of little advantage, its chance for survival might be as good as that of a normal animal. Indeed, it may even be better since eyes are apt to be injured in such dark caverns. If such were true, blind amphibians would have a selective advantage over those with functional eyes, and would then increase in the population until the entire population would be sightless. This has actually happened in certain caves. The sickle-cell trait in man, mentioned earlier (p. 537) is also a case in which a mutation has an advantage in a particular environment.

Since mutations occur only very rarely it is difficult to study enough cases to understand what part they play in evolution. However, by studying large populations of organisms, it is possible to determine the frequency of genes in such populations. Such studies come under the heading of **population genetics** and have received more and more attention in recent years.

The question, why is the frequency of genes in population relatively constant from generation to generation, stimulated two men in the early part of this century to propose a theory which has become a law. They were G. H. Hardy, an English mathematician, and W. Weinberg, a German physician who, working independently, formulated the basic law of population genetics now known as the **Hardy–Weinberg Law.** This law states that **the proportion of genes and genotypes in any large biparental population will remain constant generation after generation providing there is random mating, equal fertility, no mutations, no selection and no environmental change.** It fails with small populations, but in large populations it works regardless of whether the genes are dominant or recessive. This, then, gives us a base from which to determine the effect of mutations, selection, or gene flow, because they are all superimposed upon this equilibrium.

At the outset one might wonder why recessive blue eyes in human populations do not give way to dominant brown eyes with the passing of time. We know they do not, and this is not because of any association of the genes within the chromosome. It is because there is normally no selection for either eye color and so long as this happens the genes for both will combine so as to give the same proportion of brown and blue eyes in each succeeding generation. Let us see why this is so.

In any population the distribution of a pair of genes, A and *a*, either of which can occupy a given locus, each individual will possess one of the following genotypes and no other: AA, A*a*, or *aa*. Let us suppose

TABLE 26–1. Hardy-Weinberg Law: Gene Frequencies Remain Constant

Mating		Frequency		Percentage Offspring		
MALE	FEMALE	SPERM	EGGS	AA	Aa	aa
AA × AA		All A	All a	100		
AA × Aa		All A	$\frac{1}{2}$A : $\frac{1}{2}$a	50	50	
AA × aa		All A	All a		100	
Aa × AA		$\frac{1}{2}$A : $\frac{1}{2}$a	All A	50	50	
Aa × Aa		$\frac{1}{2}$A : $\frac{1}{2}$a	$\frac{1}{2}$A : $\frac{1}{2}$a	25	50	25
Aa × aa		$\frac{1}{2}$A : $\frac{1}{2}$a	All a		50	50
aa × AA		All a	All A		100	
aa × Aa		All a	$\frac{1}{2}$A : $\frac{1}{2}$a		50	50
aa × aa		All a	All a			100
			Totals	225 :	450 :	225
			Ratio	$\frac{1}{4}$:	$\frac{1}{2}$:	$\frac{1}{4}$

these exist in the population in the typical Mendelian ratio of $\frac{1}{4}$ AA : $\frac{1}{2}$ Aa : $\frac{1}{4}$ aa. Now assuming that random mating occurs and comparable offspring are produced, the next generation consists of the same proportion of genotypes, as is seen by recording all of the possible combinations in the next generation (Table 26–1).

In any population consisting of diploid, cross-fertilizing individuals the frequency of any pair of genes, A and a, must appear in each individual as one of the three combinations, AA, Aa, and aa. The number of times each combination appears in a population can be determined in the following fashion: the number of times the gene A appears in the population (frequency) can be defined as the frequency of the homozygous dominant individuals AA plus $\frac{1}{2}$ of the frequency of the heterozygous individuals Aa, and this can be designated p. Likewise, the frequency of the gene a may be defined as the frequency of the homozygous recessive individuals aa plus $\frac{1}{2}$ the heterozygous individuals, Aa; this can be designated q. We know that the sum of the three genotypes, AA, Aa, aa, equal 1.00; therefore, $p + q = 1.00$. In random mating the sperm cells carrying gene A or a unite in accordance with their relative frequencies, As you will see from Table 26–2, the three

TABLE 26–2. Calculation of Genotype Frequencies

	SPERM CELLS	
	A (p)	a (q)
Eggs		
A (p)	AA (p^2)	Aa (pq)
a (q)	Aa (pq)	aa (q^2)
Results:		
p^2 (AA) + 2 pq (Aa) + q^2 (aa)		

genotypes have the frequencies, $p^2 + pq = p$ and $q^2 + pq = q$, and we know that $p + q = 1.00$; then it becomes obvious that the frequencies of the genotypes AA, Aa, and aa in the second and all subsequent generations will be the same as in the original population.

Observe that Mendel's Laws represent the special example of the Hardy–Weinberg Law where p and q are equal, each being 0.5. If p is dominant, the phenotypic ratio is 3:1. Actually Mendel used the binomial theorem $(a + b)^2 = a^2 + 2ab + b^2$ to express his famous F_2 generations.

It is possible with this information to compute the frequencies of genotypes in the population providing one of the frequencies is known. It is usually easier to obtain figures for recessive genes, aa (q^2), because they are more obvious. With this information it is possible to compute the frequencies of the other genotypes in the population. An excellent example of this comes from humans.

The trait albinism (p. 541) occurs in white populations in the ratio of 1 in 20,000. It is recessive and is inherited as a single pair of genes; therefore, an albino has the genotype, *aa*. In order to determine how many individuals in the population are carriers for the trait, that is, how many are heterozygous (*Aa*), it is possible to use the Hardy–Weinberg Law to find the answer. The frequency of *aa*, represented by q^2, is 1/20,000. By taking the square root of this figure we get about 1/141 which represents q. We know that if $p + q = 1.00$, it follows that $p = 1 - q$. Substituting 1/141 for q, $1 - 1/141 = 140/141$, which is the value of p. Now we know the value of both q and p and from this information we can calculate the heterozygotes in the population because we know the number is represented by $2pq$, thus, $2 \times 140/141 \times 1/141 = 1/70$. This means that heterozygous condition *Aa* is present in one person in every 70, even though only 1 person in 20,000 is homozygous and has the defect.

Another example of how the Hardy–Weinberg law operates may be taken from studies of the blood antigens M and N, again in human populations. Because of usage we employ the letters M and N, although this is misleading because only a single pair of alleles are involved. (Correctly, we should use M and m or N and n). Neither allele is dominant and there is no differential selective value associated with them. Most people do not know whether they possess M, N, or MN. The genotype can be distinguished by simple serological tests.

Studies of 6,129 white people in the United States have demonstrated the following proportion of phenotypes (also given in percentages): $M = 1787$ (29.16%); $N = 1303$ (21.26%); and $MN = 3039$ (49.58%). By using the Hardy–Weinberg Law we can calculate that $q = 0.46$, which

when subtracted from 1.00 gives us $p = 0.54$. Then $p^2 = 0.29$ and $2\ pq = 0.49$, rather close to the proportions established by actual serology tests. The agreement between theoretical calculations and observations establishes the likelihood that the blood antigens are under the control of a single pair of genes in which neither is dominant to the other.

From these examples we have seen the usefulness of the Hardy–Weinberg Law. We must remember, however, that it is entirely theoretical. It assumes random mating, the lack of migration of organisms in and out of a population, the absence of differential mortality among the various genotypes, and an infinitely large population without chance gene frequency fluctuations. In spite of these exceptions the law is most important to an understanding of evolution. It demonstrates that there is nothing built into Mendelian inheritance that can alter gene frequencies in populations. If alterations do occur, they must be ascribed to other factors such as natural selection, migration, chance and mutations, both forward and backward. Evolution, then, is a result of these factors.

Natural Selection. Darwin's thoughts on the part natural selection plays in evolution have been largely retained and enlarged upon as a result of new information that has become available since his time. One of the most interesting cases of the operation of natural selection has been observed in the last one hundred years in Darwin's own England. It concerns moths belonging to the species *Biston betularia*, and the phenomenon is called **industrial melanism.**

These moths occurred in two different forms in Europe, a normal light **peppered** form and a dark **melanic** form (Fig. 26–3). The first dark moth was found near Manchester, England, in 1848, and in the years that followed it increased in numbers until today it is the only form present in areas where a great deal of heavy industry exists. What is the relationship between the indus-

Fig. 26–3. Industrial melanism illustrated by the moth, *Biston betularia*. The melanic and peppered forms are shown in both photographs. The one above is of a tree trunk covered with lichens. Find the two peppered moths. The dark form is very conspicuous. The lower photograph shows the blackened tree trunk of an industrial area. Note how conspicuous the peppered moth is, whereas the dark form blends into the background.

trialization and the predominance of the dark form?

A careful study of the tree trunks by H. B. D. Kettlewell reveals that as the gaseous by-products in industry increase, the mottled lichens are poisoned and disappear from the tree trunks leaving them dark as a result of the accumulated soot from the factories. Thus the peppered moth is clearly visible on the dark trunk, which was not the case when the moth rested on lichens. Consequently, the light-colored moths were easily observed and eaten by birds, whereas the dark colored moths were protected because they matched the dark background of the blackened trunks and were thus inconspicuous. As a result the light moths disappeared in the industrial regions, though they have been able to maintain their numbers in the rural areas of England.

We can explain these observations utilizing our knowledge of genetics and applying the rules of natural selection. Gene mutations must have always occurred in the population of these moths producing an occasional dark form, but these were immediately preyed upon owing to their conspicuous coloration. When metropolitan areas became industrialized, the gases and soot belching forth from the factory chimneys killed the lichens, thus changing the background upon which the moths rested. Now the dark color became an advantage and the light color a disadvantage, hence the dark-colored moths were selected for and preserved in the population. The light moths, on the other hand, were selected against and disappeared from the population. This demonstrates natural selection in action and results from man's efforts to change the face of the earth.

Attempts have been made to classify natural and artificial selection into several types. The first type of selection has been called **directional**; that is, those individuals which vary in one general direction are favored whereas those that vary in the opposite direction are eliminated. The classical case of this type of selection is horse evolution which is described later (p. 578).

The second type is known as **stabilizing selection** which tends to keep down the numbers of deleterious mutant genes and gene combinations. This stabilizes and protects the normal development pattern of the species. It is antievolutionary in its effects because it tends to stop change. Examples of this type of selection are the most common and are responsible for the stability of a species. Buck deer with too large antlers would be as disadvantageous as one with small antlers, although for different reasons. In any environment there is an optimum antler size that promotes success of its owner. Mutants either way would tend to be eliminated and the population would maintain the optimal antler size. This demonstrates the fact that most mutations are harmful.

A third type is **dynamic selection** which permits the species to maintain itself in its changing environment and to take advantage of new ecological opportunities as they become available. Alterations of climate, contact with new parasites, predators, or competitors, destruction of food sources, all bring pressure to bear on the species. If new genotypes are not available the species fails to survive. Therefore dynamic natural selection involves the establishment of new genotypes that provide the animal with the means to adapt to the new environmental conditions. Even though we know that new genotypes are a result of mutation and recombination, it is reasonable to say that new genotypes are *produced* by selection. Over a long period of time selection is the principal force which directs evolution because it determines which genotypes survive, and these are the only ones in which mutations can occur.

Genetic Drift. Chance also plays an important part in evolution. Not all populations are large; some small segment of a larger population becomes isolated frequently by accident, and of necessity interbreeds. Thus by chance heterozygous gene pairs become homozygous for one allele or the other. This is called **genetic drift** and is an

exception to the Hardy–Weinberg Law which holds only for large populations. The differences noted within small populations of the same species may be explained by genetic drift.

Large populations possess a great many genes in their **gene pool** and no small group would possess all of them in the same proportion as they exist in the large population. Therefore, if a small group, by accident, became transported to some distant land their composite gene pool more than likely would not be representative of the entire population. Interbreeding through subsequent generations would tend to emphasize the genes they possess and in the proportion that existed in the original small population.

An example of genetic drift can be drawn from *Homo sapiens*. A small group (28 persons) of Germans, called Dunkers, migrated to Pennsylvania in the early part of the eighteenth century. Owing to strict religious beliefs they have remained segregated up to the present time. A number of characteristics of this group were studied and it was found, for example, that there was a high percentage of the ABO blood groups among them. The results were compared with those taken from the original stock from which the Dunkers came, still living in Germany, and the general American population. The outcome demonstrated evidence for genetic drift as regarding the percentage of type A blood. The Dunkers show 59.3 percent, the original stock, 44.6, and the American population as a whole, 39.2 percent. Obviously, the early migrants possessed a higher percentage of genes for blood type A than the German population from which they came, and have maintained it since. There seems to be no evidence that selection has played a role, since there is no advantage of one type of blood over another. It seems that pure chance was the only factor involved.

Geographic and Genetic. Darwin realized, as have most biologists since, that isolation played a very important part in the origin of new species, but it was not until inheritance was thoroughly understood that the full meaning of isolation became apparent. It is known now that in addition to **geographic isolation**, **genetic isolation** is also essential for the origin of new varieties and eventually new species.

Darwin and many other naturalists noted very early that animals living on an island were frequently quite different from those on the neighboring mainland. Other barriers, such as a mountain range, a deep canyon, a wide body of water, or a desert, have been known to separate members of a species long enough for differences to be detected between the two groups.

It is understandable that when a barrier occurs in the range of a given species so that groups of individuals of that species are completely separated, the continual occurrence of mutations and the slightly different environments acting selectively upon them eventually brings about divergence of the two groups. If the barrier continues for a sufficiently long period of time, the differences may become sufficient to constitute two species. If the barrier is then removed so that the two groups once again intermingle they will probably maintain their identity as different species. However, if the barrier is removed at a time when they have not changed enough, that is, while they are still varieties, it is possible that the differences may be lost through interbreeding. The latter instance is illustrated by the human species, *Homo sapiens*. Although he has occupied nearly all portions of the earth's surface and does show some marked physical variations, the various races have not lost their ability to interbreed. The barriers that have been set up between races have not been effective sufficiently long to bring about infertility. With the intervention of modern means of transportation those original geographic barriers can be overcome, so theoretically man should lose what changes have been produced by this isolation, providing there is a free mingling of all races. But such, as we know, is not the case. Society places barriers that can be as effective as infertility or vast geographical obstacles.

It is interesting to note that under natural conditions hybridization is only rarely observed among closely related species of animals. To be sure, it has been possible to effect it under laboratory conditions—the biscow, for example, as a result of the mating of a bison and a cow—but such cases are unusual and do not normally occur in nature. Since the animals occupy the same environment and are therefore subject to the same conditions, why do they not mate freely and thus reverse the differences established earlier through isolation, inducing reverse evolution in this way? This question can best be answered from a study of genetics.

Genetic. When two species or even varieties mate, the genes are not as perfectly compatible as they were in the common ancestor. This results in low fertility or in most cases complete sterility. Thus such crosses under natural conditions would be unfruitful if they did occur and the hybrid would have little chance of establishing itself. Also, the hybrid itself is often sterile, as is true of the mule (p. 515). Furthermore, slight differences in the mating season might act as a barrier even though the animals were fertile. These and many other slight differences act as barriers to keep the two species distinct and separate.

Many kinds of gene changes occur, from slight modifications in the gene itself to an alteration in the arrangement of the genes on the chromosomes, any of which may make fertility impossible. Under laboratory conditions these changes sometimes occur spontaneously in a species that in all other respects is like the parent form. Such genetic changes are as effective in isolating animals as any physical barrier could be. Many biologists agree that interfertility is the acid test of a species. No matter what other characteristics two species may have in common, they are distinct species if they lack interfertility. The problem of establishing species on such a basis is almost hopeless at the present time because of our lack of information concerning the genetic constitution of all but a very few species of animals. These are the animals extensively used in the laboratory, such as *Drosophila,* mice, rats, and paramecium.

Neoteny. A long-known observation among different groups of animals may well be one of the important forces that brings about evolutionary change. This is the curious phenomenon of **neoteny** where sexual maturity occurs during the larval or juvenile stage of the animal. It occurs among insects, the early vertebrates, and may explain how man came to be the strange animal he is. Great steps in evolution, called **macroevolution,** may be explained by neoteny.

The classical case among vertebrates is the Mexican axolotyl (p. 256). Normally a salamander breathes by means of external gills during the larval stage but develops lungs in the adult. The axolotyl becomes sexually mature during the larval stage and never does acquire air-breathing lungs. Thus an animal has drastically changed by the simple procedure of becoming sexually mature at a time in its life cycle when its body is quite unlike the normal adult. This could account for a major step in evolution, perhaps in a very short time and without the usual microevolutionary steps by which it is believed most evolution takes place. This observation becomes quite startling when applied to man.

Adult man in physical appearance is more similar in many ways to a young ape than to the adult ape. Such physical characteristics as the relatively huge brain, the form of the teeth, the flatness of the face, hairlessness, angle of the head with the trunk, and the sutures between the bones of the skull, all resemble those in the baby ape. These observations led Bolk in 1926 to call man a neotenous primate. Perhaps man has simply lost the capacity to mature and remains, in part, in the juvenile stage.

Such phrases as "struggle for existence" and "survival of the fittest" appealed to the emotions of many people of Victorian England and spurred them on to provide such other phrases as "eat or be eaten" and "nature red in tooth and claw." This idea of rugged competition seemed to reflect the attitude of industrialists of this period and helped justify their methods. However, in the current century, with the advent of genetics and a better perspective of the problem of evolution, it became clear not only that biological theories could not be used to explain (much less justify) economic or social phenomena but that the fierceness of the struggle for existence leading to natural selection was greatly overemphasized as well. It became apparent that the **fit** survived, and not the **fittest,** which implied that only one survived. And this survival could be the result of cooperation as well as competition. Fighting, pugnacity, and aggressiveness are often less conducive to biological success than cooperation. To "live and let live" promotes the welfare of species as much as mortal combat. Indeed, success depends on both cooperation and competition; at one time struggle may be important, at another cooperation. Survival of the offspring of a species may depend on parental care as well as fierce defense. Today we are inclined to interpret natural selection as including both cooperation and competition.

DIRECTION OF EVOLUTION

At the beginning of life there must have first existed only one species. After a time this must have varied as a result of mutations, thus giving natural selection an opportunity to initiate evolution. It is conceivable that evolution might have proceeded in any or all directions, and perhaps it did at first, but after a time it seems to have headed in a certain rather general but definite direction, which has continued more or less uninterrupted ever since. This idea is often referred to as **orthogenesis** or **straight-line evolution.** One of the classic evidences for straight-line evolution is that of the horse.

Horse Evolution

The modern horse has had an interesting evolution and one that can be followed better than most because of the abundance of fossils left from the Eocene to the present.

Eohippus, the earliest known ancestor of the horse, occurred in early Eocene times and was anatomically quite remote from *Equus,* our modern horse (Fig. 26–4). It was about the size of a jackrabbit and possessed functional toes on each appendage, four in front and three behind. There were rudiments of others which, when accounted for, demonstrated that the animal had descended from an ancestor with five digits, the usual number among vertebrates. In the forefeet only the inside or "thumb" was rudimentary while in the hind feet

Fig. 26–4. The evolution of the horse illustrates orthogenesis. Probably these animals did not progress along a direct path from the tiny *Eohippus* to *Equus* but rather there were many deviations along the way. Progress was in a general way toward the present-day horse, but there were many paths that led to extinction.

Eocene   Oligocene   Miocene   Pliocene   Pleistocene

Eohippus   Miohippus   Merychippus   Pliohippus   Equus (modern)

both the thumb and the "little finger" were missing but could be seen as functionless splints along the next foot bone. These tiny mammals apparently lived in seclusion, feeding on succulent leaves and avoiding contact with the huge mammals of this period.

In the Oligocene we find a somewhat larger animal, *Miohippus,* with only three toes on the forefeet, although a rudiment of a fourth remained. The middle toe on each foot apparently bore the brunt of the weight of the animal because it was much larger than those on either side.

During the Miocene *Merychippus* came into prominence, still with three toes but with added emphasis on the middle one which bore all of the body weight. The other two toes, while externally visible, appear to have been of little or no use to the animal. *Merychippus* was larger than any of the horses up to that time, standing about 3 or 4 feet in height.

This gradual tendency toward greater size and reduction of the number of toes continued into the Pliocene, where *Pliohippus* was without external lateral toes, although retaining splints indicating the historic loss.

Horses that appeared in the Pleistocene had acquired essentially the same proportions as the modern horse. There is a single hoof on each foot, although paired lateral splints give evidence of the ancient toes of its forebears. The size is larger than any of the earlier horses, and by selective breeding man has produced the kinds of horses he desires, from Shetland ponies to draft horses.

Through a period of 60 million years, the horse has evolved from a tiny animal living in the forest for protection and feeding on succulent leaves to a huge grass-eating beast roaming the prairies where fleet-footedness is essential for survival.

Opportunity in Evolution

Another feature that is probably important in **oriented evolution,** which also involves the relation of organisms and environment, is the relative time or co-incidence with which these impinge upon one another. When the environment is just right for an animal to evolve in a certain direction, the animal must not only be present but also must be sufficiently adaptable (have sufficient mutations available) to take advantage of the opportunity. Simpson called this "Opportunity of Evolution." Because of these environmental opportunities, animals follow within certain restricted limits, progressing toward an apparent goal, as in the case of the horse. As the environment changes, the organisms change with it. Living things themselves constitute more opportunities by providing a new environment; the body of a vertebrate as a home for a blood parasite is an illustration. Thus new environments are constantly being created (and others destroyed) which open up more and more opportunities for a greater variety of living things. This has produced an untold number of kinds of animals, of which probably over a million are alive today. Evolution constantly operates with materials available. Frequently a structure is produced that may not be perfect but serves a purpose more or less satisfactorily. Perhaps ¡there is no structure in the body of any animal that performs its function to a perfect degree. Why, then, are not all of the organs of animals perfect, or more broadly, why is the animal itself more or less imperfect in every detail? There is constant selection among physical variations after they appear and a constant process of adaptation, but complete and absolute fitness to the environment in every conceivable way seems quite impossible to attain.

To illustrate this point, any one of hundreds of examples might be chosen. The matter of photoreception is a good illustration of how all groups of animals have devised some method or receiving light rays. Among all of the photoreceptors there is none that we might call a perfect photoreceptor. The primate eye is one of the best and we are all aware of its imperfections.

There are various types of photoreceptors among animals from the very simple eye spots in the protozoa to the complex molluscan and vertebrate eyes. From the planarian eye we see steady improvement concomitant with the development of organ systems until finally among the mollusks, arthropods, and vertebrates we see the best eyes of all. It is interesting to note that each of these types has faced the same problem but has solved it differently. The arthropod eye consists of many tiny units aggregated into a compound structure which forms many images, giving a composite picture. Thus, the problem of receiving light has been satisfactorily solved but in an entirely different manner from that of the mollusk or vertebrate. On the other hand, the likeness of the vertebrate and squid or cuttlefish eye (Fig. 26–5) is amazing. The fact that two distantly related groups of animals could solve photoreception in almost identincal ways seems incredible. Each animal had different materials at hand to build a photoreceptor and each built it differently, but both came out with essentially the same organ, varying only in the placement of the sensory cells of the retina (the rods and cones in the vertebrate are directed away from the light source, whereas in the molluscan eye they are reversed). The pathways by which these eyes were formed are entirely different in the two groups, which merely proves the earlier contention that animals solve similar problems in different ways.

Fig. 26–5. There is a remarkable similarity between the molluscan (cuttlefish, left) and vertebrate (human, right) eyes, yet they have evolved along two quite different pathways. This type of evolution is called convergent evolution and is abundantly illustrated among animals.

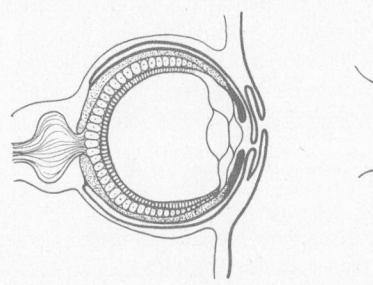
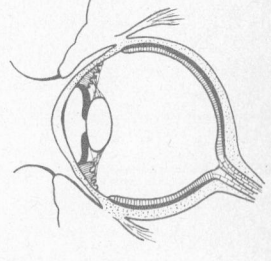

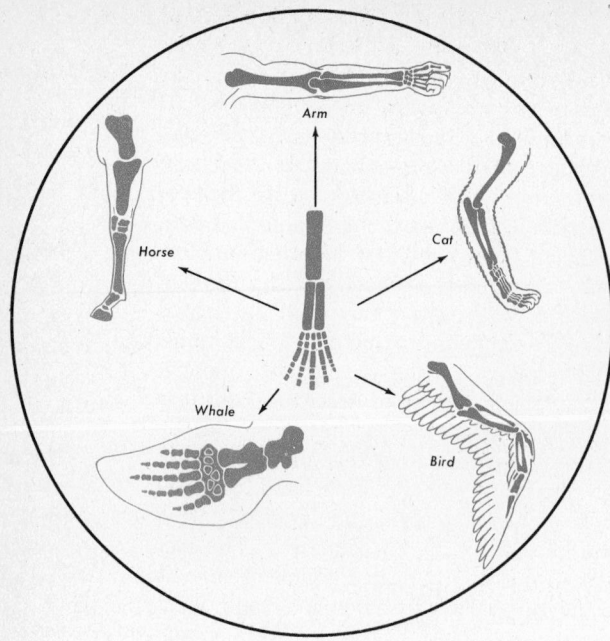

Fig. 26–6. The front appendage of vertebrates illustrates divergent evolution. The hypothetical vertebrate appendage was a five-toed structure which has been modified in many ways as represented by the arm of man, the front leg of the cat, the wing of the bird, the flipper of the whale, and the front leg of the horse. Each modification satisfies the needs of the particular animal living in an environment that requires such an appendage for survival.

Divergence and Convergence in Evolution

As opportunities have presented themselves, animals have evolved in many directions with respect to certain organs in order to take advantage of all possible environments. This is called **divergent evolution** or **adaptive radiation**. At some later time some members of various groups have evolved toward life in a particular environment; this is called **convergent evolution**. The manner in which these two types of evolution operate can best be understood by following a specific structure or organ through its evolution in several groups of animals. A good example is the forefoot of land vertebrates.

The ancestral land vertebrate appendage was a five-toed structure designed for surface locomotion. As new environments became available the forefoot underwent **divergent evolution**, becoming highly modified for locomotion in many ways. For example, it became a single-toed leg in horses, a five-toed clawed leg in cats, a flipper in whales, an arm in man, and a wing in birds (Fig. 26–6). This fundamental five-digit structure has become an effective instrument for locomotion in water, on land, and in the air. This is the situation in which we find vertebrates today. But long before vertebrates reached their present state, ancestral forms, even after they became adapted to land life, had the opportunity of taking to the air which has some obvious advantages over locomotion on land. Members of three groups took advantage of this opportunity and underwent **convergent evolution** toward locomotion in the air. These were: first, the reptiles (pterodactyls), then the birds, and finally the mammals (bats) (Fig. 26–7). They all produced wings which were essentially the same, although varying slightly in minor details. None of them are perfect instruments of flight, yet they are sufficiently good to permit both the birds and the bats very real success, the former more than the latter.

Another case of convergent evolution is illustrated by still other members of these same three groups (reptiles, birds, and mammals) which returned to the water and became modified in many ways for an aquatic life (Fig. 26–7). The primitive reptile *Ichthyosaurus* took on the shape of a modern fish and its appendages become fin-like. The whale, a mammal, developed fishlike appendages and body form. The penguin, a bird, lost its ability to fly but became a proficient swimmer, using its wings as flippers.

Finally divergent evolution is nicely illustrated in Australia where the ancient marsupial stock gave rise to a variety of animals that closely parallel the placental mammals on other continents (Fig. 26–8). For example, there are shrewlike marsupials, squirrel-like marsupials, doglike marsupials and many others. It appears that, given only marsupials to begin with and no carnivores to destroy them, divergence has been commensurate with the possible niches, paralleling similar situations where placental mammals were the starting point. There are many similar cases of divergent and convergent evolution among animals that could be cited. They confirm one of the fundamental features of evolution, namely, that all opportunities are exploited with the materials at hand.

EVOLUTION THROUGH EXPERIMENTATION

Under natural conditions, animal hybridization probably is of little significance in evolution. Domestication among animals was carefully studied by Darwin and he used the results of these studies as a potent argument in favor of natural selection. He observed, as we all have, that under the influence of selective breeding a wide variety of animals can be produced and if such forms appeared in nature, they certainly would be assigned different species names. Darwin argued that new species could come into being by substituting man as the selective agent. Since the environment was not endowed with reasoning powers, the tendency would be to select in many different directions with the ultimate goal always the adaptation to the particular environment in which the animal found itself.

Under the guiding hand of man a prodigious variety of animals have been produced to satisfy his whims. He has done a notable job in the chicken, for example, which varies from the cocky Bantam to the tremendous Brahmin and is a far cry from the Oriental jungle fowl from which present-day chickens are largely or wholly descended. This ancestor laid perhaps two dozen eggs per year, whereas some present-day breeds equal that number in two weeks. Domesticated animals seem to preserve their interfertility, quite contrary to conditions found in nature. The reason why they do is that man plans it. Sterile animals are of no

use to him, so he has selected them for high fertility along with other traits that he wishes to accentuate. Even with some of his domestic animals the beginnings of infertility have been evident. For example, some crosses between dogs are all but impossible, as between a very large one and a very small one. Furthermore, some crosses often produce malformed animals that soon die, or again the number of offspring may

Fig. 26–7. Divergent and convergent evolution can be illustrated by a study of the anterior appendages of three different groups of vertebrates. The ancestral reptile appendages diverged for locomotion (walking and flying) at a very early time. Some members of each class (reptiles, birds, mammals) subsequently underwent convergence toward flight and life in the water by modifying their anterior appendages to function more effectively in these environments.

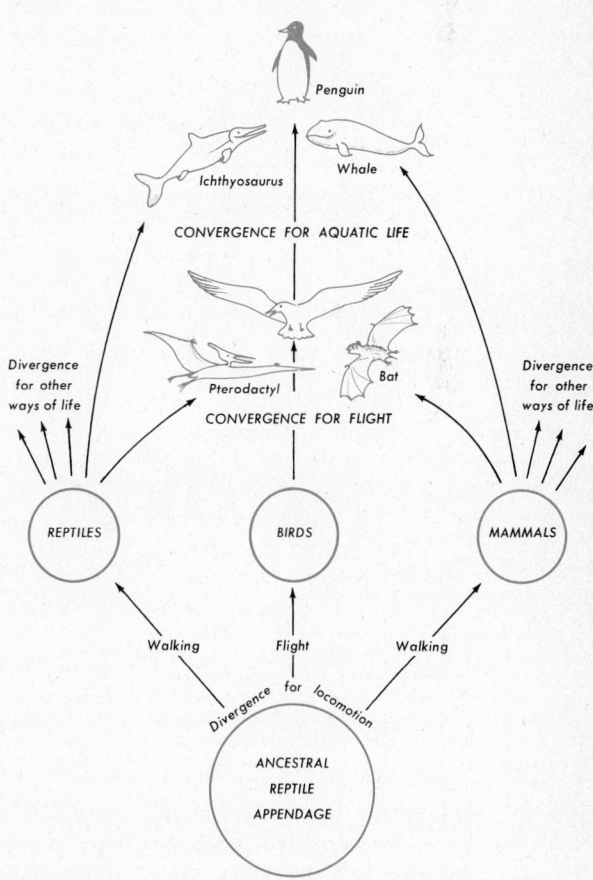

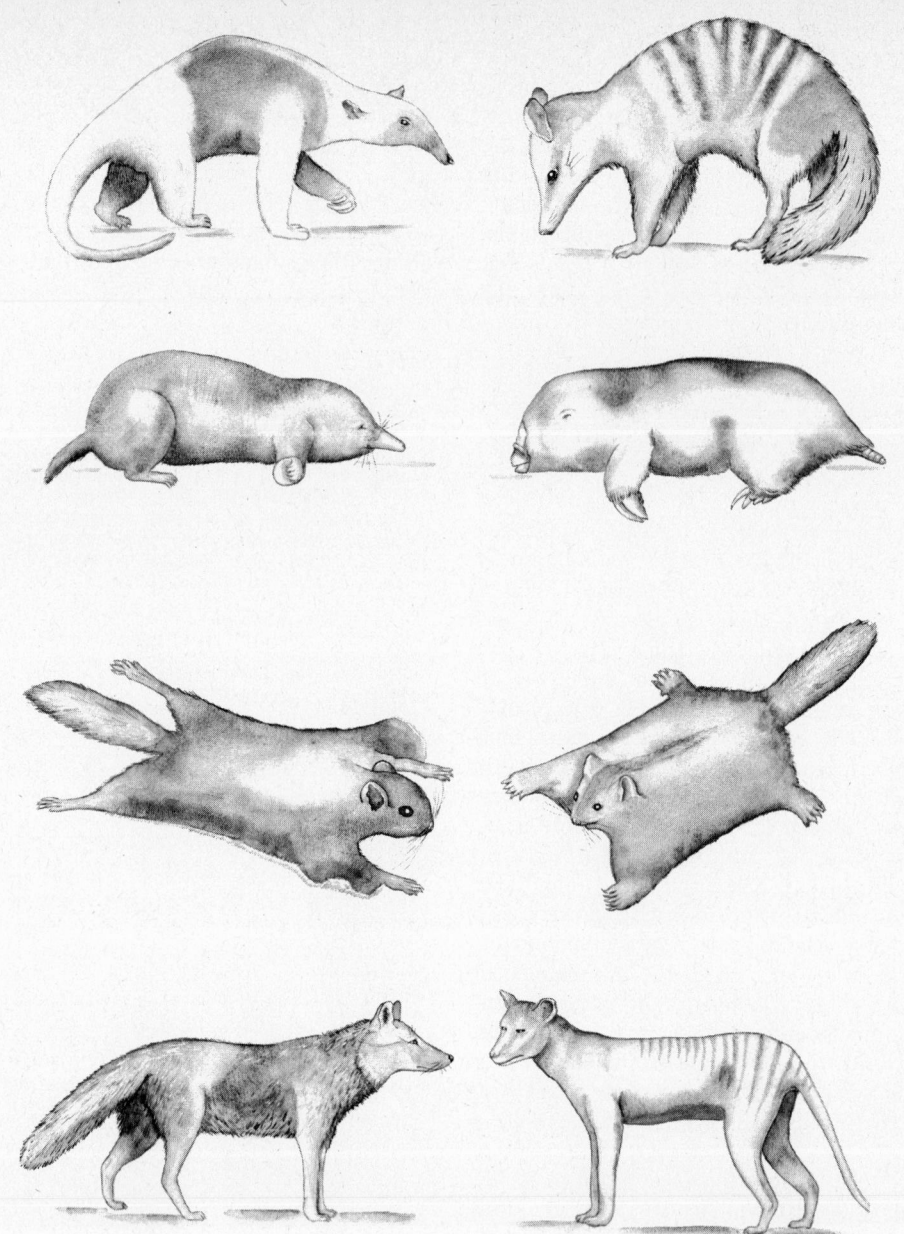

Fig. 26–8. A comparison of placental (left) and marsupial (right) mammals illustrating divergent evolution in both groups based on their modes of life. (Top to bottom) Those that are adapted for feeding on ants (*Tamandua tetradactyla* and *Myrmecobius fasciatus*), the molelike forms that are adapted for burrowing (*Parascalops breweri* and *Motoryctes typhlops*), squirrels that are modified for soaring (*Glaucomys volans* and *Petaurus sciureus*), and wolflike carnivorous animals (*Vulpes vulpes* and *Thylacinus cynocephalus*).

be small. The incompatibilities noted now, after several hundred generations, may, perhaps, become so great that a strict genetic barrier will be placed between certain varieties of dogs alive today, inaugurating a new species. If that happens, man will have

Fig. 26–9. A case of protective coloration. Find the chipmunk.

created a new species and thus controlled evolution under his own guidance. We must conclude, therefore, that in the domestication process, man has within his grasp the very mechanism by which evolution proceeds.

THE END RESULT OF EVOLUTION— ADAPTATION

Out of all this should come animals remarkably adapted to their environment but, as we have already seen, this is not completely so. In spite of the rather careless way in which most animals are keyed to their environments, certain examples of adaptation forever intrigue the naturalist and the layman. Among these, color adaptations are particularly interesting.

The matter of camouflage is an important factor in survival and is used almost universally by animals. Such concealing or **protective** coloration has been selected for, and the outer coverings of some animals match their background astoundingly well (Fig. 26–9). Most animals, particularly those that fly or swim, are dark on the dorsal side and white or lighter in color on

the ventral side. Thus potential predators from beneath looking up will be less apt to see them because they blend with the sky and those from above would experience the same difficulty because of blending with the darker earth or stream bottom. Other animals, such as fish, frogs, and lizards, can change their skin color to match the background upon which they happen to find themselves. The positive survival value of such colorations has been shown by experimentation in which the prey with colors most in contrast to the background is always picked up first by the predators.

Some animals have found survival value in **mimicking** another species that carries a potent defensive or offensive weapon, such as a venomous snake, for example (Fig. 26–10). Such well-equipped animals are frequently highly colored; this advertises their danger and warns unsuspecting predators that might take them by mistake. Some animals mimic living or dead leaves, and certain butterflies are so faithful in such resemblance that they may go unobserved even by a careful naturalist (Fig. 26–11).

Fig. 26–10. These two snakes demonstrate mimicry in animals. The poisonous coral snake (*Micrurus fulvius tenere*), in the top picture, is mimicked by the false coral snake (*Oxyrhopus baileyi*). Note the remarkable resemblance between the two.

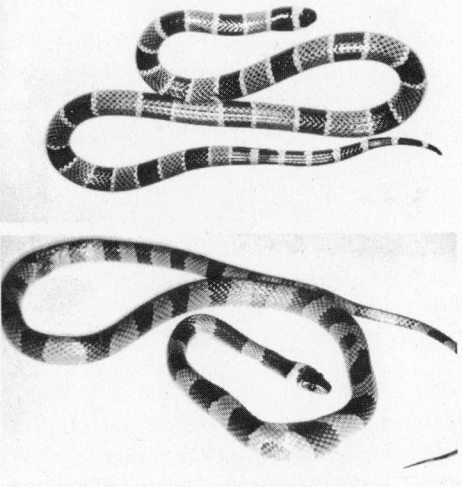

Fig. 26–11. Certain insects resemble leaves upon which they rest, a circumstance which obscures them from their enemies.

These adaptations and many others undoubtedly have a marked effect on survival, but every such case is open to question and should be interpreted in the light of actual observations.

Things of much greater importance than color, body shape, horns, claws, and similar characters are such physiological qualities as ability to resist disease, high or low temperatures, desiccation, and so on. Fitness for a specific environment involves many qualities, and it is the product of all of them that makes for survival. It follows that most of them must be favorable, although an animal may survive and be rather successful while carrying along some unfavorable traits. When the weight of the adverse qualities exceeds that of the favorable ones the animal becomes extinct. This has happened to millions of species in the distant past and has happened in recent times, as in the case of the passenger pigeon.

Adaptations tend toward better and better fitness for the environment which results in specialization (Fig. 26–12). When this has been carried to the extreme, the animal is said to be "overspecialized," a fact which biologists have long considered important in the extinction of certain species and even large groups of animals. Specialization is inevitable, and all animals are specialized, some much more than others. Land animals, for instance, are adapted for obtaining oxygen from air, and if the earth were suddenly covered with water they would all be eliminated. This is specialization in a broad sense. As adaptation proceeds, animals become adapted to narrower and narrower limits of their environments and some, such as ruminants, have become adapted to eating grass and probably could not survive if all grasses of the earth were destroyed. Thus survival is intricately linked with consistency of specific environments.

The more an animal becomes specialized, the more closely it is forced to live in the environment for which it is fitted. Therefore, the more generalized types are less apt to become extinct. For example, omnivores are more apt to survive under conditions where the food is likely to change than strict herbivores or carnivores. If food becomes scarce and the omnivorous animal can no longer exist as a predator it can subsist on vegetable matter, or it may devour some vegetable matter and some meat as the situation demands. Obviously, then, as animals become more specialized they are endangering their possibility of survival.

EVOLUTION AND THE FUTURE

Some believe that evolution has run its course and that all of the niches of the world have been filled so that no further change is possible. Furthermore, others contend that the apotheosis of evolution, namely, man, has been attained and there is now "no place to go." One might be inclined to agree with these contentions if there were no evidence of evolution going on around us and if man had become perfect in every respect, leaving no room for improvement. Most of us would not subscribe to either of these statements. From what we know of the past it seems likely that as long as there is life on earth there will be continuing change or evolution. The directions it will take are unpredictable to us now, but that it will continue seems assured.

To be sure, the abundance of life in existence now might suggest that all of the possible environments capable of supporting life have been taken, leaving no room for any more forms. A similar situation might have seemed to be true even in the Cambrian, where all of the present phyla had their representatives. There were fewer niches then but all of them were probably filled, and as ages passed new niches appeared and new species filled them. There is no reason to believe that a similar process is not going on at the present time. New environments are constantly being created and those animals with an ample supply of mutations move into them, with the aid of natural selection. There may come a time,

Fig. 26-12. Two cases of adaptation. (Left) The attachment organ of the shark sucker, located on the dorsal side of the head, makes it possible for this bony fish to adhere to the ventral surface of sharks, thus getting a free ride and profiting by feeding on the bits of food left over when the shark kills its prey. (Right) The Pallas' murre nests on narrow ledges of vertical cliffs in the Pribilof Islands. When disturbed, the sharply pointed egg rolls in tight circles, thus retaining its position. This adaptation has great survival value for the species.

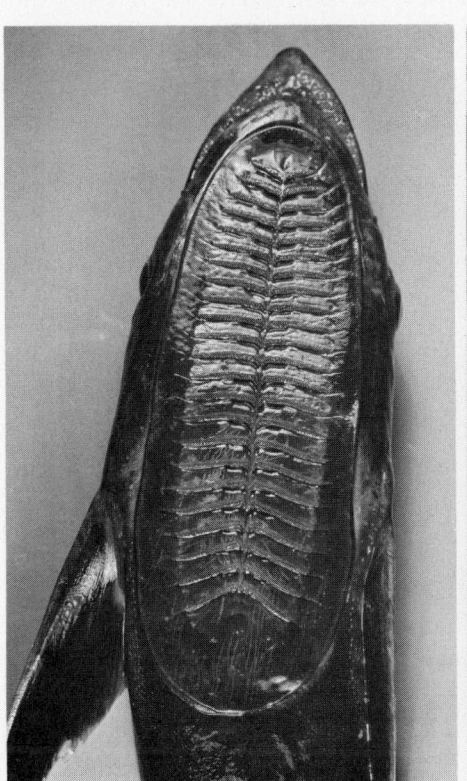

perhaps a billion or so years in the future, when all life will cease, but up to the end living things will undoubtedly attempt to adapt.

It is highly probable that the evolution of certain groups could be repeated if the ones living today were wiped out. Many of the "immortals" such as *Sphenodon* and the opossum are still with us and perhaps they could radiate once again, giving rise to new groups of reptiles and mammals. Undoubtedly the new forms would not be exactly like the ones we see today, but they might be as diversified. One might ask why the opossum has not done this very thing today, why it is still essentially the same creature that lived in the Cretaceous. The most obvious reason is that placental mammals have occupied nearly all of the environments that are available to it and the competition gives it little opportunity to crowd in. Hence it has been able only to maintain itself up to the present. But if the placental mammals were suddenly removed, the descendants of the opossum would soon fill the spaces left and there would be rapid evolution among the various forms, just as there was in Australia.

Likewise, if the human species came to an abrupt end, the spaces left would be filled by other forms, perhaps not exact duplicates but certainly some possessing the unique qualities of man. The entire trend in evolution has been toward a more intricate coordinating system, and there is no reason to believe that such a trend would cease if there were no animals possessing those qualities on the earth. To be sure, no other groups are likely to usurp that position among animals so long as man maintains his present place. Unless he retrogresses, it is highly improbable that any other group could match or excel him in reasoning power.

What does future evolution hold for man? Can he profit by what he knows about this grand drama of evolution, can he control its future trends with respect to the animals around him, and most important, can he control it with respect to his own destiny? These are the really important problems that scientists and laymen alike wish to understand and, of course, they are the most difficult to answer.

The evolution of man has placed him in a unique position among animals in that he is the first and only animal who can control his environment to a greater or lesser extent. He builds shelters, clothes his body, plans his food needs and provides for them, and in general improves his physical environment wherever his own ingenuity and available materials will permit. He knows enough about evolution to take advantage of it in controlling animal and plant life around him, which enhances his own chances for survival. In other words, he can give evolution direction with a goal. He is rapidly gaining greater and greater control over his environment in all respects, so that his continued existence is more assured from that point than it ever has been. However, there is a sinister aspect to all of this because, while he has acquired sufficient knowledge to direct evolution to a certain extent in the organic world about him, he has done almost nothing in directing his own evolution, either physical or social. Further confusing the issue, he has at his disposal means by which he can annihilate himself, in which respect he is unique among animals. Moreover, he seems to be little concerned about either the quality or the quantities of his members. It seems that a creature with such a remarkable nervous system; who is able to control his earthly environment, and initiate extraplanetary excursions, should be able to manage his own development, the facts for which he has before him. But there is the matter of time, even if the earnest intention is there. Will we act soon enough? If we do, *Homo sapiens* could evolve into a better adapted man both physically and socially; if we do not, our place will be taken by some lesser creature.

Books

BATES, M., and HUMPHREY, P. S., *The Darwin Reader*. New York: Scribner's, 1956.

BERRILL, N. J., *The Origin of Vertebrates*. London: Oxford University Press, 1955.

CARTER, G. S., *Animal Evolution: A Study of Recent Views of Its Cause*. London: Judgwick and Jackson, 1951.

*COLBERT, E. H., *Evolution of the Vertebrates*. New York: Wiley, 1955.

*DARWIN, C., *On the Origin of Species by Means of Natural Selection, or the Preservation of Favored Races in the Struggle for Life*. New York: D. Appleton and Company, 1875.

*DOBZHANSKY, T., *Evolution, Genetics and Man*. New York: Wiley, 1955.

DODSON, E. O., *Evolution: Process and Product*. New York: Reinhold Publishing Co., 1960.

EHRLICH, P. R., and HOLM, R. W., *The Process of Evolution*. New York: McGraw-Hill, 1963.

MAYR, E., *Animal Species and Evolution*. Cambridge, Mass.: The Belknap Press of Harvard University Press, 1963.

MERRILL, D. J., *Evolution and Genetics*. New York: Holt, Rinehart & Winston, 1962.

MOODY, P. A., *Introduction to Evolution*. New York: Harper & Row, 1962.

*SIMPSON, G. G., *The Meaning of Evolution*. New Haven, Conn.: Yale University Press, 1951.

SIMPSON, G. G., *This View of Life: The World of an Evolutionist*. New York: Harcourt, Brace & World, 1964.

STEBBINS, G. L., *Processes of Organic Evolution*. Englewood Cliffs, N.J.: Prentice-Hall, 1966.

WALD, G., "Biochemical Evolution," in *Trends in Physiology and Biochemistry*, E. S. G. Barron, ed., New York: Academic Press, 1952.

* Available in paperback.

Articles

CROW, J. F., "Ionizing Radiation and Evolution." *Scientific American*, **201**, 138, September, 1959.

DOBZHANSKY, T., "The Genetic Basis of Evolution." *Scientific American*, **182**, 32, January, 1950.

GLAESSNER, M. F., "Pre-Cambrian Animals." *Scientific American*, **204**, 72, March, 1961.

LACK, D., "Darwin's Finches." *Scientific American*, **188**, 166, April, 1953.

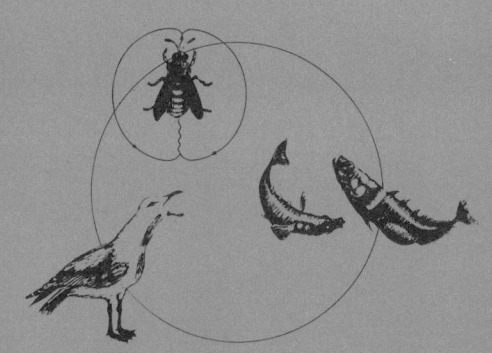

VIII

ANIMALS
IN THEIR
ENVIRONMENTS

27

ANIMAL BEHAVIOR

With special attention devoted to the vertebrates, let us turn to a discussion of **how animals do what they do**—that is, their **behavior**—a field sometimes called **ethology.** Ethology is such a vast subject, involving both psychology and the social sciences, that only those aspects of greatest interest to zoologists are dealt with here. Essentially all of the disciplines related to the biological sciences apply to animal behavior. The refined techniques of the biochemist, neurophysiologist, endocrinologist, ecologist, and many others are brought to bear on problems of behavior. While the study of the behavior of animals has a long history, it is only recently, with the advances and technical refinements of cognate sciences, that great strides have been made in this field. Behavioral information has become so important to our understanding of animals that it seems appropriate to attempt to consolidate it into a single chapter rather than interweave it in the chapters of the various animal groups.

It is extremely difficult to interpret what one sees in the activity of animals. We are so apt to describe the behavior of animals in anthropomorphic terms; that is, we attribute human characteristics to animals lower than man. This should be avoided because it is highly unlikely that any other animal has the same characteristics as man. However, it is difficult not to use some terms that have human implications because we are limited by our language which is replete with words that express human emotions and motivations.

A similar pitfall in attempting to describe the behavior of animals is the tendency to attach some purpose to their activity, such as: birds fly south in the fall because they do not want to perish from the rigors of the northern winters. Migration can be explained in terms of responses to stimuli resulting from the physiology of the bird in the fall of the year. It is neither necessary nor accurate to express the behavior in teleological (purposeful) terms. It is highly unlikely that the individual bird has foresight and anticipates what events will take place in the future. Through millions of years of evolution the behavior patterns built around all of the factors that cause migration have had survival value and consequently have been preserved and elaborated. Explanations must fit the facts of observation, and teleological explanations are unwarranted unless purpose can be demonstrated.

One method of studying behavior in any animal is to record it in great detail, employing whatever instrumentation is necessary. Some of these instruments are complicated and very expensive; others are simple and inexpensive. It is necessary in most studies to be able to recognize the individual animal from others which may look very much like it. Sometimes natural markings are adequate but more often this is not the case. If an individual can be identified without handling, so much the better because molestation of the animal may influence its behavior. Marking the animal may be necessary and there are many ways of doing this. Cellulose paint has been used satisfactorily in marking insects. The colors and the type of mark can be varied almost infinitely. Birds are commonly marked by colored leg-bands.

Cinematography is most helpful in analyzing behavior because movies can be observed again and again and in slow motion in order to decipher minute details. Tape recordings are important in studying sounds of animals. These can be converted to spectrograms in which the sounds are visualized on paper. This permits a high degree of accuracy when analyzing the various phases of the sound, particularly the songs of birds and insects.

Models that mimic some feature of the animal are often used effectively to determine just which factors elicit a response. Often some special feature such as a colored spot is the identifying characteristic that is recognized, whereas shape, size, etc., are ignored. Models of various shapes and colors

can detect this feature in the behavior of an animal. Varying intensities of light and lengths of exposure are useful in the study of rhythms.

It is best to study behavior under natural conditions, that is, in the animal's own environment. More critical aspects may be studied in the laboratory but one must always be conscious of the artificiality of laboratory conditions and interpret his findings in that light.

ORIENTATION

Animals orient themselves in their environment as a result of external stimuli coming to them from many sources. Some stimuli are simple and the response seems to be straightforward. This may be true in the simplest of animals (protozoa) but other factors, discussed later, are involved in metazoa, and particularly in the higher forms.

The study of orientation in animals has resulted in a system of classification which is convenient though it does not necessarily explain how the behavior occurs. Animals which move but do not orient in response to a stimulus are exhibiting a kind of behavior called **kinesis.** Kineses can be demonstrated in choice chambers which are round containers, divided in half (Fig. 27–1). To study reactions to light, one-half of the floor can be covered with black paper, the other with white paper. For humidity studies a dehydrating agent (H_2SO_4) is placed beneath the floor. Other choices, such as those resulting from olfactory stimulants, can be used. Under experimental conditions the movements of animals can be observed and recorded.

Sowbugs, which normally live in a damp place, when placed in a choice chamber of high and low humidity, at first, move rapidly over the entire area. Later they begin to slow down in the moist half of the chamber (Fig. 27–1) and, finally, they stop and congregate in the region with the most favorable humidity. Thus, there is an inverse correlation be-

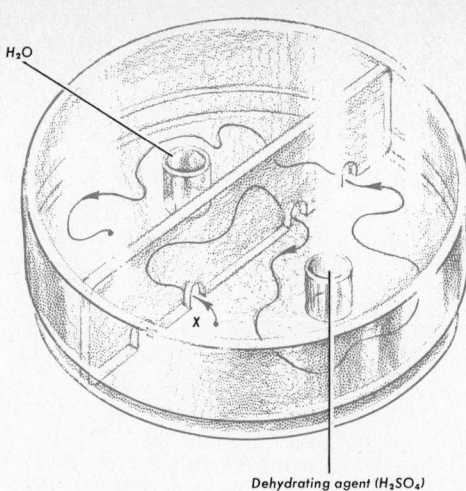

H₂O

X

Dehydrating agent (H₂SO₄)

Fig. 27-1. A choice chamber designed for experiments with relative humidity. The humidity is low on the right side and high on the left. A sowbug, released at X, moves through a circuitous path, passing from one side to the other but finally stops on the side with high relative humidity.

tween the speed of the sowbugs and the relative humidity, and this accounts for their eventual distribution. This type of behavior is called **orthokinesis.**

Another category is **klinokinesis.** Some animals distribute themselves in favorable environments by changing the frequency or amount of turning. Planarians, which normally avoid bright light, when placed in a choice chamber, one-half of which is dark and the other half light, will move into the most favorable illumination. They do this by turning more frequently when in the light area and less frequently in the dark region. By frequently changing direction the animal is more likely to reach a favorable environment. However, if its frequent turning keeps it in the bright area, after a time it begins moving in a straight line, until a favorable light intensity is reached.

Other types of orientation behavior called **taxes** (singular, **taxis**) involve movement directly toward the stimulus. A prefix can be added to describe oriented movements toward particular kinds of stimuli, such as, **phototaxis** (light), **chemotaxis** (chemicals), etc. The best examples of this type of orientation are **phototaxes.** Using planaria again, if an animal is given a choice of responding to two equally bright lights placed at equal distances away, it moves toward a point midway between them. It orients so that the light impinges on both eyes with equal intensity. Similarly, grayling butterflies fly directly toward the sun when pursued by predators. They orient so that they receive equal illumination in each eye. The bright light partially blinds its pursuer, thus giving the butterfly a better chance to escape. If one eye is blinded, the animal flies in circles because it no longer receives equal stimulation in both eyes.

Jacques Loeb, a noted American physiologist, in 1918 explained this orientation quite simply. He attributed such responses to quantitative differences in the stimulation of the bilateral photoreceptors and effectors (muscles). The eye which receives brightest light sends more impulses to the leg muscles on the same side causing them to exert greater tension, thus turning the head and body of the animal toward the light. Both eyes then receive equal illumination. Some years later, Mast (1938) presented evidence that the processes of orientation differ among the various groups of animals and cannot be explained by Loeb's simple model. Among some insects, such as dragonflies, if one eye is blinded the animal can still pursue and capture insects. If the legs of a fly (*Eristalis*) are removed on one side, it walks toward a light, moving in a sideways manner. Thus, if the insect is prevented from approaching the light in the usual manner, it compensates by changing its responses and gets there in another way. Orientation abilities are obviously adaptive, and many different orienting mechanisms are involved, differing among the various animal groups.

In spite of much experimental work on simple orientations, we are still a long way from explaining completely how various animals find their way about. It is a well-known fact that pets such as cats and dogs return home from great distances. Even lower animals such as certain insects return to their homes after wandering, often many miles.

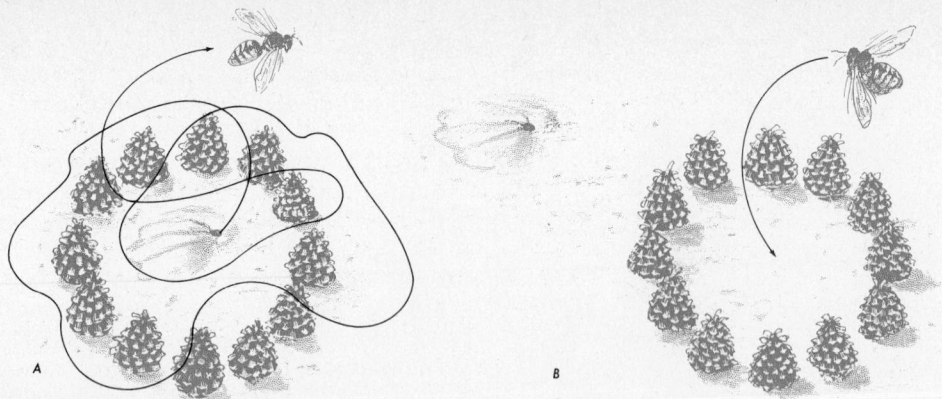

Fig. 27–2. An experiment to show that the solitary female digger wasp locates its nest utilizing landmarks. (A) Fir cones are placed around the nest entrance while she is inside; upon emerging she flies around the nest a few times then leaves. (B) The cones are then moved a short distance to one side. When she returns she flies to the center of the circle of cones and searches for her nest.

How this is done is fairly well understood in some animals, but not in others.

Landmarks are utilized by animals, from insects to mammals, as guideposts. For example, the female bee-killer wasp, *Philanthus triangulum*, digs a hole in the ground where she places a bee she has paralyzed with her sting. She deposits an egg on the bee which serves as food for her developing young. Before leaving the nest on her next sortie, she circles it for about 30 seconds, apparently getting her bearings. If the observer places a ring of objects around the nest while she is in her nest and these are moved to a new location nearby when she is away, she returns and searches in the middle of the circle of objects (Fig. 27–2). She does the same if the objects are changed from pine cones to blocks of wood. Obviously the configuration of the design, in this case a circle, is recognized rather than the objects themselves. If the number of objects is reduced by one-half (same diameter), the wasp still searches in the center of them for her nest.

The migration of birds over thousands of miles from their place of hatching in the north to their southern wintering grounds has amazed naturalists for centuries. The routes taken are characteristic for the species and cannot be accounted for by random wandering. Some species even make nonstop flights over several thousand miles. The golden plovers are good examples. The adult Pacific coast form, which cannot land on water, each fall flies from Alaska to the Hawaiian and Marquesas Islands where it overwinters and returns the next spring going over the same route (Fig. 27–3). It traverses the open Pacific Ocean where there are no landmarks which could be used as guideposts. The Atlantic population nests in northern Canada and in the fall takes an easterly route flying over several thousand miles of the open Atlantic Ocean to its winter range in central South America. The return trip the next spring is mostly over land. The Arctic tern follows a very circuitous route in its migrations, which are the longest of any bird known. Its nesting range is very extensive in the circumpolar regions of the earth. It follows several migration routes to its southern range in Antarctica which is 11,000 miles away, but owing to its circuitous path, it probably flies over 25,000 miles each year. Somehow the map of the route which has been followed by these birds for centuries is engraved in their nervous systems which makes it possible for each generation to follow these very precise routes.

Many birds also have remarkable homing abilities, a fact recognized by pigeon fanciers. Homing pigeons will return to their home lofts when released as far as 600 miles away, averaging 50 or 60 miles per hour. One of the most remarkable cases is that of the manx shearwater, a marine fish-eating bird, which was taken from England to Boston by plane where it was released. Twelve days later it was back home after flying 3000 miles over the Atlantic Ocean. This type of navigation differs from seasonal migration, although the ability to determine direction may be functionally the same.

There are several theories to account for orientational abilities of birds, a promising one being celestial navigation. With the approach of the time of migration, caged starlings fly to the side of the cage that is in the direction of their normal migratory route, providing they can see the sun. In cloudy weather they lose their orientation. Moreover, if the position of the sun is changed with mirrors, they change their direction of flight accordingly.

Many birds migrate at night and there is some evidence that they use star patterns for orientation. European warblers reared in large cages with glass tops fly in the direction of their normal migration routes. If the birds cannot see the sky they are completely disoriented. The most convincing evidence that birds can orient by star patterns is from studies of the indigo bunting. When placed in a planetarium during the fall migration period, the birds fly in a southerly direction according to the star patterns even though the sky has been altered so that it is not true south.

It seems that birds can orient by celestial signals, which partly accounts for their migration patterns. What stimuli they use to find their way home, as in the case of the manx shearwater, remains a mystery.

Other environmental cues, such as the earth's magnetic field, have been studied intensely but so far the results indicate that birds cannot detect gradients in the magnetic field of the earth. However, recent experiments show that tiny magnets placed around the head of homing pigeons indicate that under overcast skies they often lose their orientation. They seem to respond to the magnetic field, contrary to earlier experiments.

Animals may also utilize odors for orientation, that is, they follow a trail of scent laid down by another of their species or some other species. Almost every object has a

Fig. 27–3. The migration routes of the golden plover and Arctic tern are indicated on this map. The Pacific golden plover nests in Alaska and in the fall flies over the open ocean to its winter home on several southern islands, returning the next spring. The Atlantic population takes an easterly route over great stretches of ocean to its winter home in central South America, but returns north the next spring by another route which is mostly over land. The Arctic terns nest in their southern range in the Antarctic, a distance of 11,000 miles. They probably return over similar routes.

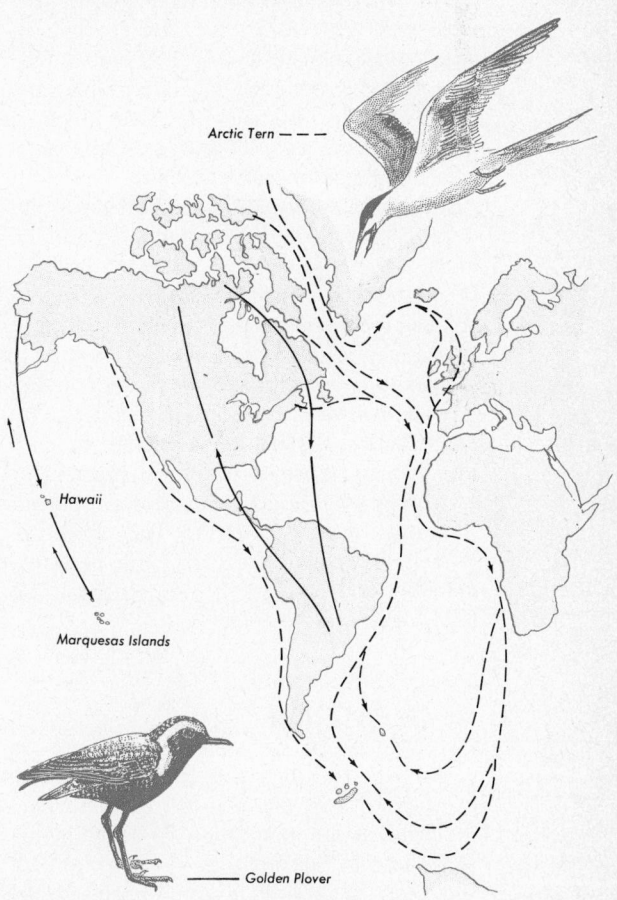

Arctic Tern — — —

Hawaii

Marquesas Islands

———— Golden Plover

specific odor, though humans cannot detect differences in many cases. Many other animals, however, have evolved highly refined olfactory organs which give them greater awareness of odors in their environment. Excellent examples of this occur among the insects where some such as the male silkworm moth can detect the odor of the female two or more miles away. Some species of ants (leaf cutting, p. 230) lay a chemical trail that is followed by others of the same species. We have more to say about these substances (pheromones) later.

In recent years another most spectacular way in which animals utilize chemical signals for orientation and navigation was discovered in fish. It has been known for a long time that salmon return to spawn in the same stream in which they hatched several years earlier. The young salmon swims down the fresh water stream to the ocean where it grows to adulthood. During this time it forages great distances, but when spawning time comes, it finds the exact stream in which it hatched and makes its way to the source where the eggs are laid. Of the thousands of streams that empty into the ocean, it is indeed remarkable that the salmon can find its home stream. The answer seems to be that it possesses remarkably sensitive olfactory organs (chemoreceptors). This has been studied by a number of investigators. Destroying their olfactory organs causes salmon, as well as other fish, to lose the capacity to detect odors, making them unable to find the proper spawning stream. Apparently, each stream has a characteristic spectrum of chemicals dissolved in it. The receptor-response system of the young fish becomes indelibly impressed with this spectrum of odors, and it never forgets.

COMMUNICATION

Probably all but the simplest animals communicate in one way or another. The evolution of communication has taken many paths so that the methods are highly varied and some are bizarre. The most remarkable

occur among social animals, although those that lead solitary lives most of the time do communicate at certain times of the year—particularly during the breeding season. The aspects of the physical environment that are most commonly used in communication are chemicals (smell), sound (hearing), and light (sight), although others may be used that we are not aware of.

Chemical Communication

Chemical cues are utilized by many animals in communication between members of the same species. Substances are secreted externally which are recognized by all members of the species. These are called **pheromones** and they may function much like hormones except they are secreted to the outside of the body. When released by one member of a species, a pheromone may influence the behavior of another of the same species in two ways. It may initiate an immediate response in the recipient or it may condition the recipient so that at some later time its behavior will be influenced by certain environmental stimuli.

Many workers in this field have concentrated on pheromones in insects, where they seem to be almost universal. Some species of ants, leaf cutting ants, for example, release a pheromone from the anus on their return from a successful foraging venture. This provides a trail for other members of the species to follow to the food source. The trail is not laid if no food is found and once the food is consumed and ants stop making trips, the trail soon volatilizes thus preventing others from making a futile trip. The pheromone, being species-specific, assures that only members of one species recognize the odor.

The so-called sex-attractants (pheromones) are prevalent among insects as well as other animals. The well-known attraction of male moths (silkworm, cecropia, etc.) to the female over great distances (two or more miles) has fascinated biologists and school children alike. It is now known that the female releases powerful pheromones in minute quantities which the male detects in fantastically great dilution. He responds only to the sex attractant and not to visual stimuli. If the female is shielded from his vision he still moves toward her, providing the pheromone is present in the air. If he can see her in a closed glass container, he loses all interest in her.

Mammals also release pheromones that have profound effects on their behavior. Dogs are notorious for staking out their territory by releasing a small amount of urine on trees and other convenient objects. The intense interest shown by a second dog sniffing at these objects must mean that he obtains information from this experience. Monkeys when released in a new cage gently touch their anal region to various areas of the enclosure. Pheromones are apparently left on these areas which will be recognized by other monkeys. In nature, this act may have significance in denoting the home territory of the monkey. Female mammals when in heat release pheromones that are recognized by the male at some distances. The importance of this behavior has obvious survival value, particularly in those mammals that are solitary most of the time.

Pheromones that have a delayed action in the behavior of the recipient are common among insects. For example, population density of bees seems to be controlled through a system of pheromones. The workers regulate the production of new queens by altering the food of the larvae. So long as there is a healthy queen in the hive and the population is not too great, the workers will not provide the proper diet to induce a new queen. If the colony is separated in half by a gauze partition, the queen is in contact with only those workers on her side. After a while the workers on the other side will build queen cells indicating that they are not receiving the inhibiting stimulus from the queen. It is known that the queen releases a pheromone that is spread over her body when she cleans herself. As the workers tend her they lap up the pheromone which is then passed from one to another during a peculiar kind of food exchange system. The workers make oral contact and by a process of regurgitation and feeding, materials are exchanged, including the pheromone from the queen. Thus, the workers are constantly aware of a healthy queen in their midst. In the separated hive, the workers on the queenless side do not receive the hormone and respond by making preparations for producing a new queen. This pheromone apparently is absorbed through the digestive tract and has a physiological effect which alters the workers' behavior. However, the same substance acts externally as a pheromone because it attracts drones to the virgin queen during her nuptial flight.

Mammals also produce long-term pheromones which help to regulate population densities. Under crowded conditions some mammals have a decreased birth rate. For example, if female mice are confined to close quarters and overcrowded, estrus is blocked. If the olfactory organs are removed, the cycle returns to normal. Apparently, pheromones in high concentration, as they would be under these crowded conditions, influence the breeding capabilities of mice.

Sound Communication

Sound is perhaps the most obvious, to us at least, means of communication in animals. Sounds vary widely, from the mechanism producing the sound to the receptors to the interpretation of the sounds. Sounds have evolved to different extents in different groups. In some groups, particularly insects, sounds have become so characteristic that it is sometimes easier to distinguish between species by slight variations in their song patterns than from their morphology.

Male mosquitoes recognize the presence of the female by being able to receive the sounds set up by the fluttering of her wings. When a male is sexually mature, his antennae are fringed with numerous extensions resembling hairs. These are sensitive to sound vibrations resulting from the beating of the female's wings. When he senses these pulses he flies directly to the female and mates with her. It is quite remarkable that the male mosquito can detect, among the thousands of other sound signals coming through the air, the pattern that is characteristic of his own species and no other.

Among vertebrates sound communication is well developed, indeed, in one of them, man, it is much more complex than that of any other animal. Even the most advanced primates cannot be taught to speak beyond a very few simple words.

Many ethologists have studied the acoustical behavior of amphibians, particularly frogs. The familiar croaking of male frogs in early spring is pleasant to our ears, but it has more meaning to the female. Shortly after coming out of hibernation male frogs begin emitting their call which is strikingly characteristic for each species. This sound attracts the females which move closer to him, providing they are ready to lay their eggs. Immature females ignore the call. Amplexus (p. 257) follows, during which the female lays her eggs and the male fertilizes them.

The songs of birds have been analyzed more than any other animal sounds. Whereas most birds make some sort of sound, there is wide variation in their ability to sing. What does the song mean to the bird? Singing is particularly pronounced preceding the nesting period, hence it must be associated with sexual behavior. But it has other meanings as well. It is now generally agreed that bird songs function in maintaining a territory, as a display for recognizing others of the same species, and for attracting females. It induces synchronization of the sexual cycles in both sexes by stimulating the secretion of sex hormones.

Mammals, exclusive of man, produce a variety of sounds that are identified by other members of the species. The growl, whine, and bark of the dog communicate its emotions. We are familiar with these signs and can interpret them owing to our familiarity with the animal. Wild mammals have a similar spectrum of sounds which function in communication. In recent decades one group, the porpoises (small whales) have been carefully studied and some remarkable results have been obtained. These are intelligent creatures, certainly approaching the

brightest land animals, and they produce a wide range of sounds that convey many types of information. Most of our information comes from experiments performed in large tanks of sea water where electronic equipment is installed which records the sounds. The porpoise has an extensive "vocabulary" of whistles, grunts, squeals, and clicking sounds. These were first identified during the World War II when equipment became sufficiently refined to record them. Whereas many of these sounds are used for communication between members of their own species, others function in locating objects in the water. This is called **echolocation** or **sonar.** This is not a form of communication in the strict sense because it involves only one individual. Echolocation is extremely efficient in porpoises because it can not only enable location of the smallest object in total darkness, but also recognition of food (a fifteen-centimeter fish) at considerable distance. Frequent high-pitched whistling sounds are emitted and the echoes detected. The vocal apparatus is designed to make these high-pitched sounds and the ears can receive vibrations over a wide range (150–150,000 vibrations per second), far beyond those to which we are sensitive. Porpoises can swim through a maze in very murky water and avoid all objects. If clear plastic barriers are placed in the tank, they never touch them. If blind-folded, they can navigate a complex maze to find their food. This system for navigating and locating objects, together with the ability to communicate with one another, requires a highly developed nervous system. This is true of porpoises. Their cerebral hemispheres equal ours in size and complexity and the brain areas where sounds are recorded rival ours. Moreover, the auditory equipment is even superior. These animals seem to the mental giants of the seas, as man is of the land.

Echolocation is utilized by many aquatic forms, including invertebrates such as arthropods. Among land forms, the bats possess remarkable sonar equipment about which we have a good deal of information because they have been studied more than any other mammal. The first careful investigations were made by the Italian, L. Spallanzani, in 1793. He observed that bats flew through totally dark caves without colliding with one another or with the walls, yet they were absolutely silent. He then set up experiments to prove how they did this. His first thought was that they had extremely keen night vision, but when he blinded them they flew about the dark room as efficiently as bats with functional eyes. However, if he covered their heads with hoods or plugged their ears they could not navigate in the dark. He concluded that sound was somehow used in navigation, but since they were silent (to Spallanzani) the secret was left unsolved for nearly 150 years when Professor Donald R. Griffin of Rockefeller University discovered the answer to this riddle. By employing highly sensitive sound-recording equipment, he discovered that bats do indeed emit sounds but because their high frequency is beyond the limits of our ears, we cannot hear them. Like the porpoises, they use sonar, this time in air, for detecting objects, whether it be food (insects) or obstacles in their path. They fly rapidly in a dark room never touching wires 0.5 mm in diameter strung one wing-span apart. They are able to capture insects in flight, employing sonar only.

The bat has a typical mammalian larynx, but it is highly developed for producing sounds at extremely high frequencies (Fig. 27–4). We are sensitive to sounds not exceeding 20,000 cycles per second, whereas some bats recognize sounds of frequencies over 200,000. Likewise, the sound-receiving apparatus is specialized to record these high-pitched sounds. The auditory regions of the brain are also highly developed.

The bat's refined hearing is utilized to detect echoes of its own voice. The emitted sound bounces off objects and returns to its ears where it is conducted to the acoustic center in the brain. It is here that it recognizes its own voice. Experiments have shown that even when background sound (noise) of all frequencies is 30,000 times the intensities of the echoes it still is able to avoid tiny wires and can locate insects. It obviously disregards all other sounds except its own—in other words, it can concentrate on one particular sound just as we can, in the midst of meaningless sound signals.

Bats have been pursuing and capturing insects for a very long time and it is interesting to note that some insects have evolved sensing equipment to detect the bat. For example, the night-flying owlet moths have a pair of ears located laterally between the abdomen and the thorax which are sensitive to sounds emitted by the bat. The moth can detect the bat's chirp at a distance of 20 feet vertically and 100 feet horizontally. Upon hearing these sounds, the moth immediately flies in a zigzag manner, greatly increasing its chance of evading its pursuer. Some moths even produce high-frequency sounds that may "jam" the sonar equipment of the bat, thus decreasing its chances of capturing the moth. Thus, we see that as the predator improves its hunting techniques, the prey increases the efficiency of its escaping devices. Such selective pressure on prey and predator have resulted in highly refined sensing devices, many of which we do not understand.

COURTSHIP

The behavior of animals during courtship has attracted a great deal of attention from ethologists. This type of behavior is usually species-specific and involves activities that are striking and easily observed. Courtship involves mutual responses and results in mat-

Fig. 27–4. A bat emitting high-pitched sonic waves while in flight.

ing. The partners must first find each other, which demands that certain signals (sounds, odors, display patterns, etc.) be sufficiently obvious to be recognized at a distance. These are usually species-specific also so that mating between different species does not occur. Through selection these have often become very elaborate, varying only in subtle details between closely related species. The slight variations, however, are adequate to separate the species. Among the various species of *Drosophila*, for example, courtship patterns have certain elements in common but the way they are carried out identifies each species. The behavior pattern during courtship in *D. melanogaster* illustrates the basic pattern for the genus.

By placing a pair of *Drosophila* in a small transparent box, one can observe their behavior during courtship. After the third day from emergence the courtship begins; earlier than this period the female is unreceptive and nothing happens. The male moves toward the female and gently touches her with the tarsi on his first pair of legs. He then faces her side, at the same time raising and lowering his wings. One wing is then extended and fluttered which creates a flow of air stimulating the antennae of the female. During this maneuver he licks the female's abdomen. The female then opens her genital plates, if she is receptive, and the male mounts and copulates with her. These are the elements of courtship behavior in the genus *Drosophila* but there are many variations among the numerous species.

Animals that travel by themselves most of the year, avoiding the opposite sex, must reverse this behavior during the breeding season. Hormones play an important role in initiating courtship in such animals. Once they reach a certain level in the animal, courtship begins, but not otherwise, even if the two sexes are forced to live in close contact with one another.

One of the best known examples of courtship in vertebrates occurs in the three-spined stickleback fish, described by N. Tinbergen of England. In the spring the mature male is bright red underneath and the female is drab with a distended abdomen caused by the contained eggs (Fig. 27–5). The male builds an elaborate nest which consists of a chamber with openings at both ends. He remains near the nest until a female approaches. The sight of her excites him to initiate a sequential series of signals that then occur alternately in the male and female, resulting finally in fertilization of the eggs. He first performs a zigzag "dance" which consists of a series of short dashes around her. She responds by curving her body dorsally exposing her swollen abdomen. The male then swims down to his nest, leading her. She will not follow if she is not sexually mature. If she follows him to the nest he points his nose toward one of the openings. The female then pushes past him and enters the nest which is just large enough to accommodate her body with head and tail protruding. The male then nudges her in the tail region; this signal stimulates the female to release her eggs. When she finishes she swims off never to bother any further about caring for the eggs or young. The male immediately enters the nest and fertilizes the eggs. From this time on he has the responsibility of caring for the eggs during their incubation. Here we see a series of signals and responses by both sexes which culminates in the fertilization of the eggs. Such elaborate behavior has come about by natural selection through millions of years of evolution.

Birds are famous for their behavior during courtship. They use both sound (song) and visual signals (displays). The bird's song functions as a warning to rival males and as an attraction for females. Before courtship actually begins the male marks out a **territory** which is an area that is defended against other males of the same species. Other species are usually permitted to enter or leave the territory without molestation. This behavior can be readily observed in robins. In the early spring the male picks out an area and from a perch located somewhere in the territory, he sings a specific song. Other male robins entering the area are immediately attacked and driven off. The song advertises that the territory is occupied and that cruising males had better look elsewhere. It also has the function of inviting the female to share the territory. During the early stages of staking-out territories, the boundaries are not well defined, but later as competition increases and nesting begins, they become firmly fixed. The extent of the territory is often determined by the vigor of the male's song. The loudest singers usually occupy the largest territories, indeed, they may even extend them into the areas of weaker singers.

The function of the territory varies with different birds. It may serve as a source of food as well as an area for breeding. Robins and eagles, for example, obtain most of their food from their territories whereas sea birds get their food from the sea and consequently require only a very small area for breeding activities. In the first cases, the territory may function to reserve feeding space or good nest sites, and keep other males away. In the second case, small territories encourage concentration of great numbers of birds which has the advantage of protection against predators.

All birds respond much the same when an intruder enters their territory. The owner takes on an aggressive attitude, first moving toward the intruder, which is usually sufficient to put him to flight. If the intruder does not leave, the owner runs toward him with head lowered. Instead of attacking, he passes the intruder and promptly raises his head. This aggressive act usually prompts the intruder to leave the territory. The intruder, being in foreign territory, usually adopts a submissive attitude and retreats without a fight. Birds, as well as other animals, rarely make bodily contact and thus do not injure or kill one another. The survival value for the species of this behavior is obvious.

Among some mammals territorialism is pronounced. All of us have seen a dog defend the home of his master against the invasion of another dog of considerably greater size. The moment the property boundary is

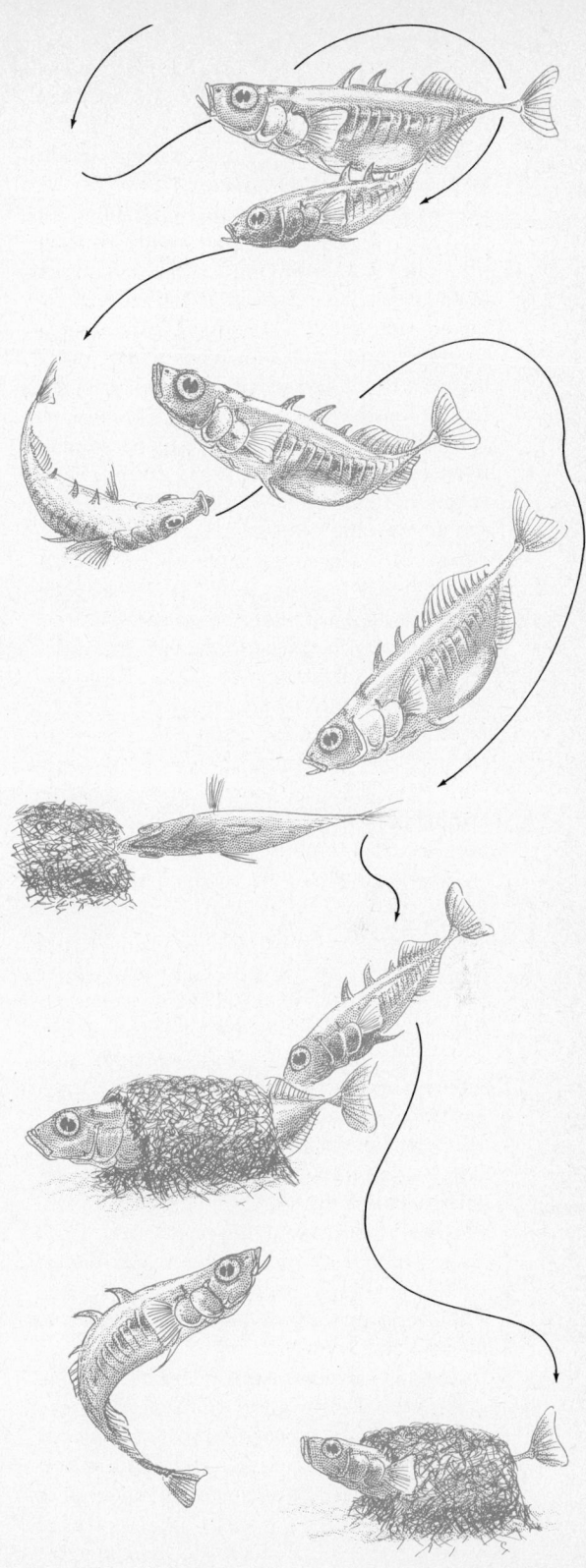

Fig. 27–5. The various stages in the courtship of a pair of three-spined sticklebacks. See text for explanation.

approached the size difference between the two animals becomes the deciding factor in reference to any further encounters. The larger dog respects the home territory of the first dog and gives ground; however, on neutral territory all dogs fight members of their own species to the extent of their physical capabilities. Once within his own boundaries a dog, as well as most other animals, defends his territory with super strength and determination. Just what happens to an animal physiologically, under these conditions, is not very clear. Since the reaction occurs so universally among animals it seems quite clear that it has great survival value and must have been selected for throughout the evolution of each species. Those with the most aggressive nature regarding their territory would be most apt to maintain an area sufficient to provide food for the mates themselves, together with their subsequent offspring. The less aggressive would be unlikely, or less likely, to hold their feeding area in the first place and if they did any offspring that might be produced would have less chance of survival. Consequently, throughout time the tendency for aggressiveness has had high selective value so that we see it displayed in many, but not all, mammals.

One of the more interesting cases of territorialism occurs among the Uganda (Africa) kob, a small beautiful antelope. Professor H. K. Buechner describes the special type of territorialism centered around the breeding behavior of these animals. Each population of kob, consisting of from 800 to 1000 animals, has a special region, called the stamping ground, where the dominant males gather. This is the breeding arena and it is in continuous operation the year round. Each male "stakes out" his territory, consisting of an area about 50 feet in diameter, which he defends against all other males. A dozen to fifteen males succeed in possessing such territories; all others are kept in the background either by fear or by a lack of interest. Consequently, each new generation is a product of only a few aggressive males; thus aggressiveness is retained in the genetic makeup of

Fig. 27–6. Some of the courtship displays of the male mallard duck; (1) bill shake; (2) head flick; (3) grunt whistle; (4) head-up–tail-up; and (5) down-up.

these animals and is selected for generation after generation.

Each male fights for his stamping ground and retains it with all his might until he moves out for food or water. It is then taken over by another male. Upon the return of the original owner, a fight ensues and the winner retains the stamping ground. A hierarchy exists also among the stamping grounds; that is, the most dominant male occupies the most desirable stamping ground which is the one on elevated ground usually near the center. This one has a view!

The females, when in estrus, move into the breeding area and make directly for the one occupied by the most dominant male, and accept any lesser area with reluctance. Once in the stamping ground the male goes through a simple display of his neck colors and then attempts to mount the female. If he is not successful she may wander into the stamping ground of another male. Once this happens, the previous male loses all interest in her and returns contentedly to feeding within his own territory. The important point in this observation is that the male is maintaining his own territory and seems to have little interest in the female, until she enters his territory. Males evidently compete for females by competing for real estate. They never fight dangerously for their territory, that is, they gesture with sternness but have very little actual contact, which has the advantage of preserving the species. When off the stamping ground the males display no antagonism toward one another. Copulation is also very rare off the stamping grounds. Here then we see attachment to a piece of ground, an instinctive trait, which has the secondary effect of stimulating sexual desire in both males and females.

K. Z. Lorenz of Bavaria has described courtship in many birds, among them the mallard duck (Fig. 27–6). Shortly following their return to northern waters in the spring mallards congregate, along with many other species of ducks, on ponds where they begin their courtship. The male periodically raises his wings, head, tail, and emits a sharp whistle, then settles back in the water. At other times he dips up some water in his bill, twists his head, quickly tossing it to one side. This act is immediately followed by his rising in the water, arching his neck, and emitting a whistle and a grunt. A variation on this pattern is to raise his rear end as high as possible and whistle. This courtship behavior has a profound effect on the female as well as the male. It stimulates the production of sexual hormones in both sexes so that fruitful mating follows. Moreover, as the male goes through his dances, certain identifying markings, usually colored patches, are conspicuously displayed so that they can be no mistake about the species, thus preventing hybridization.

According to N. Tinbergen nearly all species of gulls show eight different postures and movements during courtship (Fig.

Fig. 27–7. Display postures of the herring gull during courtship: (1) oblique long call; (2) choking (up and down movement of the head); (3) facing away; and (4) upright. These postures are similar in closely related species as shown by the examples of the long call: (A) grea skua; (B) herring gull; (C) kittiwake; and (D) black-headed gull.

27–7). These are so precise, though modified among the different species, that they can be useful in taxonomy.

Most animals, including man, experience emotional conflicts when threatened and their behavior is remarkably similar. When an intruder moves into the home territory of another animal of the same species, the latter is simultaneously stimulated by two drives, one to attack and the other to flee. These two drives are registered in its behavior. For example, when a pair of ducks move into the territory of another pair, the female in her home territory attacks the couple by running at them in an aggressive manner. When she gets close to them she immediately retreats to the male for protection. Her escape reaction suddenly dominates aggressive behavior. She is torn between fighting and fleeing. This creates strong activation of two antagonistic drives and causes tension. Under these conditions so-called **displacement activities** appear which are registered in behavior that is associated with other activities of the animal, such as those engaged in during courtship. Many examples are known (Fig. 27–8). When a gull's territory is threatened, it begins vigorously pulling grass which is the type of activity the gull undergoes during nest building. This seems to serve as an outlet for the tensions incurred by the encounter. Avocets respond by going to sleep, while domestic roosters peck vigorously at imag-

inary food during brief intervals between attacks. Even man shows displacement activities when angry. He may scratch his head or clench his fists when unable to express his feelings directly. Or he may even pound the desk with his fists to relieve the tension built up through conflict. Displacement activities seem to be basic responses to tense situations which are registered in specific behavior of animals. Moreover they are inherited. If ducks with slightly different displacement activities are hybridized, the offspring exhibits combinations of both types, indicating that the behavior patterns are inherited.

Whereas the courtship patterns mentioned so far involve primarily visual displays and sounds, certainly other factors such as chemicals (mentioned earlier), tactile stimuli, and perhaps others we are not aware of, also function. These signals, sometimes called **releasers**, function as a battery of stimuli which evoke the final mating act.

SIGN STIMULI

Animals respond to only a few of the many stimuli they receive from their environment. Through evolution, response to these special stimuli has become a part of the animal's make-up so that it reacts instinctively to them. In other words, the response to certain stimuli is genetically fixed. The stim-

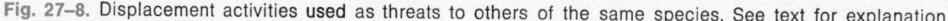

Fig. 27–8. Displacement activities **used** as threats to others of the same species. See text for explanation.

uli that are effective in initiating this behavior are referred to by ethologists as **sign stimuli** and the factors (sounds, actions, structures, etc.) that produce sign stimuli are called **releasers**. The animal itself possesses **releasing mechanisms** (neural, hormonal) that are selectively sensitive to sign stimuli coming to it. To sum up then, the releasers in the animal's environment produce sign stimuli to which the animal is sensitive. The releasing mechanisms then function to bring about the behavior pattern that is observed. It is not as simple as it sounds because other factors such as motivation, previous learning experience, and recent sensory stimulation may alter the behavior pattern. Let us consider some experiments that have been performed by a number of biologists to analyze behavior. Much valuable information has been gained by exposing animals to models which emphasize certain characteristics such as shape, color, size, etc., and noting their response. It often turns out that the animals respond to only one or two cues (sign stimuli) and ignore all others. For example, Tinbergen showed that it was the red belly of the male stickleback that stimulated other males to attack (Fig. 27–9). He did this by exposing males to models that resemble the fish closely but without the red belly and to others that had very unfishlike anatomy but with red undersides (Fig. 27–9). The males always attacked those with red coloration and ignored the others. This demonstrates that the sign stimulus came only from the red coloration.

Another example of sign stimuli is the red spot near the end of the yellow beak of the harbor gull, *Larus* (Fig. 27–9). The significance of this red spot became apparent when

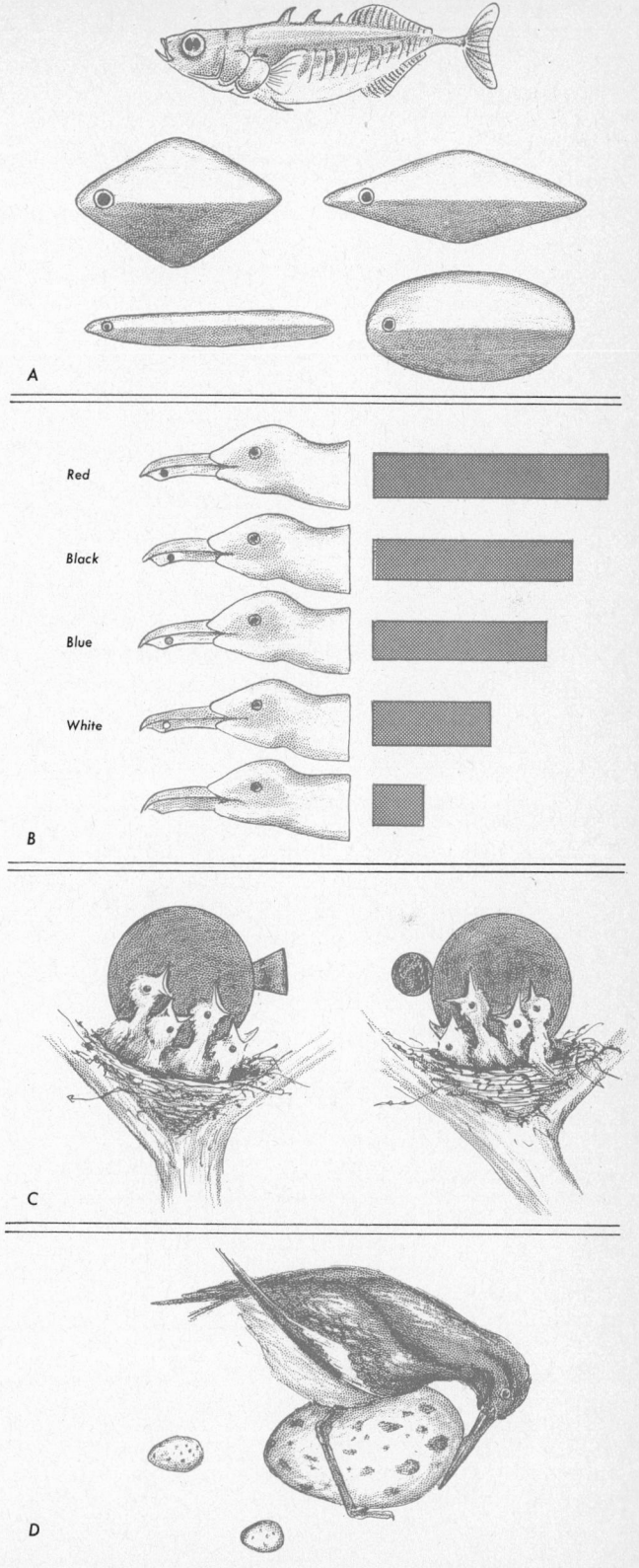

Fig. 27–9. Several sign stimuli. (A) Models of Sticklebacks. One is identical to the male fish but without a red belly. The others are unfishlike but have red undersides. (B) Gull beak models with different colored spots and no spot. The bars at the right indicate the relative frequency of responses of the newly hatched young. (C) Nestling thrushes responding to model parents with different shaped "heads." (D) Oystercatcher trying to incubate an oversized egg, even though its own egg lies close by. It also ignores the slightly larger than its own but similarly marked egg of the herring gull.

nestling gulls were presented with model bills which had spots painted different colors or without a spot at all. The young gulls under normal conditions peck at the red spot when begging for food. They were completely puzzled by the lack of a spot or colored spots (other than red) on the models and failed to peck at them. The red spot is the sign stimulus for gulls when they are nestlings and hungry.

Another example is the response to nestling thrushes to sign stimuli. When the young hatch, they are blind. They respond instantly when the nest is touched by opening their mouths and stretching their heads straight upward. When their eyes open, the head is directed toward the head of the parent sitting on the edge of the nest. When models made of disks of about the same size as the adult were presented to the young birds, they always pointed their open mouths toward the "head" of the model (Fig. 27–9). It did not matter what the configuration of the "head" was. The sign stimulus was the over-all design, that is, a large "body" and a small "head." Other experiments demonstrated that the sign stimulus was the relative size of the body and head of the simulated parent. When the model had two "heads," one large and the other small (about the same as that of the parent in relation to the body), the young pointed their heads toward the latter. When the "body" of the model was doubled in size and the two "heads" remained the same as before, the young pointed at the larger "head." Obviously it is the ratio of head to body size that constitutes the sign stimulus.

When the male English robin in its own territory is presented with a bundle of red feathers and a stuffed male robin with its breast painted gray, it attacks the red feather model and ignores the stuffed robin. The red color is the sign stimulus which precipitates the attack. However, this occurs only in the robin's territory during the nesting season; in other places and at other times of the year no such response is evoked by the model. The releasing mechanism operates only under specific conditions of time, space, and hormonal level.

Tinbergen and others have shown that it is possible to accentuate the response of an animal to an object by exaggerating the sign stimuli. The female oystercatcher, for example, will attempt to incubate an egg that has markings like her own but is much larger, even when her own eggs are close by (Fig. 27–9).

SOCIAL ANIMALS

Animals that live in permanent groups have evolved special behavior patterns that insure the cohesion of the group. However, there is wide variation in the degree to which animals form groups, from those that aggregate periodically for some external reason, such as snakes in winter, to those where the young remain with the parents forming a true society. Even among those that aggregate for short periods of time, certain stimuli are responsible for keeping them in a group. For example, the individuals in a swarm of grasshoppers move back into the group if they accidentally wander outside the swarm. Apparently, they react positively to the sound and visual stimuli of the beating wings of members of the swarm. Woodlice aggregate under the bark of trees initially as a mutual response to the relative humidity. Recently it has been shown that the individuals give off an odor which attracts them to one another. There are gradations from this type of aggregation to the so-called social insects which have evolved a very compact organization based on morphologically and physiologically differentiated members of the group (castes), each performing specific functions. This type of social life has reached its peak among hymenopteran insects, and it is best known in the honeybees. Karl von Frisch of Germany initiated a long series of experiments with honeybees, the results of which have become the basis for much of what we know about behavior in these social insects.

As we learned earlier (p. 228), castes of honeybees differ in morphology and physiology. The worker has anatomical features that make pollen-collecting efficient. The queen does not have these features but has a gigantic egg-producing apparatus. The behavior of each is characteristic of the function it performs in the colony. The worker does several jobs in a sequential order during her life. Her first assignment is to work in the hive where she cleans cells for three days. Then she feeds the larvae (4–10 days) during which time her nursing glands are highly developed. From the tenth to the sixteenth day she builds combs. During this time her wax glands reach their greatest development. For the next few days she receives nectar and pollen from other workers and stores them. For about one day (20th) she stands guard and from this time on to the end of her life, she forages for food. This is the usual pattern, but it is not absolutely rigid. The behavior of the worker is thus correlated with its morphological and physiological state, all for the successful maintenance of the colony as a whole.

Since all members of the hive share the same food, and this is exchanged among individuals (p. 228), they all acquire an odor distinctive for the hive. This distinguishes members of the hive from enemies or nonmembers. When a bee lands at the entrance of the hive, it is inspected by the guard bees which are always on the alert. If the bee has the proper odor it is admitted but those with the wrong odor are securely held by a leg and carefully inspected by the guards. If the intruder is from another hive and arrived at this one by mistake, it is submissive during the inspection and is released unharmed by the guards. However, if the stranger shows signs of fighting, as it would if its mission is to rob food, the guards promptly sting it to death. Obviously the odor prompts the specific behavior of the worker bees.

One of the essentials of life in the bee colony is the communal effort in foraging for food. It is in this aspect of behavior that von Frisch made his classic contribution. He demonstrated that bees have an elaborate communication system, which he called the "language of the bees," that facilitates the search and acquiring of food

Obviously the food-getting process would be more efficient if when a foraging bee came across a good source of nectar, it could direct others to the source. By setting out sugar-water scented with honey some distance from the hive, von Frisch found that when a scout bee discovered it, in a matter of minutes other bees came, even though many hours may have elapsed before the first one found the food. It was very possible that the successful forager had communicated its good fortune to others in the hive. To examine this possibility von Frisch placed glass in the sides of the hives so he could observe the activities of the bees. He painted the thorax of the scout bee while she was feeding at the sugar-water so that he could distinguish her from other workers when she returned to the hive. The moment she returned to the hive she first gave food to several other workers, then began "dancing" vigorously on the vertical surface of the honeycomb. She walked in circles, alternating right and left, and then repeating many times (Fig. 27–10). This "round dance," as von Frisch called it, excited the other workers. They placed their antennae near her and followed her around for a short time, and then left the hive and flew directly to the sugar-water. Apparently the information about the source of the food was conveyed by the first worker to the others. The directions as to the location must have been very accurate. Von Frisch continued his experiments to determine the kinds of information that were transmitted between bees. He put sugar-water dishes 10 meters in all directions from the hive and, by carefully observing the marked bees as they returned, he demonstrated that the round dance merely indicated that there was food nearby but it did not indicate either direction or distance. He then placed sugar-water at different distances from the hive, one at 10 meters and another at 300 meters. Whereas the bees returning from the 10 meter dish executed the round dance, those from the 300 meter dish danced differently. They ran a short distance in a straight line and wagged their abdomens rapidly, then circled and repeated the wagging, always going in the same direction. They turned alternately right and left with each cycle of the dance. Von Frisch called this the "wagging

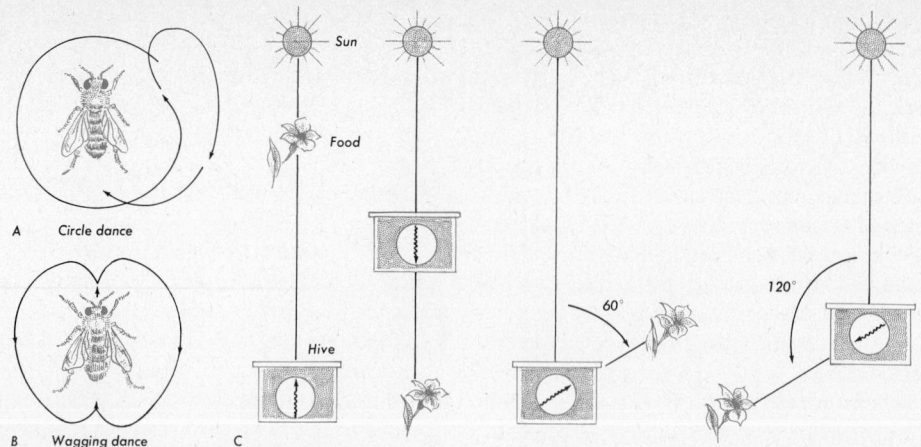

Fig. 27–10. (A) The circle dance. (B) The wagging dance. (C) The relationship of the sun, food, hive, and angle of the bees' wagging dance oriented in a vertical direction. The wavy line in the hive indicates the direction taken by the bee.

dance." By comparing the number of turns made with the distance, von Frisch concluded that there was a direct relationship (Fig. 27–11). In addition, he discovered that that part of the dance where the bee runs in a straight line indicated the direction of the food in relation to the sun. In other words, she "pointed." She did this by executing the wagging dance on the vertical surface of the comb in the dark hive. She substituted gravity for the position of the sun. Where the straight run of her dance was directly vertical (moving against gravity) she meant that the food was in the direction of the sun. Moving toward gravity (straight down) meant the food was directly opposite the sun. Running at angles between up and down indicated that the food would be found at corresponding angles from the sun. Workers performed their dances and gave accurate information even on a cloudy day if they could see a patch of blue sky now and then. They have the remarkable ability to analyze polarized light with their compound eyes. When light is polarized it travels in one plane and is reflected from particles in the air. As the sun moves, the angle of polarized light has a direct relationship between the sun and the observer. Thus, the bee can locate the sun and orient its dance accordingly.

Bees can calculate the movement of the sun by a "clock" mechanism (p. 609). It has been shown that they frequently continue to dance long after the sun has set. They indicate the direction of the food in relation to the nonexistent sun as if it were continuing across the sky instead of being below the horizon.

These experiments seemed to explain all that was necessary to understand communication among bees. Recently two workers, A. M. Wenner at the University of California and H. Esch at the University of Notre Dame investigated bee sounds during various activities in the hive. By placing a small microphone near the returning forager bee during her wagging dance, they recorded a whirring sound. Some workers promptly flew off to the food. After a careful analysis of these sounds it was concluded that the returning bee emitted the whirring sound during the wagging dance, and the duration of the bursts of sound indicated the distance to the food source. It appears, then, that visual, sound, and probably chemical signals are used in conveying information involved in food gathering.

Other behavior patterns in social bees are related to reproduction, protection, temperature regulation, constructing combs, etc.

These are essential for the operation of colonies of large numbers of individuals. They have evolved from simple behavior patterns to their present complexity over millions of years.

The structure of social vertebrate behavior patterns is quite different from that in social insects. In many vertebrate societies there is a dominance hierarchy, that is, a linear "peck order," well illustrated by birds. There is usually one bird in the flock (limited to one sex) that can peck any other and never be pecked itself. Below this α-bird is the β-bird who can peck all others except the α-bird. This behavior passes linearly down from bird to bird until the last one is pecked by all others. The α-bird has certain rights because of his rank in the group. He has the first choice of the food, nesting places, and available females. He maintains this position of dominance until he grows too old or becomes ill, either of which forces him to a lower level and another bird takes his place. Administering male hormones to a bird can raise its position in the hierarchy. Under normal circumstances the levels of dominance remain rather stable. This has survival value for each individual because it prevents fighting which is a useless waste of time and may even bring injury and death to individual members of the group.

Among social mammals there is wide variation in group structure. Among carnivores, such as wolves, the social group may consist of an old male, one or two adult females, and the young which may be full grown. The large herbivores, on the other hand, aggregate into herds of many individuals. The dominance hierarchy is simple in the wolf family but becomes more complex in a herd of herbivores such as bison or elk. A hierarchy can be established only if all individuals recognize one another or at least recognize classes of individuals. In a large herd this is difficult and there is usually no well defined peck order.

In some groups, such as elephant seals, contests to acquire the α-position may be bloody affairs. The bull fights savagely by rearing up and crashing down upon the rival, attempting to stab him with his tusks. This creates disorder and often death among the young. Since belligerence assures the male a large harem, this behavior has the selective advantage that only the strongest and most aggressive males father the next generation, thus maintaining and improving the vigor of the group.

This type of aggressive behavior to acquire the top rank in the group is unusual. Male deer engage in combat by locking their horns to test one another's strength. After a time one gives up by disengaging his antlers and making a hasty retreat, maintaining as much composure as the situation allows. The dominant male retains his position and no one is injured.

Among primates, such as baboons, a rigid dominance hierarchy exists. The α-male carries his tail erect as he walks through the group. If a male of lower rank is occupying his favorite resting place the mere attitude of the α-male is sufficient to cause the intruder to retreat without a fight.

One must be conservative about drawing parallels between social dominance in lower

Fig. 27–11. This graph shows the inverse relationship between the distance of the food source and the rapidity of turning per unit time in the wagging dance.

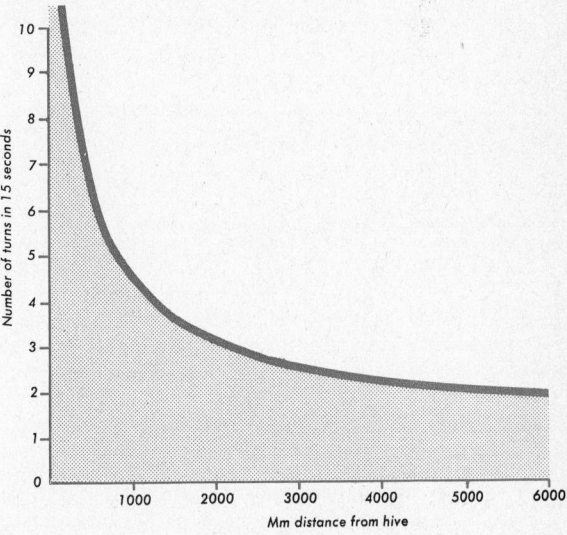

animals, even primates, and human societies. Whereas there may be some basic behavior patterns that resemble peck orders in humans, there are so many other factors that influence man's behavior that it is unlikely that meaningful comparisons can be made.

BIOLOGICAL RHYTHMS AND CLOCKS

It has been known for a long time that plant and animal activities are rhythmic in nature; that is, they occur at regular intervals. Those that occur the same time each day are called **circadian** (circa—about, dieus —a day) **rhythms.** Other rhythms that occur monthly or yearly are also known. Let us discuss circadian rhythms.

Since plants and animals evolved on a planet that turns on its axis approximately every 24 hours, it is not surprising that they have geared their activities to this cycle. Changes in temperature, light, radiation, etc., all of which have a profound effect on living things, occur with each revolution of the earth. Many animals possess a rhythmic pattern in their activities. Mice, for example, are active at night and quiet during the day; *Drosophila* emerge at a specific time in the morning no matter what their age or under what circumstances they have been reared; certain crabs are dark in the morning and lighter in the afternoon. Even among protozoa a striking circadian rhythm exists. *Chlamydomonas* divides at a specific hour during the night. Some species of paramecia conjugate only at night.

This rhythmicity of behavior has been attributed to some kind of internal timing device called a biological "clock," by which the organism tells time. No evidence has been found for a morphological structure that could be ascribed such a function. Nevertheless, there is abundant evidence that animals can tell time. This means that they must be able to compensate for the changing position of the sun. We have already discussed how birds and bees navigate by noting the position of the sun. They must somehow correlate position of the sun with time of day. If a caged bird is trained to seek its food in a certain direction, say west, it always searches in a westerly direction no matter what time of day it is put in the cage. In other words, it knows which way is west by checking the position of the sun, which of course varies constantly. Bees tell time accurately as indicated by the fact that they know at what time of day each species of flower contains the most nectar and visit it only at that time. They do this even when placed in a continuous light source and constant temperature. Obviously their time signals come from an internal clock.

Such experiments as cited above suggest that biological clocks are rigid and are not influenced by the environment. This is not altogether true. Under normal conditions, stimuli from the environment set the clock, and it is repeatedly set throughout the life of the organism. Because this is so, it is possible to experimentally manipulate the environment so as to change the setting of the clock. For example, if starlings, placed in an open cage so they can see the sun, are trained to orient toward the north and then are transferred to a room for several weeks in which the period of light and darkness is shifted six hours ahead of the natural day, when replaced in the original cage they will orient toward the east. Their clocks have been reset six hours ahead so that they now interpret the position of the sun 90 degrees off; they recognize east as north.

Under natural conditions, the clock in animals is being constantly reset so that all of its physiological mechanisms are synchronized into a harmoniously operating organism and everything goes along smoothly. When the clocks are experimentally tampered with, serious consequences may result. It has been demonstrated that the various organs of the animal do not respond to a shift in environmental conditions synchronously, that is, some may take longer than others to adjust to the new situation. If, for example, the protein-synthesizing system adjusts in three days so that a particular enzyme, say adenosine triphosphatase, is produced in quantity, but the mechanism for the synthesis of the enzyme phosphorylase does not adjust for a week, the animal would have a deficiency of the second enzyme, hence would not be able to convert ATP to ADP and would be lacking sufficient energy to perform normally. Likewise, the levels of hormones and the responsiveness of the target organs could be thrown out of synchrony by the failure of each to respond synchronously to the shift in the environment.

A thorough knowledge of the various clocks in man is important because in modern living we are constantly shifting our environment. For example, pilots who fly between continents at regular intervals are exposing themselves to radical shifts in their environments. Space travel is another factor in modern life which undoubtedly influences man's behavior by divorcing him from the rhythmic environment which he evolved in. This is a critical field of research that demands close attention in the years ahead.

LEARNING

In the preceding pages we have often referred to unlearned (instinctive) behavior and have said it was inherited. Now we want to discuss learned behavior. Is there a clearcut distinction between the two? Attempts were made a few decades ago to regard the behavior of some animals as purely instinctive and of others as learned. Psychologists emphasized learning as the most important factor in behavior and biologists felt that instincts were equally important. As more information has become available, it has become clear to both groups of scientists that instincts and learning are basic to an understanding of animal behavior, particularly that of higher animals.

Behavior shows a mixture of inheritance and learning in some animals which has been beautifully demonstrated by W. H. Thorpe of England who studied the male English chaffinch which has a very distinctive song under natural conditions. When males are raised artificially from egg to adult without hearing the songs of other males, they sing, but the song sounds monotonous, even to us. An analysis of the sound track (spectrogram) shows that it retains some of the quality of the normal chaffinch song, but it is not nearly so elaborate. If nestling males are exposed to tape-recordings of singing adult males, when the young males begin to sing the next season, their song resembles that of the recorded song. Young males actually learn songs that are not quite like the normal chaffinch song, but this is possible only within the first three years; later the birds are unable to learn a new song. Apparently, chaffinch males inherit the basic song pattern but the parts of the song that give it quality are learned. This is a good example of combining instinct and learning in animal behavior.

Kinds of Learning

Learning may be defined as relatively permanent changes in behavior as a result of experience. It does not involve those changes that occur during maturation or those associated with transient physiological changes such as those resulting from hunger, fatigue, and sensory adaptation. The types of learning have been classified in a variety of ways by both biologists and psychologists. We confine our discussion to a few of the most commonly recognized types.

Habituation. In this type of learning, perhaps the most elementary, the animal responds less frequently and finally not at all to sign stimuli that have no significance. In other words, it learns to separate the meaningful from the meaningless stimuli in its environment, thus saving much of its energies. This can well be illustrated in young birds.

Young chicks, pheasants, and turkeys crouch or show some other alarm behavior when an object moves overhead (Fig. 27–12). This innate behavior is faithfully executed when the chick is very young. After a time the crouching behavior gradually wanes and they do not respond to **all** objects or birds moving overhead. They become habituated only to objects they see fre-

quently, such as falling leaves and common small birds. However, they still crouch whenever they see an unfamiliar object in the sky. This response to unfamiliar objects has survival value because birds of prey such as hawks and owls are much less common than sparrows or robins, consequently the chick rarely sees one. When it does it crouches, thus making it less conspicuous to the predator. As much as we would like to say that the chick recognizes the **shape** of the predator, it apparently does not. Actually the chick has never experienced being snatched up by an owl so it does not even know that the owl is dangerous. It has merely learned the shape of familiar objects and responds only to unfamiliar shapes. This behavior is permanently fixed so that the chick responds similarly throughout its life.

Imprinting. This is a highly specialized type of learning which is most easily observed in birds shortly after hatching. K. Lorenz, for example, showed that a duckling or gosling follows the first large, moving object it sees within a few hours after hatching. Ordinarily this is its mother, but it follows almost any object, a man or even a mechanical model. For example, it will follow a large green box, with a ticking clock, designed to move along a wire, with persistence and endurance. If the model is drawn over an area where the terrain is irregular, the duckling becomes more quickly and firmly imprinted. It seems that if greater effort is needed to follow the object, the duckling learns faster. Imprinting occurs at a short, critical period, about the first 36 hours for a duckling. Once imprinted on some artificial model during this period,

Fig. 27–12. Newly hatched chicks respond to any object that passes overhead by crouching and sitting very still in a protective reaction. By habituation it learns to ignore birds it sees repeatedly, in this case the gull (top sketch). It crouches when unfamiliar objects appear in the sky such as the hawk (middle sketch). When the model (lower sketch) is moved to the right the chick crouches because it resembles the unfamiliar hawk; when it is moved to the left the chick ignores the model because it resembles the familiar gull.

it will not be imprinted later, even on its own mother.

Imprinting occurs in many vertebrates, but how widespread it is among other animals is not known. The case of the salmon finding its birthplace, mentioned earlier, may be a case of imprinting in fishes. Dog trainers know that puppies must be in contact with people during the 6th and 7th week of life if they are to be good pets. Puppies in isolation until after the 14th week rarely ever react with friendliness toward man. Professor H. Harlow of the University of Wisconsin demonstrated that young monkeys raised in isolation from their parents became extremely unsocial and emotionally disturbed. When adults, they refused to mate. Apparently among all the animals studied, from birds to primates, normal social behavior is imprinted during early life. It must depend on some special condition of the nervous system during early development.

Conditioning. This is a simple form of learning first described by the Russian physiologist, Ivan Pavlov. It involves the association of a simple reflex with a stimulus with which it was not previously associated. Rewards or punishments (reinforcement) are used to reinforce the stimulus. Pavlov's original experiments were with dogs which he fed meat and recorded the amount of saliva they produced. Then he rang a bell each time he fed them; after many trials the dogs salivated when they heard the bell even though they had no contact with food. Pavlov called this reaction to the bell a **conditioned reflex.** The animal establishes a new reflex which is somehow impressed on its nervous system. An association between the new stimulus and the usual stimuli of taste, sight, and odor of food has become fixed in the dog's nervous system so that now all give the same response. This type of experiment has been repeated many times by other workers using other animals, different stimuli, and other reflexes. It is well established that conditioned reflexes are a simple type of learning.

Other, more complicated behavior, involv-

ing more than simple reflexes, can be ascribed to conditioning. In this type of learning, sometimes called **trial and error** learning, the animal must do something. The experiment is centered around the animal's desire for food. The animal is starved until its motivation for securing food is high, and then it is placed in a maze of which there are several types. One type, the Thorndike puzzle box, presents the animal with several choices of pathways to the food. Variations on this scheme include doors to be opened, lids to be lifted, etc., which must be manipulated in sequence to acquire the food. Another type is the Skinner box which resembles a slot machine. The animal has several choices of levers or buttons to operate in order to obtain food. The frequency of trials made by the animal can be recorded automatically, providing a permanent record of the animal's activity.

In both types of boxes the animal must at first spend some time exploring, that is, he tries and fails, tries again repeatedly until at last he succeeds and obtains the food. With time his errors become less and less until finally he always succeeds and thus has learned by trial and error. This is a type of learning which we all recognize in ourselves. We learn to like or dislike certain foods by trying them. A child learns very quickly to avoid hot fluids or objects by first having been burned by them. Trial and error learning occurs in all vertebrates and probably most invertebrates. Factors that alter this type of learning are studied in great detail by biologists and psychologists. Obviously anything that speeds up learning is important to man and beast alike.

An extension of trial and error learning is abstract **generalization.** Birds, pigeons, for example, when trained to peck at a certain colored button to receive food will peck at a button of similar color if the original colored button is not there. Even more remarkable, canaries can be taught to find a food pellet under an unusual object placed in a series of like objects. For example, if a series of screws are placed in a row and one aspirin is placed anywhere in the series, the canary will look under the aspirin. If the experiment is reversed, it will look under the lone screw. Whatever single unusual object appears in the series the bird always looks under the unusual object first.

Insight Learning. This type of learning, often referred to as **reasoning**, reaches its peak development in higher primates, especially in man. It is distinguished from other types of learning by the fact that the animal does not require a period of trial and error, but responds to a new situation correctly the first time. By merely surveying the new situation it solves the problem mentally, making use of past experiences, and then makes the appropriate moves to get the right answer on the first try. In other words, insight learning involves the capacity to respond correctly to stimuli which are totally different from anything the animal has experienced. The animal may be familiar with some fragments of the problem, that is, he may recognize the items from past experience, but in a different context. He can utilize this information to manipulate the items to solve the new problem. This can best be illustrated with chimpanzees. Köhler long ago showed that these animals obtain bananas hung out of reach in their cage by correctly stacking boxes that were originally scattered about the floor. If given two separate rods that fit together, they will place them end-to-end in order to reach food that is out of reach when using either stick alone.

There are many other types of learning such as concept formation, principle learning, and symbolic processes which are very complex, particularly those in man, but these are covered in psychology courses and we do not discuss them.

NEUROLOGICAL BASIS
OF BEHAVIOR

It is quite obvious that much of behavior is linked with the nervous system from planaria to man. We understand quite a bit about neurons, nerve pathways, parts of the brain and broadly how they function, what the nerve impulse is, and many other things. But there is a great void in our knowledge of how all these function to bring about the behavior we see in animals. Is instinctive behavior and learning localized in the brain? How do hormones alter an animal's behavior? What is memory and where is it stored? Is there a memory coding system similar to the genetic code? All of these are profound questions for which we have no good answers in spite of the tremendous effort being made by scientists all over the world.

The Brain

Highly refined surgery has made it possible to localize centers which govern anger, sleep, hunger, thirst, sex drive, and pleasure, primarily in mammals. Most of our information has come from experiments on rats and monkeys, although more and more is being learned about these centers in man through head injuries as well as surgery.

These centers were discovered by inserting fine needlelike metal electrodes into specific regions of the brain, of a rat, for example, and keeping them there permanently by firmly fixing them to the skull. This is done under anaesthesia and once healed the rat feels no pain. Very mild electric shocks can be given and the effect noted on the animal's behavior. One of the most interesting results of such experiments was the localization of the **pleasure center** in the hypothalamus. When a rat has an electrode in its pleasure center and is given the opportunity to stimulate itself by pressing a lever in a Skinner box, it does so repeatedly, as much as several thousand times an hour. If he is given a choice between refraining from pressing the lever and suffering severe pain (electric shock) he makes the sacrifice and presses the lever. He even chooses the pleasure from the electrodes over sex. Rats with similar electrodes in other parts of the brain will depress the lever only a few times an hour.

By employing the same technique, other centers have been localized in the hypothalamus, not only in rats but in mice, goats, monkeys, and even birds. The antagonistic actions of these centers have been demonstrated in the same animal. By placing electrodes in the "angry" and "friendly" centers of a monkey and connecting them with tiny transistor radios attached to the animal, its behavior can be controlled from a transmitting station located some distance away. The animal can be alternately thrown into fits of anger or friendly gestures.

Food intake is also controlled by centers in the hypothalamus. They are the **satiety center** and the **hunger center** located close together. If the satiety center is destroyed with an electric cautery the rat eats voraciously, doubling its weight in a few weeks. Its weight then levels off and the rat eats just enough to maintain this abnormal weight. If the hunger center is similarly destroyed in another rat, it stops eating and finally starves to death. Whether one or the other of these centers is stimulated depends on the glucose level in the blood. Fasting causes a very slight drop in blood glucose and this difference stimulates the hunger center; the opposite results following a meal. The satiety center cells have a high affinity for glucose as shown by feeding rats glucose "tagged" with a toxic gold compound. The compound accumulates in these cells and destroys them. Such animals become obese just as the ones did that had their satiety centers destroyed with an electric cautery.

The presence of these centers in the hypothalamus has been amply demonstrated in many animals. Impulses sent out from these centers to those parts of the body that are involved in the specific action are controlled by the cerebral cortex. This final expression of these combined actions constitutes the animal's behavior.

Memory

For a long time it was thought that memory somehow left a record, called an "engram," in the brain. Attempts to find such a morphological or physiological trace have never been successful. A new approach

to studying memory came about through studies of transfer of learning from one animal to another. Planaria has been used in many studies because it has the most primitive complete nervous system consisting of a brain and two lateral nerve cords, and it reproduces by fission which is not true of higher forms. The experimental procedure is to place the worm in a shallow plastic trough, one-half inch in diameter and one foot long, filled with water. The animal moves from one end of the trough to the other. Training consists of suddenly flashing a bright light which is followed by an electric shock. The light usually causes no reaction in the untrained animal, but the shock causes a strong contraction. After a number of trials the planarian is conditioned to the light; it contracts more frequently when the light is flashed.

Conditioned animals are then cut in half and the length of time that both "head" and "tail" retain the conditioned reflex when fully regenerated, compared to the trained controls, is measured. It turns out that both heads and tails retain the reflex the same as the uncut controls. By cutting a trained animal in half, letting it grow a new tail, then cutting off the head and letting it grow a new head, an animal resulted which had none of its original tissues, yet a significant amount of the training was retained. Apparently, retention involves the whole animal, not just the nervous system. Perhaps some sort of blueprinting or coding occurs in all of the cells. The most likely substances would be DNA or RNA. Experiments have been done with the latter and its hydrolytic enzyme, RNAse. Trained worms that were exposed to a weak solution of this enzyme failed to retain what they had learned, indicating that RNA was somehow linked with learning.

Variations of these experiments have been extended to mice and rats. RNA extracted from the brains of trained animals and injected into untrained animals improve the latter's ability to learn. Also, some workers have shown that antimetabolites, such as azoguanine (blocks RNA synthesis) and puromycin (blocks protein synthesis) interfere with memory in several animals, including fish. Such results imply an active role of RNA in learning and memory.

There has been much controversy over these experiments; some investigators swear by the results, others have been unable to repeat them. Based on what we know about

DNA as the molecule that permits coding of the most intricate information, both in replication and transcription, it is not unreasonable to assume that RNA could code memory. This is a very exciting field which requires much more research before meaningful conclusions can be made.

SUGGESTED SUPPLEMENTARY READINGS

Books

CARTHY, J. D., *An Introduction to the Behavior of Invertebrates*. New York: Macmillan, 1958.

*CARTHY, J. D., *The Study of Behavior*. New York: St. Martin's, 1966.

*DETHIER, V. G., and STELLAR, E., *Animal Behavior*, 3rd ed. Englewood Cliffs, N.J.: Prentice-Hall, 1970.

DORST, J., *The Migration of Birds*. Boston: Houghton Mifflin, 1961.

W. ETKIN, ed., *Social Behavior and Organization among Vertebrates*. Chicago: University of Chicago Press, 1964.

*FRISCH, K. VON, *Bees: Their Vision, Chemical Senses, and Language*. Ithaca, N.Y.: Cornell University Press, 1964.

*GRIFFIN, D. R., *Bird Migration*. Garden City, New York: Natural History Press, 1964.

GRIFFIN, D. R., *Echoes of Bats and Men*. Garden City, N.Y.: Doubleday Anchor Books, 1959.

GRIFFIN, D. R., *Listening in the Dark*. New Haven, Conn.: Yale University Press, 1958.

HINDE, R. A., *Animal Behavior*, 2nd ed., New York: McGraw-Hill, 1970.

*LORENZ, K. Z., *King Solomon's Ring*. New York: Crowell, 1952.

LORENZ, K. Z., *On Aggression*. New York: Harcourt, Brace & World, 1966.

NATIONAL GEOGRAPHIC SOCIETY, *The Marvels of Animal Behavior*. Washington, D.C.: National Geographic Society, 1972.

SCHMIDT-KOENIG, K., "Current Problems in Bird Orientation," in *Advances in the Study of Behavior*, Vol. 1. New York: Academic Press, 1965.

THORPE, W. H., "The Ontogeny of Behavior," in *Ideas in Modern Biology*, J. A. Moore, ed. Garden City, N.Y.: Natural History Press, 1965.

TINBERGEN, N., *The Study of Instinct*. New York: Oxford University Press, 1951.

TINBERGEN, N., *Social Behavior in Animals*. New York: Wiley, 1953.

TINBERGEN, N., "Behavior and Natural Selection," in *Ideas in Modern Biology*, J. A. Moore, ed. Garden City, N.Y.: Natural History Press, 1965.

YOUNG, W. C., "The Hormones and Mating Behavior," in *Sex and Internal Secretions*, W. C. Young, ed., 3rd ed., Vol. 2. Baltimore: Williams & Wilkins, 1961.

* Available in paperback.

Articles

CARR, A., "The Navigation of the Green Turtle." *Scientific American*, May, 1965.

DILGER, W. C., "The Behavior of Lovebirds." *Scientific American*, January, 1962.

EIBL-EIBESFELDT, I., "The Fighting Behavior of Animals." *Scientific American*, December, 1961.

ESCH, H., "The Evolution of Bee Language." *Scientific American*, April, 1967.

FRISCH, K. VON, "Dialects in the Language of the Bees." *Scientific American*, August, 1962.

HASLER, A. D., and LARSEN, J. A., "The Homing Salmon." *Scientific American*, August, 1955.

HESS, E. H., "Imprinting in Animals." *Scientific American*, March, 1958.

HESS, E. H., "Imprinting in Birds." *Science*, 146, 1128, 1964.

KORTLANDT, A., "Chimpanzees in the Wild." *Scientific American*, May, 1962.

LORENZ, K. Z., "The Evolution of Behavior." *Scientific American*, December, 1958.

SHAW, E., "The Schooling of Fishes." *Scientific American*, June, 1962.

THORPE, W. H., "The Language of Birds." *Scientific American*, October, 1956.

TINBERGEN, N., "The Curious Behavior of the Stickleback." *Scientific American*, December, 1952.

TINBERGEN, N., "The Evolution of Behavior in Gulls." *Scientific American*, December, 1960.

TODD, J. H., "The Chemical Languages of Fishes." *Scientific American*, 224, 98, 1971.

WENNER, A. M., "Sound Communication in Honeybees." *Scientific American*, April, 1964.

WILSON, E. O., "Pheromones." *Scientific American*, May, 1963.

28

ANIMAL ECOLOGY

The relationship of animals to their environment and to one another is the science of **ecology.** Particular attention is here devoted to animal ecology although no ecological study can ignore the role played by plants.

Great variations in the environments of the world are caused by such physical factors as light, temperature, and moisture, all of which have a profound effect upon the physical and physiological characteristics of animals. These physical factors not only determine the kinds of animals that are able to survive in certain regions but are also instrumental in building up associations between animals and plants. Thus the problem in ecology is twofold: first, to consider the individual animal in terms of certain physical factors in its environment; second, to study the relationship between organisms living together.

PHYSICAL AND CHEMICAL FACTORS IN THE ENVIRONMENT

Temperature

Everyone is fully aware of his own sensitivity to change in temperature. We usually want our houses at a relatively constant temperature of about 23°C and experience discomfort if it deviates a few degrees one way or the other. Our internal environment is even more critical—there a rise of a few degrees indicates sickness of a serious sort. What is true of man in this respect is equally true of many animals. When we consider that the temperatures known to us range from 273°C below zero to several thousand degrees above zero, it is rather remarkable that life exists in that extremely narrow range of a few degrees above freezing to about 45°C. Even within these narrow limits the physiological processes do their best work at an **optimal point** around the middle, on either side of which the rate of physiological reaction falls off. Animals tend to seek out a temperature that, at least most of the time, will permit their bodily activities to proceed at an optimal rate.

Since animals are found in all parts of the earth, even the polar regions, they must find ways of surviving extremes of temperature with the least amount of discomfort to themselves. Those living in colder regions either have a constant body temperature (**homiothermal**) or else have developed a hardiness to cold that permits them to survive. The internal environment of the warm-blooded animals—birds and mammals—is relatively constant and always maintains the temperature at which physiological activities can proceed at an optimal rate. Cold-blooded animals (**poikilothermal**), on the other hand, vary their internal temperature and rate of reaction in accordance with the external environment. When the temperature drops, the animal becomes sluggish, even to the point of complete inactivity. Some can stand freezing for short periods of time. On a chilly morning in the fall of the year it is simple to capture a cold-blooded animal, from a common housefly to a rattlesnake, but the task becomes more difficult on a hot summer day when the temperature approaches 35°C. Only at the higher temperature are most activities at their maximum.

During cold seasons some mammals undergo a period of inactivity called **hibernation,** when their temperature drops and metabolic processes are reduced to a minimum. Hibernating rodents, such as the jumping mouse (Fig. 28–1), pass into almost complete inactivity, their heart and breathing rates slowing down markedly. Indeed metabolism is just enough to keep the animal alive. The energy to maintain life is derived from stored fat, hence the fat woodchuck in the fall and the lean woodchuck in the spring of the year.

Some cold-blooded animals put forth communal effort to prevent too great a drop in temperature. Bees, for example, become very active on cold winter days, beating their wings almost continuously. This

Fig. 28–1. Jumping mouse (*Zapus hudsonius*) in active condition (above) and hibernating (below).

keeps the temperature in the hive above freezing even though the outside temperature may be several degrees below zero. Snakes frequently aggregate in dens in the fall of the year for the apparent purpose of keeping warm. Even though they are cold-blooded, their temperature stays slightly above that of the external environment. By coiling about one another in large masses, the whole group stays a little warmer because the individual heat loss is reduced.

Moisture

We are already familiar with the importance of water in relation to life processes (p. 41); here we need to consider it as an essential part of the environment. Getting the proper amount of water at the right time is one of the basic problems that confronts animals. This is sometimes very difficult and, as a result, animals are equipped with various means for maintaining their water supply at a constant level. Although too much water is as detrimental to some animals as too little is to others, probably the greatest problem for most animals is the conservation of water. Some animals survive periods of intense heat by going into an inactive state called **estivation**. This is strikingly demonstrated by the African lungfish which lives in regions that are apt to dry up during the summer months (Fig. 28–2). With the approach of hot weather and desiccation, the fish burrows in the mud and secretes a capsule in which it passes the warm dry months. When the temperature drops and moisture returns, it resumes its active life once more.

Some large animals, desert turtles and lizards for example, never require water in the liquid state; they manage very well on that which is taken in with their food. Camels are notorious for their ability to work long periods without water. They can exist a week or more on dry food, and if green plants are available it is not uncommon for them to go without water for a month. Jack rabbits, mountain goats,

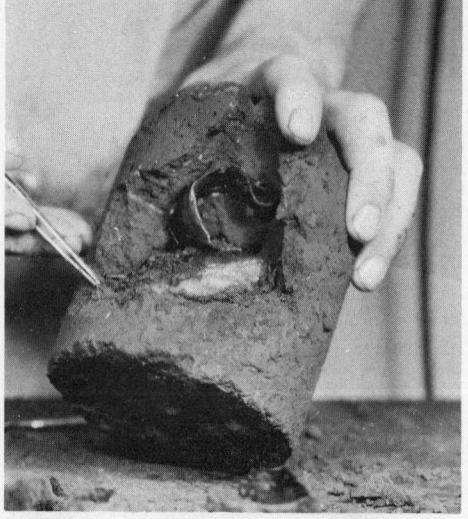

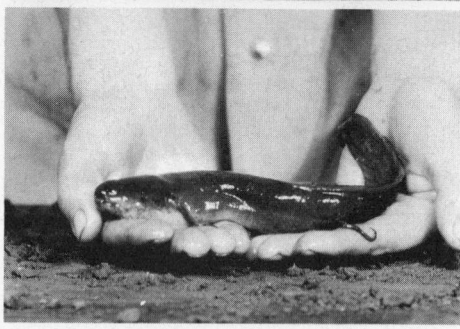

Fig. 28–2. The African lungfish (*Protopterus*) undergoes estivation during periods of drought when the water disappears from its normal habitat. If placed in a container filled with mud it will form its capsule and remain dormant for many months in this condition. Pictured here is such a fish being released from a can of mud. When placed in water the animal immediately breathes by means of its gills like any other fish.

jumping mice, and other mammals living in arid regions are very well fitted to conserve their water intake, which is usually only that provided in the food. Most mammals, however, require a great abundance of water, especially those that perspire, such as man and the horse. Excessive moisture is fatal for some animals. The earthworm, for example, is driven from its burrows after heavy rains because it cannot get enough oxygen from the water.

Light

The sun emits electromagnetic radiations which are characterized by a definite **wavelength** and **energy content** (Fig. 28–3). The wave lengths extend from less than one angstrom (gamma rays) to more than a thousand meters (Hertzian or radio waves). This entire extension is referred to as the radiation spectrum. The energy content is inversely proportional to the wavelength, that is, the longer the wave length the less the energy. Waves that are utilized in radio, television, and radar signals are some of the longest waves and they are also some of the weakest as far as energy is concerned. As the waves become progressively shorter their energy content increases; X-rays and gamma rays contain a great deal of energy and are also penetrating, hence their use in "looking at" the inside of bodies that light does not penetrate.

Human beings are sensitive to only a very small portion of the radiation spectrum, that is, only to the **light waves.** Our photoreceptor, namely the human eye, is the recording instrument by which we measure light waves based on colors. The eye in combination with the brain perceives the longer light wave lengths as red and the shorter ones as violet. Intermediate wave lengths are recorded as orange, yellow, green and blue (from longer to shorter). Combined, the entire spectrum of waves

visible to man (visible spectrum) appears as "white" light.

Most animals seem to be sensitive to the visible spectrum, although some respond to wavelengths to which we are insensitive. On the other hand, we see wave lengths that some animals are unaware of. Our receptors pick up only about 1/125 of the total radiation spectrum. Even though light waves make up such a small segment of this spectrum, all animals are profoundly affected by them.

Most animals from protozoa to mammals orient themselves in relation to light intensity. Single-celled organisms seek out the light intensity that suits their needs, usually neither too bright nor too dim. Moths fly toward a light and pillbugs avoid bright light. The advantage to the moth may be questionable, but the advantage to the pillbug is related to the fact that it breathes by means of gills and must, therefore, seek out damp places which are more likely to be found in dark places than in well-lighted regions.

The reproductive cycle of some animals, particularly birds, is definitely influenced by light. If daylight is supplemented by artificial illumination the reproductive organs are stimulated to work longer, hence more eggs and young. Some of the lower vertebrates, particularly fish and amphibians, have the ability to change color. They usually attempt to match the background upon which they are resting, obtaining the obvious advantage of camouflage (Fig. 28–4).

Fig. 28–3. Radiation spectrum of the sun. See text for explanation.

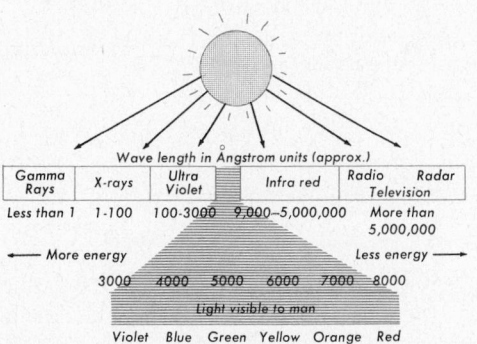

Environmental Cycles

The very nature of the movements of the earth brings about the recurrence of conditions in rhythmic fashion. The earth spins with remarkable precision and this periodism impresses itself on the physical as well as the biological matter existing on it.

The Water Cycle. The abundance of water on the earth is striking; over 73 per-

Fig. 28–4. A case of concealment by acquiring the color and position of the surrounding environment. Note how the upper part of this swamp eel (*Fluta alba*) resembles the surrounding eelgrass.

Fig. 28–5. The water cycle.

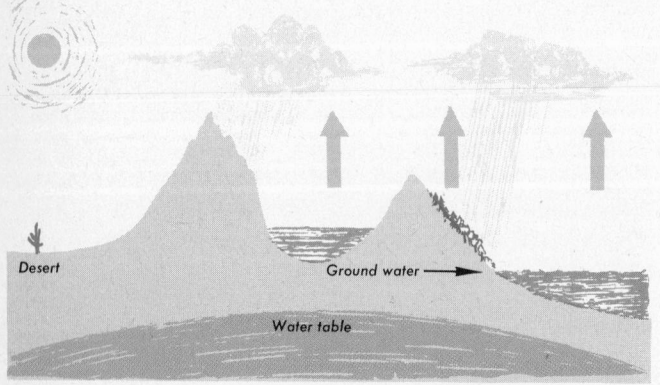

Desert

Ground water ➝

Water table

cent of the surface is covered by it, and all exposed bodies and the atmosphere above the earth contain water. The organisms that live on the planet are composed primarily of water. Its temperature has a profound effect on all objects near it or on it. Much of the sun's heat energy is stored in the oceans, thus moderating the temperature variation from day to night. Equally important is the screening effect atmospheric water has upon high-energy radiations coming from space. Without water, planet Earth would be uninhabitable, at least with life as we know it. Water follows well established cycles which play their principal role in establishing climate in the various areas of the earth. (Fig. 28–5).

Water constantly **evaporates** from the earth's surface and this is very significant over the oceans, particularly in the tropical regions of the earth because increased temperature hastens evaporation. Water vapor condenses as it rises to form clouds and with subsequent cooling falls as rain or snow. This frequently takes place over the land, particularly on the windward side of mountains. This results in a high rainfall on the oceanside of a mountain range, with very little on the opposite side. The result is fertile valleys on one side of a mountain range and desert on the other. Obviously this movement of water profoundly affects the organisms living in these environments. Most of the rain falling on land eventually makes it way back to the oceans. In mountainous areas the flow is rapid and the resulting torrential streams cut deeply into the earth constantly transforming its surface. As a result of the eroding effect of water new habitats are provided for living things while others are destroyed. Water that seeps into the soil ultimately reaches impervious rock formations which in general conform to the exposed surface of the earth. The underground water, like surface water, finally reaches the ocean only to evaporate again and repeat the cycle.

Within the large bodies of water, such as oceans, massive up and down movements of water, caused by the temperature differential between the hot tropics and the polar regions, bring about a constant shifting of water between the poles and the equator. This results when the cold water of the polar region sinks and the warm water in the tropics rises. These forces, together with the rotation of the earth, as well as wind patterns, create **oceanic currents** which influence the climate, not only in the sea itself but also on the land masses nearly everywhere on the earth. Again plant and animal life is drastically influenced by such water movements.

The tremendous mass of protoplasm that exists on earth also has some influence on global climate, and as this mass increases the effect will become more pronounced. Plants and animals release carbon dioxide to the atmosphere in ever-increasing amounts, and this is accentuated by the quantity added as waste from industry. Also, atomic explosions and supersonic aircraft tend to destroy the ozone in the upper atmosphere which protects us from fatal ultraviolet rays.

Chemical Cycles

The elements of which all organisms are composed come from the environment and return to the environment upon the death and subsequent decomposition of the organism. There is, then, a constant cycle of the elements. An atom of carbon residing in a protein molecule that goes to make up one of our muscle cells, let us say, may have been incorporated into any carbon-containing molecule of thousands of plants and animals before us, and will become a part of thousands of living things following us. It might be thought of as a kind of "reincarnation," so to speak, but not the variety that usually comes to mind when this word is mentioned. All elements found in protoplasm follow specific cycles, three of which—the **carbon, nitrogen,** and **phosphate** cycles—are discussed briefly.

Carbon, being the core element of protoplasm, is conspicuously present in all living things and, like all elements, follows a cyclic pattern (Fig. 28–6). Plants utilize the carbon in carbon dioxide to manufacture fats, carbohydrates, and proteins, as well as many other essential food products. These foods are eaten, digested, and absorbed by animals, and the carbon becomes a part of the body of the animal. During metabolism carbohydrates are burned, releasing carbon dioxide into the air again. Similarly, carbon dioxide is released at night by plants as they oxidize carbohydrates to obtain energy. It must be pointed out, however, that plants also produce carbon dioxide during the daytime, but because it is utilized immediately in the process of photosynthesis, its release is obscured. The burning of organic matter and decaying of dead plants and animals also release carbon dioxide into the air.

The nitrogen cycle (Fig. 28–7) is somewhat more complicated than the carbon cycle, primarily because plants cannot utilize atmospheric nitrogen. Nitrogen in the air must be converted first to **nitrites** (NO_2) and then to **nitrates** (NO_3) before the plants can make use of it in producing proteins. This conversion is brought about by nitrogen-fixing bacteria some of which are in the nodules that protrude from the roots of certain leguminous plants, such as beans and clover. Again the plant proteins are consumed by animals and converted into their own proteins, or reduced to urea

Fig. 28–6. The carbon cycle.

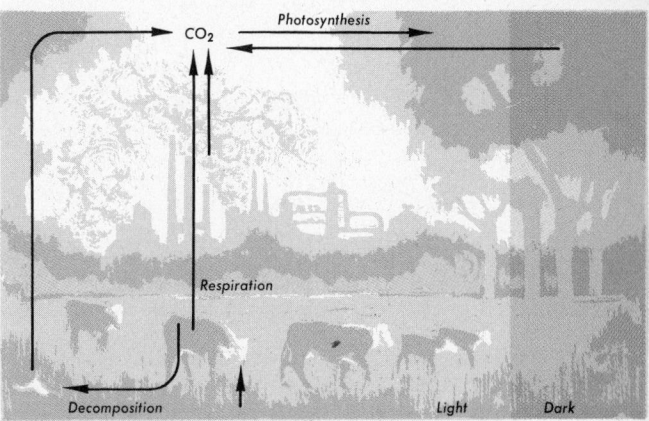

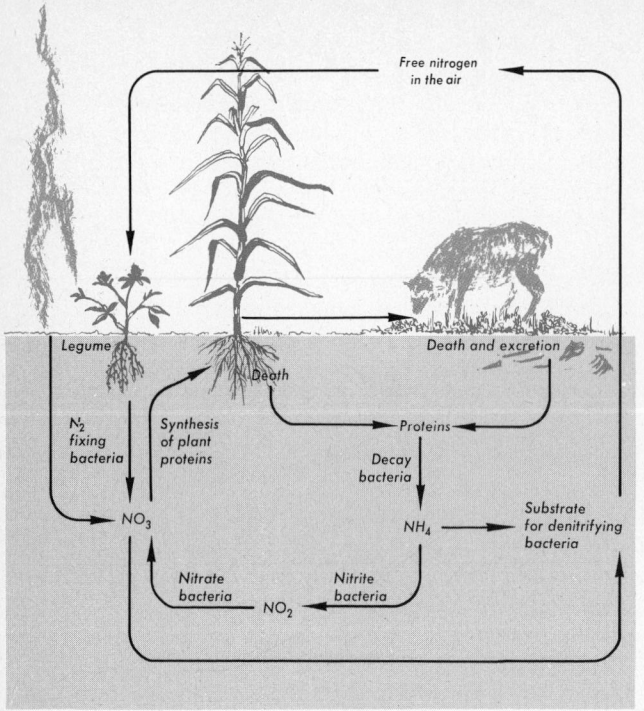

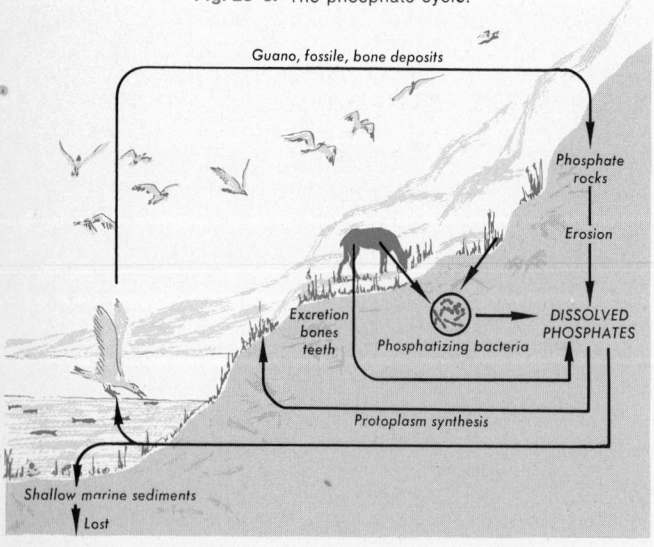

and lost from the body. Urea, as well as the dead body of an animal or plant, decomposes to form either atmospheric nitrogen or nitrates which are then used by the plants. Bacteria play an important part in the nitrogen cycle. If all bacteria suddenly disappeared from the earth, we would soon be short of nitrates and eventually of the basic materials that produce protoplasm.

The phosphate cycle (Fig. 28–8) is important in the economy of nature. Fossils, phosphate-containing rocks, bones, and guano deposits are the chief sources of phosphates. These are dissolved during rain storms, stream action, etc., then are taken up by plants, and eventually become incorporated into the protoplasm of animals, particularly herbivorous animals. Carnivores obtain their phosphates from the animals they feed upon. Decomposing bacteria add phosphates to the environment. Some phosphates are lost by becoming incorporated into the sediment at the bottom of bodies of water. Eventually, however, they form rocks which again enter the cycle at some future time.

Fig. 28–7. The nitrogen cycle.

Fig. 28–8. The phosphate cycle.

Nutrition—Food Chains

All animals with the exception of a few protozoa depend ultimately on plants for their food. The plant manufactures fats, carbohydrates, and proteins, and the animal breaks these down for its own use. The plants, therefore, are continually building up the organic world while animals are constantly tearing it down.

Solar energy that has been stored by the plant passes to the animal that eats the plant. However, it does not always expend itself completely in a passage that involves only two organisms. Often there are many intermediates which transfer the energy through a **food chain,** in which one animal after another is eaten until the energy can be released only by the death of the last animal in the chain. All of the food chains in a given community constitute a **food cycle.**

In an abundantly populated fresh-water pond, plants and animals are constantly dying, falling to the bottom, and decomposing. This disintegrating organic material forms a source of energy for the growth of bacteria. In addition, many algae (simple plants) grow by the utilization of simpler substances, just as all plants do. These two then, bacteria and unicellular plants, form the basis of food for tiny organisms such as protozoa. Small protozoa are eaten by larger ones, these in turn are eaten by rotifers, then crustacea, aquatic insects, and finally by fishes—first smaller fish, than larger ones. The latter either die or are eaten by fish-eating mammals such as mink, bear, or man. In the first case, the chain ends with the death of the fish; in the second, by the death of the mammal (Fig. 28–9).

On land, a food chain may be illustrated as follows (Fig. 28–9): grasshoppers feed on plants and are then eaten by a frog, which in turn is devoured by a snake. Finally, a hawk feeds on the snake. As the hawk grows older and loses some of its faculties, sooner or later it falls prey to another carnivore (flesh eater). This transfer of energy may go on almost endlessly.

Ecologists are very much interested in the "ecological efficiency" of food chains, that it, just how much utilizable energy is transferred from one stage in the food chain to the next. This can be determined in its simplest form by measuring the number of calories in the plants that serve as food for the herbivore as well as the number of calories in the herbivore and carnivore, in a simple plant–herbivore–carnivore food chain. This can be expressed as follows:

Fig. 28–9. Food chains: The chain always starts with plants which are eaten by herbivores. These in turn are eaten by a series of carnivores, in one case ending with the hawk and in the other with the large fish.

food eaten by the carnivore and the carnivore converts about 10 percent of the herbivore tissue into carnivore tissue. If we take a four-stage food chain; plants–herbivores–carnivores–secondary carnivores, and calculate the energy flow through it, we

$$\frac{\text{Cal. of herbivore eaten by carnivore per unit time}}{\text{Cal. of plants eaten by herbivore per unit time}}$$

This has been done both in the laboratory and in the field by several workers and the ecological efficiency comes out to be about 10 percent, that is, the herbivore converts about 10 percent of the plant food into

come out with some revealing figures. The secondary carnivores would receive about $1/100$ ($\frac{1}{10} \times \frac{1}{10} \times \frac{1}{10}$) of the energy in the plants. This illustrates why a carnivore is a very expensive animal to feed. In human

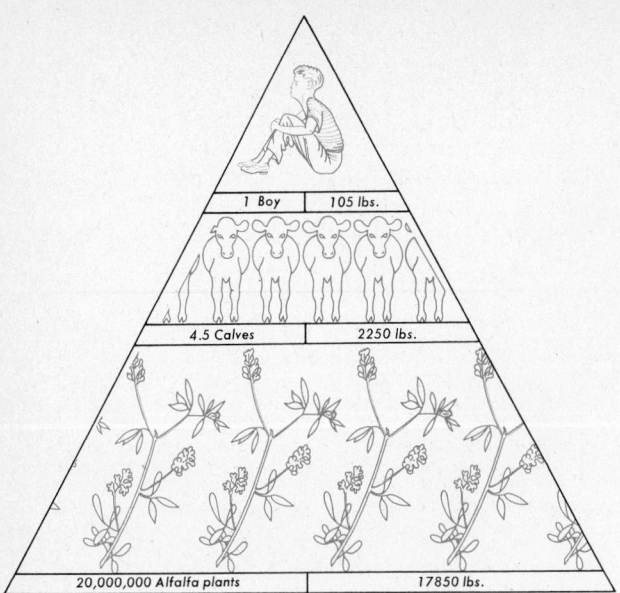

1 Boy	105 lbs.
4.5 Calves	2250 lbs.
20,000,000 Alfalfa plants	17850 lbs.

Fig. 28–10. A schematic representation to illustrate ecological pyramids of numbers of organisms and protoplasmic mass (biomass). A large amount of plant material is required to support a few calves and these, in turn, are more numerous and massive than the boy at the top of the pyramid who is supported by their protoplasm.

populations only the affluent can afford to eat much meat.

Ecological Pyramids

In any group of organisms, whether they are in the ocean, on land, in the forest, in the soil, or elsewhere, there are usually a great many small organisms associated with a few large ones. This natural arrangement is often thought of as taking the shape of a pyramid, those organisms at the base representing the large number of smaller ones and those toward the apex indicating the few larger ones (Fig. 28–10). The smallest organisms at the base supply the food for those just above them, and these in turn are fed upon by the ones higher in the pyramid and so on to the animals at the top. It is also quite clear that the actual protoplasmic mass (biomass) as well as the numbers decrease as we ascend the pyramid

—that is, to produce a given amount of human protoplasm, a great deal more cow protoplasm is required; similarly, a unit weight of calf protoplasm depends on a much greater weight of alfalfa. One can also see the same principle if one measures the energy involved in each of these layers. Just as in the food chains, useful energy is lost in the transfer through each step from the smaller organisms to the larger. The pyramids of biomass describe graphically the over-all effect of the food-chain relations for the entire ecological group.

Relationships Among Organisms

Up to this point we have considered the physical environment and the energy cycles involving organisms in that environment. Now we must think about the relationships among organisms. Organisms do not live alone; they are always associated with members of their own kind and other species associated together in areas known as **biotic communities.** Such communities are more or less definite units held together by the interdependence of their individual members. Within communities, smaller groups whose members are closely associated with each other form units which are called **populations.** A population may be made up of one species or it may consist of one to several species, and these need not be closely related. There is no hard and fast separation between population and the community, but they may be considered levels of organization above that of the individual that can be studied more or less separately.

Two other terms used by ecologists, **habitat** and **niche,** should become a part of our vocabulary before we go into this complex study of ecology. The habitat of an organism is the place where it lives. For example, the habitat of the earthworm is moist humus soil. This is where you would go to find it. The habitat includes other organisms as well as the nonliving environment. The ecological niche, on the other hand, is the "status of an organism in its community." Biologically speaking, one may say that the habitat is the "address" of the organism, whereas the niche is its "profession." Continuing with this analogy, if it is desired to become acquainted with a person, it would be a simple matter to look up his address. This would be insufficient information to really know the person. To do this we would need to know his occupation, his interests, and his function in the community. A similar situation exists with organisms; we usually want to know more than just where they can be found. We must discover what their activities are in the community, what they eat, their movements, and their effect on other associated organisms. Referring again to the earthworm, it is a simple matter to find the earthworm in moist humus soil, but with this information we know very little about this organism. To really understand the animal we should know something about its diet, the parasites that live within the body, and how it may affect other organisms in the community, what effect it has upon the soil, and many other things. The description of ecological niches of organisms is one of the most important aspects of ecology, because it is from such knowledge that an understanding of the community can be had.

Another term used by ecologists and which is important as a concept is the ecological system or **ecosystem.** It is defined as a natural unit of living organisms which, together with the organisms' nonliving environment, forms a stable system through interaction; the exchange of materials between organisms and environment follows a circular path. The ecosystem is the most complex functional unit in ecology, since it includes both the nonliving environment and the organisms, each contributing to the other and both essential for the continuance of life as we find it on earth.

Populations of Single or Similar Species

The study of population alone has proved valuable not only in the control of insect pests and predators, in the increase of game and fish, and in other redistributions of animal life, but it has also been very important in business and government, as for example in the formulation of insurance and retirement plans. During every census more and more information is gathered about people in order to learn what is happening to our population. This makes it possible to predict future trends and also sheds some light on what might be done to influence the ultimate outcome.

When organisms of a single or similar species are numerous, they take on characteristics of the population which are not those of single individuals of that population. Some of these characteristics are **population density, birth rate, death rate, age distribution,** and **reproductive potential.** Both birth and death are characteristics of the individual, but birth and death rates are not; for they are found only in populations.

The success or failure of a population is dependent on **population density,** which means the number of organisms per unit of area or volume—the protozoa in a liter of lake water or the meadow mice per acre of land. This is often very important to know, especially when one is trying to determine the capacity of a certain area to support certain species of plants or animals, or to know the effect of one species on another. For example, one grasshopper per acre of a farmer's land would have little effect on his cereal crop, but one million per acre would.

It is also important to know not only the density of a population but also whether or not it is changing and at what rate. This is possible by counting the number of fish that swim past a certain point per hour or the number of animals caught in a trap in a certain specified time. Another method is to catch a certain number of animals in a trap, say 100, tagging them by clipping the tail, for example, and then releasing them. This may be repeated again at some later time and the number of tagged animals noted in the traps. If in the second sample of 100, 10 were found with tags, then the population (P) in this area would be

$100/P = 10/100$, or $P = 1,000$. This is a rather accurate method of estimating populations and has been extensively used.

Populations also grow in a characteristic manner, and when the numbers are plotted against time we arrive at a so-called *population growth curve.* We might start by considering how growth occurs in single cells, protozoa, for example. If one or only a few protozoa are placed in a flask containing complete nutrients for this particular cell, in a short time there are millons. Since they are single cells, we may count them at regular intervals during this period of increase and from this information construct a curve which will tell us something about how growth occurs (Fig. 28–11). Such a curve is said to be a **sigmoid,** because it is S-shaped. At the beginning the cells fail to divide for a time, as indicated on the curve by the so-called **lag phase;** just why this occurs is not known. They then begin to divide at a rapid and uniform rate. This is called the **logarithmic phase** of growth because the cells increase in a geometric manner, that is, 2, 4, 8, 16, and so on. Once they reach this phase they continue dividing at a uniform rate, so the curve is straight. As the limits of food in the culture are reached or the

Fig. 28–11. Growth of cells, such as protozoa in a flask of nutrients, follows very precise stages. During the initial stage there is little increase in numbers (lag phase). This is followed by a rapid uniform rate of increase (logarithmic phase) that continues until the food becomes exhausted or the accumulated wastes become toxic, or both factors operate simultaneously. Growth then remains at a plateau for a time (stationary phase). This cycle can be repeated any number of times by transplanting some of the cells to a fresh medium.

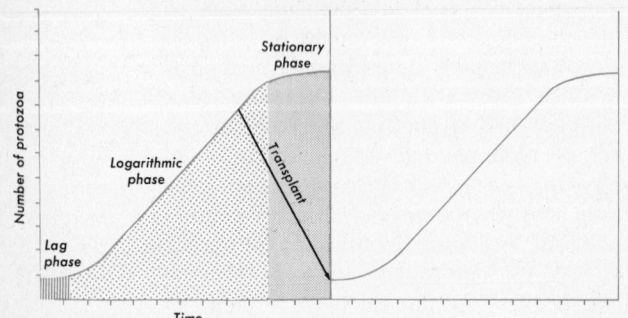

accumulation of wastes inhibits further divisions, the growth rate gradually falls off until the population finally reaches a plateau. Either the organisms no longer divide, or if they do, they die as fast as they are produced, for the total number remains the same. This is called the **stationary phase.** This curve can be repeated again and again by simply taking some of the organisms in the stationary phase and placing them in fresh medium. If the medium is continually changed, the culture can be kept in the logarithmic phase indefinitely. Apparently, these cells can continue to grow and divide at a uniform rate with no sign of aging. Aging then may be only the exhaustion of food or the accumulation of wastes in any community of cells.

Whether one studies the growth of protozoa or human beings, the curve is essentially the same. Such curves have been invaluable in predicting the outcome of certain populations in environments where conditions are favorable to unlimited growth.

Natality or **theoretical birth rate** is the innate capacity of a population to increase. This capacity usually far exceeds the actual birth rate because adverse environmental conditions drastically influence the actual numbers of offspring produced. The theoretical birth rate can be determined under optimal conditions rather accurately, and it is usually found to be rather constant for each species. This capacity to produce young is often helpful in making predictions about the possibility of a species overrunning a new environment once released in it. For all practical purposes, the actual birth rate is more important in any given situation because it determines the ultimate numbers of any species in any constant environment.

Mortality is the antithesis of natality, that is, it refers to the death of organisms per unit of time in a population. Here again there is a minimum mortality which refers to the number of deaths that would occur owing to the ravages of old age. This is an ideal figure which is constant for a population but in actuality rarely occurs.

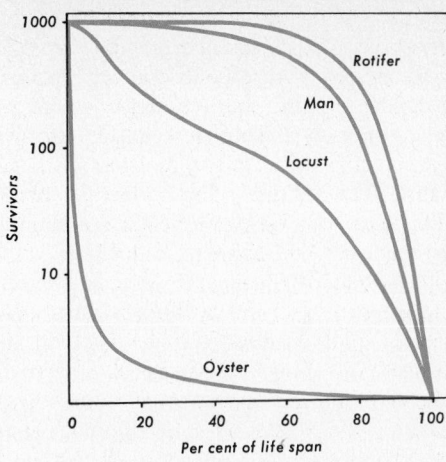

Fig. 28–12. Several types of survival curves indicating the numbers of individuals out of 1,000 that survive in a community. Numbers of survivors are plotted on a logarithmic scale and ages as percentage of total life span.

Environmental factors, including the composition and size of the population, greatly influence this ideal figure. It is possible to plot a so-called **survival curve** to indicate graphically the number of individuals that actually survive in a community (Fig. 28–12). By plotting the number of survivors out of 1,000 against percentage of total life span, one can compare the survival of organisms with quite different life spans. Rotifers live a short time, then nearly all die at once. Oyster larvae, on the other hand, die in great numbers very early in life, and only those that become attached to a rock or shell survive and mature. The locust is typical of most plants and animals, since a relatively constant number die throughout the life span. Even though mortality in man is more frequent in the first five days of life than during the following several years, it is still relatively low. Survival is high until the mid-forties when once again the death rate increases until nearly all are gone at the hundredth year. The curve plotted for man applies only to civilized man and is influenced by modern medicine. It would be somewhat modified for primitive populations.

Another characteristic of a population is **age distribution**, which influences both natality and mortality. Reproduction is usually restricted to the middle-age groups of animals, and mortality varies with age, some dying earlier than others. For this reason, the ratio of the various age groups in a population more or less determines the capacity of the population to reproduce itself. In general, a rapidly expanding population contains a greater proportion of young members, whereas a declining population has a higher percentage of old individuals. A stationary population consists of a more or less even distribution of animals of different ages. The progress of a population can sometimes be determined by observing the age of its members. The age of fish, for example, can be determined by counting the growth rings on the scales, thus making it possible to know the number of fish in each age group. This information is important to commercial fishermen in deciding whether or not a fish population is expanding or declining. They can then control their fishing practices accordingly.

One of the most important pieces of information one needs to know about any population is its power to increase its numbers when the age ratio is stable and the environment is optimal. Ecologists call this characteristic of a population its **reproductive potential**. Obviously this power must be kept in check, and the factors that do that are collectively spoken of as **environmental resistance**. These two antagonistic forces maintain a balance in nature which is observed in any community of organisms.

Although each species of animal has the potentiality of overrunning the earth, it actually never does. Populations grow in size and die out just as individual organisms do. Indeed, the number of individuals in a given population rarely remains constant but *fluctuates* in time. The many interrelations that exist between a given population and the population of other species may act as a regulatory mechanism and serve as a system of checks and balances to keep numbers of individuals within relatively narrow limits. Such controlling mechanisms may be competition for food, infectious disease, predation, and many others. Moreover, a single species may fluctuate widely from season to season. Grasshoppers may be very numerous one year, devouring all vegetation over large areas, whereas the next year there may be few. Barring man's intervention, this may be caused by unseasonal weather during the young stages when the organism is sensitive to adverse conditions. Indeed, this rise and fall in the population of certain species is so regular that it can be predicted. Sometimes animals reach tremendous numbers, then go into a decline from which they never recover, and eventually become extinct. The passenger pigeon is a good example. In other cases, like the American bison and the whooping crane, an attempt is being made to save them from extinction by the animal who nearly caused it in the first place, namely, man.

If the reproductive potential of a species is permitted to express itself to any great degree, the results are sometimes unfortunate. One example will suffice. The English sparrow was first introduced in Brooklyn, New York, in 1850 and 1852 for the purpose of controlling certain insect pests that were destroying the shade trees of the city. In England the bird was desirable and because of its natural enemies existed in modest numbers. In America, however, it was free from its predators and the full powers of its reproductive capabilities came into play. Within a few years it became a pest. Instead of eating the insect pests, it fed on garden produce and cereal grains, man's own food, and in addition destroyed other insect-eating birds. By 1886 it had spread to Salt Lake City and today its distribution is continent-wide.

Such mistakes as the one just described have been made on numerous occasions by man. Sometimes injurious animals are imported into new regions because they have escaped border inspections and have subsequently become established later to become serious pests. Great care is now taken to prevent this from happening. Many states have inspections on railroads and highways to keep any injurious pests out. Airplanes must be carefully inspected when they fly from one region to another, particularly when the two are great distances apart. The danger of introducing certain disease-carrying insects, such as mosquitos, into a new environment is obvious.

Man has intentionally introduced some animals to prey upon others with excellent success. One good illustration is that of the ladybird beetle introduced from Australia a few years ago to destroy the cottony scale which was playing havoc with the citrus crops of California. This required the research efforts of an entomologist who studied the enemies of the cottony scale in its native Australia. The ladybird was finally decided upon and when brought to this country proved very successful in partially controlling the pest. It has not, as yet, become a pest itself.

Our economic zoologists are well aware of the great precaution that must be exercised in interrupting the delicate balance of life in any environment. For example, when the insecticide DDT was first available for public consumption, many enthusiastic laymen wanted it spread from airplanes over wide swamp areas in order to destroy mosquitos as well as other insects that have mostly a nuisance value. If our practical zoologists had not prevented that procedure, the damage might have been so tremendous that it would have taken several generations and perhaps many millions of dollars to repair. Think of the destruction to honeybees as one example. Besides destroying them as a source of honey, there would be untold damage to orchards due to unpollinated flowers. Birds, amphibians, and reptiles feed largely on insects, to say nothing of the aquatic life that subsists on these very numerous little animals. We can tamper with the environment only in small areas for specific purposes; any large-scale operation must be carried out with utmost caution so that the balanced plan of the community of animals is not disturbed.

Populations of two different species may interact with one another in a number of ways. Broadly speaking they may (a) compete for food, space, or other common needs, or (b) live in close association, either benefiting or harming one another (symbiosis). These are not clear-cut distinctions and one is sometimes hard put to be certain what the relationship is between two closely associated populations.

Competition. When two populations of interacting species occupy the same habitat, competition and a concomitant decrease in number of both populations result. The two populations compete primarily for food and space; hence, something must happen in a relatively short time. One possibility is that the less endowed competitor dies or is forced to move to another niche leaving the other to occupy the area. The second possibility is that both populations survive but with a reduction in numbers in both groups. Obviously, since populations are bound to overlap at their boundaries, competition to a greater or lesser extent occurs in these regions. It becomes extremely difficult to determine at what point two populations are in equilibrium or in competition. This problem has stimulated considerable research and some pertinent observations have been made.

The Russian ecologist, Gause, concluded on the basis of field laboratory observations, that each niche is occupied by only one species. Even though niches overlap, each one is filled by a single species. One of Gause's own experiments with protozoa illustrates the principle. When two species of paramecium, *P. caudatum* and *P. aurelia*, are grown under identical cultural conditions in separate containers their growth rates are quite different—*P. aurelia* grows much faster than *P. caudatum*. When the two species are grown together, *P. aurelia* outstrips *P. caudatum*, indeed completely eliminates it after about two weeks. Apparently, *P. aurelia* utilizes food effi-

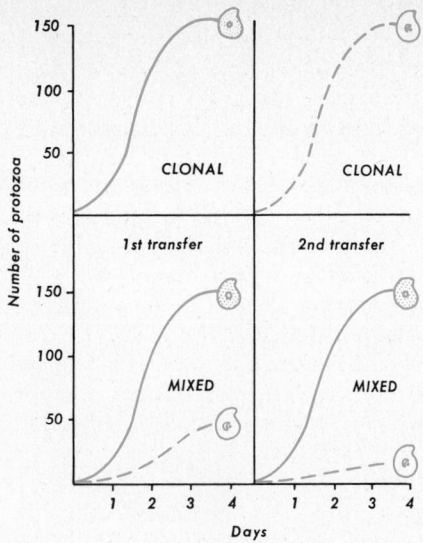

Fig. 28–13. Gause's principle illustrated by the protozoan, *Tetrahymena pyriformis*. Competition between micronucleate and amicronucleate in the same environment. The latter always overgrows the former.

ciently, hence is the more successful competitor when both occupy the same niche. Thus *P. aurelia* and *P. caudatum* occupy different niches.

Again using protozoa, the presence or absence of a micronucleus seems to determine the success or failure of ciliates of the same species (Fig. 28–13). When equal numbers of two strains of *Tetrahymena pyriformis*, one with a micronucleus and the other without one, are inoculated into a tube of culture medium and permitted to grow for several days, a marked difference in their relative numbers is noted. The amicronucleate cells are far more numerous than those with micronuclei. If a small sample of this culture is transferred to a second tube and then permitted to grow for several more days, the difference in numbers of the two strains is even more striking. When this is done a third time, it is difficult to find even a single micronucleate cell among thousands of amicronucleate protozoa. In this case, even though these two strains have identical morphology, other than the presence or absence of a micro-

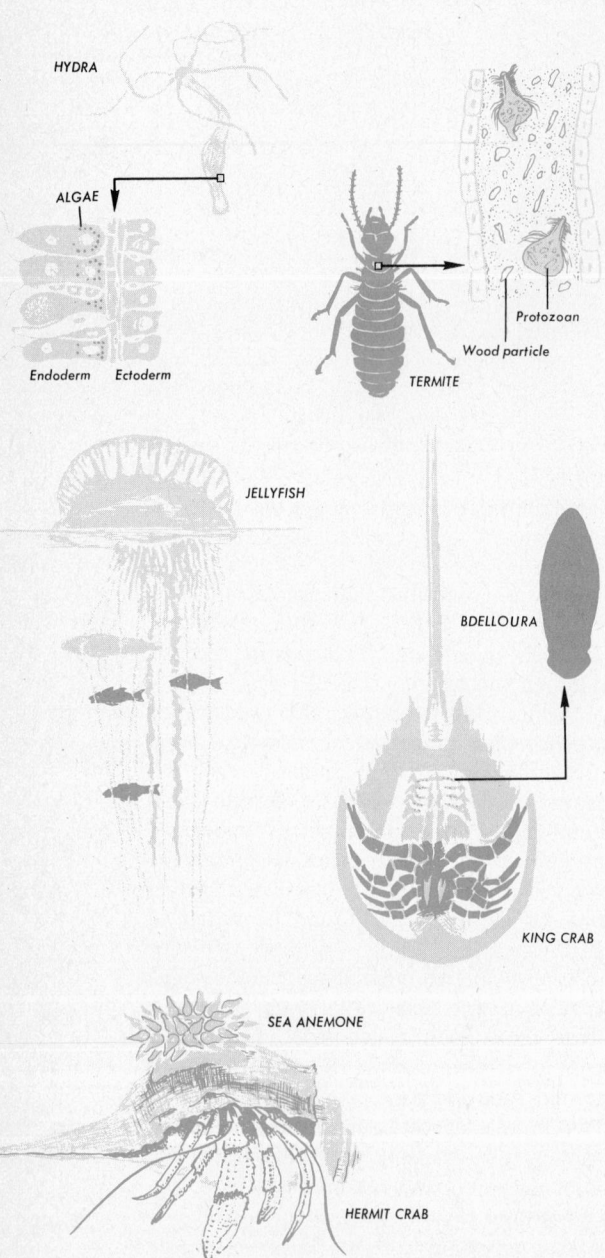

HYDRA

ALGAE

Endoderm   Ectoderm

TERMITE

Protozoan

Wood particle

JELLYFISH

BDELLOURA

KING CRAB

SEA ANEMONE

HERMIT CRAB

Fig. 28–14. Plants and animals became associated intimately in many ways. Here are some illustrations. See text for explanation.

nucleus, they seem to occupy different niches, at least in a test tube. Whether or not this situation occurs in nature is not known.

Gause's principle has been examined with many competing populations both in the laboratory and in the field. Flour beetles have been used in laboratory studies. When *Tribolium* and *Oryzaephilus* are grown together the former always exterminates the latter owing to the fact that it is more aggressive in destroying the immature stages of the former. Both populations survive, however, if glass tubes are provided in the flour because the larvae of *Oryzaephilus* have a place to hide. Therefore, by introducing a second niche both populations survive owing to the reduction in competition. If this can be carried over to natural conditions it becomes clear that the more the niches the greater the number of species that can survive.

Many examples can be drawn from field studies. At first these may seem to be exceptions to Gause's principle but upon more careful scrutiny, the principle holds. One example illustrates the point. In Africa, two species of flamingos feed side by side in shallow lakes and would certainly appear to be in direct competition with one another for food. However, upon closer examination of the anatomy of the screening mechanism in the mouth region, it becomes quite clear that one feeds only on microscopic algae whereas the other can take larger food particles such as small animals. These birds then, though mingling in the same habitat, occupy different niches.

A great deal of evidence has accumulated from both laboratory and field studies which seems to confirm Gause's contention that only one species can occupy any one niche.

Symbiosis. Some very interesting interrelationships between organisms have been established, primarily for the purpose of obtaining energy, although some seem to have other functions. Collectively, these relationships are spoken of as **symbiosis**. They range all the way from a loose, more or less haphazard association to a closely knit relationship in which the two or more organisms are forced to live together. These may be associations between plants, between animals, or between plants and animals. The antithesis of symbiosis is **antibiosis,** which is also common among certain organisms, particularly the lower plants. Instead of living together, some organisms produce substances (antibiotics) which insure that others, at least certain others, remain at a safe distance. A famous example is the mold *Penicillium notatum* which secretes a substance that has the property of preventing growth of a large variety of bacteria and, because of its nontoxic effect when taken into the body, has become a very important medicine in the treatment of infectious diseases. There are many other examples of this type of antibiosis.

Commensalism. This is a loose association of two animals in which one derives benefit from the combination whereas the other does not. These associations vary greatly in degree; indeed it is often difficult to be certain of the exact effect of the relationship on the members involved. One example illustrates this type of association. A flatworm, *Bdelloura,* can usually be found on the gills of the king crab (Fig. 28–14). From this association the flatworm is able to pick up bits of food which are dispersed into the sea water as the crab tears up its prey. It is doubtful that the crab receives any particular benefit from this association, although it seems to suffer no inconvenience from the presence of the worms. There are many examples of commensalism among the colorful animals on coral reefs (Pl. 17).

Mutualism. A situation where animals live together with benefit to both is called **mutualism,** and this arrangement seems to be far more common than commensalism. There are all gradations of this association, from those who only casually meet and become associated to those that are always found together and, indeed, cannot live apart. One interesting example is that of

certain jellyfish (*Physalia,* Fig. 28–14) and several species of small fish. The fish live among the tentacles of the jellyfish which offer protection by their stinging cells. On the other hand, the jellyfish benefits by the fact that the tiny fish act as lures by attracting larger fish to come within shooting distance of the deadly stinging darts. Thus both cooperate in the kill and both benefit. A similar relationship exists between certain large sea anemones and several species of small fish (Pl. 17). Whereas these associations frequently occur, they may be only temporary; neither party is forced to live with the other.

Another illustration of a temporary association is that of the hermit crab and sea anemones (Fig. 28–14). In this case, the sea anemone, attaching itself to the shell occupied by the crab, gets free transportation to areas which the crab finds attractive because of an abundance of food. In return for the ride, the sea anemone acts as a camouflage, making the shell resemble the rest of the ocean floor. In addition, because of its powerful battery of stinging cells, it functions as a line of defense against possible enemies of the crab. Some primitive chordates (prochordates) have adopted the same association with hermit crabs.

Another interesting association, where the relationship is more or less compulsory, is the case of the metazoan hydra and a unicellular plant, an alga. The algae live in the hydra's inner layer of cells (endoderm) where they carry on photosynthesis, releasing oxygen which is utilized by the hydra. The hydra, in turn, releases carbon dioxide which is used by the algae. While in nature this situation usually exists, it has been possible to separate them in the laboratory and each can survive without the other.

In some cases of mutualism the association of two animals is so intimate that neither can live without the other. The best illustration of this is among the termites, or white ants (Fig. 28–14). This association came to the attention of biologists when it was thought that termites could survive indefinitely on pure carbohydrate. They feed on wood alone, receiving only the small amount of nitrogenous compounds that are present in wood. Upon investigation it was found that great hordes of complex protozoa inhabit the termite's intestines. If the termite was warmed up a bit the protozoa died and such defaunated termites lived only a short time. Likewise, the protozoa could not survive outside the body of the termite. Apparently the protozoa produce enzymes that digest cellulose to sugars which are utilized by the termite. The termite, on the other hand, provides a good abundant home for the protozoa.

Parasitism. This is also a forced relationship between two animals, but it is a one-way proposition. The **parasite** lives at the expense of the **host,** taking all and giving nothing in return, possibly even causing injury to the host. An ideal parasite withdraws just enough nourishment from its host to maintain itself in good health and in reasonable numbers. If the parasite removes too much from its host, so that the latter becomes sick and dies, then the parasite too is destroyed. Many parasites reach a satisfactory balance with their host in which the latter merely contributes a home for the parasite and is not apparently injured by it.

Parasitism probably arose shortly after life originated on the earth. Some animals soon found that they could live to advantage either in or on the body of another. Perhaps at first the relationship was a perfectly harmless one, something like commensalism, but as the association persisted the parasite became more and more dependent on the host for its existence. It modified its body both morphologically and physiologically in accordance with its parasitic habit. In earlier relationships the parasite probably clung to the outside of its host, later going into the shallow cavities such as the mouth and cloaca. Some, of course, were inadvertently swallowed with the food, and these after a time became adapted to life in the gut where all of their food requirements were provided for. Others that learned to live in the bodies of insects, such as mosquitoes, also learned to thrive in the blood of vertebrates because

they were dumped into that environment every time the mosquito feasted on a blood meal.

Through similar food chains, parasites must have learned to live first in one host, then in another, and sometimes in a third, all in sequence, in order to complete their life cycle. This arrangement had the advantage of spreading the species but it also had the serious disadvantage of depending not only on three hosts but also on the necessity that the hosts be sequentially arranged in time and space. Should any one of the hosts die out, the parasite would likewise be destroyed because it could not complete its cycle. In fact, this is a most effective way to control certain dangerous parasites. Killing mosquitoes in order to control malaria is a familiar example.

Parasites have been so intimately tied to their hosts for so many millions of years that ecologists today can often trace the history of certain species by comparing their parasites. In the earlier chapters we studied several different kinds of parasites as they occur in the various animal groups.

Predation. Parasitism should be distinguished from **predation,** in which one animal also lives at the expense of another. A predatory animal feeds upon another by eating its entire body, frequently at one sitting. A parasite might just as surely destroy the host, but it does so in an entirely different manner. A cat kills and eats a mouse; the cat is living at the expense of the mouse. The cat has a tapeworm inside its intestine; the tapeworm is living at the expense of the cat. The first case is predation, the second, parasitism. "The difference between a predator and a parasite is simply the difference between living upon capital and income, between the burglar and the blackmailer. The general results are the same although the methods employed are different." [C. Elton, *Animal Ecology* (New York: Macmillan, 1939)].

Biotic Communities

A biotic community, as defined earlier, is any collection of populations existing in a defined area or habitat. It is not a closely organized unit, but it is sufficiently well defined to possess characteristics of its own which are different from those of its individual and population components. It consists of the living organisms of the ecosystem and it may be large or small. The concept of the community implies that widely diverse organisms live together in an orderly fashion, not as independent self-sufficient organisms. Thus it is possible to study the different types of communities and determine their characteristics, which can be very important when one wishes to alter them in such a way as to encourage or discourage the growth of one particular species. For example, mosquitoes can be effectively controlled simply by changing the water levels where they breed at the proper time of year. This is much more economical than using expensive poisons to kill the larvae.

There is wide variation in the importance of the many species of plants and animals that inhabit a community, some having much greater influence than others. Usually a relatively few species exert the major controlling influence owing to their numbers, size, or activities. For example, in a pasture, bluegrass and cattle may be the dominant organisms, therefore the controlling influence in this community. In all communities there are usually a very few dominant species and these are responsible for the principal characteristics of that community.

Ecological Succession. In nearly all communities there is an orderly process of change in which one community replaces another in a given area. In a typical ecosystem, community development begins with a **pioneer stage** which is gradually replaced by several sequential stages, each more mature than the preceding one, until a **climax** is reached where the community becomes relatively stable. An interesting point about ecological succession is that it is directional; a careful examination of a community reveals the stage it is in at any one time, and a reasonably accurate prediction can be made as to what will happen to it in the future. This type of prognostication is extremely helpful in mak-

ing plans for the development of areas. The early development of pioneer stages is often very gradual but the intermediate stages may progress rapidly; indeed each year may bring different organisms into the community. Gradually the process slows down until the climax is finally reached. We usually think of the organisms as changing, but they also bring about changes in the environment. As a matter of fact, the action of the community on the habitat tends to make the area less favorable for itself and better suited for other groups of organisms, which follow in sequential order. When the climax is attained, the community changes but little until some catastrophic event occurs which greatly alters the whole environment, as in forest fires, the draining of lakes, or flooding of the land.

Succession can be clearly seen in many aquatic environments such as lakes. "Young" lakes contain very little plant and animal life, but later as they begin to fill in with sediment, increasing numbers and kinds of living things appear and disappear in succession. Lakes which are relatively "old" and not chemically polluted are characterized by the presence of rooted vegetation, bluegills, and frogs. Cattails and arrowhead constitute the principal vegetation around the edges of the water. A few years later such a lake would contain more and more rooted vegetation and it would also grow smaller and smaller in size as the sediment produced by the plants and animals filled up the basin. Bluegills would be replaced by catfish and golden shiners. Finally the lake or pond would become a swamp and then dry land on which a continued sequence of vegetation and animals would follow with time. The swampy regions would support willows, tamarack, spruce, and cedar. Later these would be replaced by aspen and birch, and finally the climax of this particular region would consist of maples and basswood trees. Along with this plant succession occur succeeding associations of animals, the invertebrates, fish, amphibia, birds, deer, and squirrels.

A recently exposed pioneer area such as a rocky cliff is suitable only for primitive plants such as algae and lichens, later to be replaced by mosses and ferns. These plants, together with a few animals, gradually change the rocky region to one containing a thin soil. Erosion washes most of this soil into the valleys, beginning the slow process of filling them. Farther down the mountainside the soil supports shrubs; still farther down where the soil is deep, trees such as pines thrive. With time these communities gradually change, and this change is always toward the climax vegetation and the associated animals.

Another characteristic of biotic communities is **vertical stratification** of the organisms of the community. This is owing to the vertical differences in such physical factors as temperature, light, oxygen, and so on, as well as to the composition of the underlying soil. Stratification of organisms makes possible many more niches in a community, thus reducing interspecific competition and permitting a more effective use of the solar energy reaching the area. A good illustration of stratification is found in many lakes of the temperate regions of the earth. During the

summer the water is stratified temperature-wise, the circulating surface layer being oxygen-rich and warm, the deeper layers of cold water having a lower oxygen content. Fish such as the large-mouth bass inhabit the surface waters, the walleye and sauger the deeper layers. By measuring the depth, oxygen, and temperature, one can predict where the fish may be found, a very helpful bit of information for the fisherman.

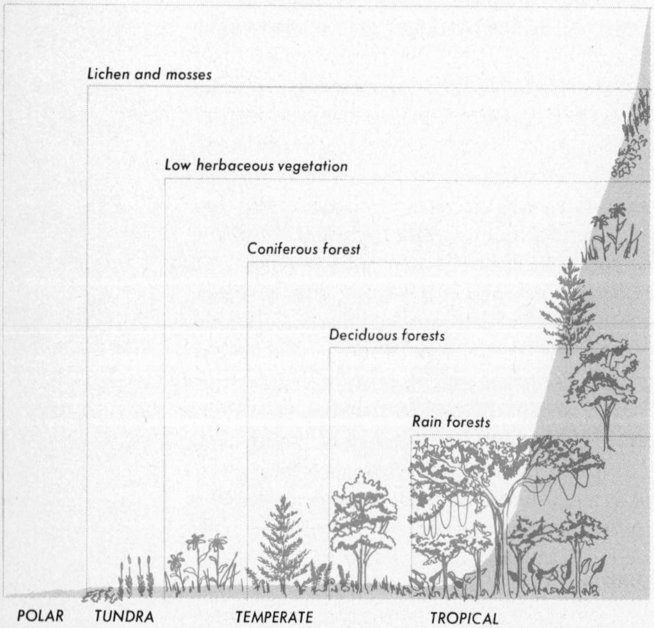

Latitude and Elevation. As one moves north from the equator a striking difference is noted in the plant and animal life. The same observation is made when climbing a mountain (Fig. 28–15). Temperature and precipitation more or less correlate between these two axes and consequently the habitats are bound to be similar. At the base of a mountain in the tropics one finds typical tropical rain forests, deserts, or grasslands, and as one moves up the slope one encounters in succession, deciduous forests and temperate zone grasslands, then coniferous forests, then low shrubby growths, and finally nothing but mosses and lichens. Beyond this point there is no plant or animal life, merely snow and ice. One finds essentially the same sequence when moving from the equator to the poles.

The mountains in the temperate zone and farther north have vegetation at their bases similar to that generally found in the latitude where the mountain is located, and on their slopes, vegetation that exists still farther north. We see then that the farther north a mountain is located the fewer habitats it provides for plants and animals. Consequently, in the extreme north where polar ice exists, there is almost no life at all.

The wrinkling of the earth, its turning on its axis, and its spatial relationship with the sun create a wide variety of habitats at its surface. Plants and animals have invaded these rather generally, and ecosystems have been established. These may be as large as the great oceans, plains, deserts, or as small as ponds or brooks. The limits of space preclude a coverage of each of these but we can take a brief look at two of them.

Fig. 28–15. A sketch demonstrating the life zones at different altitudes and latitudes. Note that the successively higher habitats support the same types of vegetation as are found at successively greater distances from the equator.

Snow and ice

Lichen and mosses

Low herbaceous vegetation

Coniferous forest

Deciduous forests

Rain forests

POLAR   TUNDRA   TEMPERATE   TROPICAL

Fig. 28–16. A typical freshwater pond. The association of plants and animals in such an enviroment is extremely complex.

ECOSYSTEMS

A Freshwater Pond

A pond is defined as a small body of fresh water, usually not more than two or three meters in depth, its temperature being approximately the same throughout (Fig. 28–16). Many animals and plants live in such a limited environment, and even within its confines there are definite regions which support specific animals. The open water is largely devoid of both fish and rooted plants, but the shores support a variety of animal and plant life, depending on the relative amounts of mud, sand, or rocks.

If the bottom is muddy, many plants, such as water lilies, grow in profusion. In protected places around the edge there may be several varieties of fish, principally bass and pickerel. Crayfish and small fish may be seen darting here and there in search of food. Tadpoles can be found near the bottom. The water teems with tiny crustacea and larvae of insects like midges (small gnatlike flies in the adult stage), which form the basic food for young fish. By scooping up some of the

mud in a fine mesh net, many other animals can be noted, including snails of various sizes and shapes and perhaps a few leeches.

Many different kinds of flying insects make their home around the edge of the pond. The dragonfly and May fly larvae can be found. An occasional diving beetle (Fig. 28–17) may be picked up. This is an interesting insect because it is so well adapted to aquatic life even though it must breathe air. It carries a film of air under its wings which acts as a reservoir for underwater maneuvering. The hind legs are large and beautifully designed for swimming under water.

The pond may include a sandy shore where animals of a different kind live. Snails, different from those found on the mud bottom, crawl over the sand from which they remove the small plants growing there. Frogs, toads, and turtles may live around the edge of the pond, and birds such as the redwinged blackbird may inhabit the vegetation along the shore. Although these animals do not live in the water, they do contribute to the combined interrelationships of the community. They seek at least some of their food in the water and when they die their bodies may fall into the water where they are eaten by animals living there.

It is obvious that there must exist many complex food chains in such a well-defined community. The chief occupation of each living thing is to nourish itself, a need that results in a severe struggle for existence. Rarely does an animal die a natural death, for the moment it wavers it is pounced upon and destroyed by another, thus becoming a part of a long or short food chain. There is, however, a complete food cycle for the entire community which involves certain general groups of plants and animals. The green plants always provide the beginning of such a food chain. In the case of the pond, the water plants extract their simple needs from the water and manufacture food which is consumed by **plant feeders**, such as the tiny crustacea that feed on algae and the snails that eat larger plants. These animals are pursued and eaten by a large variety of carnivores such as the predaceous diving beetles. These

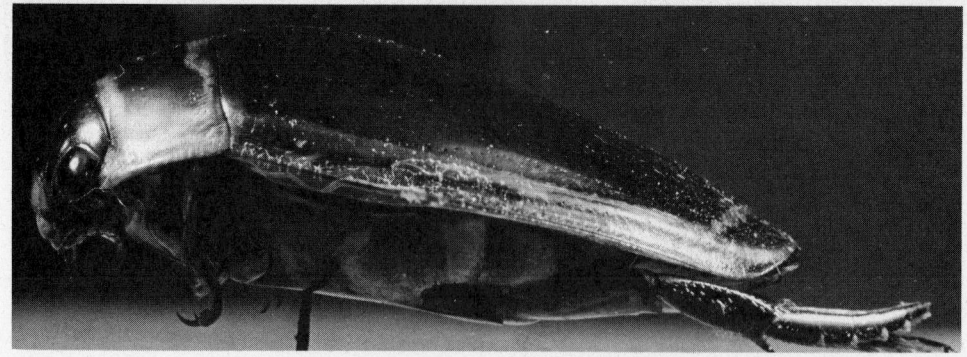

Fig. 28–17. Aquatic insects are modified in many ways for life in the water. The predaceous diving beetle (*Dytiscus*) shown here has its hind legs fringed with hairlike bristles which serve to increase the effectiveness of these appendages when used in swimming. As a result, it is an excellent swimmer. The insect also stores reserve air under its wings for use while submerged.

predators are eaten by larger carnivores such as fish. Some, like the midge larvae and May fly larvae, are content with the dead bodies of plants and animals, and are known as **scavengers.** Finally, the organic matter that remains after all animals are through with it is decomposed by bacteria, known as **decomposers,** so that the inorganic compounds that are needed by the plants are restored to the water. Thus the cycle is completed.

Seasonal changes occur in the pond community and the animals living there must adjust to them. Usually there is an abundance of water in the pond during the spring of the year, but as fall approaches there may be very little water left. Consequently animals must migrate to other ponds or go into a resting stage until conditions improve. The sheet of ice which covers the pond during the winter excludes most of the light, which results in less photosynthesis, hence less oxygen. If animals are to survive the winter they must be able to get along on minimum quantities of oxygen as well as withstand low temperatures. Each animal living permanently in such a pond community is able to meet these situations and is found year after year in the same locality.

Fig. 28–18. Because of the exacting environmental conditions of a desert plain, few plants and animals subsist there. But even here there is an intricate relationship between them.

A Desert Plain

In contrast to the pond, the desert environment supports only those plants and animals that can survive on very little or no water, except that taken with the food. If we study the life existing on the plains of our own Southwest, we find a strange group of plants and animals, all of which are adapted in one way or another to life in a hot dry climate (Fig. 28–18). This region is not truly desert; it is about half-way between grassy plains and true desert. The vegetation is composed chiefly of cacti, yuccas, and other plants particularly well adapted to prevent loss of water through transpiration. They also possess water-holding devices which permit storing of water during the brief wet seasons to be used when moisture is reduced to minimal levels. They are protected from the marauding attacks of hungry and thirsty animals by their sharp stiff spines and tough outer coverings.

Desert animals possess some very interesting adaptations to life in this dry climate some of which were discussed earlier (p. 443). The tough impervious integuments of reptiles such as snakes and lizards reduce water loss to a minimum. Moreover, they excrete **uric acid** (birds do likewise) which is essentially dry and in the form of crystals. Mammals, on the other hand, excrete **urea**

which requires considerable quantities of water to flush it out of the body. In spite of this "fault" some mammals, such as the kangaroo rat and the pocket mouse, can live indefinitely on dry seeds and do not require drinking water. Moreover they do not use water for temperature regulation and their urine is extremely concentrated. By day they remain in their burrows where the relative humidity is 30–50 percent as compared to 0–15 percent above ground. They are abroad at night when the relative humidity is about the same as it is in the burrows. Those desert animals that require drinking water and cannot survive on dry food manage to quench their thirst by eating the succulent cacti and other desert plants that have the capacity to store water. The largest carnivores of the desert are foxes, coyotes, and bobcats, all of which feed on the lesser creatures, which are primarily vegetarians.

Compared to the pond situation, the desert food cycle is rather simple, as would be expected where there are fewer organisms. Like the food cycle in any ecosystem, the desert cycle begins with the plants. These are preyed upon by the various herbivorous animals, which in turn are eaten by the carnivores, which probably also eat one another. Eventually death overtakes them and the elements of which each is composed return to the soil to be used again by the plants.

Fig. 28–19. Life on Heron Island, one of the thousands that make up the Greater Barrier Reef. (Top) A scene from the reef looking back on the exposed island at low tide. The scattered corals appearing above the water are the dead skeletons; living corals are below the water. (Bottom) A mutton bird (*Puffinis pacificus*), one of the thousands that nest on the island. This confused bird laid its egg on the ground; normally they are laid in burrows in the coral sand.

Other Ecosystems

There are a great many other ecosystems in a wide range of habitats. Some of these are ocean, grassy plains, tundra, forest, and mountains. One of the most interesting communities of the ocean ecosystem is the coral reef (p. 163), the largest of which is the Great Barrier Reef off the northeast coast of Australia. Of the thousands of islands which make up this 1,200-mile long reef, one— Heron Island located at the southern tip— has been studied quite extensively due to the fact that a biological station is located there. The vegetation consists of a few plants and the Pisonia tree (Fig. 28–19). The noddy tern nests in these trees and the mutton bird or wedgetailed shearwater burrows in the coral sand where it lays its single large egg. The enormous green and loggerhead turtles seasonally lay their eggs in the coral sand (Fig. 28–20). Whereas the exposed island itself is only several hundred feet across, the reef extends about 5 miles and is covered by shoal waters that vary in depth with the tides. At low tide the biologists can have a most delightful experience studying the vast array of coral life. There is probably no greater concentration of life in any other ecosystem and the variety in color, form, and activity is fantastic.

In the ecosystems of the world many

Fig. 28–20. Heron Island. A giant female loggerhead turtle (*Thallassochelys caretta*) laying eggs. She digs a hole with her hind flippers, deposits her eggs, and then covers them so expertly that they cannot be detected. The warm sun incubates the eggs.

niches have developed, inhabited by certain species of plants and animals which are similar as regards specific needs. Together they constitute the complex life pattern of the earth.

THE ECOSPHERE AND THE FUTURE

La Mont Cole introduced the term **eco-sphere** which is defined as "the sum total of life on earth together with the global environment and the earth's total resources." The word is a hybrid of ecosystem and biosphere (all living organisms on earth). By treating our planet from this point of view, some estimates can be made on how much life it can support.

According to Cole's estimates, the earth can produce a maximum of 410 billion tons of protoplasm (all types) per year. The available energy places an upper limit on production of living things. Just how close we are to this maximum is not known but with the rate of energy utilization increasing exponentially, there is no doubt but that the limit will be reached at some future time. In spite of present and future accomplishments in improving productivity, we will someday reach a plateau in which the amount of available energy will be converted to the maximal amount of protoplasm. In the meantime, what is happening to the energy available to us?

To answer this question we must consider what influence man is having on the ecosphere. This most populous of all animals demands a large fraction of the total productivity in food. If he eats nothing but plants, it has been estimated that he consumes one percent of the total productivity of the earth. This is remarkable when we consider that there are well over one million species of animals. If he ate only animal protein he would consume four percent of the total resources. Such a tremendous consumer must have a drastic effect on the physiology of the ecosphere, and indeed he does.

Man is utilizing fossil fuels and minerals

at an alarming rate. The supply is not inexhaustible. Energy can be obtained from cleaving atoms, but here too the supply is not unlimited. More distressing is the waste of natural resources. Streams cause the erosion of valuable topsoil which is essential for plant growth. This diminishes the arable land of the world. Moreover, the streams are being polluted at an unprecedented rate which destroys both plants and animals in them, to say nothing of rendering the water itself unsuitable for human consumption. Even more discouraging is the rapid pollution of the air we breathe. This has markedly increased in industrial areas in recent decades. It has already reached a point where the health of people living in these areas is affected. Fortunately this sad state of affairs has come to public attention, particularly in the United States, and laws are being enacted to curb further degradation and in some instances reverse the damage already done.

The deleterious effect on the ecosphere brought about by man would not take place if his numbers were kept within bounds. Unfortunately, this is not the case. Man, like all animals, has the capacity to reproduce himself far beyond his food supply, if no checks intercede. In all parts of the world populations are "exploding," primarily as a result of a drop in death rates. At the present time approximately 270,000 babies are born each day and only 142,000 people die, leaving a daily increase of 128,000. This means that each year approximately 47,000,000 individuals are added to the total world population. Observe from Figure 28–21 that by the year 2000 the population will reach between six and eight billion as compared to 3.9 billion in 1975. Since this is a logarithmic curve (p. 628), the situation will be compounded with each year thereafter until human protoplasm will crowd everything else off the earth. It has been estimated that if the present rate continues by the year 2500 there would be five square yards of the earth's surface for each person. The heart of New York City has one person for every 35 square yards today, and most people would agree that even

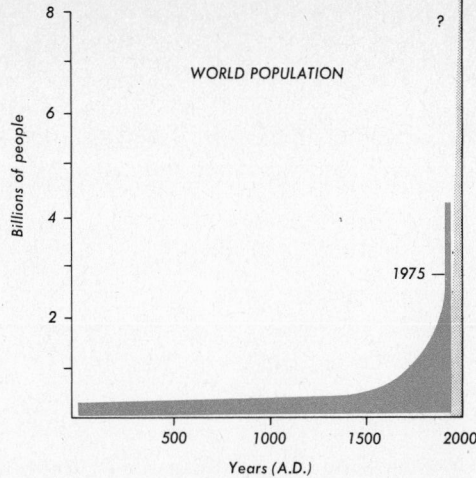

Fig. 28–21. Human populations follow the logarithmic growth curve just as other organisms do. This graph shows the curve for the entire world population from the beginning of the Christian era to the year 2000. The extrapolated curve beyond 1975 indicates a population of over eight billion by the year 2000.

this concentration covering deserts and mountain peaks would hardly be compatible with the "good life." There seems little doubt that the "population problem" must be dealt with as hastily as possible. Even now sociologists, as well as biologists, are deeply concerned and many already fear it is too late to bring about effective controls.

Even though there are a number of ways to improve food production, and these are being explored, their combined effect does little more than delay the problem of overpopulation. Indeed, with increased food supply there is always increased numbers of offspring. The only remedy for overpopulation is a lower birth rate. This has religious, social, national, and biological implications which means that it is a very complex problem. Nevertheless, it is one about which all responsible people must be thinking. College students, who will become the leaders of tomorrow, should take an active part in bringing about a satisfactory solution to this problem. If it is solved we can look forward to a better world for all mankind.

Books

ALLEE, W. C., EMERSON, A. E., PARK, O., PARK, T., and SCHMIDT, K. P., *Principles of Animal Ecology*. Philadelphia: Saunders, 1949.

BATES, M., *The Forest and the Sea*. New York: Random House, 1960.

*CARSON, R., *The Sea Around Us*. New York: Oxford University Press, 1961.

*CARSON, R., *Silent Spring*. Boston: Houghton Mifflin, 1962.

CHANDLER, A. V., and READ, C. P., *Introduction to Parasitology*. New York: Wiley, 1961.

ELTON, C., *Animal Ecology*. New York: Macmillan, 1939.

GILBETT, K., and McNEIL, F., *The Great Barrier Reef and Adjacent Islands*. Sydney, Australia: The Coral Press Pty., 1959.

NOBEL, R. D., and NOBLE, G. A., *The Biology of Animal Parasites*. Philadelphia: Lea and Febiger, 1961.

ODUM, E. P., *Ecology*. New York: Holt, Rinehart & Winston, 1968.

RICKLEFS, R. E., *Ecology*. Newton, Mass.: Chiron Press, 1973.

* Available in paperback.

Articles

BARTHOLOMEW, G. A., and HUDSON, J. R., "Desert Round Squirrels." *Scientific American*, **205**, 107, November, 1961.

BROWER, L. P., "Ecological Chemistry." *Scientific American*, **220**(2), 22, February, 1969.

COLE, L. C., "The Ecosphere." *Scientific American*, **198**, 83, April, 1958.

DEEVEY, E. S., "Life in the Depths of a Pond." *Scientific American*, **185**, 68, October, 1951.

DIETZ, R. S., "The Sea's Deep Scattering Layers." *Scientific American*, **207**, 44, August, 1962.

FREJKA, T., "The Prospects for a Stationary World Population." *Scientific American*, **228**, 15, March, 1973.

ISAACS, J. P., "The Nature of Ocean Life." *Scientific American*, **221**(3), 146, September, 1969.

NICHOLAS, G., "Life in Caves." *Scientific American*, **192**, 98, May, 1955.

RYTHER, F. H., "The Sargasso Sea." *Scientific American*, **194**, 98, January, 1950.

WENT, F. W., "The Ecology of Desert Plains." *Scientific American*, **192**, 68, April, 1955.

WOODWELL, G. M., "Toxic Substances and Ecological Cycles." *Scientific American*, **216**, 24, March, 1967.

29

CLASSIFICATION OF THE ANIMAL KINGDOM

Subkingdom Protozoa

PHYLUM PROTOZOA.   Protozoa. Mostly unicellular, some colonial. Habitat moist or aquatic, freshwater and marine. Free-living or parasitic. Sexual and asexual reproduction.

SUBPHYLUM SARCOMASTIGOPHORA.   Possess flagella, pseudopodia, or both locomotor organelles. Single nucleus is usual. Sexual stages, when present, are syngamous.

SUPERCLASS I. MASTIGOPHORA.   One or more flagella present in trophozoites; singles or colonies; asexual reproduction by longitudinal division; no sex in many groups. May be autotrophs, heterotrophs, or both.

Class 1. Phytamastigophorea.   Plantlike; most have chromatophores; one or two flagella; amoeboid forms present in some groups; sexual reproduction in some orders; mostly free-living. There are 10 orders; the most significant are listed below.

Order 1. Chrysomonadida.   One to three flagella; yellow to brown chromatophores. *Ochromonas.*

Order 2. Dinoflagellida.   Two flagella, one transverse, one trailing; transverse and longitudinal groove; green, yellow, or brown. *Peridinium, Gonyaulax, Noctiluca.*

Order 3. Euglenida.   Two flagella arising from reservoir; usually green; changes shape when moving. *Euglena, Phacus, Peranema.*

Order 4. Volvocida.   Two to four apical flagella, grass green, cup-shaped chromatophores; some form colonies. *Chlamydomonas, Haematococcus, Gonium, Volvox.*

Class 2. Zoomastigophorea.   Colorless, animal-like with variable structures. Have one to many flagella; no chromatophores; some are amoeboid. Mostly parasites. Three orders listed as examples below.

Order 1. Kinetoplastida.   One to four flagella; have a kinetoplast which contains DNA. *Trypanosoma, Leishmania.*

Order 2. Diplomonadida.   Bilaterally symmetrical. *Giardia.*

Order 3. Trichomonadida.   Four to six flagella, one directed posteriorly; no sexuality. *Trichomonas.*

SUPERCLASS II. OPALINATA.   Numerous cilia or flagella, no mouth, numerous nuclei. Parasites of amphibians. *Opalina.*

SUPERCLASS III. SARCODINA.   Move by means of pseudopodia; body naked or covered by a test. Mostly free-living. Numerous subgroups of which only three orders are included below.

Order 1. Amoebida.   All move by lobopodia, a special type of pseudopod. *Amoeba, Entamoeba.*

Order 2. Foraminferida.   Possess a test with one or more chambers and reticulopodia, a special type of pseudopod. *Globigerina.*

Order 3. Actinophryida.   No skeletons; axopodia, a special type of pseudopod. *Actinosphaerium, Actinophrys.*

SUBPHYLUM SPOROZOA.   Produce spores; single nucleus; cilia and flagella absent except in microgametes. All parasitic. Sample subtaxonomy below.

Class Teleosporea. Reproduce both sexually and asexually, move by gliding; microgametes are flagellated.

Order Eugregarinida.   Parasites of annelids and arthropods. *Monocystis.*

Order Eucoccida.   Have schizogony and sexual stages in life cycle. Live in cells of both vertebrates and invertebrates. *Eimeria.*

Suborder Haemosporina.   Gametes develop separately; zygote usually motile; sporozoites naked; schizogony in vertebrate host and sporogony in invertebrate host. *Plasmodium.*

SUBPHYLUM CILIOPHORA.   Possess cilia with subpellicular infraciliature; two types of nuclei; have sex. Mostly free-living.

Subclass Holotrichia.   Ciliature usually uniform. *Paramecium, Tetrahymena.*

Subclass Peritrichia.   No body cilia; attached by stalk; bell-shaped; many colonial. *Vorticella.*

Subclass Suctoria.   No external cilia; attached by stalk; have sucking tentacles. *Tokophrya.*

Subclass Spirotrichia.   Sparse cilia, cirri in some. *Blepharisma, Stentor, Euplotes.*

Subkingdom Metazoa

645

CLASSIFICATION
OF THE
ANIMAL
KINGDOM

PHYLUM MESOZOA. Simplest of all metazoa. Consist of outer cellular layer, commonly ciliated, which encloses one or more reproductive cells. All parasitic. *Dicyemida.*

PHYLUM PORIFERA. Sponges. Adults are sedentary and the larvae flagellated. Loosely diploblastic but not true ectoderm and endoderm. Choanocytes. Spicules give support to body. Various canal systems.

    Class Calcarea. Spicules composed of calcium carbonate; shallow water. Ascon, sycon, and rhagon canal types. *Sycon, Leucosolenia, Scypha.*

    Class Hexactinellida. Spicules composed of silicic acid (silica), hence known as "glass sponges." Deep water, marine. *Euplectella, Hyalonema.*

    Class Demospongia. Soft sponges. Skeletons of siliceous spicules and spongin fibers. Spicules absent in some species. Rhagon canal system. Freshwater and marine. Value as sponges for man. *Spongia, Spongilla.*

PHYLUM COELENTERATA. Coelenterates. Basically two-layered animals (diploblastic), some show mesoglea cells. Single gastrovascular cavity with one opening. All are aquatic, mostly marine. Radially symmetrical. Nematocysts. Colonial forms demonstrate polymorphism.

    Class Hydrozoa. Both polyps and medusae in life cycle. Velum in medusae. Variable size. Fresh- and salt-water. *Hydra, Obelia, Gonionemus, Physalia.*

    Class Scyphomedusae. Jellyfishes. Polyps reduced or absent. Umbrella margin of medusae notched, no velum. Cells in the mesoglea. *Pelagia, Cyanea.*

    Class Anthozoa. Sea pens, sea anemones, corals. Polyps only, individual or colonial. Septa, mesenteries, and tentacles conspicuous. *Astrangia, Metridium, Cribrina.*

PHYLUM CTENOPHORA. Ctenophores (comb jellies, sea walnuts, cat's-eyes). Pelagic (floaters) and marine. Biradial symmetry. Comblike structures a typical characteristic. All are free-swimming. *Cestum, Beroë.*

PHYLUM PLATYHELMINTHES. Flatworms. Dorsoventrally flattened and bilaterally symmetrical. Well-developed mesoderm, hence triploblastic. Mostly parasitic, some terrestrial, freshwater, and marine. Mostly hermaphroditic (monoecious). Excretory system composed of flame-cells.

    Class Turbellaria. Free-living flatworms. Externally ciliated. Mouth ventral. Freshwater, marine, and terrestrial. *Planaria, Phagocata, Dugesia.*

    Class Trematoda. Flukes. Cuticle forms external covering. Suckers present and mouth usually anterior. Parasitic. *Opisthorchis, Fasciola, Schistosoma.*

    Class Cestoda. Tapeworms. Cuticle forms external covering. Scolex with suckers and (usually) hooks. Few to many proglottids. Parisitic. *Taenia, Echinococcus.*

PHYLUM NEMERTINEA. The nemertines. Possess a circulatory system, an anus, and an eversible proboscis enclosed in a tubular cavity called the rhynchocoel. Body form slender, flat and soft. Mostly marine. *Amphiporus.*

PHYLUM ASCHELMINTHES. Diverse group of animals. All possess pseudocoel. Simple digestive tract with muscular pharynx.

Class Rotifera. Wheel animals. Sexes separate (dioecious). Bilaterally symmetrical. Primitive flame-cells. Mostly free-living, some parasitic. Grinding organ, the mastax with well-developed jaws and wheel-like cilia on anterior end. *Epiphanes, Asplanchna.*

Class Nematoda. Round worms. External covering smooth. Cilia absent. Bilaterally symmetrical. Mostly dioecious. Complete digestive tract, mouth to anus. Tube-within-a-tube body plan. *Ascaris, Trichinella, Necator, Wuchereria.*

Class Acanthocephala. Spiny-headed worms. Possess a pseudocoel. Slightly flattened. Proboscis present, usually bearing hooks. Dioecious. Parasitic in the intestines of vertebrates. *Gigantorhynchus.*

PHYLUM ENTOPROCTA. Small group of animals, mostly colonial, attached by a stalk. Ciliated tentacles. Pseudocoel is present. Mostly marine. *Pedicellina.*

PHYLUM ECTOPROCTA. Permanently attached animals which form arborescent or encrusting colonies. Possesses a special structure the lophophore. *Bugula.*

PHYLUM BRACHIOPODA. Lamp shells. Possess bivalve shell and stalks. Resemble mollusks. Marine. *Magellania.*

PHYLUM CHAETOGNATHA. Arrow worms. Mouth contains stout, chitinous teeth, which gives it its name. Relation to other phyla doubtful. Marine. *Sagitta.*

PHYLUM MOLLUSCA. Mollusks. Snails, clams, squids, octopuses, and others. Shell or shells usually present. A ventral muscular foot characteristic. Segmentation absent. Mantle in some forms. 90,000 species, the third largest phylum.

Class Amphineura. Chitons. Eight dorsal body plates in most species. Body bilaterally symmetrical. Large flat foot, strongly adhesive. Marine, on rocks near shore. *Cryptochiton, Mopalia.*

Class Scaphopoda. Tooth shells. Foot modified for burrowing. Shell shaped like a tooth and open at both ends. Fine, threadlike tentacles. Marine and bottom dwellers. *Dentalium.*

Class Gastropoda. Snails, slugs, and whelks. Single, strong, flat ventral foot. Special feeding organ, the radula. Head faintly distinct, tentacles prominent. Trochophore larvae in some species. Both fresh- and salt-water. Economically important to man. *Haliotis, Littorina, Helix.*

Class Pelecypoda. Clams. No radulas or tentacles. Functional mantle present. Foot spade-shaped adapted for digging in sand and mud. Fresh- and salt-water. *Venus, Ostrea, Anodonta.*

Class Cephalopoda. Squids, octopuses, and nautiluses. Foot divided into arms or tentacles which vary in number among the different species and which possess adhesive sucking cups or disks. Some possess internal or external shell. Radula and siphon present. All marine. *Loligo, Nautilus, Sepia.*

PHYLUM ANNELIDA. Segmented worms. Serially arranged segments or metameres. Bilaterally symmetrical. Well-developed digestive, circulatory, nervous, muscular, and reproductive systems. Tube-within-a-tube body plan well developed. Coelum present. Some marine forms possess a trochophore larva. Marine, freshwater, terrestrial, a few are parasitic.

Class Archiannelida. Small simple marine annelids. *Polygordius.*

Class Polychaeta. Clamworms. Parapodia on each segment and bearing setae. Tentacles in head region. Usually dioecious. Mostly marine. *Nereis, Aphrodite, Chaetopterus.*

Class Oligochaeta. Earthworms. Parapodia and head absent. Setae small. Monoecious. Damp soil and freshwater. *Lumbricus, Tubifex, Enchytraeus.*

Class Hirudinea. Leeches. Parapodia, setae, and head absent. One or two suckers. Segments subdivided into annuli. Freshwater and parasitic. *Hirudo, Placobdella, Hemopis.*

PHYLUM ONYCHOPHORA. Peripatus. Rare group showing both annelid and arthropod characteristics. Short unjointed appendages. Body annelidlike. Breathing and circulatory systems resemble arthropods, also coelom. Probably similar to arthropod ancestors. *Peripatus.*

PHYLUM ARTHROPODA. Arthropods. Most distinguishing characters are the jointed appendages and chitinous or limy exoskeleton. Basic body regions are head, thorax, and abdomen (modified in some). Heart dorsal, nervous system ventral. Bilateral symmetry. Largest number of species of all phyla—750,000.

SUBPHYLUM TRILOBITOMORPHA. Trilobites. All extinct.

SUBPHYLUM CHELICERATA. Without antennae and first pair of appendages modified into pincerlike chelicerae. Six pairs of appendages.

Class Eurypterida. Sea scorpions. All extinct.

Class Xiphosurida. Horseshoe or king crab. Very ancient forms with few living representatives. Marine. *Limulus.*

Class Pycnogonida. Sea spiders. Shortened body and very long legs. Marine. *Nymphon.*

Class Arachnida. Spiders, scorpions, ticks, and mites. Breathing organs are book lungs in scorpions and spiders. *Lactrodectes, Centrurus, Eutrombicula, Dermacentor.*

SUBPHYLUM MANDIBULATA. Appendages near mouth modified into mandibles or jaws.

Class Crustacea. Crustaceans (crabs, lobsters, crayfish, barnacles, sowbugs). Limy skeleton. Breathe with gills. Two pairs of antennae. Mostly freshwater or marine; head and thorax fused into a cephalothorax. *Cambarus, Callinectes, Balanus.*

Class Diplopoda. Millipedes. Cylindrical bodies of many segments, each with 2 pairs of legs (50 to 200 pairs). Short antennae. No poison glands. *Julus, Spirobolus.*

Class Chilopoda. Centipedes. Many segments, each bearing a pair of legs, which range in number from 15 to 170 in different species. Prominent segmented antennae. Possess poison glands. *Scolopendra, Lithobius.*

Class Insecta. Insects. Three pairs of walking legs and usually wings at some stage. A pair of antennae. Three body regions: head, thorax, and abdomen. Graded metamorphosis (complete, incomplete, gradual, or none). Many orders, some of which are listed below.

Order Thysanura. Bristletails or silverfish. No metamorphosis. Wingless. Chewing mouth parts absent. Long caudal bristles. *Lepisma, Thermobia.*

Order Neuroptera. Ant lions, lacewings, doodlebugs. Metamorphosis complete. Biting mouth parts. Wings with cross veins. Adults and larvae terrestrial and carnivorous. *Chrysopa, Myrmeleon.*

Order Orthoptera. Grasshoppers, locusts, crickets, cockroaches, and others. Gradual metamorphosis. Biting mouth parts. Usually two pairs of wings, many with leaping legs. Called the singing insects. *Mantis, Romalea.*

Order Odonata. Dragonflies and damsel flies. Metamorphosis incomplete. Two pairs of membranous wings and large compound eyes. Name derived from toothlike biting mouth parts. Larval forms are aquatic and breathe by means of rectal gills. *Gomphus.*

Order Ephemeroptera. May flies. Incomplete metamorphosis. Two pairs of wings, hind ones small. Rudimentary chewing mouth parts. Adults live very short periods. *Ephemera, Polymitarcys.*

Order Anoplura. Sucking lice. Gradual metamorphosis. Wingless. Sucking and piercing mouth parts. Ectoparasites of the hairy regions of mammals. *Pediculus, Phthirius.*

Order Hemiptera. True bugs. Gradual metamorphosis. Piercing and sucking mouth parts. Two pairs of wings in some, others are wingless. *Cimex, Lygus.*

Order Homoptera. Cicadas, aphids, leaf hoppers, and others. Metamorphosis gradual. Sucking and piercing mouth parts. Mostly with two pairs of wings held dorsally. Parasitic on trees and shrubs. *Aphis, Magicicada.*

Order Coleoptera. Beetles, fireflies, weevils, and others. Metamorphosis complete. Chewing mouth parts. Forewings usually heavy and protective. Exoskeleton horny or leathery. Largest order: 250,000 species. *Leptinotarsa, Dytiscus.*

Order Lepidoptera. Butterflies and moths. Complete metamorphosis. Larvae with chewing mouth parts, adults with sucking. Usually four membranous scaly wings. Compound and simple eyes. Proboscis long and coiled in most. Larvae spin cocoons from silk and make other provisions for pupae. Antennae are feathery in adult moths and knobbed in adult butterflies. Moths hold wings horizontally when at rest, whereas butterflies hold wings vertically. Some larvae do serious damage to crops. *Carpocapsa, Hibernia, Anosia, Papilio.*

Order Siphonaptera. Fleas. Metamorphosis complete. Piercing and sucking mouth parts. Small head, body flattened laterally, and leaping legs. Ectoparasites on birds and mammals. *Ctenocephalides, Pulex.*

Order Hymenoptera. Bees, ants, wasps, and others. Referred to as the social insects. Complete metamorphosis. Chewing and sucking mouth parts. Four membranous wings. Some possess a sting, saw, or a piercing organ. Of great economic significance. *Apis, Bombus, Pheidole.*

Order Diptera. Flies, mosquitos, and others. Metamorphosis complete. Two wings. *Culex, Musca.*

PHYLUM ECHINODERMATA.
Echinoderms. Starfish, sea urchins, and others. Endoskeleton and water vascula system characteristic. Also pentamerous body plan. Adults radially symmetrical. All marine.

SUBPHYLUM PELMATOZOA.
Primitive stalked forms. Only one living class.

Class Crinoidea. Sea lilies. Stalked forms attached to ocean floor. Adults usually pentamerously branched to form arms with pinnules. Madreporic plate absent. Tube-feet without suckers. *Pentacrinus, Antedon.*

SUBPHYLUM ELEUTHEROZOA.
Free-moving unattached forms. All marine.

Class Asteroidea. Starfishes. Most are prominently pentamerous. Madreporic plate and pedicellariae present. Numerous tube-feet lying in ambulacral grooves. Most possess powers of autonomy and regeneration of lost arms. *Asterias, Solaster.*

Class Ophiuroidea. Brittle stars. Arms markedly flexible and set off from central body. No ambulacral grooves. Autonomy and regeneration marked. No pedicellariae. *Ophioderma, Aphiothrix.*

Class Echinoidea. Sea urchins, sand dollars, and others. Body rounded or extremely flattened. Madreporic plate aboral. Arms absent. Tube-feet, spines, and pedicellariae present. Skeletal tests present. *Arbacia, Strongylocentrotus.*

Class Holothuroidea. Sea cucumbers. Soft cylindrical bodies. Much reduced skeleton, spicules only embedded in body wall. Anal respiration by means of respiratory trees. Arms, pedicellariae, and spine absent. Madreporic plate internal. Muscular body wall. Autonomy and regeneration marked. *Thyone, Cucumaria.*

PHYLUM HEMICHORDATA.
Acorn worms. Well-developed enterocoelum and gill slits. Primitive dorsal nervous system and internal skeleton in form of a notochord. Prominent proboscis. Wormlike. All marine. *Saccoglossus.*

PHYLUM CHORDATA. Chordates. All possess, at some stage in their life cycle, well developed gills or gill slits, dorsal tubular nerve cords and a notochord.

SUBPHYLUM UROCHORDATA. Tunicates (sea squirts). Larvae show chordate characteristics whereas the adults become sedentary and lose the neural tube and notochord. Adults possess primitive circulatory system. All marine. *Styela, Amaroucium.*

SUBPHYLUM CEPHALOCHORDATA. Amphioxus. Fusiform body shape showing prominent muscle segments (myotomes). Coelom well developed. Circulatory system without discrete heart. Marine. *Branchiostoma.*

SUBPHYLUM VERTEBRATA. Vertebrates. Internal skeleton of cartilage or bone (both in some), the main axis being the spinal column composed of overlapping vertebrae. Nervous system with brain dorsal to the digestive tract. Two pairs of appendages. Red blood.

Class Agnatha. Lampreys and hag fish. Scales, jaws, and appendages absent. Mostly ectoparasitic on bodies of other vertebrates. Both monoecious and dioecious in different species. Some have rasping tongue and more than seven pairs of gills. Most primitive of vertebrates. Marine and freshwater. *Petromyzon, Bdellostoma.*

Class Chondrichthyes. Sharks, rays, skates, and chimaeras. Cartilaginous endoskeleton. Placoid scales in most species. Rudimentary notochord. Spiral valve in intestine. No opercula over gills in most. Mostly marine. *Squalus, Raja.*

Class Osteichthyes. Bony fishes. Scales are never placoid. Endoskeleton of bone and cartilage. Caudal fin effective in locomotion. Swim bladder present, becomes lungs in some species. Paired pectoral and pelvic fins, highly modified in some. Operculum present. Marine and freshwater.

Subclass Choanichthyes. Lobe-finned and lungfishes. Nostrils open into mouth cavity. Paired fins possess enlarged median lobes. Most species fossils. *Protopterus, Latimeria.*

Subclass Actinopterygii. Ray-finned fishes. Nostrils not connected with mouth cavity. Variable body form, mostly streamlined. *Acipenser, Amia, Perca.*

Class Amphibia. Frogs, toads, salamanders, and others. Larvae usually aquatic, adults usually terrestrial. Metamorphosis in most. Adult skin glandular and soft. Usually four walking legs on adults. Eyelids usually present. Adult heart with two auricles and one ventricle; larvae possess one auricle and one ventricle. None marine. Poikilothermal.

Order Apoda. Caecilians. Legless, wormlike. Short tail, or completely lacking. Tiny scales in some. *Ichthyophsis.*

Order Caudata. Newts, salamanders, hellbenders, mud puppies. Tailed forms. Front and hind legs about equal in size. No scales. *Triturus, Ambystoma, Necturus.*

Order Salienta. Frogs and toads. Tailless forms. Front legs small, hind legs large for leaping. Eggs laid in jelly masses and strings. Metamorphosis clearly indicated. No scales. *Rana, Bufo.*

Class Reptilia.   Lizards, snakes, turtles, alligators, and others. Scales or bony plates form external covering. Poikilothermal. The heart ventricle partially separated in most species. Air-breathing throughout life.

Order Squamata.   Lizards and snakes. Scales smooth on snakes, horny and thick on lizards. Cloacal opening transverse. Copulatory organs paired. *Agkistrodon, Anolis.*

Order Testudinata.   Turtles, terrapins, and tortoises. Beaks, bony plates to form boxlike shell, and short legs are the chief characteristics. Turtles live in either fresh or salt water, terrapins live in freshwater, and tortoises live on land. Some reach great size and age. *Chelonia, Terrapene.*

Order Crocodilia.   Alligators, crocodiles, and others. Body form lizardlike and large. Crocodiles have elongated pointed snouts whereas those of alligators are rounded. Two complete ventricles present. Bony plates overlying heavy skin. Valves in nostrils and ears which are closed when under water. *Crocodylus, Alligator.*

Order Rhynchocephalia.   A "living fossil" possessing characteristics of primitive reptiles. Found in New Zealand. Possesses a parietal eye in roof of skull. *Sphenodon.*

Class Aves.   Birds. Feathers characteristic. Wings usually present for flight. Light skeleton. Beak present. Homiothermal. Heart consisting of two auricles and two ventricles. Ubiquitous in distribution. 30,000 species and subspecies. Many orders. The student is referred to bird books for classification. *Pelecanus, Gallus.*

Class Mammalia.   Mammals. Mammary glands which secrete milk for nourishing the young. Hair present on at least part of the body. Sweat and sebaceous glands, red blood, young usually born alive (viviparous), and homiothermal. Over 17,000 species and subspecies.

Subclass Prototheria.   Monotremes. Egg-laying and possessing reptilian characteristics, particularly in the urogenital system and the intestinal openings into the cloaca. *Echidna, Ornithorhynchus.*

Subclass Theria.   Marsupials and placental mammals.

Infraclass Metatheria.   Opossum, kangaroo, and others. Possess primitive placenta and abdominal pouches which house mammary glands. *Didelphis, Macropus.*

Infraclass Eutheria.   Placental mammals. Embryos develop in uterus of female. All viviparous.

Order Insectivora.   Shrews, moles, and others. Shrews smallest living mammals. Nocturnal, usually terrestrial, and insect-eating. *Sorex, Scalopus.*

Order Chiroptera.   Bats. Forelimbs modified into wings for flight. Sharp teeth and claws. Usually nocturnal. Variable diets, from fruit to blood of vertebrates including man (Vampire bats). *Myotis, Desmodus.*

Order Primates. Monkeys, lemurs, gibbons, and many others including man. Large hands and feet with opposable "thumb." High intelligence. Usually arboreal. *Ateles, Gorilla, Homo.*

Order Carnivora. Mostly flesh-eating. Dogs, cats, bears, mink, seals, and others. Canine teeth large. Usually five-clawed toes. Terrestrial and widespread. *Canis, Felis, Procyon.*

Order Perissodactyla. Horse, rhinoceros, tapir, and others. Hoofed mammals with odd number of toes. Long legs and large body size. *Equus, Tapirella.*

Order Artiodactyla. Cow, camel, deer, and others. Hoofed mammals with even number of toes. Ruminants have four-chambered stomachs. Hoofs on two or four toes. *Bos, Camelus.*

Order Proboscidea. Elephants. Extension of upper lip and nose into a "trunk." Upper incisor teeth greatly enlarged to form tusks. Massive animals with thick skull and skin. *Elephas.*

Order Sirenia. Sea cows—manatee and dugong. Aquatic mammals. Hind legs are lost and front ones developed into flippers. Transverse tail. *Manatus.*

Order Cetacea. Whales, porpoises, and others. Aquatic, primarily marine. Anterior appendages modified into paddles for swimming. Neck absent, small ear-openings, smooth body, and very little hair (modified in some to form whalebone). Usually large, one (the blue whale) is largest animal that has ever lived. *Physeter, Mesoplodon.*

Order Edentata. Anteaters, sloths, and others. Teeth absent in some (anteaters) and others (armadillos and sloths) possess no incisors and canines and no enamel on any teeth. All members highly specialized. Many extinct forms, such as the giant sloth and armadillo. *Bradypus, Dasypus.*

Order Rodentia. Gnawing mammals. Squirrel, hamster, and others. Incisor teeth well developed and chisel-like, canine teeth absent. *Mus, Rattus.*

Order Lagomorphia. Rabbits, conies, and others. Chisel-like incisor teeth, canines absent. *Lepus, Ochotona.*

APPENDICES

GLOSSARY

The glossary is included as an aid in not only defining biological terms, but also in helping the student pronounce them. Their origin is indicated as follows: Gr., Greek; L., Latin; M.E., Middle English; A.S., Anglo-Saxon; Fr., French. Words that do not appear here may be found by referring to the index.

Long words with two accented syllables contain both the primary accent (″) and the secondary accent (′).

Ab-do′men (L. *abdere*, to hide). The belly region of animals, specifically in mammals that part which extends from the diaphragm to the pelvis.

Ab-duc′tor (L. *ab*, from; *ducere*, to lead). A muscle which pulls a structure away from the median line of the animal.

Ab′i-o-gen″e-sis (L. *a*, not; *bio*, life; *genesis*, birth). The doctrine whereby life originates from the nonliving world.

Ab-o′ral (L. *ab*, from; *oris*, mouth). The side opposite the mouth.

Ab-sorp′tion (L. *ab*, away; *sorbere*, to suck in). The taking up of fluids or other substances by living systems.

A-byss′al (L. *abyssus*, bottomless). The bottom waters of the deep sea, marked by the absence of light and hence of plants; where carnivores live.

Ac-cre′tion (L. *accrescere*, to increase). The addition of material to an inert mass of the same material.

Ac′e-tyl-cho″line. A chemical substance produced at or near parasympathetic nerve endings which is thought to be involved in conduction of nerve impulses across synapses.

A-coe′lo-mate (Gr. *a*, not; *koilos*, hollow). Animals without a coelom.

Ac′ro-meg″a-ly (Gr. *akron*, point, peak; *megas*, big). An organic disease caused by the hypersecretion of the anterior pituitary after the long bones have reached maturity.

ACTH. Ad-re″no-cor″ti-co-troph′ic hormone. A hormone from the anterior pituitary.

Ad-ap-ta′tion (L. *ad*, to; *aptare*, to fit). A process by which an organism becomes better suited to its environment.

Ad-duc′tor (L. *ad*, to; *duco*, lead). A muscle which pulls a structure toward the median line of the animal.

Ad′e-nine. A pyrimidine component of nucleic acids and nucleotides.

Ad-he′sion (L. *adhaerere*). A molecular force at surfaces causing unlike molecules to cling together.

Ad′i-pose (L. *adeps, adipis*, fat). Pertaining to fat.

Ad-re′nal gland (L. *ad*, to; *renes*, kidneys; *glans*, acorn). A double endocrine gland located on or near the kidney; consists of the cortex and the medulla, each secreting different hormones.

Ad-re′nal-in. A hormone produced by the medullary portion of the adrenal glands.

Ad-sorp′tion (L. *ad*, to; *sorbere*, to suck in). The attachment of molecules of one substance to the surface of other substances.

A′er-o″bic (Gr. *aer*, air; *bios*, life). Growing only in air or free oxygen. Aerobic respiration involves the Kreb's cycle with oxygen acting as the terminal hydrogen acceptor to form water.

Af′fer-ent (L. *affere*, to bring). Carrying toward. An afferent artery carries blood to the lungs.

Ag-glu′ti-nin (L. *ad*, to; *glutinare*, to glue). A substance in the blood which causes cells or microörganisms to clump.

Al′bi-nism (L. *albus*, white). A condition in which the normal pigment of the skin, hair, and eyes is lacking.

Al-i-men′tary (L. *alimentum*, from; *alere*, to nourish). Pertaining to the digestive tract.

Al-lan′to-is (Gr. *allas*, sausage; *eidos*, form). An extra embryonic membrane arising as an outgrowth of the cloaca in reptiles, birds, and mammals.

Al-lele′ (**allelomorph**) (Gr. *allelon*, of one another). The alternative forms of a gene having the same locus in homologous chromosomes which influence "alternative" characters.

Al′ler-gy (Gr. *all, ergy, ergon*, work). A condition in which substances that are normally harmless cause a marked hypersensitivity or reaction.

Al-ve′o-lus (L. dim. of *alveus*, pit). A small cavity, such as the alveoli of the lungs or the socket of a tooth.

A-mi′no ac′id (from amine NH_2). An organic compound resulting from protein breakdown; it must contain at least one amino group (NH_2) and one acid group (COOH).

Am′i-to″sis (Gr. *a*, without; *mitos*, thread). Direct nuclear division without formation of condensed chromosomes or spindle fibers.

Am′ni-on (Gr. dim. of *amnos*, lamb). A transparent membrane surrounding the embryos of reptiles, birds, and mammals.

Am′ni-o″ta (Gr. dim. of *amnos*, lamb). A group of vertebrates possessing an amnion and an allantois—reptiles, birds, and mammals.

A-moe′boid (Gr. *amoebe*, change). Cell movements resembling those of the amoeba.

Am′phi-as″ter (Gr. *amphi*, both sides of; *aster*, star). A figure formed by the spindle fibers and the two asters in the dividing cell.

Am-phib′i-a (Gr. *amphi*, of both kinds; *bios*, life). Class of vertebrates including frogs, toads, and salamanders which usually spend their larval life in water (water breathers) and their adult life on land (air breathers).

Am′y-lase (L. *anylum*, starch; Gr. *opsis*, appearance). A carbohydrate-splitting enzyme produced by the pancreas.

An-ab′o-lism (Gr. *ana*, up; *bole*, stroke). Constructive metabolism.

An′aer-o″bic (Gr. *an*, negative; *aer*, air; *bios*, life). Growing only in the absence of air or free oxygen.

A-nal′o-gous (Gr. *ana*, up; *logos*, ratio, proportion). Body parts with similar functions but usually genetically dissimilar.

An-am′ni-o″ta (Gr. *a*, not; *amnion*, inner membrane around the fetus). A group of vertebrates without an amnion—cyclostomes, fishes, and amphibia.

An′a-phase (Gr. *ana*, up; *phasis*, appearance, aspect). A stage in mitosis following metaphase when chromosomes migrate from the equatorial plate to the poles of the cell.

A-nas′to-mo″sis (Gr. *ana*, up; *stoma*, mouth). The union of two or more blood vessels, nerves, or other structures.

A-nat′o-my (Gr. *ana*, up; *temnein*, to cut). The study of the structure of animals and plants.

An′i-on (Gr. *ana* up + *iōn* going). A negatively charged ion.

An-nel′i-da (L. *annulus*, ring; Gr. *eidos*, resemblance). A phylum of animals.

An-ten′na (L. a sailyard). Tactile sense organs on the heads of arthropods.

An-ten′nules (L. dim. of antenna). Tactile sense organs located near the antenna of many arthropods.

An-te′ri-or (L. comp. of *ante*). Toward the head end of an animal.

An″ti-bod′y (Gr. *anti*, against; M.E. *bodi*, body). A substance produced in the body as a result of the presence of a foreign body (antigen).

An′ti-gen (Gr. *anti*, against; *genes*, born). A substance which, when introduced into the body, stimulates the production of antibodies.

An′ti-tox″in (Gr. *anti*, against; L. *toxicum*, poison). A substance in the body fluids naturally occurring or induced which neutralizes specific poisons or toxins.

A′nus (L. *anus*). The exit of the alimentary canal opposite the mouth.

A-or′ta (Gr. *aeirein*, to lift). The large artery carrying blood away from the heart.

Ap′er-ture (L. *aperire*, to uncover). An opening.

Ap′i-cal (L. *apex*, summit). The top or apex.

Ap-pend′age (L. *ad*, to; *pendere*, to hang). A part of the body that extends some distance, such as an arm or leg.

Ar-ach′nid-a (Gr. *arachne*, spider). A class of the Arthropoda, including the spiders, ticks, mites, and scorpions.

Ar-bo′re-al (L. *arbor*, tree). Tree-living.

Ar-chen′ter-on (Gr. *arche*, beginning; *enteron*, gut). The primitive gut of an embryo produced by gastrulation.

Ar′te-ry (L. *arteria*, windpipe, artery). A blood vessel carrying blood away from the heart.

Ar′thro-pod-a (Gr. *arthron*, joint; *pous*, foot). A phylum of animals.

Ar-tic′u-la-tion (L. dim. of *articulus*, joint). A joint.

A-sex′u-al re-pro-duc′tion (L. *a*, not; *sexus*, sex; L. *re*, again; *productio*, production). Reproduction without involving sex cells.

As-sim′i-la″tion (L. *ad*, to; *similis*, like). The conversion of the end products of digestion into protoplasm.

As′ter (Gr. *aster*, star). The star-shaped structure fomed round the centriole in mitosis.

A-sym′me-try (Gr. *a*, without; *syn*, with; *metron*, measure). Without symmetry.

At′a-vism (L. *atavus*, ancestor). Resemblance to a remote ancestor.

ATP. Ad-e-no″sine-tri-phos′phate. The major energy-bearing molecule in cells.

At′ro-phy (Gr. *a*, not; *trephein*, to nourish). The gradual regression of the whole body or of its parts.

Au′di-to-ry (L. *audire*, to hear). Pertaining to hearing or ears.

Au′ric-le (L. *auris*, ear). Earlike projection, as on the head of planaria, or a heart chamber into which the veins empty.

Au-to-nom′ic (Gr. *autos*, self; *nomos*, law). Self-operating.

Au′top-sy (Gr. *autos*, self; *optos*, seen). The dissection of a body (post mortem) to determine the cause of death.

Au′to-some (Gr. *autos*, self; *some*, body). The chromosomes other than the sex chromosomes.

Au-tot′o-my (Gr. *autos*, self; *tomos*, cutting). The automatic breaking off of a part of the body, such as in arthropods.

A′ves (L. *avis*, bird). A class of Vertebrata to which the birds belong.

A-xen′ic (*a* neg. + Gr. *xenos* a guest-friend, stranger). A culture containing one species of living organism.

Ax′is (L. *axis*, axle). An imaginary line through an animal, about which the organism is oriented; also, it refers to the anterior second vertebra.

Ax′on (Gr. *axon*, axle). The nerve fiber that conducts the impulse away from the cell body.

Bar′na-cle. A crustacean.

Ber′i-ber′i. A disease resulting from thiamine deficiency.

Bi′ceps (L. *bi*, two; *caput*, head). A muscle with two parts at one end, at the flexor muscle of the human arm.

Bi-cus′pid (L. *bis*, twice; *cuspis*, point). A structure with two cusps or points, as a bicuspid tooth or biscuspid valve between the left auricle and ventricle of the mammalian heart.

Bi-lat′er-al sym′me-try (L. *bis*, twice; *latus*, side; Gr. *syn*, with; *metron*, measure). The arrangement of body parts so that the right and left halves are mirror images of each other.

Bile pig′ments (L. *bilis*, bile; *pingere*, to paint). The colored pigments of bile (bilirubin, biliverdin) resulting from hemoglobin breakdown.

Bi′na-ry fis′sion (L. *bi*, two; *fissus*, cleft). The type of reproduction (asexual) in which division into two parts is approximately equal.

Bi-no′mi-al no′men-cla-ture (L. *bis*, twice; *nomen*, name; *calare*, to call). The international system of naming animals whereby two names are used. The first is generic, the second specific.

Bi-o-gen′e-sis (Gr. *bios*, life; *genesis*, birth). The doctrine that life comes only from preëxisting life.

Bi-ol′o-gy (Gr. *bios*, mode of life; *logos*, discourse).The science of plant and animal life.

Bi-o-lu′mi-nes′cence (Gr. *bios*, mode of life; L. *lumen*, light). The emission of light by living organisms.

Bi′o-mass. Total weight of living organisms in any area or volume.

Blas′to-coel (Gr. *blastos*, bud; *koilus*, hollow). Cavity of the blastula.

Blas′to-pore (Gr. *blastos*, bud; *poros*, passage). The opening into the archenteron of the gastrula.

Blas′tu-la (Gr. *blastos*, bud). The early embryo in which the cells form a hollow ball.

Bow′man's cap′sule (named after Sir Wm. Bowman, English physician. L. *capsula*, a little box). The cuplike structure at the end of the kidney tubule which surrounds the glomerulus.

Bra′chi-al (L. *brachium*, arm). Pertaining to the arm of a vertebrate.

Bra′chi-ate (L. *brachium*, arm). A manner of swinging from limb to limb in trees with the hands, such as apes employ.

Brack′ish. Water intermediate in salt content between sea water and freshwater.

Bron′chi-ole (Gr. *bronchos*, windpipe). A tiny air tube in the lung.

Bron′chus (Gr. *bronchos*, windpipe). One of the two main air passages of the trachea.

Bruise. A blood clot in the tissue.

Buc′cal (L. *bucca*, cheek or mouth cavity). The mouth cavity.

Bud′ding (M.E. *budde*). Reproduction by the splitting off and subsequent development of a small portion of the original animal.

Cae′cum (L. *caecus*, blind gut). A blind sac.

Cal-ca′re-ous (L. *calcarius*, limestone, limy). Containing lime or calcium carbonate.

Cal′or-ie (L. *calor*, heat). The amount of heat energy equal to that used in raising a kilogram of water from 14.5 to 15.5 degrees centigrade. A calorie, when written with a small *c*, is one thousandth of a Calorie.

Can′cer (L. crab, ulcer, sign of the zodiac). One of a variety of malignant tumors.

Ca′nine. (L. *canis*, dog). Pertaining to the pointed tooth next to the incisors, and so named because of its doglike characteristics.

Cap′il-la-ry (L. *capillus*, hair). Tiny blood vessel with walls of one cell layer which connects arterioles to venules.

Cap′sule (L. *capsula*, little box). Any fibrous or membranous envelope, covering a part or the whole of an organism.

Car′bo-hy″drate (L. *carbo*, coal; Gr. *hydor*, water). An organic compound (starch, sugar) containing carbon, hydrogen, and oxygen in which the last two atoms are in the same proportions as in water (H_2O).

Car′di-ac (Gr. *kardia*, the heart). Pertaining to the heart.

Car-niv′o-ra (L. *caro*, flesh; *vorare*, to devour). Flesh-eaters.

Car′ti-lage (L. *cartilago*, gristle). A white elastic secretion forming certain skeletal structures of vertebrates, particularly of embryos and the very young.

Cas-tra′tion (L. *castratio*). The act of removing the sex glands, usually from males.

Ca-tab′o-lism (Gr. *kata*, down; *ballein*, to throw). Destructive metabolism.

Cat′a-lyst (Gr. *kata*, down; *lysis*, a loosing). A substance that alters the rate of a chemical reaction, which would go very slowly or not at all without it.

Cat′i-on (Gr. *kata* down + *iōn* going). A positively charged ion.

Cau′dal (L. *cauda*, tail). Pertaining to the tail.

Cen′tral dog′ma (L. *centralis*, center; Gr. *dogma*, opinion, decree). The concept that hereditary information is encoded in DNA, which can be transcribed into RNA, and in turn translated into protein. New and identical copies of DNA are made by replication.

Cen′tri-ole (Gr. *kentron*, a center). A small body in the centrosome around which the asters form during mitosis.

Cen′tro-some (Gr. *kentron*, a center; *soma*, body). A tiny differentiated area of a cell near the nucleus containing the centriole. Also called cell center.

Ce-phal′ic (Gr. *kephate*, the head). Pertaining to or near the head.

Cer′e-bel″lum (L. diminutive of *cerebrum*, brain). The part of the brain in higher vertebrates which controls muscular coordination.

Cer′e-brum (L. *cerebrum*, brain). The anterior part of the brain, which is conspicuous in man, and is the seat of thought, memory, and reasoning.

Cer′vi-cal (L. *cervix*, neck). Pertaining to the neck.

Char′ac-ter (Gr. *charassein*, to engrave). Used specifically in biology to designate any trait, function, or structure of an organism.

Chi′tin (Gr. *chiton*, a kind of garment, tunic). A horny substance forming the skeleton of arthropods and other animals.

Chlo′ro-phyll (Gr. *chloros*, green; *phyllon*, leaf). A green pigment in plants essential for photosynthesis.

Cho′a-no-cyte (Gr. *choane*, funnel; *kytos*, hollow). Flagellated cells with collars, typical of porifera.

Chor′date (L. *chorda*, cord). An animal possessing a notochord which subsequently may be replaced by backbone (tunicates, vertebrates, etc.).

Chro′ma-tin (Gr. *chroma*, color). The dark, staining material of the chromosomes.

Chro′ma-to-phore (Gr. *chroma*, color; *pherein*, to bear). Pigment-containing cells or bodies in the skin of certain animals, such as the frog and some reptiles which are able to change their skin color.

Chro'mo-mere (Gr. *chroma*, color; *meros*, part). One of the many linearly arranged bead-like structures found on a chromosome.

Chro-mo-nem'a (Gr. *chroma*, color; *nema*, thread). Thread-like strands of chromatin visible in the nucleus during mitosis and in some cells during interphase.

Chro'mo-somes (Gr. *chroma*, color; *soma*, body). Deeply staining rod-shaped bodies within the nucleus and conspicuously visible during cell division. They contain the genes.

Cil'i-a (L. *cilium*, eyelid). Microscopic, hairlike projections from certain cells which vibrate, causing movement.

Cir"ca-di'an (L. *circa* about + *dies* a day). Approximately 24 hours.

Class (L. *classis*, collection). Principal subdivision of a phylum.

Cleav'age (A.S. *cleofan*, to cut). Divisions of the fertilized egg.

Cli'to-ris (Gr. *chleio*, close). An organ of the female mammal homologous with the penis of the male.

Clo-a'ca (L. *cloaca*, sewer). A common receptacle for digestive and excretory wastes and the reproductive cells of lower vertebrates.

Clone (Gr. *klon*, twig). A group of animals produced by asexual reproduction from a single individual.

Co-ag'u-la"tion (L. *cogere*, to drive together). A process of changing a sol into a gel.

Co-coon' (Fr. *cocon*, shell). A protective covering for eggs, larvae, pupae, or adult animals.

Coe'lom (Gr. *koilus*, hollow). The body cavity lined with tissue of mesodermal origin in which the digestive and other organs lie.

Co-en'zyme. A relatively simple substance which is involved in the transfer of hydrogen atoms during oxidative reactions in protoplasm.

Col'loid (Gr. *kolla*, glue; *eidos*, form). A system in which particles larger than molecules of one substance are suspended throughout a second substance.

Co'lon (Gr. *kolon*, the colon). The part of the large intestine which extends from the caecum to the rectum.

Com-men'sal-ism (L. *cum*, together; *mensa*, table). The intimate association of two species in which one is benefited and the other is neither helped nor harmed.

Con'ju-ga"tion (L. *cum*, together; *jugare*, to join, marry). A sexual process in unicellular organisms where two individuals unite temporarily and exchange their nuclear material, later dividing.

Con-trac'tile vac'u-ole (L. *cum*, together; *trahere*, to draw; *vacuus*, empty). A space in the cytoplasm of certain species of protozoa where fluids collect before being periodically discharged to the outside.

Cop'u-la"tion (L. *copulare*, to couple). The union of two individuals in which spermatozoa are transferred from the male to female.

Cor'ne-a (L. *cornu*, horn). The transparent outer covering of the anterior part of the eye.

Cor'o-na-ry (L. *coronarius*, crown). Encircling in the manner of a crown; applied to nerves, vessels, etc.

Cor'pus (L. *corpus*, body). A body.

Cor'tex (L. *cortex*, bark). The outer layer.

Cra'ni-um (Gr. *kranion*, the head). The portion of the skull which envelops the brain.

Cross'ing o'ver. The process in which homologous chromosomes break and exchange corresponding parts.

Cu-ta'ne-ous (L. *cutis*, the skin). Pertaining to the skin.

Cyst (Gr. *kystis*, a bladder). The stage of an organism where it is encased in a resistant wall.

Cy-tol'o-gy (Gr. *kytos*, hollow; *logos*, discourse). The science that deals with cell structure.

Cy'to-plasm (Gr. *kytos*, hollow; *plasma*, something molded). The protoplasm of a cell, exclusive of the nucleus.

Dac'tyl (Gr. *daktylos*, finger). Pertaining to the finger.

De-am'i-na"tion (L. *de*, from; *amino*, amino group). The process by which the amino group is removed from the amino acid.

Def-e-ca'tion (L. *de*, from; *faecis*, dregs). The evacuation of the bowels.

De'glu-ti"tion (L. *de*, down; *glutire*, to swallow). The act of swallowing.

Den'drite (Gr. *dendron*, tree). The branching protoplasmic outgrowth of a nerve cell.

De-ox"y-ri'bose. A five-carbon sugar with less oxygen than the formula Cx $(H_2O)_4$ indicates.

Der'mal (Gr. *derma*, skin). Pertaining to the skin, especially the deeper layers.

Der'mis (Gr. *derma*, skin). The inner layer of the skin.

Di-al'y-sis (L. *dialysis*, separation). The process by which crystalloids and colloids are separated in solution. This involves a natural or artificial membrane through which unequal diffusion takes place.

Di'a-phragm (Gr. *diaphragma*, midriff). The internal muscular layer found between the mammalian thoracic and abdominal cavities.

Di-as'to-le (Gr. *dia*, through; *stellein*, to set, place). The relaxation phase of the cardiac cycle.

Di-en-ceph'a-lon (Gr. *dia*, between; *engkephalon*, brain). An area of the vertebrate brain immediately posterior to the cerebrum.

Dif'fer-en'ti-a"tion (L. *differre*, to carry apart). Specialization of cells and tissues as a result of growth and development.

Dif-fu'sion (L. *diffundere*, to pour). The movement of molecules as a consequence of their kinetic energy.

Di-ges'tion (L. *digestio*, digestion). The conversion of food into materials that can be absorbed and assimilated.

Dig'it (L. *digitus*, finger). A finger or a toe.

Di-hy'brid (Gr. *dis*, twice; L. *hibrida*, mixed offspring). The offspring of parents differing in two characters.

Di-mor'phism (Gr. *dis*, twice; *morphe*, shape). Having two forms.

Di-oe'cious (**diecious**) (Gr. *dis*, twice; *oikos*, house). Having the sexes in separate individuals.

Dip'lo-blas"tic (Gr. *diploos*, double; *blastos*, bud). Having two germ layers.

Dip'loid (Gr. *diploos*, double). The number of chromosomes in somatic cells.

Dis'tal (L. *dis*, apart; *stare*, to stand). Most distant from the point of attachment.

Di-ur'nal (L. *dies*, day). Associated with daylight.

Di'ver-tic"u-lum (L. *de*, away; *vertere*, to turn). A pouch or pocket leading out from a tube.

DNA. Abbreviation for **desoxyribose nucleic acid.**

Dom'i-nant. A gene which expresses its effect even in the presence of a different gene for the same character.

Dor'sal (L. *dorsum*, back). The upper surface of any animal.

DPN. Abbreviation for **diphosphopyridine nucleotide.** See NAD.

Duct (L. *ducere*, to lead). A tube for the passage of metabolic products.

Duct'less gland. A gland which pours its secretion directly into the bloodstream.

Du'o-de"num (L. *duodeni*, 12 each). The first part of the small intestine posterior to the stomach (about as long as the width of 12 fingers).

Du'ra ma'ter (L. *dura*, hard; *mater*, mother). The tough outer covering of the brain and nerve cord.

Dys-gen'ic (Gr. *dys*, hard, ill; *gignesthai*, to be born). Tending to bring about genetic degradation of a species.

Ec-cen'tric (L. *ex*, without; *centrum*, center). Away from center.

Ec'dy-sis (Gr. *ek*, out; *dyein*, to enter). To shed the exoskeleton, as in arthropods.

E-col'o-gy (Gr. *oikos*, house; *logos*, discourse). The division of biology dealing with the relation of plants and animals to their environment.

Ec'to-derm (Gr. *ektos*, outside; *derma*, skin). The outermost germ layer of the gastrula which gives rise to the nervous system.

Ec'to-par"a-site (Gr. *ektos*, outside; *para*, beside; *sitos*, food). A parasite which lives on the outermost surface of its host.

Ec'to-plasm (Gr. *ektos*, outside; *plasma*,

something molded). The outer layer of cytoplasm in a cell.

Ec'to-sarc (Gr. *ektos*, outside; *sarx*, flesh). The outer layer of certain protozoa.

Ef-fect'or (L. *effectus*, to effect). A nerve end organ which serves to distribute impulses which activate muscle contraction and gland secretion.

Ef-fer'ent (L. *efferre*, to bear out). Conveying away from, as motor nerves that conduct impulses away from the cord.

Egg (L. *ovum*, egg). The animal gamete or ovum formed by the female.

E-jac'u-la-tion (L. *ejaculatus*, to throw out). The emission or discharge, as of sperm cells.

E-lec'tro-lyte (Gr. *electron*, amber; *lytos*, dissolved, dissoluble). A substance which, when in solution, dissociates into ions and thus conducts an electric current.

El-ec'tron. The part of an atom having a negative charge.

El'e-ment (L. *elementum*, of obscure origin). A simple substance which cannot be decomposed by ordinary chemical means and consisting of one kind of atom.

Em'bry-o (Gr. *embryon*, embryo). The organism in an early stage of development.

Em'bry-ol"o-gy (Gr. *embryon*, embryo; *logos*, discourse). The science of the development of the organism.

E-mul'si-fi-ca"tion (L. *e*, out; *mulgere*, to milk). The process of dividing fat into particles of very small size.

En-cyst' (Gr. *en*, in; *kystis*, bladder). To become enclosed in a cyst.

En-der-gon'ic. A reaction that requires energy.

En'do-crine (Gr. *endon*, within; *krinein*, to separate). Pertaining to a ductless gland.

En'do-derm (Gr. *endon*, within; *derma*, skin). The innermost germ layer of the gastrula which gives rise to the lining of the digestive tract and its derivatives.

En'do-par"a-site (Gr. *endon*, within; *para*, beside; *sitos*, food). A parasite living within the body of its host.

En'do-plasm (Gr. *endon*, within; *plasma*, something molded). Inner cytoplasm surrounded by estoplasm.

En'do-plasmic re-tic'u-lum (Gr. *endon*, within; *plasma*, something molded. L. dim. of *rete*, net). Cytoplasmic double membranes. Those with clinging ribosomes are called "rough," those without ribosomes are called "smooth."

En'do-sarc (Gr. *endon*, within; *sarx*, flesh). Inner mass of protoplasm in a protozoan.

En'do-skel"e-ton (Gr. *endon*, within; *skeletos*, hard). Internal bony and cartilaginous structure of animals.

En'do-the"li-um (Gr. *endon*, within; *thele*, nipple). Epithelial lining of the circulatory organs.

En'er-gy (Gr. *energein*, to be active). The capacity to do work.

En-ter-o-ki'nase (Gr. *enteron*, gut; *kinein*, to move). Intestinal enzyme which activates pancreatic trypsinogen.

En'ter-on (Gr. *enteron*, gut). That part of the digestive tract derived from endoderm.

En'to-derm. See Endoderm.

En'to-mol"o-gy (Gr. *entomon*, insect; *logos*, discourse). The study of insects.

En-vi'ron-ment (Fr. *environ*, about, thereabouts). External or internal surroundings.

En'zyme (Gr. *en*, in; *zyme*, leaven). Organic catalyst.

Ep-i-der'mis (Gr. *epi*, upon; *derma*, skin). The outermost layer of the skin.

Ep'i-gen"e-sis (Gr. *epi*, upon; *gignesthai*, to be born). The concept that development begins with an undifferentiated cell.

Ep-iph'y-sis (Gr. *epi*, upon; *phyein*, to grow). The tip of a bone separated in early development by cartilage but later becoming a part of the bone.

Ep'i-the"li-um (Gr. *epi*, upon; *thele*, nipple). A sheet of cells covering either external or internal parts of body surfaces.

E'qua-to"ri-al plate (L. *aequator*, one who equalizes). The platelike arrangement of chromosomes formed at the equator during cell division.

E-rep'sin (L. *eripere*, to set free). A mixture of peptone and proteose-splitting enzymes produced by the intestinal mucosa.

E-soph'a-gus (Gr. *oisophagos*, gullet). The tube extending from the pharynx to the stomach.

Es'ti-va"tion (L. *aestivus*, pertaining to summer). Inactivity brought about by extreme dryness and heat.

Es'tro-gen (L. *aestus*, fire, glow). A hormone produced by the ovarian follicle which, together with one of the pituitary hormones, influences estrus.

Es'trus (L. *aestus*, fire, glow). The mating period in female mammals marked by intensified sexual urge.

Eu-gen'ics (Gr. *eu*, well; *genos*, birth). The science of applying genetic knowledge for the improvement of the human species.

Eu-sta'chi-an tube (named after *Eustachi*, Italian physician. L. *tuba*, pipe). A tube leading from the pharynx to the middle ear.

Eu-then'ics (Gr. *euthenein*, to thrive). The science of improving the human race by improving the environment.

E-vag'i-na"tion (L. *e*, out; *vagina*, sheath). An out-pocketing of some part or organ.

E-vis'cer-ate (L. *ex*, out; *viscera*, entrails). To remove the internal organs.

Ev'o-lu"tion, (L. *evolvere*, to unroll; Gr. *organon*, instrument, tool). Descent with modification.

Ex-cre'tion (L. *ex*, out; *cernere*, to sift). Discharge of wastes of metabolism.

Ex-er-gon'ic. A reaction which yields energy.

Ex'o-skel"e-ton (Gr. *exo*, outside; *skeletos*, hard). The hardened external structure of animals.

Ex'pi-ra"tion (L. *ex*, out; *spirare*, to breathe). To breathe out.

Ex-ter'nal res'pi-ra"tion. The exchange of gases between the alveoli of the lungs and the blood.

Fac'et (L. *facies*, face). A subdivision of the compound eye in arthropods.

FAD.
A hydrogen acceptor in the respiratory chain.

Fam'i-ly (L. *familia*, from *famulus*, servant). The main subdivision of an order.

Fas'ci-a (L. a band). A sheet of connective tissue which covers and binds parts together.

Fau'na (L. *faunus*, a god of the woods). The animal life characteristic of a region.

Fe'ces or fae'ces (L. *faeces*, dregs). Undigested, unabsorbed food residue.

Feedback. Stimuli passing from the effector to the receptor indicating that some action has taken place.

Fer'ti-li-za"tion (L. *fertilis*, from *ferre*, to bear). The fusion of the sperm with the egg to produce a zygote.

Fe'tus (L. a bringing forth). The unborn young of any viviparous animal.

Fi'brin (L. *fibra*, band). The essential insoluble protein found in the blood clot.

Fi-brin'o-gen (L., *fibra*, band; Gr. *gignesthai*, to produce). The soluble protein material which is converted to fibrin during clotting.

Fis'sion (L. *fissus*, cleft). Asexual reproduction in which the cell divides into two parts.

Fis'sure (L. *fissus*, cleft). Any groove, furrow, cleft, or slit.

Fla-gel'lum (L. *flagellum*, whip). A mobile, whiplike process.

Fol'li-cle (L. *folliculus*, small sac). A small excretory or secretory sac or gland.

Fo-ra'men (L. *foramen*, an opening). A hole in a bone or membrane.

Fos'sa (L. *fossa*, ditch). A pit or depression found in bone.

Fos'sil (L. *fossilis*, from *fodere*, to dig). Any naturally preserved record of prehistoric life.

FSH. Follicle stimulating hormone from the anterior pituitary gland.

Func'tion (L. *fungi*, to perform). Plant or animal action.

Gam'ete (Gr. *gametes*, spouse). A mature germ cell.

Gam'e-to-gen"e-sis (Gr. *gametes*, spouse; *genesis*, birth). The development of gametes.

Gan'gli-on (Gr. *ganglion*, enlargement). A mass of nerve cell bodies.

Gas'tric (Gr. *gaster*, belly). Pertaining to the stomach.

Gas'tro-vas"cu-lar (Gr. *gaster*, belly; L. *vasculum*, vessel). A cavity used both for digestion and circulation.

Gas'tru-la (Gr. *gaster*, belly). The early embryonic stage following the blastula in which the embryo consists of two germ layers.

Gas'tru-la"tion (Gr. *gaster*, belly). The process of invagination of the blastula to form the gastrula.

Gene. Hereditary units located on the chromosomes.

Ge-net'ics (Gr. *genesthai*, to be born). The science of heredity.

Gen'i-tal (L. *genere*, to beget). Taxonomic ing to the reproductive organs.

Gen'o-type (Gr. *genesthai*, to be produced; *typos*, impression). Genic constitution.

Ge'nus (L. *genere*, to beget). Taxonomic subdivision of a family.

Ge-ot'ro-pism (Gr. *ge*, earth; *repein*, to turn). Response to gravity.

Germ lay'er (L. *germen*, germ). An embryonic primary cell layer.

Germ plasm (Gr. *plasma*, something molded). Reproductive and hereditary substance of individuals which is transmitted in direct continuity to the germ cells of succeeding generations.

Ges-ta'tion (L. *gerere*, to carry). The period of pregnancy.

Gill (Gr. *cheilos*, lip). An aquatic respiratory organ.

Gill arches (Gr. *cheilos*, lip; *arcus*, bow). the walls bearing the gills.

Gill slit (pharyngeal cleft) (A.S. *slitan*). Paired openings in the wall of the pharynx of chordates which permits the water that entered through the mouth to escape externally during breathing.

Giz'zard (Fr. *giser*, gizzard). An enlarged muscular part of the digestive tract.

Gland (L. *glans*, acorn). A cell or collection of cells which produces a specific product.

Glo-mer'u-lus (L. *glomare*, to make a ball). A tuft or cluster of blood vessels projecting into the capsule of each uriniferous tubule.

Glot'tis (Gr. *glotta*, tongue). Opening from the pharynx into the larynx.

Gly'co-gen (Gr. *glykus*, sweet; *gen*, producing). A carbohydrate stored in many parts of the body; also known as "animal starch."

Goi'ter (L. *gutter*, throat). An enlargement of the thyroid gland, usually resulting from a deficiency of iodine in the diet.

Golgi body. A cytoplasmic particle which probably functions in the storing of secretory products.

Go'nad (Gr. *gonos*, reproduction). A gamete-producing reproductive organ.

Go-nan'gi-um (Gr. *gone*, seed; *angeion*, vessel). The reproductive individual of a hydroid colony.

Gre-gar'i-ous (L. *gregarius*, from *grex*, a herd). The property of animals of flocking together.

Gul'let (L. *gula*, gullet). Esophagus.

Hab'i-tat (L. *habitare*, to dwell). Environment of an organism.

Hap'loid (Gr. *haploos*, single; *eidos*, form). The reduced number of chromosomes found in the gametes; half the diploid number.

He'mo-coel (Gr. *haima*, blood; *koilus*, hollow). A body cavity functioning as a part of the circulatory system.

He'mo-cy"a-nin (Gr. *haima*, blood; *kyanos*, a dark-blue substance). An oxygen-carrying pigment, copper-bearing and blue in color, found in the blood of mollusks and arthropods.

He'mo-glo"bin (Gr. *haima*, blood; L. *globus*, globe). Oxygen-carrying red pigment of blood.

He'mo-phil"i-a (Gr. *haima*, blood; *phil*, to love). A hereditary condition in man in which blood clotting is abnormally slow.

He-pat'ic (Gr. *hepar*, liver). Pertaining to the liver. Hepatocyte—liver cell.

Her-biv'o-rous (L. *herba*, herb; *vorare*, to devour). Subsisting on plants.

Her-ed'i-ty (L. *hereditas*, heirship). Organic resemblance based on descent.

Her-maph'ro-ditic (Gr. *Hermes*, *Aphrodite*). Possessing both male and female reproductive organs.

Het'er-o-zy"gote (Gr. *heteros*, different; *zeugon*, yolk). A hybrid formed from gametes having different genes for the same trait.

Hex"o-ki'nase. An enzyme that initiates the phosphorylation of glucose.

Hi'ber-na"tion (L. *hiems*, winter). The dormant state in which some animals spend the winter.

His'ta-mine. An organic compound which is a powerful dilator of the capillaries and a stimulator of gastric secretions; found in all plants and animals.

His'to-gen"e-sis (Gr. *histos*, tissue; *gignesthai*, to be born). The origin, development, and differentiation of tissues from undifferentiated cells of the embryonic germ layers.

His-tol'o-gy (Gr. *histos*, tissue; *logos*, study). The study of the microscopic structure of tissues and organs.

Hol'o-zo"ic (Gr. *holós*, whole, *zoion*, animal). Nutrition involving the ingestion and digestion of organic material.

Ho-mol'o-gy (Gr. *homos*, same; *logos*, study). The study of organs which result from common embryonic origin. They may or may not have the same function.

Ho'mo-zy"gote (Gr. *homos*, same; *zeugon*, yolk). An organism formed from gametes containing like genes for a given character.

Hor'mone (Gr. *hormon*, from *hormaein*, to arouse or excite). Chemical substance produced by an endocrine gland which, when transported to another area, produces a specific effect.

Host (L. *hostis*, stranger). The organism upon which a parasite lives.

Hy'a-line (Gr. *hyalos*, clear). A translucent, albumenoid material.

Hy'brid (L. *hybrida*, mongrel). An organism formed from the union of gametes differing in one or more genes; a heterozygote.

Hy'dranth (Gr. *hydra*, water serpent; *anthos*, flower). A vegetative branch of a hydroid colony.

Hy'dro-car"bon (Gr. *hydor*, water; L.

carbo, coal). An organic compound formed only of hydrogen and carbon.

Hy-drol'y-sis (Gr. *hydor*, water; *lysis*, a loosing). Chemical decomposition by reaction with water.

Hy'men (Gr. *umen*, membrane). A membranous fold which usually partially or wholly occludes the external orifice or vagina during virginity.

Hy'oid (Gr. *hyoides*, Y-shaped). A Y-shaped group of bones at the base of the tongue.

Hy'per-ton"ic (Gr. *huper*, beyond; *tonikos*, strength). High osmotic pressure in reference to another solution.

Hy-poph'y-sis (Gr. *hupo*, under; *physis*, growth). The pituitary gland.

Hy-poth'e-sis (Gr. proposal). Tentative solution or proposal concerning a problem.

Hy'po-ton"ic (Gr. *hupo*, under; *tonikos*, strength). A lesser osmotic pressure in reference to another solution.

I-den'ti-cal twins. Twins arising from the same fertilized egg and therefore having the same genetic constitution.

Il'e-um (L. groin). The most posterior portion of the small intestine.

Il'i-um (L. flank). The dorsal bone of the pelvic girdle.

Im-mu'ni-ty (L. *immunis*, free). The power which a living organism possesses to resist and overcome infection.

In"breed'ing. The crossing of closely related animals.

In-fun-dib'u-lum (L. *infundibulum*, funnel). A stalklike evagination of the diencephalon.

In-gest' (L. *ingestus*, from *ingerere*, to put in). To take in food.

In'or-gan"ic (L. *in*, not; Gr. *organikos*, instrument). Pertaining to substances of non-organic origin.

In-ser'tion (L. *insertus*, from *inserere*, to connect, insert). Point of attachment of a muscle to the movable part.

In'su-lin (L. *insula*, island). Hormone secreted by the pancreatic islets of Langerhans.

In-teg'u-ment (L. *integumentum*, covering). The outermost covering of an organism; skin.

In'ter-cel''lu-lar (L. *inter*, between; *cellula*, cells). Between cells.

In-ter'nal se-cre'tion. Secretion into the bloodstream.

In'ter-sti''tial (L. *inter*, between; *sistere*, to set). Pertaining to intercellular spaces.

In-tes'tine (L. *intestinus*, internal). Part of the digestive tract posterior to the stomach.

In'tra-cel''lu-lar (L. *intra*, within; *cellula*, cells). Within cells.

In'tus-sus-cep''tion (L. *intus*, within; *suscipere*, to receive). Growth by the addition of new materials within protoplasm.

In-vag'i-nate (L. *in*, in; *vagina*, sheath). An in-pushing of a cellular layer into a cavity.

In-ver'te-brate (L. *in*, not; *vertebra*, joint). An animal without a vertebral column.

In'vo-lu''tion (L. *in*, in; *volvere*, to roll). A rolling or turning inward of cells over a rim.

I'on (Gr. *ion*, going). A portion of a molecule in solution carrying a charge.

Ir'ri-ta-bil''i-ty (L. *irrito*, excite). Ability to respond to a stimulus.

Is'chi-um (Gr. *ischion*, hip). The posterior ventral bone of the pelvic girdle.

I'so-ton''ic (Gr. *isos*, equal; *tonikos*, strength). Having the same osmotic pressure.

Je-ju'num (L. *jejunus*, empty). The digestive tract lying between the duodenum and the ileum.

Jug'u-lar (L. *jugulum*, collar bone). Pertaining to the neck (jugular) vein.

Kid'ney. The major excretory organ in vertebrates.

Ki-ne'to-some (Gr. *kinētos*, movement + *sōma*, body). The base piece from which the flagellum or cilium arises. Similar to a centriole.

La'bi-al (L. *labium*, lip). Pertaining to lips.

Lab'y-rinth (Gr. *labrys*, double ax). Part of the inner ear in higher vertebrates, com-posed of semicircular canals, utriculus, sacculus, and cochlea.

Lac'ri-mal (L. *lacrima*, tear). Pertaining to tears.

Lac'te-al (L. *lac*, milk). Pertaining to milk; also referring to lymph vessels in the small intestine.

La-cu'na (L. *lacuna*, cavity). A small pit, hollow, or cavity, as in bone or cartilage.

La-mel'la (L. small plate). A thin layer.

La-nu'go (L. *lanuginosus*, wool). The soft woolly hair covering the human fetus, usually shed shortly before birth.

Lar'va (L. *larva*, ghost). An immature, free-living stage of an animal.

Lar'ynx (Gr. *larynx*, larynx). A structure containing the vocal cords located at the top of the trachea and below the root of the tongue in all vertebrates except birds.

Lat'er-al (L. *latus*, side). Toward the side.

Le'thal gene (L. *lethum*, death). A gene capable of bringing about the death of the organism.

Leu'co-cyte (Gr. *leucos*, white; *kytos*, cell). A white blood cell.

LH. Luteinizing hormone from the anterior pituitary gland.

Lig'a-ment (L. *ligamentum*, bandage). A tough, fibrous band of connective tissue.

Lin'gual (L. *lingua*, tongue). Pertaining to the tongue.

Link'age (M.E. *linke*). Phenomenon occurring when a series of genes are passed on as a unit.

Li'pase (Gr. *lipos*, fat). Fat-splitting enzyme.

Lip'id (Gr. *lipos*, fat). Pertaining to fat.

Lum'bar (L. *lumbus*, loin). Pertaining to the region of the back between the thorax and the pelvis.

Lu'men (L. cavity). The internal cavity within a structure.

Lymph (L. *lympha*, water). Fluid found in the lymph vessels containing fat, white blood cells, and plasma.

Mac'ro-nu''cle-us (Gr. *makros*, large; L. *nucleus*, kernel). The large nucleus of a ciliate (such as paramecium) as distinguished from the micronucleus.

Ma-la'ri-a (L. *mal*, bad; *aria*, air). An

infectious disease caused by protozoa (sporozoa) and transmitted by certain mosquitos.

Man′di-ble (L. *mandere*, to chew). The lower jawbone in vertebrates of either jaw in arthropods.

Man′tle (L. *mantellum*, cloak). A sheetlike tissue enclosing soft structures of an animal, such as a mollusk.

Ma-nu′bri-um (L. *manas*, hand). The uppermost part of the sternum; also the structure bearing the mouth in the medusa.

Ma-rine′ (L. *marinus*, from *mare*, the sea). Pertaining to the sea.

Mas′ti-ca″tion (L. *masticare*, to chew). Referring to chewing.

Ma-ter′nal (L. *maternus*, of a mother). Referring to the mother.

Mat′u-ra″tion (L. *maturus*, ripe). The final stages in the production of gametes in which the number of chromosomes is reduced to one-half (diploid to haploid) the number characteristic of the species.

Me′di-an (L. *medius*, middle). Pertaining to the midline.

Me-dul′la (L. *medulla*, marrow). The distinct inner portion of a structure.

Me-dul′la ob-lon-ga′ta (L. oblong medulla). The most posterior portion of the brain.

Med′ul-la-ry plate, groove, or tube (L. *medullaris*, narrow). Neural plate, groove, or tube found in the embryonic development of the vertebrate nervous system.

Med′ul-lat-ed. A nerve fiber with a myelin covering or sheath.

Mei-o′sis (Gr. to make smaller). See Maturation.

Mem′brane (L. *membrana*, membrane). Any thin cellular sheet or layer.

Me-nin′ges (Gr. *meninx*, membrane). The three membranes enveloping the brain and the spinal cord.

Men′o-pause (Gr. *men*, month; *pausis*, cessation). The period when menstruation normally ceases; change of life.

Mes′en-chyme (Gr. *mesos*, middle; *engchein*, to pour in). A loose embryonic connective tissue derived chiefly from mesoderm.

Mes′en-ter-y (Gr. *mesos*, middle; *enteron*, intestine). A membrane supporting an organ in the abdominal cavity; also a partition found in certain coelenterates.

Mes′o-derm (Gr. *mesos*, middle; *derma*, skin). The midlayer of embryonic cells found between the ectoderm and the endoderm.

Mes′o-gle″a (**mesogloea**) (Gr. *mesos*, middle; *gloia*, glue). A noncellular, jellylike substance found in coelenterates.

Mes′o-neph″ros (Gr. *meso*, middle; *nephros*, kidney). A kind of vertebrate kidney found in all embryos and in certain adult fish and amphibians.

Mes′o-some (Gr. *mesos*, middle; *soma*, body). Thickened infoldings of the cell membrane in certain bacteria, believed to be involved with division and respiration.

Me-tab′o-lism (Gr. *meta*, beyond; *ballein*, to throw). The sum total of the chemical and physical processes ocurring in protoplasm.

Met-a′chron-y (Gr. *metachronos*, done afterwards). One acting after the other, applied to the rhythm of cília.

Met′a-gen″e-sis (Gr. *meta*, over; *genesis*, origin). Alternation of sexual and asexual generations.

Met′a-mere (Gr. *meta*, over; *meros*, part). Homologous segment of the body.

Me-tam′er-ism. The possession of a succession of homologous parts.

Met′a-mor″pho-sis (Gr. *meta*, over; *morphe*, form). The structural changes taking place in the transformation of a larva to an adult.

Met′a-phase (Gr. *meta*, after; *phasis*, appearance). The midstage of mitosis during which there is a lengthwise separation of chromosomes at the equatorial plate.

Met′a-zo″a (Gr. *meta*, over; *zoion*, animal). The multicellular animals in which there is a differentiation of the somatic cells.

Mi′cron (Gr. *mikros*, small). One thousandth part of a millimeter.

Mi′cro-nu″cle-us (Gr. *mikros*, small; L. *nucleus*, kernel). The small reproductive nucleus of certain protozoa.

Mi-gra'tion (L. *migratio*). Moving from place to place.

Mil'li-li''ter (L. *mille*, one thousand; Fr. *litre*, liter). Thousandth part of a liter.

Mit'o-chon''dri-a (Gr. *mitos*, thread; *chondros*, grain). Small granules or rod-shaped structures found in the cytoplasm, containing oxidative enzymes.

Mi-to'sis (Gr. *mitos*, thread). Nuclear division.

Molt (M.E. *mouten*, from L. *mutare*, to change). To shed an outer covering.

Mo-noe'cious. Containing both sexes in the same organism, hermaphroditic.

Mon'o-hy''brid (Gr. *monos*, single; L. *hybrida*, mongrel). The offspring of parents differing in one character.

Mor-phol'o-gy (Gr. *morphe*, form; *logos*, discourse). The study of form and structure.

Mu'cous (L. *mucus*, slime). Pertaining to mucus.

Mu'cus (L. *mucus*, slime). A watery secretion which covers mucous membranes.

Mu'tant (L. *mutare*, to change). A variation which breeds true.

Mu-ta'tion (L. *mutare*, to change). A permanent transmissible change in the character of an offspring.

My'e-lin (Gr. *myelos*, marrow). A fatlike substance forming a sheath around the axis of a medullated nerve.

My'o-fi''bril (Gr. *mys*, muscle; L. *fibrilla*, a small fiber). One of the slender, protoplasmic threads found in the muscle fiber which runs parallel with the long axis.

NAD. Nicotinamide adenine dinucleotide (formerly called DPN). A hydrogen acceptor in the respiratory chain.

NADP. Nictinamide adenine dinucleotide phosphate (once called TPN).

Na'res (L. *naris*, nostril). Openings of the air passages in vertebrates, both external and internal.

Na'sal (L. *nasus*, nose). Pertaining to the nose.

Nem'a-to-cyst (Gr. *nema*, thread; *kystis*, a bladder). A stinging body found in coelenterates.

Ne-ot'e-ny (*neo* + Gr. *teinein*, to extend). Sexually mature larval or juvenile animals.

Ne-phrid'i-o-pore (Gr. *nephros*, kidney; *poros*, passage). The external opening of the invertebrate excretory organ.

Ne-phrid'i-um (Gr. *nephros*, kidney). An invertebrate excretory organ.

Neph'ro-stome (Gr. *nephros*, kidney; *stoma*, mouth). The funnel-shaped opening at the inner end of the nephridium.

Neu'ral (Gr. *neuron*, nerve). Pertaining to the nervous system.

Neu'ri-lem''ma (Gr. *neuron*, nerve; *lemma*, covering). The outermost nerve fiber sheath.

Neu'ron, or neu'rone (Gr. *neuron*, nerve). The nerve cell.

Niche. The role and location of an animal in its habitat.

No'to-chord (Gr. *notos*, back; *chorde*, string). The cylindrical rod of supportive tissue found in chordates, dorsal to the digestive tract and ventral to the nerve cord.

Nu-cle'ic acid. A group of molecules composed of nucleotide complexes, the important types being desoxyribose nucleic acid (DNA) and ribose nucleic acid (RNA).

Nu-cle'ol-us (L. diminutive of *nucleus*). A round, conspicuous body found within the nucleus of most cells.

Nu'cleo protein. A giant molecule containing nucleic acid and protein.

Nu'cle-o-tide. A molecule containing phosphate, five-carbon sugar (ribose or desoxyribose), and a purine or a pyrimidine.

Nu'cle-us (L. kernel). A dense spheroid body, containing chromatin, found within the cell.

Nu-tri'tion (L. *nutrimentum*, nourishment). Sum total of the processes involved in food assimilation.

Oc-cip'i-tal (L. *occiput*, back of the head). Pertaining to the back of the head.

O-cel'lus (L. a little eye). The simple eye found in invertebrates.

Oc′u-lar (L. *oculus,* an eye). Pertaining to the eye; also the eyepiece of a microscope.

Ol-fac′to-ry (L. *olfacere,* to smell). Pertaining to the sense of smell.

Om-ma-tid′i-um (Gr. *omna,* eye). A small rod-like unit in the invertebrate compound eye.

Om-niv′o-rous (L. *ominis,* all; *vorare,* to devour). Subsisting on food of all types.

On-tog′e-ny (Gr. *onto,* being; *genos,* birth). The evolution of developmental history of an organism.

O′o-cyte (Gr. *oon,* egg; *kytos,* cell). The original cell of the ovarian egg before the formation of polar bodies.

O′o-gen″e-sis (Gr. *oon,* egg; *genesis,* origin). The origin and development of the ovum.

O′o-gon′i-um (Gr. *oon,* egg; *gonos,* offspring). The primordial cell which gives rise to the ovarian egg.

Oph-thal′mic (L. *opthalmia,* the eye). Pertaining to the eye.

Op′tic (Gr. *optikos,* sight). Pertaining to the eye or vision.

O′ral (L. *os,* mouth). Pertaining to the mouth.

Or′bit (L. *orbis,* circle). The bony eye socket.

Or′gan (Gr. *organon,* an instrument). A group of tissues associated to perform one or more functions.

Or′gan-elle″. A minute organ found in protozoa.

Or-gan′ic com′pound (Gr. *organon,* an implement; L. *componere,* to put together). A carbon-containing molecule.

Or′gan-ism (Fr. *organisme*). Any living thing.

Or′ga-nog″e-ny (Gr. *organon,* an instrument, implement; *genesis,* birth). The developmental processes involved in the formation of specialized tissue and organ systems.

Or′i-gin (L. *orior,* rise, become visible). Part of the muscle attached to an immovable structure.

Or′tho-gen″e-sis (Gr. *orthos,* straight; *genesis,* descent). Progressive evolution in a given direction.

Os-mo′sis (Gr. *osmos,* pushing). Diffusion through a semi-permeable membrane.

O′to-lith (Gr. *ous,* ear; *lithos,* stone). A small calcareous mass found in the auditory organ of many animals.

O′va-ry (L. *ovarium,* ovary). The primary female sex gland in which the eggs are formed.

O′vi-duct (L. *ovum,* egg; *ducere,* to lead). The ibe through which the eggs are carried from the ovary to the uterus or outside the body.

O-vip′a-rous (L. *ovum,* egg; *pario,* to produce). Producing eggs which hatch outside the body.

O′vi-pos″i-tor (L. *ovum,* egg; *ponere,* to place). An organ found in female insects for depositing eggs.

O′vo-vi-vip″a-rous (L. *ovum,* egg; *vivus,* alive; *parere,* to bear). Producing eggs which hatch within the body.

O-vu-la′tion (L. *ovum,* egg). The discharge of the unimpregnated ovum from the ovary.

O′vum (L. egg). The mature female gamete.

Ox′i-da″tion (Gr. *oxys,* acid). An increase of positive charges on an atom, or the loss of negative charges.

Ox″y-to′cin. A hormone from the posterior pituitary which stimulates uterine contraction and release of milk.

Pae′do-gen″i-sis (Gr. *pais,* a child; *genesis,* origin). Reproduction while in the immature or larval stage.

Pa′le-on-tol″o-gy (Gr. *palaios,* old; *ons,* being; *logos,* discourse). Study of fossils.

Palp (L. *palpare,* to feel). One of the pointed sense organs attached to the mouth of some invertebrates.

Pan′cre-as (Gr. *pan,* all; *kreas,* flesh). A digestive gland located behind the stomach and opening by a duct into the duodenum.

Par′a-site (Gr. *para,* beside; *sitos,* food; or *parasitos,* eating beside another). An organism which lives on or in another living organism from which it receives an advantage without compensation.

Par′a-thy″roid (Gr. *para,* near; *thyreoeides,*

shield-shaped). A small endocrine gland located near the thyroid.

Pa-ren'chy-ma (Gr. *para*, beside; *enchyma*, infusion). Loose, spongy connective tissue.

Pa-ri'e-tal (L. *parietis*, a wall). Pertaining to the coelomic wall.

Par'the-no-gen''e-sis (Gr. *parthenos*, virgin; *genesis*, origin). Reproduction by development of the egg without its being fertilized by the male element.

Pa-ter'nal (L. *paternus*, from *pater*, father). Pertaining to a father.

Path'o-gen''ic (Gr. *pathos*, suffering; *genesis*, to produce). Disease-producing.

Pa-thol'o-gy (Gr. *pathos*, disease; *logos*, study). The study of abnormal structures and processes.

Ped'al (L. *pes*, foot). Pertaining to the foot or feet.

Pe-lag'ic (L. *pelagicus*, from *pelagus*, sea). Inhabiting the open sea.

Pel'li-cle (L. *pellicula*, small skin). A thin skin, or film.

Pe'nis (L.). The male organ of copulation.

Pen'ta-dac''tyl (Gr. *pente*, five; *daktylos*, finger). Having five digits.

Pep'sin (Gr. *pepsis*, a cooking, digesting). The protein-splitting enzyme of the gastric juice.

Per'i-car''di-um (Gr. *peri*, around; *cardia*, heart). The membranous sac surrounding the heart.

Per'i-os''te-um (Gr. *peri*, around; *osteon*, bone). The tough, fibrous membrane surrounding bone.

Pe-riph'er-al (Gr. *periphereia*, from *peri*, around; *pherein*, to bear, carry). Pertaining to the surface.

Per'i-stal''sis (Gr. *peri*, around; *stalsis*, constriction). The wave of contraction by which the alimentary canal propels its contents.

Per'i-to-ne''um (Gr. *peri*, around; *tenein*, to stretch). The thin epithelial membrane which lines the coelom and invests the viscera.

pH. Symbol which represents the concentration of hydrogen ions in a solution.

Phag'o-cyte (Gr. *phagein*, to eat; *kytos*, cell). A type of white blood cell which ingests foreign substances.

Pha-lan'ges (Gr. *phalanx*, long line of battle). Digital bones.

Phar'ynx (Gr. *pharynx*). The portion of the digestive tract located between the mouth and esophagus.

Phe'no-type (Gr. *phaino*, show; *typto*, strike). The total of visible and physiological traits common to a group of individuals.

Phos'phor-y-la'tion. The addition of a phosphate group to a compound.

Pho'to-syn''the-sis (Gr. *phos*, light; *synthesis*, a putting together). The chemical combination, in the green plant, of carbon dioxide and water, to form carbohydrate, in the presence of light and chlorophyll.

Phy-log'e-ny (Gr. *phylon*, race, branch; *geny*, become). The evolutionary history of a race or group.

Phy'lum (Gr. *phylon*, tribe). One of the main divisions of the animal or plant kingdom.

Phys'i-ol''o-gy (Gr. *physis*, nature; *logos*, study). The study of functions of organisms.

Pig'ment (L. *pingere*, to paint). Coloring matter.

Pin'e-al (L. *pinea*, pine cone). An evaginated structure on the roof of the brain.

Pi-tu''i-tar'y bod'y (L. *pituita*, phlegm; A.S. *bodig*). An endocrine gland.

Pla-cen'ta (L. *placenta*, a flat cake). The round, flat organ within the uterus of mammals formed during embryonic development and attached to the embryo by means of an umbilical cord.

Plan'u-la (L. *planus*, flat). The ciliated, free-swimming larva found in the development of certain invertebrates.

Plas'ma (Gr. something molded). The liquid part of the blood or lymph.

Plas'ma mem'brane (L. *membrana*, skin covering). The external cytoplasmic membrane of a cell.

Plas'tid (Gr. *plastides*, to form). A pigmented cytoplasmic inclusion.

Plat'y-hel-min''thes (Gr. *platy*, flat; *helmins*, worm). A phylum of animals.

Pleur'al (Gr. *pleura*, side). Pertaining to the

cavity in which the lungs are contained.

Plex'us (L. interwoven). A nervous or vascular network.

Poi'kil-o-therm'al (Gr. *poikilos*, variegated). Having a variable body temperature. Opposed to *homiothermal*.

Po'lar bod'y (L. *polaris*, axis). The nonfunctional cell formed during the maturation of the egg cell.

Pol'y-mor"phism (Gr. *polys*, many; *morphe*, form). The occurrence of more than two types of individuals in the same species.

Pol'yp (Gr. *polypous*, many-footed). The attached form of a coelenterate usually possessing a mouth and tentacles at the free end.

Pos-te'ri-or (L. latter). Situated behind.

Pre-co'cious (L. *praecox*, ripe before its time). Characterized by early maturity.

Pre-da'ceous (L. *praedo*, prey). Capturing of live animals for food.

Pri-mor'di-al (L. *primordium*, beginning). Original or primitive.

Prin'ci-ple (L. *principium*, beginning). Scientific fact, theory, or law.

Pro-bos'cis (Gr. *proboskis*, trunk). Tubular extension of the lips, nose, or pharynx.

Proc'to-de"um (Gr. *proktos*, anus; *hodos*, way). The most posterior part of the digestive tract near the anus, lined with ectoderm.

Pro-ges'ter-one. A hormone produced by the corpus luteum.

Pro-lac'tin (*pro* + L. *lac*, milk). A hormone produced by the pituitary which stimulates the production of milk in mammals.

Pro-neph'ros (Gr. *pro*, before; *nephros*, kidney). A primordial kidney.

Pro-nu'cle-us (Gr. *pro*, before; L. *nucleus*, kernel). The nucleus of either the egg or sperm cell during the interval existing between the penetration of the sperm into the egg and the subsequent union to form the germinal nucleus.

Pro'phase (Gr. *pro*, before; *phasis*, appearance). The first stage of mitotic division during which the chromosomes become visible.

Pros'tate gland (Gr. *prostates*, one who stands before; L. *glans*, an acorn). An accessory sex gland found in the male

surrounding the neck of the bladder and urethra.

Pro'te-in (Gr. *protos*, first). An organic compound composed of amino acids forming an essential part of the protoplasm.

Pro'to-plasm (Gr. *protos*, first; *plasma*, something molded). The colloidal living substance constituting the physical basis of life.

Pro'to-zo"a (Gr. *protos*, first; *zoion*, animal). A phylum of animals.

Prox'i-mal (L. *proximus*, next). Near the point of attachment of an organ.

Pseu'do-coel (Gr. *pseudo*, false; *koilia*, body cavity). A body cavity not completely lined with mesoderm, as found in the round worm.

Pseu'do-po"di-a (Gr. *pseudo*, false; *pous*, foot). A temporary protoplasmic projection found in amoeba or in some amoeba-like cells.

Pty'a-lin (Gr. *ptyalon*, spittle). Salivary enzyme which acts on starch.

Pu'bis (L. *pubes*, mature). The pubic bone, one of the three bones forming the pelvis.

Pul'mo-nar-y (L. *pulmo*, lung). Pertaining to the lung.

Pu'rine. A characteristic substance in nucleic acids. The two common ones are adenine and guanine.

Py-lor'ic (Gr. *pylorus*, gate). Pertaining to the posterior portion of the stomach or the pylorus.

Py'rim-i-dine. A characteristic substance in nucleic acids. The three common ones are thymine, cytosine, and uracil.

Quad'ru-ped (L. *quattuor*, four; *pes*, foot). Four-footed animal.

Ra'di-al sym'me-try (L. *radius*, way). The condition in which similar parts are regularly arranged around a central axis, for example, as seen in starfish.

Ra'mus (L.). A branch, as of a blood vessel, bone, or nerve.

Re-cep'tor (L. *receptor*, receiver). A sensory end-organ.

Rec'tum (L. *rectus*, straight). The most posterior portion of the large intestine.

Re-gen'er-a"tion (L. *re*, again; *generare*,

to beget). The renewal or repair of a structure.

Re′nal (L. *renes*, kidney). Pertaining to the kidney.

Ren′nin (A.S. *gerinnan*, to curdle, coagulate). Enzyme secreted by the stomach wall which causes the coagulation of casein in milk.

Re′pro-duc″tion (L. *re*, again; *pro*, forth; *ducere*, to lead). The production of offspring.

Res′pi-ra″tion (L. *re*, again; *spirare*, to breathe). The gaseous metabolism at the cellular level.

Re-sponse′ (L. *re*, again; *spondere*, to promise). Reaction to stimulus.

Ret′i-na (L. *rete*, net). The innermost layer of the eye containing the light sensitive receptors.

Rh fac′tor (the Rhesus factor). A chemical factor found in the blood of some individuals which has genetic significance.

Rho-dop′sin (Gr. *rhodon*, rose; *opsis*, sight). Visual purple. A substance found in the retina and associated with vision.

RNA. Abbreviation for **ribose nucleic acid.**

Ru-di-men′ta-ry (L. *rudis*, unwrought, rude). Imperfectly developed, or having no function.

Ru′mi-nant (L. *rumen*, throat). A group of cud-chewing animals including the cow, goat, and deer, which have a stomach with four complete cavities.

Sap′ro-zo″ic (Gr. *sapros*, rotten; *zoion*, living being). Living on dead or decaying organic matter.

Sar′co-lem″ma (Gr. *sarx*, flesh; *lemma*, covering). The elastic sheath investing each striated muscle fiber.

Sar′co-plasm (Gr. *sarx*, flesh; *plasma*, liquid). The fluid protoplasmic matter of striated muscles.

Se-ba′ceous glands (L. *sebum*, tallow, grease; L. *glans*, an acorn). Small skin glands associated with hair follicles which produce a waxy secretion for lubrication of the skin.

Se-cre′tin. A hormone produced by the intestinal wall which controls the secretion of pancreatic juice.

Se-cre′tion (L. *secretio*, from *secernere*, to separate). The process of separating various substances from the blood for use by the organism; also the material produced.

Sed′en-tar-y (L. *sedere*, to sit). Permanently attached form.

Seg′ment (L. *segmentum*, a piece cut off). A portion of a metameric organism.

Self′-fer-ti-li-za′tion. Fertilization of an egg by a sperm from the same individual.

Se′men (L. *serere*, to sow). The sperm-containing fluid of male animals.

Sem′i-cir″cu-lar ca-nals′ (L. *semi*, half; *circulus*, circle). Canals in the inner ear of the vertebrate which make up an essential part of the sense organ of equilibrium.

Sem′i-nal (L. *semen*, seed). Pertaining to the semen.

Sem′i-nal re-cep′ta-cles (L. *semen*, seed; L. *recipere*, to receive). Saclike receptacles in the female used for the storage of sperm after copulation.

Sem′in-al ves′i-cles (L. *semen*, seed; L. *vesica*, bladder). Saclike structures in the male used to store sperm.

Sem′i-nif″er-ous tu′bule (L. *semen*, seed; *ferro*, to carry; *tubules*, small tube). A small duct used to convey seminal fluid.

Sen′so-ry cell (L. *sensus*, sense). Any receptor cell.

Sep′tum (L. *septum*, partition). A partition between adjoining cavities.

Ser′i-al ho-mol′o-gy (L. *series*, join; Gr. *homos*, same; *logos*, discourse). The serial repetition of structures having the same embryonic origin.

Se-rol′o-gy (L. *serum*, liquid). The study of serums and their actions.

Ses′sile (L. *sedere*, to sit). Attached, as opposed to free-living.

Se′tae (L. *seta*, bristle). Stiff bristle, as, for example, the setae in the parapodia of *Nereis*.

Si′nus (L. cavity). A cavity or hollow space.

Sol′ute (L. *solvere*, to loosen). The dissolved substance in a solution.

Sol′vent (L. *solvere*, to loosen). The substance in which the solute is dissolved. It is the continuous liquid portion of a solution.

So′mite. A serial segment or metamere of an

animal.

Spe′cies (L. appearance). A primary subdivision of a genus.

Sperm (Gr. seed). Male sex cell.

Sper′ma-ry. Male reproductive gland, testis.

Sper′ma-tid. The male germ cell just prior to the formation of the sperm.

Sper′ma-to-gen″e-sis (Gr. *sperma*, seed; *genesis*, origin). The process by which spermatozoa are formed.

Sper′ma-to-zo″on (Gr. *sperma*, seed; *zoion*, animal). A mature male reproductive cell, sperm.

Sphinc′ter (Gr. *sphingein*, to bind tightly). A ringlike muscle surrounding a natural orifice; the opening is closed by its contraction.

Spic′ule (L. *spiculum*, a little point). A sharp, needlelike body, characteristic of sponges.

Spi′nal col′umn (L. *spina*, thorn, spine). A continuous series of vertebrae in vertebrates which houses the spinal cord.

Spi′nal cord. The tubular part of the central nervous system which extends posteriorly from the brain throughout the length of the spinal column.

Spin′dle (A.S. *spinnan*, to spin). The spindle-shaped threads or fibers associated with the chromosomes in mitosis, radiating out from the centrosomes.

Spi′ra-cle (L. *spiraculum*, air hole). A breathing orifice of insects and the first gill slit in cartilaginous fishes.

Spleen (Gr. *splen*, spleen). A large organ situated in the upper part of the abdomen on the left side and lateral to the stomach having to do with red blood cell disintegration.

Spore (Gr. *spora*, seed). A special encapsulated reproductive body of one of the lower organisms.

Spor′u-la″tion (Gr. *sporo*, seed). Reproduction by multiple fission, forming spores.

Stat′o-cyst (Gr. *statos*, stationary; *kystis*, sac). An organ of equilibrium found in invertebrates.

Stat′o-lith (Gr. *statos*, standing; *lithos*, stone). The solid body within a statocyst.

Ster′num (L. breastbone). Breastbone.

Stig′ma (Gr. a pricked mark). A light-sensitive pigment spot in certain protozoa.

Stim′u-lus (L. *stimulare*, to incite). An environmental change which causes a response.

Strat′i-fied (L. *stratum*, a covering). Arranged in layers.

Stri′at-ed (L. *stria*, channel). Cross-striped, such as in skeletal muscle.

Sub-cu-ta′ne-ous (L. *sub*, under; *cutis*, skin). The region immediately beneath the skin.

Sub′mu-co″sa (L. *sub*, under; *mucosus*, mucus). One of the layers of tissue in the wall of the digestive tract.

Su′ture (L. *sutura*, from *suere*, *sutum*, to sew). The point of fusion between two bones.

Swim′mer-et (A.S. *swimman*). Paired biramous appendage of the crayfish, just posterior to the walking legs.

Sym′bi-o″sis (Gr. *syn*, together; *bios*, living). Living together of two species.

Sym″me′try (Gr. *syn*, together; *meton*, measure). The regular or reversed disposition of parts around a common axis, or on each side of any plane of the body.

Syn′apse (Gr. *syn*, together; *hapto*, unite). The region of contact of two adjacent neurons.

Syn-ap′sis. Temporary union of maternal and paternal homologous chromosomes prior to the first maturation division.

Syn-cyt′i-um (Gr. *syn*, together; *kytos*, cell). A multinucleate mass of protoplasm.

Sys′tem (Gr. *syn*, together; *histanai*, to place). A group of organs having the same general function.

Sys′to-le (Gr. *syn*, with; *stellein*, to set, place). The contraction phase of the cardiac cycle.

Tac′tile (L. *tangere*, to touch). Pertaining to the sense of touch.

Tax-on′o-my (Gr. *taxis*, arrangement; *nomos*, law). The science of the classification of living things.

Tel′o-phase (Gr. *telos*, end; *phasis*, aspect). The last stage in mitotic division.

Ten′don (L. *tendere*, to stretch). A fibrous cord of connective tissue which binds a muscle to a bone or to other structures.

Ten'ta-cle (L. *tentare*, to touch, feel). A slender, whiplike organ for feeling or motion found in invertebrates.

Ter-res'tri-al (L. *terra*, earth). Living on the ground.

Tes'tis (L. *testis*). The sperm-forming male gonad.

Tes-tos'ter-one. Male sex hormones.

Tet'rad (Gr. *tetra*, four). The four chromosomes which arise during maturation from the pairing and splitting of a homologous pair of chromosomes.

The'o-ry (Gr. *theoria*, a beholding, specula- . A formulated hypothesis.

Tho-rac'ic (Gr. *thorax*, chest). Pertaining to the chest or thorax.

Tho'rax (Gr. chest). The chest.

Throm'bin (Gr. *thrombos*, clot). A substance produced in shed blood when prothombin comes in contact with thromboplastin. It is essential in clot formation.

Throm'bo-plas''tin (Gr. *thrombos*, clot; *plastikos*, to form, mold). A substance released from injured cells or platelets that initiates blood clotting.

Thy'mus (Gr. *thymos*, thymus). A ductless gland-like body situated ventral and anterior to the heart and below the thyroid.

Thy'roid (Gr. *thyreos*, shield; *eidos*, resemble). An endocrine gland located in the neck of vertebrates which secretes thyroxine.

Thy-rox'ine. The hormone secreted by the thyroid.

Tis'sue (L. *texere*, to weave). A collection of cells, usually similar, organized for the performance of a specific function.

Tox'in (Gr. *toxicon*, poison). A poisonous substance of plant or animal origin.

Tra'che-a (Gr. *tracheia*, windpipe). The breathing tube or windpipe of vertebrates and the tiny air tubes of certain arthropods.

Trait (L. *tractus*, a drawing). An inherited character.

Trans-duc'tion. The passage of genetic material from one bacterium to another by means of a virus.

Trip'lo-blas''tic (Gr. *triplax*, triple; *blastos*, bud). Having three germ layers.

Tryp'sin (Gr. *truein*, rubbing down; *pepsis*, digesting). A protein-splitting pancreatic enzyme.

Tym-pan'ic mem'brane (Gr. *tympanon*, eardrum; L. *membrana*, skin covering). The eardrum.

Um-bil'i-cal cord (L. *umbilicus*, navel). The cord-like connection between the placenta and the fetus of mammals.

U-re'a (Gr. *ouron*, urine). A nitrogenous metabolic waste of mammals found in urine.

U're'ter (Gr. *oureter*, ureter). The tube which carries urine from the kidney to the bladder or to the cloaca.

U-re'thra (Gr. *oureter*, ureter). The tube that carries urine from the bladder to the surface, and in the male conveys the seminal fluids.

U'rine (L. *urina*, urine). The fluid secreted by the kidney.

U'rin-if''er-ous tu'bule (L. *urina*, urine; *ferre*, to bear; *tubulus*, any small tube). One of the excretory tubules in the kidney of higher vertebrates, consisting of a coiled tube and a capsule.

U'ro-gen''i-tal (u'ri-no-gen''i-tal sys'tem) (Gr. *ouron*, urine; *gignesthai*, to produce). Pertaining to both the excretory and reproductive systems.

U'ter-us (L. womb). The hollow muscular posterior end of the oviduct where the developing animal is contained and nourished.

Vac'u-ole (L. *vacuum*, empty). Small space in the cytoplasm.

Va-gi'na (L. sheath). The canal in the female reproductive tract which receives the penis in mating.

Va'gus (L. wandering). The tenth cranial nerve.

Va-ri-a'tion (L. *variare*, to change). Difference in structure or function exhibited by individuals of the same species.

Va-ri'e-ty (L. *varietas*, difference). One of the subdivisions of the species.

Va'sa ef'fer-en''ti-a (L. vessels; L. *efferens*, bringing out). Small ducts which carry sperm from the testis to the epididymis in higher vertebrates.

Vas'cu-lar (L. *vasculum*, little vessel). Pertaining to vessels.

Vas de'fer-ens (L. vessel; L. carrying down). The tube or duct which conveys sperm away from the testis.

Vein (L. *vena*, vein). A blood vessel that carries blood to the heart.

Ven'tral (L. *venter*, belly). The lower surface, opposite to dorsal.

Ven'tri-cle (L. *ventriculus*, little belly). A lower muscular pumping chamber of the heart, also a brain cavity.

Ver'mi-form ap-pen'dix (L. *vermis*, worm; *forma*, form; *ad*, to; *pendo*, hand). A small, blind pouch projecting from the caecum in some mammals.

Ver'te-bral col'umn (L. *vertere*, to turn). See Spinal column.

Ver'te-brate (L. *vertebratus*). Animal with a backbone.

Ves-tig'i-al (L. *vestigium*, footstep). A rudimentary or degenerate structure which at one time was functional or better developed.

Vil'lus (L. *villus*, hair). A very small, finger-like projection, especially found in the vertebrate intestinal lining for increasing absorption.

Vi'rus (L. slimy liquid, poison). The simplest living organism composed almost entirely of nucleoprotein. They are capable of reproduction only within cells and are visible only under the electron microscope.

Vis'cer-a (L. internal organs). Internal organs of the body cavity.

Vi'ta-min (L. *vita*, life; M.E. *amine*, a chemical radical). A general term for a number of unrelated organic substances found in many foods and necessary, in small quantities, for normal metabolic activity.

Vi-tel'line (L. *vitellus*, yolk). Pertaining to egg yolk.

Vi-vip'a-rous (L. *vivus*, alive; *parere*, to bear). Producing young that receive nourishment for their development from the uterine wall of the mother.

Warm-blood'ed. Animals possessing a temperature regulating device which enables them to maintain a constant body temperature.

Zo'o-ge-og"ra-phy (Gr. *zoion*, animal; *ge*, earth; *graphein*, to write). A branch of zoology dealing with the geographic distribution of animals.

Zy'gote (Gr. *zygotos*, united). The fertilized egg.

‖ ACKNOWLEDGMENTS FOR ILLUSTRATIONS AND COLOR PLATES

Figure no.	
2–9	adapted from Urey and Miller
2–13	Virus Laboratory, University of California, Berkeley; courtesy of Robley C. Williams
2–14	from *Cells and Organelles* by Alex B. Novikoff and Eric Holtzman. Copyright © 1970 by Holt, Rinehart, and Winston, Publishers. Reproduced by permission of Holt, Rinehart, and Winston, Publishers. Micrograph courtesy of A. Ryter
2–28	redrawn from Novikoff and Holtzman, 1970
3–1	redrawn from E. B. Wilson, *The Cell in Development and Heredity*
3–3	courtesy of G. Meek and Wiley Interscience. Redrawn with permission
3–4	courtesy of George G. Rose
3–5	photographic arrangement from: Robert Dyson, *Cell Biology, A Molecular Approach*. Allyn & Bacon, Inc., Boston, 1974.

676

(a) courtesy of W. Bloom and D. W. Fawcett. A *Textbook of Histology*, 9th Edition, Philadelphia: W. B. Saunders Co., 1968, p. 117.

(b) courtesy of Murray Barr

(c) courtesy of Thomas L. Hayes and the Lawrence-Berkeley Laboratory, University of California and *Experimental and Molecular Pathology*, **10**, 186, 1969

(d) courtesy of J. A. Clarke and A. J. Salsbury, *Nature*, **215**, 402, 1967

677

ACKNOWLEDGE-
MENTS FOR
ILLUSTRATIONS
AND COLOR
PLATES

3–8   courtesy of Norman E. Kemp

3–11   courtesy of K. Porter, D. Prescott, J. Frye and *Journal of Cell Biology*, **57**, 815, 1973

3–15   courtesy of David E. Green, John H. Young and *American Scientist*, **59**, 96, 98, 1971

3–16   courtesy of Keith R. Porter

3–17   courtesy of Keith R. Porter

3–18   courtesy of Keith R. Porter

3–22   *top:* courtesy of Jean André

3–23   Y. Nonomura, G. Blobel and D. Sabatini, *Journal of Molecular Biology*, **60**, 303, 1971

3–30   George Lower

5–1   courtesy of *Smithsonian Report*, E. L. Greene, 1907

6–4   diagram from F. Warner, 1972. Macromolecular organization of eukaryotic cilia and flagella. In *Advances in Cell and Molecular Biology* (E. J. DuPraw, ed.) p. 427, Academic Press, New York

6–5   courtesy of F. Seydel and D. Outka

6–20   courtesy of Turtox

6–24   courtesy of C. P. Gilmore and New York Graphic Society; micrograph taken by Mrs. Marianna Wilke and Mr. Fred Behnke, Kodak Industrial Laboratory, Rochester, New York. Courtesy of Eastman Kodak Co.

6–35   courtesy of the University of Michigan Library

7–3   modified after Hyman

7–5   partly after Schulze

7–12   courtesy of Keith Gillett, Australian Museum, Sidney

8–2   partly after Wilhelmi and Ude, and from Steinmann and Breslau

8–5   redrawn after Belding

8–9   modified after Borradaile and Potts

9–2   redrawn and modified after Goodchild and Wilson

10–3   after R. Hartwig

10–6   based on Lefevre and Curtis

10–7   George Lower

10–9   *right:* courtesy of Keith Gillett

10–12   modified after G. A. Drew

10–13   photo by M. Woodbridge Williams

11–6   photo by J. T. Darby, courtesy of Bertha Allison

12–2   courtesy of L. B. Kellum

12–5   courtesy of Keith Gillett

12–8   redrawn partly after Herrick

12–10   based on Herrick

12–11   redrawn partly after Wigglesworth

12–16   redrawn mostly after Snodgrass

12–22   partly after Snodgrass

678

ACKNOWLEDGE-
MENTS FOR
ILLUSTRATIONS
AND COLOR
PLATES

679

ACKNOWLEDGE-
MENTS FOR
ILLUSTRATIONS
AND COLOR
PLATES

15–3   redrawn with modifications from Neal and Rand
15–12   courtesy of Hugh Huxley
16–1   redrawn with modifications after Kahn
16–16   partly after Arey
16–24   courtesy of Paul A. Wright
16–27   specimens supplied by Paul A. Wright
16–31   courtesy of Hans Selye, *Textbook of Endocrinology*, Acta, Inc.
16–32   courtesy of Hans Selye, *Textbook of Endocrinology*, Acta, Inc.
16–33   courtesy of Hans Selye, *Textbook of Endocrinology*, Acta, Inc.
17–6   after Kavanov from Moment
17–10   after Kahn
17–11   courtesy of L. E. Hanson, University of Minnesota
19–1   redrawn in part from Carlson and Johnson
19–4   after various sources
21–3   New York Zoological Society
21–6   courtesy of I. Muul and John Alley
21–9   "courtesy of Carnegie Institution of Washington" with credit to C. F. Reather,
            RBP, FBPA as photographer
21–12   models from Cleveland Health Clinic, University Museums, University of
            Michigan
22–6   redrawn in part from Etkin
22–9   redrawn from Spemann
22–11   redrawn from Spemann
22–12   courtesy of William Cristinelli, Michigan Department of Conservation
22–14   modified after Briggs and King
22–15   redrawn after Barth
22–16   redrawn after Patten after Shannes
22–17   redrawn after Patten after Shannes
23–3   based on a model by Crick and Watson
23–6   courtesy of K. Porter, D. Prescott, J. Frye and *Journal of Cell Biology*, 57, 815,
            1973
23–7   *left:* courtesy of Oscar Miller and Barbara Beatty, Visualization of Nucleolar
            Genes, *Science*, 164, 955, 1969
            right: courtesy of C. P. Gilmore and New York Graphic Society; micrograph
            taken by Dr. K. M. Marimuthu
23–8   courtesy Turtox
23–10   courtesy of Daniel Mazia
23–12   courtesy of Bruce Voeller
24–2   courtesy of Central Scientific Co.
24–21   modified from Emmens
24–22   courtesy Turtox
24–27   redrawn from Dent and Harris
24–30   modified from Pauling, Itano, Singer, and Wells
24–32   redrawn from Etkin
24–34   courtesy of James V. Neel
24–35   courtesy of Jan Powers
24–36   *top:* courtesy of Dr. P. Fogh-Anderson, University of Human Genetics, Copen-
            hagen, Denmark
            bottom: courtesy of Arne Birch-Jensen, Department of Surgery, Sygehuset,
            Skive, Denmark

680

**ACKNOWLEDGE-
MENTS FOR
ILLUSTRATIONS
AND COLOR
PLATES**

24–37   percentages taken from Cyril Burt, *Intelligence and Fertility*, Eugenics Society (London, 1946)

25–1   age estimates from Simpson, The Meaning of Evolution, Yale University Press

25–2   courtesy of I. G. Reiman, University Museums, University of Michigan

25–3   from specimens of the University Museums, University of Michigan

25–5   after Simpson, *The Meaning of Evolution*, Yale University Press

25–6   courtesy of I. G. Reiman, University Museums, University of Michigan

25–7   after Simpson, *The Meaning of Evolution*, Yale University Press

25–12   partly after Patten

26–1   courtesy of the University of Michigan Library

26–3   from the experiments of H. B. D. Kettlewell

26–4   mostly from specimens from the University Museums, University of Michigan

26–5   after Hesse and others

26–7   adapted in part from Simpson, *The Meaning of Evolution*, Yale University Press

26–10   courtesy of Dr. William Duellman

26–11   courtesy Turtox

26–12   *left:* courtesy of Keith Gillett
           right: courtesy of O. Wilford Olson

27–2   redrawn from N. Tinbergen, *The Study of Instinct*

27–3   modified from *Migration of Birds*, U.S. Fish and Wildlife Service, 1950

27–4   courtesy of William Burt

27–5   redrawn from N. Tinbergen, *The Study of Instinct*

27–6   adapted from K. Z. Lorenz, "The Evolution of Behavior." Copyright © 1958 by Scientific American, Inc. All rights reserved.

27–7   adapted from N. Tinbergen, "The Evolution of Behavior in Gulls." Copyright © 1960 by Scientific American, Inc. All rights reserved.

27–8   redrawn from N. Tinbergen, *The Study of Instinct*

27–9   redrawn from N. Tinbergen, *The Study of Instinct*

27–10   adapted from K. von Frisch

27–11   adapted from K. von Frisch

27–12   redrawn from N. Tinbergen, *The Study of Instinct*

28–1   courtesy of C. W. Schwartz

28–2   courtesy Turtox

28–4   courtesy of M. Woodbridge Williams

28–10   data from Odum

28–12   data from several sources

28–16   courtesy of Charles Ray, Jr.

28–18   courtesy of Greater Arizona, Inc., Phoenix, Arizona; photo by Mclaughlin Co.

28–21   estimates from World Health Organization

Plate 1   George Lower

Plate 2A   George Lower

Plate 2B   George Lower

Plate 3A   courtesy of Edward Slater

Plate 4   courtesy of Edward Slater

Plate 5A   courtesy of Edward Slater

Plate 5B	courtesy of Edward Slater
Plate 6A	courtesy of Edward Slater
Plate 6B	courtesy of Edward Slater
Plate 7	courtesy of Edward Slater
Plate 8A	George Lower
Plate 8B	George Lower
Plate 9A	George Lower
Plate 9B	courtesy of Edward Slater
Plate 10	courtesy of Edward Slater
Plate 11	courtesy of Edward Slater
Plate 13A	George Lower
Plate 13B	courtesy of Edward Slater
Plate 15A	courtesy of Harold Trapido
Plate 15B	courtesy of Harold Trapido
Plate 16A	courtesy of Harold Trapido
Plate 16B	George Lower
Plate 17A	courtesy of Edward Slater

ACKNOWLEDGE-
MENTS FOR
ILLUSTRATIONS
AND COLOR
PLATES

INDEX

Bold-face numbers refer to pages on which illustrations relating to the subject occur. Numbers in italics refer to glossary definitions of terms.